FINITE ELEMENT ANALYSIS
Theory and Programming
Second Edition

FINITE ELEMENT ANALYSIS
Theory and Programming

Second Edition

C S Krishnamoorthy
Indian Institute of Technology
Madras

McGraw Hill Education (India) Private Limited

FINITE ELEMENT ANALYSIS: Theory and Programming, Second Edition

Copyright © 1994, 1987, by McGraw Hill Education (India) Private Limited.

39th reprint 2019
RDBYRDLTDDACB

This edition can be exported from India only by the publishers,
McGraw Hill Education (India) Private Limited.

ISBN (13 digits): 978-0-07-462210-0
ISBN (10 digits): 0-07-462210-2

40 41 42 43 44 ADA 23 22 21 20 19

Published by McGraw Hill Education (India) Private Limited,
444/1, Sri Ekambara Naicker Industrial Estate, Alapakkam, Porur, Chennai 600116,
Tamil Nadu, India.

Printed and Bound in India at Adarsh Printers, Jhilmil Colony Industiral Area,
Delhi 110095.

Visit us at: www.mheducation.co.in ; (Toll free in India): **1800 103 5875**
Write to us at: info.india@mheducation.com

CIN: U80302TN2010PTC111532

Dedicated

to

MY PARENTS

Preface
to the Second Edition

It has been quite gratifying to the author that in the last six years since its publication, the book has found wide acceptance amongst students and teachers of the subject. In order to make it a more effective text book, the author has felt the necessity of providing exercises under each chapter and also including additional material and references wherever appropriate. The second edition is aimed to fulfil these requirements and the salient additional features are explained below.

In the brief review of the principles of structural mechanics in Chapter 2, expressions for the consideration of initial stress, strain and temperature effects are included.

Use of Lagrange interpolation function for the derivation of shape functions is described in Chapter 3 under the new section Lagrange and Serendipity Elements.

In some cases of discretization, triangular elements are needed. A new section on Triangular Elements is included in Chapter 4 on Isoparametric Elements. Computational aspects of element stiffness matrix are also discussed.

The five basic steps to be followed in finite element analysis are illustrated through an example in Chapter 5.

The patch test provides a simple means of testing the validity of element formulation and its implementation in a computer program. A brief introduction to the procedure is added in Chapter 8.

In the last few years considerable amount of research has been carried out on the analysis of thin plates. The shear deformation in plates, the Reissner-Mindlin theory (or referred to as Mindlin's theory) and its application to the analysis of thin plates are described in detail in Chapter 10. The shear locking problems and the selective/reduced integration techniques are also discussed therein.

Though the majority of finite element application in engineering design is in the area of structural mechanics, it has been found in many situations that stress analysis due to tempeature requires computation of distribution of temperature in a body. A new Chapter (12) on Conduction Heat Transfer has been contributed by Prof.K.N.Seetharamu of Mechanical Engineering Department, I.I.T., Madras. Following weighted residual technique, the element properties are derived and application to one-, two- and three-dimensional heat conduction problems is illustrated through examples. The approach presented in this Chapter would also enable the student to extend the application of finite element technique to solve other field problems characterized by the governing differential equations.

In order to focus the attention of the readers, a separate Chapter (14) on Finite Element Analysis Software is devoted to highlight the capabilities of and current trends in FEA packages. Finite element modelling and assessment of reliability of the solution have been a problem to the analysts. For the last few years, intensive research is being carried out in the areas of *error estimates* and *adaptive meshing* and these developments are discussed in this Chapter.

Finite element analysis is a powerful numerical tool and in order to help students master the techniques, 190 problems involving derivations, hand calculations, computer programming and use of PASSFEM are given under various Chapters of the book.

The author would like to express his deep sense of gratitude to Professor K.N.Seetharamu, Department of Mechanical Engineering, I.I.T., Madras for contributing the Chapter 12 on Conduction Heat Transfer which makes wider use of the book possible. The exercises given at the end of each Chapter are based on the courses on Finite Element Analysis conducted by the author, at I.I.T., Madras. The author is grateful to the research scholars and graduate students at I.I.T., Madras, and to the participants of short courses conducted by the author at various organisations, for their helpful suggestions and feed-back information on various topics.

The author is thankful to his research scholars, Mr G.Ramesh and Mr Benny Raphael for proof reading the various Chapters of the second edition. The author records his deep appreciation and thanks to Mr S.Rajeev, Lecturer, Department of Civil Engineering, to Mrs S.Srividhya, Project staff, for typesetting using LaTex, to Mr R.Muthusamy for his excellent support in the preparation of drawings and to Mrs S.Sankari of the Departmental Computer Facility for her help.

Thanks are due to Dr N.Subrahmanyam, Mr N.V.Seshan and Mr Rajiv Banerji and the production staff of the Tata McGraw-Hill Publishing Company for their enthusiastic support in bringing out this second edition of the book.

C S KRISHNAMOORTHY

Preface
to the First Edition

The stress analysis in the fields of civil, mechanical and aero-space engineering, naval architecture, off-shore engineering and nuclear engineering is invariably complex, and for many of the problems it is extremely difficult and tedious to obtain analytical solutions. In these situations, engineers usually resort to numerical methods to solve the problems. With the advent of computers, one of the most powerful techniques that has been developed in the realm of engineering analysis is the finite element method, and the method, being general, can be used for the analysis of structures / solids of complex shapes and complicated boundary conditions.

With the rapid development of the method a number of books have been published on this subject during the last ten years. Many of the books are concerned with the description of the method and development of elements for various applications. Also, the analysis of framed structures is usually treated separately under matrix methods of structural analysis.

The finite element method is a product of the computer age, and the application of the method to solve practical problems requires use of computer programs for analysis. The method is presented here in a computer-oriented manner so that the student or engineer can thoroughly understand the theory and the connected programming aspects of the various elements. The analysis of framed structures is formulated using the finite element model which makes the coverage unified and widely applicable to the analysis of all types of structural systems.

An essential feature of the book is the presentation of a general three-dimensional finite element analysis program (PASSFEM) around a main core program. Following the description of the theory for each type of element, the computational aspects of the element are illustrated through the subroutines. This approach will help acquire knowledge and experience in the development and use of finite element analysis programs.

After the introduction of the method in Chapter 1, the basic structural mechanics background required for learning the method is presented in Chapter 2 through a brief review of the basic equations of elasticity and the principles of virtual displacement and potential energy.

The displacement model which is widely used in package programs is introduced in Chapter 3 along with the derivation of the element properties. One of the significant developments in 'finite elements' is the concept of isoparametric elements, and the basic theory is presented in detail in Chapter 4.

For computer programming, it has been found that the direct stiffness method of assemblage of elements is efficient. The method and efficient solution technique are discussed in Chapter 5.

It has been my aim to bridge the educational gap between the theory and programming aspects of the finite element method. In an attempt to achieve this goal, the main program of PASSFEM is presented in Chapter 6. The salient features of the main program are described in this Chapter and the computer code is given in Appendix I. The program is designed in a modular form so that the various routines that will be presented in the subsequent Chapters dealing with different types of elements could be integrated to form a general-purpose three-dimensional finite element analysis program. The computer code is written in standard FORTRAN IV so that it can be used on most of the machines.

A unified treatment is adopted for the analysis of framed structures and presented in Chapter 7. The stiffness properties of truss and beam members are derived using the displacement model and the approach is consistent with other types of elements. Computer routines are presented for the analysis of truss and framed structures.

In Chapters 8 to 11, plane stress/strain, axisymmetric solid, three-dimensional solid, plate bending and shell elements are described in detail. Computer program listings for the routines are presented in Appendices 2 to 8 for one or two shapes of elements for each type of analysis. It may be pointed out that provision is made in the element library of PASSFEM for three shapes for each element type. This would provide the student an opportunity to gain more experience in finite element programming by developing the routines for shapes other than the ones presented here. A number of examples are included to illustrate the use of various types of elements.

For the analysis of large structural systems, 'substructuring' also known as 'super element' technique has been found to be quite efficient. This technique along with programming aspects for multi-level substructuring is well explained in Chapter 12. The finite element formulation for dynamic analysis is also described in this Chapter.

In practice, one encounters a large number of package programs for finite element analysis. A discussion on some of the widely used package programs and the software techniques used are presented in the section on finite element software. There has been significant advancement in the use of computer graphics for pre- and post-processing in the input and output stages of finite element programs. Also, the recent developments in software engineering like the use of problem-oriented language, data base management and artificial intelligence leading to the development of an efficient finite element software system giving more flexibility and facility to the user are highlighted in this Chapter. Having discussed the theory and programming aspects in the earlier Chapters, the purpose of this Chapter is to present the techniques and advances for large-scale finite element analysis of structural systems so that the student will acquire the necessary background and information to solve large-sized problems in professional practice.

The main aim of this book is to provide a comprehensive text for a first level course on finite element analysis with detailed presentation of the theory and programming of elements used in various types of stress analysis. The material presented here is based on the course of lectures which I have offered during the last ten years at IIT Madras and at other industrial organisations. I am indebted to various research workers and authors of textbooks and papers on this subject and selected references to their works are given at the end of each Chapter. A number of research scholars and graduate students have helped in the development of various types of elements for the program, PASSFEM. I would particularly like to express my gratitude and appreciation to Drs A.Panneerselvam and M.Inbasakaran and to Messrs Dani Thomas, S.M.Vaidya, Mahalinga Gowda, H.Narendra, Nagaraju, M.N.Ravi and Ravindra Babu for their contribution to the finite element library. I would also like to thank Profs K.S.Sankaran and L.N.Ramamurthy, former Heads of Department of Civil Engineering, IIT Madras, for their constant encouragement while writing this book.

C S KRISHNAMOORTHY

Contents

Preface to the Second Edition   *vii*

Preface to the First Edition   *ix*

1 Introduction   **1**
  1.1   Field conditions   1
  1.2   Boundary conditions   2
  1.3   Approximate solutions   4
       References   18
       Exercises   19

2 Basic Principles of Structural Mechanics   **22**
  2.1   Equilibrium conditions   22
  2.2   Strain-displacement relations   26
  2.3   Linear constitutive relations   28
  2.4   Principle of virtual work   34
  2.5   Energy principles   43
  2.6   Application to finite element method   56
       References   56
       Exercises   57

3 Element Properties   **63**
  3.1   Displacement models   64
  3.2   Relation between the nodal degrees of freedom and generalised coordinates   66
  3.3   Convergence requirements   67
  3.4   Natural coordinate systems   69
  3.5   Shape functions (Interpolation functions)   75
  3.6   Element strains and stresses   96
  3.7   Element stiffness matrix   97
  3.8   Static condensation   103
       References   105
       Exercises   106

4 Isoparametric Elements   **109**
  4.1 Two-dimensional isoparametric elements 111
  4.2 Computation of stiffness matrix for
      isoparametric elements 118
  4.3 Convergence criteria for
      isoparametric element 139
      References 140
      Exercises 141

**5 Direct Stiffness Method of Analysis and
Solution Technique**   **145**
  5.1 Assemblage of elements -
      Direct stiffness method 146
  5.2 Gauss elimination and matrix decomposition 155
      References 171
      Exercises 172

**6 Computer Program for Finite Element Analysis
(PASSFEM)**   **175**
  6.1 Main routine 176
  6.2 Subroutine PASSIN 178
  6.3 Subroutine FELIB 182
  6.4 Subroutine COLUMH 184
  6.5 Subroutine CADNUM 186
  6.6 Subroutine PASSEM 188
  6.7 Subroutine PASOLV 190
  6.8 Subroutine PASLOD 191
  6.9 Subroutine DISP 192
  6.10 Guidelines for adding element routines 192
  6.11 Input details 198
      References 201
      Exercises 201

7 Analysis of Framed Structures   **204**
  7.1 Two-dimensional truss element 204
  7.2 Three-dimensional truss element 212
  7.3 Two-dimensional beam element 214
  7.4 Three-dimensional beam element 237
  7.5 Shear deformation in beams 250
  7.6 Offset beam element (BEAM 2) 263
  7.7 Subroutine for 3D truss element 267
  7.8 Subroutine for 3D beam element 270
  7.9 Subroutine for boundary element 274
  7.10 Examples 278
      References 287
      Exercises 288

8 Plane Stress and Plane Strain Analysis **296**

  8.1 Triangular elements 297
  8.2 Rectangular elements 303
  8.3 Isoparametric elements 308
  8.4 Incompatible displacement models 315
  8.5 The Patch test 322
  8.6 Reinforced concrete element 324
  8.7 Axisymmetric solid element 327
  8.8 Subroutine PLANE 330
  8.9 Examples 335
      References 341
      Exercises 343

9 Three-Dimensional Stress Analysis **350**

  9.1 Three-dimensional solid elements 350
  9.2 Eight noded isoparametric solid element 355
  9.3 Twenty noded isoparametric solid element 366
  9.4 Properties of element faces 368
  9.5 Element load vector 370
  9.6 Evaluation of stresses 374
  9.7 Subroutine THREDS 377
  9.8 Examples 383
      References 398
      Exercises 399

10 Analysis of Plate Bending **402**

  10.1 Basic theory of plate bending 402
  10.2 Displacement functions 406
  10.3 Plate bending elements 407
  10.4 Shear deformation in plates 412
  10.5 Four noded isoparametric element (Plate4) 420
  10.6 Eight noded isoparametric element (Plate8) 430
  10.7 Subroutine PLATE 437
  10.8 Examples 442
      References 456
      Exercises 459

11 Analysis of Shells **465**

  11.1 Thin shell theory 465
  11.2 Review of shell elements 467
  11.3 Bilinear degenerated shell element 468
  11.4 Eight noded shell element 486
  11.5 Subroutine SHELL 495
  11.6 Examples 499
      References 511
      Exercises 512

12 Conduction Heat Transfer   **515**
 12.1 Basic equations of heat transfer 516
 12.2 Governing differential equation
   for heat conduction 517
 12.3 Discrete system 521
 12.4 Formulation of the finite element method
   for heat conduction 523
 12.5 Galerkin's method for Quasi harmonic
   - equation Heat conduction equation 525
 12.6 Field of application of
   quasi-harmonic equation 528
 12.7 One-dimensional heat conduction 530
 12.8 Two-dimensional conduction heat transfer 535
 12.9 Three-dimensional heat conduction 542
 12.10 Transient heat conduction problems 547
 12.11 Modifications to plane stress/strain
   program to determine temperature
   distributions in a body 553
   References 554
   Exercises 555

13 Substructuring Technique   **559**
 13.1 Multilevel substructuring technique 559
 13.2 Basic approach to substructuring technique 560
   References 574

14 Finite Element Analysis Software   **575**
 14.1 Pre- and post-processors 576
 14.2 Finite element analysis software 580
 14.3 Error estimates and adaptive meshing 586
   References 591

   Computer Program (PASSFEM) 593
   Index 704

Chapter 1

Introduction

Most of the problems of solid mechanics in various branches of engineering are boundary value problems. Any solution to these problems must satisfy the following conditions.

1.1 Field Conditions

The field variables, displacements (strains) and stresses (or stress-resultants) must satisfy the governing condition which can be mathematically expressed in any one of basic forms [1]

1.1.1 Differential Equations

One or more partial differential equations are usually obtained by vectorial treatment of the principles of mechanics. These equations form the governing equations that the field variables must satisfy for equilibrium and compatibility conditions of the solid or structural system. These differential equations can also be obtained by using the Euler-Lagrange equations of the variational principle as explained in Chapter 2.

For example, it is well known from elementary strength of materials [2] that the fourth order differential equation for beam problems is given by

$$EI \frac{d^4 w}{dx^4} = p \qquad (1.1)$$

Any solution to the variable w must satisfy the above equation.

In the case of isotropic thin plates, the governing partial differential equation that must be satisfied by the variable w, the vertical displacement of the plate, is given in well known texts on the subject [3].

$$\frac{\partial^4 w}{\partial x^4} + 2\frac{\partial^4 w}{\partial x^2 \partial y^2} + \frac{\partial^4 w}{\partial y^4} = \frac{p}{D} \tag{1.2}$$

where

$$D = \frac{Eh^3}{12(1-\mu^2)}$$

1.1.2 Variational Formulation

In this approach the extremum value of the functional is sought as the solution to the problem. The governing functional may be the total potential energy or total complementary energy of the system. Using the calculus of variations it can be shown that the extremum or stationary value of the functional is obtained by equating the first variation of the functional to zero. In problems of structural mechanics, the condition leads to the equilibrium or compatibility equation for the problem, and the field variable must satisfy these equations. For example, the potential energy of an isotropic plate subjected to uniformly distributed load of intensity p is given by the functional [3]

$$\Pi = \frac{Eh^3}{24(1-\mu^2)} \int\int \left\{ \left(\frac{\partial^2 w}{\partial x^2} + \frac{\partial^2 w}{\partial y^2}\right)^2 \right.$$

$$-2\,(1-\mu)\left[\frac{\partial^2 w}{\partial x^2}\frac{\partial^2 w}{\partial y^2} - \left(\frac{\partial^2 w}{\partial x \partial y}\right)^2\right]\right\} dxdy$$

$$-\int\int w\,p\,dxdy \tag{1.3}$$

The solution to the field variable, w, must yield stationary (extremum) value for the potential functional Π and also should satisfy all the prescribed kinematic boundary conditions. A brief summary of the principles used in this type of formulation is given in Chapter.2. The strain energy principle used for the analysis of beams, framed structures and plate bending problems treated in texts [2, 3, 4, 5] fall under this category.

1.2 Boundary Conditions

For structural mechanics problems the boundary conditions may be kinematic, i.e., where the displacements (and slope, i.e., derivative of displacement), may be prescribed, or static i.e. where forces (and moments) may be prescribed. In problems where time is involved the initial values may have to be specified.

Figure 1.1 shows a cantilever beam AB subjected to a uniformly distributed load. If the vertical deflection w at any point is taken as a field variable, it must satisfy the differential equation Eq.(1.1), which is an equilibrium condition,

$$EI \frac{d^4 w}{dx^4} = p$$

And the solution to the above equation must also satisfy the boundary conditions at A and B as follows.

Kinematic boundary conditions at A: Displacement $w = 0$ and slope

$$\frac{dw}{dx} = 0$$

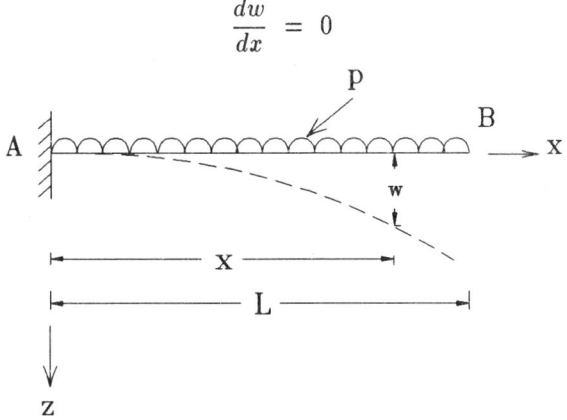

Fig. 1.1 A cantilever beam

static or force boundary conditions at B: shear force,

$$EI \frac{d^3 w}{dx^3} = 0$$

and bending moment

$$EI \frac{d^2 w}{dx^2} = 0$$

In subsection 2.3.2 the variational formulation of the above problem is discussed and the application of Euler-Lagrange equation to obtain the governing differential equation is illustrated for the above cantilever beam problem.

1.3   Approximate Solutions

It is not possible to obtain analytical solution for many engineering
problems. An analytical solution is a mathematical expression that
gives the value of the field variable at any location in the body. For
problems involving complex shapes, material properties and complicated boundary conditions, it is difficult and in many cases in-
tractable to obtain analytical solution that satisfies the governing
differential equations or gives extremum value to the governing functional. Hence, for most of the practical problems the engineer resorts
to numerical methods that provide approximate but acceptable solutions. The three methods that are used are as follows:

1. Functional approximation

2. Finite difference method

3. Finite element method

A brief description of the first two methods is given in the subsequent sections and then the finite element method is introduced as
a powerful numerical method, widely used in practice.

1.3.1   Functional Approximation

A set of independent functions satisfying the boundary conditions is
chosen and a linear combination of a finite number of them is taken to
approximately specify the field variable at any point. The unknown
parameters that combine the functions are found out in such a way
to achieve at best the field condition, which is represented through
variational formulation. The well known classical methods such as
Rayleigh-Ritz, Galerkin and collocation are based on functional approximation but vary in their procedure for evaluating the unknown
parameters [6]. The Rayleigh-Ritz method is briefly described below.

Consider a simply supported beam, shown in Fig.1.2 subjected
to a central concentrated load P and a uniformly distributed load
of intensity p_o. In this problem if the deflected shape of the beam
is known, the bending moment and shear force at any cross-section
can be determined. Consider the following approximation to the
deflection curve that satisfies the boundary condition.

$$w = a_1 \sin\frac{\pi x}{L} + a_2 \sin\frac{3\pi x}{L} \qquad (a)$$

where a_1 and a_2 are unknown parameters.

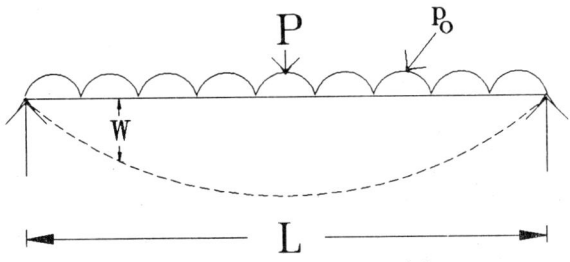

Fig. 1.2 Simply supported beam

It is known from the elementary strength of materials that the strain energy, U, of the beam due to bending is

$$U = \frac{EI}{2} \int_o^L \left(\frac{d^2w}{dx^2}\right)^2 dx \qquad (b)$$

Substituting for w from equation (a) into equation (b),

$$U = \frac{EI}{2} \int_o^L \left(\frac{-\pi^2 a_1}{L^2}\sin\frac{\pi x}{L} - \frac{9\pi^2 a_2}{L^2}\sin\frac{3\pi x}{L}\right)^2 dx$$

$$= \frac{EI\pi^4}{4L^3}(a_1^2 + 81a_2^2) \qquad (c)$$

The potential energy due to loads is given by,

$$H = -\int_o^L p_o\, w\, dx - P\, w_{max}$$

$$= -\int_o^L p_o \left(a_1\sin\frac{\pi x}{L} + a_2\sin\frac{3\pi x}{L}\right) dx - P(a_1 - a_2)$$

$$= -\frac{2p_o L}{\pi}\left(a_1 + \frac{a_2}{3}\right) - P(a_1 - a_2) \qquad (d)$$

The total potential energy, Π, of the beam is

$$\Pi = U + H$$

$$= \frac{EI\pi^4}{4L^3}(a_1^2 + 81a_2^2) - \frac{2p_o L}{\pi}\left(a_1 + \frac{a_2}{3}\right)$$

$$- P(a_1 - a_2) \qquad (e)$$

It will be shown in subsection 2.3.2 that for stable equilibrium of the body the potential energy attains stationary value. It can be seen from equation (e) that the total potential energy is now expressed in terms of the parameters a_1 and a_2. Hence, for stationary value of Π the following conditions must be satisfied.

$$\frac{\partial \Pi}{\partial a_1} = 0 \qquad \frac{\partial \Pi}{\partial a_2} = 0 \qquad (f)$$

Applying the above conditions to equation (e) we get,

$$\frac{EI\pi^4}{2L^3} a_1 - \frac{2p_oL}{\pi} - P = 0 \qquad (g)$$

$$\frac{81EI\pi^4}{2L^3} a_2 - \frac{2p_oL}{3\pi} + P = 0 \qquad (h)$$

Solving the above two equations we get,

$$a_1 = \frac{2L^3}{EI\pi^4}\left(\frac{2p_oL}{\pi} + P\right) \qquad (i)$$

$$a_2 = \frac{2L^3}{81EI\pi^4}\left(\frac{2p_oL}{3\pi} - P\right) \qquad (j)$$

Thus the maximum deflection at the centre of the beam is,

$$w_{max} = a_1 - a_2$$

$$w_{max} = \frac{PL^3}{48.11EI} + \frac{p_oL^4}{76.82EI} \qquad (k)$$

The maximum deflection practically coincides with the exact value of

$$\frac{PL^3}{48EI} + \frac{p_oL^4}{76.8EI}$$

The bending moment at any point along the beam is given by

$$M_x = EI\frac{d^2w}{dx^2} \qquad (l)$$

After substituting for the values of a_1 and a_2 from equation(i) and (j) into equation (a) and differentiating, the bending moment M_c at the centre of the beam, x = L/2, can be shown to be

$$M_c = \frac{PL}{4.44} + \frac{p_oL^2}{8.05} \qquad (m)$$

It may be noted that the error in the first term is 9.92 per cent and that in the second term is 0.62 per cent. This error can be reduced by adding more terms to the approximate (or trial) function for w, i.e, into equation (a).

The above procedure can be extended to the analysis of a three-dimensional solid. In general a deformable body consists of infinite material points and, therefore, it has infinitely many degrees of freedom. By the Rayleigh-Ritz method such a continuous system is reduced to a system of finite degrees of freedom. For the case of three-dimensional solid, the variation of the field variables, displacements u, v and w can approximately be represented by the following trial functions.

$$u = a_1\phi_1(x,y,z) + a_2\phi_2(x,y,z) + \ldots + a_n\phi_n(x,y,z)$$

$$v = b_1\beta_1(x,y,z) + b_2\beta_2(x,y,z) + \ldots + b_n\beta_n(x,y,z)$$

$$w = c_1\psi_1(x,y,z) + c_2\psi_2(x,y,z) + \ldots + c_n\psi_n(x,y,z)$$

$$(1.4)$$

where a_i, b_i and c_i are linearly independent unknown parameters and $\phi_i(x,y,z), \beta_i(x,y,z)$ and $\psi_i(x,y,z)$ where i = 1,2...,n are continuous functions in x, y and z that satisfy all the kinematic boundary conditions. By this approximation the body is reduced to have 3n degrees of freedom. Now the potential energy of the body can be expressed by a functional in terms of these parameters.

$$\Pi(u,v,w) = \Pi(x,y,z,a_1 \ldots a_n, b_1 \ldots b_n, c_1 \ldots c_n) \qquad (1.5)$$

As was indicated earlier that for stable equilibrium of the body the potential energy attains a stationary value and as the potential energy functional is in terms of the parameters a_i, b_i and c_i the following equations must be satisfied.

$$\frac{\partial\Pi}{\partial a_1} = 0 \quad \frac{\partial\Pi}{\partial a_2} = 0 \quad \cdots \quad \frac{\partial\Pi}{\partial a_n} = 0$$

$$\frac{\partial\Pi}{\partial b_1} = 0 \quad \frac{\partial\Pi}{\partial b_2} = 0 \quad \cdots \quad \frac{\partial\Pi}{\partial b_n} = 0$$

$$\frac{\partial\Pi}{\partial c_1} = 0 \quad \frac{\partial\Pi}{\partial c_2} = 0 \quad \cdots \quad \frac{\partial\Pi}{\partial c_n} = 0 \qquad (1.6)$$

Thus from Eq.(1.6), we get 3n linear algebraic equations to solve the unknown parameters a_i, b_i and c_i.

It may be noted that the assumed trial functions must be continuous and satisfy all the prescribed boundary conditions, and no simple guidelines are available to select such functions. Also the classical approach of arriving at the equations of the type (1.6) is quite cumbersome. Hence, except in simple situations, this approach could not be used to solve practical problems. However, the concepts used in Rayleigh-Ritz method, i.e., representing the variation of the field variable by trial function and finding the unknown parameters through minimization of potential energy, are well exploited in the finite element method.

1.3.2   Finite Difference Method

Prior to the introduction of finite element method, finite difference method has been used for solving numerically some of the difficult problems of structural mechanics. In this approach the body or the system is 'discretized' by a mesh of nodal points as shown in Fig.1.3

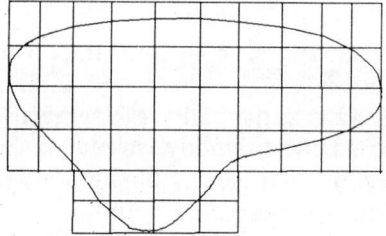

Fig. 1.3 Illustration of discretization for finite difference method

The field variable is represented by the discrete values of the variable at the 'nodes'. For example, in the case of plate bending problem the normal displacement w will be taken as unknown variable at each one of the nodes. The governing differential equation and the boundary conditions are converted to finite difference form. In the case of isotropic thin plates, the finite difference form of the governing differential Eq.1.2 applied to one of the nodal point (of a square mesh) in Fig.1.4 can be shown to be,

$$20w_1 - 8(w_2 + w_3 + w_4 + w_5) + 2(w_6 + w_7 + w_8 + w_9)$$

$$+w_{10} + w_{11} + w_{12} + w_{13} = \frac{p}{D}\lambda^4 \tag{1.7}$$

where λ is the step size and

$$D = \frac{Eh^3}{12(1 - \mu^2)}$$

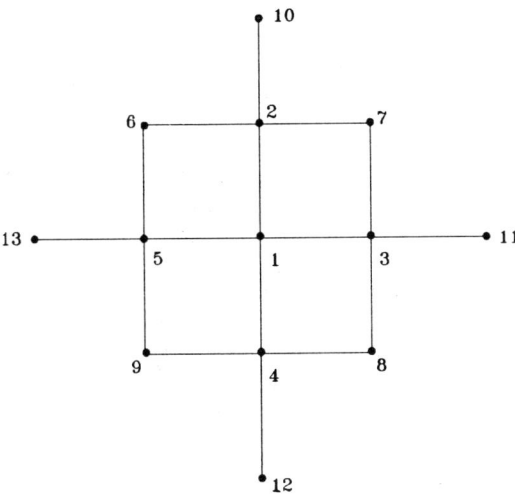

Fig. 1.4 Finite difference pattern

The finite difference form of the governing differential equation and boundary conditions are then applied to each of the nodes in turn and this will yield a set of linear algebraic equations in the discrete field variable, w. The resulting equations are then solved for the nodal values of this variable. The procedure is illustrated through an example given below.

Consider a simply supported square plate of size 4 m × 4 m subjected to a pressure of 6 kN/m^2 including its selfweight. The thickness of the plate is 12 cm. Assume $E = 2 \times 10^7$ kN/m^2 and $\mu = 0.15$.

The discretization of the plate with 25 nodal points is shown in Fig.1.5. It may be noted that the additional nodal points outside the plate boundaries 26 to 37 are required for analysis as will be shown below.

Since the loading and boundary conditions are symmetric about the two axes, only one quadrant of the plate is considered for analysis. The following boundary conditions are applied.

(i) Displacement, w, is zero along the supports

$$w_1 = w_2 = w_3 = w_4 = w_5 = w_6 = w_{10} = w_{11} = w_{15}$$
$$= w_{16} = w_{20} = w_{21} = w_{22} = w_{23} = w_{24} = w_{25} = 0 \qquad (a)$$

(ii) Bending moment is zero along the supports

$$\frac{\partial^2 w}{\partial x^2} = 0 \quad or \quad \frac{\partial^2 w}{\partial y^2} = 0$$

Using the central difference, $\partial^2 w/\partial x^2$ or $\partial^2 w/\partial y^2 = 0$ at a point 'n'

$$= \frac{w_{n+1} - 2w_n + w_{n-1}}{2\lambda^2} \tag{b}$$

Applying the condition (b) at the points 2, 3, 4, 6, 10, 11, 15, 16, 20, 22, 23 and 24 we get the following conditions.

$$w_{26} = -w_7 \quad w_{27} = -w_8 \quad w_{28} = -w_9$$

$$w_{29} = -w_7 \quad w_{30} = -w_9 \quad w_{31} = -w_{12}$$

$$w_{32} = -w_{14} \quad w_{33} = -w_{17} \quad w_{34} = -w_{19}$$

$$w_{35} = -w_{17} \quad w_{36} = -w_{18} \quad w_{37} = -w_{19} \tag{c}$$

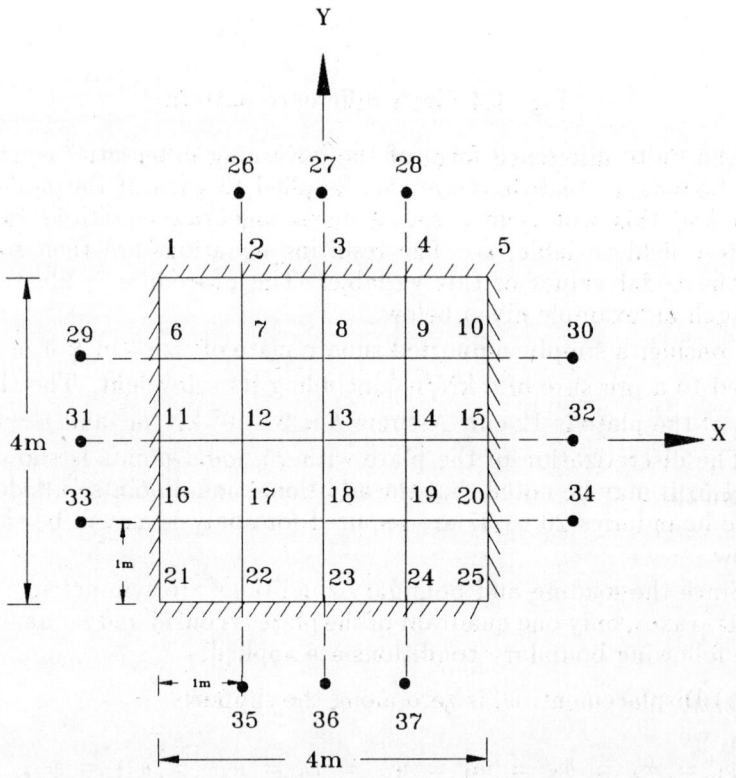

Fig. 1.5 Discretization of a simply supported plate

Applying the finite difference pattern of Eq.1.7 to the nodal points 8, 9 and 13 and making use of the conditions (a) and (c) and symmetry conditions we get,

$$20w_8 - 8(w_9 + w_9 + w_{13}) + 2(2w_8) + (-w_8 + w_8) = \frac{p\lambda^4}{D}$$

$$20w_9 - 8(w_8 + w_8) + 2(w_{13}) + (-w_9 - w_9 + w_9 + w_9) = \frac{p\lambda^4}{D}$$

$$20w_{13} - 8(4w_8) + 2(4w_9) + 0 = \frac{p\lambda^4}{D} \qquad (d)$$

Simplifying the above equations (d) we get

$$24w_8 - 16w_9 - 8w_{13} = \frac{p\lambda^4}{D}$$

$$-16w_8 + 20w_9 + 2w_{13} = \frac{p\lambda^4}{D}$$

$$-32w_8 + 8w_9 - 20w_{13} = \frac{p\lambda^4}{D} \qquad (e)$$

Solving the above equations, we get

$$w_{13} = 1.03125\frac{p\lambda^4}{D}$$

$$w_8 = 0.75\frac{p\lambda^4}{D}$$

$$w_9 = 0.5468\frac{p\lambda^4}{D} \qquad (f)$$

From the given data we get,

$$D = \frac{Eh^3}{12(1 - \mu^2)} = \frac{2 \times 10^7 \times 0.12^3}{12(1 - 0.15)^2} = 2946.3 \text{ kN } \text{ m}$$

$$p = 6.0 \quad \text{and} \quad \lambda = 1.0$$

Substituting the above numerical values in equation (f) for $w_{max} = w_{13}$, we get,

$$w_{max} = 1.03125\frac{p\lambda^4}{D}$$

$$= 1.03125 \times \frac{6.0 \times 1.0^4}{2946.3}$$

$$= 2.1 \times 10^{-3}m = 2.1 \text{ mm}$$

The bending moment,

$$M_x = -D\left[\frac{\partial^2 w}{\partial x^2} + \mu \frac{\partial^2 w}{\partial y^2}\right]$$

Applying the central difference for M_x at the point 13 we get

$$
\begin{aligned}
M_{x\ max} &= -D(1+\mu)\frac{2w_8 - 2w_{13}}{\lambda^2} \\
&= -2946.3(1+0.16)\left[\frac{2 \times 0.75 - 2 \times 1.03125}{1.0^2}\right] \times \frac{6 \times 1.0^4}{2946.3}r \\
&= +3.8813 \text{ kN m/m}
\end{aligned}
$$

From the Theory of Plates [3] solution we get the solution as,

$$w_{max} = 0.00406\frac{pa^4}{D} = \frac{0.00406 \times 6 \times 4^4}{2946.3}$$

$$= 2.117 \times 10^{-3}m = 2.117 \text{ mm}$$

Taking only the first two terms of the series solution for M_x given in reference [3] and substituting for $\mu = 0.15$,

$$M_x = 0.042192pa^2 = 0.042192 \times 6 \times 4^2 = 4.0504 \text{ kN m/m}$$

Comparing the values obtained by finite difference method with the standard solution the displacement is very close to the theoretical solution and the maximum moment differs by 4.18 per cent. It can be seen that for a discretization of $\lambda = 1.0$ the accuracy obtained is sufficient from practical point of view. However, in a general case a finer discretization may be necessary to get the solution closer to the theoretical values.

The above example illustrated the method of solving a problem when the finite difference form of the governing differential equation is known. However, it may be noted that the finite difference form of the differential equations can be obtained easily for meshes of regular shapes, but it becomes quite involved to derive it for irregular meshes. Moreover, the governing differential equation itself will become complex in the case of materials other than isotropic and setting up the difference pattern for irregular shapes will be complicated. Also the procedure does not lend itself to computerisation for solving many classes of problems in general. However, the concept of 'discretization' used in this method formed the basis of the finite element method.

1.3.3 Finite Element Method

A key contribution to the development of matrix methods for structural analysis was made by Argyris and Kelsey [7]. In their contribution they presented matrix formulation for force and displacement methods of analysis using energy theorems of structural mechanics. It was the work of Turner, Clough, Martin and Topp [8] that led to the discovery of the finite element method. Clough in his subsequent work [9] gave the physical interpretation to the method and it appears that he was the first to use the terminology 'finite element'. Since then, tremendous advances have been made in the last 25 years both on the mathematical foundations and generalisation of the method to solve field problems in various areas of engineering analysis [10, 11, 12]. During the same period due to rapid development in computer technology, large number of package programs have been developed for finite element analysis which made it possible for wider use of this technique in practice. A brief summary of the method is given below and detailed description of the method for stress analysis is presented in subsequent Chapters.

The finite element method combines in an elegant way the best features of the two approximate methods of analysis discussed above. In particular the method can be explained through physical concept and hence is most appealing to the engineer. And the method is amenable to systematic computer programming and offers scope for application to a wide range of analysis problems.

The basic concept is that a body or a structure may be divided into smaller elements of finite dimensions called 'finite elements'. The original body or the structure is then considered as an assemblage of these elements connected at a finite number of joints called 'Nodes' or 'Nodal Points'. The concept of discretization used in finite difference method is adopted here. Figures 1.6 to 1.9 illustrate the finite element discretization of some of the structural systems.

The properties of the elements are formulated and combined to obtain the solution for the entire body or structure. For example, in the displacement formulation widely adopted in finite element analysis, simple functions known as 'shape functions' are chosen to approximate the variation of displacement within an element in terms of the displacement at the nodes of the element. This follows the concept used in the Rayleigh-Ritz procedure of functional approximation method but the difference is that the approximation to field variable is made at the element level. The strains and stresses within an element will also be expressed in terms of the nodal displacements. Then the *principle of virtual displacement or minimum potential energy* is used to derive the equation of equilibrium for the element and the nodal displacements will be the unknowns in the equations.

The equations of equilibrium for the entire structure or body are then obtained by combining the equilibrium equation of each element such that the continuity of displacement is ensured at each node where the elements are connected. The necessary boundary conditions are imposed and the equations of equilibrium are solved for the nodal displacements. Having thus obtained the values of displacements at the nodes of each element, the strains and stresses are evaluated using the element properties derived earlier.

Thus, instead of solving the problem for the entire structure or body in one operation, in this method attention is mainly devoted to the formulation of properties of the constituent elements. The procedure for combining the elements, solution of equations and evaluation of element strains and stresses are the same for any type of structural system or body. Hence, the finite element method offers scope for developing general purpose programs with the properties of various types of elements forming an 'element library' and the other procedures of analysis forming the common core segments. This modular structure of the program organisation is well exploited in the large number of program packages, now available for practical application to various disciplines of engineering.

In the subsequent chapters of the book, the properties of various types of elements, procedures for assemblage of elements, solution technique and development of computer programs for various types of problems will be presented.

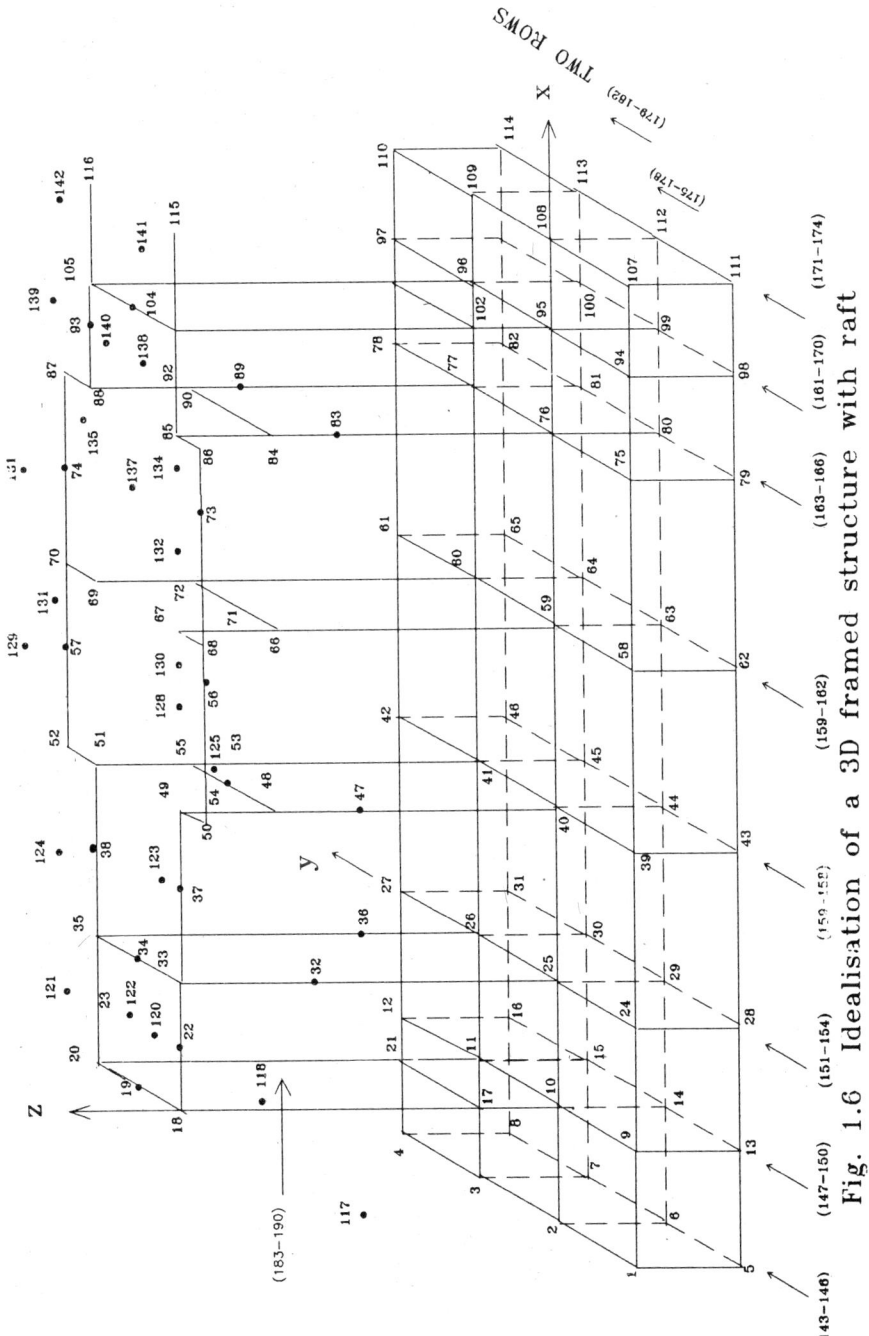

Fig. 1.6 Idealisation of a 3D framed structure with raft

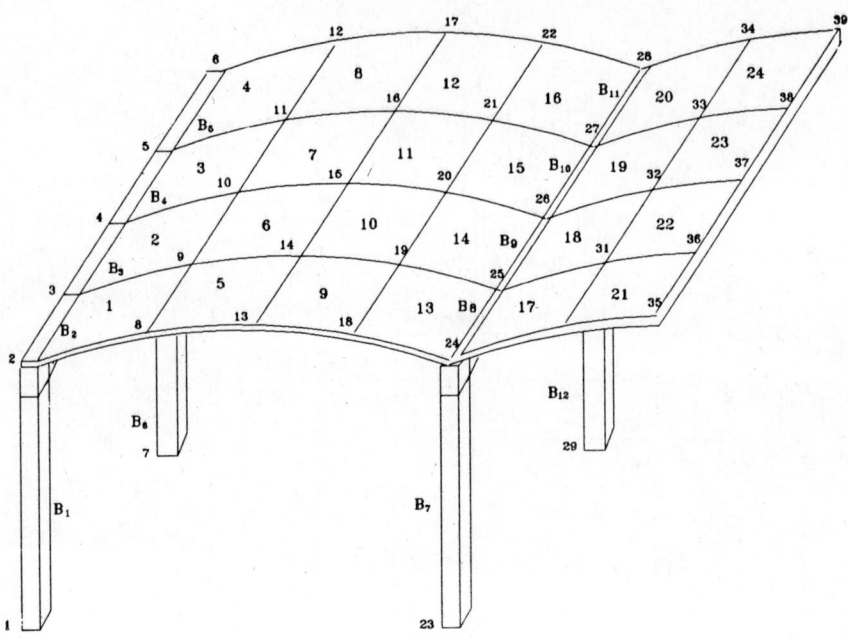

Fig. 1.7 Analysis model of a shell roof

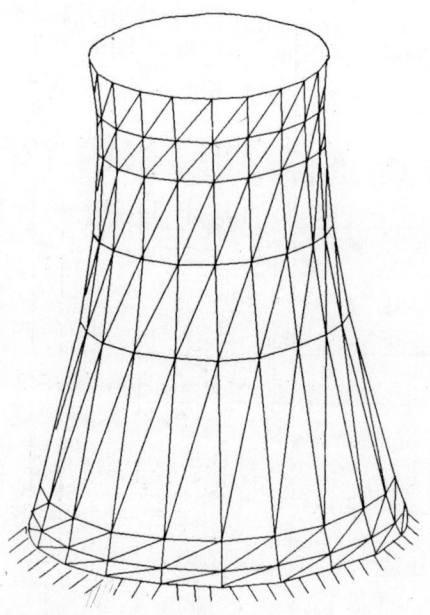

Fig. 1.8 Idealisation of a Cooling tower

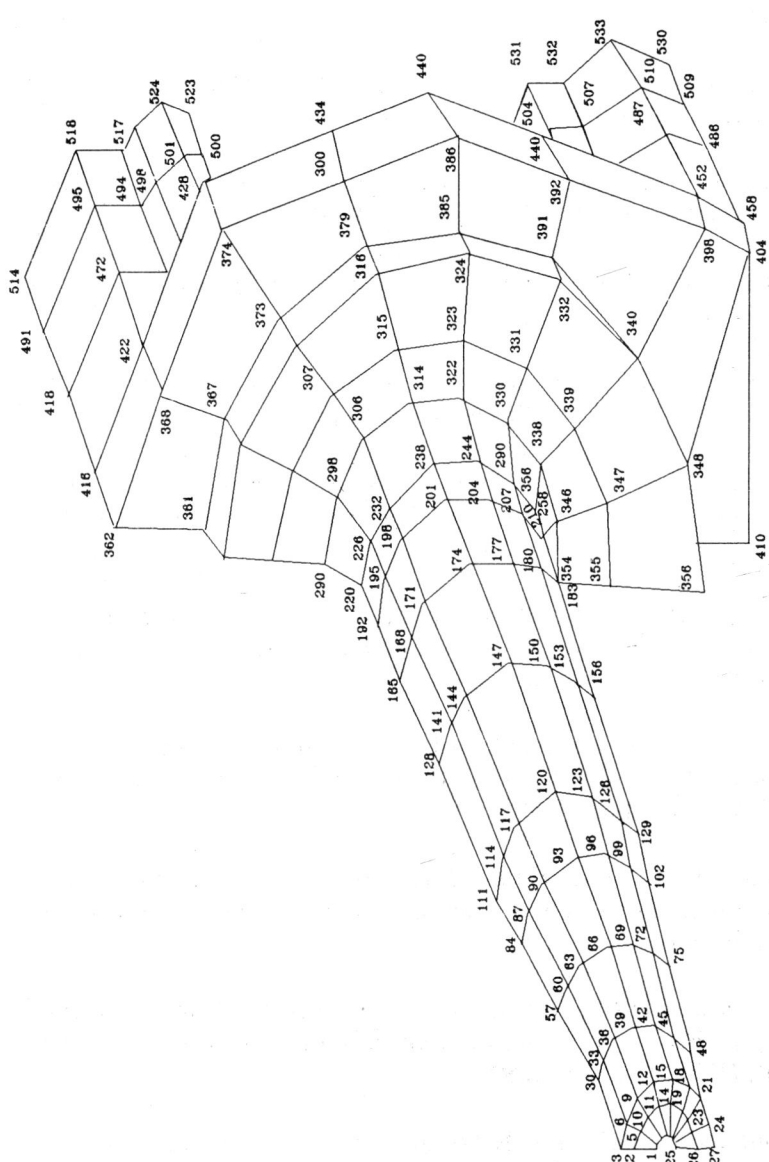

Fig. 1.9   A finite element model for axle arm

REFERENCES

1. Crandall,S.H., *Engineering Analysis*, McGraw-Hill, New York, 1956.

2. Timoshenko,S., *Elements of Strength of Materials*, Parts I and II, Van Nostrand, New Jersey, 1956.

3. Timoshenko,S. and S.Woinowsky-Krieger, *Theory of Plates and Shells*, McGraw-Hill, New York, 1959.

4. Timoshenko,S. and J.N.Goodier, *Theory of Elasticity*, McGraw-Hill, New York, 1970.

5. Dym,C.L. and I.H.Shames, *Solid Mechanics- A Variational Approach*, McGraw-Hill, New York, 1973.

6. Richards,T.H., *Energy Methods in Structural Analysis with an Introduction to Finite Element Techniques*, Ellis Harwood Ltd., Chichester, 1977.

7. Argyris,J.H. and S.Kelsey, *Energy Theorems and Structural Analysis*, Butterworth, London, 1960 (reprinted from Aircraft Engineering, 1954-55).

8. Turner,M.J., R.W.Clough, H.C.Martin and L.J.Topp, *Stiffness and Deflection Analysis of Complex Structures*, J.Aeron.Sci. Vol.23, No.9, pp.805-824, 1956.

9. Clough,R.W., *The Finite Element in Plane Stress Analysis*, Proc. 2nd ASCE Conference on Electronic Computation, Pittsburgh, pa., September 1960.

10. Zienkiewicz,O.C. and R.L.Taylor, *The Finite Element Method Vol I, Basic Formulation and Linear Problems*, McGraw-Hill, (U.K) Limited, 1989.

11. Zienkiewicz, O.C. and K.Morgan, *Finite Elements and Approximation*, John Wiley and Sons, Inc., New York, 1983.

12. Reddy, J.N., *An Introduction to the Finite Element Method*, McGraw Hill Book Co., Singapore, International Student Edition, 1985.

EXERCISES

1.1 A Cantilever beam is shown in Fig. E1.1. It is subjected to an uniformly distributed load, p and, concentrated load P_o and moment M_o at the free end as shown.

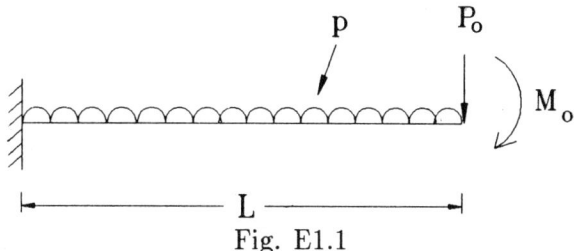

Fig. E1.1

Solve the problem by Rayleigh-Ritz procedure by using

a. algebraic polynomials, and
b. trignometric functions

Try one and two parameter functions and compare the results of deflection and bending moment with the exact solution.

1.2 For the simply supported beam shown in Figure 1.2 in this Chapter, obtain the solution by Rayleigh-Ritz procedure using algebraic functions. Try one and two parameter functions and compare the results with the exact solution.

1.3 A Cantilever beam is shown in Fig. E1.3.

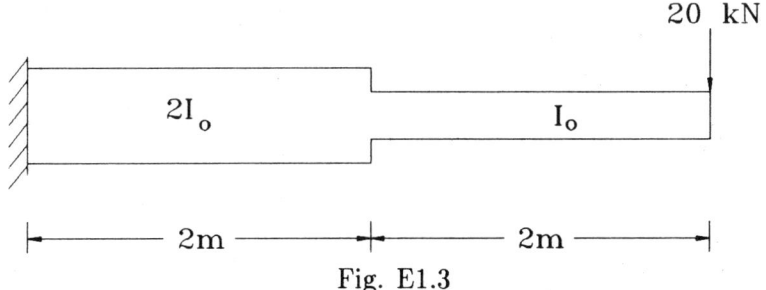

Fig. E1.3

Compute the deflection at the free end by using Rayleigh-Ritz procedure. Compare the results with one, two and three parameter solutions by using,

a. algebraic polynomials, and
b. trignometric functions.

1.4 A bar fixed at one end and free at the other end is loaded as shown in Fig.E1.4. Calculate the displacement and stresses using Rayleigh-Ritz procedure.

Compare the solution with exact results using, one, two and three terms in the polynomials.

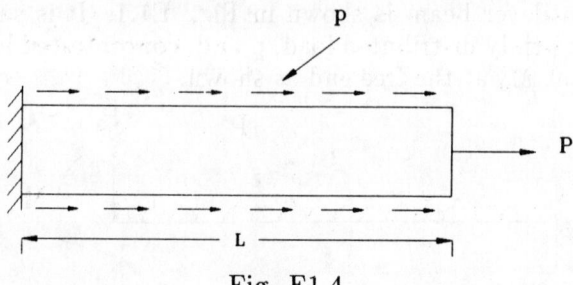

Fig. E1.4

1.5 A fixed ended beam subjected to uniformly distributed load of intensity p is shown in Fig.E1.5. Using the Rayleigh-Ritz procedure, compute the maximum deflection and bending moment and shear force at the fixed ends. Compare the results with analytical solution using one and two parameters in the approximate function.

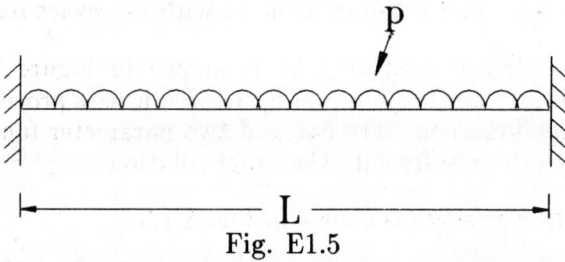

Fig. E1.5

1.6 Solve the problem 1.3 by using finite difference method. Use meshes with 2 and 4 nodes and compare the results.

1.7 Analyse the simply supported beam shown in Fig.E1.7 by finite difference method.

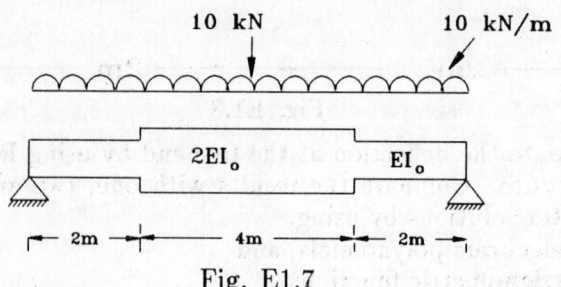

Fig. E1.7

EI_o = 400 units.
Use discretization with three and seven nodes. Compare the results with the exact solution.

1.8 The beam shown in Fig.E1.8 is subjected to a vertical load of 20 kN. Compute the bending moment and shear force at the fixed end by finite difference approach and compare them with the analytical solution.

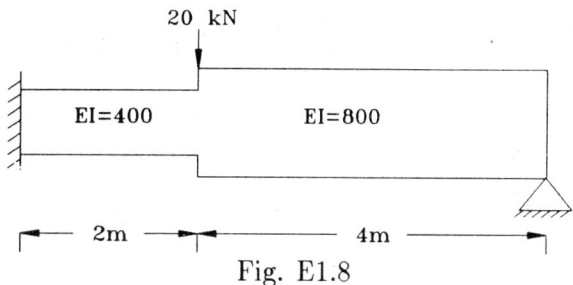

Fig. E1.8

1.9 Solve the problem of a simply supported square plate described in page 9 when it is subjected to a concentrated load P distributed over an area 1m × 1m at the centre. Use finite difference method and compare the results with the analytical solution.

1.10 Solve the plate problem described in page 9, when the edges are fixed. Compare the results with the analytical solution.

Chapter 2

Basic Principles of Structural Mechanics

The study and application of finite element technique requires a sound knowledge of the basic equations of elasticity and a thorough understanding of the principle of virtual work and energy theorems. A number of text books are available on the subject [1,2,3,4]. In this chapter, a brief review is made of those principles of structural mechanics which are essential for the derivation of the finite element properties.

2.1 Equilibrium Conditions

Consider the case of a deformable body which is in equilibrium. The body is subjected to external forces and a set of internal forces developed due to deformation of the solid. The external force may be due to (i) body forces due to weight of the body, centrifugal forces, etc. that act within the boundaries of the body and expressed as force per unit volume, and (ii) surface forces which act on the boundary surface of the body expressed as force per unit area of the surface. Let us examine the conditions for equilibrium at a point within the body.

Figure 2.1 shows a differential element of sides dx, dy and dz within the boundaries of the deformed body and the general state of stress and the variation from point A to A' are also shown therein. The forces acting on the element parallel to the x-axis, are shown in Fig.2.1(b). As $\Sigma F_x = 0$ for equilibrium we get

$$\frac{\partial \sigma_x}{\partial x}\,dxdydz + \frac{\partial \tau_{yx}}{\partial y}dxdydz + \frac{\partial \tau_{zx}}{\partial z}dxdydz + X_b dxdydz = 0$$

where X_b is the component of the body force in the x direction.

Dividing by $dxdydz$ we get the following equilibrium condition

$$\frac{\partial \sigma_x}{\partial x} + \frac{\partial \tau_{yx}}{\partial y} + \frac{\partial \tau_{zx}}{\partial z} + X_b = 0$$

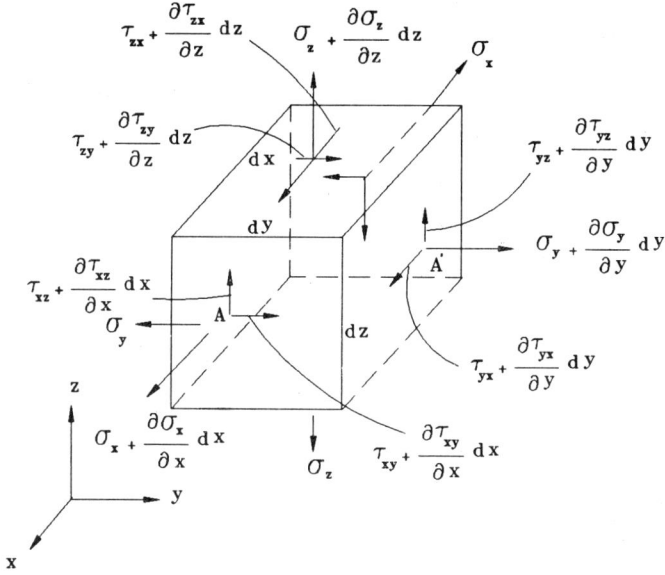

Fig. 2.1a General state of stress on a
differential element

Similarly we can apply the conditions of static equilibrium for forces along the directions of y and z axes, as $\Sigma F_y = 0$ and $\Sigma F_z = 0$. Thus we get the following differential equations of equilibrium in terms of stresses and body forces

$$\frac{\partial \sigma_x}{\partial x} + \frac{\partial \tau_{yx}}{\partial y} + \frac{\partial \tau_{zx}}{\partial z} + X_b = 0$$

$$\frac{\partial \tau_{xy}}{\partial x} + \frac{\partial \sigma_y}{\partial y} + \frac{\partial \tau_{zy}}{\partial z} + Y_b = 0$$

$$\frac{\partial \tau_{xz}}{\partial x} + \frac{\partial \tau_{yz}}{\partial y} + \frac{\partial \sigma_z}{\partial z} + Z_b = 0 \qquad (2.1)$$

where Y_b and Z_b are body force components along y and z directions respectively.

The three remaining equations of statics for moments about the axes x, y and z are $\Sigma M_x = 0, \Sigma M_y = 0$, and $\Sigma M_z = 0$. Now taking moments of forces about the x, y and z axes, we can show that

$$\tau_{xy} = \tau_{yx}, \quad \tau_{xz} = \tau_{zx}, \quad \tau_{yz} = \tau_{zy} \tag{2.2}$$

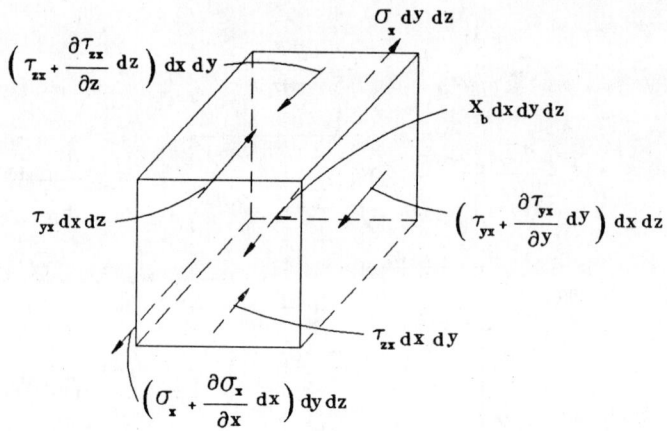

Fig. 2.1b Forces acting on the element in the x direction

Thus, the state of stress at any point can be defined by six components of stress $\sigma_x, \sigma_y, \sigma_z, \tau_{xy}, \tau_{yz}$ and τ_{zx}. It may be represented as

$$\{\sigma\}^T = [\sigma_x \quad \sigma_y \quad \sigma_z \quad \tau_{xy} \quad \tau_{yz} \quad \tau_{zx}] \tag{2.3}$$

Using the Eq.2.2, the equilibrium Eq.2.1 can be rewritten as

$$\frac{\partial \sigma_x}{\partial x} + \frac{\partial \tau_{xy}}{\partial y} + \frac{\partial \tau_{zx}}{\partial z} + X_b = 0$$

$$\frac{\partial \tau_{xy}}{\partial x} + \frac{\partial \sigma_y}{\partial y} + \frac{\partial \tau_{yz}}{\partial z} + Y_b = 0$$

$$\frac{\partial \tau_{zx}}{\partial x} + \frac{\partial \tau_{yz}}{\partial y} + \frac{\partial \sigma_z}{\partial z} + Z_b = 0 \tag{2.1a}$$

Consider an element of area ΔS on the surface (boundary) of a solid in equilibrium and let X_s, Y_s and Z_s be the components of external forces (per unit area) acting on the surface (Fig.2.1 (c)). For equilibrium it can be shown [2, 3] that the following equations must be satisfied by the internal stresses and surface forces

$$\sigma_x l + \tau_{xy} m + \tau_{zx} n = X_s$$

$$\tau_{xy} l + \sigma_y m + \tau_{yz} n = Y_s$$

$$\tau_{zx} l + \tau_{yz} m + \sigma_z n = Z_s \qquad (2.4)$$

where l, m and n are the direction cosines of the normal to the boundary surface. These Eq. 2.4 are often termed as static boundary conditions.

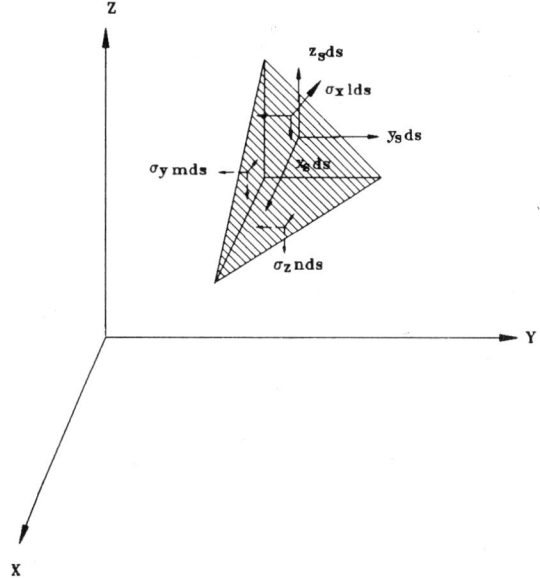

· Fig.2.1 (c) Forces acting on an element on the boundary

2.1.1 Equilibrium Conditions for Two-Dimensional Stress Distribution

The equilibrium conditions for two-dimensional stress distribution can be deduced from Eq.2.1a and 2.4, and are given below.

$$\frac{\partial \sigma_x}{\partial x} + \frac{\partial \tau_{xy}}{\partial y} + X_b = 0$$

$$\frac{\partial \tau_{xy}}{\partial x} + \frac{\partial \sigma_y}{\partial y} + Y_b = 0 \qquad (2.5)$$

and the static boundary conditions are as follows:

$$\sigma_x l + \tau_{xy} m = X_s$$

$$\tau_{xy} l + \sigma_y m = Y_s \qquad (2.6)$$

2.2   Strain-Displacement Relations

The displacement at a point of a deformable body can be described by the components $u, v,$ and w parallel to the cartesian coordinates axes. These components of the displacement can, in general, be represented as functions of x, y and z. The strains in the deformed body can be expressed as partial derivatives of the displacements u, v and w. The strain-displacement relations are given by the following expressions [2].

$$\epsilon_x = \frac{\partial u}{\partial x} + \frac{1}{2}\left[\left(\frac{\partial u}{\partial x}\right)^2 + \left(\frac{\partial v}{\partial x}\right)^2 + \left(\frac{\partial w}{\partial x}\right)^2\right]$$

$$\epsilon_y = \frac{\partial v}{\partial y} + \frac{1}{2}\left[\left(\frac{\partial u}{\partial y}\right)^2 + \left(\frac{\partial v}{\partial y}\right)^2 + \left(\frac{\partial w}{\partial y}\right)^2\right]$$

$$\epsilon_z = \frac{\partial w}{\partial z} + \frac{1}{2}\left[\left(\frac{\partial u}{\partial z}\right)^2 + \left(\frac{\partial v}{\partial z}\right)^2 + \left(\frac{\partial w}{\partial z}\right)^2\right]$$

$$\gamma_{xy} = \frac{\partial v}{\partial x} + \frac{\partial u}{\partial y} + \left[\frac{\partial u}{\partial x}\frac{\partial u}{\partial y} + \frac{\partial v}{\partial x}\frac{\partial v}{\partial y} + \frac{\partial w}{\partial x}\frac{\partial w}{\partial y}\right]$$

$$\gamma_{yz} = \frac{\partial w}{\partial y} + \frac{\partial v}{\partial z} + \left[\frac{\partial u}{\partial y}\frac{\partial u}{\partial z} + \frac{\partial v}{\partial y}\frac{\partial v}{\partial z} + \frac{\partial w}{\partial y}\frac{\partial w}{\partial z}\right]$$

$$\gamma_{zx} = \frac{\partial w}{\partial x} + \frac{\partial u}{\partial z} + \left[\frac{\partial u}{\partial x}\frac{\partial u}{\partial z} + \frac{\partial v}{\partial x}\frac{\partial v}{\partial z} + \frac{\partial w}{\partial x}\frac{\partial w}{\partial z}\right] \qquad (2.7)$$

The components of the strain $\epsilon_x, \epsilon_y, \epsilon_z, \gamma_{xy}, \gamma_{yz}$ and γ_{zx} define the state of strains in the deformed body, and can be written as

$$\{\epsilon\}^T = [\epsilon_x \quad \epsilon_y \quad \epsilon_z \quad \gamma_{xy} \quad \gamma_{yz} \quad \gamma_{zx}] \qquad (2.8)$$

The relations given in Eq.2.7 are non-linear partial differential equations in the unknown components of displacement. However, in many problems of stress analysis when the deformations are small, we can assusme that the products and squares of the first derivatives are negligible compared with the derivatives themselves. By this assumption we considerably reduce the complexity in the analysis of deformable solids and the strain-displacement relations in Eq.2.7 reduce to linear relations as given below.

$$\epsilon_x = \frac{\partial u}{\partial x} \qquad \epsilon_y = \frac{\partial v}{\partial y} \qquad \epsilon_z = \frac{\partial w}{\partial z}$$

$$\gamma_{xy} = \frac{\partial v}{\partial x} + \frac{\partial u}{\partial y} \qquad \gamma_{yz} = \frac{\partial w}{\partial y} + \frac{\partial v}{\partial z} \qquad \gamma_{zx} = \frac{\partial u}{\partial z} + \frac{\partial w}{\partial x} \qquad (2.9)$$

The above equation can be written in a matrix form as

$$\begin{Bmatrix} \epsilon_x \\ \epsilon_y \\ \epsilon_z \\ \gamma_{xy} \\ \gamma_{yz} \\ \gamma_{zx} \end{Bmatrix} = \begin{bmatrix} \frac{\partial}{\partial x} & 0 & 0 \\ 0 & \frac{\partial}{\partial y} & 0 \\ 0 & 0 & \frac{\partial}{\partial z} \\ \frac{\partial}{\partial y} & \frac{\partial}{\partial x} & 0 \\ 0 & \frac{\partial}{\partial z} & \frac{\partial}{\partial y} \\ \frac{\partial}{\partial z} & 0 & \frac{\partial}{\partial x} \end{bmatrix} \begin{Bmatrix} u \\ v \\ w \end{Bmatrix} \qquad (2.10)$$

If the deformation is not small, the above assumption cannot be made and the resulting problem has to be considered as geometrically nonlinear.

In section 2.1 the equilibrium conditions are expressed as differential equations and are to be satisfied by the stress components. Similarly, it must be noted that the displacements u, v and w and the strain components must also satisfy certain conditions. First of all, the displacement components must be consistent with the kinematic boundary conditions of the solid or the structure. Then the conditions to be satisfied by the strain components are briefly discussed below.

If Eqs.2.9 are examined, we would notice that if the displacement components are known as continuous functions of x, y and z (satisfying also the kinematic boundary conditions) the strain components can be easily evaluated. However, if strain components are specified then we have six equations to determine three displacement components, viz. u, v and w. It, therefore, follows that the u, v and w cannot be arbitrarily determined by specifying the strain components. Hence, additional conditions are to be imposed on the strain components in order to establish unique values of u, v and w ensuring continuity, throughout the body during deformation. These conditions are known as compatibility conditions. These conditions can be obtained by eliminating u, v and w by appropriate differentiation of Eqs. 2.9. The six compatibility conditions are given below and a more detailed derivation is given in references 2 and 3.

$$\frac{\partial^2 \epsilon_x}{\partial y^2} + \frac{\partial^2 \epsilon_y}{\partial x^2} = \frac{\partial^2 \gamma_{xy}}{\partial x \partial y}$$

$$\frac{\partial^2 \epsilon_y}{\partial z^2} + \frac{\partial^2 \epsilon_z}{\partial y^2} = \frac{\partial^2 \gamma_{yz}}{\partial y \partial z}$$

$$\frac{\partial^2 \epsilon_z}{\partial x^2} + \frac{\partial^2 \epsilon_x}{\partial z^2} = \frac{\partial^2 \gamma_{xz}}{\partial x \partial z}$$

$$2\frac{\partial^2 \epsilon_x}{\partial y \partial z} = \frac{\partial}{\partial x}\left(-\frac{\partial \gamma_{yz}}{\partial x} + \frac{\partial \gamma_{xz}}{\partial y} + \frac{\partial \gamma_{xy}}{\partial z}\right)$$

$$2\frac{\partial^2 \epsilon_y}{\partial z \partial x} = \frac{\partial}{\partial y}\left(\frac{\partial \gamma_{yz}}{\partial x} - \frac{\partial \gamma_{xz}}{\partial y} + \frac{\partial \gamma_{xy}}{\partial z}\right)$$

$$2\frac{\partial^2 \epsilon_z}{\partial x \partial y} = \frac{\partial}{\partial z}\left(\frac{\partial \gamma_{yz}}{\partial x} + \frac{\partial \gamma_{xz}}{\partial y} - \frac{\partial \gamma_{xy}}{\partial z}\right) \qquad (2.11)$$

2.3   Linear Constitutive Relations

To determine the stresses in the members of a structure or in a deformable solid, it is necessary to know the components of stress as functions of the components of strain and vice versa. We assume that the structures or bodies under consideration are made of elastic material that obeys Hooke's law. In its general form, Hooke's law states that the six components of stress may be expressed as linear functions of the six components of strain. Thus for a linear elastic, anisotropic and homogeneous material the relation is given below.

$$\begin{Bmatrix} \sigma_x \\ \sigma_y \\ \sigma_z \\ \tau_{xy} \\ \tau_{yz} \\ \tau_{zx} \end{Bmatrix} = \begin{bmatrix} C_{11} & C_{12} & \dots & C_{16} \\ C_{21} & C_{22} & \dots & C_{26} \\ \dots & \dots & \dots & \\ \dots & \dots & \dots & \\ \dots & \dots & \dots & \\ C_{61} & C_{62} & \dots & C_{66} \end{bmatrix} \begin{Bmatrix} \epsilon_x \\ \epsilon_y \\ \epsilon_z \\ \gamma_{xy} \\ \gamma_{yz} \\ \gamma_{zx} \end{Bmatrix}$$

or

$$\{\sigma\} = [C]\{\epsilon\} \qquad (2.12a)$$

where [C] is called the constitutive matrix. And the inverse relation for strains and stresses can be expressed as

$$\begin{Bmatrix} \epsilon_x \\ \epsilon_y \\ \epsilon_z \\ \gamma_{xy} \\ \gamma_{yz} \\ \gamma_{zx} \end{Bmatrix} = \begin{bmatrix} d_{11} & d_{12} & \dots & \dots & d_{16} \\ d_{21} & d_{22} & \dots & \dots & d_{26} \\ \dots & \dots & \dots & \dots & \\ \dots & \dots & \dots & \dots & \\ \dots & \dots & \dots & \dots & \\ d_{61} & d_{62} & \dots & \dots & d_{66} \end{bmatrix} \begin{Bmatrix} \sigma_x \\ \sigma_y \\ \sigma_z \\ \tau_{xy} \\ \tau_{yz} \\ \tau_{zx} \end{Bmatrix}$$

$$\text{or} \qquad \{\epsilon\} = [D]\{\sigma\} \tag{2.12b}$$

The matrices [C] and [D] are symmetric and it can be seen that 21 elastic constants are required to describe the constitutive relation for a general anisotropic solid.

If the material has three orthogonal planes of symmetry it is said to be orthotropic and only nine constants are required to describe the constitutive relations given below.

$$\begin{Bmatrix} \sigma_x \\ \sigma_y \\ \sigma_z \\ \gamma_{xy} \\ \gamma_{yz} \\ \gamma_{zx} \end{Bmatrix} = \begin{bmatrix} C_{11} & C_{12} & C_{13} & 0 & 0 & 0 \\ & C_{22} & C_{23} & 0 & 0 & 0 \\ & & C_{33} & 0 & 0 & 0 \\ & sym. & & C_{44} & 0 & 0 \\ & & & & C_{55} & 0 \\ & & & & & C_{66} \end{bmatrix} \begin{Bmatrix} \epsilon_x \\ \epsilon_y \\ \epsilon_z \\ \gamma_{xy} \\ \gamma_{yz} \\ \gamma_{zx} \end{Bmatrix} \tag{2.12c}$$

where $x - y, y - z$ and $z - x$ are material symmetry planes.

2.3.1 Initial Stress and Strain, Temperature Effects

In some situations, there may be initial residual stress $\{\sigma_o\}$ present in the body before deformations take place. This 'initial stress' $\{\sigma_o\}$ must be superposed to the stress caused by deformation.

The body may also be subjected to 'initial strains', $\{\epsilon_o\}$, which may be due to temperature changes, shrinkage, crystal growth and so on. The stresses will be caused by the difference between the actual and initial strains.

Thus, in the presence of initial stresses, $\{\sigma_o\}$, and 'initial strains' $\{\epsilon_o\}$, the Eq. 2.12a becomes,

$$\{\sigma\} = [C] \{\{\epsilon\} - \{\epsilon_o\}\} + \{\sigma_o\} \tag{2.13}$$

As indicated above, the temperature effects can be treated as initial strains $\{\epsilon_o\}$ caused in the body. For example, if T is the change in temperature the initial strains produced in an orthotropic material are,

$$\{\epsilon_o\}^T = [\alpha_x T \quad \alpha_y T \quad \alpha_z T \quad 0 \quad 0 \quad 0] \tag{2.14}$$

where α's are the coefficients of thermal expansion in the principal directions. No shear strains are caused by thermal expansion. Then, to calculate the effects due to temperature changes, Eq. 2.14 is substituted in Eq. 2.13 and setting $\{\sigma_o\} = \{0\}$.

2.3.2   Linear Elastic Isotropic Material

An isotropic material is one for which every plane is a plane of symmetry of material behaviour and only two constants (Youngs Modulus, E and Poisson ratio μ) are required to describe the constitutive relation. The following equation includes the effect due to temperature changes as may be necessary in certain cases of stress analysis.

$$
\begin{Bmatrix} \epsilon_x \\ \epsilon_y \\ \epsilon_z \\ \gamma_{xy} \\ \gamma_{yz} \\ \gamma_{zx} \end{Bmatrix} = \frac{1}{E}
\begin{bmatrix}
1 & -\mu & -\mu & 0 & 0 & 0 \\
 & 1 & -\mu & 0 & 0 & 0 \\
 & & 1 & 0 & 0 & 0 \\
 & Sym. & & 2(1+\mu) & 0 & 0 \\
 & & & & 2(1+\mu) & 0 \\
 & & & & & 2(1+\mu)
\end{bmatrix}
\begin{Bmatrix} \sigma_x \\ \sigma_y \\ \sigma_z \\ \tau_{xy} \\ \tau_{yz} \\ \tau_{zx} \end{Bmatrix}
$$

$$
+\alpha T \begin{Bmatrix} 1 \\ 1 \\ 1 \\ 0 \\ 0 \\ 0 \end{Bmatrix} \tag{2.15}
$$

The inverse relation of stresses in terms of strain components can be expressed as

$$
\begin{Bmatrix} \sigma_x \\ \sigma_y \\ \sigma_z \\ \tau_{xy} \\ \tau_{yz} \\ \tau_{zx} \end{Bmatrix} = \overline{E}
\begin{bmatrix}
(1-\mu) & \mu & \mu & 0 & 0 & 0 \\
 & (1-\mu) & \mu & 0 & 0 & 0 \\
 & & (1-\mu) & 0 & 0 & 0 \\
 & Sym. & & \frac{1-2\mu}{2} & 0 & 0 \\
 & & & & \frac{1-2\mu}{2} & 0 \\
 & & & & & \frac{1-2\mu}{2}
\end{bmatrix} \times
$$

$$
\times \begin{Bmatrix} \epsilon_x \\ \epsilon_y \\ \epsilon_z \\ \gamma_{xy} \\ \gamma_{yz} \\ \gamma_{zx} \end{Bmatrix} - \frac{E\alpha T}{1-2\mu} \begin{Bmatrix} 1 \\ 1 \\ 1 \\ 0 \\ 0 \\ 0 \end{Bmatrix} \tag{2.16}
$$

where

$$\overline{E} \;\; = \;\; \frac{E}{(1+\mu)\,(1-2\mu)}$$

α = coefficient of thermal expansion

T = change in temperature

2.3.3   Two-Dimensional Stress Distributions

The problems of solid mechanics can be formulated as three- dimensional problems and finite element technique can be used to solve them. However, such an approach may prove prohibitively costly. In many practical situations, the geometry and loading will be such that the problems can be reduced to two or one-dimensional problems without much loss of accuracy. The two-dimensional idealizations in stress analysis are described below.

Plane Stress: The plane stress condition is characterised by very small dimensions in one of the normal directions. Fig.2.2(a) shows an example of plane stress conditions. In these cases the stress components σ_z, τ_{xz} and τ_{yz} are zero and it is assumed that no stress component varies across the thickness. The state of stress is then specified by σ_x, σ_y and τ_{xy} only (functions of x and y) and is called plane stress.

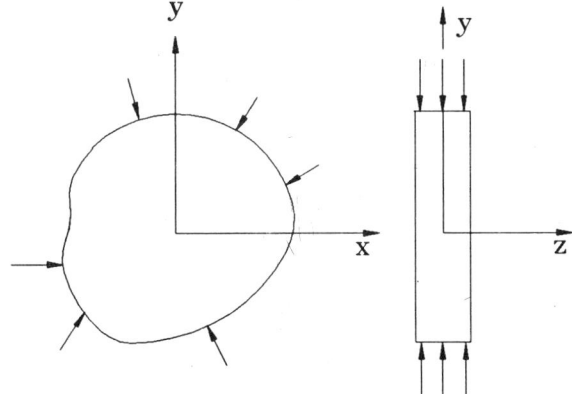

Fig.2.2 (a) Plane stress example: Thin plate
with in plane loading

The constitutive relation is given by

$$\begin{Bmatrix} \sigma_x \\ \sigma_y \\ \tau_{xy} \end{Bmatrix} = \frac{E}{1-\mu^2} \begin{bmatrix} 1 & \mu & 0 \\ \mu & 1 & 0 \\ 0 & 0 & \frac{1-\mu}{2} \end{bmatrix} \begin{Bmatrix} \epsilon_x \\ \epsilon_y \\ \gamma_{xy} \end{Bmatrix} - \frac{E\alpha T}{1-\mu} \begin{Bmatrix} 1 \\ 1 \\ 0 \end{Bmatrix} \qquad (2.17)$$

Also the strain components can be expressed in terms of the stress and are given below:

$$\begin{Bmatrix} \epsilon_x \\ \epsilon_y \\ \gamma_{xy} \end{Bmatrix} = \frac{1}{E} \begin{bmatrix} 1 & -\mu & 0 \\ -\mu & 1 & 0 \\ 0 & 0 & 2(1+\mu) \end{bmatrix} \begin{Bmatrix} \sigma_x \\ \sigma_y \\ \tau_{xy} \end{Bmatrix} + \alpha T \begin{Bmatrix} 1 \\ 1 \\ 0 \end{Bmatrix} \qquad (2.18)$$

Also $\qquad \epsilon_z = \dfrac{-\mu}{1-\mu}(\epsilon_x + \epsilon_y) + \dfrac{1+\mu}{1-\mu}\alpha T$

and $\qquad \gamma_{yz} = \gamma_{zx} = 0 \qquad\qquad\qquad (2.19)$

Plane Strain: Problems involving a long body whose geometry and loading do not vary significantly in the longitudinal direction are referred to as plane strain problems. Some typical examples are given in Fig.2.2(b). In these situations, a constant longitudinal displacement corresponding to a rigid body translation and displacements linear in z corresponding to rigid body rotation do not result in strain. Hence, if we consider a cross section away from the ends, it can be assumed that $w = 0$ and displacements u and v are functions of x and y but are independent of z. It follows that,

$$\epsilon_z = \gamma_{zx} = \gamma_{yz} = 0 \qquad\qquad\qquad (2.20)$$

The constitutive relation for elastic istropic material is given by,

$$\begin{Bmatrix} \sigma_x \\ \sigma_y \\ \tau_{xy} \end{Bmatrix} = \frac{E}{(1+\mu)(1-2\mu)} \begin{bmatrix} (1-\mu) & \mu & 0 \\ \mu & (1-\mu) & 0 \\ 0 & 0 & \frac{(1-2\mu)}{2} \end{bmatrix} \begin{Bmatrix} \epsilon_x \\ \epsilon_y \\ \gamma_{xy} \end{Bmatrix}$$

$$- \frac{E\alpha T}{(1-2\mu)} \begin{Bmatrix} 1 \\ 1 \\ 0 \end{Bmatrix} \qquad\qquad (2.21)$$

It can also be shown that

$$\sigma_z = \mu(\sigma_x + \sigma_y) - E\alpha T \qquad\qquad\qquad (2.22a)$$

$$\tau_{yz} = \tau_{zx} = 0 \qquad\qquad\qquad (2.22b)$$

Solving Eqs. 2.21 for the strains ϵ_x, ϵ_y and γ_{xy} gives

$$\left\{ \begin{array}{c} \epsilon_x \\ \epsilon_y \\ \gamma_{xy} \end{array} \right\} = \frac{(1+\mu)}{E} \begin{bmatrix} (1-\mu) & -\mu & 0 \\ -\mu & 1-\mu & 0 \\ 0 & 0 & 2 \end{bmatrix} \left\{ \begin{array}{c} \sigma_x \\ \sigma_y \\ \tau_{xy} \end{array} \right\} + (1+\mu)\alpha T \left\{ \begin{array}{c} 1 \\ 1 \\ 0 \end{array} \right\}$$

$$(2.23)$$

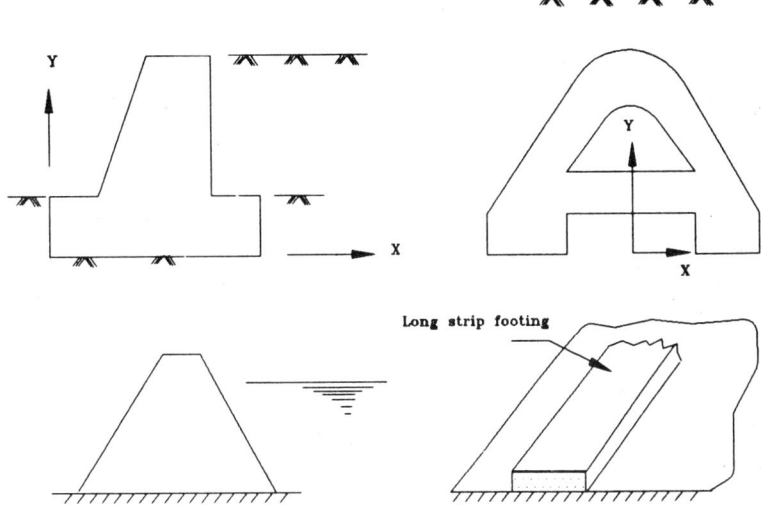

Fig.2.2 (b) Plane strain examples

Axisymmetric Stress Distribution: Many problems in stress analysis which are of practical interest involve solids of revolution (axisymmetric solids) subject to axially symmetric loading. The simplest examples are a circular cylinder loaded by a uniform internal or external pressure, a circular footing resting on a soil mass (Fig.2.2(c)), pressure vessels, rotating wheels, flywheels etc. The deformation being symmetrical with respect to the y-axis, the stress components are independent of the angle θ and all derivatives with respect to θ vanish. The components w, $\gamma_{x\theta}$, $\gamma_{\theta y}$ $\tau_{x\theta}$ and $\tau_{\theta y}$ are zero. The strain-displacement relations are given by [3],

$$\epsilon_x = \frac{\partial u}{\partial x} \quad \epsilon_\theta = \frac{u}{x} \quad \epsilon_y = \frac{\partial v}{\partial y} \quad \gamma_{xy} = \frac{\partial u}{\partial y} + \frac{\partial v}{\partial x} \qquad (2.24)$$

Thus, the two components of displacements in any plane section of the body along its axis of symmetry define completely the state of strain and therefore, the state of stress. The constitutive relation is

$$\left\{\begin{array}{c} \sigma_x \\ \sigma_y \\ \sigma_\theta \\ \tau_{xy} \end{array}\right\} = \frac{E}{(1+\mu)(1-2\mu)} \begin{bmatrix} (1-\mu) & \mu & \mu & 0 \\ \mu & 1-\mu & \mu & 0 \\ \mu & \mu & 1-\mu & 0 \\ 0 & 0 & 0 & \frac{1-2\mu}{2} \end{bmatrix} \left\{\begin{array}{c} \epsilon_x \\ \epsilon_y \\ \epsilon_\theta \\ \gamma_{xy} \end{array}\right\}$$

$$(2.25)$$

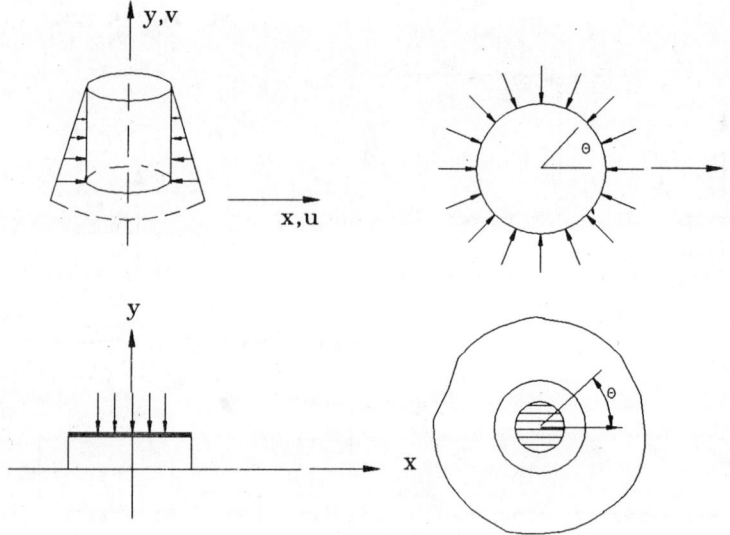

Fig. 2.2(c) Axisymmetric Problems

2.4   Principle of Virtual Work

A force P is said to do work if it gives rise to a displacement r. Work is the product of the force and the corresponding displacement, ie. $W = Pr$. When the force P and displacement r are unrelated by the cause-effect relation, the work $W = Pr$ is called the virtual work. Hence, the virtual work may be caused by true forces moving through imaginary displacements or vice versa. Accordingly the principle of virtual work can be divided into two parts: (i) the principle of virtual displacements and (ii) the principle of virtual forces.

The principle of virtual displacements is based on the virtual work done by the true forces moving through the virtual displacements. It will be shown later that this principle establishes the conditions of equilibrium and is used in the 'displacement models' of the finite element method. On the other hand, the principle of virtual forces uses the virtual work done by the virtual forces in moving through the true displacements and establishes the compatibility conditions, and is used in the 'equilibrium models'.

In the following discussion, attention will be confined only to the principle of virtual displacements which will be used in the derivation of element properties in Chap. 3. However, for a detailed discussion on the subject, readers may refer to excellent books on this topic [1,2].

2.4.1 Principle of Virtual Displacements

In the case of a three-dimensional solid or a structure, the displacement at any point can be expressed by its components u, v and w parallel to the cartesian coordinate axes x, y and z. The displacements u, v and w are continuous functions of x, y and z. Any virtual displacement will also be continuous functions and in addition that they should satisfy the kinematic boundary conditions.

Consider a two-dimensional body shown in Fig.2.2(a). The equilibrium conditions to be satisfied by the stresses within the body and on the boundary are given in Eqs.2.5 and 2.6 and are as follows:

$$\frac{\partial \sigma_x}{\partial x} + \frac{\partial \tau_{xy}}{\partial y} + X_b = 0$$

$$\frac{\partial \tau_{xy}}{\partial x} + \frac{\partial \sigma_y}{\partial y} + Y_b = 0 \tag{2.5}$$

and

$$\sigma_x l + \tau_{xy} m = X_s$$

$$\tau_{xy} l + \sigma_y m = Y_s \tag{2.6}$$

Multiply the equilibrium Eqs 2.5 by δu and δv respectively and integrate over the area of the solid.

$$\int \int \left[\left(\frac{\partial \sigma_x}{\partial x} + \frac{\partial \tau_{xy}}{\partial y} + X_b \right) \delta u + \left(\frac{\partial \tau_{xy}}{\partial x} + \frac{\partial \sigma_y}{\partial y} + Y_b \right) \delta v \right] dx dy = 0 \tag{2.26}$$

In order to give a physical interpretation to the equation we can consider δu and δv as virtual displacements and also expand each one of the terms by using the Green's theorem in two-dimensions. If $\phi(x, y)$ and $\psi(x, y)$ are continuous functions with their first and second partial derivatives are also continuous, then according to Green's theorem,

$$\int\int\left(\frac{\partial\phi}{\partial x}\cdot\frac{\partial\psi}{\partial x}+\frac{\partial\phi}{\partial y}\cdot\frac{\partial\psi}{\partial y}\right)dxdy$$

$$=-\int\int\phi\left(\frac{\partial^2\psi}{\partial x^2}+\frac{\partial^2\psi}{\partial y^2}\right)dxdy$$

$$+\int\phi\left(\frac{\partial\psi}{\partial x}l+\frac{\partial\psi}{\partial y}m\right)dS \qquad (2.27)$$

Consider the first term of the integral in Eq.2.26 Let $\phi=\sigma_x$ and assume ψ such that

$$\frac{\partial\psi}{\partial x}=\delta u \quad \text{and} \quad \frac{\partial\psi}{\partial y}=0$$

Then

$$\int\int\frac{\partial\sigma_x}{\partial x}\delta u\;dxdy=-\int\int\sigma_x\frac{\partial\delta u}{\partial x}\;dxdy+\int\sigma_x l\delta u\;dS \qquad (2.28)$$

Similarly it can be shown that

$$\int\int\frac{\partial\sigma_y}{\partial y}\delta v\;dxdy=-\int\int\sigma_y\frac{\partial\delta v}{\partial y}dxdy+\int\sigma_y\;m\;\delta v\;dS \qquad (2.29)$$

and

$$\int\int\left(\frac{\partial\tau_{xy}}{\partial x}\delta v+\frac{\partial\tau_{xy}}{\partial y}\delta u\right)dxdy$$

$$=-\int\int\tau_{xy}\left(\frac{\partial\delta v}{\partial x}+\frac{\partial\delta u}{\partial y}\right)dxdy$$

$$+\int(\tau_{xy}m\delta u+\tau_{xy}l\delta v)dS \qquad (2.30)$$

Thus Eq.2.26 is transformed to

$$-\int\int\left[\sigma_x\frac{\partial\delta u}{\partial x}+\sigma_y\frac{\partial\delta v}{\partial y}+\tau_{xy}\left(\frac{\partial\delta v}{\partial x}+\frac{\partial\delta u}{\partial y}\right)\right]dxdy$$

$$+\int\int(X_b\delta u+Y_b\delta v)dxdy+\int[(\sigma_x l+\tau_{xy}m)\delta u$$

$$+(\tau_{xy}l+\sigma_y m)\delta v]dS=0 \qquad (2.31)$$

Each integral in the above equation has physical significance and is explained below.

The second integral on the left side of Eq.2.31 represents the virtual work done by the body forces due to virtual displacements δu and δv. The line integral can be simplified using Eq.2.6 as

$$\int [(\sigma_x l + \tau_{xy} m)\delta u + (\tau_{xy} l + \sigma_y m)\delta v]dS$$

$$= \int (X_s \delta u + Y_s \delta v)dS \qquad (2.32)$$

Thus, the total external virtual work done δW_e is given by

$$\delta W_e = \int\int (X_b \delta u + Y_b \delta v)dxdy$$

$$+ \int (X_s \delta u + Y_s \delta v)dS \qquad (2.33)$$

To bring out the physical significance of the first integral on the left-hand side of Eq.2.31, consider an element dxdy of unit thickness subject to a stress σ_x, as shown in Fig.2.3.

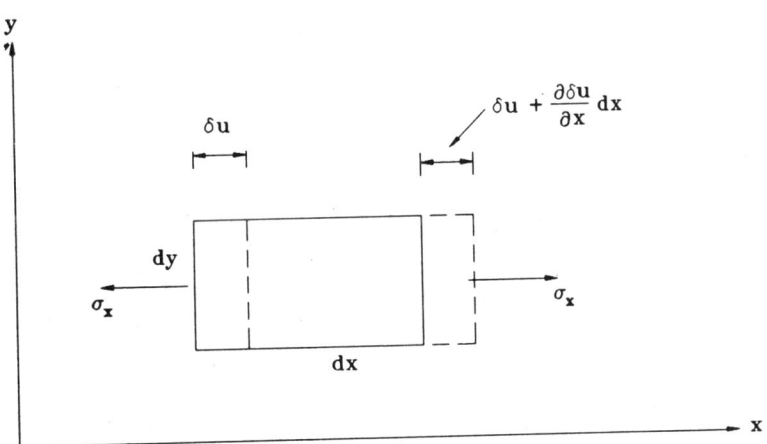

Fig. 2.3 Differential element subject to stress

Let a virtual displacement δu be given to the element and the displaced position is shown in Fig.2.3. This results in an internal virtual work done by the stress σ_x as

$$\text{internal virtual work} = \sigma_x dy\left(\delta u + \frac{\partial \delta u}{\partial x}dx\right) - \sigma_y\, dy\, \delta u$$

$$= \sigma_x \frac{\partial \delta u}{\partial x}\, dxdy \qquad (2.34)$$

Extending this concept it can thus be seen that the first integral of Eq.2.31 represents the total internal virtual work done by the stress system σ_x, σ_y and τ_{xy}.

It may be observed that due to the virtual displacement δu there will be change of strain $\delta \epsilon_x$

$$\delta \epsilon_x = \frac{\left(\delta u + \frac{\partial \delta u}{\partial x} dx\right) - \delta u}{dx}$$

i.e.

$$\partial \epsilon_x = \frac{\partial \delta u}{\partial x} \tag{2.35}$$

Substituting in Eq.2.34, the internal virtual work done by the stress σ_x can be expressed as

$$\text{internal virtual work} = \sigma_x \, \delta \epsilon_x \, dx \, dy \tag{2.36}$$

Thus we can write the total internal virtual work given by the first integral of Eq.2.31 as

$$\delta U = \int \int (\sigma_x \delta \epsilon_x + \sigma_y \delta \epsilon_y + \tau_{xy} \delta \gamma_{xy}) \, dx \, dy \tag{2.37}$$

The above derivations can be extended to a three-dimensional stress system and the total external and internal virtual work can be expressed by the following equations:

$$\delta W_e = \int \int \int_V (X_b \delta u + Y_b \delta v + Z_b \delta w) \, dV$$

$$+ \int \int_{S1} (X_s \delta u + Y_s \delta v + Z_s \delta w) \, dS \tag{2.38}$$

where S_1 is the portion of the surface on which forces are prescribed. The above equation can be expressed concisely using matrix notation as,

$$\delta W_e = \int \int \int_V \{\delta u\}^T \{X\} \, dV + \int \int_{S1} \{\delta u\}^T \{p\} \, dS \tag{2.38a}$$

where

$$\{\delta u\}^T = [\delta u \quad \delta v \quad \delta w]$$

$$\{X\}^T = [X_b \quad Y_b \quad Z_b] \quad \text{and} \quad \{p\}^T = [X_s \quad Y_s \quad Z_s] \tag{2.38b}$$

and

$$\delta U = \int \int \int_V (\sigma_x \delta \epsilon_x + \sigma_y \delta \epsilon_y + \sigma_z \delta \epsilon_z + \tau_{xy} \delta \gamma_{xy} + \tau_{yz} \delta \gamma_{yz} + \tau_{zx} \delta \gamma_{zx}) dV$$

$$(2.39)$$

or in concise matrix form,

$$\delta U = \int \int \int_V \{\delta \epsilon\}^T \{\sigma\} dV \qquad (2.39a)$$

With the above explanation, the Eq.2.31 becomes

$$-\delta U + \delta W_e = 0 \qquad (2.40)$$

i.e.,

$$\delta W_e = \delta U \qquad (2.41)$$

Thus, it can be said from Eq.2.41 that for equilibrium to exist the total external virtual work is equal to the total internal virtual work. This mathematical statement is a necessary condition for equilibrium. Again if we apply Green's theorem to the Eq.2.40 in the opposite direction, it can be shown that we can get following equation,

$$\int \int \left\{ \left(\frac{\partial \sigma_x}{\partial x} + \frac{\partial \tau_{xy}}{\partial y} + X_b \right) \delta u + \left(\frac{\partial \sigma_y}{\partial y} + \frac{\partial \tau_{xy}}{\partial x} + Y_b \right) \delta v \right\} dx dy$$

$$= \int \{ (\sigma_x l + \tau_{xy} m - X_s) \delta u + (\tau_{xy} l + \sigma_y m - Y_s) \delta v \} dS \qquad (2.42)$$

For any value of virtual displacement δu and δv, the above relation can be true only if each of the terms in the parenthesis vanishes and this once again leads to the Eqs. 2.5 and 2.6 of equilibrium.

We, therefore, conclude the $\delta W_e = \delta U$ is a necessary and sufficient condition of equilibrium.

Thus, the principle of virtual displacement can be stated as that a deformable system is in equilibrium if the total external virtual work is equal to the total internal virtual work for every virtual displacement satisfying the kinematic boundary conditions.

It may be noted here that the principle of virtual displacement is an equilibrium requirement and is independent of the material behaviour.

2.4.2 The Unit Dummy Displacement Method

Consider a deformable body or structural system under a stress condition $\sigma_x, \sigma_y, \sigma_z, \tau_{xy}, \tau_{yz}, \tau_{zx}$. Let a virtual displacement δr be given at the point and in the direction of the external load P, while keeping the position of all other loads fixed, ie., zero virtual displacement to all the other external loads. The total external virtual work is $P\delta r$.

This virtual displacement δr, causes virtual strains

$$\delta\epsilon_x^r \quad \delta\epsilon_y^r \quad \delta\epsilon_z^r \quad \delta\gamma_{xy}^r \quad \delta\gamma_{yz}^r \quad \delta\gamma_{zx}^r$$

From the principle of virtual displacement we get,

$$P\delta r = \int\int\int_V (\sigma_x\delta\epsilon_x^r + \sigma_y\delta\epsilon_y^r + \sigma_z\delta\epsilon_z^r + \tau_{xy}\delta\gamma_{xy}^r + \tau_{yz}\delta\gamma_{yz}^r + \tau_{zx}\delta\gamma_{zx}^r)\ dV$$

$$(2.43)$$

Assuming δr is unity, we get,

$$P = \int\int\int_V (\sigma_x\delta\epsilon_x' + \sigma_y\delta\epsilon_y' + \ldots + \tau_{zx}\delta\gamma_{zx}')dV \qquad (2.44)$$

where $\delta\epsilon_x', \delta\epsilon_y' \ldots \delta\gamma_{zx}'$ are the virtual strains due to unit virtual displacement at the position and in the direction of P or in compact form it can be expressed as

$$P = \int\int\int_V \{\delta\epsilon'\}^T\{\sigma\}dV \qquad (2.44a)$$

It may be noted here that $\{\sigma\}$ is the vector containing the components of true (actual) stress and $\{\delta\epsilon'\}$ is the vector of virtual strains due to unit virtual displacement.

Physically P can be interpreted as a measure of the stiffness of the structure i.e. P is the force developed at some point i on the deformable system due to unit displacement at i in the direction of P. The use of the principle of virtual displacement which forms the basis for the displacement method of analysis is illustrated by the following example. Figure 2.4 shows a three bar truss subjected to a vertical load of 50 kN.

The components of the actual displacement of node 1 are u and v as shown in Fig.2.4a. Due to this deformation the elongation and strains ϵ_1, ϵ_2 and ϵ_3 of the members can be determined for this simple truss structure and are given below.

$$\epsilon_1 = \frac{u\cos 45^\circ + v\sin 45^\circ}{2\sqrt{2}} = \frac{(u+v)}{4}$$

$$\epsilon_2 = \frac{u}{2}$$

$$\epsilon_3 = \frac{u\cos 45^\circ - v\sin 45^\circ}{2\sqrt{2}} = \frac{u-v}{4}$$

(a)

Let a unit virtual displacement along the u direction be given. The virtual strain in the members can be determined from Eq.(a).

$$\delta\epsilon_1' = 1/4$$
$$\delta\epsilon_2' = 1/2$$
$$\delta\epsilon_3' = 1/4 \qquad\qquad (b)$$

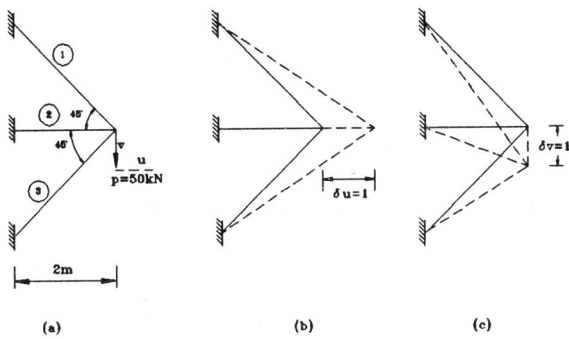

(a) (b) (c)

Fig. 2.4 Three bar truss

The internal virtual work done in a member

$$= \int\int\int_V \sigma_x \delta\epsilon' dV = EAL\epsilon\delta\epsilon'$$

Using equations (a) and (b) the total internal virtual work due to $\delta u = 1$ is given below

$$\delta U = EA \cdot 2\sqrt{2}\frac{(u+v)}{4}\frac{1}{4} + EA2\frac{u}{2}\frac{1}{2} + EA2\sqrt{2}\frac{(u-v)}{4}\frac{1}{4}$$

$$= EAu\left\{\frac{\sqrt{2}}{8} + \frac{1}{2} + \frac{\sqrt{2}}{8}\right\} + EAu\left\{\frac{\sqrt{2}}{8} - \frac{\sqrt{2}}{8}\right\}$$

$$= EAu\left\{\frac{2\sqrt{2}+4}{8}\right\}$$

The external virtual work

$$\delta W_e = 0 \times 1 = 0$$

Using the principle of virtual displacement,

$$\delta W_e = \delta U$$

$$EAu\left\{\frac{2\sqrt{2}+4}{8}\right\} = 0$$

i.e.

$$u = 0 \qquad\qquad (c)$$

Following the same procedure, the virtual strains caused due to $\delta v = 1$ are given by,

$$\delta\epsilon_1' = \frac{1}{4}$$

$$\delta\epsilon_2' = 0$$

$$\delta\epsilon_3' = -\frac{1}{4}$$

The total internal virtual work due to $\delta v = 1$ is,

$$\delta U = EA2\sqrt{2}\frac{(u+v)}{4}\frac{1}{4} + EA2\sqrt{2}\frac{(u-v)}{4}\left(-\frac{1}{4}\right)$$

substituting $u = 0$

$$\delta U = EAv\left\{\frac{\sqrt{2}}{8} + \frac{\sqrt{2}}{8}\right\} = \frac{EAv\sqrt{2}}{4}$$

The external virtual work is

$$\delta W_e = 50 \times 1$$

Equating $\delta W_e = \delta U$, we get

$$EAv\frac{\sqrt{2}}{4} = 50$$

$$v = \frac{200}{\sqrt{2}AE} \qquad\qquad (d)$$

Substituting the solution for

$$u = 0 \quad \text{and} \quad v = \frac{200}{\sqrt{2}AE}$$

in Eq. (a), the strains in the members can be calculated as,

$$\epsilon_1 = \frac{200}{4\sqrt{2}AE} = \frac{50}{\sqrt{2}AE}$$

$$\epsilon_2 = 0$$

$$\epsilon_3 = \frac{-200}{4\sqrt{2}AE} = \frac{-50}{\sqrt{2}AE} \qquad (e)$$

The forces in the members are,

$$
\begin{aligned}
F_1 &= AE\frac{50}{\sqrt{2}AE} = \frac{50}{\sqrt{2}} = 25\sqrt{2} \text{ kN} \quad (\text{tension}) \\
F_2 &= 0 \\
F_3 &= AE\left(-\frac{50}{\sqrt{2}AE}\right) \\
&= -25\sqrt{2} \text{ kN} \quad (\text{compression})
\end{aligned}
$$

$$(f)$$

This example illustrates the use of the principle as it forms the basis for the displacement method of analysis. The following steps may be noted:

1. Displacements are taken as unknowns.

2. Equilibrium equations are established using the principle of virtual displacement.

3. The equilibrium equations are solved to determine the actual displacements.

4. The strains in the members are calculated knowing the displacements.

5. Stresses and forces are then computed from the values of strains.

It may be noted here that the above procedure is followed in the finite element analysis using displacement model and is described in the subsequent chapters.

2.5 Energy Principles

In the area of structural mechanics, principles based on potential and complementary energy have been used in solving problems such as beams, trusses, frames, plate bending etc.The two important principles are (i) principle of minimum potential energy and (ii) principle of minimum complementary energy. It can be shown that the principle of virtual displacement can be formulated using variational principle for total potential energy and the principle of virtual forces can be formulated using variational principle for total complementary energy. Such a formulation requires a basic knowledge of calculus of variations and an introduction to the subject is given in the subsequent section.

The principle of minimum potential energy forms the basis of the displacement formulation for finite element analysis and hence is discussed in detail here. The principle of minimum complementary energy can be used to develop equilibrium model and as this is not widely used in programs it is not described here. References [2, 4, 5] contain detailed presentation of the energy principle and other connected theorems used in structural analysis.

2.5.1 Potential Energy

Potential energy is the capacity of the forces to do work. For structural systems or deformable bodies the forces are classified as external forces due to loads and internal forces due to stresses (or stress resultants). Thus the total potential energy is the sum of the potential energy of the external and internal forces.

Consider a spring of stiffness k subjected to a force P as shown in Fig.2.5. Let the displacement of the spring be r. While moving through the displacement r, P loses some of the capacity for doing work when it displaces in the direction it acts. Thus, the change in potential energy of the load P is $-Pr$. It may be noted here that the change in the potential energy is $-Pr$ instead of $-Pr/2$ since this potential energy arises from the magnitude of the force and its capacity to displace and it is not in any way dependent on the linear properties of the spring.

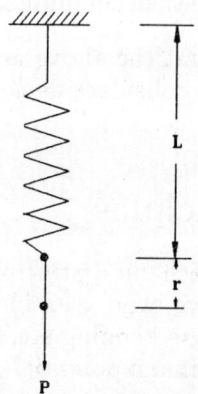

Fig. 2.5 A spring system

The internal forces developed in a deformable body or members of structural system also have a capacity to perform work. As the body deforms, stresses are developed which result in internal forces.

These forces perform work in moving through internal displacements until the final configuration is attained. If the strained body were allowed to slowly return to its unstrained state, it would be capable of returning at least a portion of the work done by the external forces. This capacity of the internal forces to perform work by virtue of the strained state of the body is called strain energy.

For the spring shown in Fig.2.5, as the deformation in the spring changes from 0 to r, the strain energy of the spring is $1/2kr^2$. Thus, the total potential energy of the spring is

$$\Pi = \frac{1}{2}kr^2 - Pr \qquad (2.45)$$

The above simple example has helped to illustrate the potential energy calculation for a deformable system. We can now generalise and get the relations for a general three-dimensional body subjected to external and body forces. Let u, v and w be the final displacement components. Following the concept used in the example, the total potential energy of the external forces can be expressed by

$$H = -\int\int\int_V (X_b u + Y_b v + Z_b w)dV$$

$$-\int\int_{S1}(X_s u + Y_s v + Z_s w)dS \qquad (2.46)$$

$$\text{i.e.,} \quad H = -W_e \qquad (2.46a)$$

where W_e is the external work done by the external and body forces. The Eq.2.46 is given in matrix form as

$$H = -\int\int\int_V \{u\}^T\{X\}dV - \int\int_{S1}\{u\}^T\{p\}dS \qquad (2.47a)$$

If concentrated loads F_i, are present on the body or structure, the potential energy due to these loads have to be included as,

$$H = -\int\int\int_V \{u\}^T\{X\}dV - \int\int_{S1}\{u\}^T\{p\}dS - \sum_{i=1}^{n} F_i\ r_i \qquad (2.47b)$$

where r_i is the displacement corresponding to F_i and the summation includes all such loads.

Next, we would derive an expression for the internal strain energy for a three-dimensional solid subjected to stress system. Consider a differential element shown in Fig.2.6 subject to a stress of σ_x.

The differential or incremental work done due to increment in displacement is

$$dU = \sigma_y \ dy \ dz d\left(u + \frac{\partial u}{\partial x}dx\right) - \sigma_x \ dy \ dz \ du$$

$$= \sigma_x d\left(\frac{\partial u}{\partial x}\right) \ dxdydz \qquad (2.48)$$

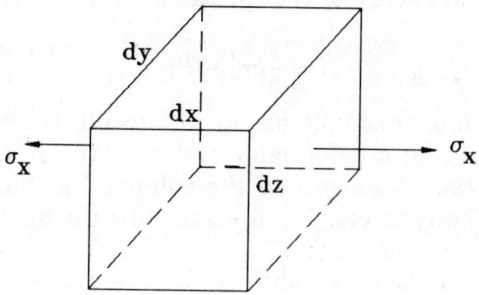

Fig. 2.6 Three dimensional differential element

Since,

$$\frac{\partial u}{\partial x} = \epsilon_x \quad \text{we get,}$$

$$dU = \sigma_x \ d\epsilon_x \ dxdydz \qquad (2.49)$$

As the strains increase from zero to the final value ϵ_x, the differential work done which is the strain energy developed in the element is

$$\int_0^{\epsilon_x} \sigma_x \ d\epsilon_x \ dV \qquad (2.50)$$

If the element is subjected to all the six stress components, the total strain energy developed in the element is given by

$$\left[\int_0^{\epsilon_x} \sigma_x d\epsilon_x + \int_0^{\epsilon_y} \sigma_y d\epsilon_y + \int_0^{\epsilon_z} \sigma_z d\epsilon_z \right.$$

$$+ \int_0^{\gamma_{xy}} \tau_{xy} d\gamma_{xy} + \int_0^{\gamma_{yz}} \tau_{yz} d\gamma_{yz} +$$

$$\left. \int_0^{\gamma_{zx}} \tau_{zx} d\gamma_{zx} \right] dV \qquad (2.51)$$

The integration is performed from initial to the final state of strain.

Let U_0 be the strain energy density, i.e., strain energy per unit volume. Then generalizing Eq.2.49 we can express the increment in strain energy density as

$$dU_o = \sigma_x d\epsilon_x + \sigma_y d\epsilon_y + \sigma_z d\epsilon_z + \tau_{xy} d\gamma_{xy} + \tau_{yz} d\gamma_{yz} + \tau_{zx} d\gamma_{zx} \qquad (2.52)$$

From differential calculus we know that dU_o can be expressed as

$$dU_o = \frac{\partial U_o}{\partial \epsilon_x} d\epsilon_x + \frac{\partial U_o}{\partial \epsilon_y} d\epsilon_y + \frac{\partial U_o}{\partial \epsilon_z} d\epsilon_z +$$

$$\frac{\partial U_o}{\partial \gamma_{xy}} d\gamma_{xy} + \frac{\partial U_o}{\partial \gamma_{yz}} d\gamma_{yz} + \frac{\partial U_o}{\partial \gamma_{zx}} d\gamma_{zx} \qquad (2.53)$$

Comparing Eqs. 2.52 and 2.53, we get

$$\frac{\partial U_o}{\partial \epsilon_x} = \sigma_x \qquad \frac{\partial U_o}{\partial \epsilon_y} = \sigma_y \qquad \frac{\partial U_o}{\partial \gamma_{zx}} = \tau_{zx} \qquad (2.54)$$

Thus, the strain energy density has the property that its partial derivative with regard to any strain component is the corresponding stress component.

Eq.2.54 can be expressed in matrix form as,

$$\{\frac{\partial U_o}{\partial \epsilon}\} = \{\sigma\} \qquad (2.55)$$

Substituting $\{\sigma\} = [C]\{\epsilon\}$ from Eq. 2.12 into Eq.2.55, we get

$$\{\frac{\partial U_o}{\partial \epsilon}\} = [C]\{\epsilon\} \qquad (2.56)$$

Integration of Eq.2.56 with respect to the strains yields

$$U_o = \frac{1}{2}\{\epsilon\}^T [C]\{\epsilon\} \qquad (2.57)$$

$$\text{or} \qquad U_o = \frac{1}{2}\{\epsilon\}^T \{\sigma\} \qquad (2.57a)$$

In the case of linearly elastic isotropic body, substituting for [C] from Eq.2.16, assuming T = 0 and simplifying we get,

$$U_o = \frac{1}{2E}(\sigma_x^2 + \sigma_y^2 + \sigma_z^2) - \frac{2\mu}{E}(\sigma_x\sigma_y + \sigma_y\sigma_z + \sigma_z\sigma_x) +$$

$$\frac{(1+\mu)}{E}(\tau_{xy}^2 + \tau_{yz}^2 + \tau_{zx}^2)$$

or, $U_o = \frac{1}{2}(\sigma_x \epsilon_x + \sigma_y \epsilon_y + \sigma_z \epsilon_z + \tau_{xy} \gamma_{xy} + \tau_{yz} \gamma_{yz} + \tau_{zx} \gamma_{zx})$ (2.28)

The total strain energy, U, can be obtained by integrating U_o over the volume of the solid. Thus,

$$U = \int \int \int_V U_o \, dV \qquad (2.59)$$

And substituting for U_o from Eq.2.57,

$$U = \frac{1}{2} \int \int \int_V \{\epsilon\}^T [C] \{\epsilon\} dV \qquad (2.60a)$$

In case initial stress and strain, and temperature effects are to be considered, the stress-strain relation given by Eq. 2.13 is to be used for the evaluation of strain energy. It can be shown,

$$U = \int \int \int_V (\frac{1}{2}\{\epsilon\}^T [C] \{\epsilon\} - \{\epsilon\}^T [C] \{\epsilon_o\} + \{\epsilon\}^T \{\sigma_o\}) \, dV \quad (2.60b)$$

In the above equation, a term, $\frac{1}{2} \int \int \int_V \{\epsilon_o\}^T [C] \{\epsilon_o\} \, dV$ is ommitted since it is a constant and has no effect in the variational process to arrive at equilibrium equation described in section 2.5.3.

Thus the total potential energy Π is defined as the sum of the external potential energy and strain energy

$$\Pi = H + U \qquad (2.61)$$

Substituting from Eqs.2.47 and 2.60a,

$$\Pi = \frac{1}{2} \int \int \int_V \{\epsilon\}^T [C] \{\epsilon\} dV - \int \int \int_V \{u\}^T \{X\} dV$$

$$- \int \int_{S1} \{u\}^T \{p\} dS \qquad (2.62)$$

2.5.2   Variational Approach to Solid Mechanics

There has been considerable interest shown in the recent past in the application of calculus of variations to the formulation of finite element models. Such an approach makes it possible to apply finite element technique to solve, in general, field problems. Also it helps to study mathematically in a rigorous manner the convergence and other properties of elements and assemblage.

In this section a brief introduction is given to the calculus of variations and also its application in the area of solid mechanics [5, 6]. The theory presented here will be used in the next section to derive the *principle of minimum potential energy*.

In the calculus of variations we are concerned with functionals and their extremum values. A functional, in simple terms, is a function of functions. More specifically a functional is defined as an expression that takes on a particular value which is dependent on the function used in the functional. In many areas of applied mathematics we are concerned with functionals in the integral form involving first or higher-order derivatives, for example

$$I = \int_{x_1}^{x_2} F(x, y, y', y'')dx \qquad (2.63)$$

where the variables y, and the first and second derivatives y' and y'' are functions of x. The integral is defined between the two points x_1 and x_2 and the corresponding values of y will be,

$$y(x_1) = y_1$$

$$y(x_2) = y_2$$

In variational procedure we seek the function $y = y(x)$ that makes I stationary (i.e. to be a maximum or minimum). Let an approximate solution be represented as $\bar{y}$. Then the variation from the exact solution y is expressed as

$$\delta[y(x)] = \bar{y}(x) - y(x) \qquad (2.64)$$

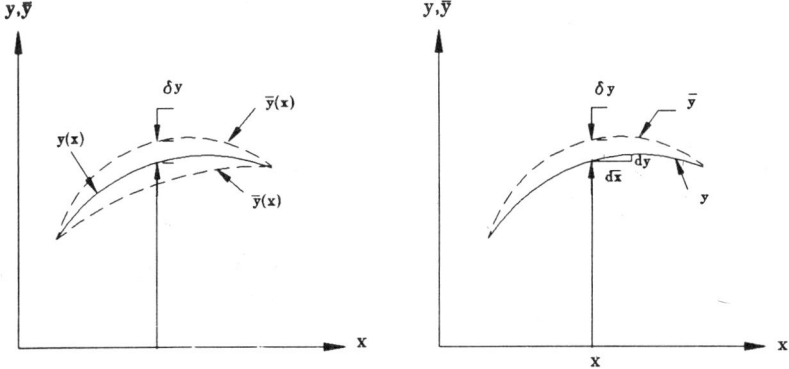

Fig. 2.7 Illustration of δy in functionals

In Fig.2.7 the extremum solution y and other trial functions $\bar{y}$ are indicated. It must be noted that δy, the first variation in y is defined as an infinitesimal change in y for a fixed value of the independent variable x. This is in contrast to the differentiation process wherein dy is associated with dx. Fig. 2.7b shows clearly the difference between δy and dy.

Having understood the difference between δy and dy, it can be shown that δ, termed delta operator used in variational calculus is similar to d, the differential operator used in differential calculus [7, 8]. Thus,

$$\delta\left(\frac{dy}{dx}\right) = \frac{d(\delta y)}{dx} \qquad (2.65)$$

$$\delta \int F \, dx = \int \delta \, F \, dx \qquad (2.66)$$

$$\delta F = \frac{\partial F}{\partial y}\delta y + \frac{\partial F}{\partial y'}\delta y' + \frac{\partial F}{\partial y''}\delta y'' \qquad (2.67)$$

We know from differential calculus that the necessary condition for a function to be a minimum or maximum is that its first derivative must be zero. Similarly it can be shown that for the functional I to attain a stationary value its first variation must be zero [8],

$$\delta I = \dot{0} \qquad (2.68)$$

In solid mechanics the functional usually has a physical meaning, for example, the potential energy of a deformable solid. And finding the first variation leads to minimum potential energy of the system which corresponds to the equilibrium condition. Before we derive such conditions in the next section, we shall study the first variation of the functional given by Eq.2.63. Thus,

$$\delta I = \int_{x_1}^{x_2} \left(\frac{\partial F}{\partial y}\delta y + \frac{\partial F}{\partial y'}\delta y' + \frac{\partial F}{\partial y''}\delta y''\right) dx = 0 \qquad (2.69)$$

In order to eliminate the variations $\delta y'$ and $\delta y''$ from Eq.2.69, the second and third terms are integrated by parts,

$$\int_{x_1}^{x_2} \frac{\partial F}{\partial y'}\delta y' dx = \left[\frac{\partial F}{\partial y'}\delta y\right]_{x_1}^{x_2} - \int_{x_1}^{x_2} \frac{d}{dx}\left(\frac{\partial F}{\partial y'}\right)\delta y \, dx \qquad (2.70)$$

$$\int_{x_1}^{x_2} \frac{\partial F}{\partial y''}\delta y'' dx = \left[\frac{\partial F}{\partial y''}\delta y'\right]_{x_1}^{x_2} - \int \frac{d}{dx}\left(\frac{\partial F}{\partial y''}\right)\delta y' \, dx \qquad (2.71(a))$$

$$= \left[\frac{\partial F}{\partial y''}\delta y'\right]_{x_1}^{x_2} - \left[\frac{d}{dx}\left(\frac{\partial F}{\partial y''}\right)\delta y\right]_{x_1}^{x_2} + \int_{x_1}^{x_2}\frac{d^2}{dx^2}\left(\frac{\partial F}{\partial y''}\right)(\delta y)dx \qquad (2.71(b))$$

Therefore, δI becomes,

$$\delta I = \int_{x_1}^{x_2}\left[\frac{\partial F}{\partial y} - \frac{d}{dx}\left(\frac{\partial F}{\partial y'}\right) + \frac{d^2}{dx^2}\left(\frac{\partial F}{\partial y''}\right)\right]\delta y dx$$

$$+ \left[\left\{\frac{\partial F}{\partial y'} - \frac{d}{dx}\left(\frac{\partial F}{\partial y''}\right)\right\}\delta y\right]_{x_1}^{x_2} + \left[\frac{\partial F}{\partial y''}\delta y'\right]_{x_1}^{x_2} = 0 \qquad (2.72)$$

Since the variation δy is arbitrary, the terms must vanish individually. Thus, the following equations are to be satisfied for the functional I to attain an extremum value.

$$\frac{\partial F}{\partial y} - \frac{d}{dx}\left(\frac{\partial F}{\partial y'}\right) + \frac{d^2}{dx^2}\left(\frac{\partial F}{\partial y''}\right) = 0 \qquad (2.73)$$

and

$$\left[\left\{\frac{\partial F}{\partial y'} - \frac{d}{dx}\left(\frac{\partial F}{\partial y''}\right)\right\}\delta y\right]_{x_1}^{x_2} = 0 \qquad (2.74(a))$$

$$\left[\left\{\left(\frac{\partial F}{\partial y''}\right)\delta y'\right\}\right]_{x_1}^{x_2} = 0 \qquad (2.74(b))$$

Equation 2.73 is known as Euler-Lagrange equation corresponding to the functional defined by Eq.(2.63). In the case of solid mechanics, the Euler-Lagrange equation yields the governing differential equation for the problem [5].

Equations 2.74 give the boundary conditions for the problem. It may be noted that these conditions can be satisfied in combination as follows:

$$\delta y(x_1) = 0, \quad \delta y(x_2) = 0$$

$$\delta y'(x_1) = 0, \quad \delta y'(x_2) = 0 \qquad (2.75(a))$$

or

$$\frac{\partial F}{\partial y'} - \frac{d}{dx}\left(\frac{\partial F}{\partial y''}\right)\Big|_{x_1}^{x_2} = 0$$

$$\frac{\partial F}{\partial y''}\Big|_{x_1}^{x_2} = 0 \qquad (2.75(b))$$

The conditions expressed in Eq.2.75(a) are called kinematic boundary conditions and conditions of Eq.2.75(b) are known as natural boundary conditions. In solid mechanics problems, the kinematic boundary conditions specify displacement conditions of the boundary and the natural boundary conditions specify force conditions of the boundary.

Thus the first variation of the functional, for extremum value, has yielded Euler-Lagrange equation and kinematic and natural boundary conditions. The physical significance of these conditions are illustrated in the following example.

Example: Consider a cantilever beam shown in Fig.2.8 subjected to an uniformly distributed load. The flexural rigidity of the beam is EI.

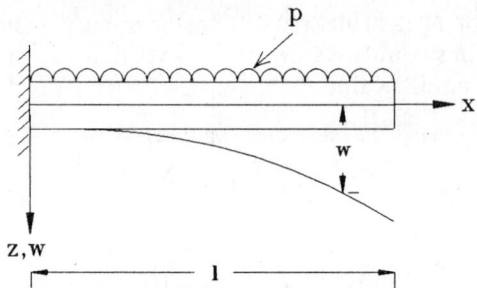

Fig. 2.8 A Cantilever Beam

From simple theory of bending it can be shown that the strain energy due to bending of the beam

$$= \frac{EI}{2} \int_o^l \left(\frac{d^2 w}{dx^2}\right)^2 dx$$

Potential energy due to load

$$= - \int_o^l pw dx$$

Total potential energy of the system

$$\Pi = \frac{EI}{2} \int_o^l \left(\frac{d^2 w}{dx^2}\right)^2 dx - \int_o^l pw \, dx$$

i.e. $\Pi = \int_o^l \left[\frac{EI}{2}\left(\frac{d^2 w}{dx^2}\right)^2 - pw\right] dx$ (a)

The Euler-Lagrange equation for this case can be formed from Eq.2.73 as

$$\frac{\partial F}{\partial w} - \frac{d}{dx}\left(\frac{\partial F}{\partial w'}\right) + \frac{d^2}{dx^2}\left(\frac{\partial F}{\partial w''}\right) = 0 \qquad (b)$$

$$\text{where} \quad F = \frac{EI}{2}(w'')^2 - pw \qquad (c)$$

$$\text{and} \quad \frac{\partial F}{\partial w} = -p; \quad \frac{\partial F}{\partial w'} = 0; \quad \frac{\partial F}{\partial w''} = EIw'' \qquad (d)$$

Substituting the values from Eq.(d) into the Euler-Lagrange equation. (b) yields

$$EIw^{iv} - p = 0 \qquad (e)$$

It is well known that Eq.(e) is the differential equation of equilibrium of beams. Thus, it is clear that the Euler-Lagrange equation yields the governing differential equation for the problem.

The boundary conditions necessary for obtaining the extremum value of II are given by the Eqs.2.75. At the clamped end $x_1 = 0$, have $w(0) = w'(0) = 0$ which satisfy the kinematic boundary condition given in Eq.2.75.

At the free end $x_2 = l$, the natural boundary condition of Eq.2.75b must be satisfied. Thus,

$$\frac{\partial F}{\partial w'} - \frac{d}{dx}\left(\frac{\partial F}{\partial w''}\right) = 0 \quad \text{i.e.} \quad EIw''' = 0 \qquad (f)$$

$$\text{and} \quad \frac{\partial F}{\partial w''} = 0, \quad \text{i.e.} \quad EIw'' = 0 \qquad (g)$$

We know from elementary strength of materials that Eqs.(f) and (g) express shear force and bending moment and they are zero at the free end. It is also noted that the natural boundary conditions specify force conditions at the boundary.

2.5.3 Principle of Stationary Potential Energy

In the last section we have shown the application of calculus of variation in the area of structural mechanics wherein the functional concerned is the total potential energy. We shall now establish the properties of the potential energy and formalize the theorems associated with it.

The total potential energy of the deformable system is expressed by Eq.2.61

$$\text{II} = U + H$$

where U is the strain energy of the system and H is the potential energy due to loads and body forces.

The total potential energy is given by Eq.2.62 as

$$\Pi = \frac{1}{2} \int \int \int_V \{\epsilon\}^T [C]\{\epsilon\} dV - \int \int \int_V \{u\}^T \{X\} dV$$

$$- \int \int_{S_1} \{u\}^T \{p\} dS$$

Taking the first variation of Eq.2.62 and substituting $\{\sigma\} = [C] \{\epsilon\}$ we get

$$\delta\Pi = \int \int \int_V \{\delta\epsilon\}^T \{\sigma\} dV - \int \int \int_V \{\delta u\}^T \{X\} dV$$

$$- \int \int_{S_1} \{\delta u\}^T \{p\} dS \qquad (2.76)$$

The first term on the right hand side of Eq.2.76 is the first variation of the strain energy U and the expression is the same as the internal virtual work, δU, given by Eq.2.39(a). Similarly, the second and third terms of Eq.2.76 are the first variation of the potential energy due to body forces and external loads respectively and these expressions are the same as given by Eq.2.38a for the external virtual work done δW_e. Hence, the first variation in U and W_e can be interpreted as the corresponding virtual work done, and

$$\delta\Pi = \delta U - \delta W_e \qquad (2.77)$$

But from the principle of virtual displacement we know that

$$\delta W_e = \delta U \quad \text{and hence,} \quad \delta\Pi = 0 \qquad (2.78)$$

Thus using the concept of the principle of virtual displacement we can interpret the condition that $\delta\Pi = 0$ as follows: 'A deformable body or structural system is in equilibrium if the first variation of the total potential energy is zero for every virtual displacement satisfying the boundary conditions'.

It has been shown in the previous section dealing with the calculus of variations, that the first variation of the functional expresses the condition of the functional to assume a stationary value. Thus the condition that $\delta\Pi = 0$ may be stated as:

'If a deformable body or structural system is in equilibrium the total potential energy of the system has a stationary value'.

It is known in ordinary differential calculus that for a function of single variable to attain a minimum value the second derivative, d^2v/dx^2 must be positive. Similarly, using the calculus of variations it can be shown that the second variation of the potential energy $\delta^2\Pi$ is positive and that the total potential energy is a minimum for equilibrium state. More detailed derivation is given in reference [2,5].

We have shown in the case of principle of virtual displacement that it is a necessary and sufficient condition for equilibrium. Now if we apply the calculus of variation for extremizing the total potential energy functional (Eq.2.76), we can arrive at the differential equations of equilibrium and boundary conditions.

The variational formulation as expressed in Eq. 2.78 is called the *weak* formulation. For example, consider the potential energy functional Π for the beam problem given by Eq.(a) in section 2.5.2. The continuity requirement is weakened since w need to be only twice differentiable compared to the requirement of fourth order by the governing differential equation (e).

Now the use of principle of stationary potential energy is illustrated through the example of three bar truss shown in Fig.2.4. The strains in the three members are given in Eq.(a) of section 2.4.2. As the members are subject to uniform axial strains only, the strain energy for each member is,

$$U = \frac{1}{2} A L_i E \epsilon_i^2 \qquad (a)$$

Thus, the total strain energy, U, for the three bars is

$$U = EA\left[\frac{(u+v)^2}{8\sqrt{2}} + \frac{u^2}{4} + \frac{(u-v)^2}{8\sqrt{2}}\right]$$

The potential energy due to loads $= -50v$
The total potential energy,

$$\Pi = EA\left[\frac{(u+v)^2}{8\sqrt{2}} + \frac{u^2}{4} + \frac{(u-v)^2}{8\sqrt{2}}\right] - 50v \qquad (b)$$

Using the principle of stationary potential energy, for equilibrium $\delta\Pi = 0$. In this case Π is a function of u and v. Using the calculus of variations, we can show that

$$\delta\Pi = \frac{\partial\Pi}{\partial u}\delta u + \frac{\partial\Pi}{\partial v}\delta v$$

Since the variations δu and δv are arbitrary, for $\delta\Pi = 0$,

$$\frac{\partial\Pi}{\partial u} = 0 \quad \text{and} \quad \frac{\partial\Pi}{\partial v} = 0 \qquad (c)$$

Applying the above conditions,

$$\frac{\partial\Pi}{\partial u} = EAu\left[\frac{2\sqrt{2}+4}{8}\right] = 0 \qquad (d)$$

i.e. $u = 0$

$$\frac{\partial \Pi}{\partial v} = EA\left[\frac{(u+v)}{4\sqrt{2}} + \frac{(u-v)}{4\sqrt{2}}(-1)\right] - 50 = 0$$

Substituting for $u = 0$, in the above equation we get

$$EAv\frac{\sqrt{2}}{4} = 50$$

$$\text{or} \qquad v = \frac{200}{\sqrt{2}AE} \qquad\qquad (e)$$

It may be noted that the Eqs (d) and (e) are identical to the Eqs. (c) and (d) of the example in subsection 2.4.2 obtained using the principle of virtual displacement.

Having solved for the displacements, the strains can be calculated using Eq.(a) and forces in the members are obtained as in Eq.(f) of section 2.4.2.

2.6   Application to Finite Element Method

For most of the problems in structural mechanics it is difficult to obtain the exact solution directly by using the energy principles. Approximate (or trial) functions are chosen for the unknown variables and by extremizing the functional, we obtain the necessary equation to get the approximate solution.

It will be shown in the next chapter that in finite element method, we approximate the displacement variation through trial functions, and minimizing the total potential energy functional we obtain the approximate solution satisfying the boundary conditions.

The strain-displacement and constitutive relations for various problems and the principle of minimum potential energy formulations described here are used in the finite element formulations described in subsequent chapters.

REFERENCES

1. Argyris, J.H. and S.Kelsey, *Energy Theorems and Structural Analysis*, Butterworths, London, (1960).

2. Oden, J.T., *Mechanics of Elastic Structures*, McGraw-Hill, New York, 1967.

3. Timoshenko, S.J. and J.N.Goodier, *Theory of Elasticity*, 3rd Edition, McGraw-Hill, New York, 1970.

4. Tauchert, T.R., *Energy Principles in Structural Mechanics*, McGraw-Hill, New York, 1974.

5. Dym, C.L. and I.H.Shames, *Solid Mechanics - A Variational Approach*, International Student Edition, McGraw-Hill Kogahusha Ltd., 1973.

6. Washizu, K. *Variational Methods in Elasticity and Plasticity*, 2nd Edition, Pergamon Press, 1965.

7. Richards, T.H., *Energy Methods in Structural Analysis with an Introduction to Finite Element Techniques*, Ellis Horwood Ltd., England, 1977.

8. Forry, M., *Variational Calculus in Science and Engineering*, McGraw-Hill, New York, 1968.

EXERCISES

2.1 Fig. E2.1 shows a truss subjected to loads P and Q at Joint 1.

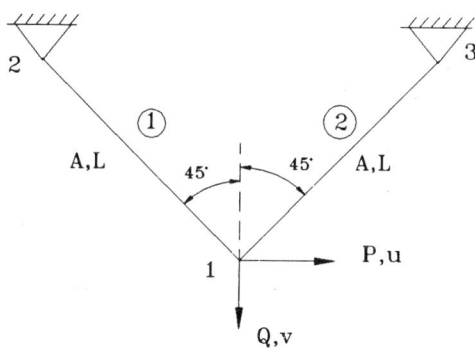

Fig. E2.1

Using the principle of virtual displacement derive the equilibrium conditions for the joint 1. Compute the deflections u and v, and also the forces in the members 1 and 2.

2.2 Using the unit dummy-displacement method, compute th stiffness due to unit rotation at one end of the beam, the farthe end being fixed as shown in Fig. E2.2.

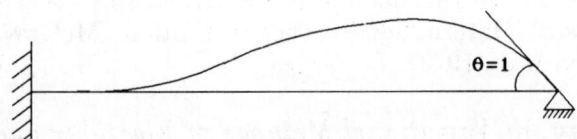

Fig. E2.2

2.3 A truss is shown in Fig. E2.3. The area of cross-section of all the members is the same.

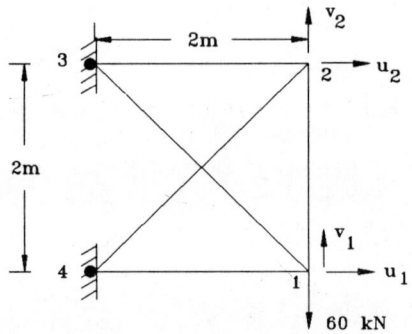

Fig. E2.3

Establish the condition of equilibrium for the structure in terms of nodal displacements u_1, v_1, u_2 and v_2 by the principle of virtual displacement.

2.4 Derive from first principles, the strain energy of an axial bar in terms of strain ϵ and an initial strain, due to temperature $\epsilon_o(\alpha T)$. Assume the material obeys Hooke's law.

2.5 The axial bar shown in Fig. E2.5 is tapered and the area of cross-section at the ends are A_1 and A_2. Calculate the strain energy of the bar assuming the material obeys Hooke's law.

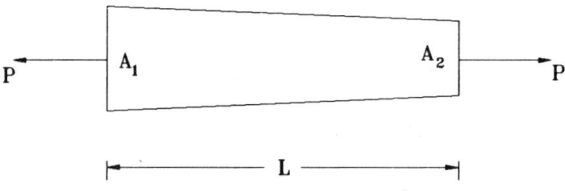

Fig. E2.5

2.6 The stresses developed in a thin plate are:

$$\sigma_x = \frac{Ez}{1-\mu^2} \left(\frac{\partial^2 w}{\partial x^2} + \mu \frac{\partial^2 w}{\partial y^2}\right)$$

$$\sigma_y = \frac{Ez}{1-\mu^2} \left(\frac{\partial^2 w}{\partial y^2} + \mu \frac{\partial^2 w}{\partial x^2}\right)$$

$$\tau_{xy} = \frac{Ez}{1+\mu} \frac{\partial^2 w}{\partial x \partial y}$$

where w is the component of displacement normal to the plane of the plate and z is the coordinate measured from the middle plane of the plate. The energy contributed by the remaining components is assumed to be negligible. Show that the strain energy developed in the plate is given by,

$$U = \frac{Eh^3}{12(1-\mu^2)} \int \int \left[\left(\frac{\partial^2 w}{\partial x^2} + \frac{\partial^2 w}{\partial y^2}\right)^2 - 2(1-\mu)\right.$$

$$\left. \left[\frac{\partial^2 w}{\partial x^2} \frac{\partial^2 w}{\partial y^2} - \left(\frac{\partial^2 w}{\partial x \partial y}\right)^2\right]\right] \, dx dy$$

where h is the thickness of the plate.

2.7 An axial bar shown in Fig. E2.7 is fixed at one end. It is subjected to a distributed axial force p/unit length and a force P at the free end. Calculate the total potential energy of the bar. Using the variational principles derive the governing differential equation for the bar.

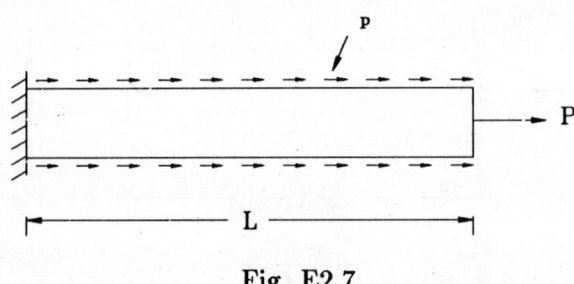

Fig. E2.7

2.8 A cantilever beam shown in Fig.E2.8 is subjected to an uniformly distributed load p/unit length and end forces M_o and P_o.

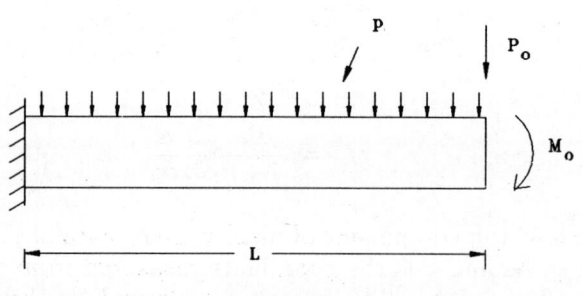

Fig. E2.8

Derive the governing functional for this problem using the differential equation for beam bending problems.

$$EI \frac{\partial^4 w}{\partial x^4} = p$$

2.9 A beam shown in Fig. E2.9 is subjected to a load of intensity p per unit length and an axial force Q_1 at the left end. Using the variational approach derive the governing differential equations and, the natural and kinematic boundary conditions for the problem. Show the uncoupling of bending and axial deformations, and hence the governing conditions can be treated independently for each mode.

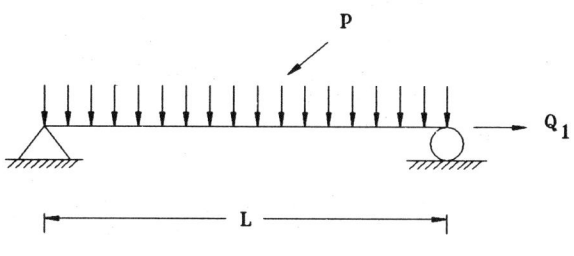

Fig. E2.9

2.10 The axial bar shown in Fig. E2.10 is subjected to a load P at a distance of $\frac{L}{2}$ from one of the fixed ends. Use the principle of stationary potential energy to calculate the reactions R_1 and R_2.

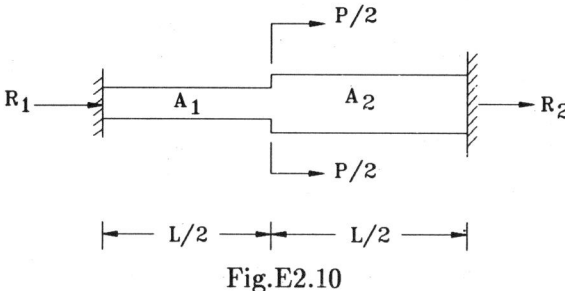

Fig.E2.10

2.11 Determine the total potential energy functional for a beam resting on an elastic foundation, with one end rigidly fixed and the other end propped on an elastic support as shown in Fig. E2.11. Assume K_f is the subgrade reaction-spring constant per unit length and K_s is the spring constant for the flexible end support. Using the variational principles, obtain the governing differential equation and boundary conditions for this beam.

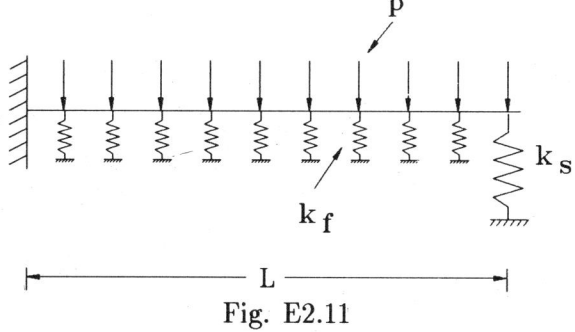

Fig. E2.11

2.12 Solve the problem 2.1 by the principle of stationary potential energy.

2.13 Solve the problem 2.3 by the principle of stationary potential energy.

2.14 Derive the virtual work and stationary potential energy equations if the body is also subjected to a temperature raise of T.

2.15 A beam is subjected to a uniformly distributed load, p and end forces Q_1, M_1 and Q_2, M_2 as shown in Fig. E2.15. Derive the governing differential equation and the boundary conditions using the principle of stationary potential energy.

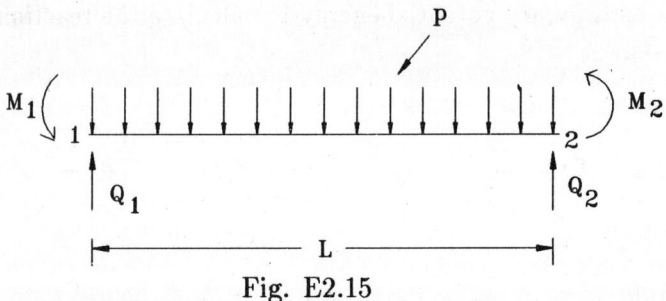

Fig. E2.15

Chapter 3

Element Properties

As described earlier, the discretization of the structure or body into finite elements forms the first step in the analysis of a complicated structural system. Figure 3.1 shows some typical elements for various types of structures. Triangular or quadrilateral elements are employed for plane stress or plane strain or plate bending problems. Shell structures are discretized with either flat or curved elements. Axisymmetric solids are discretized using ring type elements. For problems which require a strict three-dimensional analysis, hexahedral or tetrahedral elements are used for discretization. One of the widely used formulations for finite element analysis is based on the assumption for the variation of displacement in the element and such models are called displacement models or displacement formulation. In this chapter the properties of the element required for performing the analysis are described for such a formulation.

Plane Stress/strain and Axisymmetric Analysis

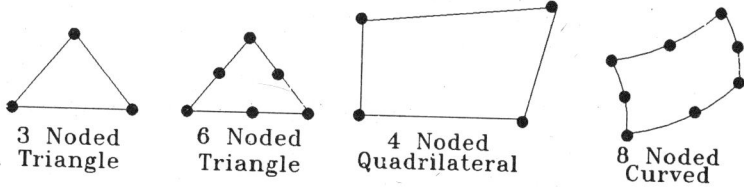

| 3 Noded Triangle | 6 Noded Triangle | 4 Noded Quadrilateral | 8 Noded Curved |

Plate bending and shells

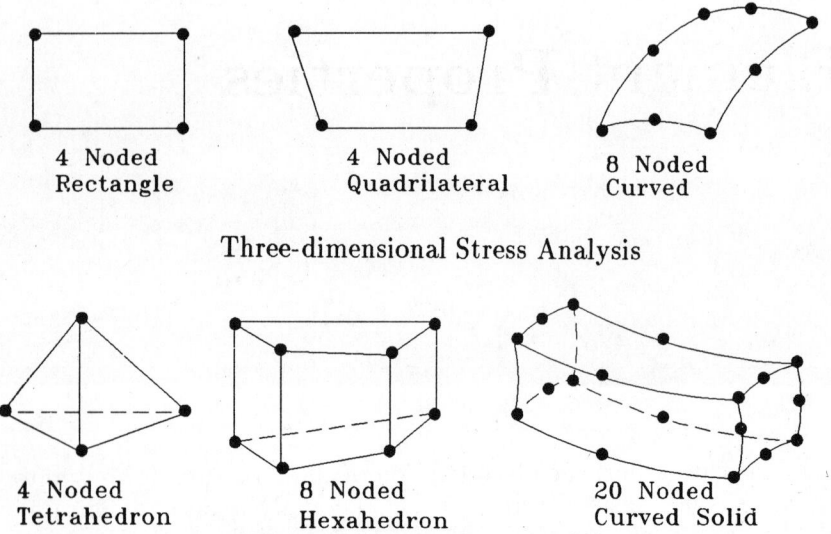

4 Noded
Rectangle

4 Noded
Quadrilateral

8 Noded
Curved

Three-dimensional Stress Analysis

4 Noded
Tetrahedron

8 Noded
Hexahedron

20 Noded
Curved Solid

Fig. 3.1 Typical finite elements

3.1   Displacement Models

For the displacement formulation, the essential step consists of approximating the variation of displacement in the element by suitable functions. It has generally been found that polynomials are much easier for mathematical operations and provide the student a feel for the degree of approximation. Hence, they are widely used in finite element analysis.

Consider a triangular element shown in Fig.3.2. This element is used for plane stress or plane strain analysis. The displacement variation within the element can be represented by the following function.

$$u = \alpha_1 + \alpha_2 x + \alpha_3 y$$

$$v = \alpha_4 + \alpha_5 x + \alpha_6 y \qquad (3.1)$$

where $\alpha_1, \alpha_2, ..., \alpha_6$ are unknown coefficients and are also called as generalised coordinates. This Eq.3.1 can be represented in a matrix form as

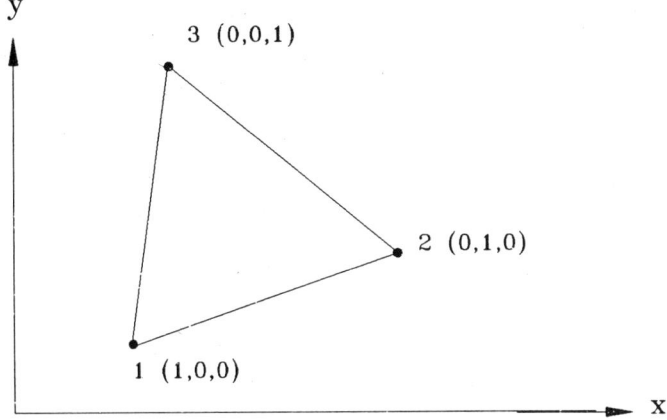

$$\{u\} = \begin{Bmatrix} u \\ v \end{Bmatrix} = \begin{bmatrix} 1 & x & y & 0 & 0 & 0 \\ 0 & 0 & 0 & 1 & x & y \end{bmatrix} \begin{Bmatrix} \alpha_1 \\ \alpha_2 \\ \alpha_3 \\ \alpha_4 \\ \alpha_5 \\ \alpha_6 \end{Bmatrix} \qquad (3.2a)$$

i.e. $\{u\} = [\phi]\{\alpha\}$ $\qquad$ (3.2b)

Fig. 3.2

In general, if a polynomial of degree n is assumed, the displacements for a two-dimensional element can be expressed as

$$
\begin{aligned}
u &= \alpha_1 + \alpha_2 x + \alpha_3 y + \alpha_4 x^2 + \alpha_5 xy + \alpha_6 y^2 + \ldots + \alpha_m y^n \\
v &= \alpha_{m+1} + \alpha_{m+2} x + \alpha_{m+3} y + \alpha_{m+4} x^2 + \alpha_{m+5} xy
\end{aligned}
$$

$$+\alpha_{m+6} y^2 + \ldots + \alpha_{2m} y^n \qquad (3.3)$$

The above equations can be written in matrix form as

$$\{u\} = \begin{Bmatrix} u \\ v \end{Bmatrix} = \begin{bmatrix} \{\phi_2\}^T & \{0\}^T \\ \{0\}^T & \{\phi_2\}^T \end{bmatrix} \{\alpha\}$$

i.e. $\{u\} = [\phi]\{\alpha\}$ $\qquad$ (3.4)

where $\{\phi_2\}^T = \begin{bmatrix} 1 & x & y & x^2 & xy & y^2 & \ldots & y^n \end{bmatrix}$ $\qquad$ (3.4a)

$$\{\alpha\}^T = [\alpha_1 \quad \alpha_2 \quad \alpha_3 \quad \ldots \quad \alpha_{2m}] \tag{3.4b}$$

In the case of three-dimensional elements the displacement functions are given by

$$
\begin{aligned}
u &= \alpha_1 + \alpha_2 x + \alpha_3 y + \alpha_4 z + \alpha_5 xz + \ldots + \alpha_m z^n \\
v &= \alpha_{m+1} + \alpha_{m+2} x + \alpha_{m+3} y + \ldots + \alpha_{2m} z^n \\
w &= \alpha_{2m+1} + \alpha_{2m+2} x + \alpha_{2m+3} y + \ldots + \alpha_{3m} z^n
\end{aligned}
$$

$$\tag{3.5}$$

Expressing it in matrix form,

$$
\{u\} = \left\{ \begin{array}{c} u \\ v \\ w \end{array} \right\} = \begin{bmatrix} \{\phi_3\}^T & \{0\}^T & \{0\}^T \\ \{0\}^T & \{\phi_3\}^T & \{0\}^T \\ \{0\}^T & \{0\}^T & \{\phi_3\}^T \end{bmatrix} \{\alpha\}
$$

$$\text{i.e.} \quad \{u\} = [\phi]\{\alpha\} \tag{3.6}$$

where $\quad \{\phi_3\}^T = [1 \quad x \quad y \quad z \quad xz \quad \ldots \quad z^n]$

$$\{\alpha\}^T = [\alpha_1 \quad \alpha_2 \quad \ldots \quad \alpha_{3m}] \tag{3.6a}$$

3.2   Relation between the Nodal Degrees of Freedom and Generalised Coordinates

The unknown coefficients α's of the polynomials can be expressed in terms of the displacements of the nodes of the element, called as nodal degrees of freedom. It thus facilitates to specify the displacements within an element by nodal displacements. This is illustrated with reference to the triangular element shown in Fig.3.2. Substituting the coordinates of nodes in Eq.3.2

$$
\{d\} = \begin{bmatrix} u_1 \\ u_2 \\ u_3 \\ v_1 \\ v_2 \\ v_3 \end{bmatrix} = \begin{bmatrix} 1 & x_1 & y_1 & 0 & 0 & 0 \\ 1 & x_2 & y_2 & 0 & 0 & 0 \\ 1 & x_3 & y_3 & 0 & 0 & 0 \\ 0 & 0 & 0 & 1 & x_1 & y_1 \\ 0 & 0 & 0 & 1 & x_2 & y_2 \\ 0 & 0 & 0 & 1 & x_3 & y_3 \end{bmatrix} \begin{bmatrix} \alpha_1 \\ \alpha_2 \\ \alpha_3 \\ \alpha_4 \\ \alpha_5 \\ \alpha_6 \end{bmatrix} \tag{3.7}
$$

or in general

$$\{d\} = [A]\{\alpha\} \tag{3.8}$$

The above equation expresses the relation between the nodal degrees of freedom and generalised coordinates. The elements of the matrix [A] are computed by substituting the coordinates of the nodes of the element in the polynomial displacement function. In general, the matrix [A] is a square matrix and hence the total number of generalised coordinates equals the total number of nodal, external and internal, degrees of freedom.

Solving the Eq.3.8 for $[\alpha]$ we get

$$[\alpha] = [A]^{-1}\{d\} \tag{3.9}$$

Substituting in Eq.3.2,

$$\{u\} = [\phi][A]^{-1}\{d\} \tag{3.10}$$

$$\text{or} \qquad \{u\} = [N]\{d\} \tag{3.11}$$

where [N] is called the interpolation or shape function and the Eq.3.11 directly expresses the displacement $\{u\}$ at any point within the element in terms of nodal displacements $\{d\}$.

3.3 Convergence Requirements

The finite element method provides a numerical solution to a complex problem. It may, therefore, be expected that the solution must converge to the exact solution under certain circumstances. It can be shown that the displacement formulation of the method leads to an upper bound to the actual stiffness of the structure. Hence as the mesh is made finer the solution should converge to the correct result and this would be achieved if the following three conditions are satisfied by the assumed displacement function (model) [1, 2, 3, 5, 7, 10, 11].

1. The displacement function must be continuous within the element. This condition can easily be satisfied by choosing polynomials for the displacement model.

2. The displacement function must be capable of representing rigid body displacements of the element. That is when the nodes are given such displacements corresponding to a rigid body motion, the element should not experience any strain and hence leads

to zero nodal forces. The constant terms in the polynomials used for the displacement models would usually ensure this condition.

3. The displacement function must be capable of representing constant strain states within the element. The reason for this requirement can be understood if we imagine the condition when the body or structure is divided into smaller and smaller elements. As these elements approach infinitesimal size, the strains in each element also approach constant values. Hence the assumed displacement function should include terms for representing constant strain states. For one, two and three-dimensional elasticity problems, the linear terms present in the polynomial satisfy the requirement. However, in the case of beam, plate and shell elements, this condition will be referred to as 'constant curvature' instead of 'constant strains'.The difficulty of satisfying this condition in these situations would be discussed later.

3.3.1   Compatibility Requirement

The displacements must be compatible between adjacent elements. That is when the elements deform there must not be any discontinuity between elements, i.e. elements must not overlap or separate. In the case of beam, plate and shell elements, this requirement would also include that there should be no sudden changes in slope across the interelement boundaries.

Elements which satisfy all the three convergence requirements and compatibility condition are called 'Compatible or Conforming Elements'. And elements which violate the compatibility requirement are known as 'incompatible elements'. There are many instances reported in the literature wherein incompatible elements have been successfully used. They have been found to give good results and are used to improve certain basic properties of the elements which are not possible otherwise. In Chapter 8, a detailed description is given for one such element used for solving plane stress problems. However, the one drawback of such elements is with respect to convergence as it may not be monotonic and may be from above or below.

3.3.2   Geometric Invariance

Besides the convergence and compatibility requirements, one of the important considerations in choosing proper terms in the polynomial expansion is that the element should have no preferred direction. That is the displacement shapes will not change with a change in local coordinate system. This property is known as geometric isotropy, or geometric invariance [8].

Geometric invariance is achieved if the polynomial includes all the terms, i.e., the polynomial is a complete one. However, invariance may be achieved if the polynomial is 'balanced' in case all the terms cannot be included. This 'balanced' representation can be illustrated with respect to Pascal triangle for two-dimensional polynomials:

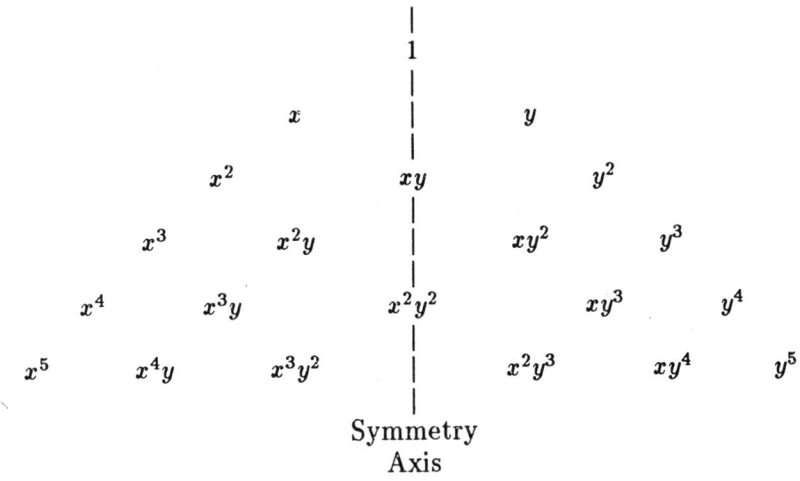

For example, if we would like to construct a polynomial with four terms, invariance is achieved by selecting

$$u = \alpha_1 + \alpha_2 x + \alpha_3 y + \alpha_4 xy$$

Thus, geometric invariance can be achieved by selecting the corresponding order of terms on either side of the axis of symmetry.

3.4 Natural Coordinate Systems

In finite element formulation, the natural coordinate system has been found to be quite effective in formulating the element properties. The natural coordinate system is a local system in which a point within an element will be expressed by dimensionless set of numbers whose magnitude never exceed unity. Moreover, these systems will be so defined that the nodal points will have unit magnitude or zero, or a convenient set of fractions. This type of expressing the coordinates also facilitates the integration to compute element stiffness.

3.4.1   One-Dimensional Line Element

Figure 3.3 shows one-dimensional line element. The natural coordinates of any point P is defined as follows

$$L_1 = 1 - \frac{x}{l}$$

$$\text{and} \quad L_2 = \frac{x}{l}$$

(3.12)

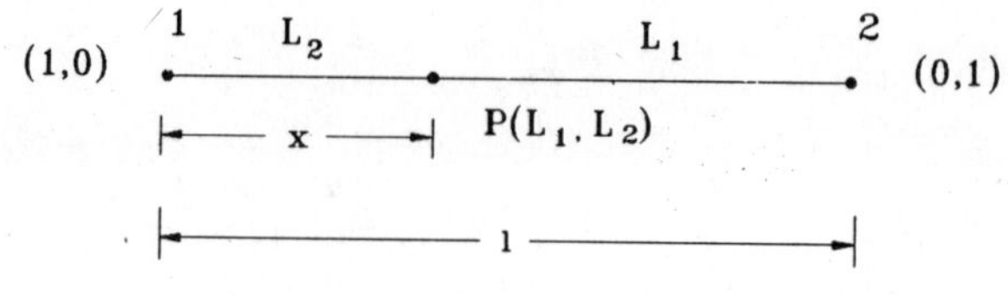

Fig. 3.3 Line element

Thus any point P is referred to by the new coordinates L_1, L_2. The relation between the cartesian and the natural coordinate can be expressed in matrix form as,

$$\left\{\begin{matrix} 1 \\ x \end{matrix}\right\} = \begin{bmatrix} 1 & 1 \\ 0 & l \end{bmatrix} \left\{\begin{matrix} L_1 \\ L_2 \end{matrix}\right\}$$

(3.13)

and the inverse relation between the natural coordinate and cartesian coordinate can be computed as,

$$\left\{\begin{matrix} L_1 \\ L_2 \end{matrix}\right\} = \frac{1}{l} \begin{bmatrix} l & -1 \\ 0 & 1 \end{bmatrix} \left\{\begin{matrix} 1 \\ x \end{matrix}\right\}$$

(3.14)

The differentiation with respect to x can be written as

$$\frac{d}{dx} = \frac{\partial}{\partial L_1} \cdot \frac{\partial L_1}{\partial x} + \frac{\partial}{\partial L_2} \cdot \frac{\partial L_2}{\partial x}$$

From Eq.3.14,

$$\frac{\partial L_1}{\partial x} = -\frac{1}{l} \quad \text{and} \quad \frac{\partial L_2}{\partial x} = \frac{1}{l}$$

Hence,

$$\frac{d}{dx} = \frac{1}{l}\left(\frac{\partial}{\partial L_2} - \frac{\partial}{\partial L_1}\right)$$

(3.15)

And the integration over the length l is given by

$$\int_l L_1^p \, L_2^q \, dl = \frac{p! \, q!}{(p+q+1)!} \, l \qquad (3.16)$$

where 0! is defined as equal to unity.

3.4.2   Two-Dimensional Triangular Element

Consider a triangular element shown in Fig.3.4.

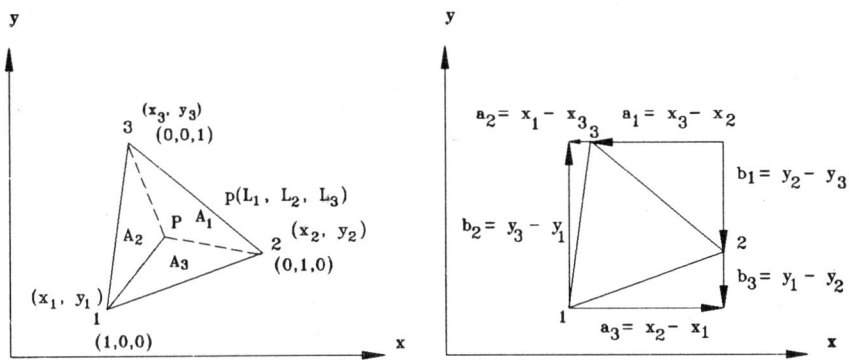

Fig. 3.4 Triangular element-local coordinates

A point P will be defined by the following set of natural coordinates as,

$$L_1 = \frac{A_1}{A}; \quad L_2 = \frac{A_2}{A}; \quad L_3 = \frac{A_3}{A} \qquad (3.17)$$

where A_1, A_2 and A_3 are the areas of the three subtriangles, subtended by the point P and the nodes as shown in Fig.3.4 and A is the total area of the triangle. Thus,

$$A_1 + A_2 + A_3 = A$$

and $\quad L_1 + L_2 + L_3 = 1 \qquad (3.18)$

The natural coordinates of nodes 1, 2 and 3 are as follows:

Node $1 - (1, 0, 0)$

Node $2 - (0, 1, 0)$

Node $3 - (0, 0, 1)$

And the coordinates of the nodes at midsides are (Fig.3.8)

$$\text{Node} \ \ 4 - (1/2, 1/2, 0)$$

$$\text{Node} \ \ 5 - (0, 1/2, 1/2)$$

$$\text{Node} \ \ 6 - (1/2, 0, 1/2)$$

The area of the triangle can be expressed as function of the x, y coordinates of the nodes 1, 2 and 3.

$$A = \frac{1}{2} \begin{vmatrix} 1 & x_1 & y_1 \\ 1 & x_2 & y_2 \\ 1 & x_3 & y_3 \end{vmatrix} \tag{3.19}$$

The area of the subtriangles are

$$A_1 = \frac{1}{2} \begin{vmatrix} 1 & x & y \\ 1 & x_2 & y_2 \\ 1 & x_3 & y_3 \end{vmatrix}$$

$$A_2 = \frac{1}{2} \begin{vmatrix} 1 & x_1 & y_1 \\ 1 & x & y \\ 1 & x_3 & y_3 \end{vmatrix}$$

$$A_3 = \frac{1}{2} \begin{vmatrix} 1 & x_1 & y_1 \\ 1 & x_2 & y_2 \\ 1 & x & y \end{vmatrix} \tag{3.20}$$

where (x, y) are the coordinates of an arbitrary point P inside or on the boundaries of the element.

The relationship between cartesian and natural coordinate can be shown as follows [4].

$$\begin{Bmatrix} 1 \\ x \\ y \end{Bmatrix} = \begin{bmatrix} 1 & 1 & 1 \\ x_1 & x_2 & x_3 \\ y_1 & y_2 & y_3 \end{bmatrix} \begin{Bmatrix} L_1 \\ L_2 \\ L_3 \end{Bmatrix} \tag{3.21}$$

and the inverse relation between natural and cartesian coordinates is given by

$$\begin{Bmatrix} L_1 \\ L_2 \\ L_3 \end{Bmatrix} = \frac{1}{2A} \begin{bmatrix} (x_2 y_3 - x_3 y_2) & (y_2 - y_3) & (x_3 - x_2) \\ (x_3 y_1 - x_1 y_3) & (y_3 - y_1) & (x_1 - x_3) \\ (x_1 y_2 - x_2 y_1) & (y_1 - y_2) & (x_2 - x_1) \end{bmatrix} \begin{Bmatrix} 1 \\ x \\ y \end{Bmatrix} \tag{3.22}$$

For the calculation of element properties, the derivatives with respect to the global coordinates are required. These can be evaluated by using the chain rule of partial differentiation as,

$$\frac{\partial f}{\partial x} = \frac{\partial f}{\partial L_1} \cdot \frac{\partial L_1}{\partial x} + \frac{\partial f}{\partial L_2} \cdot \frac{\partial L_2}{\partial x} + \frac{\partial f}{\partial L_3} \cdot \frac{\partial L_3}{\partial x}$$

$$= \frac{b_1}{2A} \cdot \frac{\partial f}{\partial L_1} + \frac{b_2}{2A} \cdot \frac{\partial f}{\partial L_2} + \frac{b_3}{2A} \cdot \frac{\partial f}{\partial L_3}$$

$$= \sum_{i=1}^{3} \frac{b_i}{2A} \cdot \frac{\partial f}{\partial L_i} \qquad (3.23a)$$

where $b_1 = y_2 - y_3$; $b_2 = y_3 - y_1$ and $b_3 = y_1 - y_2$ as indicated in Fig. 3.4.
Similarly it can be shown that,

$$\frac{\partial f}{\partial y} = \sum_{i=1}^{3} \frac{a_i}{2A} \frac{\partial f}{\partial L_i} \qquad (3.23b)$$

where $a_1 = x_3 - x_2$; $a_2 = x_1 - x_3$ and $a_3 = x_2 - x_1$ as indicated in Fig.3.4. The area integral can be computed as [12]

$$\int_A L_1^p \, L_2^q \, L_3^r \, dA = \frac{p! \, q! \, r!}{(p+q+r+2)!} 2A \qquad (3.24)$$

where $0!$ is defined as equal to unity.

3.4.3   Three-Dimensional Tetrahedron Element

Consider an arbitrary tetrahedron shown in Fig. 3.5. Any point P within the element may be defined by a set of natural coordinates as

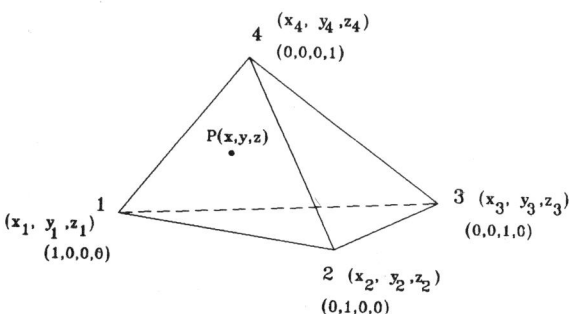

Fig. 3.5 Tetrahedron element

$$L_i = \frac{V_i}{V} \tag{3.25}$$

where V_i is the volume of the subelement bound by point P and face i and V is the total volume of the element. For example L_i may be interpreted as the ratio of the volume of the subelement P234 to the total volume of element 1234. The volume of the element V is given by the determinant of the nodal coordinates as follows:

$$V = \frac{1}{6} \begin{vmatrix} 1 & 1 & 1 & 1 \\ x_1 & x_2 & x_3 & x_4 \\ y_1 & y_2 & y_3 & y_4 \\ z_1 & z_2 & z_3 & z_4 \end{vmatrix} \tag{3.26}$$

The relationship between the cartesian and natural coordinates of point P may be expressed as [13]

$$\begin{Bmatrix} 1 \\ x \\ y \\ z \end{Bmatrix} = \begin{bmatrix} 1 & 1 & 1 & 1 \\ x_1 & x_2 & x_3 & x_4 \\ y_1 & y_2 & y_3 & y_4 \\ z_1 & z_2 & z_3 & z_4 \end{bmatrix} \begin{Bmatrix} L_1 \\ L_2 \\ L_3 \\ L_4 \end{Bmatrix} \tag{3.27}$$

and it may be noted that the identity included in the first row,

$$L_1 + L_2 + L_3 + L_4 = 1$$

render the matrix invertible. The inverse relation is given by

$$\begin{Bmatrix} L_1 \\ L_2 \\ L_3 \\ L_4 \end{Bmatrix} = \frac{1}{6V} \begin{bmatrix} V_1 & a_1 & b_1 & c_1 \\ V_2 & a_2 & b_2 & c_2 \\ V_3 & a_3 & b_3 & c_3 \\ V_4 & a_4 & b_4 & c_4 \end{bmatrix} \begin{Bmatrix} 1 \\ x \\ y \\ z \end{Bmatrix} \tag{3.28}$$

where V_i is the volume subtended from face i and terms a_i, b_i and c_i represent the projected area of face i on the x, y, z coordinate planes respectively and are given as follows:

$$a_i = (z_j y_k - z_k y_j) + (z_k y_l - z_l y_k) + (z_l y_j - z_j y_l)$$

$$b_i = (z_j x_k - z_k x_j) + (z_k x_l - z_l x_k) + (z_l x_j - z_j x_l)$$

$$c_i = (y_j x_k - y_k x_j) + (y_k x_l - y_l x_k) + (y_l x_j - y_j x_l)$$

$$i, j, k, l \quad \text{in cyclic order} \quad (1 \to 2 \to 3 \to 4 \to 1) \tag{3.29}$$

The partial derivatives of the natural coordinates with respect to the cartesian coordinates are given by

$$\frac{\partial L_i}{\partial x} = \frac{a_i}{6V}, \quad \frac{\partial L_i}{\partial y} = \frac{b_i}{6V}, \quad \frac{\partial L_i}{\partial z} = \frac{c_i}{6V} \qquad (3.30)$$

The general integral taken over the volume of the element is given by,

$$\int_v L_1^p \ L_2^q \ L_3^r \ L_4^s \ dV = \frac{p! \ q! \ r! \ s!}{(p+q+r+s+3)!} \cdot 6V \qquad (3.31)$$

3.5 Shape Functions (Interpolation Functions)

In finite element analysis using the displacement model, we assume the variation of displacement within an element since the 'true' variation of displacement is not known. In general, in higher mathematics, it is necessary in many situations to deal with functions whose analytical form is either totally unknown or else is of such a nature that the function cannot easily be subjected to such operations as may be required. In either case, it is desirable to replace the given function by another function which can be more easily handled. This operation of replacing or representing a given function by simpler one is known as interpolation in a broad sense. In finite element literature it is also referred to as 'Shape Function'.

There are two types of interpolation functions: (i) Lagrange interpolation and (ii) Hermitian interpolation. In the Lagrange interpolation, which is widely used in practice, the assumed function takes on the same values as the given function at specified points. In the Hermitian type of function, the slopes of the function also takes the same value as the given function at specified points. In this section attention will be restricted to Lagrange interpolation or shape function.

The Lagrange type shape function can be derived using the Eqs. 3.10 and 3.11. It can also be constructed more directly by using the Lagrange's interpolation formula. Both the approaches are illustrated in the following sections. In the derivation of these functions, it has been found to be quite advantageous to use the natural coordinate system as will be shown later. Experienced analyst would find it easier to derive these functions by intuition as observed by Zienkiewicz [1]. However, it has been the experience of the author that the students prefer a formal path to be consistent and straightforward in the derivation of the functions, which are illustrated in the following examples.

3.5.1   Shape Function for a Linear Model for Triangular Element

The linear displacement variation of a triangular element shown in Fig. 3.6 can be expressed in natural coordinates as,

$$u = \alpha_1 L_1 + \alpha_2 L_2 + \alpha_3 L_3 \tag{3.32}$$

$$\text{i.e.} \quad u = \{\phi_2\}^T \{\alpha\} \tag{3.33}$$

where $\{\phi_2\}^T = [L_1 \; L_2 \; L_3]$ and $\{\alpha\}^T = [\alpha_1 \; \alpha_2 \; \alpha_3]$

Substituting the natural coordinates for nodes 1, 2 and 3, the 'u' components of nodal displacements $\{d_u\}$ are given by

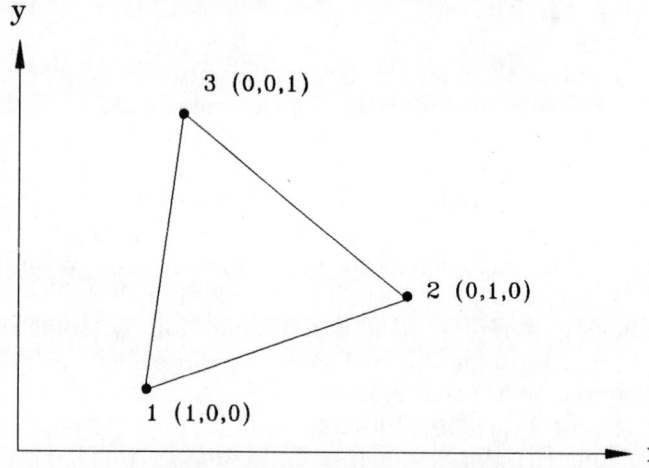

Fig. 3.6 Three noded triangular element (Constant Strain Triangular Element)

$$\{d_u\} = \begin{Bmatrix} u_1 \\ u_2 \\ u_3 \end{Bmatrix} = \begin{bmatrix} 1 & 0 & 0 \\ 0 & 1 & 0 \\ 0 & 0 & 1 \end{bmatrix} \{\alpha\} \tag{3.34}$$

$$\text{Hence,} \quad \{\alpha\} = \begin{bmatrix} 1 & 0 & 0 \\ 0 & 1 & 0 \\ 0 & 0 & 1 \end{bmatrix} \begin{Bmatrix} u_1 \\ u_2 \\ u_3 \end{Bmatrix} \tag{3.35}$$

and from Eq.3.33,

$$u = \{\phi_2\}^T \begin{bmatrix} 1 & 0 & 0 \\ 0 & 1 & 0 \\ 0 & 0 & 1 \end{bmatrix} \{d_u\}$$

or can be expressed in terms of interpolation function as

$$u = \{N_2\}^T \{d_u\}$$

where $\{N_2\}^T$ can be worked out as,

$$\{N_2\}^T = [L_1 \ L_2 \ L_3] \begin{bmatrix} 1 & 0 & 0 \\ 0 & 1 & 0 \\ 0 & 0 & 1 \end{bmatrix} = [L_1 \ L_2 \ L_3] \tag{3.36}$$

Similarly, the displacement v can be expressed as

$$v = \{N_2\}^T \{d_v\} \tag{3.37}$$

The two displacement components u and v of any point inside an element can now be defined as,

$$\{u\} = \begin{Bmatrix} u \\ v \end{Bmatrix} = \begin{bmatrix} \{N_2\}^T & \{0\}^T \\ \{0\}^T & \{N_2\}^T \end{bmatrix} \{d\} \tag{3.38}$$

i.e.

$$\{u\} = [N]\{d\}$$

where

$$\{d\} = \begin{Bmatrix} \{d_u\} \\ \{d_v\} \end{Bmatrix} \tag{3.38a}$$

and

$$[N] = \begin{bmatrix} L_1 & L_2 & L_3 & 0 & 0 & 0 \\ 0 & 0 & 0 & L_1 & L_2 & L_3 \end{bmatrix} \tag{3.39}$$

The elements of the shape function vector $\{N_2\}$ can in general be denoted as N_i. For the three noded triangular element the values of N_i are given by Eq.3.36, i.e., $N_1 = L_1$; $N_2 = L_2$; $N_3 = L_3$. The shape function N_i gives a unit value at the node i and zero value at the other nodes of the element. Fig.3.7 shows the assumed variation of the displacement u plotted normal to the element and the shape functions N_1, N_2 and N_3. Thus as given in Eq.3.38, the sum of the three functions each multiplied by the corresponding nodal displacement describes the variation of displacement in the element, i.e.,

$$u = N_1 u_1 + N_2 u_2 + N_3 u_3$$

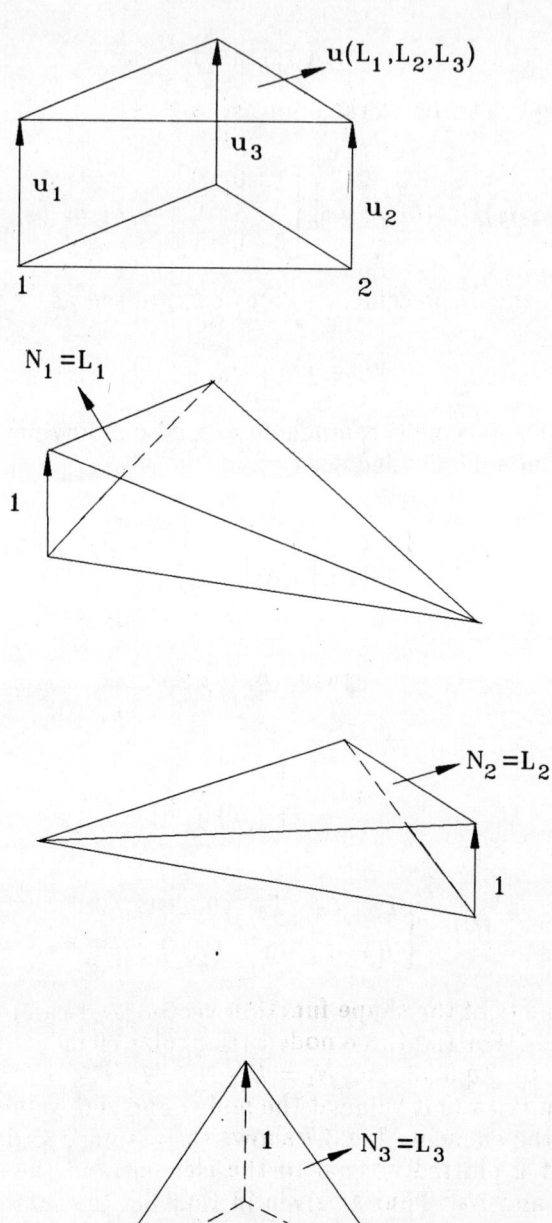

Fig. 3.7 Three noded triangular element
shape function

3.5.2 Shape Function for a Quadratic Model for Triangular Element

The quadratic variation of displacements in a six noded triangular element can be expressed by,

$$u = \alpha_1 L_1^2 + \alpha_2 L_2^2 + \alpha_3 L_3^2 + \alpha_4 L_1 L_2 + \alpha_5 L_2 L_3 + \alpha_6 L_3 L_1 \qquad (3.40)$$

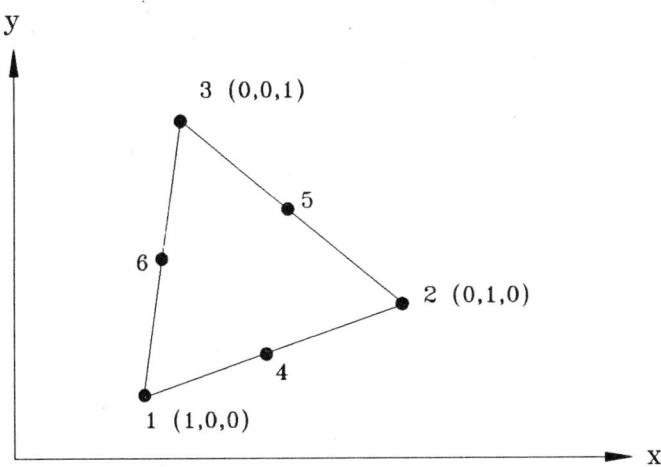

Fig. 3.8 Six noded triangular element (Linear Strain Triangular Element)

It may be noted that all the terms in the above polynomial are quadratic since any term which is not quadratic can be made so by proper multiplication of $L_1 + L_2 + L_3$ whose magnitude is equal to unity. Substituting for nodal coordinates the 'u' component of nodal displacement is given by

$$\{d_u\} = \begin{Bmatrix} u_1 \\ u_2 \\ u_3 \\ u_4 \\ u_5 \\ u_6 \end{Bmatrix} = \begin{bmatrix} 1 & 0 & 0 & 0 & 0 & 0 \\ 0 & 1 & 0 & 0 & 0 & 0 \\ 0 & 0 & 1 & 0 & 0 & 0 \\ 1/4 & 1/4 & 0 & 1/4 & 0 & 0 \\ 0 & 1/4 & 1/4 & 0 & 1/4 & 0 \\ 1/4 & 0 & 1/4 & 0 & 0 & 1/4 \end{bmatrix} \begin{Bmatrix} \alpha_1 \\ \alpha_2 \\ \alpha_3 \\ \alpha_4 \\ \alpha_5 \\ \alpha_6 \end{Bmatrix} \qquad (3.41)$$

i.e.

$$\{d_u\} = [A]\{\alpha\} \qquad (3.42)$$

Finite Element Analysis-Theory and Programming

and

$$\{\alpha\} = [A]^{-1} \{d_u\} \qquad (3.43)$$

The inverse of the matrix [A] can be shown to be,

$$[A]^{-1} = \begin{bmatrix} 1 & 0 & 0 & 0 & 0 & 0 \\ 0 & 1 & 0 & 0 & 0 & 0 \\ 0 & 0 & 1 & 0 & 0 & 0 \\ -1 & -1 & 0 & 4 & 0 & 0 \\ 0 & -1 & -1 & 0 & 4 & 0 \\ -1 & 0 & -1 & 0 & 0 & 4 \end{bmatrix} \qquad (3.44)$$

Thus proceeding along the same lines of linear model we get

$$\{N_2\}^T = \{\phi_2\}^T [A]^{-1} \qquad (3.45)$$

$$= [L_1^2 \quad L_2^2 \quad L_3^2 \quad L_1 L_2 \quad L_2 L_3 \quad L_3 L_1]$$

$$\times \begin{bmatrix} 1 & 0 & 0 & 0 & 0 & 0 \\ 0 & 1 & 0 & 0 & 0 & 0 \\ 0 & 0 & 1 & 0 & 0 & 0 \\ -1 & -1 & 0 & 4 & 0 & 0 \\ 0 & -1 & -1 & 0 & 4 & 0 \\ -1 & 0 & -1 & 0 & 0 & 4 \end{bmatrix}$$

$$= [(L_1^2 - L_1 L_2 - L_3 L_1) \ (L_2^2 - L_1 L_2 - L_2 L_3)$$

$$(L_3^2 - L_2 L_3 - L_3 L_1) \ 4L_1 L_2 \ 4L_2 L_3 \ 4L_3 L_1]$$

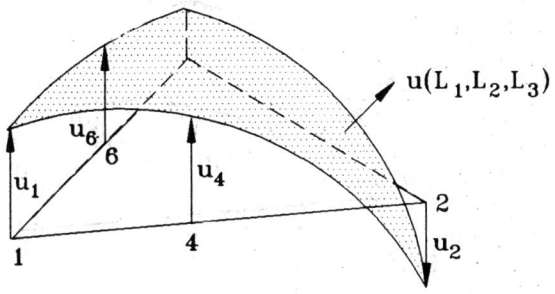

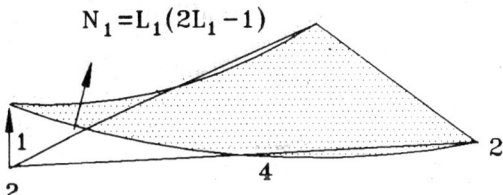

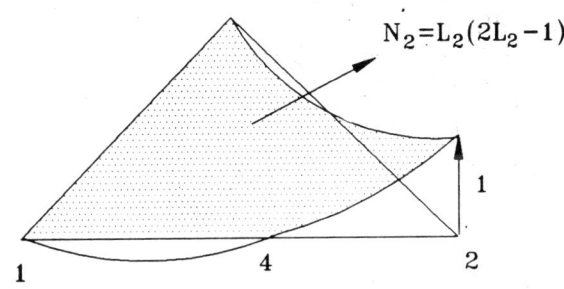

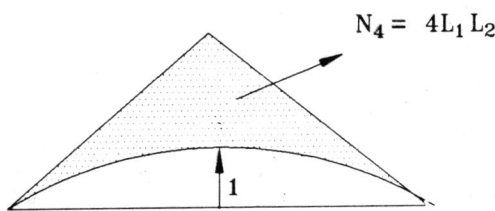

Fig. 3.9 Six noded triangular element shape functions

Simplifying,

$$\{N_2\}^T = [L_1(2L_1 - 1)\ L_2(2L_2 - 1)\ L_3(2L_3 - 1)$$

$$4L_1L_2\ 4L_2L_3\ 4L_3L_1] \tag{3.46a}$$

or can be expressed in concise form as

$$\{N_2\}^T = [N_1\quad N_2\quad N_3\quad N_4\quad N_5\quad N_6] \tag{3.46b}$$

where each component N_i is given in Eq.3.46a. Figure 3.9 shows the plot of the shape function for $N_1\ N_2$ and N_4. Thus the displacements u and v at any point inside an element can be given by,

$$\{u\} = \begin{Bmatrix} u \\ v \end{Bmatrix} = \begin{bmatrix} \{N_2\}^T & \{0\}^T \\ \{0\}^T & \{N_2\}^T \end{bmatrix} \{d\} \tag{3.47a}$$

i.e

$$\{u\} = [N]\ \{d\} \tag{3.47b}$$

where $\{N_2\}^T$ is given by the Eq.3.46 and

$$\{d\}^T = [u_1\ u_2\ u_3\ u_4\ u_5\ u_6\ v_1\ v_2\ v_3\ v_4\ v_5\ v_6] \tag{3.47c}$$

3.5.3   Shape Function for First Order Rectangular Element

The natural coordinates for the rectangular element shown in Fig.3.10 are defined by

$$r = \frac{x - x_c}{a} \qquad s = \frac{y - y_c}{b} \tag{3.48}$$

where x_c and y_c are the coordinates of the centre of the element. Assuming the polynomial function in natural coordinates, the displacement u can be expressed by

$$u = \alpha_1 + \alpha_2 r + \alpha_3 s + \alpha_4 rs \tag{3.48a}$$

The nodal displacements $\{d_u\}$ can be obtained by substituting the coordinates for the nodes as

$$\{d\} = \begin{Bmatrix} u_1 \\ u_2 \\ u_3 \\ u_4 \end{Bmatrix} = \begin{bmatrix} 1 & -1 & -1 & 1 \\ 1 & 1 & -1 & -1 \\ 1 & 1 & 1 & 1 \\ 1 & -1 & 1 & -1 \end{bmatrix} \begin{Bmatrix} \alpha_1 \\ \alpha_2 \\ \alpha_3 \\ \alpha_4 \end{Bmatrix} \tag{3.49}$$

and

$$\{\alpha\} = [A]^{-1}\{d_u\}$$

where $[A]^{-1}$ can be shown to be equal to

$$[A]^{-1} = \begin{bmatrix} 1/4 & 1/4 & 1/4 & 1/4 \\ -1/4 & 1/4 & 1/4 & -1/4 \\ -1/4 & -1/4 & 1/4 & 1/4 \\ 1/4 & -1/4 & 1/4 & -1/4 \end{bmatrix} \tag{3.50}$$

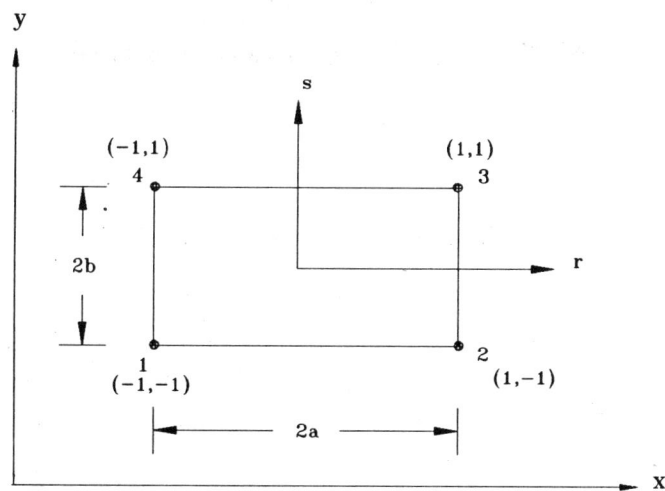

Fig. 3.10 Four noded rectangular element

Thus

$$\{N_2\}^T = \{\phi_2\}^T[A]^{-1}$$

$$= [1 \ r \ s \ r \ s] \begin{bmatrix} 1/4 & 1/4 & 1/4 & 1/4 \\ -1/4 & 1/4 & 1/4 & -1/4 \\ -1/4 & -1/4 & 1/4 & 1/4 \\ 1/4 & -1/4 & 1/4 & -1/4 \end{bmatrix} \tag{3.51}$$

and

$$\{N_2\}^T = \left[\frac{(1-r)(1-s)}{4} \quad \frac{(1+r)(1-s)}{4} \quad \frac{(1+r)(1+s)}{4} \quad \frac{(1-r)(1+s)}{4}\right]$$

$$\tag{3.52a}$$

or can be expressed in concise form as

$$\{N_2\}^T = [N_1 \; N_2 \; N_3 \; N_4] \tag{3.52b}$$

where each component N_i is given in Eq.3.52(a) Fig.3.11 shows the plot of the shape functions for N_1, N_2, N_3 and N_4. The shape function for the element is thus given by,

$$[N] = \begin{bmatrix} \{N_2\}^T & \{0\}^T \\ \{0\}^T & \{N_2\}^T \end{bmatrix} \tag{3.53}$$

where $\{N_2\}^T$ is given by Eq.(3.52). And,

$$\{u\} = [N]\{d\}$$

where $\qquad \{d\}^T = [u_1 \; u_2 \; u_3 \; u_4 \; v_1 \; v_2 \; v_3 \; v_4] \tag{3.53a}$

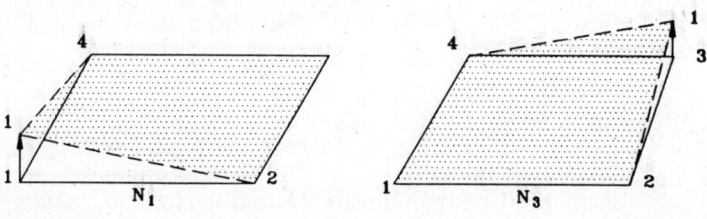

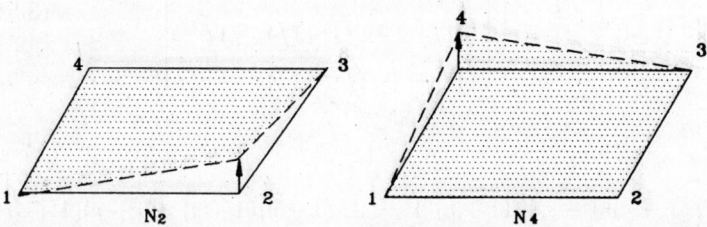

Fig. 3.11 Four noded rectangular element shape functions

3.5.4 Shape Function for Second Order Rectangular Element

The variation of displacement u can be expressed by the following polynomial in natural coordinates.

$$u = \alpha_1 + \alpha_2 r + \alpha_3 s + \alpha_4 r^2 + \alpha_5 rs + \alpha_6 s^2 + \alpha_7 r^2 s + \alpha_8 rs^2 \quad (3.54)$$

In the above expression the cubic terms r^3 and s^3 are omitted and 'geometric invariance', described in Sec.3.3.2, is maintained by the above choice of the terms.

The nodal displacements $\{d_u\}$ can be obtained by substituting the coordinates for the nodes as

$$\{d_u\} = \begin{Bmatrix} u_1 \\ u_2 \\ u_3 \\ u_4 \\ u_5 \\ u_6 \\ u_7 \\ u_8 \end{Bmatrix} = \begin{bmatrix} 1 & -1 & -1 & 1 & 1 & 1 & -1 & -1 \\ 1 & 1 & -1 & 1 & -1 & 1 & -1 & 1 \\ 1 & 1 & 1 & 1 & 1 & 1 & 1 & 1 \\ 1 & -1 & 1 & 1 & -1 & 1 & 1 & -1 \\ 1 & 0 & -1 & 0 & 0 & 1 & 0 & 0 \\ 1 & 1 & 0 & 1 & 0 & 0 & 0 & 0 \\ 1 & 0 & 1 & 0 & 0 & 1 & 0 & 0 \\ 1 & -1 & 0 & 1 & 0 & 0 & 0 & 0 \end{bmatrix} \begin{Bmatrix} \alpha_1 \\ \alpha_2 \\ \alpha_3 \\ \alpha_4 \\ \alpha_5 \\ \alpha_6 \\ \alpha_7 \\ \alpha_8 \end{Bmatrix}$$

$$(3.55)$$

and

$$\{\alpha\} = [A]^{-1}\{d_u\} \quad (3.56)$$

where $[A]^{-1}$ can be shown to be equal to

$$[A]^{-1} = \frac{1}{4} \begin{bmatrix} -1 & -1 & -1 & -1 & 2 & 2 & 2 & 2 \\ 0 & 0 & 0 & 0 & 0 & 2 & 0 & -2 \\ 0 & 0 & 0 & 0 & -2 & 0 & 2 & 0 \\ 1 & 1 & 1 & 1 & -2 & 0 & -2 & 0 \\ 1 & -1 & 1 & -1 & 0 & 0 & 0 & 0 \\ 1 & 1 & 1 & 1 & 0 & -2 & 0 & -2 \\ -1 & -1 & 1 & 1 & 2 & 0 & -2 & 0 \\ -1 & 1 & 1 & -1 & 0 & -2 & 0 & 2 \end{bmatrix} \quad (3.57)$$

Thus,

$$\{N_2\}^T = \{\phi_2\}^T [A]^{-1} \qquad (3.58)$$

where

$$\{\phi_2\}^T = [1 \quad r \quad s \quad r^2 \quad rs \quad s^2 \quad r^2s \quad rs^2] \qquad (3.58a)$$

and $[A]^{-1}$ is given by Eq. 3.57.

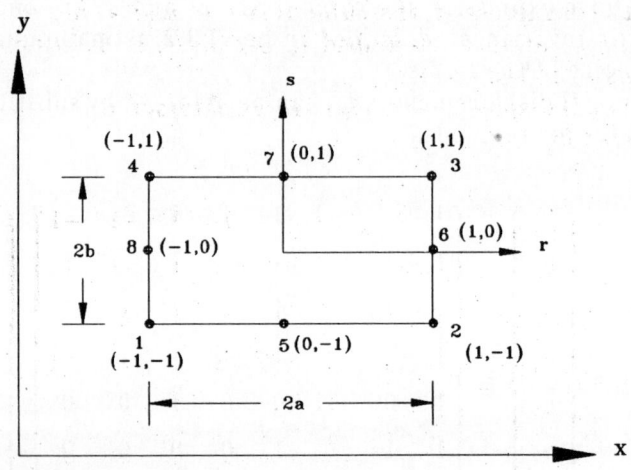

Fig. 3.12 Eight noded rectangular element

After substitution in Eq.3.58 and simplifying,

$$
\begin{aligned}
\{N_2\}^T = & [\frac{1}{4}(1-r)(1-s)(-r-s-1), \frac{1}{4}(1+r)(1-s)(r-s-1) \\
& \frac{1}{4}(1+r)(1+s)(r+s-1), \frac{1}{4}(1-r)(1+s)(-r+s-1) \\
& \frac{1}{2}(1+r)(1-r)(1-s), \frac{1}{2}(1+r)(1+s)(1-s) \\
& \frac{1}{2}(1+r)(1-r)(1+s), \frac{1}{2}(1-r)(1+s)(1-s)]
\end{aligned}
$$

$$(3.59a)$$

or can be expressed in concise form as

$$\{N_2\}^T = [N_1 \quad N_2 \quad N_3 \quad N_4 \quad N_5 \quad N_6 \quad N_7 \quad N_8] \qquad (3.59b)$$

where each element of N_i is given in Eq. 3.59(a).

The shape function for the element is thus given by

$$[N] = \begin{bmatrix} \{N_2\}^T & \{0\} \\ \{0\}^T & \{N_2\}^T \end{bmatrix} \tag{3.60}$$

where $\{N_2\}^T$ is given by Eq.3.59.

3.5.5 Lagrange and Serendipity Elements

In the earlier sections, the shape functions have been derived using explicit inverse of the $[A]$ matrix. As indicated in section 3.5. Lagrange interpolation formula or polynomials can be used to derive the shape functions and this approach is particularly simpler to derive higher order elements.

Lagrange interpolation function (or Lagrange polynomials)

The Lagrange interpolation function associated with node i is defined by,

$$f_i(r) = \frac{(r - r_1)(r - r_2)\ldots\ldots(r - r_{i-1})(r - r_{i+1})\ldots\ldots(r - r_n)}{(r_i - r_1)(r_i - r_2)\ldots\ldots(r_i - r_{i-1})(r_i - r_{i+1})\ldots\ldots(r_i - r_n)} \tag{3.61}$$

The functions f_i satisfy the property,

$$f_i(r_j) = \begin{cases} 1 & \text{if } i = j \\ 0 & \text{if } i \neq j \end{cases} \tag{3.62}$$

where r_j denotes the r coordinate of the j th node in the element. For an element with n nodes, f_i $(i = 1, 2, ...n)$ are functions of degree $n-1$. We shall define the shape function for an n noded one-dimensional element as.

$$N_i = f_i \tag{3.63}$$

Consider the two noded one-dimensional element discussed in section 3.4.1. In order to extend the derivation of shape functions to two-and three-dimensional elements, we now define the natural coordinate system r which has the origin at the centre of the element and the natural (or normalised) coordinates of nodes 1 and 2 are $r = -1$ and $r = +1$ respectively.

The natural coordinate r can be expressed as,

$$r = \frac{2(x - x_1)}{l} - 1 \tag{3.64}$$

where $x_1 = x$ coordinate of node 1.

The shape functions for the two noded one-dimensional element shown in Fig. 3.13, can be derived using Eq. 3.61.

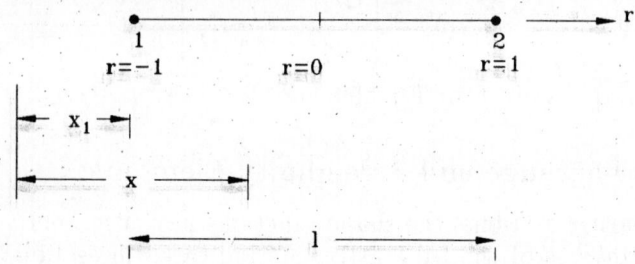

Fig. 3.13 Natural coordinates of one-dimensional element

$$N_1 = f_1(r) = \frac{(r - r_2)}{(r_1 - r_2)} = \frac{(r - 1)}{-1 - (1)} = \frac{1}{2}(1 - r) \qquad (3.65a)$$

$$N_2 = f_1(r) = \frac{(r - r_1)}{(r_2 - r_1)} = \frac{(r + 1)}{1 - (-1)} = \frac{1}{2}(1 + r) \qquad (3.65b)$$

These functions are linear as shown in Fig. 3.14.

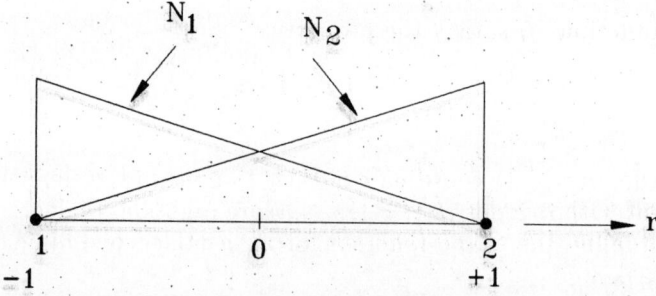

Fig. 3.14 Shape functions for two noded one-dimensional element

Consider now a three noded one-dimensional element shown in Fig. 3.15. The shape functions are given by

$$N_1(r) = f_1(r) = \frac{(r - r_2)(r - r_3)}{(r_1 - r_2)(r_1 - r_3)} = \frac{r(r - 1)}{(-1)(-2)} = \frac{1}{2}r(r - 1)$$
$$(3.66a)$$

$$N_2(r) = f_2(r) = \frac{(r - r_1)(r - r_3)}{(r_2 - r_1)(r_2 - r_3)} = \frac{(r + 1)(r - 1)}{(1)(-1)} = (1 - r^2)$$
$$(3.66b)$$

$$N_3(r) = f_3(r) = \frac{(r - r_1)\,(r - r_2)}{(r_3 - r_1)\,(r_3 - r_2)} = \frac{(r + 1)\,r}{(2)\,(1)} = \frac{1}{2}r(r+1) \quad (3.66c)$$

Figure 3.15 shows the shape functions which are quadratic.

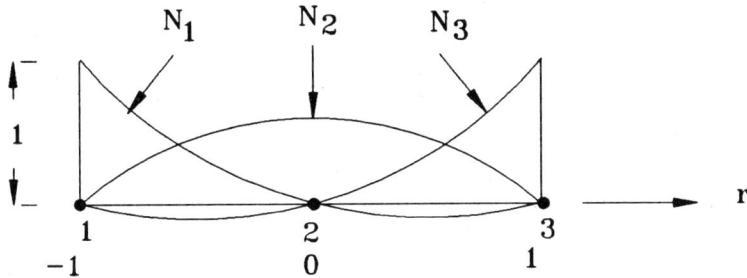

Fig. 3.15 Quadratic shape functions for one-dimensional element

The Lagrange interpolation functions for the rectangular elements can be obtained from the one-dimensional Lagrange interpolation functions presented above by taking the tensor product of the r-direction one-dimensional interpolation functions ($f_a(r)$) with the s-direction one-dimensional interpolation functions ($f_b(r)$). The formula is,

$$N_i(r,\ s) = f_a(r)\ f_b(s) \qquad (3.67)$$

where the indices have to be appropriately related to the nodes of the element and one-dimensional shape functions.

Four Noded Rectangular Element

Consider the four noded rectangular element discussed in section 3.5.3. The procedure for deriving the shape functions using Eq. 3.67 is schematically illustrated in Fig. 3.16.

Using Eqs. 3.67 and 3.65 the shape functions for the four noded element are obtained as follows,

$$N_1(r,s) = f_1(r)\ f_1(s) = \frac{1}{4}\,(1 - r)\,(1 - s)$$

$$N_2(r,s) = f_2(r)\ f_1(s) = \frac{1}{4}\,(1 + r)\,(1 - s)$$

$$N_3(r,s) = f_2(r)\ f_2(s) = \frac{1}{4}\,(1 + r)\,(1 + s)$$

$$N_4(r,s) = f_1(r) \, f_2(s) = \frac{1}{4}(1-r)(1+s) \qquad (3.68)$$

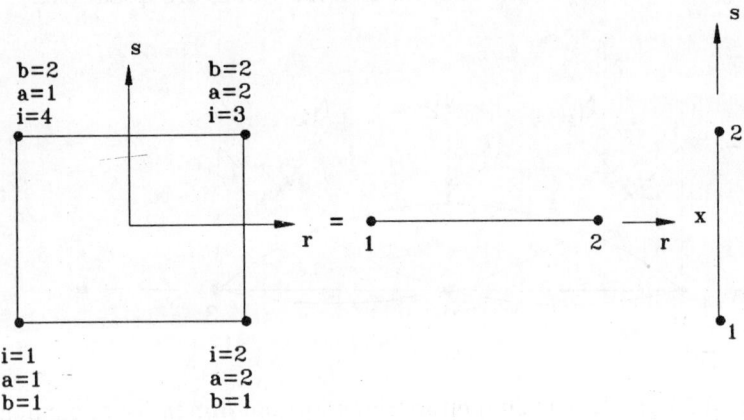

Fig. 3.16 Four noded two-dimensional element-relationships
between shape function indices

The shape functions given by Eq. 3.68 are exactly the same of Eq. 3.52a derived in section 3.5.3.

The derivation of shape function is further illustrated through a nine noded two-dimensional element.

Nine noded two-dimensional element

Figure 3.17 shows the nine noded Lagrange element.

The shape functions are obtained using Eqs. 3.67 and 3.66 and the relationship between indices given in Fig. 3.17. For example,

$$N_1(r,s) = f_1(r) \, f_1(s) = \frac{1}{4} \, rs(r-1)(s-1)$$

$$N_5(r,s) = f_2(r) \, f_1(s) = \frac{1}{2} \, s(1-r^2)(s-1)$$

Similarly, the other shape functions are derived and the shape functions for the element are given below.

$$N_1 = \frac{1}{4} \, rs \, (r-1)(s-1), \qquad N_2 = \frac{1}{4} \, rs(r+1)(s-1)$$

$$N_3 = \frac{1}{4} \, rs \, (r+1)(s+1), \qquad N_4 = \frac{1}{4} \, rs(r-1)(s+1)$$

$$N_5 = \frac{1}{2} s (1 - r^2) (s - 1), \qquad N_6 = \frac{1}{2} r(r + 1) (1 - s^2)$$

$$N_7 = \frac{1}{2} s (1 - r^2) (s + 1), \qquad N_8 = \frac{1}{2} r(r - 1) (1 - s^2)$$

$$N_9 = (1 - r^2) (1 - s^2) \qquad (3.69)$$

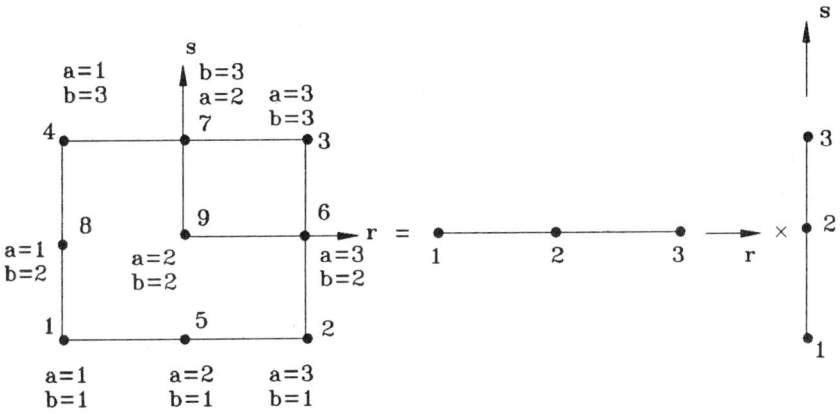

Fig. 3.17 Nine noded two-dimensional Lagrange element

The polynomial associated with Eq. 3.69 is given by,

$$\{\phi_2\}^T = [1 \quad r \quad s \quad r^2 \quad rs \quad s^2 \quad r^2 s \quad r s^2 \quad r^2 s^2] \qquad (3.70)$$

The above procedure can be extended to derive the shape functions for eight and twenty seven noded three-dimensional elements. Elements whose shape functions are formed from the products of one-dimensional Lagrange interpolation functions are called Lagrange elements.

Serendipity Elements

The higher order Lagrange elements do have internal nodes as may be seen for the nine-noded element shown in Fig. 3.17. These internal nodes do not connect with adjoining elements and hence can be condensed out at the element level using the procedure given in section 3.8. Alternatively, we can develop elements with nodes present only on the boundaries (i.e with no internal nodes) and these elements are called *serendipity elements*. A typical example is the eight noded rectangular element described in section 3.5.4.

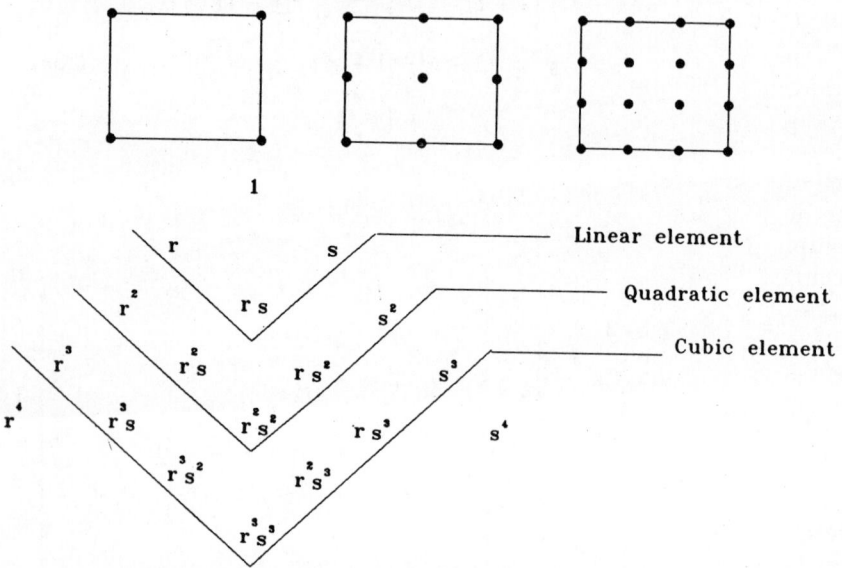

Fig. 3.18a Lagrange Elements and Pascal Triangle

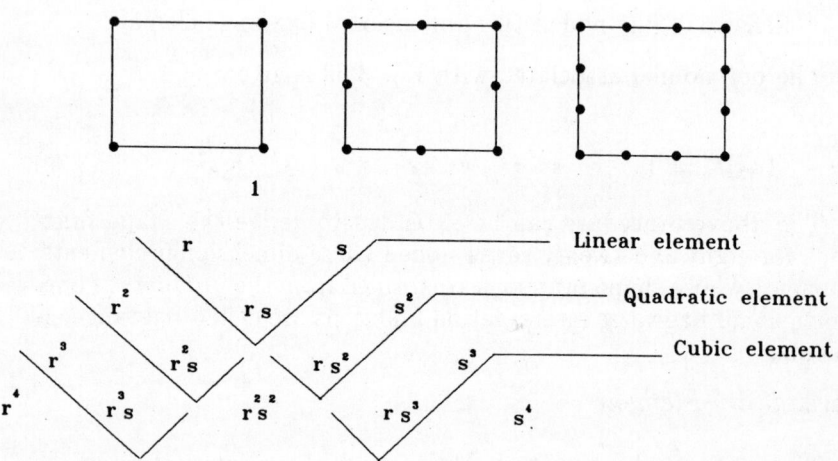

Fig. 3.18b Serendipity elements and Pascal Triangle

The shape functions for these elements can not be obtained using tensor products of one-dimensional Lagrange interpolation functions. The procedure followed for the eight noded serendipity element given in section 3.5.4 can be used. However, alternative techniques are suggested in reference 1.

Figure 3.18a shows the Pascal triangles for two-dimensional Lagrange rectangular elements. Examination of the polynomials given in Fig. 3.18b shows that the serendipity elements leave out the middle terms of the Pascal triangle. The Lagrange elements (which have one or more internal nodes) have these terms and hence, have a better degree of completeness in polynomial function. The degree of completeness is generally considered to be the important measure of accuracy of an element. Thus, it may be seen that Lagrange elements have better accuracy.

3.5.6 Elements with Variable Number of Nodes

In some cases of finite element analysis we may have to mix elements with different number of nodes. This situation usually arises in discretization of stress concentration areas where a higher order element will be used and regions away from the critical zones a lower order element can be used.

Suppose for example (Fig.3.19) that a six noded triangular element is used with an element like 2, which has five nodes, two of which lie along the interelement boundary 2-3. Similarly the element 4 has only four nodes, three of which lie along the interelement boundary 2-6-7.

The shape functions for the five noded and four noded elements such as for 2 and 4 can be developed by suitably degrading the shape functions of six-noded triangular element as illustrated below.

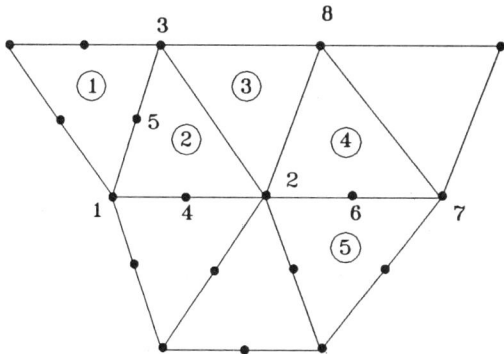

Fig. 3.19 Mixing of elements with different number of nodes

Five Noded Triangular Element

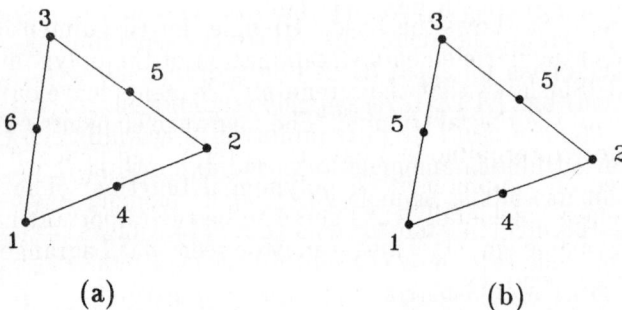

(a)   (b)

Fig. 3.20 Degrading for five noded element

The shape function for a six noded triangular element is given by Eq.3.47, i.e,

$$u = N_1 u_1 + N_2 u_2 + N_3 u_3 + N_4 u_4 + N_5 u_5 + N_6 u_6 \qquad (3.71)$$

i.e., expressed as the components of $\{N_2\}^T$ and are given by Eq.3.46

$$N_1 = L_1(2L_1 - 1), \quad N_2 = L_2(2L_2 - 1), \quad N_3 = L_3(2L_3 - 1),$$

$$N_4 = 4L_1 L_2, \quad N_5 = 4L_2 L_3, \quad N_6 = 4L_3 L_1$$

Referring to Fig.3.20b for the five noded triangular element the displacement along the line 2-3 is constrained to vary linearly and hence the displacement at 5′ may be expressed as

$$u_{5'} = \frac{u_2 + u_3}{2} \qquad (3.72)$$

Substituting the values of $u_{5'}$ for u_5 in Eq.3.71 we get

$$u = N_1 u_1 + N_2 u_2 + N_3 u_3 + N_4 u_4 + N_5 \left(\frac{u_2 + u_3}{2}\right) + N_6 u_6$$

$$= N_1 u_1 + \left(N_2 + \frac{N_5}{2}\right) u_2 + \left(N_3 + \frac{N_5}{2}\right) u_3 + N_4 u_4 + N_6 u_6 \qquad (3.73)$$

The shape function for the five noded triangular element can be expressed as

$$u = N_1' u_1 + N_2' u_2 + N_3' u_3 + N_4' u_4 + N_5' u_5$$

Comparing Eqs.3.71 and 3.73 and noting that node 6 of six noded triangle corresponds to node 5 of five noded triangle, we get

$$N_1' = N_1, \quad N_2' = N_2 + \frac{N_5}{2}, \quad N_3' = N_3 + \frac{N_5}{2}, \quad N_4' = N_4, \quad N_5' = N_6 \quad (3.74a)$$

Thus

$$
\begin{aligned}
N_1' &= L_1(2L_1 - 1) \\
N_2' &= L_2(2L_2 - 1) + \frac{4L_2 L_3}{2} = L_2(1 - 2L_1) \\
N_3' &= N_3 + \frac{N_5}{2} = L_3(2L_3 - 1) + \frac{4L_2 L_3}{2} = L_3(1 - 2L_1) \\
N_4' &= 4L_1 L_2 \\
N_5' &= 4L_3 L_1
\end{aligned}
$$

$$(3.74b)$$

Four Noded Triangular Element

The shape function for the four noded triangular element can be derived following the same procedure as indicated above. Thus,

$$u_{5'} = \frac{u_2 + u_3}{2} \quad \text{and} \quad u_{6'} = \frac{u_3 + u_1}{2} \quad (3.75)$$

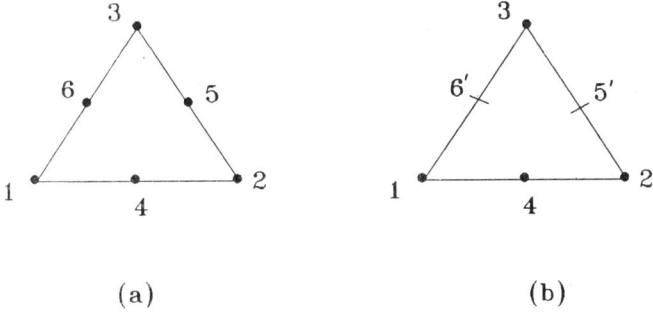

(a)   (b)

Fig. 3.21 Degrading for four noded element

Substituting the values of $u_{5'}$ and $u_{6'}$ in Eq.3.71 we get

$$u = N_1 u_1 + N_2 u_2 + N_3 u_3 + N_4 u_4 + N_5 \left(\frac{u_2 + u_3}{2}\right) + N_6 \left(\frac{u_3 + u_1}{2}\right)$$

$$= \left(N_1 + \frac{N_6}{2}\right) u_1 + \left(N_2 + \frac{N_5}{2}\right) u_2 + \left(N_3 + \frac{N_5 + N_6}{2}\right) u_3 + N_4 u_4 \quad (3.76)$$

The shape function for the four noded element can be expressed as,

$$u = N_1' u_1 + N_2' u_2 + N_3' u_3 + N_4' u_4 \quad (3.77)$$

Comparing Eqs.3.76 and 3.77, we get,

$$N_1' = N_1 + \frac{N_6}{2} = L_1(2L_1 - 1) + \frac{4L_3 L_1}{2} = L_1(1 - 2L_2)$$

$$N_2' = N_2 + \frac{N_5}{2} = L_2(2L_2 - 1) + \frac{4L_2 L_3}{2} = L_1(1 - 2L_1)$$

$$N_3' = N_3 + \frac{N_5 + N_6}{2} = L_3(2L_3 - 1) + \frac{4L_2 L_3 + 4L_3 L_1}{2} = L_3$$

$$N_4' = N_4 = 4L_1 L_2 \quad (3.78)$$

3.6   Element Strains and Stresses

The strain displacement relations of the theory of elasticity described in Chapter 2 could be used to derive the relation between the strains at any point inside an element in terms of nodal displacements. If $\{\epsilon\}$ is the vector of strains at any point inside an element, the relation can be expressed as,

$$\{\epsilon\} = [B]\{d\} \quad (3.79)$$

where [B] is called the strain-displacement matrix and will be functions of the derivatives of shape function [N].

If $\{\sigma\}$ is the vector of stresses corresponding to the strains $\{\epsilon\}$, the element stresses can be expressed in the form,

$$\{\sigma\} = [C][B]\{d\} \quad (3.30)$$

where [C] is the constitutive matrix.

Example: For the constant strain triangle shown in Fig.3.6, the [B] matrix can be worked out as follows: The displacement model for this element is given by Eq.3.39 as

$$\{u\} = \begin{bmatrix} L_1 & L_2 & L_3 & 0 & 0 & 0 \\ 0 & 0 & 0 & L_1 & L_2 & L_3 \end{bmatrix} \{d\}$$

where

$$\{d\}^T = \begin{bmatrix} u_1 & u_2 & u_3 & v_1 & v_2 & v_3 \end{bmatrix}$$

From Eq.2.9

$$\epsilon_x = \frac{\partial u}{\partial x} \quad \epsilon_y = \frac{\partial v}{\partial y} \quad \gamma_{xy} = \frac{\partial u}{\partial y} + \frac{\partial v}{\partial x}$$

Using Eqs.3.23a and 3.23b

$$\frac{\partial u}{\partial x} = \frac{b_1}{2A} u_1 + \frac{b_2}{2A} u_2 + \frac{b_3}{2A} u_3 = \frac{1}{2A}(b_1 u_1 + b_2 u_2 + b_3 u_3) \tag{3.81}$$

Similarly,

$$\frac{\partial v}{\partial y} = \frac{1}{2A}(a_1 v_1 + a_2 v_2 + a_3 v_3) \tag{3.82}$$

and

$$\frac{\partial u}{\partial y} + \frac{\partial v}{\partial x} = \frac{1}{2A}(a_1 u_1 + a_2 u_2 + a_3 u_3 + b_1 v_1 + b_2 v_2 + b_3 v_3) \tag{3.83}$$

Thus,

$$\{\epsilon\} = \begin{Bmatrix} \epsilon_x \\ \epsilon_y \\ \gamma_{xy} \end{Bmatrix} = \frac{1}{2A} \begin{bmatrix} b_1 & b_2 & b_3 & 0 & 0 & 0 \\ 0 & 0 & 0 & a_1 & a_2 & a_3 \\ a_1 & a_2 & a_3 & b_1 & b_2 & b_3 \end{bmatrix} \{d\}$$

$$\tag{3.84}$$

i.e.

$$\{\epsilon\} = [B]\{d\}$$

where

$$[B] = \frac{1}{2A} \begin{bmatrix} b_1 & b_2 & b_3 & 0 & 0 & 0 \\ 0 & 0 & 0 & a_1 & a_2 & a_3 \\ a_1 & a_2 & a_3 & b_1 & b_2 & b_3 \end{bmatrix} \tag{3.85}$$

3.7 Element Stiffness Matrix

In the previous section the element strains and stresses have been expressed in terms of the nodal displacements through the use of strain-displacement matrix. It can also be noted that once the shape function for an element is chosen the [B] matrix can be evaluated. Now it is necessary to derive the equilibrium condition for an element which would help us to build up the condition for the assemblage of elements constituting the body or structure.

We have described in Chapter 2, the principle of virtual displacement and principle of stationary potential energy which establish the condition of equilibrium for displacement formulation. It has been established therein that both the principles lead to identical conditions and are illustrated in the following derivation. However, it must be noted the principle of stationary potential energy helps us also to establish the bounds and convergence conditions on the solution and these are described in references [1, 11].

Following the Eqs. 2.38a and 2.39a, the principle of virtual displacements yields,

$$\int\int\int_V \{\delta\epsilon\}^T\{\sigma\}dV = \int\int\int_V \{\delta u\}^T\{X\}dV + \int\int_{S1} \{\delta u\}^T\{p\}dS \quad (3.86)$$

Substituting from Eq.(3.80)

$$\{\sigma\} = [C][B]\{d\}$$

into Eq.3.86 above,

$$\int\int\int_V \{\delta\epsilon\}^T[C][B]\{d\}dV = \int\int\int_V \{\delta u\}^T\{X\}dV + \int\int_{S1} \{\delta u\}^T\{p\}dS$$

Applying strain-displacement relationships,

$$\{\epsilon\} = [B]\{d\}$$

Since [B] is independent of displacements, the virtual strain $\{\delta\epsilon\}$ due to virtual displacement $\{\delta d\}$ is given by

$$\{\delta\epsilon\} = [B]\{\delta d\} \quad (3.87)$$

Also from Eq.3.11, $\{u\} = [N]\{d\}$ it can be observed that,

$$\{\delta u\} = [N]\{\delta d\} \quad (3.88)$$

Substituting the above relationships in Eq.3.86 we get

$$\{\delta d\}^T \int\int\int_V [B]^T[C][B]dV\{d\} = \{\delta d\}^T \int\int\int_V [N]^T\{X\}dV$$

$$+\{\delta d\}^T \int\int_{S1} [N^S]^T\{p\}dS \quad (3.89)$$

where $[N^S]$ is the shape function along the boundary S_1 where forces are prescribed. Since the variations of nodal displacements are arbitrary the above equilibrium equation can be written as

$$\int \int \int_V [B]^T[C][B]dV\{d\} = \int \int \int_V [N]^T\{X\}dV + \int \int_{S1} [N^S]^T\{p\}dS$$

(3.90)

i.e.

$$[k]\{d\} = \{Q\}$$

(3.90a)

where $[k]$ is called the stiffness matrix of the element and $\{Q\}$ is the nodal load vector, and are given by

$$[k] = \int \int \int_V [B]^T[C][B]dV$$

(3.91)

$$\{Q\} = \int \int \int_V [N]^T\{X\}dV + \int \int_{S1} [N^S]^T\{p\}dS$$

(3.92)

The Eq.3.90 of equilibrium can be also derived by using the principle of stationary potential energy. The total potential energy for an element is expressed by using Eq.2.62 as

$$\Pi = \frac{1}{2} \int \int \int_V \{\epsilon\}^T[C]\{\epsilon\}dV - \int \int \int_V \{u\}^T\{X\}dV - \int \int_{S1} \{u\}^T\{p\}dS$$

Substituting for $\{u\}$ and $\{\epsilon\}$ from Eqs. 3.11 and 3.79, we get

$$\Pi = \frac{1}{2} \int \int \int_V \{d\}^T[B]^T[C][B]\{d\}dV - \int \int \int_V \{d\}^T[N]^T\{\dot{X}\}dV$$

$$- \int \int_{S1} \{d\}^T[N^S]^T\{p\}dS$$

(3.93)

where V is the volume of the element and S_1 is the portion of the boundary where forces are prescribed.

According to the principle of stationary potential energy, the first variation of Π must be zero for equilibrium condition. Thus taking the first variation of Eq.3.93, we get

$$\delta\Pi = \{\delta d\}^T \left(\int \int \int_V [B]^T[C][B]dV\{d\} - \int \int \int_V [N]^T\{X\}dV \right.$$

$$\left. - \int \int_{S1} [N^S]^T\{p\}dS \right) = 0$$

(3.94)

Since Eq.3.94 should hold good for any variation of $\{\delta d\}$

$$\int\int\int_V [B]^T [C][B] dV \{d\} - \int\int\int_V [N]^T \{X\} dV - \int\int_{S1} [N^S]^T \{p\} dS = 0$$

(3.95)

The above equation is exactly the same equilibrium condition expressed by Eq.3.90. Defining the stiffness matrix and nodal load vector by Eqs.3.91 and 3.92, the equilibrium condition is expressed by

$$[k]\{d\} = \{Q\}$$

(3.90a)

In case the element is subjected to initial strains and stresses, the strain energy U, given in Eq. 2.60b is to be used in the calculation of the potential energy, Π. Following the same procedure as above, the equilibrium equation for the element can be obtained as,

$$\int\int\int_V [B]^T [C] [B] \, dV \{d\} - \int\int\int_V [B]^T [C] \{\epsilon_o\} \, dV + \int\int\int_V [B]^T \{\sigma_o\} \, dV$$

$$-\int\int\int_V [N]^T \{X\} \, dV - \int\int_{S_1} [N^S]^T \{p\} \, dS = 0$$

(3.96)

Thus the equilibrium equation is reduced to the same form,

$$[k] \{d\} = \{Q\}$$

(3.96a)

where

$$[k] = \int\int\int_V [B]^T [C] [B] \, dV$$

(3.97a)

$$\{Q\} = \int\int\int_V [B]^T [C] \{\epsilon_o\} \, dV - \int\int\int_V [B]^T \{\sigma_o\} \, dV$$

$$+ \int\int\int_V [N]^T \{X\} \, dV + \int\int_{S_1} [N^S]^T \{p\} \, dS$$

(3.97b)

The evaluation of the stiffness matrix and nodal load vector are illustrated by a simple example given below. Also in chapters dealing with the applications to various analyses, examples are included for computation of the above properties for specific type of elements.

Example: Constant Strain Triangular Stiffness Matrix

The shape function for constant strain triangular element is given by Eq.3.39

$$[N] = \begin{bmatrix} L_1 & L_2 & L_3 & 0 & 0 & 0 \\ 0 & 0 & 0 & L_1 & L_2 & L_3 \end{bmatrix}$$

The strain-displacement relation is given by Eq.3.85

$$[B] = \frac{1}{2A} \begin{bmatrix} b_1 & b_2 & b_3 & 0 & 0 & 0 \\ 0 & 0 & 0 & a_1 & a_2 & a_3 \\ a_1 & a_2 & a_3 & b_1 & b_2 & b_3 \end{bmatrix}$$

If the thickness of the element is constant of value 'h' the stiffness matrix for the element is given by

$$[k] = h \int \int [B]^T [C][B] \, dx dy \qquad (3.98)$$

It can be observed that the matrix [B] is constant and the constitutive matrix [C] depends on the type of stress analysis, i.e., plane stress or plane strain and can be obtained from Eqs. 2.17 and 2.21. Hence, the stiffness matrix of the element can be worked out as

$$[k] = h[B]^T [C][B] \int \int dx dy$$

$$[k] = Ah \, [B]^T \, [C] \, [B] \qquad (3.99)$$

Thus, in this case the computation of stiffness matrix is only an evaluation of triple product of the above matrices.

3.7.1 Nodal Load Vector

The nodal load vector is given by

$$\{Q\} = h \int \int [N]^T \{X\} dA + h \int [N^S]^T \{p\} ds \qquad (3.100)$$

where $\{X\}$ is the vector of body forces and
$\{p\}$ is the vector of surface traction.
For a two-dimensional plane stress or strain element considered in the present example, the body force components can be given as

$$\{X\}^T = [\bar{X} \quad \bar{Y}] \qquad (3.101)$$

For gravity loads $\bar{X} = 0$ and $\bar{Y} = -\rho$ g where ρ is the mass density. The inplane surface forces are

$$\{p\}^T = [p_x \quad p_y] \qquad (3.102)$$

where p_x and p_y are the prescribed load intensities which are positive if applied in the positive coordinate directions.

The nodal load vector due to body forces can be computed as

$$h \int \int [N]^T \{X\} dA = h \int \int \begin{bmatrix} L_1 & 0 \\ L_2 & 0 \\ L_3 & 0 \\ 0 & L_1 \\ 0 & L_2 \\ 0 & L_3 \end{bmatrix} \begin{Bmatrix} \bar{X} \\ \bar{Y} \end{Bmatrix} dA$$

Applying the integration rule, vide Eq.3.24, we get

$$h \int \int [N]^T \{X\} dA = \frac{hA}{3} \begin{Bmatrix} \bar{X} \\ \bar{X} \\ \bar{X} \\ \bar{Y} \\ \bar{Y} \\ \bar{Y} \end{Bmatrix} = \frac{-\rho g h A}{3} \begin{Bmatrix} 0 \\ 0 \\ 0 \\ 1 \\ 1 \\ 1 \end{Bmatrix} \tag{3.103}$$

if the body forces are only due to gravity loads as mentioned above.

To illustrate the details of working of nodal load vector due to surface traction, consider a load intensity of $p_x^{(1)}$ uniformly distributed along side 1 shown in Fig.3.22.

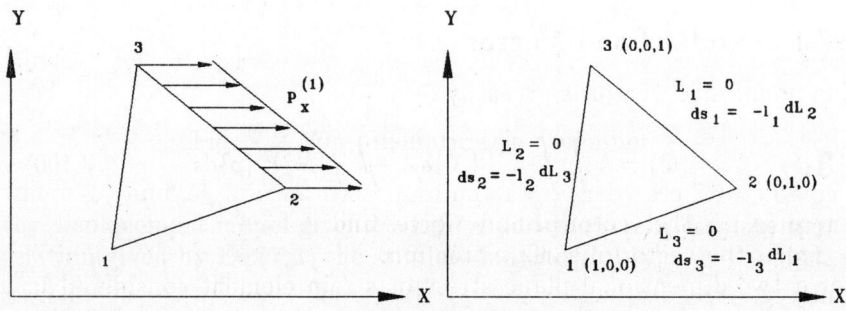

Fig. 3.22 Triangular element subjected to surface traction

Along the side '1' of the element, $L_1 = 0$ and $L_3 = 1 - L_2$ and it can be shown that

$$ds_1 = -l_1 \, dL_2$$

where l_1 is the length of the side 1.

The nodal load vector $\{Q\}$ corresponding to $p_x^{(1)}$ is given by

$$h \int [N^S]^T \{p\} ds_1 = h l_1 \int_0^1 \begin{bmatrix} 0 & 0 \\ L_2 & 0 \\ (1-L_2) & 0 \\ 0 & 0 \\ 0 & L_2 \\ 0 & (1-L_2) \end{bmatrix} \left\{ \begin{matrix} p_x^{(1)} \\ 0 \end{matrix} \right\} dL_2$$

$$= \frac{h p_x^{(1)} l_1}{2} \left\{ \begin{matrix} 0 \\ 1 \\ 1 \\ 0 \\ 0 \\ 0 \end{matrix} \right\} \tag{3.104}$$

Proceeding on similar lines the nodal load vector due to gravity loads and surface traction of intensities $p_x^{(i)}$ and $p_y^{(i)}$ acting on a side i can be shown to be

$$\{Q\} = -\frac{pghA}{3} \left\{ \begin{matrix} 0 \\ 0 \\ 0 \\ 1 \\ 1 \\ 1 \end{matrix} \right\} + \frac{h}{2} \left\{ \begin{matrix} p_x^{(2)} \ l_2 + p_x^{(3)} \ l_3 \\ p_x^{(1)} \ l_1 + p_x^{(3)} \ l_3 \\ p_x^{(1)} \ l_1 + p_x^{(2)} \ l_2 \\ p_y^{(2)} \ l_2 + p_y^{(3)} \ l_3 \\ p_y^{(1)} \ l_1 + p_y^{(3)} \ l_3 \\ p_y^{(1)} \ l_1 + p_y^{(2)} \ l_2 \end{matrix} \right\} \tag{3.105}$$

3.8   Static Condensation

It is possible to construct an element by assembling the basic simple elements. For example, the four noded quadrilateral element shown in Fig.3.23a can be formed by assembling four CST elements. Also in some cases internal nodes maybe needed to include all the terms of the shape function, as is the case of triangular element shown in Fig.3.23b which has nine external nodes and one internal node necessary for completeness of the cubic interpolation used for the displacement model.

It may be observed that the internal nodes of such elements do not connect with the adjoining elements in the assemblage. Hence, the displacements of these internal nodes are not required to formulate the overall equilibrium equations for the structure. We may, therefore, eliminate such internal degrees of freedom and thus express the stiffness matrix and nodal load vector corresponding to displacements of external nodes of the element. The procedure is explained below.

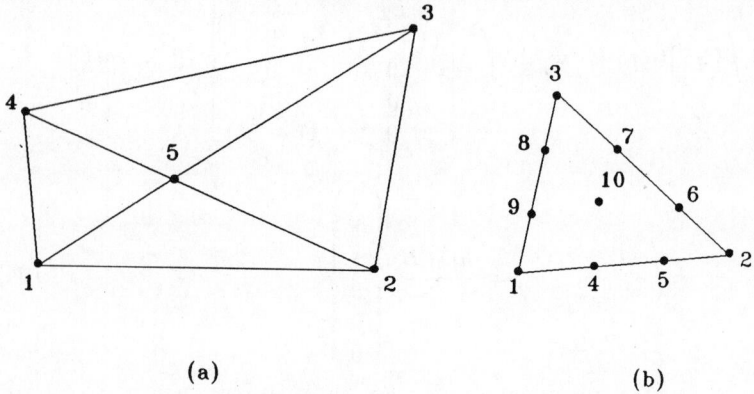

(a)   (b)

Fig. 3.23 Elements with internal nodes

The equilibrium equation for the element is given by Eq.3.90

$$[k]\{d\} = \{Q\}$$

Rearranging the above equation by partitioning the relevant terms corresponding to external and internal degrees of freedom

$$\left[\begin{array}{c|c} [k_{11}] & [k_{12}] \\ \hline [k_{21}] & [k_{22}] \end{array}\right] \left\{\begin{array}{c} \{d_1\} \\ \hline \{d_2\} \end{array}\right\} = \left\{\begin{array}{c} \{Q_1\} \\ \hline \{Q_2\} \end{array}\right\} \qquad (3.106)$$

where $\{d_2\}$ is the vector of displacements and $\{Q_2\}$ is the vector of loads corresponding to degrees of freedom of internal nodes. The above equations can be expressed as

$$[k_{11}]\{d_1\} + [k_{12}]\{d_2\} = \{Q_1\} \qquad (3.107a)$$

$$[k_{21}]\{d_1\} + [k_{22}]\{d_2\} = \{Q_2\} \qquad (3.107b)$$

Solving Eq.3.107 for $\{d_2\}$,

$$\{d_2\} = -[k_{22}]^{-1}[k_{21}]\{d_1\} + [k_{22}]^{-1}\{Q_2\} \qquad (3.108)$$

Substituting the above solution for $\{d_2\}$ into Eq.3.107a, we get

$$[k_{11}]\{d_1\} - [k_{12}][k_{22}]^{-1}[k_{21}]\{d_1\} = \{Q_1\} - [k_{12}][k_{22}]^{-1}\{Q_2\} \qquad (3.109)$$

The above Eq.3.106 can be written as

$$[\bar{k}]\{d_1\} = \{\bar{Q}\} \tag{3.110}$$

where

$$[\bar{k}] = [k_{11}] - [k_{12}][k_{22}]^{-1}[k_{21}] \tag{3.111}$$

$$\{\bar{Q}\} = \{Q_1\} - [k_{12}][k_{22}]^{-1}\{Q_2\} \tag{3.112}$$

where $[\bar{k}]$ is called effective, reduced or condensed stiffness matrix and $\{\bar{Q}\}$ is the effective nodal load vector, corresponding to external nodes of the element. This procedure is called 'Static Condensation'. However, from computational point of view the 'static condensation' is efficiently carried out by Gauss elimination procedure of solving equations. It involves only a few steps in computer programming as is illustrated in reference [2].

REFERENCES

1. Zienkiewicz, O.C. and R.L.Taylor, *The Finite Element Method, Vol.1, Basic Formulation and Linear Problems*, McGraw-Hill (U.K) Limited, 1989.

2. Desai, C.S. and J.F.Abel, *Introduction to the Finite Element Method.*, Affiliated East-West Press Pvt. Ltd., New Delhi, East-West Student Edition, 1977.

3. Dunne, P., *Complete Polynomial Displacement Fields for the Finite Element Method*, The Aeronautical Journal, Vol. 72, 1968.

4. Felippa, C.A., *Refined Finite Element Analysis of Linear and Nonlinear Two-Dimensional Structures*, Ph.D. Dissertation, University of California, Berkeley, 1966.

5. Irons, B.M. and K.J. Drapper *Inadequacy of Nodal Connections in a Stiffness Solution of Plate Bending*, AIAA J., Vol.3, No.5, May 1965.

6. Fried, I., *Some Aspects of the Natural Coordinate System in the Finite Element Method*, AIAA, J., Vol.7, No.7, July 1969.

7. Irons, B.M. and A.Razaaque, *Experience with the patch Test. The Mathematical Foundations of the Finite Element Method with Applications to Partial Differential Equations*, K.Aziz, (Ed.)., Academic Press, London, pp 557-587, 1972.

9. Scarborough, J.B., *Numerical Mathematical Analysis*, The Johns Hopkins Press, Baltimore, U.S.A., 1966.

10. Taylor, R.L., *On the Completeness of Shape Functions for Finite Element Analysis.*, International Journal for Numerical Methods in Engineering, Vol.4, No.4, pp.17-22, 1972.

11. Strang, G. and G.J. Fix, *An Analysis of the Finite Element Method*, Prentice-Hill, Englewood Cliffs, N.J., 1973.

12. Eisenberg, M.A., and L.E. Malvern, *On Finite Element Integration in Natural Coordinates*, International Journal for Numerical Methods in Engineering, vol.7 No.4. pp.574-75, 1973.

13. Clough, R.W., *Comparison of Three-Dimensional Finite Elements*, Proc. Symposium on Application of Finite Elements in Civil Engineering, Vanderbilt University, Nashville, Tenn. Nov. pp 1-26, 1969.

EXERCISES

3.1 Using the Lagrange interpolation formula construct the shape function in natural coordinates for one-dimensional axial element with 4 nodes. Sketch the shape functions.

3.2 Using the bilinear shape function for a four noded rectangular element, develop the shape function for a five noded rectangular element.

3.3 Develop the shape functions for a twelve noded rectangular element belonging to the Serendipity family. Sketch the shape functions.

3.4 Using the degeneration technique derive expressions for four through nine noded two-dimensional element. Draw a flow chart for programming purposes. Check the results for the eight noded serendipity element with shape function expressions given in equation (3.59a).

3.5 Develop the shape function for an eight noded element shown in Fig. 9.10.

3.6 Develop the shape functions for 20 noded 3D element shown in Fig. 9.10.

3.7 Derive the shape function for a wedge-shaped element shown in Fig. E3.7 by degenerating the eight noded hexahedral element.

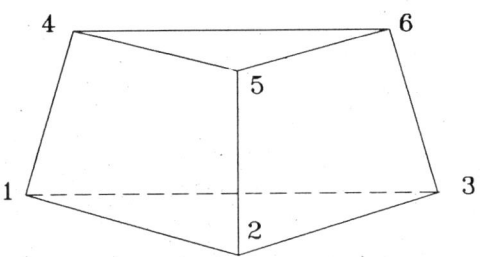

Fig. E3.7

3.8 Show that the shape functions along the interelement boundary 3-4 are the same for the two rectangular elements shown in Fig. E3.8.

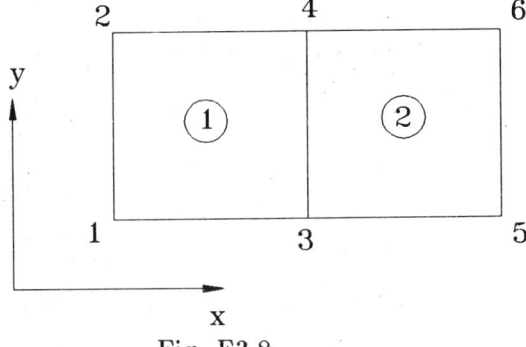

Fig. E3.8

3.9 Show that the shape function representing the variation of displacement along the inter element boundary 2-3 is the same for the two elements shown in Fig. E3.9.

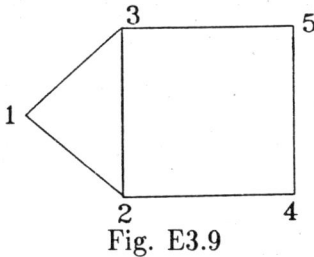

Fig. E3.9

3.10 A finite element mesh is shown in Fig. E3.10. Develop the shape function for the transition element so that continuity of displacement is maintained across the inter-element boundary.

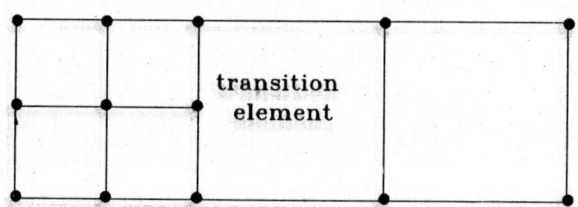

Fig. E3.10

3.11 Test the adequacy of the displacement functions used for the following elements with respect to satisfaction of equilibrium conditions within the element.

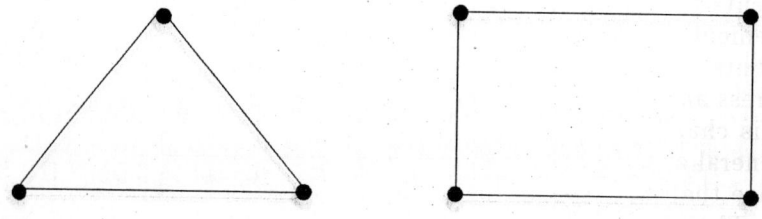

Fig. E3.11

3.12 Derive the stiffness matrix for a three noded triangular element with varying thickness. Assume thickness is known at the nodes as h_1, h_2 and h_3.

3.13 An axial bar with three nodes is subjected to a linearly varying axial force. Compute the nodal load vector.

3.14 Compute the nodal load vector for the bar in exercise 3.13 when it is subjected to a quadratically varying axial force.

3.15 Compute the nodal load vector due to self-weight for the four noded rectangular element.

Chapter 4

Isoparametric Elements

For the analysis of structural problems of complex shapes involving curved boundaries or surfaces, simple triangular or rectangular elements are no longer sufficient. This has led to the development of elements of more arbitrary shape and are called isoparametric elements. These elements are widely used in two and three-dimensional stress analysis and, plates and shell problems. It would be shown in this chapter that the formulation of isoparametric element is quite general and it becomes easy to deal with different types of elements once the basic formulation is understood.

The concept of isoparametric element is based on the transformation of the parent element in local or natural coordinate system to an arbitrary shape in the cartesian coordinate system as shown by examples in Fig.4.1. A convenient way of expressing the transformation is to make use of the shape functions of the rectilinear elements in their natural coordinate system and the nodal values of the coordinates. Thus the cartesian coordinates of a point in an element may be expressed as [1]

$$x = N_1' \, x_1 + N_2' \, x_2 + \cdots + N_n' \, x_n$$

$$y = N_1' \, y_1 + N_2' \, y_2 + \cdots + N_n' \, y_n$$

$$z = N_1' \, z_1 + N_2' \, z_2 + \cdots + N_n' \, z_n$$

or in matrix form

$$\{x\} = [N'] \, \{x_n\} \tag{4.1}$$

where $[N']$ are the shape functions of the parent rectilinear element and $\{x_n\}$ are the nodal coordinates of the element. The shape functions will be expressed through the natural coordinate system r, s and t.

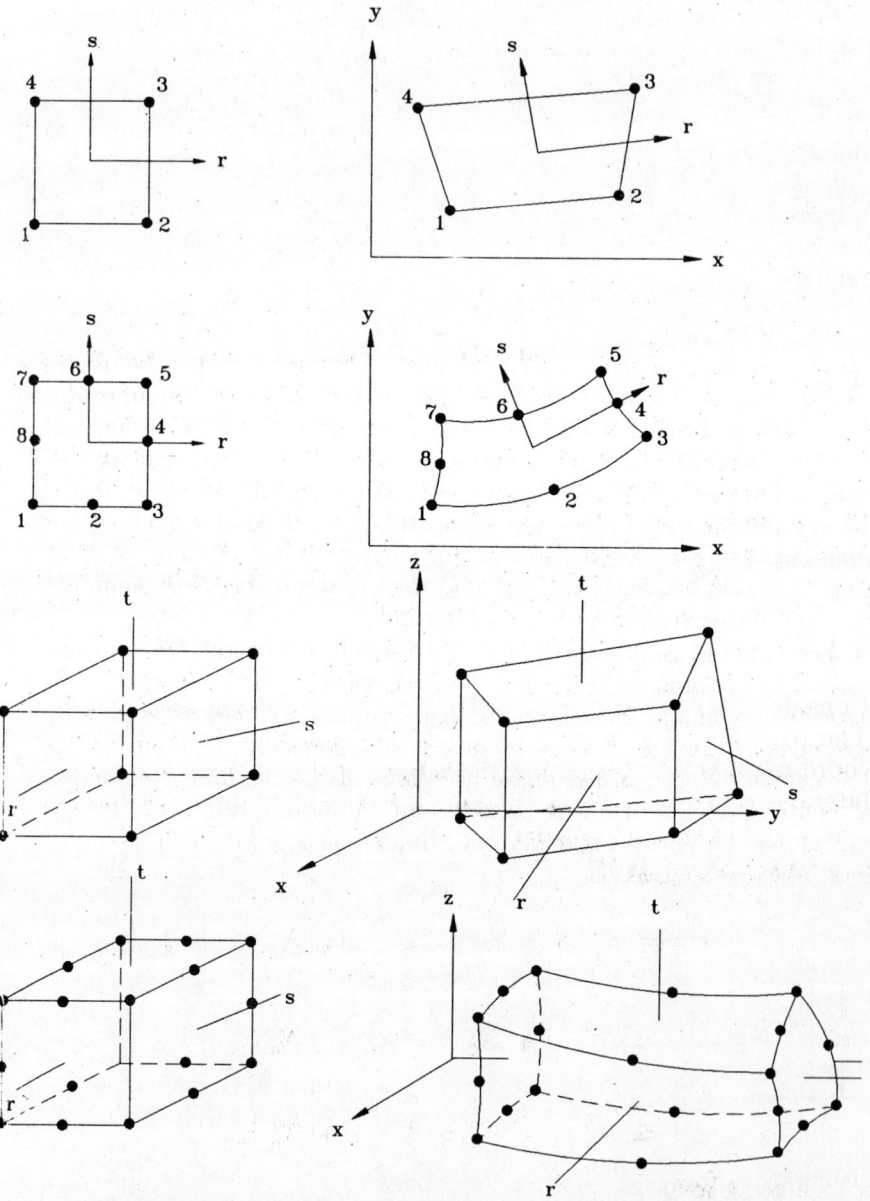

Fig. 4.1 Isoparametric elements-coordinate transformation

The shape functions $[N']$ used in the above transformation thus help us to define the geometry of the element in the cartesian coordinate system. If these shape functions $[N']$ are the same as the shape functions $[N]$ used to represent the variation of displacement in the element, these elements are called 'isoparametric' elements.

$$\text{i.e.,} \quad \{x\} = [N]\{x_n\} \tag{4.1a}$$

And in cases where the geometry of the element is defined by shape functions of order higher than that for representing the variation of displacements, the elements are called 'superparametric'. Similarly if more nodes are used to define displacement compared to the nodes used to represent the geometry of the elements, then they would be referred to as 'subparametric' elements.

4.1 Two-Dimensional Isoparametric Elements

The formulation and the computation of the element properties will be illustrated through a detailed working for a four noded plane element.

4.1.1 Four-Noded Two-Dimensional Element (PSQ4)

Consider a quadrilateral two-dimensional element shown in Fig.4.2. The parent element is a rectangle mapped into a square in natural coordinates and this in turn is transformed into an arbitrary quadrilateral element with straight boundaries.

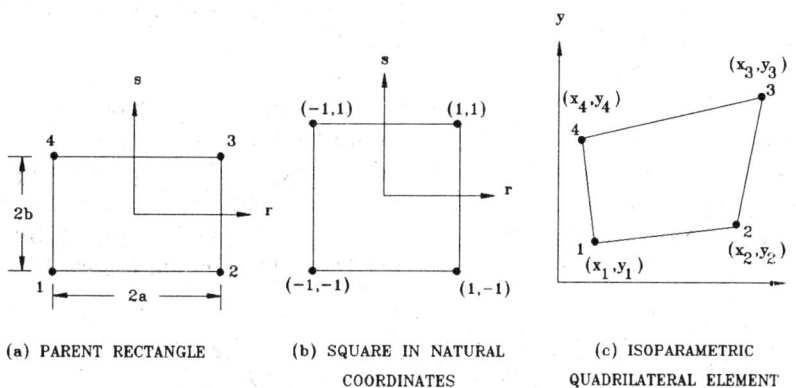

(a) PARENT RECTANGLE   (b) SQUARE IN NATURAL COORDINATES   (c) ISOPARAMETRIC QUADRILATERAL ELEMENT

Fig. 4.2 Four Noded Isoparametric Quadrilateral Element

The shape function used for representing the variation of displacement for a four noded rectangular element given by Eq.3.52 can now be used to describe the geometry of the arbitrary quadrilateral in the cartesian system. However, from computer programming point of view, the terms are rearranged to suit the description of the vector $\{x_n\}$ as

$$\left\{ \begin{matrix} x \\ y \end{matrix} \right\} = \begin{bmatrix} N_1 & 0 & N_2 & 0 & N_3 & 0 & N_4 & 0 \\ 0 & N_1 & 0 & N_2 & 0 & N_3 & 0 & N_4 \end{bmatrix} \left\{ \begin{matrix} x_1 \\ y_1 \\ x_2 \\ y_2 \\ x_3 \\ y_3 \\ x_4 \\ y_4 \end{matrix} \right\} \tag{4.2a}$$

where N_i (i = 1, 2, 3, 4) are given by Eq.3.52 as

$$N_1 = \frac{(1-r)(1-s)}{4} \quad N_3 = \frac{(1+r)(1+s)}{4}$$

$$N_2 = \frac{(1+r)(1-s)}{4} \quad N_4 = \frac{(1-r)(1+s)}{4} \tag{4.2b}$$

Thus this transformation relates a unit square in r and s coordinates to an arbitrary quadrilateral in cartesian (x, y) coordinate system whose shape and size are determined by the eight nodal coordinates x_1, y_1, x_2, y_2, x_3, y_3, x_4, y_4. The above relation also helps to determine the x, y coordinates of any point in the element when the corresponding natural coordinates r and s are given.

It can be observed from Eqs. 3.81 and 3.82 that we need to calculate the derivatives of the functions with respect to the global, i.e., cartesian coordinates. We, however, note that the shape functions used for describing the geometry of the element and displacement variation are expressed in the natural coordinates (r, s). The relationship between the two coordinate systems can be computed by using the chain rule of partial differentiation and is given below:

$$\left\{ \begin{matrix} \frac{\partial}{\partial r} \\ \frac{\partial}{\partial s} \end{matrix} \right\} = \begin{bmatrix} \frac{\partial x}{\partial r} & \frac{\partial y}{\partial r} \\ \frac{\partial x}{\partial s} & \frac{\partial y}{\partial s} \end{bmatrix} \left\{ \begin{matrix} \frac{\partial}{\partial x} \\ \frac{\partial}{\partial y} \end{matrix} \right\} = [J] \left\{ \begin{matrix} \frac{\partial}{\partial x} \\ \frac{\partial}{\partial y} \end{matrix} \right\} \tag{4.3a}$$

where $[J]$ is the Jacobian matrix. Hence, the derivatives with respect to cartesian coordinate system can be given as

$$\left\{ \begin{array}{c} \frac{\partial}{\partial x} \\ \frac{\partial}{\partial y} \end{array} \right\} = [J]^{-1} \left\{ \begin{array}{c} \frac{\partial}{\partial r} \\ \frac{\partial}{\partial s} \end{array} \right\} \tag{4.3b}$$

From Eq.4.2a

$$x = \sum_{i=1}^{4} N_i x_i \quad \text{and} \quad y = \sum_{i=1}^{4} N_i y_i$$

and noting that N_i is a function in (r, s), the Jacobian $[J]$ can be evaluated as

$$[J] = \begin{bmatrix} \frac{\partial N_1}{\partial r} & \frac{\partial N_2}{\partial r} & \frac{\partial N_3}{\partial r} & \frac{\partial N_4}{\partial r} \\ \frac{\partial N_1}{\partial s} & \frac{\partial N_2}{\partial s} & \frac{\partial N_3}{\partial s} & \frac{\partial N_4}{\partial s} \end{bmatrix} \begin{bmatrix} x_1 & y_1 \\ x_2 & y_2 \\ x_3 & y_3 \\ x_4 & y_4 \end{bmatrix} \tag{4.4}$$

Substituting from Eq.4.2b for the shape functions N_i we get,

$$[J] = \begin{bmatrix} -\frac{(1-s)}{4} & +\frac{(1-s)}{4} & \frac{(1+s)}{4} & -\frac{(1+s)}{4} \\ -\frac{(1-r)}{4} & -\frac{(1+r)}{4} & \frac{(1+r)}{4} & +\frac{(1-r)}{4} \end{bmatrix} \begin{bmatrix} x_1 & y_1 \\ x_2 & y_2 \\ x_3 & y_3 \\ x_4 & y_4 \end{bmatrix} \tag{4.5}$$

Let the inverse of $[J]$ as required in Eq.4.3b be expressed as,

$$[J]^{-1} = \begin{bmatrix} J_{11}^* & J_{12}^* \\ J_{21}^* & J_{22}^* \end{bmatrix} \tag{4.6}$$

It should be observed here that in order to transform the x and y coordinates into r and s coordinates the inverse of $[J]$ must exist. Hence, the determinant of the Jacobian $[J]$ must be non zero at every point of (r, s).

$$|J| = det[J] = \frac{\partial x}{\partial r} \frac{\partial y}{\partial s} - \frac{\partial x}{\partial s} \frac{\partial y}{\partial r} \neq 0 \tag{4.7}$$

The inverse of $[J]$ exists when there is one-to-one correspondence between the natural and cartesian coordinates as given by equations of type 4.2a. This condition imposes certain restrictions on elements, for example, the interior angle should not be too small or too large.

Also mapping a right-hand coordinate system into a left-hand coordinate system should be avoided as it leads to a negative Jacobian. For higher order elements, the location of nodes on the sides other than the corner nodes, the distances have to satisfy certain minimum requirements to satisfy the above condition (Eq. 4.7) on the determinant Jacobian. These aspects are well brought out in exercises 4.1 to 4.4 given at the end of this chapter.

Now the derivatives of displacements in global coordinates can be expressed in terms of local coordinates as

$$
\begin{Bmatrix}
\dfrac{\partial u}{\partial x} \\[2mm]
\dfrac{\partial u}{\partial y} \\[2mm]
\dfrac{\partial v}{\partial x} \\[2mm]
\dfrac{\partial v}{\partial y}
\end{Bmatrix}
=
\begin{bmatrix}
J_{11}^* & J_{12}^* & 0 & 0 \\
J_{21}^* & J_{22}^* & 0 & 0 \\
0 & 0 & J_{11}^* & J_{12}^* \\
0 & 0 & J_{21}^* & J_{22}^*
\end{bmatrix}
\begin{Bmatrix}
\dfrac{\partial u}{\partial r} \\[2mm]
\dfrac{\partial u}{\partial s} \\[2mm]
\dfrac{\partial v}{\partial r} \\[2mm]
\dfrac{\partial v}{\partial s}
\end{Bmatrix}
\tag{4.8}
$$

The strain-displacement relations given by Eq.2.9 can be expressed in matrix form as,

$$
\{\epsilon\} =
\begin{Bmatrix}
\epsilon_x \\
\epsilon_y \\
\gamma_{xy}
\end{Bmatrix}
=
\begin{bmatrix}
1 & 0 & 0 & 0 \\
0 & 0 & 0 & 1 \\
0 & 1 & 1 & 0
\end{bmatrix}
\begin{Bmatrix}
\dfrac{\partial u}{\partial x} \\[2mm]
\dfrac{\partial u}{\partial y} \\[2mm]
\dfrac{\partial v}{\partial x} \\[2mm]
\dfrac{\partial v}{\partial y}
\end{Bmatrix}
\tag{4.9}
$$

Substituting from Eq.4.8, we get

$$
\{\epsilon\} =
\begin{bmatrix}
J_{11}^* & J_{12}^* & 0 & 0 \\
0 & 0 & J_{21}^* & J_{22}^* \\
J_{21}^* & J_{22}^* & J_{11}^* & J_{12}^*
\end{bmatrix}
\begin{Bmatrix}
\dfrac{\partial u}{\partial r} \\[2mm]
\dfrac{\partial u}{\partial s} \\[2mm]
\dfrac{\partial v}{\partial r} \\[2mm]
\dfrac{\partial v}{\partial s}
\end{Bmatrix}
\tag{4.10}
$$

It can be observed from Eq.4.10 that in order to compute the **strains** or the [B] matrix we need to compute the derivatives of displacement functions for u and v with respect to the local or natural coordinate systems.

The displacements u and v are expressed through the shape functions as (Eqs. 3.52 and 3.53)

$$u = \sum_{i=1}^{4} N_i \, u_i \quad \text{and} \quad v = \sum_{i=1}^{4} N_i \, v_i \qquad (4.11)$$

where u_i and v_i are displacements of the nodes 1, 2, 3 and 4. Hence

$$\begin{Bmatrix} \dfrac{\partial u}{\partial r} \\[2mm] \dfrac{\partial u}{\partial s} \\[2mm] \dfrac{\partial v}{\partial r} \\[2mm] \dfrac{\partial v}{\partial s} \end{Bmatrix} = \begin{bmatrix} \dfrac{\partial N_1}{\partial r} & 0 & \dfrac{\partial N_2}{\partial r} & 0 \\[2mm] \dfrac{\partial N_1}{\partial s} & 0 & \dfrac{\partial N_2}{\partial s} & 0 \\[2mm] 0 & \dfrac{\partial N_1}{\partial r} & 0 & \dfrac{\partial N_2}{\partial r} \\[2mm] 0 & \dfrac{\partial N_1}{\partial s} & 0 & \dfrac{\partial N_2}{\partial s} \end{bmatrix}$$

$$\begin{bmatrix} \dfrac{\partial N_3}{\partial r} & 0 & \dfrac{\partial N_4}{\partial r} & 0 \\[2mm] \dfrac{\partial N_3}{\partial s} & 0 & \dfrac{\partial N_4}{\partial s} & 0 \\[2mm] 0 & \dfrac{\partial N_3}{\partial r} & 0 & \dfrac{\partial N_4}{\partial r} \\[2mm] 0 & \dfrac{\partial N_3}{\partial s} & 0 & \dfrac{\partial N_4}{\partial s} \end{bmatrix} \begin{Bmatrix} u_1 \\ v_1 \\ u_2 \\ v_2 \\ u_3 \\ v_3 \\ u_4 \\ v_4 \end{Bmatrix} \qquad (4.12)$$

Substituting Eq.4.12 in Eq.4.10 the strain-displacement matrix [B] can be evaluated.

The stiffness matrix for the element is given by Eq.3.81 as

$$[k] = \int\!\!\int\!\!\int [B]^T \, [C] \, [B] \; dx \; dy \; dz$$

In the case of the present example of the two-dimensional element it is given by

$$[k] = h \int\!\!\int [B]^T \, [C] \, [B] \; dx \; dy \qquad (4.13)$$

where h is the thickness of the element. It can be shown that the elemental area in cartesian coordinates $(x - y)$ can be expressed in terms of the area in the local coordinates $(r - s)$ as,

$$dx \; dy = |J| \; dr \; ds \qquad (4.14)$$

where $|J|$ is the determinant of the Jacobian.

Substituting in Eq.4.14 the stiffness matrix is given by

$$[k] = h \int \int [B]^T [C] [B] |J| \, dr \, ds \qquad (4.15)$$

It may be observed that in Eq.4.15 all the matrices can be expressed in terms of r and s. However, we cannot in general evaluate the integration exactly because of the complexity of the expressions as the determinant $|J|$ involves polynomials in r and s which appear in the denominator. Hence, the integration for computing the stiffness matrix is usually done by resorting to numerical procedures explained in the following section.

4.1.2   Triangular Elements

Following the procedure described in the previous section, the isoparametric quadrilateral and other higher order elements belonging to this family can be developed from the parent element, which is a square in natural coordinate system. In some cases of discretization, isoparametric triangular elements are also needed. It is possible to degenerate a four noded quadrilateral element to a three noded triangular element by collapsing one of the sides and assigning the same node number to two corner nodes (see Exercise 4.5). However, isoparametric triangular elements can be developed from the parent triangular elements in natural coordinates as illustrated below. Figure 4.3 shows the parent element and the isoparametric mapping for triangular elements [3, 4, 5].

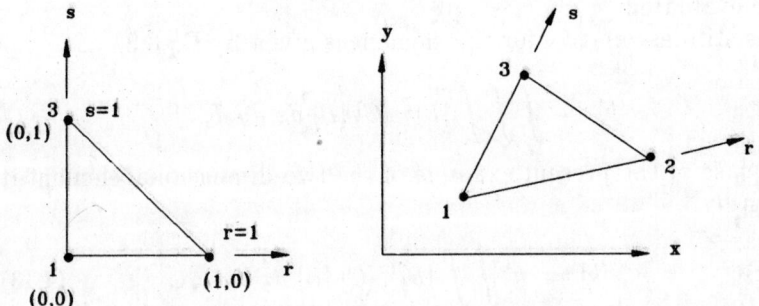

Fig. 4.3a Three noded element

The parent element is chosen as the unit isosceles right angled triangle (Fig. 4.3a). This is similar to the biunit square in $r - s$ for quadrilateral elements. Two independent natural coordinates r and s are taken for transformation to Cartesian coordinates as,

$$x = \sum_{i=1}^{3} N_i \, x_i \qquad y = \sum_{i=1}^{3} N_i \, y_i \qquad (4.16)$$

where N_i are the interpolation functions of the parent three noded triangular element (Fig. 4.3).

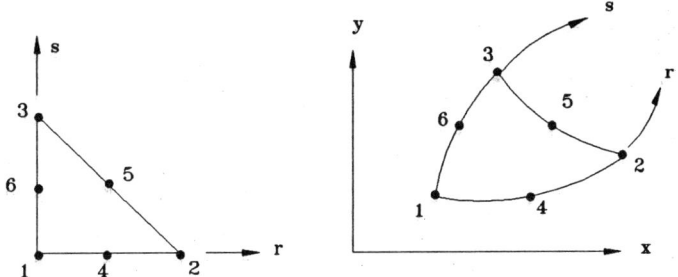

(b) Six noded element

Fig. 4.3 Mapping of triangular elements

$$N_1 = 1 - r - s \qquad N_2 = r \qquad N_3 = s \qquad (4.17)$$

It can be shown by working out the inverse transformation that the interpolation functions N_i are the name given in Eq. 3.36 for constant strain triangular element,

$$N_i = L_i \qquad i = 1, 2, 3 \qquad (4.18)$$

Thus, Eq. 4.17 reduces to,

$$L_1 = 1 - r - s \qquad L_2 = r \qquad L_3 = s \qquad (4.19)$$

where L_i are the area coordinates of the 'unit triangle' (Fig. 3.4).

The above relation helps us to use the interpolation functions for six noded element given in Eq. 3.46a to develop a six noded isoparametric triangular element.

For the evaluation of the element matrices, the Jacobian matrix has to be worked out which establishes the relationship between the two coordinate systems. For a linear triangular element the Jacobian matrix is given by,

$$[J] = \begin{bmatrix} x_2 - x_1 & y_2 - y_1 \\ x_3 - x_1 & y_3 - y_1 \end{bmatrix} \qquad (4.20)$$

4.2 Computation of Stiffness Matrix for Isoparametric Elements

It was shown in section 4.1 that the computation of strain-displacement matrix involves determination of the Jacobian matrix which for a two-dimensional element was expressed in general from of Eq.4.3. And specific expression for the $|J|$ matrix for a four noded quadrilateral element is given in Eq.4.5. Similarly in the case of three-dimensional elements, the relation between the derivatives in local (or natural) r, s, t coordinates and those in global (or cartesian) x, y, z coordinates can be shown as

$$\begin{Bmatrix} \dfrac{\partial}{\partial r} \\ \dfrac{\partial}{\partial s} \\ \dfrac{\partial}{\partial t} \end{Bmatrix} = \begin{bmatrix} \dfrac{\partial x}{\partial r} & \dfrac{\partial y}{\partial r} & \dfrac{\partial z}{\partial r} \\ \dfrac{\partial x}{\partial s} & \dfrac{\partial y}{\partial s} & \dfrac{\partial z}{\partial s} \\ \dfrac{\partial x}{\partial t} & \dfrac{\partial y}{\partial t} & \dfrac{\partial z}{\partial t} \end{bmatrix} \begin{Bmatrix} \dfrac{\partial}{\partial x} \\ \dfrac{\partial}{\partial y} \\ \dfrac{\partial}{\partial z} \end{Bmatrix} = [J] \begin{Bmatrix} \dfrac{\partial}{\partial x} \\ \dfrac{\partial}{\partial y} \\ \dfrac{\partial}{\partial z} \end{Bmatrix} \qquad (4.21)$$

Thus in the case of three-dimensional elements the Jacobian matrix $[J]$ is of the order 3×3 as given by the equation above. Hence, the derivatives with respect to cartesian coordinates can be expressed as

$$\begin{Bmatrix} \dfrac{\partial}{\partial x} \\ \dfrac{\partial}{\partial y} \\ \dfrac{\partial}{\partial z} \end{Bmatrix} = [J]^{-1} \begin{Bmatrix} \dfrac{\partial}{\partial r} \\ \dfrac{\partial}{\partial s} \\ \dfrac{\partial}{\partial t} \end{Bmatrix} \qquad (4.22)$$

Similarly the differential volume in x, y, z coordinates can be given by

$$dx \, dy \, dz = |J| \, dr \, ds \, dt \qquad (4.23)$$

where $|J|$ is the determinant of the Jacobian matrix $[J]$.

The stiffness matrix $[k]$ is given by

$$[k] = \int \int \int [B]^T \, [C] \, [B] \, |J| \, dr \, ds \, dt \qquad (4.24)$$

As was noted for two-dimensional elements the evaluation of the integral in Eq.4.24 cannot be done explicitly. The numerical integration procedure adopted widely in practice is briefly described in the following.

4.2.1 Numerical Integration

In finite element analysis we come across the following types of integrations

$$\int f(r)dr, \int \int f(r,s) \ dr \ ds, \int \int \int f(r,s,t) \ dr \ ds \ dt$$

in the one, two and three-dimensional cases, either for the computation of element stiffness or for the element nodal load vector. For the illustration of the numerical integration procedure consider first the integration of $\int_a^b f(r)dr$. This definite integral can be approximated by a properly weighted sum of particular values of the function $f(r)$ at specified points r_i, i.e. $f(r_i)$ within the limits of integration. Thus the value of the integral can be approximated as

$$\int_b^a f(r)dr = \sum_{i=1}^{n} W_i f(r_i) \qquad (4.25)$$

The specified points r_i are called sampling points and W_i are called the weights. There are several methods of numerical integration and we confine our discussion to the method known as Gauss Quadrature that is widely adopted in finite element analysis.

From Eq.4.25 it is obvious that we have to determine values for two sets of unknowns, viz., sampling points r_i and weights W_i. If the Eq.4.25 is to be exact the number of terms on the right hand side should depend on the degree m of the function $f(r)$. According to the Weierstrass theorem on polynomial approximation that any function $f(r)$ can be approximated within the given interval (a, b) by a polynomial of sufficiently high degree [6]. A polynomial of degree m is uniquely specified by $m + 1$ constants. It may be noted that the right hand side of Eq.4.25 contains $2n$ arbitrary constants, i.e. n, r_i^s and n, W_i^S and hence it is sufficient for a complete specification of the integral of a polynomial of degree $2n - 1$. Thus it is clear that if the Eq.4.25 is to be exact the degree m of the polynomial approximating $f(r)$ cannot exceed $2n-1$. If $m = 2n-1$, the constants r_i and W_i can be determined to a desired precision. This forms the basis of the Gauss quadrature. Before the evaluation of the values of the coefficients, the interval (a, b) is transformed into limits $(-1, 1)$. This can always be accomplished by suitable substitution for r. However, in the case of isoparametric elements, the integration limits are $(-1, 1)$ as can be seen from the mapping of the parent elements in natural coordinates.

The values of the sampling points r_i^S, also known as Gauss points, and the weights W_i^S are determined using Legendre polynomials. And using these values the integration can be evaluated exactly (within the given precision) for a polynomial function of degree $2n-1$ where

n is the number of sampling points chosen. For detailed derivation, readers may refer to standard texts on Numerical Analysis such as given in references [6, 7]. The values of the constants are tabulated below for the first three orders (Table 4.1)

Table 4.1 Sampling Points and Weights for Gauss Numerical Integration

n	r_i	W_i
1	$r_1 = 0.00000000$	$W_1 = 2.00000000$
2	$r_2 = -r_1 = 0.57735027$	$W_2 = W_1 = 1.00000000$
3	$r_2 = 0.00000000$	$W_2 = 0.88888889$
	$r_3 = -r_1 = 0.77459667$	$W_3 = W_1 = 0.55555556$

We note that the weights are all positive and that the weights of symmetrically placed points are the same. It may also be re-emphasised depending on the degree of the function $f(r)$ the number n of Gauss points can be chosen so that the integration is exact.

We can extend the procedure adopted above to evaluate the two and three dimensional integrals. In these cases the integral is evaluated with respect to one coordinate direction first and then with respect to other direction, and so on the procedure is followed in a similar manner. Therefore, a two-dimensional integral is evaluated as given below:

$$\int_{-1}^{+1} \int_{-1}^{+1} f(r,s) \, dr \, ds = \int_{-1}^{+1} \sum_i W_i \, f(r_i, s) \, ds$$

$$= \sum_j W_j \left[\sum_i W_i \, f(r_i, s_j) \right]$$

Hence,

$$\int_{-1}^{+1} \int_{-1}^{+1} f(r,s) \, dr \, ds = \sum_i \sum_j W_i \, W_j \, f(r_i, s_j) \qquad (4.26)$$

where W_i and W_j are the weights for one-dimensional integration.

Thus, for example using a 2×2 sampling points (i.e. four points) **for** Gauss quadrature, the integral becomes

$$\int_{-1}^{+1} \int_{-1}^{+1} f(r,s) \ dr \ ds \ = \ (1) \ (1) \ f(r_1, s_1) \ + (1) \ (1) \ f(r_1, s_2)$$

$$+ \ (1) \ (1) \ f(r_2, s_1) \ + \ (1) \ (1) \ f(r_2, s_2)$$

where the Gaussian weights W_i and W_j are equal to unity and $r_1 = s_1 = 0.57735$ and $r_2 = s_2 = -0.57735$ (Fig.4.4)

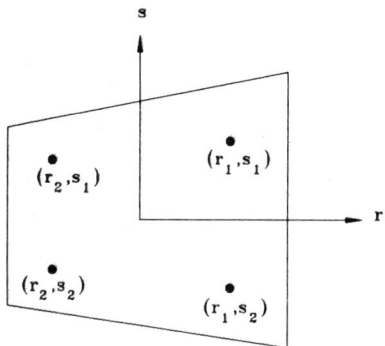

Fig. 4.4 Four point Gauss Quadrature for two-dimensional integral

The three-dimensional integral can be evaluated in exactly similar way and is given by

$$\int_{-1}^{+1} \int_{-1}^{+1} \int_{-1}^{+1} f(r,s,t) \ dr \ ds \ dt \ = \ \sum_i \sum_j \sum_k W_i \ W_j \ W_k \ f(r_i, s_j, t_k)$$

$$(4.27)$$

Triangular Elements

For triangular elements with straight edges, the formula given in Eq.3.24 can be used to evaluate the integrals. In the case of isoparametric triangular elements numerical integration needs to be used. However, the procedure described above for rectangular domain is not applicable directly since the integration limits involve the variables themselves. A great deal of work has been carried out on the development of suitable integration formula and Cowper's paper [8] can be referred to for Gauss integration formulae for triangular domains. Typical values of Gauss weights and sampling points are given in Table 4.2 [4, 5, 8].

Table 4.2 Sampling points and weights for
Gauss Quadrature over triangular domain

$$\int_0^1 \int_0^{1-r} f(r,s)\ dr\ ds \ = \ \frac{1}{2}\sum W_i\ f(r_i,s_i)$$

No. of points n	r	s	weights W_i	Geometric Location
1	$\frac{1}{3}$	$\frac{1}{3}$	1	
3	$r_1 = \frac{1}{3}$ $r_2 = \frac{1}{6}$ $r_3 = r_1$	$s_1 = r_1$ $s_2 = r_1$ $s_3 = r_2$	$W_1 = \frac{1}{3}$ $W_2 = W_1$ $W_3 = W_1$	

4.2.2   Computational Considerations in Evaluating the Integrals Numerically

For the isoparametric elements the need for adopting numerical integration procedure has been brought out earlier for the evaluation of stiffness matrix. Similar arguments can be advanced for the computation of element nodal load vector due to body forces and surface tractions. It has been observed earlier that if Gauss quadrature is adopted, the integration is exact for the polynomial of order $2n - 1$ if n sampling points are used.

The procedure for the computation of stiffness matrix given by the integral in Eq.4.24 can be summarized as follows: first of all choose the number of sampling points but a decision on this is very critical which would be discussed later in the section. Then for a sampling point i, j, k evaluate the Jacobian matrix $[J]$ and its determinant equations of type 4.5 or 4.21, in general. Using the $[J]$, compute the strain displacement matrix $[B]_{ijk}$ as the numerical values of the elements of $[B]$ are known being evaluated for the sampling point ijk. Thus the value of the function at the sampling point is given as

$$f(r_i, s_j, t_k) = [B]_{ijk}^T [C] [B]_{ijk} |J|_{ijk} \tag{4.28}$$

i.e., the product of the matrices at a sampling point. Similarly evaluating the values for all the sampling points the stiffness matrix is computed as

$$[k] = \sum_i \sum_j \sum_k W_i W_j W_k [B]_{ijk}^T [C] [B]_{ijk} |J|_{ijk} \tag{4.29}$$

Example: The numerical integration procedure for the computation of [k] is illustrated through an example given below. Consider a rectangular element shown in Fig.4.5. This element has been chosen since the values of the stiffness coefficients calculated by numerical integration procedure (i.e using Gauss Quadrature) can be compared with the exact values calculated by explicitly evaluating the integrals. The thickness of the element can be taken as 20 cm and, E = 2 × 10^3kN/cm^2 and $\mu = 0$.

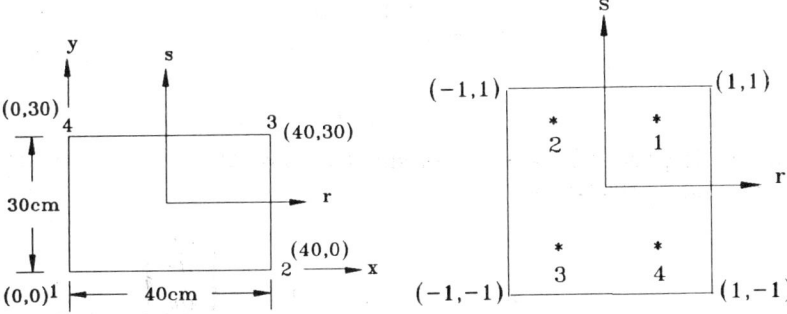

Fig. 4.5a Rectangular Element

Fig. 4.5b Element in Natural Coordinates & Gauss points

The natural coordinates of the sampling points shown in Fig.4.5(b) are given below:

$$
\begin{array}{ccc}
1 & +0.57735, & +0.57735 \\
2 & -0.57735, & +0.57735 \\
3 & -0.57735. & -0.57735 \\
4 & +0.57735 & -0.57735
\end{array}
$$

Consider the sampling point 1(0.57735, 0.57735).

The Jacobian $[J]$ at the point is evaluated by substituting for $r = 0.57735$ and $s = 0.57735$ in Eq.4.5. Thus

$$
[J] = \begin{bmatrix}
-\dfrac{(1-0.57735)}{4} & +\dfrac{(1-0.57735)}{4} & \dfrac{(1+0.57735)}{4} & \dfrac{-(1+0.57735)}{4} \\[2mm]
-\dfrac{(1-0.57735)}{4} & -\dfrac{(1+0.57735)}{4} & \dfrac{(1+0.57735)}{4} & \dfrac{(1-0.57735)}{4}
\end{bmatrix}
$$

$$
\times \begin{bmatrix}
0 & 0 \\
40 & 0 \\
40 & 30 \\
0 & 30
\end{bmatrix}
$$

$$
= \begin{bmatrix}
-0.1057 & +0.1057 & 0.3943 & -0.3943 \\
-0.1057 & -0.3943 & 0.3943 & +0.1057
\end{bmatrix}
\begin{bmatrix}
0 & 0 \\
40 & 0 \\
40 & 30 \\
0 & 30
\end{bmatrix}
\qquad (a)
$$

$$
= \begin{bmatrix}
20 & 0 \\
0 & 15
\end{bmatrix}
\qquad (b)
$$

$$
[J]^{-1} = \begin{bmatrix}
1/20 & 0 \\
0 & 1/15
\end{bmatrix}
\qquad (c)
$$

$$
|J| = 300 \qquad (d)
$$

Substituting the values of the element of the inverse Jacobian from (c) into Eq.(4.10) we get

$$
\{\epsilon\} = \begin{bmatrix}
1/20 & 0 & 0 & 0 \\
0 & 0 & 0 & 1/15 \\
0 & 1/15 & 1/20 & 0
\end{bmatrix}
\begin{Bmatrix}
\dfrac{\partial u}{\partial r} \\[2mm]
\dfrac{\partial u}{\partial s} \\[2mm]
\dfrac{\partial v}{\partial r} \\[2mm]
\dfrac{\partial v}{\partial s}
\end{Bmatrix}
\qquad (\epsilon)
$$

Evaluating $\frac{\partial N_i}{\partial r}$ and $\frac{\partial N_i}{\partial s}$ at the sampling point 1 ($r = s = 0.57735$) and substituting in Eq.4.12 we get

$$\begin{Bmatrix}\dfrac{\partial u}{\partial r}\\[4pt]\dfrac{\partial u}{\partial s}\\[4pt]\dfrac{\partial v}{\partial r}\\[4pt]\dfrac{\partial v}{\partial s}\end{Bmatrix}=
\begin{bmatrix}
-\frac{(1-0.57735)}{4} & 0 & +\frac{(1-0.57735)}{4} & 0 & +\frac{(1+0.57735)}{4} & 0 & -\frac{(1+0.57735)}{4} & 0\\[6pt]
-\frac{(1-0.57735)}{4} & 0 & -\frac{(1+0.57735)}{4} & 0 & +\frac{(1+0.57735)}{4} & 0 & +\frac{(1-0.57735)}{4} & 0\\[6pt]
0 & -\frac{(1-0.57735)}{4} & 0 & +\frac{(1-0.57735)}{4} & 0 & +\frac{(1+0.57735)}{4} & 0 & -\frac{(1+0.57735)}{4}\\[6pt]
0 & -\frac{(1-0.57735)}{4} & 0 & -\frac{(1+0.57735)}{4} & 0 & +\frac{(1+0.57735)}{4} & 0 & +\frac{(1-0.57735)}{4}
\end{bmatrix}
\underbrace{\begin{Bmatrix}u_1\\v_1\\u_2\\v_2\\u_3\\v_3\\u_4\\v_4\end{Bmatrix}}_{\{d\}}$$

$$=\begin{bmatrix}
-0.1057 & 0 & +0.1057 & 0 & +0.3943 & 0 & -0.3943 & 0\\
-0.1057 & 0 & -0.3943 & 0 & +0.3943 & 0 & +0.1057 & 0\\
0 & -0.1057 & 0 & +0.1057 & 0 & 0.3943 & 0 & -0.3943\\
0 & -0.1057 & 0 & -0.3943 & 0 & 0.3943 & 0 & +0.1057
\end{bmatrix}\{d\} \tag{f}$$

$$\{\epsilon\} = \begin{bmatrix} -0.0053 & 0 & +0.0053 & 0 & 0.0197 & 0 & -0.0197 & 0 \\ 0 & -0.0070 & 0 & -0.0263 & 0 & 0.0263 & 0 & 0.0070 \\ -0.0070 & -0.0053 & -0.0053 & +0.0053 & 0.0263 & 0.0197 & +0.0070 & -0.0197 \end{bmatrix} \{d\} \qquad (g)$$

Having thus obtained [B] matrix for the sampling point 1 from (g), the stress-displacement matrix [C] [B] can now be computed as,

$$[C][B] = 2\times10^3 \begin{bmatrix} 1 & 0 & 0 \\ 0 & 1 & 0 \\ 0 & 0 & 1/2 \end{bmatrix} \times \begin{bmatrix} -0.0053 & 0 & +0.0053 & 0 & 0.0197 & 0 & -0.0197 & 0 \\ 0 & +0.0070 & 0 & 0.0263 & 0 & 0.0263 & 0 & +0.0070 \\ -0.0070 & -0.0053 & -0.0053 & -0.0263 & 0.0053 & 0.0263 & +0.0070 & -0.0197 \end{bmatrix}$$

$$= 2\times10^3 \begin{bmatrix} -0.0053 & 0 & 0.0053 & 0 & 0.0197 & 0 & -0.0197 & 0 \\ 0 & -0.0070 & 0 & -0.0263 & 0 & 0.0263 & 0 & 0.0070 \\ -0.0035 & -0.0027 & -0.0132 & +0.0132 & 0.0035 & 0.0098 & 0.0035 & -0.0098 \end{bmatrix} \qquad (h)$$

Using Eq.4.15 and the values of Gauss weights $w_i = w_j = 1.0$, the stiffness matrix at this sampling point [k], is given by

$$[k]_i = h\, w_i\, w_j\, [B]_{ij}^T\, [C]_{ij}\, [B]_{ij}\, |J|_{ij} \qquad (i)$$

and substituting for the thickness of the element as h ≐ 20, $|J|$ from equation (d), $[B]^T$ from equation (g) and [C] [B] from equation (h), we get

$$[k]_1 = 20 \times 1 \times 1 \times 300 \times 2 \times 10^3$$

$$\begin{bmatrix}
-0.0053 & 0 & -0.0070 \\
0 & -0.0070 & -0.0053 \\
+0.0053 & 0 & -0.0263 \\
0 & -0.0263 & +0.0053 \\
+0.0197 & 0 & +0.0263 \\
0 & +0.0263 & +0.0197 \\
-0.0197 & 0 & +0.0070 \\
0 & +0.0070 & -0.0197
\end{bmatrix}$$

$$\times \begin{bmatrix}
-0.0053 & 0 & +0.0053 & 0 & 0.0197 & 0 & -0.0197 & 0 & +0.0070 \\
0 & -0.0070 & 0 & -0.0263 & 0 & 0.0263 & 0 & 0.0035 & +0.0070 \\
-0.0035 & -0.0027 & -0.0132 & +0.0027 & 0.0132 & 0.0098 & +0.0035 & +0.0035 & -0.0098
\end{bmatrix}$$

$$[k]_1 = 10^4 \begin{bmatrix} 0.0631 & 0.0227 & 0.0772 & -0.0227 & -0.2362 & +0.0823 & +0.0959 & +0.0823 \\ & 0.0760 & 0.0839 & +0.2037 & -0.0839 & -0.2832 & -0.0223 & +0.0353 \\ & & 0.4503 & -0.0852 & -0.2913 & -0.3093 & -0.2357 & +0.3093 \\ & & & +0.8472 & +0.0839 & -0.7677 & +0.0223 & -0.2832 \\ & & & & +0.8823 & +0.3093 & -0.3552 & -0.3093 \\ & & \text{Symmetric} & & & +1.0617 & +0.0827 & -0.0107 \\ & & & & & & +0.4951 & -0.0823 \\ & & & & & & & +0.2905 \end{bmatrix} \quad (j)$$

Next, consider the sampling point 2, r = -0.57735, s = 0.57735.

Following the same procedure the [B] matrix can be worked out and the stiffness matrix $[k]_2$ at this sampling point is evaluated. For brevity, only the final step is indicated below:

$$[k]_2 = 20 \times 1 \times 1 \times 300 \times 2 \times 10^3$$

$$\begin{bmatrix}
-0.0053 & 0 & -0.0263 \\
0 & -0.0263 & -0.0053 \\
+0.0053 & 0 & -0.0070 \\
0 & -0.0070 & +0.0053 \\
+0.0197 & 0 & +0.0070 \\
0 & +0.0070 & +0.0197 \\
-0.0197 & 0 & +0.0263 \\
0 & +0.0263 & -0.0197
\end{bmatrix}
\times$$

$$\begin{bmatrix}
-0.0053 & 0 & +0.0053 & 0 & 0.0197 & 0 & -0.0197 & 0 \\
0 & -0.0263 & 0 & -0.0070 & 0 & 0.0070 & 0 & +0.0263 \\
-0.0132 & -0.0027 & -0.0035 & +0.0027 & 0.0035 & 0.0098 & +0.0132 & -0.0098
\end{bmatrix}$$

$$
[k]_2 = 10^4
\begin{bmatrix}
0.4503 & 0.0852 & 0.0767 & -0.0852 & -0.2357 & -0.3093 & -0.2913 & +0.3093 \\
 & 0.8472 & 0.0223 & +0.2037 & -0.0223 & -0.2832 & -0.0839 & -0.7677 \\
 & & 0.0631 & -0.0227 & +0.0959 & -0.0823 & -0.2362 & +0.0823 \\
 & & & +0.0760 & +0.0223 & +0.00353 & +0.0839 & -0.2832 \\
 & & & & +0.4951 & +0.0823 & -0.3548 & -0.0823 \\
 & \text{Symmetric} & & & & +0.2905 & +0.3120 & -0.0108 \\
 & & & & & & +0.8823 & -0.3093 \\
 & & & & & & & +1.0617
\end{bmatrix}
$$

(k)

Similarly at the sampling point 3, $r = -0.57735$, $s = -0.57735$, the stiffness matrix $[k]_3$ is evaluated as,

$$[k]_3 = 20 \times 1 \times 1 \times 300 \times 2 \times 10^3$$

$$\begin{bmatrix}
-0.0197 & 0 & -0.0263 \\
0 & -0.0263 & -0.0197 \\
+0.0197 & 0 & -0.0070 \\
0 & -0.0070 & +0.0197 \\
+0.0053 & 0 & +0.0070 \\
0 & +0.0070 & +0.0053 \\
-0.0053 & 0 & +0.0263 \\
0 & +0.0263 & -0.0053
\end{bmatrix}$$

$$\times \begin{bmatrix}
-0.0197 & 0 & +0.0197 & 0 & 0.0053 & 0 & -0.0053 & 0 \\
0 & -0.0263 & 0 & -0.0070 & 0 & 0.0070 & 0 & +0.0263 \\
-0.0132 & -0.0098 & -0.0035 & +0.0098 & 0.0035 & 0.0027 & 0.0132 & -0.0027
\end{bmatrix}$$

$$[k]_3 = 10^4 \begin{bmatrix}
0.8823 & 0.3093 & -0.3552 & -0.3093 & -0.2359 & -0.0852 & -0.2913 & 0.0852 \\
 & 1.0617 & +0.0827 & -0.0107 & -0.0827 & -0.2847 & -0.3120 & -0.7662 \\
 & & +0.4951 & -0.0823 & +0.0959 & -0.0227 & -0.2362 & +0.0227 \\
 & & & +0.2905 & +0.0827 & +0.0050 & +0.3120 & -0.2847 \\
 & & & & +0.0631 & +0.0227 & +0.0772 & -0.0227 \\
 & \text{Symmetric} & & & & +0.0760 & +0.0839 & +0.2037 \\
 & & & & & & +0.4503 & -0.0852 \\
 & & & & & & & +0.8472
\end{bmatrix}$$

(l)

Proceeding along the same lines at the sampling point 4, r = 0.57735, s=-0.57735, the stiffness matrix $[k]_4$ is computed as

$$[k]_4 = 20 \times 1 \times 1 \times 300 \times 2 \times 10^3$$

$$\begin{bmatrix}
-0.0197 & 0 & -0.0070 \\
0 & -0.0070 & -0.0197 \\
+0.0197 & 0 & -0.0263 \\
0 & -0.0263 & +0.0197 \\
+0.0053 & 0 & +0.0263 \\
0 & +0.0263 & +0.0053 \\
-0.0053 & 0 & +0.0070 \\
0 & +0.0070 & -0.0053
\end{bmatrix}$$

$$\times \begin{bmatrix}
-0.0197 & 0 & +0.0197 & 0 & 0.0053 & 0 & -0.0053 & 0 \\
0 & -0.0070 & 0 & -0.0263 & 0 & 0.0263 & 0 & +0.0070 \\
-0.0035 & -0.0098 & -0.0132 & +0.0098 & -0.0263 & 0.0132 & 0.0027 & +0.0035 & -0.0027
\end{bmatrix}$$

$$[k]_4 = 10^4 \begin{bmatrix} 0.4951 & 0.0823 & -0.3548 & -0.0823 & -0.2362 & -0.0227 & +0.0959 & +0.0227 \\ & 0.2905 & +0.3120 & -0.0108 & -0.3120 & -0.2847 & -0.0827 & +0.0050 \\ & & +0.8823 & -0.3093 & -0.2913 & -0.0852 & -0.2357 & +0.0852 \\ & & & +1.0617 & +0.3120 & -0.7662 & +0.0827 & -0.2847 \\ & & & & +0.4503 & +0.0852 & +0.0767 & -0.0852 \\ & & \text{Sym.} & & & +0.8472 & +0.0223 & +0.2037 \\ & & & & & & +0.0631 & -0.0227 \\ & & & & & & & +0.0760 \end{bmatrix} \quad (m)$$

Now using Eq.4.29 the stiffness matrix $[k]$ of the element can be computed as the sum of the values at the four sampling points i.e., $[k] = [k]_1 + [k]_2 + [k]_3 + [k]_4$. Thus, summing the values we get

$$[k] = 10^4 \begin{bmatrix} 1.8908 & 0.4995 & -0.5561 & -0.4995 & -0.9940 & -0.4995 & -0.3908 & +0.4995 \\ & 2.2754 & +0.5009 & +0.3859 & -0.5009 & -1.1358 & -0.5009 & -1.5244 \\ & & +1.8908 & -0.4995 & -0.3908 & -0.4995 & -0.9438 & +0.4995 \\ & & & +2.2754 & +0.5009 & -1.5254 & +0.5009 & -1.1358 \\ & & \text{Sym.} & & +1.8908 & +0.4995 & -0.5561 & -0.4995 \\ & & & & & +2.2754 & +0.5009 & +0.3859 \\ & & & & & & +1.8908 & -0.4995 \\ & & & & & & & +2.2754 \end{bmatrix} \quad (n)$$

Having studied the steps required for the numerical evaluation of the stiffness matrix, it may be noted that the number of sampling points i.e., the order of the numerical integration is important from two points of view. First of all the cost of the evaluation increases considerably with increase in the order of integration and secondly the results can be affected by choosing different orders of integration.

The integration order required to evaluate each element of the matrix can be determined by studying the order of the function to be integrated. In the case of stiffness matrix, the function to be integrated is

$$f(r, s, t) = [B]^T [C] [B] |J| \qquad (4.30)$$

Since the matrix [B] and det $|J|$ are functions of r, s, t the order of the function is known and hence following the rules that the Gauss quadrature is correct upto an order $2n - 1$, the number of sampling points can be chosen.

For example in the case of rectangular element it can be shown that $|J| = ab$ where 2a and 2b are the sides of the rectangle (Fig.4.2). Then following the Eqs.4.10 and 4.12, it can be observed that the function f to be evaluated for stiffness matrix is of the type $f(r^2, rs, s^2)$ i.e. quadratic. Hence 2 x 2 order is adequate in this case. However, for other irregular cases the Jacobian $[J]$ is not a constant, which would increase the order of integration.

There are situations in finite element analysis of plates and shells where a reduced order of integration [9, 10] is being used. If one observes that the displacement model overestimates the stiffness of the element, it may appear that by adopting a reduced order of integration better results can be obtained. Though much of work has been done in this area, great caution is necessary to adopt the technique which would be discussed in much more detail in chapters on Analysis of Plates and Shells.

There is another method suggested [11] to fix up the order of integration by examining the determinant of Jacobian which is explained as follows: As the finite mesh is refined, a constant strain condition would tend to prevail in each element. This would in turn make the strain energy function for the element as

$$U = \frac{1}{2} \int \int \int \{\epsilon\}^T [C] \{\epsilon\} \, dx \, dy \, dz = \frac{1}{2} \int \int \int (\text{Const}) |J| \, dr \, ds \, dt \qquad (4.30a)$$

Hence in the limit the strain energy of the element is correctly evaluated if the volume of the element is correctly calculated. Thus the minimum order of integration is fixed depending on the order of the determinant of Jacobian $|J|$. However, based on detailed investigation, Bathe and Wilson [2] suggest the following minimum integration order given in Table 4.3, for evaluation of element stiffness matrices for two-dimensional elements.

Table 4.3 Order of Integration Using Gauss Quadrature for Two-Dimensional Elements

Element		Suggested integration order [2]
Four noded rectangle		2 x 2
Four noded quadrilateral		2 x 2
Eight noded rectangle		2 x 2
Eight noded curved		3 x 3

It may be noted that the above suggested order can be used for one dimensional and three-dimensional cases by deducing the appropriate integration orders for each plane of the element. Also the other element mass matrices and nodal load vectors can be computed using the above order of integration.

4.2.3 Fast Element Stiffness Computation

Programming the computation of element stiffness matrix is an important step in the development of finite element software. In the conventional implementation procedure, the Eq. 4.29 can be directly used to program the computations to be done at each Gauss point. However, in many cases, especially in linear analysis with isotropic material, it is possible to exploit the special structures of [B] and [C] matrices. This results in greater computational efficiency [12].

Consider the following matrix multiplication.

$$
\begin{bmatrix} b_{11} & b_{12} & 0 \\ b_{21} & b_{22} & 0 \\ 0 & 0 & b_{33} \end{bmatrix}
\begin{bmatrix} a_{11} & 0 \\ 0 & a_{22} \\ a_{31} & a_{32} \end{bmatrix}
=
\begin{bmatrix} a_{11} b_{11} & a_{22} b_{12} \\ a_{11} b_{21} & a_{22} b_{22} \\ a_{31} b_{33} & a_{32} b_{33} \end{bmatrix}
\tag{4.31}
$$

The conventional matrix multiplication will require 18(6 × 3) multiplications to get the resultant matrix. But if we take into account the zeros in the matrices [b] and [a], we can explicitly compute the product with only 6 multiplications which leads to a saving of $66\frac{2}{3}\%$ in computation. This procedure is adopted in the fast stiffness computation.

The strain-displacement matrix [B] is similar to [a] in sparseness as can be seen from Eq. 4.32 for a n noded element for two-dimensional stress analysis.

$$
[B] = \begin{array}{c}
\\ 1 \\ 2 \\ 3
\end{array}
\begin{bmatrix}
\cdot & \cdot & \bigm| & \cdot & \cdot & \bigm| & \frac{\partial N_i}{\partial x} & 0 & \bigm| & \cdot & \bigm| & \cdot & \cdot \\
\cdot & \cdot & \bigm| & \cdot & \cdot & \bigm| & 0 & \frac{\partial N_i}{\partial y} & \bigm| & \cdot & \bigm| & \cdot & \cdot \\
\cdot & \cdot & \bigm| & \cdot & \cdot & \bigm| & \frac{\partial N_i}{\partial y} & \frac{\partial N_i}{\partial x} & \bigm| & \cdot & \bigm| & \cdot & \cdot
\end{bmatrix}
\tag{4.32}
$$

with column headings: node 1 (1, 2), node i ($2i-1$, $2i$), node n ($2n-1$, $2n$).

The matrix [b] in the Eq.4.31 is similar to constitutive matrix which for isotropic material is given by,

$$
[C] = \begin{bmatrix} E_x & E_{xy} & 0 \\ E_{xy} & E_y & 0 \\ 0 & 0 & G \end{bmatrix}
\tag{4.33}
$$

The product of [B] and [C] of Eqs. 4.32 and 4.33 gives the stress-displacement matrix, denoted by [CB] and this can be written explicitly similar to Eq. 4.31, avoiding multiplication involving zeros.

$$
\begin{array}{cccccc}
& \text{node 1} & & \text{node } i & & \text{node } n \\
& 1 \quad\quad 2 & & 2i-1 \quad\quad i & & 2n-1 \quad\quad n
\end{array}
$$

$$
[CB] = \begin{array}{c} 1 \\ 2 \\ 3 \end{array}
\left[
\begin{array}{ccc|ccc|ccc}
\cdot & \cdot & \cdot & \cdot & E_x \frac{\partial N_i}{\partial x} & E_{xy}\frac{\partial N_i}{\partial y} & \cdot & \cdot & \cdot \\
\cdot & \cdot & \cdot & \cdot & E_{xy}\frac{\partial N_i}{\partial x} & E_y \frac{\partial N_i}{\partial y} & \cdot & \cdot & \cdot \\
\cdot & \cdot & \cdot & \cdot & G\frac{\partial N_i}{\partial y} & G\frac{\partial N_i}{\partial x} & \cdot & \cdot & \cdot
\end{array}
\right]
$$

$$(4.34)$$

From the above expression it may be noted that the matrix $[CB]$ can be computed and assembled directly. This can be programmed by considering the nodes one by one.

The next step is to compute $[B]^T [CB]$ and a similar procedure is adapted.

$$[B]^T [CB] =$$

$$
\begin{array}{c} 2j-1 \\ 2j \end{array}
\left[
\begin{array}{ccc}
\cdot & \cdot & \cdot \\
\frac{\partial N_j}{\partial x} & 0 & \frac{\partial N_j}{\partial y} \\
0 & \frac{\partial N_j}{\partial y} & \frac{\partial N_j}{\partial x} \\
\cdot & \cdot & \cdot
\end{array}
\right]
\left[
\begin{array}{cc}
\cdots \vdots \; CB_{1,2i-1} & CB_{1,2i} \vdots \cdots \\
\vdots & \vdots \\
\cdots \vdots \; CB_{2,2i-1} & CB_{2,2i} \vdots \cdots \\
\vdots & \vdots \\
\cdots \vdots \; CB_{3,2i-1} & CB_{3,2i} \vdots \cdots
\end{array}
\right]
$$

$$(4.35)$$

$$
\begin{array}{cc}
& 2i-1 \quad\quad 2i
\end{array}
$$

$$
= \begin{array}{c} 2j-1 \\ \\ 2j \end{array}
\left[
\begin{array}{cc}
\vdots \; \frac{\partial N_j}{\partial x} CB_{1,2i-1} + \frac{\partial N_j}{\partial y} CB_{3,2i-1} & \vdots \; \frac{\partial N_j}{\partial x} CB_{1,2i} + \frac{\partial N_j}{\partial y} CB_{3,2i} \\
\\
\vdots \; \frac{\partial N_j}{\partial y} CB_{2,2i-1} + \frac{\partial N_j}{\partial x} CB_{3,2i-1} & \vdots \; \frac{\partial N_j}{\partial y} CB_{2,2i} + \frac{\partial N_j}{\partial y} CB_{3,2i}
\end{array}
\right]
$$

$$(4.36)$$

By the above expression it is clear that the elements of the product $[B]^T[CB]$ are directly obtained. The computed values of these elements on multiplication of the weights W_i and W_j and the determinant Jacobian $|J|$ at the Gauss points yield the elements of the stiffness matrix $k_{2j-1,2i-1}$, $k_{2j-1,2i}$, $k_{2j,2i-1}$ and $k_{2j,2i}$. The reduction in the number of multiplications can be worked out and is left as an exercise to the student.

The above method is also applicable to anisotropic material. In that case the matrix $[C]$ is fully populated and there is still saving in the computation of stiffness matrix by the above procedure. Also this procedure can be extended to the evaluation of stiffness matrix of three-dimensional solid elements.

Gupta and Mohraz [13] have suggested another computational procedure for the evaluation of element stiffness matrix.

4.3 Convergence Criteria for Isoparametric Element

In Chapter 3, we have discussed the conditions to be satisfied by the displacement functions for convergence of the solution as the mesh is refined. When these conditions are satisfied for the parent element it is necessary to study the conditions to be imposed when the element is transformed into the cartesian system to form an isoparametric element [1].

It may be noted that continuity of displacement within the element is invariably satisfied by the choice of continuous polynomial or shape function. It has been pointed out in section 3.3 that constant and linear terms should be present in the polynomial to represent rigid body mode and constant strain states of the element. Now consider the case of a three-dimensional element and without any loss of generality, the displacement u is parallel to the x-direction. The above condition can be satisfied by

$$u = \alpha_0 + \alpha_1 x + \alpha_2 y + \alpha_3 z \qquad (4.37)$$

where α_i are the constants.

The displacement component v and w in other directions can be expressed in a similar fashion. Now prescribe the displacement on any node i and the corresponding coordinates of the node i as x_i, y_i, z_i in the above Eq.4.37 and we have,

$$u_i = \alpha_0 + \alpha_1 x_i + \alpha_2 y_i + \alpha_3 z_i \qquad (4.38)$$

We know that the displacement u at any point inside an element can be expressed by the shape functions N_i and the nodal displacement u_i. Thus,

$$u = \Sigma N_i u_i \qquad (4.39)$$

Substituting in Eq.4.39 for u_i from Eq.4.38, we get

$$u = \alpha_0 \Sigma N_i + \alpha_1 \Sigma N_i x_i + \alpha_2 \Sigma N_i y_i + \alpha_3 \Sigma N_i z_i \qquad (4.40)$$

For isoparametric formulation the coordinates (x, y, z) of any point is expressed by the same shape function as,

$$x = \Sigma N_i \, x_i, \; y = \Sigma N_i \, y_i, \; z = \Sigma N_i \, z_i \qquad (4.41)$$

Thus the Eq.4.40 can be expressed as

$$u = \alpha_0 \Sigma N_i + \alpha_i \, x + \alpha_2 \, y + \alpha_3 \, z \qquad (4.42)$$

The Eq.4.42 reduces to Eq.4.37 provided that

$$\Sigma N_i = 1 \qquad (4.43)$$

Thus the condition for rigid body modes and constant strain states can be satisfied provided the shape functions satisfy the condition expressed in Eq.4.43.

REFERENCES

1. Zienkiewicz,O.C. and R.L.Taylor, *The Finite Element Method Vol I, Basic Formulation and Linear Problems*, McGraw-Hill, (U.K) Limited, 1989.

2. Bathe, K.J. and E.L. Wilson, *Numerical Method in Finite Element Analysis*, Prentice-Hill of India Private Limited, New Delhi, 1978.

3. Reddy, J.N., *An Introduction to the Finite Element Method*, Mc-Graw Hill, Singapore, International Student Edition, 1985.

4. Bathe, K.J., *Finite Element Procedures in Engineering Analysis*, Prentice-Hall of India Private Limited, New Delhi, 1990.

5. Hughes, T.J.R., R.L.Taylor, and W.Kanoknukulchai, *A Simple and Efficient Finite Element for Plate Bending*, International Journal for Numerical Methods in Engineering, Vol.11, pp. 1529-1543, 1977.

6. Kopal, Z., *Numerical Analysis*, Chapman and Hall, London, 1961.

7. Conte, S.D., *Elementary Numerical Analysis*, McGraw-Hill, New York, 1965.

8. Cowper, G.R., *Gaussian Quadrature Formulas for Triangles*, International Journal for Numerical Methods in Engineering, Vol.7, 1973, pp.405-408.

9. Zienkiewicz, O.C., R.L.Taylor, and J.M.Too, *Reduced Integration Technique in General Analysis of Plates and Shells*, International Journal for Numerical Methods in Engineering, Vol.3, pp. 275-290, 1971.

10. Pawsey, S.F. and R.W. Clough, *Improved Numerical Integration of Thick Shell Finite Elements*, International Journal for Numerical Methods in Engineering, Vol.3, pp. 545-586, 1971.

11. Cook, R.D., D.S. Malkus, and Michael E Plesha, *Concepts and Applications of Finite Element Analysis*, John Wiley, Third Edition, New York, 1989.

12. Sloan, S.W., *A Fast stiffness formulation for Finite Element Analysis of Two Dimensional Solids*, International Journal for Numerical Methods in Engineering, Vol.77, pp.1313-1323, 1981.

13. Gupta, A.K. and Mohraz, B., *A Method of Computing Numerically Integrated Stiffness Matrices*, International Journal for Numerical Methods in Engineering, Vol.5, pp. 83-89, 1972.

EXERCISES

4.1 Evaluate the Jacobian matrix [J] for the elements (a) and (b) shown in Fig. E4.1. Discuss the aspect of distortion from the parent bisquare element.

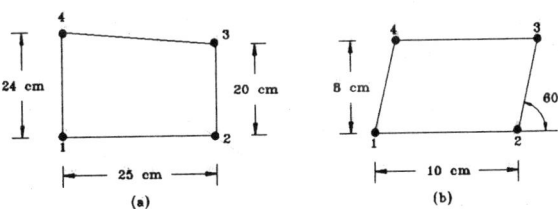

Fig. E4.1

4.2 Examine the element shown in Fig. E4.2 for validity of

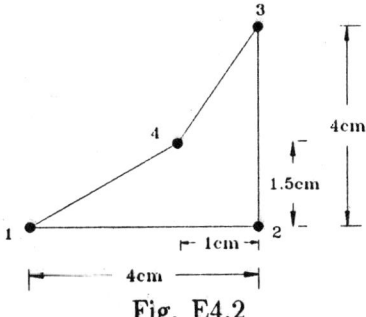

Fig. E4.2

one-to-one mapping transformation and show that for non singular $[J]$ all the interior angles must be smaller than 180 degrees.

4.3 Investigate the effect of node numbering on the transformation from the parent element to the element in Cartesian system shown in Fig. E4.3.

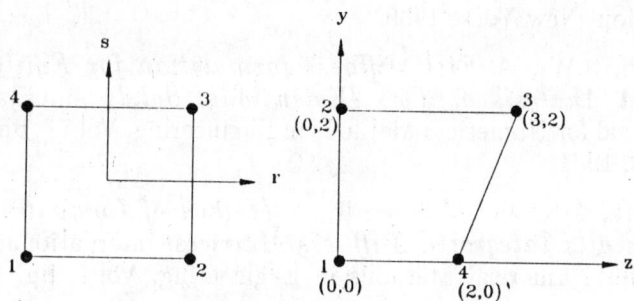

Fig. E4.3

4.4 Show that in the case of eight noded element the nodes along the sides other than the corner nodes should be placed at a distance greater than or equal to quarter of the length of the side from each corner node.

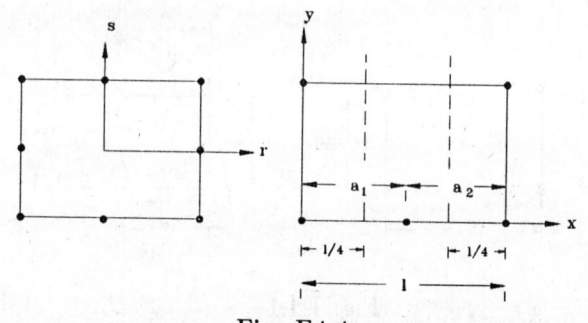

Fig. E4.4

4.5 For the element shown in Fig. E4.5 a constant strain triangular

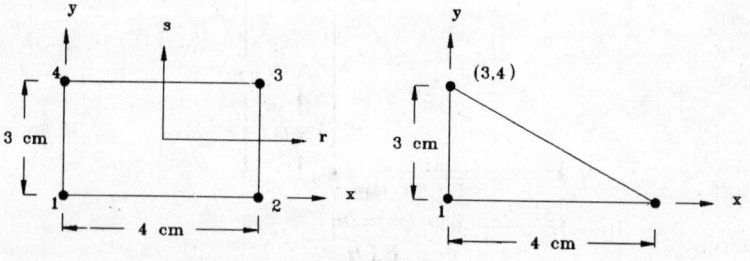

Fig. E4.5

element can be obtained by collapsing the side 3-4 of the four noded rectangular element.

4.6 Evaluate the stiffness matrix [k] using Gauss quadrature for the element 5 of the tapered cantilever beam shown in Fig. 8.15b. The origin of coordinates can be taken at the left hand corner of the beam.

4.7 The discretization of a deep beam is shown in Fig. 8.16b. Establish that 2 × 2 rule is adequate for evaluating [k]. Compute the stiffness matrix [k].

4.8 The cantilever beam shown in Fig. 8.14 is to be discretized by using four 8 noded rectangular element. Evaluate the stiffness matrix of an element using 2 × 2 Gauss rule.

4.9 Compute the stiffness matrix of the element in problem (4.8) by using 3 × 3 Gauss rule. Compare and discuss the stiffness matrices, obtained using the two different orders of integration.

4.10 For the triangular element shown in Fig. E4.10, derive the strain-displacement matrix [B] using the isoparametric transformation and compare it with the results obtained using Eq. 3.75. Compute the stiffness matrix [k] using one point quadrature rule and compare the result with the explicit integration.

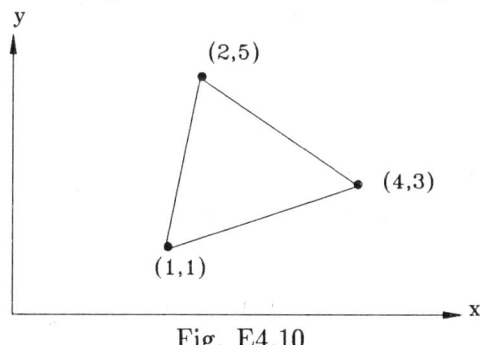

Fig. E4.10

4.11 An axial element is subjected to a constant load p_o. Compute the nodal load vector $\{Q\}$ by numerical integration for the following cases:

(a) axial element with two nodes

(b) axial element with three nodes

4.12 Compute the nodal load vector for problem (4.11) if the load p_o varies linearly.

4.13 Compute the nodal load vector for problem (4.11) if the load p_o varies as

$$p = p_o\left[1 - \left(\frac{x}{l}\right)^2\right]$$

4.14 For the rectangular element of problem 4.1, compute the nodal load vector vector due to self weight. Choose the correct order of integration by examining the integral to be evaluated.

4.15 Using the fast stiffness computation procedure, compute the stiffness matrix of the element shown in Fig. 4.5. Discuss the reduction in number of operations and computational advantages of this procedure.

4.16 Develop the necessary expressions for the fast stiffness computation of eight noded rectangular element. Establish the computational steps for evaluating [k] using Gauss Quadrature.

Chapter 5

Direct Stiffness Method of Analysis and Solution Technique

In finite element analysis, the solid or structure is considered as an assemblage of 'finite elements' connected at the nodes. After computing the stiffness matrix and nodal load vector of the elements based on the theory discussed in earlier chapter, the overall equation of equilibrium is formed by assembling these matrices using the connectivity information relating the degrees of freedom of the element and the corresponding global degrees of freedom. The congruent transformation technique or direct stiffness method can be used for constructing the equation of equilibrium of the structure [1]. However, the direct stiffness method is simple, elegant and is widely used in computer programs. The method is described in this chapter with illustrative example.

The solution of linear simultaneous equations takes significantly large percentage of total computing time in finite element static analysis. Hence, special algorithms have been developed which exploit the symmetric and banded nature of stiffness matrix. Basically all of them use the Gauss elimination procedure. The Gauss elimination procedure is described with an example.

It has been shown that following the elimination procedure the symmetric stiffness matrix [K] can be decomposed as [L] [D] $[L]^T$ and this decomposition has been found to be computationally efficient in solving linear simultaneous equations.

At the end of the chapter the basic steps to be followed for finite element stress analysis are summarised and illustrated through an example.

5.1 Assemblage of Elements — Direct Stiffness Method

Consider a continuous beam shown in Fig.5.1.

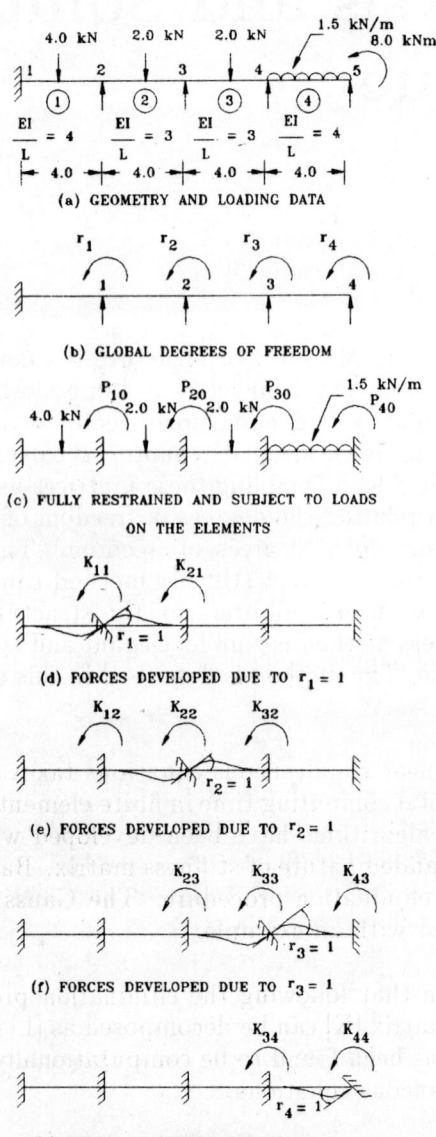

(a) GEOMETRY AND LOADING DATA

(b) GLOBAL DEGREES OF FREEDOM

(c) FULLY RESTRAINED AND SUBJECT TO LOADS ON THE ELEMENTS

(d) FORCES DEVELOPED DUE TO $r_1 = 1$

(e) FORCES DEVELOPED DUE TO $r_2 = 1$

(f) FORCES DEVELOPED DUE TO $r_3 = 1$

(g) FORCES DEVELOPED DUE TO $r_4 = 1$

Fig. 5.1 An example to illustrate direct stiffness method of assemblage of elements

The geometry and loading conditions are indicated in Fig.5.1(a) and the global or structure degrees of freedom are indicated in Fig.5.1(b). The behaviour and the stress resultants in the actual structure can be considered as the superposition of the cases (c) through (g) while the values of the displacements r_1, r_2, r_3 and r_4 are to be found out such that they satisfy the equilibrium condition corresponding to these displacements. It may be noted that the stiffness coefficient k_{ij} is the force in the displacement direction i due to unit value of displacement j, all other displacements are held zero. Here, force and displacement are used in generalized sense, i.e., includes moments and rotations. The compatibility of displacements is caused by giving unique values of displacements $(r_1, ...r_4)$ at the nodes and thus there is no discontinuity between two elements joining at a node.

Now, superposing all the cases (c) to (g), the equilibrium equation for the structure can be written as

$$
\begin{array}{llllll}
K_{11}r_1 & + & K_{12}r_2 & & & + & P_{10} = 0 \\
K_{21}r_1 & + & K_{22}r_2 & + & K_{23}r_3 & & + & P_{20} = 0 \\
& & K_{32}r_2 & + & K_{33}r_3 & + & K_{34}r_4 & + & P_{30} = 0 \\
& & & & K_{43}r_3 & + & K_{44}r_4 & + & P_{40} = 8
\end{array} \tag{5.1}
$$

Thus, for a structure or solid which is an assemblage of finite elements, the equation of equilibrium is expressed in compact form as

$$[K]\,\{r\} \;=\; \{P\} \tag{5.2}$$

where   [K]   is the stiffness matrix of the structure or global stiffness matrix

         $\{r\}$ is the displacement vector consisting of global (structure) degrees of freedom, and

         $\{P\}$ is the load vector corresponding to degrees of freedom $\{r\}$

Now our aim is to construct $[K]$ matrix and $\{P\}$ vector from the stiffness matrix, $[k]$ and nodal load vector $\{Q\}$ of the elements. This constitutes the assemblage of the elements and the procedure is explained below.

The element forming the continuous beam example is basically a two-dimensional beam with two degrees of freedom and the stiffness coefficients are shown in Fig.5.2. The equation of equilibrium for any element is given by Eq.3.80, as

$$[k]\,\{d\} \;=\; \{Q\}$$

The stiffness coefficients for this beam element are derived in chapter 7 and also it is shown therein that the values of nodal load vector $\{Q\}$ are the negative values of the 'fixed end actions' due to external loads i.e. assuming the element is fully restrained against displacements. As the assemblage of the finite elements connected at the nodes constitute the actual structure or body, the above element equations of equilibrium can be added to form the overall (structure) equilibrium Eq.5.2 through the relation of the element and global degrees of freedom that ensure the compatibility of displacement at the nodes. The assembly rules can be established through the use of potential energy theorem [2]. However, a simple and physical interpretation to the procedure is possible and is described through this example.

It may be noted that the stiffness coefficients refer to forces developed at the degrees of freedom and hence these coefficients of element stiffness matrix and load vector have to be added directly to get global stiffness coefficient, to correspond to the global degrees of freedom. Hence, if we know the connectivity, the relation between the element degrees of freedom to the global degrees of freedom, the assemblage becomes a simple process of directly adding the values as the contribution of force from different elements corresponding to a particular degree of freedom.

The stiffness matrix and nodal load vector for each beam element can be computed using the expressions given in Fig.5.2. Using these values the equation of equilibrium can be written for each element and the relation between the element degrees of freedom and the corresponding global degrees of freedom can also be established using Fig.5.1.

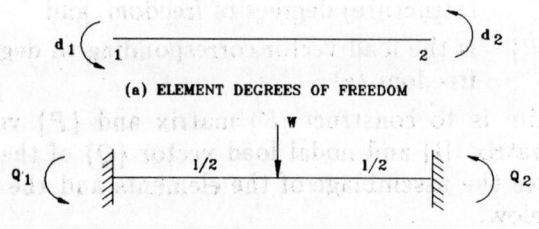

(a) ELEMENT DEGREES OF FREEDOM

(b) NODAL LOADS DUE TO A CENTRAL CONCENTRATED
LOAD ON THE ELEMENT

$$\{Q\} = \left\{ \begin{array}{c} \frac{-wl}{8} \\ \frac{wl}{8} \end{array} \right\} \quad \text{for an u.d.l.} \ \{Q\} = \left\{ \begin{array}{c} \frac{-wl^2}{12} \\ \frac{wl^2}{12} \end{array} \right\}$$

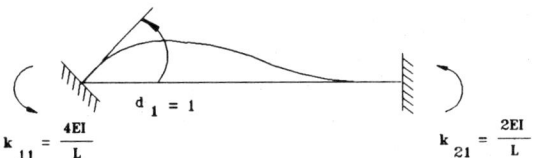

(c) FORCES (stiffness coefficients) DUE TO $d_1 = 1$

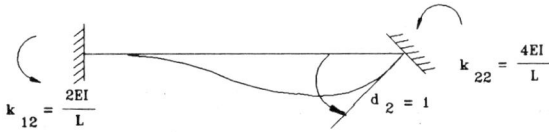

(d) FORCES (stiffness coefficients) DUE TO $d_2 = 1$

$$[k] = \begin{bmatrix} \dfrac{4EI}{L} & \dfrac{2EI}{L} \\ \dfrac{2EI}{L} & \dfrac{4EI}{L} \end{bmatrix}$$

Fig. 5.2 Stiffness matrix of the element

Element 1 The equation of equilibrium can be written as

$$\begin{bmatrix} k_{11} & k_{12} \\ k_{21} & k_{22} \end{bmatrix} \begin{Bmatrix} d_1 \\ d_2 \end{Bmatrix} = \begin{Bmatrix} Q_1 \\ Q_2 \end{Bmatrix} \tag{5.3}$$

The element degrees of freedom d_1 and d_2 correspond to 0 and r_1. For the purpose of explaining the assemblage procedure the degrees of freedom r_1, r_2, r_3 and r_4 are referred to as degrees of freedom 1, 2, 3 and 4 and in this example there are four degrees of freedom for the structure. Substituting the values of geometric and loading data, the numerical values of k_{ij} and Q_i can be computed and the above equation of equilibrium can be written as follows along with the connectivity relation

$$
\begin{array}{cc}
g.d.o.f & \quad 0 \quad\quad 1 \\
e.d.o.f & \quad 1 \quad\quad 2
\end{array}
$$

$$
\begin{array}{cc}
0 & 1 \\
1 & 2 \\
g & e \\
d & d \\
o & o \\
f & f
\end{array}
\begin{bmatrix} 16 & 8 \\ 8 & 16 \end{bmatrix} \begin{Bmatrix} d_1 \\ d_2 \end{Bmatrix} = \begin{Bmatrix} -2 \\ +2 \end{Bmatrix} \tag{5.4}
$$

Thus element stiffness coefficient k_{22} (16) refer to the force developed in the element (1) along the global displacement direction 1 due to unit value of displacement along 1. Hence the contribution from element (1) to global stiffness coefficient K_{11} is 16, which is simply obtained by identifying the row corresponding to the global d.o.f. 1 and the column corresponding to the global d.o.f. Hence, the numerical value of 16 (k_{22}) is transferred to the global stiffness coefficient K_{11} as the contribution from element (1). As the other degrees of freedom for this element are zero, the other three stiffness coefficients do not contribute to the stiffness matrix [K]. Similarly the value of Q_2 (2) is transferred as the contribution from element 1 to the element P_1 of the load vector $\{P\}$.

Element 2: For the element 2, the equation of equilibrium and the global and element degrees of freedom are given below. The computation of stiffness coefficients and the nodal load vector is carried out as before using the expressions given in Fig.5.2.

$$\begin{matrix} g.d.o.f & & 1 & 2 \\ e.d.o.f & & 1 & 2 \end{matrix}$$

$$\begin{matrix} 1 & 1 \\ 2 & 2 \end{matrix} \begin{bmatrix} 12 & 6 \\ 6 & 12 \end{bmatrix} \begin{Bmatrix} d_1 \\ d_2 \end{Bmatrix} = \begin{Bmatrix} -1 \\ 1 \end{Bmatrix} \qquad (5.5)$$

$$\begin{matrix} g & e \\ d & d \\ o & o \\ f & f \end{matrix}$$

Following the explanation given for element 1, the element stiffness coefficients k_{11}, k_{12}, k_{21} and k_{22} are the contributions from element 2 to the global stiffness coefficients as K_{11}, K_{12}, K_{21} and K_{22} since the global and element degrees of freedom are both denoted as 1 and 2. Similarly Q_1 and Q_2 are the contribution to the load vector components P_1 and P_2.

Element 3: The equation of equilibrium for element 3 is given below along with the connectivity relation.

$$\begin{matrix} g.d.o.f & & 2 & 3 \\ e.d.o.f & & 1 & 2 \end{matrix}$$

$$\begin{matrix} 2 & 1 \\ 3 & 2 \end{matrix} \begin{bmatrix} 12 & 6 \\ 6 & 12 \end{bmatrix} \begin{Bmatrix} d_1 \\ d_2 \end{Bmatrix} = \begin{Bmatrix} -1 \\ +1 \end{Bmatrix} \qquad (5.6)$$

$$\begin{matrix} g & e \\ d & d \\ o & o \\ f & f \end{matrix}$$

It can be seen that the element stiffness coefficients $k_{11}, k_{12}, k_{21}, k_{22}$ are the contributions to the global stiffness coefficients K_{22}, K_{23}, K_{32},

K_{33} and Q_1 and Q_2 to the load vector components P_2 and P_3.

Element 4: For element 4 the equation of equilibrium is given by

g.d.o.f   3   4
e.d.o.f   1   2

$$
\begin{array}{cc}
3 & 1 \\
4 & 2
\end{array}
\begin{bmatrix} 16 & 8 \\ 8 & 16 \end{bmatrix}
\begin{Bmatrix} d_1 \\ d_2 \end{Bmatrix}
= \begin{Bmatrix} -2 \\ +2 \end{Bmatrix}
\tag{5.7}
$$

$$
\begin{array}{cc}
g & e \\
d & d \\
o & o \\
f & f
\end{array}
$$

The stiffness coefficients k_{11}, k_{12}, k_{21}, k_{22} are the contributions to K_{33}, K_{34}, K_{43}, K_{44} of the global stiffness matrix and Q_1 and Q_2 are the contributions to P_3 and P_4.

5.1.1   Assemblage and overall Equation of Equilibrium

Having noted the contribution from all the elements, the global stiffness matrix and the equation of equilibrium can now be constructed. For understanding the process of transfer of contributions from different elements, the corresponding values are explicitly written down below.It may be observed that there is a moment of 8 kNm directly applied along the displacement direction r_4 which is added to P_4 along with the contribution from an element.

g.d.o.f.   1   2   3   4

$$
\begin{array}{c}
1 \\ 2 \\ 3 \\ 4
\end{array}
\begin{bmatrix}
16+12 & 6 & & \\
6 & 12+12 & 6 & \\
 & 6 & 12+16 & 8 \\
 & & 8 & 16
\end{bmatrix}
\begin{Bmatrix} r_1 \\ r_2 \\ r_3 \\ r_4 \end{Bmatrix}
= \begin{Bmatrix} 2-1 \\ 1-1 \\ 1-2 \\ 2+8 \end{Bmatrix}
\tag{5.8}
$$

Thus the overall equation of equilibrium for the structure is

$$
\begin{bmatrix}
28 & 6 & 0 & 0 \\
6 & 24 & 6 & 0 \\
0 & 6 & 28 & 8 \\
0 & 0 & 8 & 16
\end{bmatrix}
\begin{Bmatrix} r_1 \\ r_2 \\ r_3 \\ r_4 \end{Bmatrix}
= \begin{Bmatrix} 1 \\ 0 \\ -1 \\ 10 \end{Bmatrix}
\tag{5.9}
$$

The example has demonstrated the assemblage of elements by direct stiffness method which can be physically interpreted. From the computational point of view it is a process of transfer of the coefficients of the element stiffness matrix and the nodal load vector into relevant positions in the global stiffness matrix and load vector by identifying the element degrees of freedom and the corresponding global degree of freedom. This process can be easily programmed and is described in the next chapter.

5.1.2   Special Characteristics of Stiffness Matrix

The stiffness matrix of the structure is symmetric and banded. It may be noted that bandedness implies that all elements beyond the bandwidth of the matrix are zero, as Fig.5.3

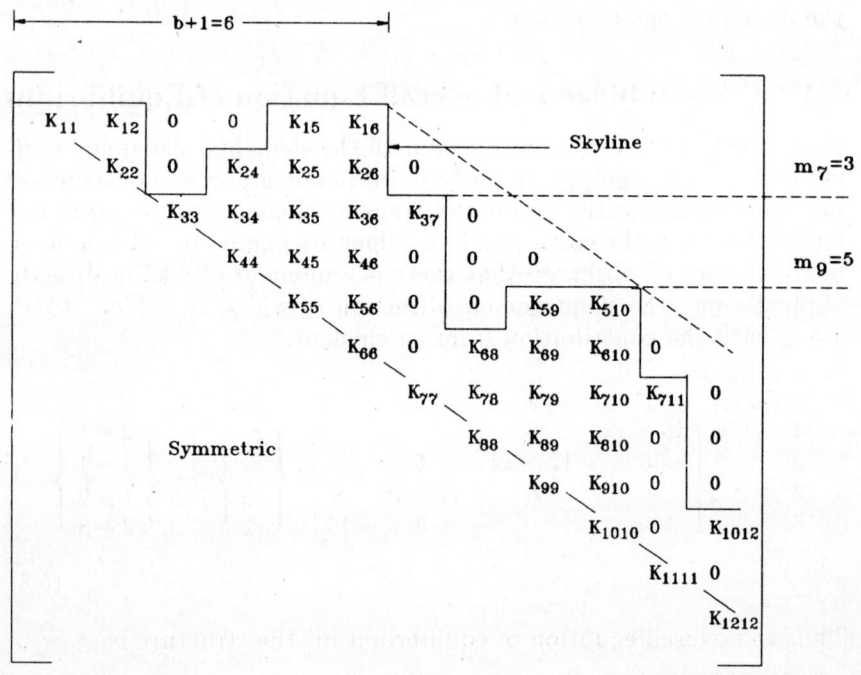

Fig. 5.3 Typical stiffness matrix of Structure-skyline
and Bandwidth Illustration

$$K_{ij} = 0 \quad j > i + b \tag{5.10}$$

where b is the half-bandwidth and $2b + 1$ is the bandwidth of the matrix $[K]$.

In the development of finite element analysis program the symmetric and banded nature of the stiffness matrix $[K]$ can be effectively made use of, for reducing the main storage requirement. One of the efficient schemes suggested is the skyline storage scheme [3, 4]. This approach is used in the program presented in the next chapter.

The skyline of the matrix defined by $m_i = 1, 2, ..., n$ is the row number of the first non-zero element in column i of the matrix. Hence, the elements above the skyline are zero. It may be observed that the half bandwidth of the stiffness matrix

$$b = \max (i - m_i), \qquad i = 1, 2, ..., n \qquad (5.11)$$

and b is equal to the maximum difference in global degrees of freedom for any one of the finite elements in the assemblage.

The computing time and the main storage requirements for solving the system of equations depends on the bandwidth of the stiffness matrix. From the above discussion it may be noted that the bandwidth depends on the numbering scheme adopted while discretizing the structure.

Figure 5.4 shows two different node numbering schemes adopted for a two-dimensional stress analysis problem. For the numbering scheme 1, the maximum difference in degree of freedom for any element works out to 9, whereas for the scheme 2 it is 23.

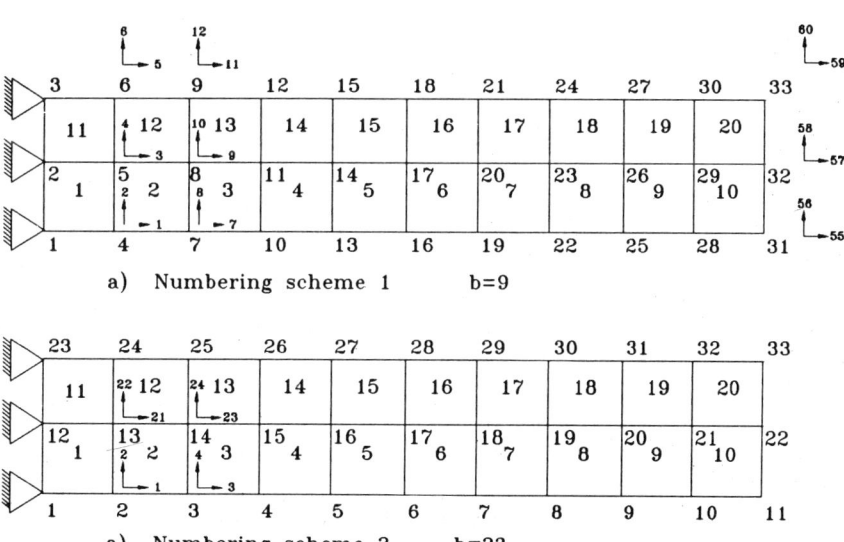

Fig. 5.4 Node numbering scheme and semi-bandwidth

The scheme 1 is preferable since the half bandwidth for this case is less than that of scheme 2. This example illustrates that it is possible to number the nodes in such a way that the maximum difference in global degrees of freedom of any element in the assemblage is as small as possible which results in a minimum bandwidth. However, an automatic bandwidth minimization procedure [5,6] is incorporated in some of the package programs.

It may be added here that there is another solution procedure called *frontal solution method* due to Irons [7]. In this method the coefficients of an equation are completely assembled from the contributions of all relevant elements and equations are reduced by Gauss elimination at the same time. The finite elements that must be considered for processing equations corresponding to one specific node define the wave front. Hence, in the frontal solution method the numbering of elements is important as it determines the length of the wave front. The frontal equation solver is well explained in reference [8].

5.1.3 Boundary Conditions and Reaction

The problems in solid or structural mechanics require sufficient kinematic boundary conditions to be specified so that the body or the structure does not experience any rigid body motion and is in equilibrium with the loads. For the problem described in the earlier section, the stiffness matrix of the structure has been constructed only for the degrees of freedom which are unrestrained, i.e., displacements which are allowed to take place. By this process the boundary conditions are imposed implicitly and equations of equilibrium are not formed corresponding to specified boundary displacements. If on the other hand it is required to compute the boundary (support) reactions or specify displacements along certain degrees of freedom the stiffness matrix of the structure ($[K_J]$) can be constructed for all the degrees of freedom including the specified boundary displacements. This stiffness matrix $[K_J]$ can be partitioned to correspond to unrestrained (unknown) displacements ($\{r\}$) and to specified displacements ($\{r_R\}$) and the equation of equilibrium can be written as

$$
\begin{bmatrix} [K] & | & [K_{DR}] \\ \text{---} & \text{---} & \text{---} \\ [K_{RD}] & | & [K_{RR}] \end{bmatrix} \begin{Bmatrix} \{r\} \\ \text{---} \\ \{r_R\} \end{Bmatrix} = \begin{Bmatrix} \{P\} \\ \text{---} \\ \{P_R\} \end{Bmatrix} + \begin{Bmatrix} \{0\} \\ \text{---} \\ \{R\} \end{Bmatrix} \quad (5.12)
$$

where $[K_{DR}]$ = stiffness matrix corresponding to unknown displacements due to unit displacements at the boundary (or support) where the displacements are specified finally.

$[K_{RD}]$ = $[K_{DR}]^T$

$[K_{RR}]$ = stiffness matrix corresponding to specified displacements due to unit displacement at these points.

$\{P_R\}$ = load vector corresponding to specified displacements

$\{R\}$ = vector of support (boundary) reactions along which displacements are specified.

The first set of Eq. 5.12 can be written as

$$[K]\{r\} = \{P\} - [K_{DR}]\{r_R\} \tag{5.13}$$

and solving which the unknown displacements $\{r\}$ can be computed. It may be noted that the second term of Eq.5.13 accounts for the equivalent load corresponding to specified displacements $\{r_R\}$ and if $\{r_R\} = \{0\}$ i.e. all boundary (support) displacements are zero, then the equation reduces to the form given earlier i.e., Eq.5.2.

After solving for $\{r\}$ the second set of Eq.5.12 can be used to compute the support (boundary) reactions as

$$\{R\} = [K_{RD}]\{r\} + [K_{RR}]\{r_R\} - \{P_R\} \tag{5.14}$$

The above theoretical procedure involves partitioning of the matrix and in actual program applications involve rearranging the stiffness matrix and also may affect the bandwidth of the matrix. A more practical way of handling this problem is to introduce a boundary element which is a spring, linear or torsional, of given stiffness, at those displacements which are specified. This technique is adopted in many programs like SAP to compute support reaction or to specify a displacement. In section 7.9 the boundary element used in PASSFEM is explained in detail.

5.2 Gauss Elimination and Matrix Decomposition

Consider a set of linear algebraic equations to be solved in the finite element analysis.

$$K_{11}r_1 + K_{12}r_2 + \cdots + K_{1n}r_n = P_1$$
$$K_{21}r_1 + K_{22}r_2 + \cdots + K_{2n}r_n = P_2$$
$$\cdots\cdots\cdots\cdots\cdots\cdots\cdots\cdots\cdots\cdots\cdots\cdots\cdots\cdots$$
$$\cdots\cdots\cdots\cdots\cdots\cdots\cdots\cdots\cdots\cdots\cdots\cdots$$
$$K_{n1}r_1 + K_{n2}r_2 + \cdots + K_{nn}r_n = P_n$$

$$\tag{5.15}$$

These equations are expressed in concise matrix form as

$$[K]\{r\} = \{P\} \tag{5.16}$$

In the Gauss elimination method, the matrix [K] is converted into an upper triangular form so that the value of the last variable can be got directly. Then back substitution yields the value of the remaining variables. An efficient way of implementing the Gauss elimination is the decomposition technique [3, 4, 9].The Gauss elimination procedure and the decomposition technique are described below.

The elimination can be carried out in the following steps.

1. Subtract l_{i1} times the first equation from the ith equation where

$$l_{i1} = \frac{K_{i1}}{K_{11}}, \quad i = 2, 3, ..., n \tag{5.17}$$

This eliminates r_1 from all but the first equation, and the equations are converted as

$$
\begin{aligned}
K_{11}r_1 \;+\; K_{12}r_2 \;+\; ... \;+\; K_{1n}r_n &= P_1 \\
K_{22}^{(2)}r_2 \;+\; ... \;+\; K_{2n}^{(2)}r_n &= P_2^{(2)} \\
... \qquad ... \qquad ... \qquad ... \qquad ... \qquad ... \\
K_{2n}^{(2)}r_2 \;+\; ... \;+\; K_{nn}^{(2)}r_n &= P_n^{(2)}
\end{aligned}
\tag{5.18}
$$

where the superscript (2) indicates that the coefficients have been modified, e.g.,

$$K_{ij}^{(2)} = K_{ij} - l_{i1}\,K_{ij} \quad j = 2, 3, ..., n \text{ for each } i = 2, 3, ..., n \tag{5.19a}$$

$$P_i^{(2)} = P_i - l_{i1}\,P_1 \quad i = 2, 3, ..., n \tag{5.19b}$$

2. Now in the set of Eq.5.18 subtract l_{i2} times the second equation from the ith equation where

$$l_{i2} = \frac{K_{i2}^{(2)}}{K_{22}^{(2)}} \quad i = 3, ..., n \tag{5.20}$$

This step eliminates r_2 from equations 3 to n and results in

$$K_{11}r_1 \quad + \quad K_{12}r_2 \quad + \quad K_{13}r_3 \quad + \quad ... \quad + \quad K_{1n}r_n \quad = \quad P_1$$

$$K_{22}^{(2)}r_2 \quad + \quad K_{23}^{(2)}r_3 \quad + \quad ... \quad + \quad K_{2n}^{(2)}r_n \quad = \quad P_2^{(2)}$$

$$K_{33}^{(3)}r_3 \quad + \quad ... \quad + \quad K_{3n}^{(3)}r_n \quad = \quad P_3^{(3)}$$

$$...\quad\quad...\quad\quad...\quad\quad...\quad\quad...\quad\quad...$$

$$...\quad\quad...\quad\quad...\quad\quad...\quad\quad...\quad\quad...$$

$$K_{n3}^{(3)}r_3 \quad + \quad ... \quad + \quad K_{nn}^{(3)}r_n \quad = \quad P_n^{(3)}$$

$$(5.21)$$

where the superscript (3) indicates that the coefficients have been modified once again.

It can be observed that while eliminating the ith variable from the set of equations all the elements of the ith column below the diagonal become zero and the elements of the ith row remain unchanged. Also the first $(i\text{-}1)$ equations are not disturbed. Let $K_{ij}^{(s)}$ be the coefficient in the i,j position at the s-th reduction step $i \geq s$ and $j \geq s$ and let $P_i^{(s)}$ represent the element P_i of the right hand side vector $\{P\}$ for $i \geq s$. The s-th step in the elimination can be given by

$$K_{ij}^{(s+1)} \; = \; 0(s \, < \, i \; \leq \; n) \tag{5.22a}$$

$$K_{ij}^{(s+1)} \; = \; K_{ij}^{(s)} \; - l_{is} \, K_{sj}^{(s)} \; (s \, < \, i \; \leq n, \, s \, < \, j \; \leq n) \tag{5.22b}$$

$$P_i^{(s+1)} = P_i^{(s)} \; - l_{is} \, P_s^{(s)} (s \; \leq i \leq n) \tag{5.22c}$$

where the multiplier 'l' at the sth step is given by

$$l_{is} \; = \; \frac{K_{is}^{(s)}}{K_{ss}^{(s)}} \quad i \; = \; s+1,...,n \tag{5.23}$$

The elements of the matrix $[K]$ and vector $\{P\}$ to be modified according to Eq.5.22b and 5.22c at the s-th step in the elimination process are indicated below.

$$\begin{bmatrix} K_{11}^{(1)} & K_{1,s-1}^{(1)} & K_{i,s}^{(1)} & K_{1,s+1}^{(1)} & K_{1,n}^{(1)} \\ ... & ... & ... & ... & ... \\ X & K_{s-1,(s-1)}^{(s-1)} & K_{s-1,s}^{(s-1)} & K_{s-1,s+1}^{(s-1)} & K_{s-1,n}^{(s-1)} \\ X & X & K_{s,s}^{(s)} & K_{s,s+1}^{(s)} & K_{s,n}^{(s)} \\ X & X & K_{s+1,s}^{(s)} & K_{s+1,s+1}^{(s)} & K_{s+1,n}^{(s)} \\ X & X & K_{n,s}^{(s)} & K_{n,s+1}^{(s)} & K_{n,n}^{(s)} \end{bmatrix} \begin{Bmatrix} P_1^{(1)} \\ P_{s-1}^{(s-1)} \\ P_s^{(s)} \\ P_{s+1}^{(s)} \\ P_n^{(s)} \end{Bmatrix} \tag{5.24}$$

Repeating the above process of elimination $n - 1$ times the $[K]$ matrix will be converted into an upper triangular form and the right hand side vector will also be correspondingly modified.

The final form of the reduced equations will be,

$$
\begin{array}{ccccccccc}
K_{11}r_1 & + & K_{12}r_2 & + & & \dots & + & K_{1n}r_n & = & P_1 \\
& & K_{22}^{(2)}r_2 & + & & \dots & + & K_{2n}^{(2)}r_n & = & P_2^{(2)} \\
\dots & \dots & \dots & \dots & & \dots & & \dots & \dots & \dots \\
& & & & & \dots & & & \dots & \dots \\
& & K_{ii}^{(i)}r_i & K_{i,i+1}^{(i)}r_{i+1} & + & & K_{in}^{(i)}r_n & = & P_i^{(i)} \\
& & & & & + & K_{nn}^{(n)}r_n & = & P_n^{(n)}
\end{array}
$$

$$(5.25)$$

The last equation gives,

$$
r_n = \frac{P_n^{(n)}}{K_{nn}^{(n)}} \tag{5.26}
$$

The value of the other variables can be evaluated by back substitution in the order $r_{n-1}, r_{n-2}, \dots, r_1$. Algebraically the back substitution can be expressed as,

$$
r_i = \frac{P_i^{(i)} - \sum_{s=i+1}^{n} K_{is}^{(i)} \, r_s}{K_{ii}^{(i)}} \qquad i = n - 1, n - 2, \dots 1 \tag{5.27}
$$

Example: The equilibrium Eq.5.9 for the continuous beam problem described in the previous section is solved below by the Gauss elimination procedure.

$$
\begin{bmatrix} 28 & 6 & 0 & 0 \\ 6 & 24 & 6 & 0 \\ 0 & 6 & 28 & 8 \\ 0 & 0 & 8 & 16 \end{bmatrix} \begin{Bmatrix} r_1 \\ r_2 \\ r_3 \\ r_4 \end{Bmatrix} = \begin{Bmatrix} +1 \\ 0 \\ -1 \\ 10 \end{Bmatrix} \tag{a}
$$

1. Elimination of r_1 from all but the first equation by subtracting l_{i1} times the first equation from ith equation.

From Eq.5.17, $l_{i1} = \frac{K_{i1}}{K_{11}} = \frac{K_{i1}}{28}$ $i = 2, 3, 4$

$$\begin{bmatrix} 28 & 6 & 0 & 0 \\ 0 & 22.714^{(2)} & 6^{(2)} & 0^{(2)} \\ 0 & 6^{(2)} & 28^{(2)} & 8^{(2)} \\ 0 & 0^{(2)} & 8^{(2)} & 16^{(2)} \end{bmatrix} \begin{Bmatrix} r_1 \\ r_2 \\ r_3 \\ r_4 \end{Bmatrix} = \begin{Bmatrix} 1.0^{(1)} \\ -0.2143^{(2)} \\ -1.0^{(2)} \\ 10.0^{(2)} \end{Bmatrix} \qquad (b)$$

2. subtracting l_{i2} times the second equation from third and fourth equation, r_2 is eliminated. From Eq.5.20,

$$l_{i2} = \frac{K_{i2}^{(2)}}{22.712} \qquad i = 3, 4$$

$$\begin{bmatrix} 28 & 6 & 0 & 0 \\ 0 & 22.714^{(2)} & 6^{(2)} & 0^{(2)} \\ 0 & 0 & 26.415^{(3)} & 8^{(3)} \\ 0 & 0 & 8^{(3)} & 16^{(3)} \end{bmatrix} \begin{Bmatrix} r_1 \\ r_2 \\ r_3 \\ r_4 \end{Bmatrix} = \begin{Bmatrix} 1.0^{(1)} \\ -0.2143^{(2)} \\ -0.9434^{(3)} \\ 10.0^{(3)} \end{Bmatrix} \qquad (c)$$

3. Subtracting l_{i3} times the third equation from the fourth equation r_3 is eliminated. From Eq.5.23,

$$l_{i3} = \frac{K_{i3}^{(3)}}{26.415} \qquad i = 4$$

Thus

$$\begin{bmatrix} 28 & 6 & 0 & 0 \\ 0 & 22.714^{(2)} & 6^{(2)} & 0^{(2)} \\ 0 & 0 & 26.415^{(3)} & 8^{(3)} \\ 0 & 0 & 0 & 13.577^{(4)} \end{bmatrix} \begin{Bmatrix} r_1 \\ r_2 \\ r_3 \\ r_4 \end{Bmatrix} = \begin{Bmatrix} 1.0 \\ -0.2143^{(2)} \\ -0.9434^{(3)} \\ 10.2857^{(4)} \end{Bmatrix} \qquad (d)$$

Using Eq.5.26, the value of r_4 can be found

$$r_4 = \frac{10.2857}{13.577} = 0.7576$$

Using Eq.5.27 the values of other variables can be calculated. Thus,

$$r_3 = \frac{-0.9434 - (8 \times 0.7576)}{26.415}$$
$$= -0.2652$$
$$r_2 = \frac{-0.2143 - [6 \times (-0.2652) + 0]}{22.714}$$

$$= \quad 0.0606$$
$$r_1 \quad = \quad \frac{1.0 - [6 \times (0.0606) + 0 + 0]}{28.0}$$
$$= \quad 0.0227$$

5.2.1   Decomposition ($[L]\,[D]\,[L]^T$)

In Gauss elimination each step consists of adding a multiple of a row to other rows.

Now consider a matrix of the form shown below:

$$[L_1]^{-1} = \begin{bmatrix} 1 & & & & & \\ -l_{21} & 1 & & & & \\ -l_{31} & & 1 & & & \\ \cdot & & & \cdot & & \\ \cdot & & & & \cdot & \\ \cdot & & & & & \cdot \\ -l_{n1} & & & & & 1 \end{bmatrix} \tag{5.28}$$

(elements not shown are zeroes)

where l_{21}, l_{31}, etc. can be obtained from Eq.(5.17). If a matrix $[K]$ is premultiplied by the above matrix the result will be equivalent to subtracting l_{i1} times the first row of $[K]$ from its ith row. This is what is done in the first step of Gauss elimination (Eq. 5.19).

Likewise it can be observed that the second step in the Gauss elimination consists of pre-multiplying the modified matrix. $[[L_1]^{-1}[K]]$ by another matrix $[[L_2]^{-1}]$ where

$$[L_2]^{-1} = \begin{bmatrix} 1 & & & & & \\ & 1 & & & & \\ & -l_{32} & 1 & & & \\ & -l_{42} & & \cdot & & \\ & \cdot & & & \cdot & \\ & \cdot & & & & \cdot \\ & -l_{n2} & & & & 1 \end{bmatrix} \tag{5.29}$$

where the elements l_{i2} are the same as given in Eq.5.20.

If we denote the final upper triangular matrix by $[S]$ then the Gauss elimination procedure can be written as

$$[S] = [L_{n-1}]^{-1} ... [L_2]^{-1}[L_1]^{-1}[K] \tag{5.30}$$

Now it can be observed that the inverse of a matrix of the type of $[L_i]^{-1}$ can be obtained simply by changing the signs of the off-diagonal elements.

$$[L_i] = \begin{bmatrix} 1 & & & & & & \\ & 1 & & & & & \\ & & \cdot & & & & \\ & & & \cdot & & & \\ & & & & 1 & & \\ & & & & l_{i+1,i} & & \\ & & & & \cdot & \cdot & \\ & & & & \cdot & & \\ & & & & l_{ni} & & 1 \end{bmatrix} \qquad (5.31)$$

Hence Eq.(5.30) can be written as

$$[K] = [L_1][L_2] \ldots [L_{n-1}][S] \qquad (5.32)$$

Again making use of the fact that the product of the matrices having the character of the type of $[L_i]$ can be obtained by simply superposing these matrices, we get

$$[L_1][L_2] \ldots [L_{n-1}] = [L] \qquad (5.33)$$

vhere

$$[L] = \begin{bmatrix} 1 & & & & \\ l_{21} & 1 & & & \\ \cdot & l_{32} & 1 & & \\ & & \cdot & \cdot & \\ \cdot & \cdot & & 1 & \\ l_{n1} & l_{n2} & & l_{n,n-1} & 1 \end{bmatrix} \qquad (5.34)$$

Therefore, Eq.5.32 becomes

$$[K] = [L][S] \qquad (5.35)$$

where [S] is an upper triangular matrix.

Thus

$$\begin{bmatrix} K_{11} & K_{12} & \ldots & K_{1n} \\ K_{21} & K_{22} & \ldots & K_{2n} \\ \ldots & \ldots & \ldots & \ldots \\ \ldots & \ldots & \ldots & \ldots \\ K_{n1} & K_{n2} & \ldots & K_{nn} \end{bmatrix} = \begin{bmatrix} 1 & & & \\ l_{21} & 1 & & \\ \ldots & \ldots & \ldots & \\ \ldots & \ldots & \ldots & \\ l_{n1} & l_{n2} & & 1 \end{bmatrix} \begin{bmatrix} s_{11} & s_{12} & \ldots & s_{1n} \\ & s_{22} & \ldots & s_{2n} \\ & & \ldots & \ldots \\ & & \ldots & \ldots \\ & & & s_{nn} \end{bmatrix}$$
$$(5.36)$$

It is possible to factorise [S] into the product $[D][\bar{S}]$ where [D] is a diagonal matrix. Thus,

$$[S] = [D][\bar{S}] \qquad (5.37)$$

where

$$D_{ii} = S_{ii} \quad i = 1, 2, \ldots, n \tag{5.38}$$

$$D_{ij} = 0 \quad i \neq j \tag{5.39}$$

$$\overline{S}_{ii} = 1 \quad i = 1, 2, \ldots, n \tag{5.40a}$$

$$\overline{S}_{ij} = S_{ij} \tag{5.40b}$$

Now using Eqs. 5.35 and 5.37, the decomposition of [K] can be expressed as

$$[K] = [L][D][\overline{S}] \tag{5.41}$$

Since [K] is a symmetric matrix and [D] is also symmetric (being a diagonal matrix), it follows that

$$[\overline{S}] = [L]^T \tag{5.42}$$

and hence

$$[K] = [L][D][L]^T \tag{5.43}$$

The decomposition of [K] in Eq.5.36 can be written as

$$
\begin{bmatrix}
K_{11} & K_{12} & \ldots & K_{1n} \\
K_{21} & K_{22} & \ldots & K_{2n} \\
\ldots & \ldots & \ldots & \ldots \\
\ldots & \ldots & \ldots & \ldots \\
K_{n1} & K_{n2} & & K_{nn}
\end{bmatrix}
=
\begin{bmatrix}
1 & & & \\
l_{21} & 1 & & \\
\ldots & \ldots & \ldots & \ldots \\
\ldots & \ldots & \ldots & \ldots \\
l_{n1} & l_{n2} & \ldots & 1
\end{bmatrix}
\times
$$

$$
\begin{bmatrix}
D_{11} & & & \\
& D_{22} & & \\
\ldots & \ldots & \ldots & \ldots \\
& & & D_{nn}
\end{bmatrix}
\times
\begin{bmatrix}
1 & l_{21} & \ldots & l_{n1} \\
& 1 & l_{32} & \ldots & l_{n2} \\
\ldots & \ldots & \ldots & \ldots & \ldots \\
\ldots & \ldots & \ldots & \ldots & 1
\end{bmatrix}
\tag{5.44}
$$

Multiplying the matrices on the right-hand side of Eq.5.44 we get Eq.5.46.

Now comparing the terms sequentially on either side of Eqs. 5.45 we can obtain the values of D_{ij} and l_{ij}. The following general expression can then be obtained for the elements of $[L]^T$ and [D] matrices.

$$l_{ij} = K_{ij} - \sum_{s=1}^{i-1} \frac{D_{ss} \, l_{si} \, l_{sj}}{D_{ii}} \quad j = 2, \ldots, n$$

and

$$D_{11} = K_{11}$$

$$D_{jj} = K_{jj} - \sum_{s=1}^{j-1} l_{sj}^2 D_{ss} \qquad (5.45)$$

$$
\begin{bmatrix}
K_{11} & K_{12} & \cdots & K_{1n} \\
K_{21} & K_{22} & \cdots & K_{2n} \\
\vdots & & & \\
K_{n1} & K_{n2} & \cdots & K_{nn}
\end{bmatrix}
=
\begin{bmatrix}
a_{11} & a_{12} & \cdots & \cdots & a_{1n} \\
 & a_{22} & \cdots & \cdots & a_{2n} \\
 & & a_{33} & \cdots & a_{3n} \\
 & & & a_{ij} & \\
\cdots & \cdots & \cdots & \cdots & \cdots \\
 & & & & a_{nn}
\end{bmatrix}
\qquad (5.46)
$$

where

$$a_{11} = D_{11}, \quad a_{12} = D_{11}l_{21}$$

$$a_{1n} = D_{11}l_{n1}, \quad a_{22} = D_{11}l_{21}^2 + D_{22}$$

$$a_{2n} = (D_{11}l_{n1}l_{21} + D_{22}l_{n2}),$$

$$a_{33} = (D_{11}l_{31}^2 + D_{22}l_{32}^2 + D_{33})$$

$$a_{3n} = (D_{11}l_{n1}l_{31} + D_{22}l_{n2}l_{32} + D_{33}l_{n3})$$

$$a_{ij} = D_{11}l_{1j}l_{1i} + D_{22}l_{2i}l_{2j} + \ldots + D_{i-1,i-1}l_{i-1,j}l_{i-1,i} + D_{ii}l_{i,j}$$

$$a_{nn} = D_{11}l_{n1}^2 + D_{22}l_{n2}^2 + \ldots + D_{nn}$$

In the program PASSFEM the stiffness matrix [K] is stored as a one-dimensional array and the scheme is explained in the next chapter. The $[L][D][L]^T$ decomposition of the matrix discussed above can be efficiently carried out by considering each column in turn. It is necessary to store only the coefficients of $[L]^T$ and [D] and that too in the same location as that of [K].

The routine PASOLV in PASSFEM solves the equations in the main memory without resorting to blocking (or partitioning) the equations which may be necessary in the case of large number of equations [10, 11]. Also it may be noted that the coefficient outside the skyline need not be computed as they are not changed, i.e., continue to have zero values during decomposition. Thus the elements are calculated column-wise and operations outside the skyline are avoided resulting in saving of computing time [3].

Denoting the skyline of the matrix by m_i the row number of the first non-zero element in column i, the Eq.5.46 can be modified to evaluate the elements column-wise.

$$l_{ij} = \frac{K_{ij} - \sum_{s=m_m}^{i-1} D_{ss}\, l_{si}\, l_{sj}}{D_{ii}}.$$

$$j = 2, \ldots, n \quad \text{and} \quad i = m_j, \ldots, \, j-1$$

$$D_{11} = K_{11}$$

$$D_{jj} = K_{jj} - \sum_{s=m_j}^{j-1} l_{sj}^2\, D_{ss} \qquad j = 2, \ldots, n$$

where

$$m_m = \max(m_i, m_j) \tag{5.47}$$

Example The decomposition of [K] using the above equations are illustrated through the same example presented under Gauss-elimination procedure (Eq.5.9)

$$[K] = \begin{bmatrix} 28 & 6 & 0 & 0 \\ 4 & 24 & 6 & 0 \\ 0 & 6 & 28 & 8 \\ 0 & 0 & 8 & 16 \end{bmatrix} \tag{a}$$

The values of m_j are, $m_1 = 1$, $m_2 = 1$, $m_3 = 2$, $m_4 = 3$

$$D_{11} = K_{11} = 28$$

For j = 2:

$$i = m_1, \ldots, 1 \quad \text{i.e.,} \quad i = 1$$

$$l_{12} = \frac{K_{12}}{D_{11}} = \frac{6}{28} = 0.2143$$

$$D_{22} = K_{22} - \sum_{s=1}^{1} l_{12}^2\, D_{11} = 24 - 0.2143 \times 28 = 22.714$$

For j = 3:

$$i = m_3, \ldots, 2 \quad \text{i.e.} \quad i = 2$$

$$l_{23} = \frac{K_{23}}{D_{22}} = \frac{6}{22.714} = 0.2642$$

$$D_{33} = K_{33} - \sum_{s=2}^{2} l_{23}^2\, D_{22} = 28 - 0.2642^2 \times 22.714 = 26.4145$$

For j = 4:

$$i = m_4, \ldots, 3 \quad \text{i.e.} \quad i = 3$$

$$l_{34} = \frac{K_{34}}{D_{33}} = \frac{8}{26.4145} = 0.3028$$

$$D_{44} = K_{44} - \sum_{s=3}^{3} l_{34}^2 D_{33} = 16 - 0.3028^2 \times 26.4145 = 13.5776$$

Hence the final elements stored along the diagonal (D_{jj}) and the elements of $[L]^T$ replacing $K_{i,j}$ for $j > 1$ are given below:

$$\begin{bmatrix} 28 & 0.2143 & 0 & 0 \\ & 22.714 & 0.2642 & 0 \\ & & 26.4145 & 0.3028 \\ & & & 13.5776 \end{bmatrix}$$

Now expressing [K] in the form of $[L]\,[D]\,[L]^T$, the equation of equilibrium 5.26 becomes.

$$[L]\,[D]\,[L]^T\,\{r\} = \{P\} \tag{5.48}$$

Premultiplying by $[L]^{-1}$

$$[D]\,[L]^T\,\{r\} = [L]^{-1}\,\{P\} \quad \text{or} \quad [D]\,[L]^T\,\{r\} = \{P'\} \tag{5.49a}$$

where

$$\{P'\} = [L]^{-1}\,\{P\} \tag{5.49b}$$

The reduced load vector $[P']$ in Eq.5.49b can be obtained during the computation of l_{ij} coefficients as

$$\{P'\} = [L_{n-1}]^{-1}\ldots [L_2]^{-1}\,[L_1]^{-1}\,\{P\} \tag{5.50}$$

Using equations of the type 5.28, the above equation can also be written as

$$P_1' = P_1$$

and

$$P_i' = P_i - \sum_{s=m_i}^{i-1} l_{si}\,P_s' \quad \text{for } i = 2, \ldots, n \tag{5.51}$$

In the computer program P_i' replaces P_i.

Before back-substitution to compute the values of r_i, the values of $\{P'\}$ are premultiplied by $[D^{-1}]$ as

$$\{\overline{P}\} = [D]^{-1} \{P'\} \tag{5.52}$$

Defining $\{\overline{P}^{(n)}\} = \{\overline{P}\}$, we get $r_n = \overline{P}_n^{(n)}$
Then by back-substitution for $i = n, \ldots, 2$, the values of $r_{n-1} \ldots,$ r_1 can be calculated by the following equation

$$\overline{P}_s^{(i-1)} = \overline{P}_s^{(i)} - l_{si} r_i \qquad s = m_i, \ldots, i-1$$

and

$$r_{i-1} = \overline{P}_{i-1}^{(i-1)} \tag{5.53}$$

The following example illustrates the above procedure.

Example: Having done the decomposition of the matrix $[K]$ of Eq.(5.9), we can now obtain the solution for r_i using the Eqs 5.51, 5.52 and 5.53.

$$\{P\} = \begin{Bmatrix} 1 \\ 0 \\ -1 \\ 10 \end{Bmatrix}$$

$$P_1' = P_1 = 1$$

$$P_2' = P_2 - \sum_{s=1}^{1} l_{s2} P_s' = 0 - (0.2143)(1) = -0.2143$$

$$P_3' = P_3 - \sum_{s=2}^{2} l_{s3} P_s' = -1 - (0.2642)(-0.2143) = -0.9434$$

$$P_4' = P_4 - \sum_{s=3}^{3} l_{s4} P_s' = 10 - (0.3028)(-0.9434) = 10.2857$$

Thus

$$\{P'\} = \begin{Bmatrix} +1.0000 \\ -0.2143 \\ -0.9434 \\ 10.2857 \end{Bmatrix}$$

At this stage, the elements of $\{P\}$ are replaced $\{P'\}$ since $[D]$ is a diagonal matrix, $[D]^{-1}$ is also diagonal matrix with the diagonal elements being reciprocal of the original elements. Thus

$$\{\overline{P}\} = [D]^{-1} \{P\} = \{\overline{P}^{(4)}\} = \begin{Bmatrix} +0.03571 \\ -0.009435 \\ -0.03572 \\ +0.75760 \end{Bmatrix}$$

Now, back substitution can be done using Eq. (5.53).

$$r_4 = \overline{P}_4^{(4)} = 0.7576$$

$i = 4$:

$$\overline{P}_3^{(3)} = \overline{P}_3^{(4)} - l_{34}\, r_4 = -0.0.3572 - (0.3028)\,(0.7576) = -0.26512$$

and

$$\{\overline{P}^{(3)}\} = \begin{Bmatrix} +0.03571 \\ -0.009435 \\ -0.26512 \\ 0.75760 \end{Bmatrix}$$

$$r_3 = \overline{P}_3^{(3)} = -0.26512$$

$i = 3$:

$$\overline{P}_2^{(2)} = \overline{P}_2^{(3)} - l_{23}\, r_3 = -0.009435 - (0.2642)\,(-0.26512) = 0.06061$$

$$\{\overline{P}^{(2)}\} = \begin{Bmatrix} 0.03571 \\ 0.06071 \\ 0.26512 \\ 0.75760 \end{Bmatrix}$$

$$r_2 = \overline{P}_2^{(2)} = 0.06061$$

$i = 2$:

$$P_1^{(1)} = P_1^{(2)} - l_{12}\, r_2 = 0.03571 - (0.2143)\,(0.06061) = 0.02272$$

and

$$\{\overline{P}^{(1)}\} = \begin{Bmatrix} +0.02272 \\ +0.06061 \\ -0.26512 \\ +0.75760 \end{Bmatrix}$$

$$r_1 = P_1^{(1)} = 0.02272$$

It may be noted that $\{\overline{P}^{(1)}\}$ contains the solution vector $\{r\}$.

The subroutine PASOLV uses the algorithm described above for solving simultaneous equations in finite element analysis. A brief account of the subroutine and the listing of the program are presented in the next chapter.

5.2.2   Basic Steps in Finite Element Analysis

Following the material presented in the earlier chapters and, the assemblage and solution procedure explained in the present chapter, it is now possible to summarise the steps to be followed in a finite element stress analysis. The cantilver beam example given in section 8.8.1 is chosen for illustration.

1. Discretization and Pre-processing of finite element model

As a first step in the analysis the given solid or structure is to be discretized into finite elements. This step requires knowledge of the physical behaviour of the solid or structure to decide on the type of analysis and elements to be used to arrive at the finite element model. In addition decision has to be made in the shape of elements to be used (higher or lower order elements), the number of elements and the pattern of the finite element mesh.

After the discretization, the nodes are numbered keeping in view the minimum bandwidth requirements discussed in this chapter. Graphics based pre-processors are available in many package programs to automatically generate the mesh and number the nodes and elements. Fig. 5.5 shows discretization of the beam with four noded rectangular elements for plane stress analysis.

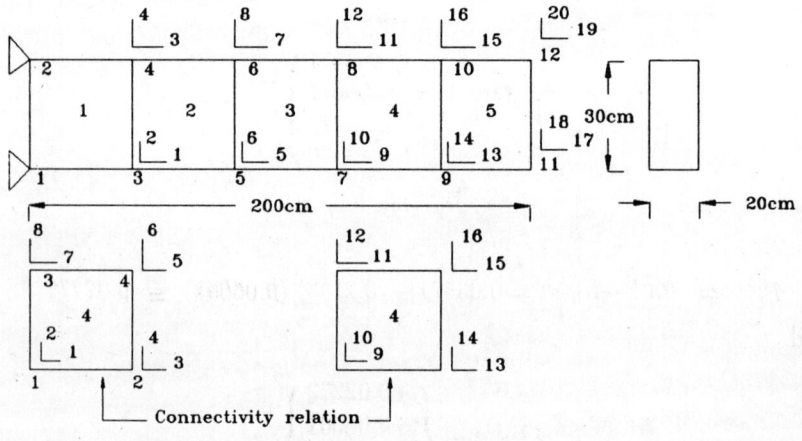

Fig. 5.5 Discretization of a cantilever beam

2. Computation of element properties

Using the expressions derived in the chapters 3 and 4, the strain-displacement matrix [B], element stiffness matrix [k] and nodal load vector {Q} are computed for each element.

3. Assemblage of elements

The direct stiffness method explained in this chapter is used to constitute the structure or global stiffness matrix [K] and nodal load vector $\{P\}$. The connectivity relation between element and global degrees of freedom is used to compute the contribution from an element to the global stiffness matrix $[K]$. The contribution from element 4 to $[K]$ matrix is given below.

$$
\begin{array}{cccccccccc}
g & d & o & f & 9 & 10 & 13 & 14 & 15 & 16 & 11 & 12 \\
e & d & o & f & 1 & 2 & 3 & 4 & 5 & 6 & 7 & 8
\end{array}
$$

$$
\begin{array}{cc}
\begin{array}{cc}
9 & 1 \\
10 & 2 \\
13 & 3 \\
14 & 4 \\
15 & 5 \\
16 & 6 \\
11 & 7 \\
12 & 8 \\
g & e
\end{array}
&
\begin{bmatrix}
k_{11} & k_{12} & k_{13} & k_{14} & k_{15} & k_{16} & k_{17} & k_{18} \\
k_{21} & k_{22} & k_{23} & k_{24} & k_{25} & k_{26} & k_{27} & k_{28} \\
k_{31} & k_{32} & k_{33} & k_{34} & k_{35} & k_{36} & k_{37} & k_{38} \\
k_{41} & k_{42} & k_{43} & k_{44} & k_{45} & k_{46} & k_{47} & k_{48} \\
k_{51} & k_{52} & k_{53} & k_{54} & k_{55} & k_{56} & k_{57} & k_{58} \\
k_{61} & k_{62} & k_{63} & k_{64} & k_{65} & k_{66} & k_{67} & k_{68} \\
k_{71} & k_{72} & k_{73} & k_{74} & k_{75} & k_{76} & k_{77} & k_{78} \\
k_{81} & k_{82} & k_{83} & k_{84} & k_{85} & k_{86} & k_{87} & k_{88}
\end{bmatrix}
\end{array}
$$

(a) Stiffness matrix [k] of element 4

$$
\begin{array}{ccccccccc}
 & 9 & 10 & 11 & 12 & 13 & 14 & 15 & 16\ldots20
\end{array}
$$

$$
\begin{array}{cc}
\begin{array}{c}
9 \\
\\
10 \\
\\
11 \\
\\
12 \\
\\
13 \\
\\
14 \\
\\
15 \\
\\
16 \\
\\
20
\end{array}
&
\begin{bmatrix}
\cdots & k_{11} & k_{12} & k_{17} & k_{18} & k_{13} & k_{14} & k_{15} & k_{16} & \cdots \\
\cdots & k_{21} & k_{22} & k_{27} & k_{28} & k_{23} & k_{24} & k_{25} & k_{26} & \cdots \\
\cdots & k_{71} & k_{72} & k_{77} & k_{78} & k_{73} & k_{74} & k_{75} & k_{76} & \cdots \\
\cdots & k_{81} & k_{82} & k_{87} & k_{88} & k_{83} & k_{84} & k_{85} & k_{86} & \cdots \\
\cdots & k_{31} & k_{32} & k_{37} & k_{38} & k_{33} & k_{34} & k_{35} & k_{36} & \cdots \\
\cdots & k_{41} & k_{42} & k_{47} & k_{48} & k_{43} & k_{44} & k_{45} & k_{46} & \cdots \\
\cdots & k_{51} & k_{52} & k_{57} & k_{58} & k_{53} & k_{54} & k_{55} & k_{56} & \cdots \\
\cdots & k_{61} & k_{62} & k_{67} & k_{68} & k_{63} & k_{64} & k_{65} & k_{66} & \cdots \\
\end{bmatrix}
\end{array}
$$

Fig. 5.6 Contribution from element 4 to the structure stiffness matrix [K]

4. Solution of equations of equilibrium

The linear simultaneous equations of equilibrium, $[K] \{r\} = \{P\}$ are solved for the nodal displacements of the structure or solid. Gauss elimination procedure or the variations of it such as the $[L] [D] [L]^T$ decomposition described in this chapter is used for this purpose. The solution for nodal displacements for the cantilever beam, with five element discretization shown in Fig. 5.5 is given below.

$$\{r\}^T = [-0.002, -0.00267, 0.002, -0.00267, -0.004,$$
$$-0.01067, 0.004, -0.01067, -0.006, -0.024,$$
$$0.006, -0.024, -0.008, -0.04267, 0.008,$$
$$-0.04267, -0.010, -0.06667, 0.010, -0.06667] \qquad (a)$$

5. Computation of stress and post-processing of results

The stresses at any point in the element can be computed using the Eq. 3.70, $\{\sigma\} = [C] [B] \{d\}$. Although one would like to get the stresses evaluated at the nodal points, they appear to be the worst sampling points. Barlow [12] has shown that for two-dimensional isoparametric elements, the Gauss points are the optimal sampling points for stress computation. These aspects are discussed further in section 10.7.2.

For the cantilever beam example, consider a typical element 4. The matrix $[C]$ $[B]$ for the element evaluated at the Gauss point 1 (Fig. 4.4b) is given by Eq.(h).

The connectivity relation shown in Fig. 5.5 can be used to obtain the nodal displacement for the element 4 from the nodal displacements $\{r\}$ of the structure in Eq. (a).

$$\{d\}^T = \{-0.006, -0.024, -0.008, -0.04267,$$
$$0.008, -0.04267, 0.006, -0.024\} \qquad (b)$$

Multiplying the [CB] matrix(Eq.h) and the displacement vector $\{d\}$ (Eq.b), the stresses at the Gauss point 1 are computed.

$$\{\sigma\} = [CB] \{d\}$$

$$\begin{Bmatrix} \sigma_x \\ \sigma_y \\ \tau_{xy} \end{Bmatrix} = \begin{Bmatrix} 0.06 \\ 0.0 \\ 0.04 \end{Bmatrix} \text{ kN/cm}^2 \qquad (c)$$

Similarly, the stresses at other sampling points can be evaluated.

Graphics based post-processors are available in all package programs that would enable plotting of the deflected shape of the structure, stress contours, variation of a particular type of stress across a given section etc.

REFERENCES

1. Gallagher, R.H., *Finite Element Analysis - Fundamentals.* Prentice-Hall, Inc., Englewood Cliffs, N.J., 1975.

2. Desai, C.S. and J.F.Abel, *Introduction to the Finite Element Method,* Affiliated East-West Press Pvt. Ltd., New Delhi, East-West Student Edition, 1977.

3. Bathe, K.J., and E.L.Wilson, *Numerical Methods in Finite Element Analysis,* Prentice-Hall of India Private Limited, New Delhi, 1978.

4. Jennings, A., *Matrix Computation for Engineers and Scientists,* John Wiley and Sons, London, 1981.

5. Cuthill, E. and J. Mckee, *Reducing the Bandwidth of Sparse Symmetric Matrices, Association of Computing Machinery,* Proceedings of the 24th National Conference, ACM Publication, New York, 1969.

6. Gibbs, N.E., and W.G.Poole, and P.K.Stockmeyer, *An Algorithm for Reducing Bandwidth and Profile of a Sparse Matrix,* SIAM, Journal for Numerical Analysis, Vol. 13, No.2, April 1976.

7. Irons, B.M., *A Frontal Solution Program,* International Journal for Numerical Methods in Engineering, Vol.2, pp.5-32, 1970.

8. Hinton, E., and D.R.J.Owen, *Finite Element Programming,* Academic Press, London, 1977.

9. Noble, B., *Applied Linear Algebra,* Prentice Hall, Inc., Englewood Cliffs N.J., 1969.

10. Wilson, E.L., K.J.Bathe, and N.P.Doherty, *Direct Solution of Large Systems of Linear Equations,* Computers and Structures, Vol.4, pp.363-372, 1974.

11. Mondkar, D.P. and G.H., Powell, *Large Capacity Equation Solver for Structural Analysis,* Computers and Structures, Vol. 4, pp. 691-728, 1974.

12. Barlow, J., *Optimal Stress Locations in Finite Element Method*, International Journal for Numerical Methods in Engineering, Vol.10, pp.243-251, 1976.

EXERCISES

5.1 The rectangular element shown in Fig. E5.1 is formed by assembling four CST elements. Compute the stiffness matrix of the element after static condensation. Compare the result with the stiffness matrix of the rectangular element given in section 4.2.2.

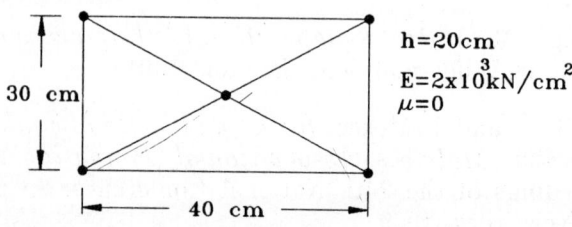

$h=20\text{cm}$
$E=2\text{x}10^3\text{kN/cm}^2$
$\mu=0$

30 cm

40 cm

Fig. E5.1

5.2 A rectangular plate shown in Fig. E5.2 is subjected to a load of 10 kN/m. Analyse the problem using CST elements.

$E=2.1\text{x}10^{11}\text{N/m}^2$ $\mu=0.3$

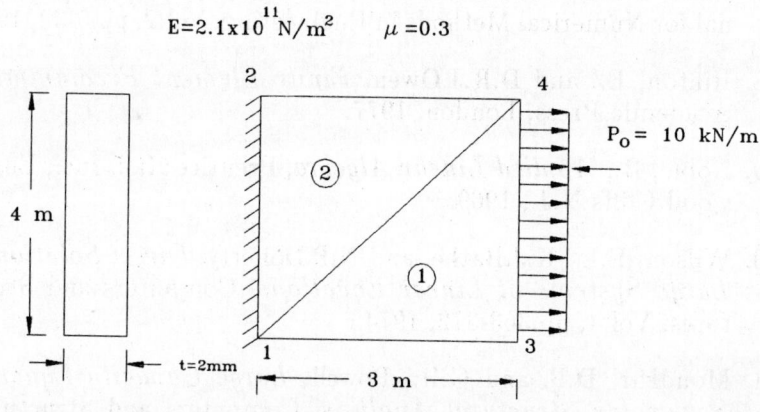

$P_0 = 10$ kN/m

4 m

t=2mm

3 m

Fig. E5.2

5.3 Using the stiffness matrix of the rectangular element given in page 102, assemble the structure stiffness matrix for the cantilever beam shown in Fig. 8.14.

5.4 Find the displacements at the nodes for the problem(5.3) when the beam is subjected to an uniformly distributed load of 4 kN/m. Plot the displacements and compare the results with the Strength of Materials solution.

5.5 Using the results of problem (5.3), compute the normal stress at the Gauss points and at the nodes. Plot the maximum stresses along the length of the beam. Discuss the finite element solution.

5.6 For the problem (5.2), using a rectangular element analyse the plate, compare the results with the solution obtained by using two triangular elements.

5.7 The equilibrium equation for a framed structure is given below. Solve for the displacements by using the decomposition, $[L]\,[D]\,[L]^T$ procedure.

$$
\begin{bmatrix}
150450 & 0 & 900 & -150000 & 0 & 0 \\
0 & 150450 & 900 & 0 & -450 & 900 \\
900 & 900 & 4800 & 0 & -900 & 1200 \\
-150000 & 0 & 0 & 200584 & -67062 & 576 \\
0 & -450 & -900 & -67062 & 90154 & -468 \\
0 & 900 & 1200 & 576 & -468 & 4800
\end{bmatrix}
\begin{Bmatrix}
r_1 \\ r_2 \\ r_3 \\ r_4 \\ r_5 \\ r_6
\end{Bmatrix}
$$

$$
=
\begin{Bmatrix}
400.0 \\
-100.0 \\
-100.0 \\
0.0 \\
-100.0 \\
100.0
\end{Bmatrix}
$$

5.8 Calculate the support reactions for a continuous beam shown in Fig. E5.8 by partitioning the stiffness matrix corresponding to unknown and known displacements.

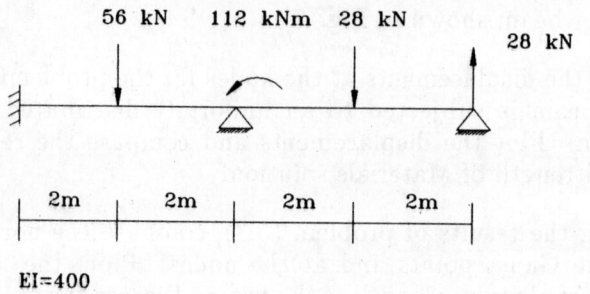

Fig. E5.8

Chapter 6

Computer Program for Finite Element Analysis - (PASSFEM)

A computer program for the analysis of structural systems by finite element method (PASSFEM) is presented here. The program is written in standard FORTRAN IV and the main objective is to provide a teaching aid to finite element programming through the development of a general purpose finite element program. With this objective the program is structured in such a way that any student or user can add a routine for any element without in any way disturbing the main program.

The program is organised to analyse structures using any of the elements or a combination of the elements in the element library. The modular element library structure is explained in section 6.3. In the subsequent chapters dealing with various types of analysis the element routines for specific shapes will be presented along with the theory. Guidelines are also given at the end of this chapter for incorporating any new element routine into the program.

In finite element analysis, the global stiffness matrix is symmetric and banded. It is seen that the zero elements outside the skyline of the stiffness matrix are not changed during the solution process. Therefore, it is required to allocate storage locations only to all the elements within the skyline but the zero elements outside the skyline need not be stored. An effective active column storage scheme taking advantage of these properties is adopted to store the elements below the skyline in a one-dimensional array 'SK'. The program implementation of this scheme is explained in this chapter.

The various routines that constitute the core of the program PASSFEM are explained in the following sections. The input details are described in the later part of the chapter. The listing of the core segments of the computer program PASSFEM (without the element routines) is presented in Appendix 1.

6.1 Main Routine

The main routine, which controls the various tasks of the program, calls a number of subroutines and its organizational structure is indicated in the flow chart in Fig.6.1. It reads the control information namely the number of structure nodes (NSN), number of element types (NET), number of material property groups (NMP) and number of load cases (NLC). The number of loaded nodes for each load case is also read by this routine.

Dynamic dimensioning is adopted for all the variables whose dimensions depend on the data sets of a particular problem. This is achieved by making use of a master array A which accommodates all the global variables and overflows only when the combination of dimensions of all the variables exceeds that of A and in such a situation the DIMENSION statement of array A only need be changed. Also, the main storage area is efficiently utilized as no excess area is provided for individual variables over their exact requirements.

The starting locations of the variables within array A are computed at this stage through statements shown in the program listing. The description for the important variables is given below:

Pointer name	Starting location	Array	Description	Dimension
N1	1	A(N1)	NDF array	6*NSN
N2	N1+6N* NSN	A(N2)	X coord.	NSN
N3	N2+NSN	A(N3)	Y coord.	NSN
N4	N3+NSN	A(N4)	Z coord.	NSN
N5	N4+NSN	A(N5)	E	NMP
N6	N5+NMP	A(N6)	PR	NMP
N7	N6+NMP	A(N7)	ND	NMP
N8	N7+NMP	A(N8)	CHT (Col.ht.)	NEQ
N9	N8+NEQ	A(N9)	ND eqn. No. of each d.o.f. of an element	No. of d.o.f of an element
N10	N9+Variable depends on element type (max. value taken)	A(N10)	NDS-Address of diagonal element of [SK]	NEQ1 (NEQ+1)

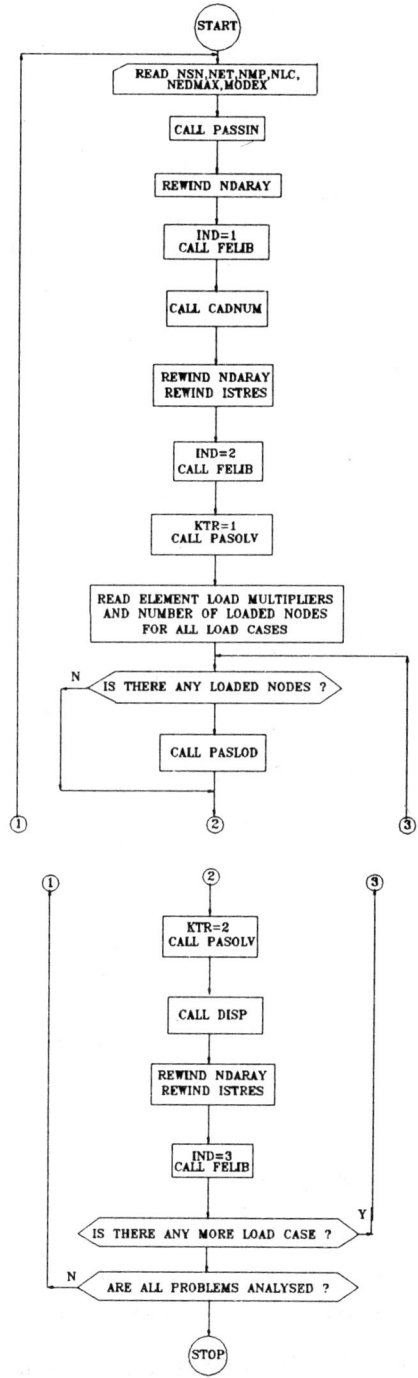

Fig. 6.1 Computer program PASSFEM-MAIN ROUTINE

N11	N10+NEQ1	A(N11)	[SK]	NSKY
N12	N11+NSKY	A(N12)	{PQ}	NEQ
N13	N12+NEQ	A(N13)	{P}	NEQ

In the listing presented here double precision is used by IMPLIC-IT REAL *8 (A-H, O-Z).

After obtaining the geometric and other data for the structure through PASSIN, the main program calls the FELIB, PASOLV and other routines at appropriate stages for the calculation of element properties, assmeblage of elements, solution of the equilibrium equations for the displacements and, finally the strains and stresses in the elements are computed and printed out.

6.2   Subroutine PASSIN

The program is set up to allow a maximum of six degrees of 'freedom at each node (Fig.6.2(a)). Corresponding to each nodal point, it must then be identified which of these degrees of freedom shall actually be used in the finite element assemblage. This is achieved by defining for each node a joint freedom array JF of dimension 6. The active degrees of freedom, those which are to be considered in the assemblage, are defined by $JF(I) = 0$ and the non-active degrees of freedom, those which are constrained to zero displacement are defined by $JF(I) = 1$.

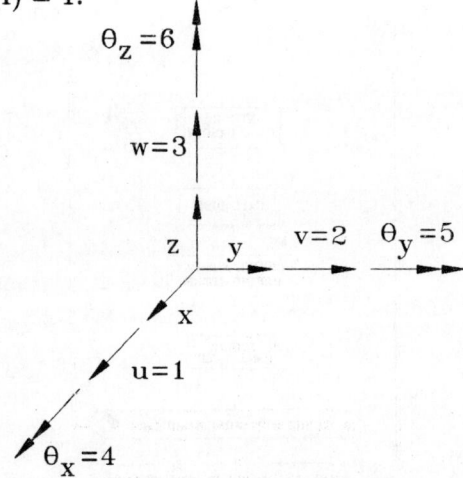

Fig. 6.2a Possible degrees of freedom at a nodal point

The input subroutine PASSIN reads, generates and prints the nodal data, viz., coordinates of nodes and the JF array. The generation of nodal data is carried out along a straight line. The JF array may be explained using the plane stress element idealization shown in Fig.6.2(b). Here each node has two degrees of freedom, the displacements u and v in x and y directions respectively as shown in Fig.6.2(c). The displacement at nodes 1, 2 and 3 are constrained to zero displacement. The JF array computed at the step of processing each node is given as follows.

$$
\begin{array}{c c c c c c c}
\text{Node No.} & & \begin{array}{c}\text{JF}\\\text{array}\end{array} & & & & \\
1 & 1 & 1 & 1 & 1 & 1 & 1 \\
2 & 1 & 1 & 1 & 1 & 1 & 1 \\
3 & 1 & 1 & 1 & 1 & 1 & 1 \\
4 & 0 & 0 & 1 & 1 & 1 & 1 \\
5 & 0 & 0 & 1 & 1 & 1 & 1 \\
6 & 0 & 0 & 1 & 1 & 1 & 1 \\
7 & 0 & 0 & 1 & 1 & 1 & 1 \\
8 & 0 & 0 & 1 & 1 & 1 & 1 \\
9 & 0 & 0 & 1 & 1 & 1 & 1 \\
10 & 0 & 0 & 1 & 1 & 1 & 1 \\
11 & 0 & 0 & 1 & 1 & 1 & 1 \\
12 & 0 & 0 & 1 & 1 & 1 & 1 \\
\end{array}
\qquad (6.1)
$$

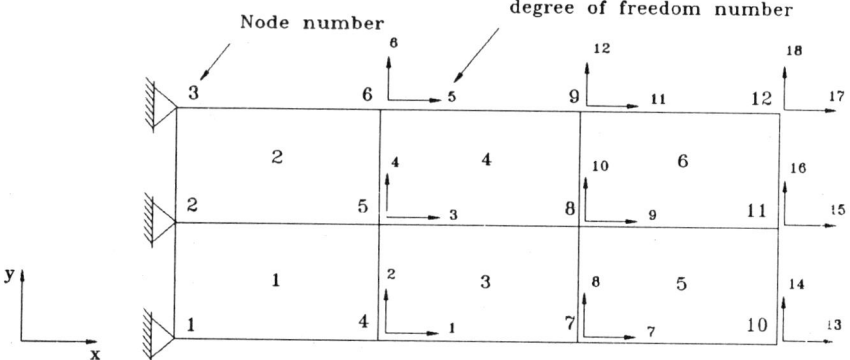

Fig. 6.2b Plane stress idealisation of a cantilever

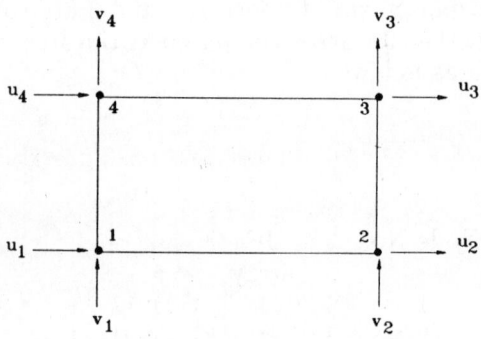

Fig. 6.2c Degrees of freedom of plane stress element

The JF array once read or generated for a node is transferred to NDF array, the element (i, j) of which corresponds to the ith degree of freedom at the node j. The NDF array thus formed for the present example is given as follows,

$$[NDF] = \begin{bmatrix} 1 & 1 & 1 & 0 & 0 & 0 & 0 & 0 & 0 & 0 & 0 & 0 \\ 1 & 1 & 1 & 0 & 0 & 0 & 0 & 0 & 0 & 0 & 0 & 0 \\ 1 & 1 & 1 & 1 & 1 & 1 & 1 & 1 & 1 & 1 & 1 & 1 \\ 1 & 1 & 1 & 1 & 1 & 1 & 1 & 1 & 1 & 1 & 1 & 1 \\ 1 & 1 & 1 & 1 & 1 & 1 & 1 & 1 & 1 & 1 & 1 & 1 \\ 1 & 1 & 1 & 1 & 1 & 1 & 1 & 1 & 1 & 1 & 1 & 1 \end{bmatrix} \qquad (6.2)$$

After defining all the active degrees of freedom by zeros in the NDF array, the equation numbers corresponding to these degrees of freedom are assigned. The procedure used is to scan each column of the NDF array and replace each zero by an equation number, which increases successively from 1 to the total number of equations. The NDF array giving equation numbers for the cantilever problem is given below.

$$[NDF] = \begin{bmatrix} 0 & 0 & 0 & 1 & 3 & 5 & 7 & 9 & 11 & 13 & 15 & 17 \\ 0 & 0 & 0 & 2 & 4 & 6 & 8 & 10 & 12 & 14 & 16 & 18 \\ 0 & 0 & 0 & 0 & 0 & 0 & 0 & 0 & 0 & 0 & 0 & 0 \\ 0 & 0 & 0 & 0 & 0 & 0 & 0 & 0 & 0 & 0 & 0 & 0 \\ 0 & 0 & 0 & 0 & 0 & 0 & 0 & 0 & 0 & 0 & 0 & 0 \\ 0 & 0 & 0 & 0 & 0 & 0 & 0 & 0 & 0 & 0 & 0 & 0 \end{bmatrix} \qquad (6.3)$$

Material properties of the various material groups viz. modulus of elasticity-E, Poisson's ratio-PR and weight density-WD are also read at this stage. A detailed flow chart of the subroutine PASSIN is given in Fig.6.3.

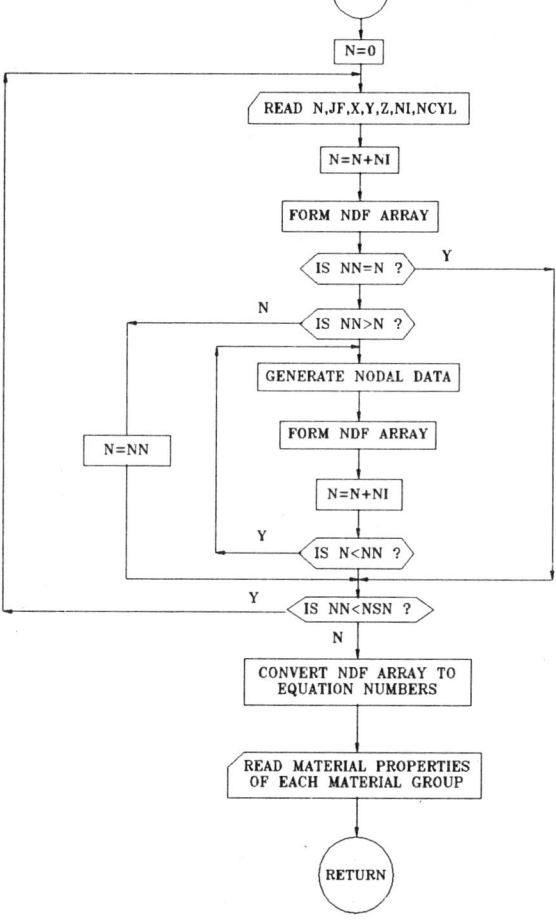

Fig. 6.3 Subroutine PASSIN

6.3   Subroutine FELIB

The program has the provision to accommodate seven element types with three shapes in each element type and it is organized in the form of a two-dimensional array called element library LLIB [7, 3] (see Table 6.1). The subroutine FELIB reads for each element type in the structure, the number of shapes (NSHAPE) and their location in the element library. If a combination of these element shapes is used for analysis, the numbering of elements should be done separately for each element shape. Any element shape can be included as a subroutine which (1) reads element data (2) computes element stiffness matrix and load vector and (3) computes element stresses. These three sets of computations are carried out in three separate stages and, they are explained with reference to the element shapes and the corresponding subroutines described in the sebsequent chapters dealing with different types of analysis.

Table 6.1 Element Library-[LLIB]

Element Type Details	L Type	Element Shapes		
3-D Truss elements	1	TRUS1	TRUS2	TRUS3
3-D Beam elements	2	BEAM1	BEAM2	BEAM3
Plane Stress/ Strain & Axi- symmetric elements	3	QUAD	RECT	CST
3-D Solid elements	4	THRD08	THRD20	THRED3
Plate bending elements	5	PLATE8	PLATE4	PLATE2
Shell elements	6	SHELL8	SHELL4	SHELL2
Boundary elements	7	BOUND1	BOUND2	BOUND3

The subroutine FELIB is called thrice by the main routine in three separate stages as shown in Fig. 6.1. At each stage it calls the respective subroutines for each element type (Fig.6.4) which in turn calls the routines for the element shapes. In the first stage, the subroutines for the element shapes read the nodal connectivity of the elements (LNC array) and compute an array ND, giving equation numbers for each degree of freedom for each element, i.e., the array ND relates each element degree of freedom to the corresponding global degree of freedom. The array ND, a one-dimensional array of dimension equal to the number of degrees of freedom of an element is stored on a scratch file which will be retrieved twice, once for the assembly of stiffness matrix and then for the computation of stresses.

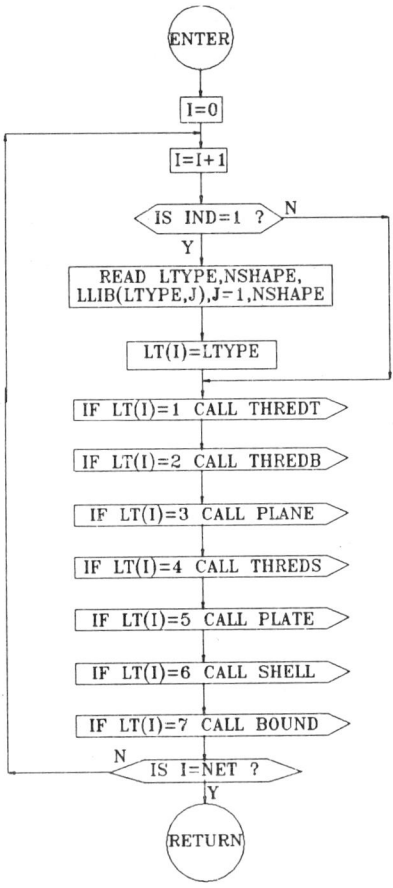

Fig. 6.4 Subroutine FELIB

The formation of the ND array may be explained through the cantilever beam example shown in Fig.6.2(b). The nodal connectivity array (LNC) for each element of the above example is given in the following:

Element No.	LNC array			
1	1	4	5	2
2	2	5	6	3
3	4	7	8	5
4	5	8	9	6
5	7	10	11	8
6	8	11	12	9

(6.4)

Using NDF array given in Eq.6.3, the equation numbers corresponding to the nodes 1, 4, 5 and 2 of the element 1 are obtained which will give the ND array for the element. The ND array for all the 6 elements of the example is given below:

Element No.			ND	array				
1	0	0	1	2	3	4	0	0
2	0	0	3	4	5	6	0	0
3	1	2	7	8	9	10	3	4
4	3	4	9	10	11	12	5	6
5	7	8	13	14	15	16	9	10
6	9	10	15	16	17	18	11	12

$$(6.5)$$

In the second stage, the element stiffness matrix is computed for each element and assembled. The computation of stresses in each element of the structure is carried out in the last stage.

6.4   Subroutine COLUMH

The subroutine COLUMH computes the height of each column of the global stiffness matrix. The column height refers to the number of elements in each column below the skyline and above the leading diagonal excluding the diagonal element of the global stiffness matrix.

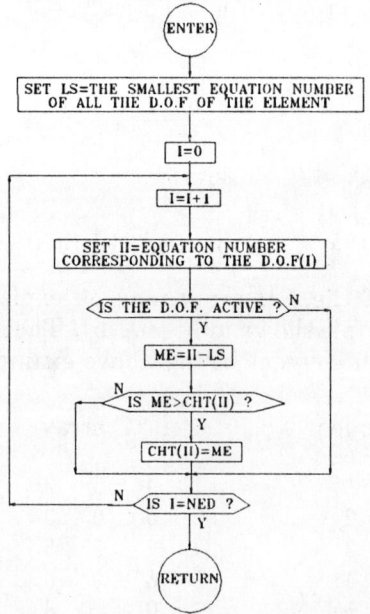

Fig. 6.6 Subroutine COLUMH

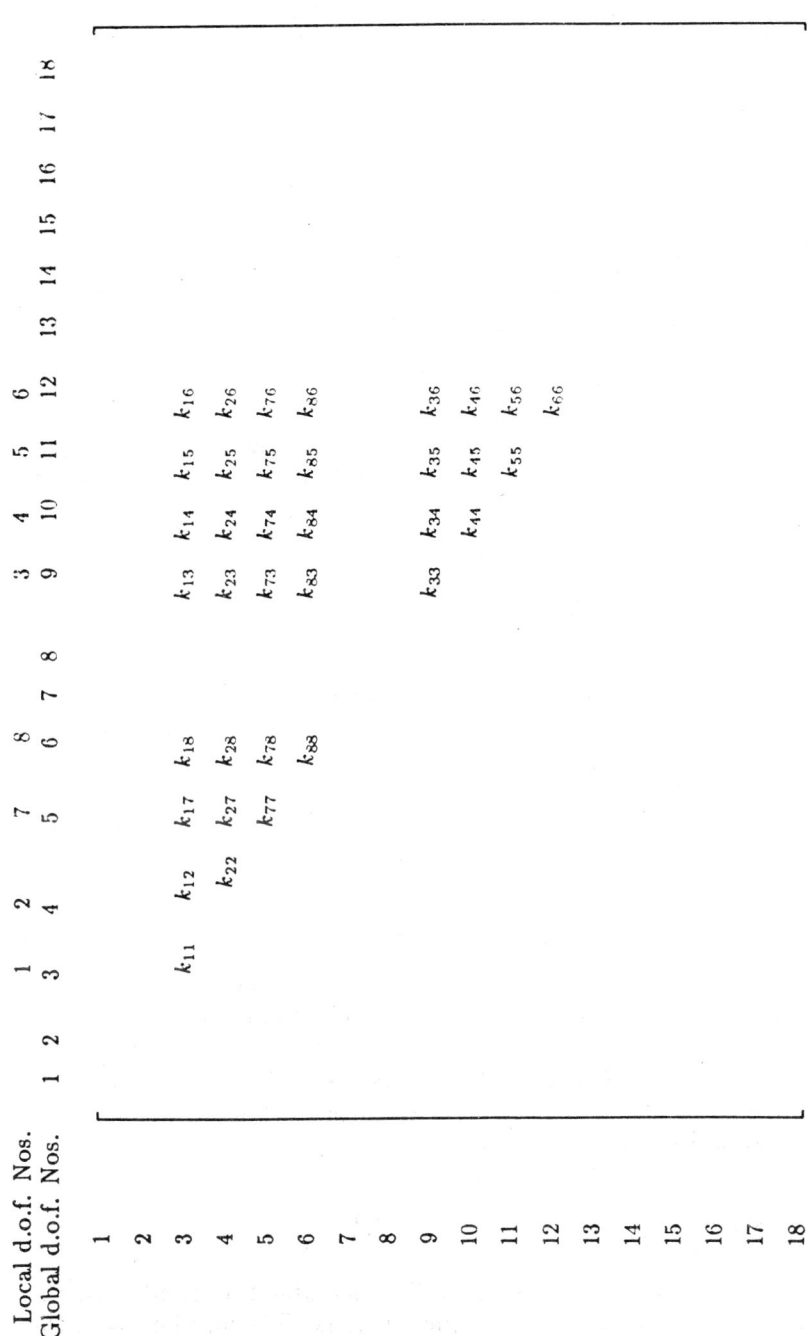

Fig. 6.5 Position of the coefficients of (EK) matrix of element 4 in global stiffness matrix.

Following the direct stiffness method of assemblage of elements, the position of the stiffness coefficients for the element 4 in the global stiffness matrix is given in Fig.6.5. It can be observed that the first non-zero coefficient in each column of the global stiffness matrix, considering the contribution from element 4 only (Fig.6.5), is along the 3rd row, which is the skyline for this stiffness matrix. Now examine the ND array for the element 4 given in the previous section.

As the ND array gives the equation number (ith row) for a particular degree of freedom, it can be observed that the row having the first non-zero element is determined by the lowest equation number in the ND array and in this case it is the 3rd row that contains the first non-zero coefficient. As the stiffness matrix is symmetric an equation number i in the ND array corresponding to ith row also refers to the ith column of the stiffness matrix, e.g., number 10 in the ND array for element 4 corresponds to 10th row and 10th column of the global stiffness matrix. Hence, while processing an element, the column heights for each equation number in the ND array of the element can be obtained by subtracting the lowest equation number in the ND array from the concerned equation number. Thus, for the element 4, the corresponding column heights for the equations 3,4,9,10,11,12,5,6 of the global stiffness matrix are 0, 1, 6, 7, 8, 9, 2, 3 respectively.

The above procedure is followed to determine the column heights for each column of the global stiffness matrix by scanning the elements sequentially. The procedure is shown in flow chart in Fig.6.6 and is also illustrated in Table 6.2 for the plane stress problem shown in Fig.6.2(b). It may be noted that while processing number of elements in the assemblage, the column heights considering the present element will be calculated first (column A in Table 6.2) and compared with the current large value of column height for each equation considering previous elements (column B in Table 6.2). If the current value for a column height is larger than the value obtained by considering the previous elements, the column height is modified as shown under column B in Table 6.2. Thus, the column heights of the global stiffness matrix are obtained after considering all the elements of the assemblage, e.g. the final values given under column B for the last element 6 in Table 6.2.

6.5   Subroutine CADNUM

The subroutine CADNUM (Fig. 6.7) calculates the number of the diagonal element of the global stiffness matrix. The numbering of the elements of the stiffness matrix for storage as one-dimensional array 'SK' is as shown in Fig.6.8 for the example problem of Fig.6.2b.

Table 6.2 Column Height Computation

Equation (Column) No.	COLUMN HEIGHTS											
	Element 1		Element 2		Element 3		Element 4		Element 5		Element 6	
	A	B	A	B	A	B	A	B	A	B	A	B
1	0	0	0	0	0	0	0	0	0	0	0	0
2	1	1	0	1	1	1	0	1	0	1	0	1
3	2	2	0	2	2	2	0	2	0	2	0	2
4	3	3	1	3	3	3	1	3	0	3	0	3
5	0	0	2	2	0	2	2	2	0	2	0	2
6			3	3	0	3	3	3	0	3	0	3
7					6	6	0	6	1	6	0	6
8					7	7	0	7	2	7	0	7
9					8	8	6	8	3	8	0	8
10					9	9	7	9	0	9	1	9
11							8	8	0	8	2	8
12							9	9	6	9	3	9
13									7	6	0	6
14									8	7	0	7
15									9	8	6	8
16									0	9	7	9
17									0	0	8	8
18									0	0	9	9

A - Column height considering the present element alone

B - Current largest values of column heights considering the preceding value and the present one.

The numbers of the diagonal elements of the global stiffness matrix are stored in a separate array called NDS. The number of a diagonal element is obtained by adding the preceding column height plus one to its diagonal number, the number of the first diagonal element being taken as 1. The number of the ith diagonal element is given by, NDS(I) = NDS(I-1) + CHT(I-1)+1. Now, the numbers of the elements in the ith column of the stiffness matrix are NDS(I), NDS(I)+1, NDS(I)+2, . . ., NDS(I+1)-1. Thus any element in 'SK' can be located by making use of this array.

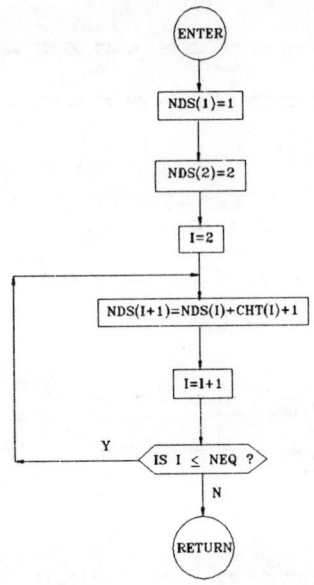

Fig. 6.7 Subroutine CADNUM

6.6   Subroutine PASSEM

Subroutine PASSEM is called at the second stage (IND=2) by each element shape routine. It assembles the element stiffness matrix to global stiffness matrix in the form of an one-dimensional array SK as shown in flow chart Fig.6.9. Consider a term EK(I,J) in the element stiffness matrix. Let the equation numbers corresponding to I and J be M and N. Only the upper triangle of the global stiffness matrix is assembled. If M or N is zero, it shows that the corresponding degree of freedom is not active and hence it is not be assembled. If M ≤ N the position of the term in SK is given by NDS (N) + N - M. For example, from Fig.6.5 and 6.8 the location of K_{24} for the element 4 in SK can be calculated as follows.

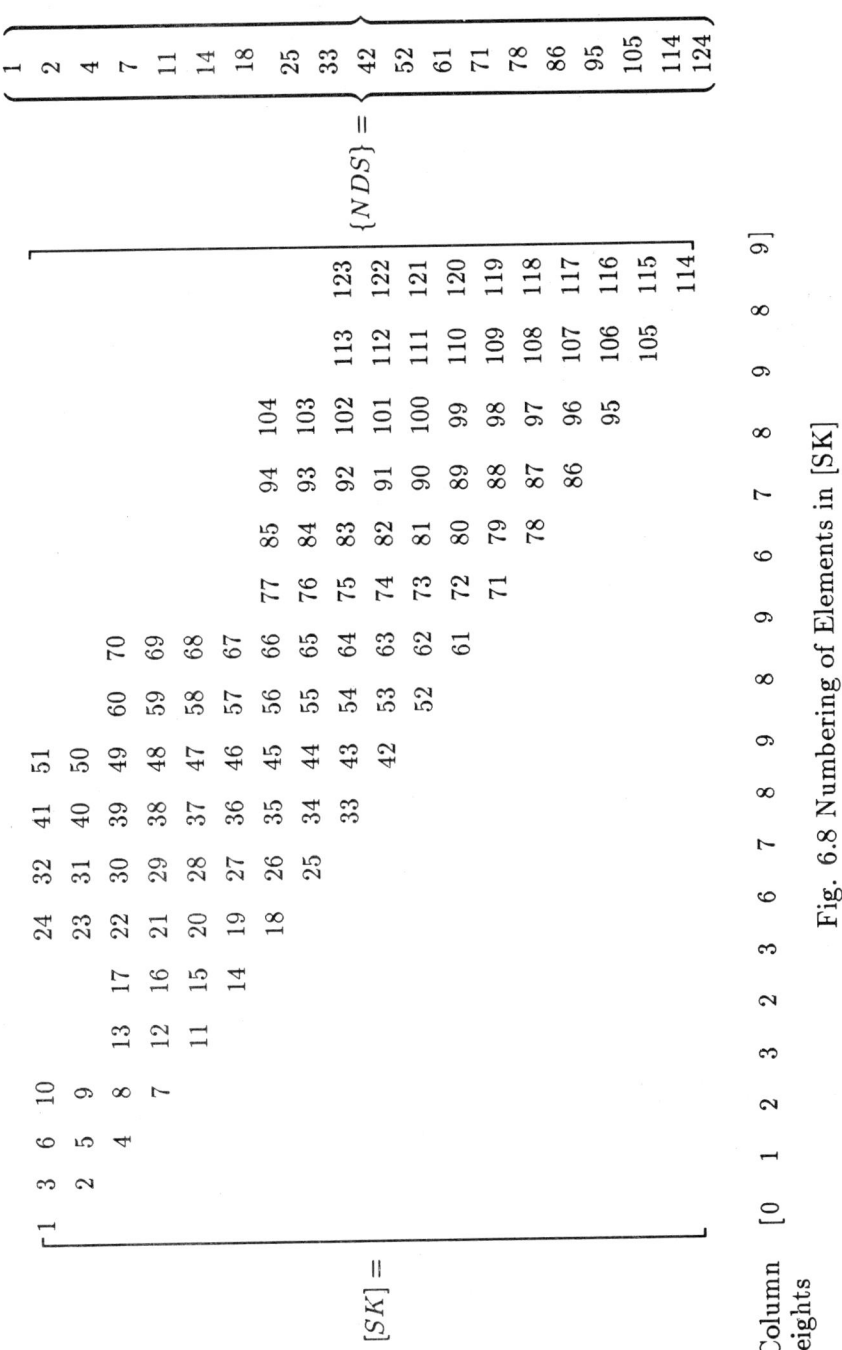

Fig. 6.8 Numbering of Elements in [SK]

From ND array of 4th element, the equation numbers correspond-ing to 2 and 4 are 4 and 10 respectively. From Fig.6.8, NDS(10) = 42 and hence the position is 42 + 10 - 4 = 48.

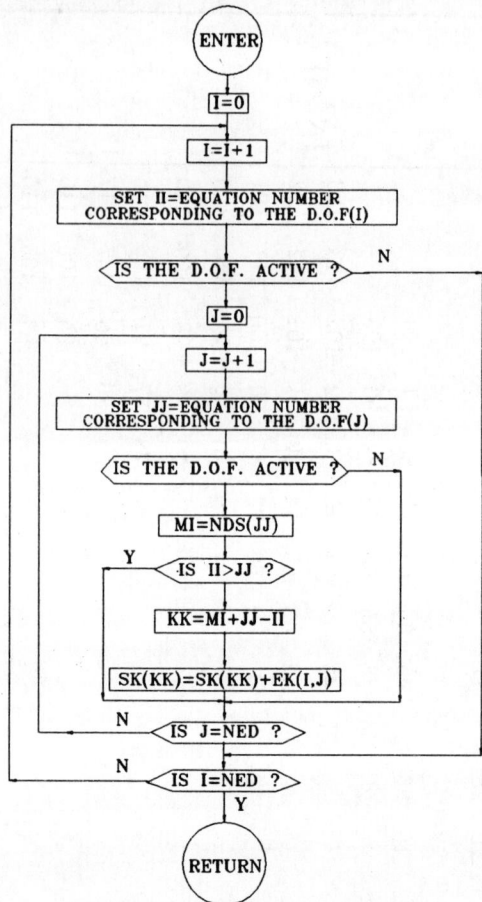

Fig. 6.9 Subroutine PASSEM

6.7   Subroutine PASOLV

In finite element analysis, the time required for the   solution   of equilibrium equations may be a large percentage of the total solution time.   The simultaneous equations in static analysis are given by Eq.5.2.

$$[K]\{r\} = \{P\}$$

where [K] is the stiffness matrix, $\{r\}$ is the displacement vector and $\{P\}$ is the load vector. The subroutine PASOLV, a direct solution routine using the algorithm based on Gauss elimination, is adopted from Reference [1]. The procedure has been explained in chapter 5. The flow chart for this routine is shown in Fig.6.10.

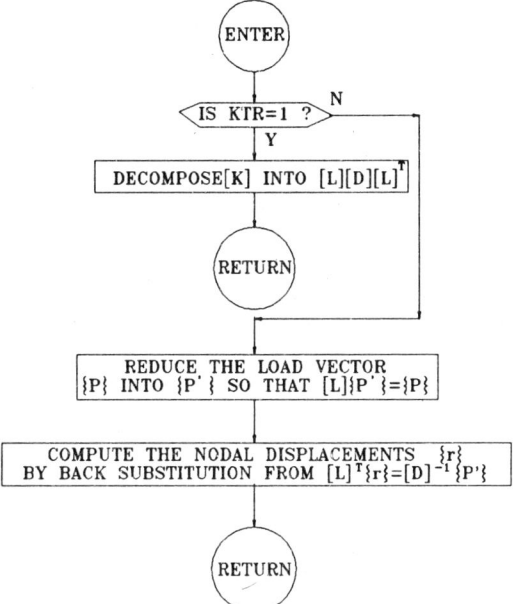

Fig. 6.10 Subroutine PASLOV

6.8 Subroutine PASLOD

The subroutine PASLOD is called by the main routine for each load case.

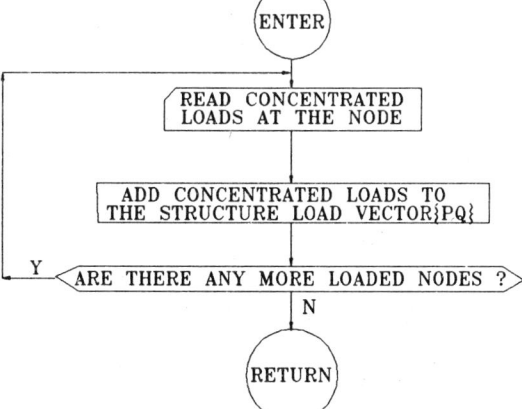

Fig. 6.11 Subroutine PASLOD

This subroutine reads the concentrated loads at the loaded nodes in the direction of each degree of freedom of the node. These loads will be added to the structure load vector from the element shapes as shown in Fig.6.11.

The concentrated load at node i in the direction j is added to the load corresponding to the equation numbers NDF (i, j).

6.9   Subroutine DISP

The subroutine DISP prints out the displacements of each node. The displacement of each node is computed from the global displacement vector as shown in Fig.6.12. The displacement corresponding to ith degree of freedom of node j is that corresponding to the equation number NDF (i, j).

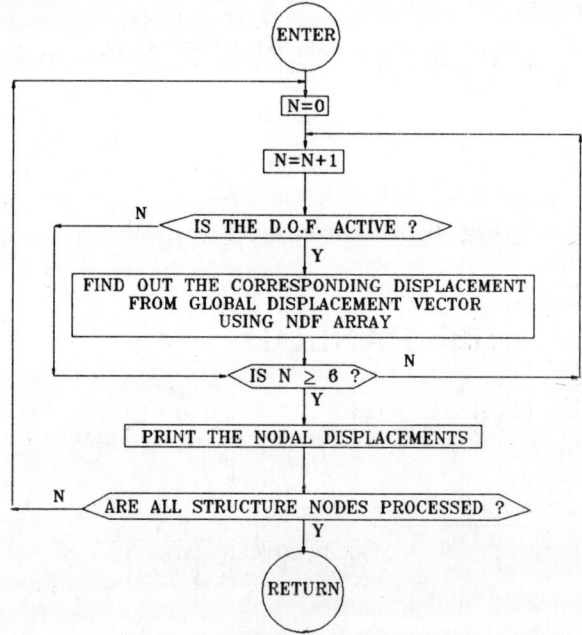

Fig. 6.12 Subroutine DISP

6.10   Guidelines for Adding Element Routines

For adding a new element routine first a control routine for the element type has to be developed. This control routine for each element type of the FELIB (Table 6.1) is given in subsequent chapters dealing with various types of analysis like plane stress/strain etc. Also the routines for one or more shapes for each element type are given therein. It may be observed that the element library in PASSFEM has provision for three shapes for each element type. Hence, in order to add additional shapes to the element library the guidelines for adding a particular shape for an element type are given below. It may be noted here that the calling statement in the control routine should be of the form [BEAM2 is the name of the shape of the element routine of type 2(Beam element)].

CALL BEAM2 (A(N2), A(N3), A(N4), A(N5), A(N6), A(N7), A(N11) A(N12), A(N1), A(N8), A(N9), A(N10), A(N13), NED).

where N1 to N13 are the starting addresses of the various global variables and NED is the variable indicating number of element degree of freedom. The details of the routine for each element shape are explained below.

6.10.1 Argument List

The argument list for a typical routine is SUBROUTINE BEAM2(X, Y, Z, E, PR, WD, SK, PQ, NDF, CHT, ND, NDS, P NED) where X, Y, Z are the arrays of the coordinates of nodes and NDF is the nodal degree of freedom array, E, PR and WD are the material properties. SK is the global stiffness matrix, PQ is the assembled element load vector (in global coordinates) and P is the final load vector. CHT and NDS are the arrays for storing column heights and diagonal addresses of stiffness matrix. The only element dependent variables are NED, which is the number of element degree of freedom, and ND, which is the degree of freedom array for the element.

6.10.2 Common Block

There are three main common blocks in the element routine namely COMMON/DIM/, COMMON/PAR/, COMMON/TAPES/.

The COMMON/DIM/ contains the starting addresses of various variables and arrays as explained in section 6.1 and is specified as

COMMON/DIM/N1, N2, N3, N4, N5, N6, N7, N8, N9, N10, N11, N12, N13.

The COMMON/PAR/ contains all the controlling parameters for a particular problem and is specified as

COMMON/PAR/IND, NET, NSN, NMP, NEQ, NSKY, NEQ1, LCOUNT where

IND   -   the flag to indicate the segment of the element routine to be executed (explained in sub-section 6.10.4)

NET   -   number of element types

NSN   -   number of structure nodes

NMP   ·   number of material property groups

NEQ   -   number of equations in the global system

NSKY   -   number of elements in the stiffness matrix

NEQ1   -   NEQ+1

LCOUNT   -   parameter to keep track of the number of element types called by FELIB and is used to recover the number of element in each element type at any stage.

The COMMON/TAPES/ contains the names of the scratch files used by the program and is specified as

COMMON/TAPES/ISTRES, NDARAY, IPR

ISTRES   -   stores the stress - displacement matrix for each element.

NDARAY   -   stores the ND and the connectivity array for the elements.

IPR   -   option for printing the nodal or element data explained in the section 6.11.

The variables in the common blocks are mainly used to dimension various global variables.

6.10.3   Local Variables

The following are the local variables used in the element routines

EK   -   element stiffness matrix (size-NED × NED)

Q   -   element load vector (size - NED)

LNC   -   element connectivity array (size-No. of element nodes × 1)

XL, YL, ZL   -   One-dimensional arrays of coordinates of nodes to which the element is connected. (size - No. of element nodes × 1)

NDEL   -   array to store number of elements in any shape.

Additional arrays which may be required for many of the routines are:

CB - stress-displacement matrix

B   - strain-displacement matrix

SIG - stress vector

All the above arrays are to be explicitly dimensioned in the individual element shape routine.

6.10.4 Routine Organisation

Each element shape routine is called thrice by the main progam and is divided into three parts. The three segments are executed depending on the value of the flag IND. The general arrangement of the subroutine for an element shape is shown in Fig. 6.13.

```
SUBROUTINE TRUSS 1   (X, Y, Z, E, PR, WD, SK,
                      PQ, NDF, CHT, ND, NDS,
                      P, NED)
IMPLICIT REAL* 8 (A-H, O-Z)
COMMON/PAR/DIM/N1,   N2, N3, N4, N5, N6, N7,
                     N8, N9, N10, N11, N12,
                     N13
COMMON/PAR/IND,   NET, NSN, NMP, NEQ, N-
                  SKY, NEQ1, LCOUNT
COMMON/TAPES/ISTRES, NDARAY, IPR
COMMON/MULT/ELMN
DIMENSION   X(NSN), Y(NSN), Z(NSN), E(NMP),
            PR(NMP), WD(NMP)
DIMENSION   SK(NSKY), PQ(NEQ), NDF(6, N-
            SN), CHT (NEQ), ND (NED), NDS
            (NEQ1), P (NEQ)
DIMENSION   EK(6,6),
            Q(6), LNC(2), XL(2), YL(2), ZL(2),
            NDEL(20)
DIMENSION   ELM (10)
SQRT (X)    = DSQRT (X)
              .  .  .  .
              .  .  .  .
            PROGRAM
              .  .  .  .
              .  .  .  .
              .  .  .  .
              .  .  .  .
```

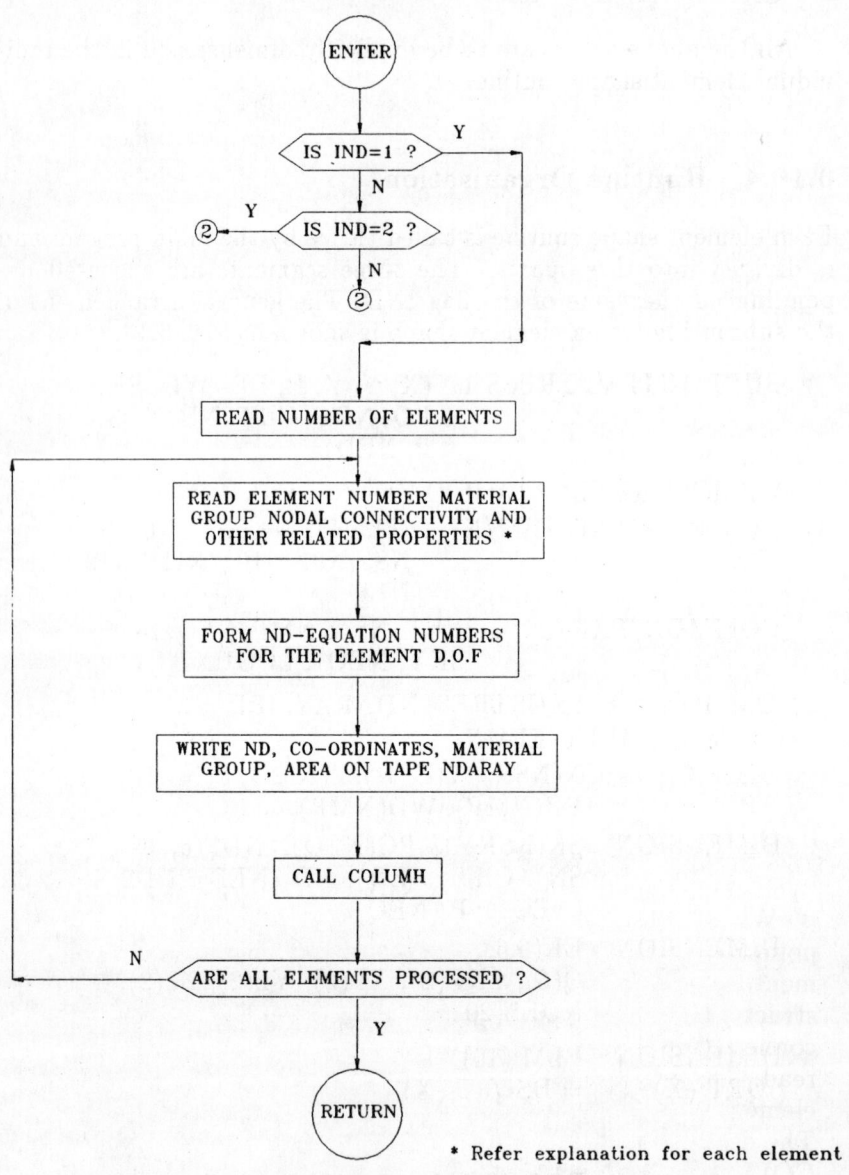

Fig. 6.13 Continued

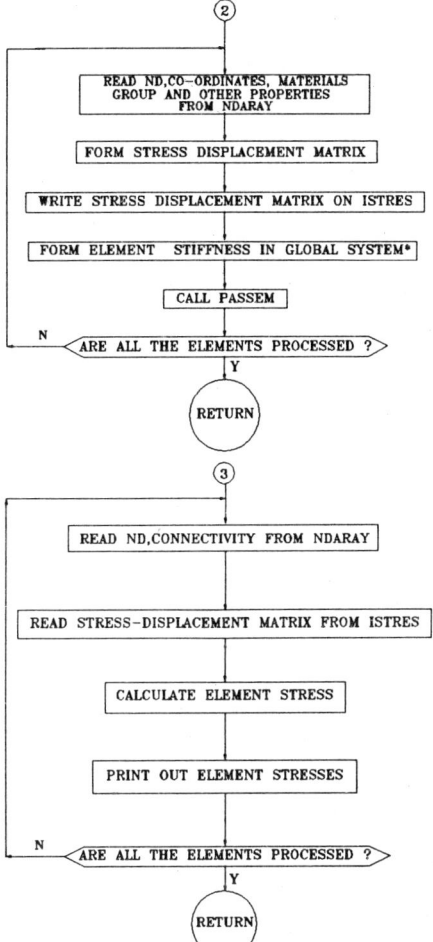

Fig. 6.13 Subroutine for typical element shape

When IND equals one the first segment is evoked. The nodal point data and material information are obtained through the argument list. Here the number of this particular type of element in the structure is to be read as input. Then for each element its nodal connectivity, material group, thickness or other relevant data are read. The nodal degree of freedom array (ND) is generated for each element and is stored in the NDARAY file in a sequential manner. The data read and generated is printed out. Finally the subroutine COLUMH is called to calculate the column heights.

When IND = 2 the control transfers to the second segment of the element routine. At this stage in addition to the nodal data and material properties, the addresses of the stiffness matrix, load vector, etc. are also obtained from the argument list. For each element ND array is read from the scratch file along with other information stored there. The stress-displacement matrix is calculated and is written on the ISTRES FILE. Next the element stiffness matrix is calculated and is passed on to the subroutine PASSEM for assembly of global stiffness matrix. The element load vector is calculated and added to the global load vector.

After the global displacements have been obtained, IND is set equal to 3 and control is transferred to the third segment of the element routine. At this stage the ND array and stress-displacement matrix are read from the NDARAY and ISTRES files respectively. Then, for each element the stresses/stress resultants at the specified points are computed and printed out.

Thus without getting involved with the main program an element routine can easily be added to the program PASSFEM. For further help, the first few statements of a typical routine (Truss element) can be referred to in the Appendix.

6.11   Input Details

The input data required for using the program PASSFEM will be explained in this section. The steps to be followed for finite element analysis using this program are given below:

1. Sketch the structure with proper dimensions (in any fixed unit system) and number the nodes and elements. While numbering the nodes care should be taken to determine the sequence that will keep the band width of the stiffness matrix as small as possible.

2. Establish the reference coordinate system (global axes) and label the nodes with proper coordinate values.

3. Prepare the input data according to the sequence described below.

1. *Problem Title* (20A4)

Columns 1 to 80 of this first *line* contain the information of the user's choice, to identify the problem/data deck and the printed output. The internal program name for this information is TITLE.

2. *Structure Data - One Line - (6I5)*

Columns		
1 - 5	Number of structure nodes (NSN)	
6 - 10	Number of element types (NET)	
11 - 15	Number of material property groups (NMP)	
16 - 20	Number of load cases (NLC)	

21 - 25   Maximum number of element degrees of freedom. (if more than one type of element is
used, the maximum number of the element degrees of freedom should be input, NEDMAX)

26 - 30   Parameter for mode of operation (MODEX)
key-in '0' for execution and any other number
for data check only.

3. (i) *Option to Supress Print out of Nodal Data* (I5)

Columns 1 - 5   Option for suppressing the print out (IPR):
key-in '0' for getting the print-out of the
nodal point, element and load data and any
other number for supressing the print out.

(ii) *Nodal Data* (I3, 1X, 6I1, 3F 10.4, 3X, I2, 3X, I2)*

One *line* per node with the following information has to be supplied.

Columns 1 - 3   Node number

5 - 10   Boundary condition code; key-in '0' if the
node is free to experience the displacement corresponding to a particular degree
of freedom and '1' if it is restrained along
that direction. Columns 5, 6 and 7 refer to
displacements along the global X, Y and
X axes respectively, and columns 8, 9 and
10 refer to rotation about the X, Y and Z
axes.

11 - 20   X coordinate of the node or r coordinate
in the cylindrical coordinate system.

21 - 30   Y coordinate of the node or θ in the cylindrical coordinate system.

31 - 40   Z coordinate of the node.

44 - 45   Increment (NI) for automatic generation of
nodal data.

Generation of nodal data will be carried out along a straight line.
The generation is done by letting two consecutive *lines* to define nodes
at the beginning and at the end of a straight line. Intermediate nodes
along this line will be automatically assigned coordinates to make
them equally spaced, and numbered at increments of NI, the

---

* Many of the present FORTRAN compilers support entry of data
separated by a comma(,) and this will be advantageous in interactive
environment compared to strict adherence to fixed format.

value of NI being taken from the *Second line* defining the node at the end of this line. If the columns 44-45 are left blank, NI will be taken as 1.

    49 - 50   Parameter (NCYL) to indicate whether the coordinates entered before are with reference to cartesian or to cylindrical coordinates system. Key-in '1' for cylindrical coordinate system and leave blank for cartesian coordinate system.

4. *Material Property Data* (I 10, E 10.3, 2F 10.4)

    One *line* for each material property group is required.

    Columns   1 - 10   Material Group Identification Number
              11 - 20   Modulus of Elasticity (E)
              21 - 30   Poisson's ratio (PR)
              31 - 40   Weight density (WD)

5. *Element Type Data*

6. *Element Data*

The data sets for 5 and 6 are element dependent and hence the input details for the element shapes are described in chapters 7, 8, 9, 10 and 11, for different element types used in various types of analysis of structural systems. The program listings are presented in Appendices 2 through 8.

7. *Element Load Multipliers* (10F5.2)

    Columns 1-5, 6-10, ..., 46-50 define the factors by which the element loads are to be multiplied for 1st, 2nd, ..., 10th load case. The multiplier (ELM) is the same for all the element loads for the concerned load case. For a particular load case the value of the element load multiplier times the load vector due to element loads is added to the load vector due to nodal loads for that load case.

8. *Nodal Load Data*

(i) Number of Loaded Nodes (10I5)
Columns 1-5, 6-10, ..., 46-50 define the number of loaded nodes for 1st, 2nd, ..., 10th load case. The provision is made for 10 load cases. If there is no concentrated load in any load case, a blank line is to be inserted.
(ii) Concentrated Loads at Nodes (I10, 6F10,4)
These data are to be entered only if the previous *line* is not a blank. Number of lines is equal to number of loaded nodes, for the 1st load case the data set is entered first, then for 2nd load case and so on.

Columns 1 - 10   Number of the node at which the load is applied.

11 - 20   21-30, . . ., 61-70, the value of concentrated loads in the direction of 1st, 2nd, ..., 6th degree of freedom.

At the end of this data set 2 blank *lines* are to be inserted to terminate the execution if this is the last problem to be analysed. If there are other problems to be analysed in the same run, enter the data set for the next problem.

REFERENCES

1. Bathe, K.J. and E.L.Wilson, *Numerical Methods in Finite Element Analysis*, Prentice-Hall of India Private Limited, New Delhi, 1978.

2. Hinton, E. and D.R.J.Owen, *Finite Element Programming*, Academic Press Inc. (London), Ltd., 1977.

3. Zienkiewicz,O.C. and R.L.Taylor, *The Finite Element Method Vol I, Basic Formulation and Linear Problems*, McGraw-Hill, (U.K) Limited, 1989.

EXERCISES

6.1 For the cantilever beam shown in Fig. 8.14, compute the ND array and column heights of the global stiffness matrix using the mesh 2 for discretization.

6.2 The cantilever beam shown in Fig. 8.14 is to be analysed using four eight noded plane stress rectangular elements. Compute the ND array and column heights of the global stiffness matrix.

6.3 Write the subroutines to compute the shape functions, their derivatives, Jacobian matrix $[J]$, $[J]^{-1}$ and $|J|$ at a given Gauss point for a four noded quadrilateral element.

6.4 Develop subroutines to compute the shape function, their derivatives, Jacobian matrix $[J]$, $[J]^{-1}$ and $|J|$ at a given Gauss point for four to eight (variable) noded two-dimensional isoparametric rectangular element.

6.5 Develop subroutines to compute the shape functions their derivatives, Jacobian matrix $[J]$, $[J]^{-1}$ and $|J|$ at a given Gauss point for three to six (variable) noded two-dimensional isoparametric triangular elements.

6.6 Using the routines of problem 6.3 develop a subroutine to compute the [B] matrix at a given Gauss point. Check the working of the routine through the example given in section 4.2.2.

6.7 Develop a subroutine to compute the contribution to stiffness matrix [k] from a Gauss point. Use the routines developed in problems 6.3 and 6.6. The routine should loop over all the chosen number of Gauss points to compute the stiffness matrix [k] of the element. The routine has to be checked for correctness of the result with the example problem given in section 4.2.2.

6.8 Modify the routine developed for problem 6.7 to include capability to compute the stiffness matrix for a three noded triangular element. Check the results by using explicit expressions.

6.9 Develop a subroutine to compute the [B] matrix for a four noded isoparametric quadrilateral element using the formulation for fast stiffness computation. Make use of the routines developed in problem 6.3.

6.10 Develop a routine to compute the stress displacement matrix, $[CB] = [C] [B]$, at a point in a quadrilateral element. This routine should make use of shape function routine developed in problem 6.3. Use the formulation for fast stiffness computation.

6.11 Making use of the subroutines of problems 6.9 and 6.10 develop a subroutine to compute the stiffness matrix [k] using numerical integration procedure. Discuss the advantages of the formulation for fast stiffness computation.

6.12 Using the flow chart given in Fig. 6.9 develop a subroutine to assemble the element stiffness matrix to global sitffness matrix using the connectivity information provided by ND array. To simplify programming efforts, the assemblage can be done assuming the element and global stiffness matrices are stored in two-dimensional arrays. However, the symmetry properties may be made use of in developing routines.

6.13 Develop a subroutine to assemble the element load vector $\{Q\}$ into the structure load vector $\{P\}$.

6.14 Develop a subroutine to compute the strains and stresses at a point in an element after the solution of equations $[K] \{r\} = \{P\}$. Make use of ND array information and assume [CB] matrix is stored in a scratch file.

6.15 A thick cylinder shown in Fig. E6.15 is to be analysed using eight noded element and the discretization is shown in the figure. Compute the ND array and column heights of the global stiffness matrix.

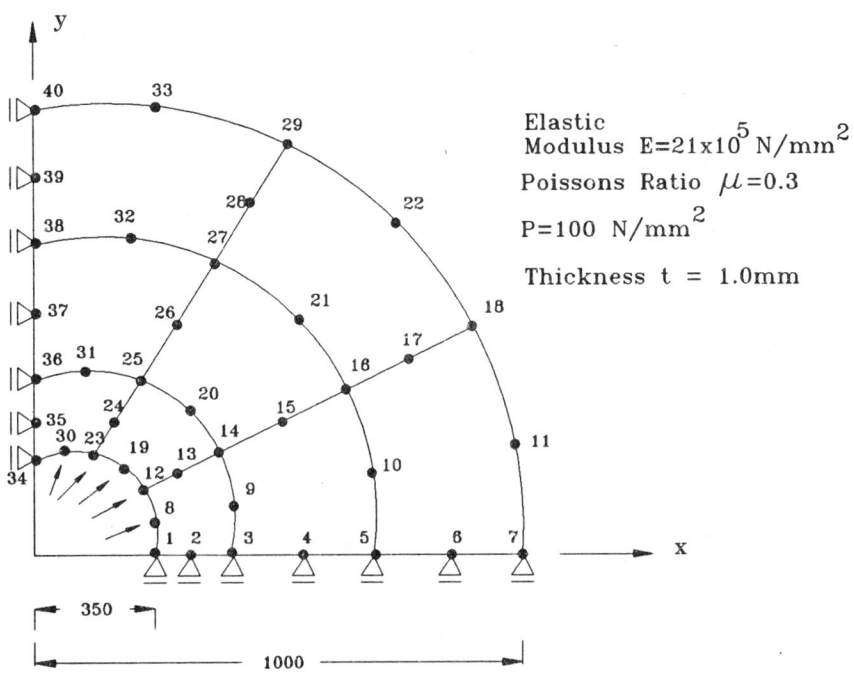

Fig. E6.15

Chapter 7

Analysis of Framed Structures

Matrix methods of analysis of framed structures have become increasingly popular and are treated in a number of books [1,2,3]. It may be noted that the finite element analysis formed a further development of the matrix methods for framed structures as applied to continuum structures. The difference in terminology is very much subtle now as it will be shown here that the finite element concepts described earlier can be used to formulate the displacement method of analysis treating the 'member' of a framed structure as an 'element'. Thus a unified formulation based on the finite element concept will be presented in this chapter for the analysis of two and three-dimensional trusses and frames.

In the simple theory of bending of beams the shear deformation is neglected. But in certain situations it becomes necessary to include the deformation due to shear. In section 7.5, the basic aspects of shear deformation in beams are explained and a simple finite element formulation is also described. The theory presented here will also be made use of in the analysis of plate bending described in Chapter 10.

Two subroutines TRUS1 and BEAM1 are presented for the analysis of three-dimensional trussess and frames, and these routines will form part of the element library of PASSFEM. The two-dimensional cases of these structures can be treated by properly specifying the nodal degrees of freedom.

7.1 Two-dimensional Truss Element

Figure 7.1(a) shows a two-dimensional truss element which is parallel to x-axis and has nodes i and j. Since it is an axial force resisting member the displacement along the x-axis will be u and its variation can be expressed as

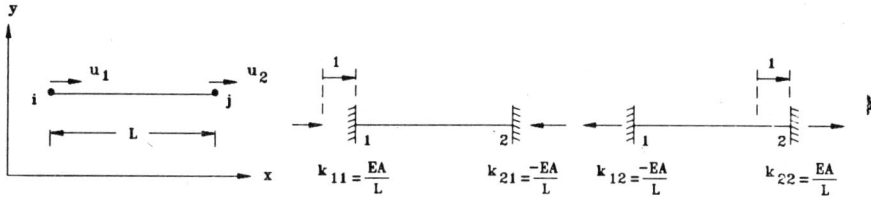

Fig. 7.1a Two-dimensional   Fig. 7.1b Truss element stiffness
    truss element   coefficients

$$u = L_1\, u_1\ +\ L_2\, u_2 \tag{7.1}$$

where L_1 and L_2 are the natural coordinates given by Eq.3.14 as

$$L_1\ =\ 1 - \frac{x}{L}\quad \text{and}\quad L_2\ =\ \frac{x}{L}$$

where L is the length of the truss element.

The strain variation can be computed using Eq.3.15 as

$$\epsilon = \frac{du}{dx} = \frac{1}{L}\left(\frac{\partial u}{\partial L_2}\ -\ \frac{\partial u}{\partial L_1}\right)$$

i.e.,

$$\epsilon_x\ =\ \frac{1}{L}\,(u_2\ -\ u_1) \tag{7.2}$$

Hence, the strain-displacement matrix [B] is given by

$$[B]\ =\ \frac{1}{L}\,[-1\quad 1] \tag{7.3}$$

The axial stress in the element is $\sigma = E\,\epsilon$, and hence

$$[C]\ =\ E \tag{7.4}$$

The stiffness matrix for the truss element, which is prismatic and having an area of cross-section A, can now be computed as

$$[k_m]\ =\ \int\int\int [B]^T\ [C]\ [B]\ dV$$

$$=\ \frac{1}{L^2}\int\int\int \begin{Bmatrix} -1 \\ 1 \end{Bmatrix}\ E\,[-1\quad 1]\ dV$$

$$= \frac{EA}{L} \begin{bmatrix} +1 & -1 \\ -1 & +1 \end{bmatrix} \qquad (7.5)$$

where $[k_m]$ is the stiffness matrix of the element/member, in the local member axes system, i.e., corresponding to the degrees of freedom at the nodes i and j in the local member axes system.

Following the description of stiffness of a structure in Chapter 3, the stiffness co-efficient k_{ij} for an element in general is defined as

k_{ij} = force developed in the degree of freedom i due to unit displacement in the degree of freedom j keeping all other displacements fixed. For a truss member the physical interpretation of the stiffness coefficients is shown in Fig.7.1(b).

If the cross-section of a truss member is not uniform, the variation can be expressed through a shape function and following the above procedure the stiffness matrix can be evaluated.' Consider a truss member shown in Fig.7.2, having area of cross-section at 1 and 2 as A_1 and A_2 respectively. The variation in the cross-sectional area A can be expressed in terms of linear function given by

$$A = \begin{bmatrix} L_1 & L_2 \end{bmatrix} \begin{Bmatrix} A_1 \\ A_2 \end{Bmatrix} \qquad (a)$$

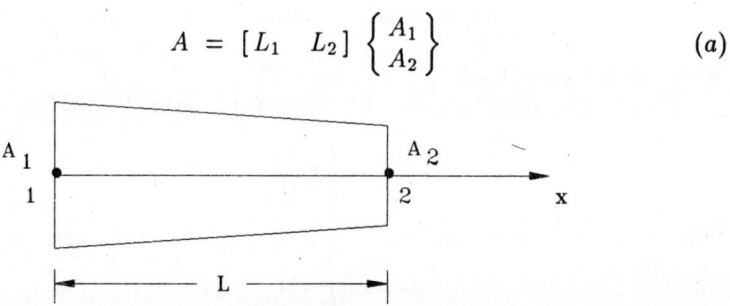

Fig. 7.2 Non-prismatic truss member

where L_1 and L_2 are natural coordinates given by Eq.3.14. Using the displacement model given by Eq.7.1 and the strain-displacement relation given by Eq.(7.3) the stiffness matrix $[k_m]$ is,

$$[k_m] = \int [B]^T E[B] \, A \, dl \qquad (b)$$

Substituting for variation of A given by Eq.(a) into (b) we get,

$$[k_m] = E \int \frac{1}{L^2} \begin{Bmatrix} -1 \\ 1 \end{Bmatrix} \begin{bmatrix} -1 & 1 \end{bmatrix} \begin{bmatrix} L_1 & L_2 \end{bmatrix} \begin{Bmatrix} A_1 \\ A_2 \end{Bmatrix} dl$$

$$= \frac{E}{L^2} \begin{bmatrix} 1 & -1 \\ -1 & 1 \end{bmatrix} \int (A_1 L_1 + A_2 L_2) \, dl$$

$$= \frac{E(A_2 + A_1)}{2L} \begin{bmatrix} 1 & -1 \\ -1 & 1 \end{bmatrix} \qquad (c)$$

Fig.7.3a shows a truss element which is inclined in $x - y$ plane. In the description to follow the stiffness matrix for this element with respect to the displacements in global $x-y$ axes will be derived by the direct approach. However, it will be shown later in this chapter that this can be evaluated by working out the rotation transformation matrix.

The axial displacements produced in the element due to unit displacements given at the nodes i and j are shown in Fig.7.3(b). The direction cosines C_x and C_y for a member i are given by,

$$C_x = \frac{x_j - x_i}{L}$$

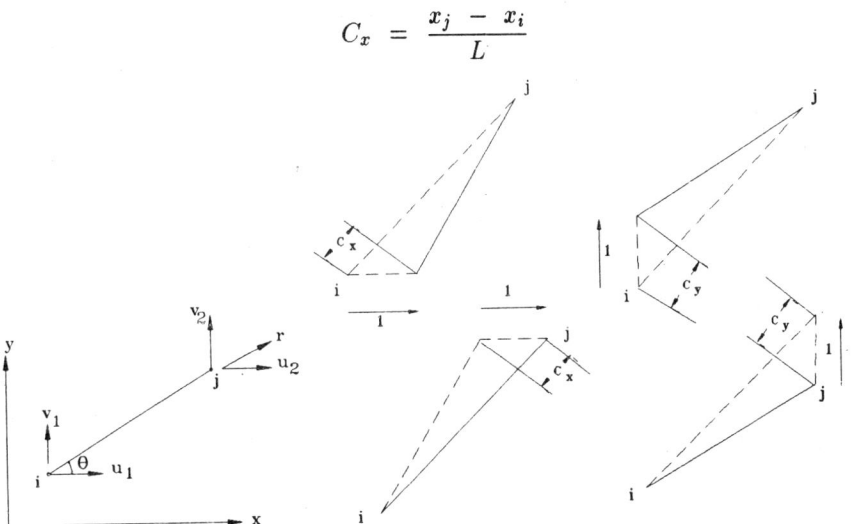

Fig. 7.3a Inclined truss member in $x - y$ plane   Fig. 7.3b Axial Displacements in 2D truss element

$$C_y = \frac{y_j - y_i}{L} \qquad (7.6)$$

and

$$L = \sqrt{(x_j - x_i)^2 + (y_j - y_i)^2}$$

where $x_i, y_i; x_j y_j$ are the coordinates of the nodes i and j.

The displacement variation in the member direction, i.e., r-direction is given by the Eq. 7.1.

$$u_r = L_i u_{1r} + L_2 u_{2r} \qquad (7.7)$$

where u_{1r} and u_{2r} are the displacements of nodes i and j in r - direction. Following the description given in Fig.7.3b these axial nodal displacements can be expressed in terms of global displacements u_1, v_1, u_2 and v_2 at the nodes i and j.

$$u_{1r} = C_x u_1 + C_y v_1$$

$$u_{2r} = C_x u_2 + C_y v_2 \tag{7.8}$$

Substituting in Eq.7.2 the axial strain ϵ_r along the member axis r of the element is given by,

$$\epsilon_r = \frac{1}{L} [C_x u_2 + C_y v_2 - C_x u_1 - C_y v_1]$$

$$= \frac{1}{L} [-C_x - C_y + C_x + C_y] \begin{Bmatrix} u_1 \\ v_1 \\ u_2 \\ v_2 \end{Bmatrix} \tag{7.9}$$

Hence, the strain-displacement matrix [B] is,

$$[B] = \frac{1}{L}[-C_x - C_y + C_x + C_y] \tag{7.10}$$

The axial stress in the element is

$$\sigma = E[B] \{d\} \tag{7.11}$$

The stiffness matrix $[k]$ in the global axes system i.e., corresponding to the degrees of freedom at the nodes i and j in the global axes system, for a prismatic element can now be evaluated as,

$$[k] = \int \int \int \frac{E}{L^2} \begin{Bmatrix} -C_x \\ -C_y \\ C_x \\ C_y \end{Bmatrix} [-C_x \quad -C_y \quad C_x \quad C_y] \, dV$$

$$= \frac{EA}{L} \begin{bmatrix} C_x^2 & C_x C_y & -C_x^2 & -C_x C_y \\ C_x C_y & C_y^2 & -C_x C_y & -C_y^2 \\ -C_x^2 & -C_x C_y & C_x^2 & C_x C_y \\ -C_x C_y & -C_y^2 & C_x C_y & C_y^2 \end{bmatrix} \tag{7.12}$$

Example 1 Consider a two-dimensional truss structure shown in Fig.7.4. The geometry and loading are symmetrical about the centre line.

Assume the area of cross section of all the members is the same. $E = 2 \times 10^4 \ kN/cm^2$. Find the force in the vertical member.

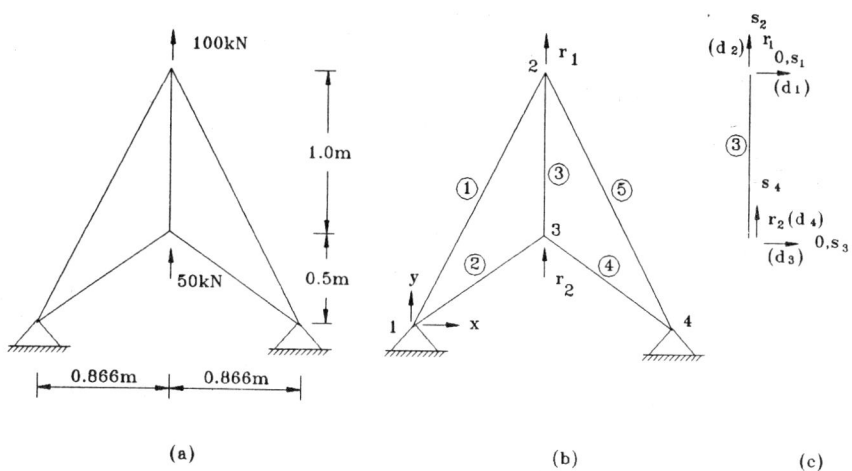

Fig. 7.4 Plane truss structure

For a two-dimensional truss element, the stiffness matrix is given by Eq.7.12 and the coefficients C_x and C_y are computed using Eq.7.6. Because of symmetry the horizontal displacements of nodes 2 and 3 are zero and thus the structure has only two degrees of freedom, viz., the vertical displacements at the joints 2 and 3. The stiffness matrix for the members 1 to 5 are computed below and the global degrees of freedom corresponding to the element degrees of freedom are indicated by the numbers outside the bracket.

Member 1: $i(0,0)$, $j(86.6, 150.0)$, L = 173.2 cm

$$C_x = 0.5, \quad C_y = 0.866$$

$$[k_1] = \frac{EA}{173.2} \begin{bmatrix} 0.25 & 0.433 & -0.25 & -0.433 \\ 0.433 & 0.75 & -0.433 & -0.75 \\ -0.25 & -0.433 & 0.25 & 0.433 \\ -0.433 & -0.75 & 0.433 & 0.75 \end{bmatrix} \quad (a)$$

The above stiffness matrix $[k_1]$ for member 1 can be written as,

$$
\begin{array}{cccc}
0 & 0 & 0 & 1 \\
1 & 2 & 3 & 4
\end{array}
\quad
\begin{array}{cccc}
g & d & o & f \\
e & d & o & f
\end{array}
$$

$$
[k_1] = \frac{EA}{1000}
\begin{bmatrix}
1.44 & 2.5 & -1.44 & -2.5 \\
2.5 & 4.33 & -2.5 & -4.33 \\
-1.44 & -2.5 & 1.44 & 2.5 \\
-2.5 & -4.33 & 2.5 & 4.33
\end{bmatrix}
\begin{array}{cc}
1 & 0 \\
2 & 0 \\
3 & 0 \\
4 & 1
\end{array}
\qquad (b)
$$

$$
\begin{array}{cc}
e & g \\
d & d \\
o & o \\
f & f
\end{array}
$$

Member 2 $i(0,0), j(86.6, 50.0), L = 100cm$

$C_x = 0.866, C_y = 0.5$

$$
\begin{array}{cccc}
0 & 0 & 0 & 2 \\
1 & 2 & 3 & 4
\end{array}
\quad
\begin{array}{cccc}
g & d & o & f \\
e & d & o & f
\end{array}
$$

$$
[k_2] = \frac{EA}{1000}
\begin{bmatrix}
7.5 & 4.33 & -7.5 & -4.33 \\
4.33 & 2.5 & -4.33 & -2.5 \\
-7.5 & -4.33 & 7.5 & 4.33 \\
-4.33 & -2.5 & 4.33 & 2.5
\end{bmatrix}
\begin{array}{cc}
1 & 0 \\
2 & 0 \\
3 & 0 \\
4 & 2
\end{array}
\qquad (c)
$$

$$
\begin{array}{cc}
e & g \\
d & d \\
o & o \\
f & f
\end{array}
$$

Member 3 $i(86.6, 150.0), j(86.6, 50.0),$

$L = 100cm, \ C_x = 0; \ C_y = -1.0$

$$
\begin{array}{cccc}
0 & 1 & 0 & 2 \\
1 & 2 & 3 & 4
\end{array}
\quad
\begin{array}{cccc}
g & d & o & f \\
e & d & o & f
\end{array}
$$

$$
[k_3] = \frac{EA}{1000}
\begin{bmatrix}
0 & 0 & 0 & 0 \\
0 & 10 & 0 & -10 \\
0 & 0 & 0 & 0 \\
0 & -10 & 0 & 10
\end{bmatrix}
\begin{array}{cc}
0 & 0 \\
2 & 1 \\
3 & 0 \\
4 & 2
\end{array}
\qquad (d)
$$

$$
\begin{array}{cc}
e & g \\
d & d \\
o & o \\
f & f
\end{array}
$$

Member 4 $i(86.6, 50.0), \ j(173.2, 0),$

$L = 100cm, \ C_x = 0.866, C_y = -0.5$

$$[k_4] = \frac{EA}{1000} \begin{array}{cccc} 0 & 2 & 0 & 0 \\ 1 & 2 & 3 & 4 \end{array} \atop \left[\begin{array}{cccc} 7.5 & -4.33 & -7.5 & 4.33 \\ -4.33 & 2.5 & 4.33 & -2.5 \\ -7.5 & 4.33 & 7.5 & -4.33 \\ 4.33 & -2.5 & -4.33 & 2.5 \end{array} \right] \begin{array}{c} 1 \\ 2 \\ 3 \\ 4 \end{array} \begin{array}{c} \\ 2 \\ 0 \\ 0 \end{array} \quad (e)$$

$$\begin{array}{cc} g & d & o & f \\ e & d & o & f \end{array}$$

$$\begin{array}{cc} e & g \\ d & d \end{array}$$

Member 5 $i(86.6, 150.0), j(173.2, 0)$;

$$\begin{array}{cc} o & o \\ f & f \end{array}$$

$L = 173.2cm.C_x = 0.5, C_y = -0.866$

$$\frac{EA}{1000} \begin{array}{cccc} 0 & 1 & 0 & 0 & g & d & o & f \\ 1 & 2 & 3 & 4 & e & d & o & f \end{array} \atop \left[\begin{array}{cccc} 1.44 & -2.50 & -1.44 & 2.50 \\ -2.50 & 4.33 & 2.50 & -4.33 \\ -1.44 & 2.50 & 1.44 & -2.50 \\ 2.50 & -4.33 & -2.50 & 4.33 \end{array} \right] \begin{array}{c} 1 \\ 2 \\ 3 \\ 4 \end{array} \begin{array}{c} 0 \\ 1 \\ 0 \\ 0 \end{array}$$

$$\begin{array}{cc} e & g \\ d & d \\ o & o \\ f & f \end{array} \quad (f)$$

Following the procedure for assembling the elements described in Chapter 5, the stiffness matrix for the structure is given by,

$$[K] = \frac{EA}{1000} \begin{bmatrix} 18.66 & -10.0 \\ -10.0 & 15.0 \end{bmatrix} \quad (g)$$

The nodal load vector is

$$\{P\} = \left\{ \begin{array}{c} 100.0 \\ 50.0 \end{array} \right\} \quad (h)$$

Hence, the equilibrium equation is,

$$\frac{EA}{1000} \begin{bmatrix} 18.66 & -10.0 \\ -10.0 & 15.0 \end{bmatrix} \left\{ \begin{array}{c} r_1 \\ r_2 \end{array} \right\} = \left\{ \begin{array}{c} 100.0 \\ 50.0 \end{array} \right\} \quad (i)$$

Solving the above equations we get,

$$r_2 = \frac{0.5372}{A}$$

$$r_1 = \frac{0.5558}{A} \quad (j)$$

The end forces in the member 3 along the global direction of displacements (Fig. 7.3c) can be found out as,

$$\{S\} = [k_3]\{d\}$$

$$= \frac{EA}{1000} \begin{bmatrix} 0 & 0 & 0 & 0 \\ 0 & 10 & 0 & -10 \\ 0 & 0 & 0 & 0 \\ 0 & -10 & 0 & 10 \end{bmatrix} \begin{Bmatrix} 0 \\ 0.5558/A \\ 0 \\ 0.5372/A \end{Bmatrix} = \begin{Bmatrix} 0 \\ 3.720 \\ 0 \\ -3.720 \end{Bmatrix} \quad (k)$$

Hence, the axial force in the vertical member is 3.720kN (tension).

7.2 Three-Dimensional Truss Element

A typical truss element arbitrarily oriented is shown in Fig.7.5.

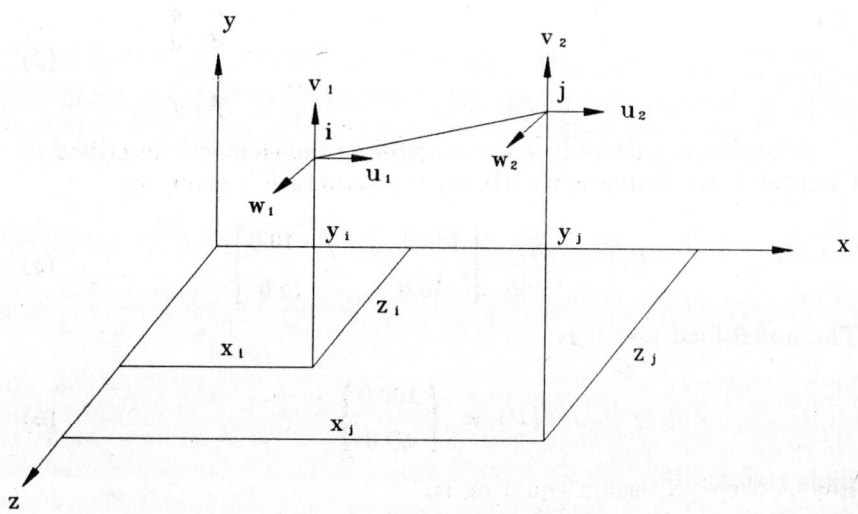

Fig. 7.5 A typical three-dimensional truss element

The direction cosines of the element are given by,

$$C_x = \frac{x_j - x_i}{L}$$

$$C_y = \frac{y_j - y_i}{L}$$

$$C_z = \frac{z_j - z_i}{L} \tag{7.13}$$

where x_i, y_i, z_i, and x_j, y_j, z_j are the coordinates of the nodes i and j respectively.

The length of the member,

$$L = \sqrt{(x_j - x_i)^2 + (y_j - y_i)^2 + (z_j - z_i)^2}$$

The variation of displacement along the member direction r is given by Eq.7.7 as

$$u_r = L_1 \, u_{1r} + L_2 \, u_{2r}$$

For a three-dimensional truss element it can be shown that,

$$u_{1r} = C_x \, u_1 + C_y \, v_1 + C_z \, w_1 \tag{7.14}$$

$$u_{2r} = C_x \, u_2 + C_y \, v_2 + C_z \, w_2$$

where $u_1, v_1, w_1, u_2, v_2, w_2$ are the displacements in global directions at nodes i and j respectively.

Substituting in Eq.7.2. the axial strain, ϵ_r along the member axis r of the element is,

$$\epsilon_r = \frac{1}{L}[C_x \, u_2 + C_y \, v_2 + C_z \, w_2 - C_x \, u_1 - C_y \, v_1 - C_z \, w_1]$$

$$= \frac{1}{L}[-C_x - C_y - C_z \ C_x \ C_y \ C_z] \begin{Bmatrix} u_1 \\ v_1 \\ w_1 \\ u_2 \\ v_2 \\ w_2 \end{Bmatrix} \tag{7.15}$$

Thus the strain displacement matrix [B] can be written as,

$$[B] = \frac{1}{L} [-C_x \ -C_y \ -C_z \ C_x \ C_y \ C_z]$$

The axial stress in the element is,

$$\sigma = E \, [B] \, \{d\} \tag{7.16}$$

The stiffness matrix [k] for a prismatic element can be evaluated as,

$$[k] = \int\int\int \frac{E}{L^2} \begin{bmatrix} -C_x \\ -C_y \\ -C_z \\ C_x \\ C_y \\ C_z \end{bmatrix} [-C_x \ -C_y \ -C_z \ C_x \ C_y \ C_z] \, dV$$

$$= \frac{EA}{L} \begin{bmatrix} C_x^2 & C_yC_x & C_zC_x & -C_x^2 & -C_yC_x & -C_zC_x \\ C_xC_y & C_y^2 & C_zC_y & -C_xC_y & -C_y^2 & -C_zC_y \\ C_xC_z & C_yC_z & C_z^2 & -C_xC_z & -C_yC_z & -C_z^2 \\ -C_x^2 & -C_yC_x & -C_zC_x & C_x^2 & C_yC_x & C_zC_x \\ -C_xC_y & -C_y^2 & -C_zC_y & C_xC_y & C_y^2 & C_zC_y \\ -C_xC_z & -C_yC_z & -C_z^2 & C_xC_z & C_yC_z & C_z^2 \end{bmatrix}$$

$$(7.17)$$

7.3   Two-Dimensional Beam Element

A typical two-dimensional beam element lying in $x - y$ plane, with the member axis parallel to x -axis is shown in Fig. 7.6a. As shown, there in, we define the axis system for the member as x_m, y_m and z_m which are local and mutually perpendicular to each other. In this case the local member axis system is parallel to the global axes.

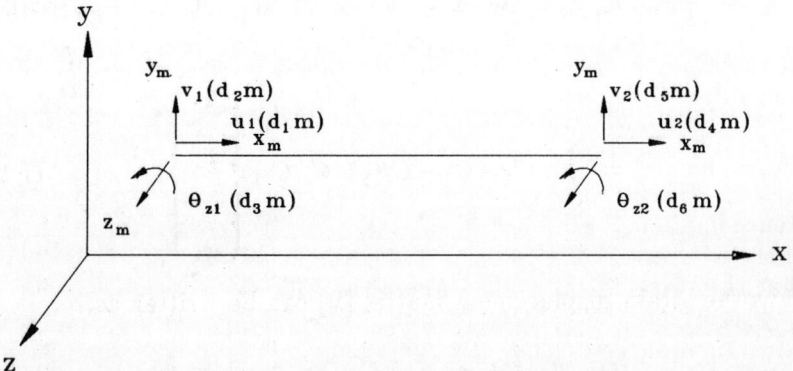

Fig. 7.6a Two-dimensional beam element with six degrees of freedom

The properties of the element will be derived below and it will be shown in Sub-section 7.3.3 how they can be evaluated if it is inclined in the $x - y$ plane.

The degrees of freedom at the nodes are

$$\{d_m\}^T \;=\; \{d\}^T \;=\; [\,u_1 \quad v_1 \quad \theta_{z1} \quad u_2 \quad v_2 \quad \theta_{z2}\,] \qquad (7.18)$$

where u_1, v_1, u_2 and v_2 are the displacements along x and y axes and θ_{z1} and θ_{z2} are rotations about z -axis at nodes i and j, and in this case the degrees of freedom referred to the member axes x_m, y_m, z_m and the global axes x, y and z are the same. It may be noted that the stiffness coefficients due to axial displacements u_1 and u_2 have been derived in section 7.1 and they do not influence the response of the member due to displacements v_1, θ_{z1}, v_2 and θ_{z2}. Hence, the stiffness coefficients due to displacements v_1, θ_{z1}, v_2 and θ_{z2} at the nodes i and j, as shown in Fig.7.6b will be derived in the following section.

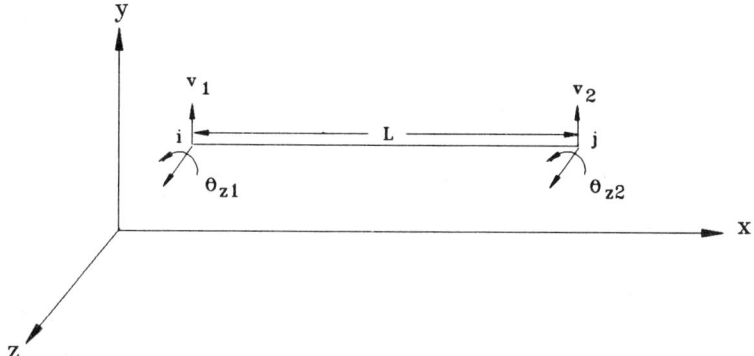

Fig. 7.6b Two-dimensional beam element with four
degrees of freedom

7.3.1 Displacement Model

In the case of a beam member shown in Fig.7.6 the rotation of the tangent to the elastic curve θ is given by

$$\theta \;=\; \frac{dv}{dx} \qquad (7.19)$$

Hence it is required to express only the variation of v along the length of the member. Following the discussion in section 3.2 it can be seen that the variation of v will be cubic and can be expressed as

$$v \;=\; \alpha_1 + \alpha_2\,x + \alpha_3\,x^2 + \alpha_4\,x^3 \qquad (7.20a)$$

or in natural coordinates as

$$v \;=\; \alpha_1\,L_1^3 + \alpha_2\,L_2^3 + \alpha_3\,L_1^2\,L_2 + \alpha_4\,L_1\,L_2^2 \qquad (7.20b)$$

It may be noted that all the terms in Eq.(7.20b) are cubic and if they are not, they can be made so by the relation $L_1 + L_2 = 1$.

Following the procedure indicated in section 3.2 the interpolation function [N] can be worked-out after evaluating the [A] matrix and using Eq.3.10. Thus,

$$v = [N_1 \quad N_2 \quad N_3 \quad N_4] \{d\} \qquad (7.21a)$$

where $\qquad \{d\}^T = [v_1 \quad \theta_{z1} \quad v_2 \quad \theta_{z2}] \qquad (7.21b)$

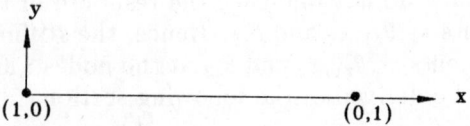

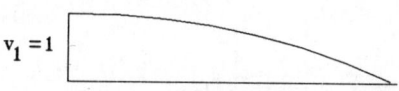
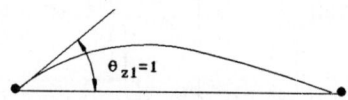

(a) Typical variation   (b) Typical variation
of v for $v = 1$ of v for $\theta_{z1} = 1$
Fig. 7.7 Displacement variation for a two-dimensional beam element

However, in this case $N_i (i = 1, ..., 4)$ can be evaluated more directly by noting that each term N_i gives the variation of v for unit value of d_i while all the other displacement components are held zero.

Evaluation of N_i (Fig.7.7a)

$$v = \alpha_1 L_1^3 + \alpha_2 L_2^3 + \alpha_3 L_1^2 L_2 + \alpha_4 L_1 L_2^2 \qquad (7.21c)$$

using Eq.3.15, $\qquad \dfrac{dv}{dx} = \dfrac{1}{L} [3\alpha_2 L_2^2 + \alpha_3 L_1^2 + 2\alpha_4 L_1 L_2$

$$-3\alpha_1 L_1^2 - 2\alpha_3 L_1 L_2 - \alpha_4 L_2^2] \qquad (7.22)$$

At node i, $v_1 = 1$, $\theta_{z1} = 0$ and at node j, $v_2 = 0$, $\theta_{z2} = 0$ (7.23)

Substituting these values in Eqs.7.20b and 7.22 we get,

$$\alpha_1 = 1, \quad \alpha_2 = 0, \quad \alpha_3 = 3, \quad \text{and} \quad \alpha_4 = 0 \qquad (7.24)$$

Hence, $\qquad\qquad\qquad v = L_1^3 + 3L_1^2 L_2$

$$= L_1^2 (3 - 2L_1)$$

Thus, $\quad N_1 = L_1^2 (3 - 2L_1)$ $\qquad$ (7.25)

Evaluation of N_2 (Fig.7.7b),

At node i, $v_1 = 0$, $\theta_{z1} = 1$ and at node j, $v_2 = 0$, $\theta_{z2} = 0$ (7.26)

Substituting these values in Eqs.7.20b and 7.22 we get,

$$\alpha_1 = 0, \ \alpha_2 = 0, \ \alpha_3 = L, \ \alpha_4 = 0 \qquad (7.27)$$

Hence, $\qquad v = L_1^2 L_2 L$

Thus, $\qquad N_2 = L_1^2 L_2 L$ $\qquad$ (7.28)

Similarly the other components, N_3 and N_4 can be evaluated.

Thus, $\quad [N] = [L_1^2(3-2L_1) \ \ L_1^2 L_2 L \ \ L_2^2(3-2L_2) \ \ -L_1 L_2^2 L]$ (7.29)

It can be shown from simple theory of bending that,

$$\epsilon_x = -y \frac{d^2 v}{dx^2} \qquad (7.30)$$

$$\sigma_x = -Ey \frac{d^2 v}{dx^2} \qquad (7.31)$$

and bending moment $\qquad M = \int_{-h/2}^{h/2} \sigma_x \, y \cdot b \cdot dy \qquad (7.32)$

Substituting Eq.7.31 in Eq.7.32, we get,

$$M = -EI_z \frac{d^2 v}{dx^2} \qquad (7.33)$$

where I_z = Moment of inertia of the section = $\int_{-h/2}^{h/2} y^2 \, b \, dy$

about the local $\ z_m \ $ axis $\qquad\qquad$ (7.34)

and I_z is assumed to be uniform for a prismatic member. However, problems involving non prismatic members can be handled by using interpolation functions to describe the geometry of the beam as indicated in reference[4]. Differentiating Eq.7.21c.

$$\frac{d^2v}{dx^2} = \frac{1}{L^2}[(6 - 12L_1) \ L \ (2L_2 - 4L_1) \ (6 - 12 \ L_2)$$

$$L(4L_2 - 2L_1)] \ \{d\} \tag{7.35}$$

Hence $M = -\dfrac{EI_z}{L^2} \ [(6 - 12L_1) \ L(2L_2 - 4L_1) \ (6 - 12L_2)$

$$L(4L_2 - 2L_1)] \ \{d\} \tag{7.36}$$

Treating the stress resultant, bending moment M, similar to the stress σ, the [B] and [C] matrices for the beam element are given by,

$$[C] = EI_z \tag{7.37a}$$

and,

$$[B] = -\frac{1}{L^2}[(6 - 12 \ L_1) \ L(2L_2 - 4L_1) \ (6 - 12L_2) \ L(4L_2 - 2L_1)] \tag{7.37b}$$

The stiffness matrix for the beam element is given by

$$[k_m] = \int [B]^T \ [C] \ [B] \ dl \tag{7.38}$$

and it may be noted that it has been reduced to a line integral as the integration across the area of cross-section has been done in Eq.7.34 and accordingly the stress resultant, M, has been considered.

Substituting Eq.7.37 in Eq.7.38, the stiffness matrix $[k_m]$ can be evaluated as

$$[k_m] = \frac{EI_z}{4} \int_0^l \begin{bmatrix} a_{11} & a_{12} & a_{13} & a_{14} \\ & a_{22} & a_{23} & a_{24} \\ \text{sym} & & a_{33} & a_{34} \\ & & & a_{44} \end{bmatrix} dl \tag{7.39}$$

where,

$$a_{11} = (6 - 12L_1)^2, \quad a_{12} = L(6 - 12L_1)(2L_2 - 4L_1)$$
$$a_{13} = (6 - 12L_1)(6 - 12L_2), \quad a_{14} = L(6 - 12L_1)(4L_2 - 2L_1)$$
$$a_{22} = L^2(2L_2 - 4L_1)^2, \quad a_{23} = L(2L_2 - 4L_1)(6 - 12L_2)$$
$$a_{24} = L^2(2L_2 - 4L_1)(4L_2 - 2L_1), \quad a_{33} = (6 - 12L_2)^2$$
$$a_{34} = L(6 - 12L_2)(4L_2 - 2L_1), \quad a_{44} = L^2(4L_2 - 2L_1)^2$$

The integration can be performed using Eq.3.16. Thus, the stiffness matrix corresponding to the degrees of freedom shown in Fig.7.6 is

$$[k_m] = \begin{bmatrix} \frac{12EI_z}{L^3} & \frac{6EI_z}{L^2} & \frac{-12EI_z}{L^3} & \frac{6EI_z}{L^2} \\ & \frac{4EI_z}{L} & \frac{-6EI_z}{L^2} & \frac{2EI_z}{L} \\ sym. & & \frac{12EI_z}{L^3} & \frac{-6EI_z}{L^2} \\ & & & \frac{4EI_z}{L} \end{bmatrix} \tag{7.40}$$

7.3.2 Stiffness Matrix for a Two-Dimensional Beam (Element with Six Degrees of Freedom)

The stiffness matrix of Eq.7.40 can be combined with the stiffness coefficients given in Eq.7.5 to form the stiffness matrix for a two-dimensional beam element shown in Fig.7.6a for a six degrees of freedom,

$$\{d_m\}^T = [d_{1m} \quad d_{2m} \quad d_{3m} \quad d_{4m} \quad d_{5m} \quad d_{6m}] \tag{7.41}$$

referred to member axes.

$$[k_m] = \begin{bmatrix} \frac{EA}{L} & 0 & 0 & \frac{-EA}{L} & 0 & 0 \\ 0 & \frac{12EI_z}{L^3} & \frac{6EI_z}{L^2} & 0 & \frac{-12EI_z}{L^3} & \frac{6EI_z}{L^2} \\ 0 & \frac{6EI_z}{L^2} & \frac{4EI_z}{L} & 0 & \frac{-6EI_z}{L^2} & \frac{2EI_z}{L} \\ \frac{-EA}{L} & 0 & 0 & \frac{EA}{L} & 0 & 0 \\ 0 & \frac{-12EI_z}{L^3} & \frac{-6EI_z}{L^2} & 0 & \frac{12EI_z}{L^3} & \frac{-6EI_z}{L^2} \\ 0 & \frac{6EI_z}{L^2} & \frac{2EI_z}{L} & 0 & \frac{-6EI_z}{L^2} & \frac{4EI_z}{L} \end{bmatrix} \tag{7.42}$$

The physical interpretation of the stiffness coefficients given by the above equation is illustrated in Fig.7.8.

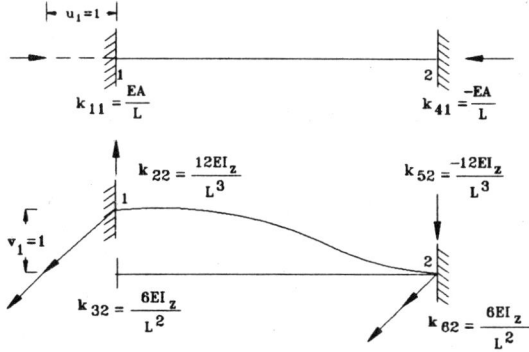

Fig. 7.8 Two-dimensional beam element stiffness (contd)

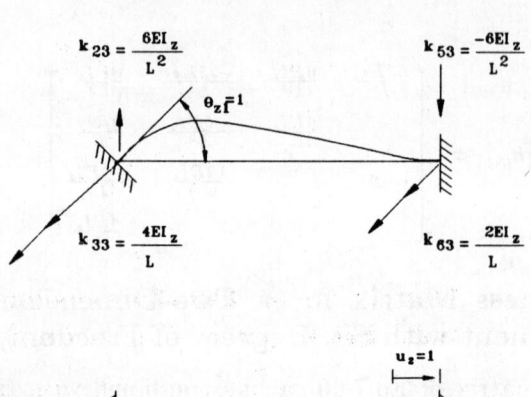

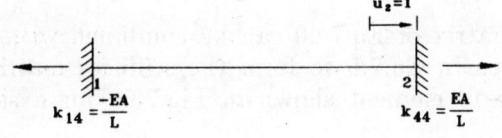

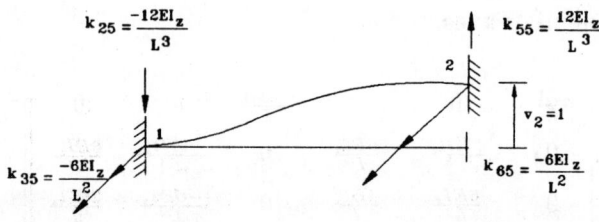

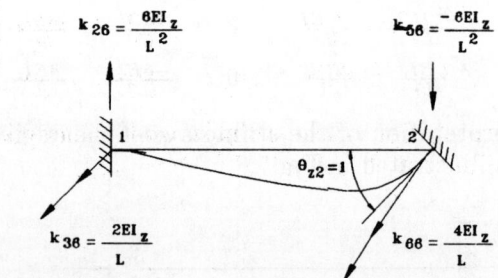

Fig. 7.8 Two-dimensional beam element stiffness

7.3.3 Computation of Element Nodal Load Vector

The computation of nodal loads due to lateral loads applied on the beam element is illustrated with reference to a beam shown in Fig.7.9. The beam is subjected to an uniformly distributed vertical load of p per unit length.

The nodal load vector $\{Q\}$ due to surface traction is given by Eq.3.82

$$\{Q\} = \int \int [N^s]^T \{p\} \, dS$$

As the load applied on the two-dimensional beam is specified as uniform intensity p per unit length the above integral reduces to

$$\{Q\} = p \int [N^s]^T \, dl \tag{7.43}$$

Substituting for $[N^s]$ from Eq.(7.29), we get the nodal load vector corresponding to the four degrees of freedom v_1, θ_{z1}, v_2 and θ_{z2}, as

$$\{Q_m\} = \{Q\} = p \int_0^L \left\{ \begin{array}{c} L_1^2(3 - 2L_1) \\ L_1^2 \, L_2 \, L \\ L_2^2(3 - 2L_2) \\ -L_1 \, L_2^2 \, 2L \end{array} \right\} dl = \left\{ \begin{array}{c} \frac{pL}{2} \\ \frac{pL^2}{12} \\ \frac{pL}{2} \\ \frac{-pL^2}{12} \end{array} \right\} \tag{7.44}$$

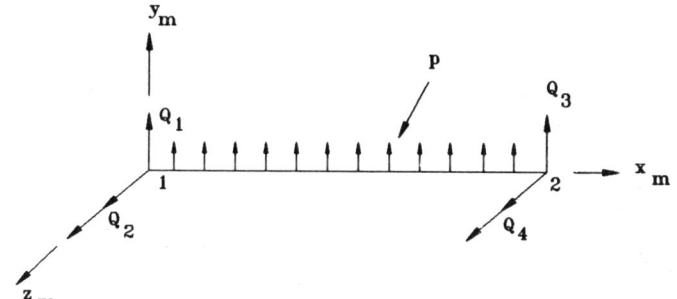

(a) Nodal load vector components

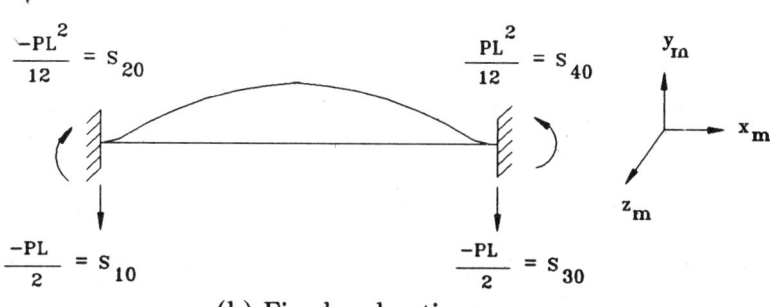

(b) Fixed end actions

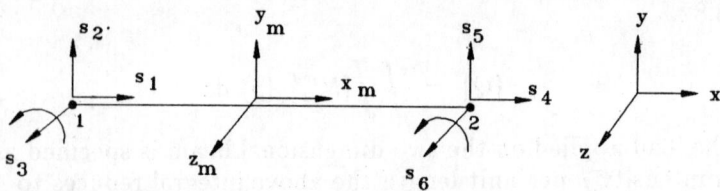

(c) Stress resultants

Fig. 7.9 Nodal load vector and stress resultant of
a two-dimensional beam member

From the above evaluation of $\{Q\}$ it can be seen that the nodal load vector components are equal to the negative values of the fixed end actions of the beam, when the ends are assumed as fixed.

7.3.4   Transformation Matrix

If the two-dimensional beam element is inclined as shown in Fig.7.10, then the stiffness matrix will first be computed in the local member axes system using the Eq.7.42.

Similarly the nodal load vector $\{Q_m\}$ due to loads normal to the member will be the negatives of the fixed end actions with respect to the member axes.

For the assemblage of elements to form the overall equations of equilibrium, transformation of these element properties is required to conform to the degrees of freedom and nodal loads expressed in the global coordinate system as shown in Fig.7.10.

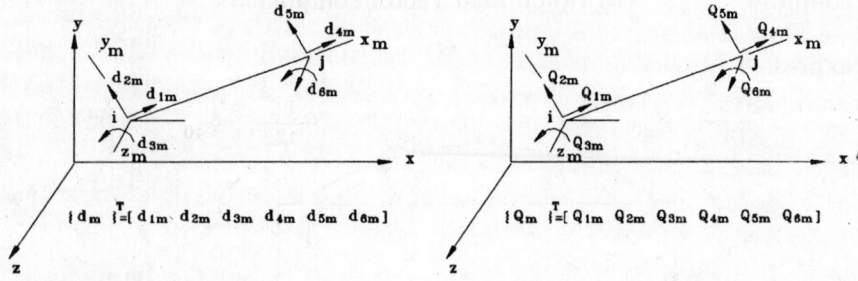

(a) Degrees of freedom
referred to member axes

(b) Nodal loads referred to
member axes

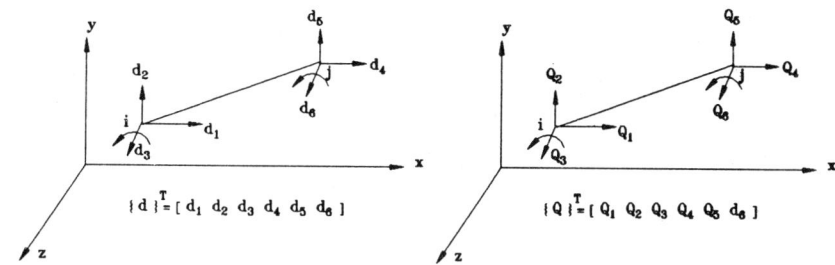

(c) Degrees of freedom   (b) Nodal loads referred to
   referred to member axes   global axes
Fig. 7.10 Two-dimensional beam element

Consider a local system of axes x_m and y_m in the $x - y$ plane ⌐otated through an angle θ as shown in Fig.7.11. The local axis z_m is the normal to the plane $x_m - y_m$ and thus coincides with the global z-axis. Let the force Q has components Q_1 and Q_2 parallel to the global axes. Resolving these force components into the axes x_m and y_m we obtain,

$$Q_{1m} = Q_1 cos\theta + Q_2 sin\theta = Q_1\, C_x + Q_2\, C_y$$

$$Q_{2m} = -Q_1 sin\theta + Q_2 cos\theta = -Q_1\, C_y + Q_2\, C_x \tag{7.45}$$

$$\text{where} \quad C_x = cos\theta \quad \text{and} \quad C_y = sin\theta \tag{7.45a}$$

If there is a moment Q_3 about the z-axis, it will have the same component about z_m axis as seen from Fig. 7.11.

Thus the components of the forces in local system of axes can be expressed in terms of their components in global system of axes as,

$$\begin{Bmatrix} Q_{1m} \\ Q_{2m} \\ Q_{3m} \end{Bmatrix} = \begin{bmatrix} C_x & C_y & 0 \\ -C_y & C_x & 0 \\ 0 & 0 & 1 \end{bmatrix} \begin{Bmatrix} Q_1 \\ Q_2 \\ Q_3 \end{Bmatrix} \tag{7.46}$$

$$\text{or} \qquad \{Q'_m\} = [T']^{-1}\, \{Q'\}$$

On further examination it can be observed that the terms in the matrix $[T']$ represent the direction cosines of the member in terms of the angle of rotation θ. It may also be seen that the transpose of $[T']$ is equal to its inverse.

$$[T']^T = [T']^{-1} \tag{7.47}$$

Hence using Eqs.7.46 and 7.47 we obtain the inverse relation for expressing the force components in global system in terms of their components in local system of axes as

$$\{Q'\} = [T']^T \{Q'_m\} \tag{7.48}$$

Since small displacements and forces can be treated as vectors the relationships obtained above for forces can be used for the displacements as well. So, we get

$$\{d'_m\} = [T'] \{d'\} \tag{7.49}$$

$$\text{and } \{d'\} = [T']^T \{d'_m\} \tag{7.50}$$

where $\{d'_m\}$ and $\{d'\}$ refer to the displacements at one node with reference to the member and global axes respectively.

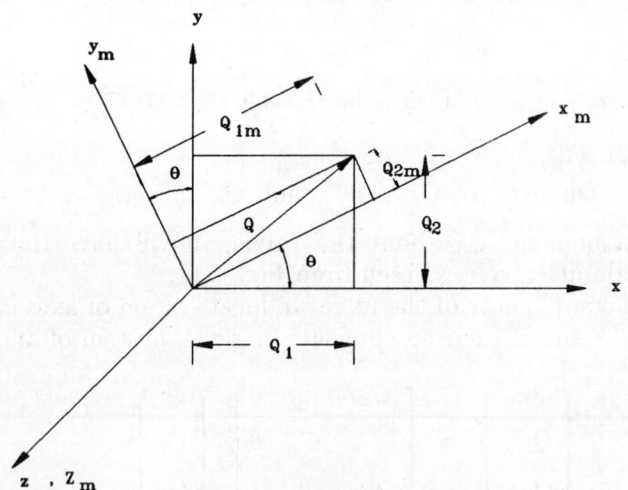

Fig. 7.11 Rotation of axes

Now considering the two-dimensional beam element shown in Fig.7.10, the forces at the two nodes with reference to the local and global axes can be related as

$$\begin{Bmatrix} Q_{1m} \\ Q_{2m} \\ Q_{3m} \\ Q_{4m} \\ Q_{5m} \\ Q_{6m} \end{Bmatrix} = \begin{bmatrix} C_x & C_y & 0 & 0 & 0 & 0 \\ -C_y & C_x & 0 & 0 & 0 & 0 \\ 0 & 0 & 1 & 0 & 0 & 0 \\ 0 & 0 & 0 & C_x & C_y & 0 \\ 0 & 0 & 0 & -C_y & C_x & 0 \\ 0 & 0 & 0 & 0 & 0 & 1 \end{bmatrix} \begin{Bmatrix} Q_1 \\ Q_2 \\ Q_3 \\ Q_4 \\ Q_5 \\ Q_6 \end{Bmatrix} \quad (7.51)$$

$$\text{or} \quad \begin{Bmatrix} \{Q'_m\}_1 \\ ---- \\ \{Q'_m\}_2 \end{Bmatrix} = \begin{bmatrix} [T'] & | & [0] \\ -- & & -- \\ [0] & | & [T'] \end{bmatrix} \begin{Bmatrix} \{Q'\}_1 \\ --- \\ \{Q'\}_2 \end{Bmatrix} \quad (7.52)$$

$$\text{i.e., } \{Q_m\} = [T]\{Q\} \quad (7.53)$$

where $[T]$ is called the rotation transformation matrix for the two-dimensional beam element.

Similarly the nodal displacements can be related as

$$\{d_m\} = [T]\{d\} \quad (7.54)$$

And the inverse relations can also be expressed as

$$\{Q\} = [T]^T\{Q_m\} \quad (7.55)$$

$$\text{and } \{d\} = [T]^T\{d_m\} \quad (7.56)$$

Now the equilibrium equation for the element in local system of axes can be expressed as

$$[k_m]\{d_m\} = \{Q_m\} \quad (7.57)$$

Substituting from Eqs.7.53 and 7.54 into Eq.7.57 we get

$$[k_m][T]\{d\} = [T]\{Q\} \quad (7.58)$$

The vectors in Eq.7.58 refer to the displacements and nodal forces in global system of axes. Premultiplying both sides of the equation by $[T]^{-1}$ which is equal to $[T]^T$, we get

$$[T]^T[k_m][T]\{d\} = \{Q\} \quad (7.59)$$

This equation relates the displacements and the nodal forces in global axes.

$$\text{Thus,} \qquad [k]\{d\} = \{Q\} \quad (7.60)$$

where $[k] = [T]^T [k_m] [T]$ is the stiffness matrix of the element in the global axes system and is obtained by transforming $[k_m]$, the stiffness matrix in the member axes system.

Similarly, the nodal loads due to loads acting on the element can be transformed to the global system using Eq.7.55 as

$$\{Q\} = [T]^T \{Q_m\} \tag{7.61}$$

7.3.5   Computation of Final Stress Resultants

In a general finite element analysis the stress is computed as

$$\{\sigma\} = [C] [B] \{d\}$$

In the case of two or three-dimensional beam elements it is necessary to compute the stress resultants at the ends of the member for the design purposes. These stress resultants correspond to the degrees of freedom as axial and shear forces and bending moments. Hence, as shown in Fig.7.8, the stiffness coefficients give the value of these actions due to unit displacements. Then after assembling the stiffness matrices and solving them and retrieving the displacements at the ends of the members by the finite element analysis procedure discussed in Chapter 5, the end actions due to end displacements for a member in the local axes system can be expressed as

$$\{S\} = [k_m] \{d_m\} \tag{7.62}$$

To the above stress resultants we should add the stress resultants due to loads on the member under fully restrained condition. Otherwise for a beam shown in Fig.7.9, the Eq.7.62 would give zero values of stress resultants which is incorrect. Thus,

$$\{S\} = [k_m] \{d_m\} + \{S_o\} \tag{7.63}$$

where $\{S_o\}$ represents the stress resultants corresponding to the nodal degrees of freedom due to loads on the member under fully restrained end conditions. These final stress resultants for a two-dimensional beam element with degrees of freedom are shown in Fig.7.9c. It may be noted that the above explanation holds good for a three-dimensional beam element and Eq.7.63 gives the final stress resultants at the ends of the member.

The above discussion suggests that in the case of analysis of other continuous structures also that the computed stresses should include those stresses arising from internal forces acting on the element under fully restrained conditions. In practice such stresses are usually ignored as they are quite tedious to compute or approximate and they tend to vanish as elements are reduced in size and thus the stresses are evaluated using Eq.3.70. However, in the case of two or three-dimensional beams, the full length of the member is considered as an element and the effects due to fully restrained conditions cannot be ignored. The standard tables, as presented in reference [1] are available to aid in such computations. Table 7.1 gives values of fixed end actions for some typical cases of loading.

Table 7.1 Fixed-end actions for beam element

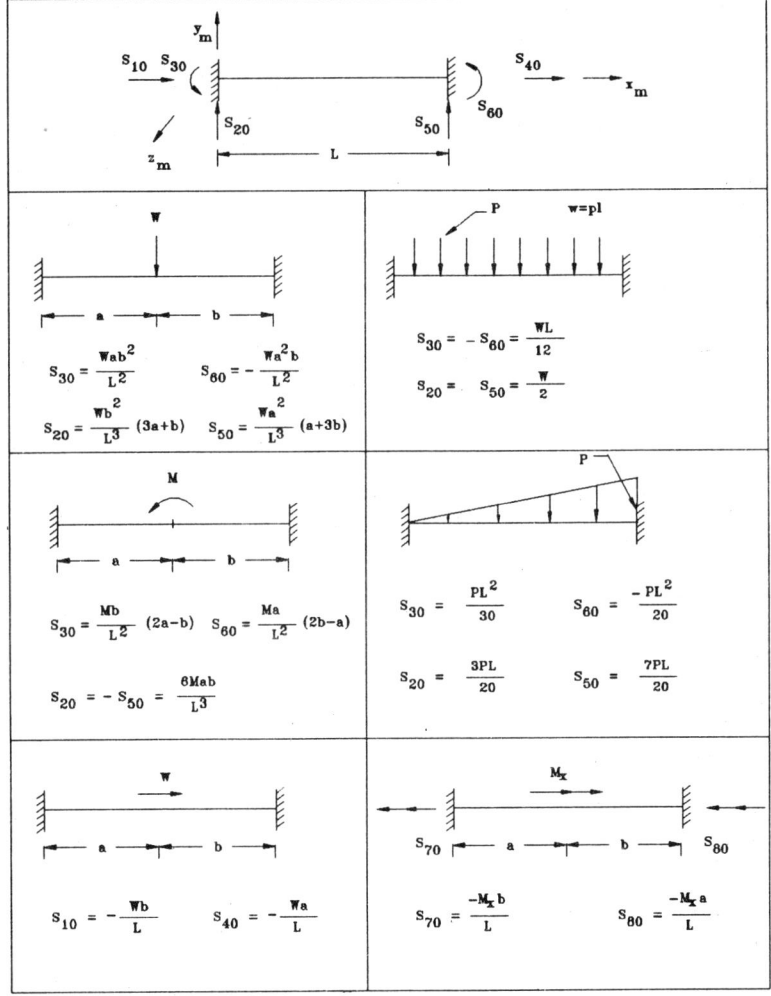

Example 2 Analyse the continuous beam shown in Fig.7.12 and draw the shear force and bending moment diagrams. The spring stiffness $= \frac{24EI}{L^3}$ and EI = 400 units.

The stiffness matrix $[k_m]$ for a two-dimensional beam with four degrees of freedom is given by Eq.7.40 and the element nodal load vector $\{Q_m\}$ due to uniformly distributed load is given by Eq.7.44. Using these equations the stiffness matrix and the element load vector for the two members of the continuous beam are calculated, and the numerical values are given below:

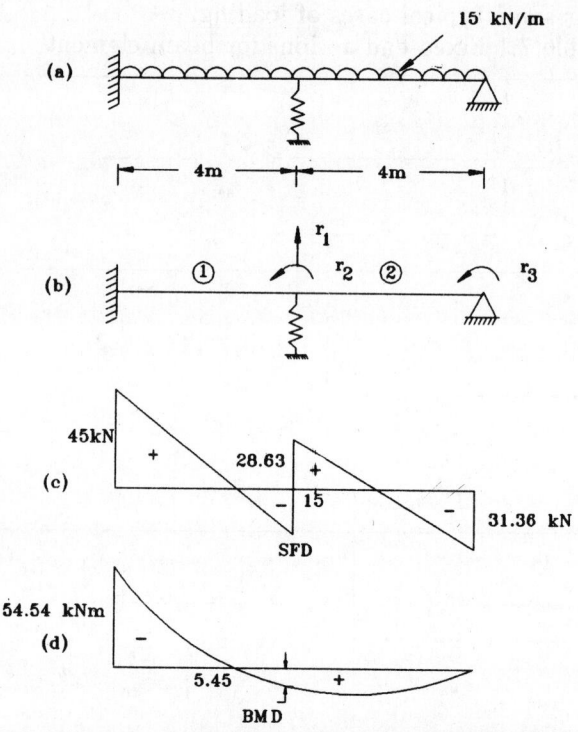

Fig. 7.12 Two span continuous beam

Member 1 $\quad \begin{array}{cccccccc} g & d & o & f & 0 & 0 & 1 & 2 \\ e & d & o & f & 1 & 2 & 3 & 4 \end{array}$

$$[k_1] = \begin{array}{c} 0 \quad 1 \\ 0 \quad 2 \\ 1 \quad 3 \\ 2 \quad 4 \end{array} \begin{bmatrix} 75 & 150 & -75 & 150 \\ & 400 & -150 & 200 \\ sym. & & 75 & -150 \\ & & & 400 \end{bmatrix} \qquad (a)$$

$$\begin{array}{cc} g & e \\ d & d \\ o & o \\ f & f \end{array}$$

$$\{Q_1\} = -\{S_{01}\} = -\left\{\begin{array}{c} \dfrac{WL}{2} \\ \dfrac{WL^2}{12} \\ \dfrac{WL}{2} \\ -\dfrac{WL^2}{12} \end{array}\right\} = \begin{array}{cc} 0 & 1 \\ 0 & 2 \\ 1 & 3 \\ 2 & 4 \end{array} \left\{\begin{array}{r} -30 \\ -20 \\ -30 \\ 20 \end{array}\right\} \qquad (b)$$

$$\begin{array}{cc} g & e \\ d & d \\ o & o \\ f & f \end{array}$$

Member 2 $\begin{array}{cccc} g & d & o & f \\ e & d & o & f \end{array}$ $\begin{array}{cccc} 1 & 2 & 0 & 3 \\ 1 & 2 & 3 & 4 \end{array}$

$$[k_2] = \begin{array}{cc} 1 & 1 \\ 2 & 2 \\ 0 & 3 \\ 3 & 4 \end{array} \left[\begin{array}{cccc} 75 & 150 & -75 & 150 \\ & 400 & -150 & 200 \\ sym. & & 75 & -150 \\ & & & 400 \end{array}\right] \qquad (c)$$

$$\begin{array}{cc} g & e \\ d & d \\ o & o \\ f & f \end{array}$$

$$\{Q_2\} = -\{S_{02}\} = \begin{array}{cc} 1 & 1 \\ 2 & 2 \\ 0 & 3 \\ 3 & 4 \end{array} \left\{\begin{array}{r} -30 \\ -20 \\ -30 \\ 20 \end{array}\right\} \qquad (d)$$

$$\begin{array}{cc} g & e \\ d & d \\ o & o \\ f & f \end{array}$$

The spring at the mid support has stiffness $= \frac{24 \times 400}{4^3} = 150$ kN/m and this resistance is added to the contribution from the two beam members to the stiffness in the displacement direction r_1. Assembling the stiffness matrix and the load vector for the structure we have the equation of equilibrium for the continuous beam as

$$\begin{bmatrix} 300 & 0 & 150 \\ & 800 & 200 \\ sym. & & 400 \end{bmatrix} \begin{Bmatrix} r_1 \\ r_2 \\ r_3 \end{Bmatrix} = \begin{Bmatrix} -60 \\ 0 \\ 20 \end{Bmatrix} \qquad (e)$$

Solving the above equations we get,

$$r_3 = \frac{2}{11}$$

$$r_2 = -\frac{1}{22}$$

$$r_1 = -\frac{16}{55} \qquad (f)$$

The final stress resultants for each member are calculated using Eq.7.63.

Member 1

$$\{S_1\} = [k_1]\{d_1\} + \{S_{o1}\}$$

$$\begin{bmatrix} 75 & 150 & -75 & 150 \\ & 400 & -150 & 200 \\ sym. & & 75 & -150 \\ & & & 400 \end{bmatrix} \begin{Bmatrix} 0 \\ 0 \\ -16/55 \\ -1/22 \end{Bmatrix} + \begin{Bmatrix} 30 \\ 20 \\ 30 \\ -20 \end{Bmatrix} = \begin{Bmatrix} 45.0 \\ 54.54 \\ 15.0 \\ 5.45 \end{Bmatrix} \qquad (g)$$

Member 2

$$\{S_2\} = \begin{bmatrix} 75 & 150 & -75 & 150 \\ & 400 & -150 & 200 \\ Sym. & & 75 & 150 \\ & & & 400 \end{bmatrix} \begin{Bmatrix} -\frac{16}{55} \\ -\frac{1}{22} \\ 0 \\ \frac{2}{11} \end{Bmatrix} + \begin{Bmatrix} 30 \\ 20 \\ 30 \\ -20 \end{Bmatrix} = \begin{Bmatrix} 28.63 \\ -5.45 \\ 31.36 \\ 0.0 \end{Bmatrix}$$

Using the above values of final end actions on the members 1 and 2, the shear force and bending moment diagrams are drawn as shown in Figs.7.12c and d.

Example 3 Analyse the plane frame shown in Fig.7.13 by using the 2D beam element. Draw the bending moment diagram.

In general the stiffness matrix $[k_m]$ of a two-dimensional beam with six degrees of freedom referred to member axes is given by the Eq. 7.42.

Using the transformation matrix $[T]$ given by Eq.7.51, the stiffness matrix $[k]$ of the member in the global axes system is computed from

Eq.7.60. Similarly the element nodal load vectors $\{Q_m\}$ and $\{Q\}$ are calculated using the procedure indicated in subsection 7.3.4 and the Eq.7.55. The numerical values for the members 1, 2 and 3 are given below.

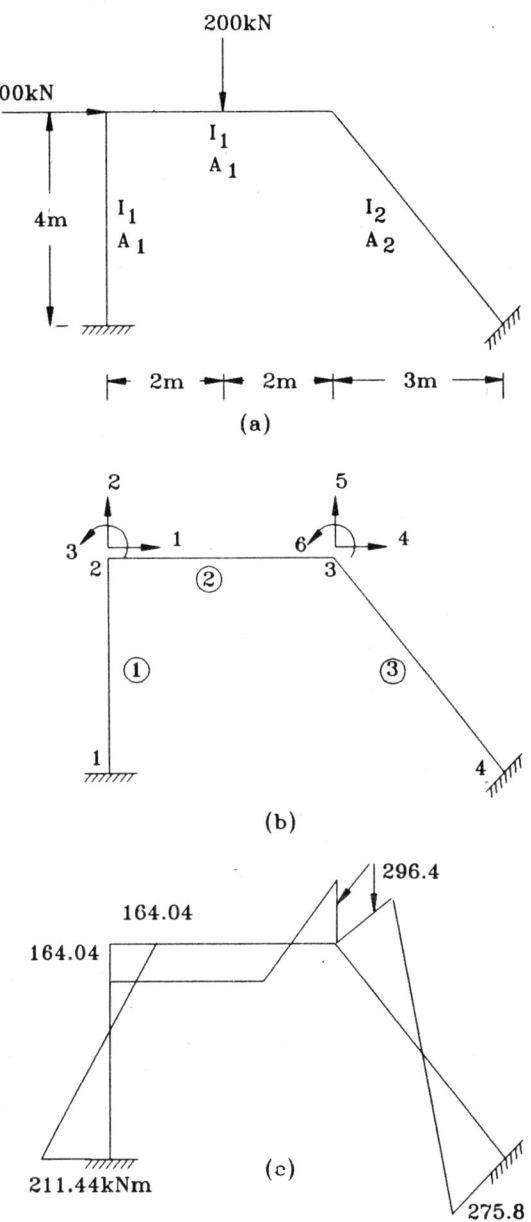

Fig. 7.13 Plane frame

$$E = 2 \times 10^7 \text{ kN/m}^2, \quad I_1 = 12 \times 10^{-5} \text{m}^4, \quad A_1 = 0.03\text{m}^2$$

$$I_2 = 15 \times 10^{-5} \text{m}^4, \quad A_2 = 0.035\text{m}^2$$

Member 1

$$[k_m] = \begin{bmatrix} 15 \times 10^4 & 0 & 0 & -15 \times 10^4 & 0 & 0 \\ 0 & 450 & 900 & 0 & -450 & 900 \\ 0 & 900 & 2400 & 0 & -900 & 1200 \\ -15 \times 10^4 & 0 & 0 & 15 \times 10^4 & 0 & 0 \\ 0 & -450 & -900 & 0 & 450 & -900 \\ 0 & 900 & 1200 & 0 & -900 & 2400 \end{bmatrix} \qquad (a)$$

$$C_x = 0, \quad C_y = 1.0$$

$$[T] = \begin{bmatrix} 0 & 1 & 0 & 0 & 0 & 0 \\ -1 & 0 & 0 & 0 & 0 & 0 \\ 0 & 0 & 1 & 0 & 0 & 0 \\ 0 & 0 & 0 & 0 & 1 & 0 \\ 0 & 0 & 0 & -1 & 0 & 0 \\ 0 & 0 & 0 & 0 & 0 & 1 \end{bmatrix} \qquad (b)$$

The stiffness matrix [k] in global axes given by,

$$[k_1] = [T]^T [k_m] [T]$$

| g | d | o | f | 0 | 0 | 0 | 1 | 2 | 3 |
| e | d | o | f | 1 | 2 | 3 | 4 | 5 | 6 |

$$[k_1] = \begin{matrix} 0 & 1 \\ 0 & 2 \\ 0 & 3 \\ 1 & 4 \\ 2 & 5 \\ 3 & 6 \end{matrix} \begin{bmatrix} 450 & 0 & -900 & -450 & 0 & -900 \\ 0 & 15 \times 10^4 & 0 & 0 & -15 \times 10^4 & 0 \\ -900 & 0 & 2400 & 900 & 0 & 1200 \\ -450 & 0 & 900 & 450 & 0 & 900 \\ 0 & -15 \times 10^4 & 0 & 0 & 15 \times 10^4 & 0 \\ -900 & 0 & 1200 & 900 & 0 & 2400 \end{bmatrix}$$

$$\begin{matrix} g & e \\ d & d \\ o & o \\ f & f \end{matrix} \qquad (c)$$

Member 2

Since the member is oriented in the global directions, no transformation is required. The member stiffness matrix in the local directions is the same as in the case of member 1.

Thus,

$$
\begin{array}{cccccccccc}
g & d & o & f & & 1 & 2 & 3 & 4 & 5 & 6 \\
e & d & o & f & & 1 & 2 & 3 & 4 & 5 & 6
\end{array}
$$

$$
[k_2] =
\begin{array}{cc}
i & 1 \\
2 & 2 \\
3 & 3 \\
4 & 4 \\
5 & 5 \\
6 & 6
\end{array}
\left[
\begin{array}{cccccc}
15 \times 10^4 & 0 & 0 & -15 \times 10^4 & 0 & 0 \\
0 & 450 & 900 & 0 & -450 & 900 \\
0 & 900 & 2400 & 0 & -900 & 1200 \\
-15 \times 10^4 & 0 & 0 & 15 \times 10^4 & 0 & 0 \\
0 & -450 & -900 & 0 & 450 & -900 \\
0 & 900 & 1200 & 0 & -900 & 2400
\end{array}
\right]
$$

$$(d)$$

$$
\begin{array}{cc}
g & e \\
d & d \\
o & o \\
f & f
\end{array}
$$

The element load vector also does not require any transformation.

$$
\{Q_2\} =
\left\{
\begin{array}{c}
0 \\
-\dfrac{WL}{2} \\
-\dfrac{WL}{8} \\
0 \\
-\dfrac{WL}{2} \\
\dfrac{WL}{8}
\end{array}
\right\}
=
\left\{
\begin{array}{c}
0 \\
-100 \\
-100 \\
0 \\
-100 \\
100
\end{array}
\right\}
$$

$$(e)$$

Member 3

$$
[k_m] =
\left[
\begin{array}{cccccc}
14 \times 10^4 & 0 & 0 & -14 \times 10^4 & 0 & 0 \\
0 & 288 & 720 & 0 & -288 & 720 \\
0 & 720 & 2400 & 0 & -720 & 1200 \\
-14 \times 10^4 & 0 & 0 & 14 \times 10^4 & 0 & 0 \\
0 & -288 & -720 & 0 & 288 & -720 \\
0 & 720 & 1200 & 0 & -720 & 2400
\end{array}
\right]
$$

$$(f)$$

$$
C_x = 0.6, \quad C_y = -0.8
$$

$$[T] = \begin{bmatrix} 0.6 & -0.8 & 0 & 0 & 0 & 0 \\ 0.8 & 0.6 & 0 & 0 & 0 & 0 \\ 0 & 0 & 1 & 0 & 0 & 0 \\ 0 & 0 & 0 & 0.6 & -0.8 & 0 \\ 0 & 0 & 0 & 0.8 & 0.6 & 0 \\ 0 & 0 & 0 & 0 & 0 & 1 \end{bmatrix} \qquad (g)$$

$$[k_3] = [T]^T [k_m] [T]$$

| g | d | o | f | 4 | 5 | 6 | 0 | 0 | 0 |
| e | d | o | f | 1 | 2 | 3 | 4 | 5 | 6 |

$$[k_3] = \begin{matrix} 4 & 1 \\ 5 & 2 \\ 6 & 3 \\ 0 & 4 \\ 0 & 5 \\ 0 & 6 \end{matrix} \begin{bmatrix} 50584 & -67062 & 576 & -50584 & 67062 & 576 \\ -67062 & 89704 & 432 & 67062 & -89704 & 432 \\ 576 & 432 & 2400 & 576 & 432 & 1200 \\ -50584 & 67062 & 576 & 50584 & -67062 & -576 \\ 67062 & -89704 & 432 & -67062 & 89704 & -432 \\ 576 & 432 & 1200 & -576 & -432 & 2400 \end{bmatrix}$$

$$(h)$$

$$\begin{matrix} g & e \\ d & d \\ o & o \\ f & f \end{matrix}$$

The stiffness contribution from each member is added to get the stiffness matrix of the structure following the procedure described in Chapter 5. The equilibrium equation for the frame is given by,

$$\begin{bmatrix} 150450 & 0 & 900 & -150000 & 0 & 0 \\ 0 & 150450 & 900 & 0 & -450 & 900 \\ 900 & 900 & 4800 & 0 & -900 & 1200 \\ -150000 & 0 & 0 & 200584 & -67062 & 576 \\ 0 & -450 & -900 & -67062 & 90154 & -468 \\ 0 & 900 & 1200 & 576 & -468 & 4800 \end{bmatrix} \begin{Bmatrix} r_1 \\ r_2 \\ r_3 \\ r_4 \\ r_5 \\ r_6 \end{Bmatrix}$$

$$= \begin{Bmatrix} 400.0 \\ -100.0 \\ -100.0 \\ 0.0 \\ -100.0 \\ 100.0 \end{Bmatrix} \qquad (i)$$

Solving the equation (i) we get,

$$\{r\} = \begin{Bmatrix} 0.2876 \\ 0.00010073 \\ -0.0395 \\ 0.2856 \\ 0.2107 \\ 0.0169 \end{Bmatrix} \tag{j}$$

Member End Forces

The stress resultants at the ends of the members are calculated using the Eq.7.63,

$$\{S\} = [k_m]\{d_m\} + \{S_o\}$$

where

$$\{d_m\} = [T]\{d\}$$

Member 1

$$\{d_m\} = \begin{bmatrix} 0 & 1 & 0 & 0 & 0 & 0 \\ -1 & 0 & 0 & 0 & 0 & 0 \\ 0 & 0 & 1 & 0 & 0 & 0 \\ 0 & 0 & 0 & 0 & 1 & 0 \\ 0 & 0 & 0 & -1 & 0 & 0 \\ 0 & 0 & 0 & 0 & 0 & 1 \end{bmatrix} \begin{Bmatrix} 0 \\ 0 \\ 0 \\ 0.2876 \\ 1.0073 \times 10^{-4} \\ -0.0395 \end{Bmatrix}$$

$$= \begin{Bmatrix} 0 \\ 0 \\ 0 \\ 1.0073 \times 10^{-4} \\ -0.2876 \\ -0.0395 \end{Bmatrix} \tag{k}$$

$$\{S\} = \begin{bmatrix} 15 \times 10^4 & 0 & 0 & -15 \times 10^4 & 0 & 0 \\ 0 & 450 & 900 & 0 & -450 & 900 \\ 0 & 900 & 2400 & 0 & -900 & 1200 \\ -15 \times 10^4 & 0 & 0 & 15 \times 10^4 & 0 & 0 \\ 0 & -450 & -900 & 0 & 450 & -900 \\ 0 & 900 & 1200 & 0 & -900 & 2400 \end{bmatrix}$$

$$\begin{Bmatrix} 0 \\ 0 \\ 0 \\ 1.0073 \times 10^{-4} \\ -0.2876 \\ -0.0395 \end{Bmatrix} + \begin{Bmatrix} 0 \\ 0 \\ 0 \\ 0 \\ 0 \\ 0 \end{Bmatrix} = \begin{Bmatrix} -15.11 \\ 93.87 \\ 211.44 \\ 15.11 \\ -93.87 \\ 164.04 \end{Bmatrix} \tag{l}$$

Member 2

Since the member axes are parallel to the global axes, $\{d_m\} = \{d\}$

$$\{S\} = \begin{bmatrix} 15 \times 10^4 & 0 & 0 & -15 \times 10^4 & 0 & 0 \\ 0 & 450 & 900 & 0 & -450 & 900 \\ 0 & 900 & 2400 & 0 & -900 & 1200 \\ -15 \times 10^4 & 0 & 0 & 15 \times 10^4 & 0 & 0 \\ 0 & -450 & -900 & 0 & 450 & -900 \\ 0 & 900 & 1200 & 0 & -900 & 2400 \end{bmatrix} \times$$

$$\begin{Bmatrix} 0.2876 \\ 1.0073 \times 10^{-4} \\ -0.0395 \\ 0.2856 \\ 0.2107 \\ 0.0169 \end{Bmatrix} + \begin{Bmatrix} 0.0 \\ 100.0 \\ 100.0 \\ 0.0 \\ 100.0 \\ -100.0 \end{Bmatrix} = \begin{Bmatrix} 300.0 \\ -15.1 \\ -164.06 \\ -300.0 \\ 215.1 \\ -296.4 \end{Bmatrix} \qquad (m)$$

Member 3

$$\{d_m\} = \begin{bmatrix} 0.6 & -0.8 & 0 & 0 & 0 & 0 \\ 0.8 & 0.6 & 0 & 0 & 0 & 0 \\ 0 & 0 & 1 & 0 & 0 & 0 \\ 0 & 0 & 0 & 0.6 & -0.8 & 0 \\ 0 & 0 & 0 & 0.8 & 0.6 & 0 \\ 0 & 0 & 0 & 0 & 0 & 1 \end{bmatrix} \begin{Bmatrix} 0.2856 \\ 0.2107 \\ 0.0169 \\ 0.0 \\ 0.0 \\ 0.0 \end{Bmatrix} = \begin{Bmatrix} 2.8 \times 10^{-3} \\ 0.3549 \\ 0.0169 \\ 0.0 \\ 0.0 \\ 0.0 \end{Bmatrix}$$

$$(n)$$

$$\{S\} = \begin{bmatrix} 14 \times 10^4 & 0 & 0 & -14 \times 10^4 & 0 & 0 \\ 0 & 288 & 720 & 0 & -288 & 720 \\ 0 & 720 & 2400 & 0 & -720 & 1200 \\ -14 \times 10^4 & 0 & 0 & 14 \times 10^4 & 0 & 0 \\ 0 & -288 & -720 & 0 & 288 & -720 \\ 0 & 720 & 1200 & 0 & -720 & 2400 \end{bmatrix} \times$$

$$\begin{Bmatrix} 2.8 \times 10^{-3} \\ 0.3549 \\ 0.0169 \\ 0.0 \\ 0.0 \\ 0.0 \end{Bmatrix} = \begin{Bmatrix} +392.0 \\ +114.4 \\ 296.1 \\ -392.0 \\ -114.4 \\ 275.8 \end{Bmatrix} \qquad (o)$$

The final bending moment diagram for the frame is shown in Fig.7.13c.

7.4 Three-Dimensional Beam Element

The stiffness coefficients due to axial deformation and due to bending in one of the planes have been described earlier. In the following section the stiffness coefficients due to torsion will be derived and it will be shown later that the stiffness matrix of the three-dimensional beam element can be evaluated using these coefficients.

7.4.1 Stiffness Coefficients due to Torsion

Consider the beam shown in Fig. 7.14.

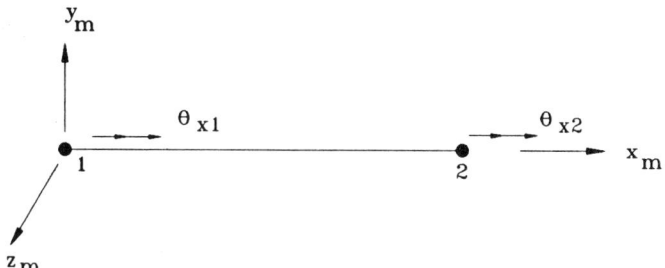

Fig. 7.14 Beam element subjected to torsion

Due to torsion there will be twists at the nodes 1 and 2 denoted as θ_{x1} and θ_{x2}. The twist along the length of the member varies linearly and is similar to the case of axial displacement described earlier. The twist of any point can be expressed through the linear interpolation function as

$$\theta_x = L_1\,\theta_{x1} + L_2\,\theta_{x2} \qquad (7.64)$$

where L_1 and L_2 are the natural coordinates of a point given by Eq.3.14. The rate of twist α, along the length of the member is given by

$$\alpha = \frac{d\theta_x}{dx} = \frac{1}{L}(\theta_{x2} - \theta_{x1}) \qquad (7.65)$$

The torsional moment M_x can be expressed as

$$M_x = GI_x\alpha = GI_x\frac{1}{L}(\theta_{x2} - \theta_{x1}) \qquad (7.66)$$

where I_x is the torsional constant and is equal to the polar moment of inertia for circular members. And G is the shear modulus.

Treating the torsional moment, M_x, as similar to the stress σ, the [B] and [C] matrices are given by,

$$[C] = GI_x$$

$$[B] = \frac{1}{L}[-1 \; +1] \tag{7.67}$$

The stiffness matrix for the beam element due to torsion only can now be evaluated as

$$[k_m] = \int_0^l [B]^T \, [C] \, [B] \, dl \tag{7.68}$$

and it may be observed that it has been reduced to a line integral as the integration across the area of cross-section has been carried out in evaluating the torsional constant I_x, and correspondingly the stress resultant M_x has been considered.

Thus substituting Eqs.7.67 into Eq.(7.68), we get,

$$[k_m] = \frac{GI_x}{L^2} \int_l \left\{ \begin{array}{c} -1 \\ 1 \end{array} \right\} [-1 \; +1] \, dl$$

$$= \frac{GI_x}{L} \begin{bmatrix} +1 & -1 \\ -1 & +1 \end{bmatrix} \tag{7.69}$$

7.4.2 Stiffness Matrix for a Three-Dimensional Beam Element

Consider a three-dimensional prismatic beam element with the member axes as shown in Fig.7.15. The x_m axis coincides with the centroidal axis of the member and is positive from i to j. The y_m and z_m axes are chosen such that $x_m - y_m$ and $x_m - z_m$ planes are principal planes of bending.

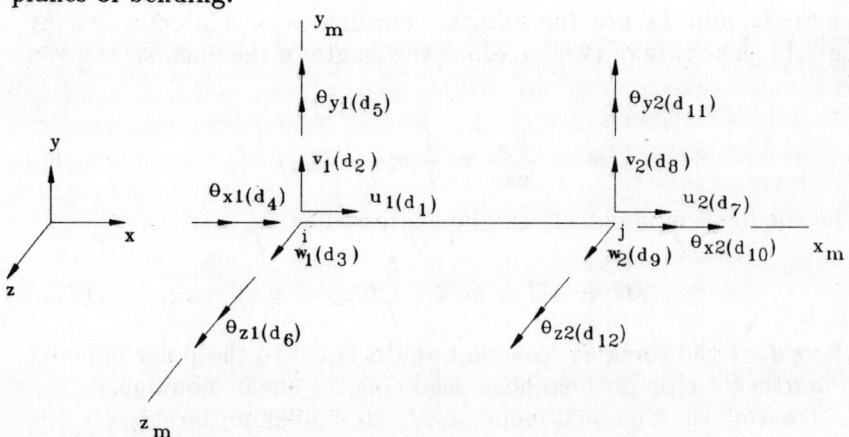

Fig. 7.15 Three-dimensional beam element

The member axes x_m, y_m and z_m are parallel to the global axes and the degrees of freedom in global and member axes system are the same as shown in Fig.7.15. But it will be shown later that for an arbitrarily inclined three-dimensional beam, the member stiffness matrix can be transformed to the global system.

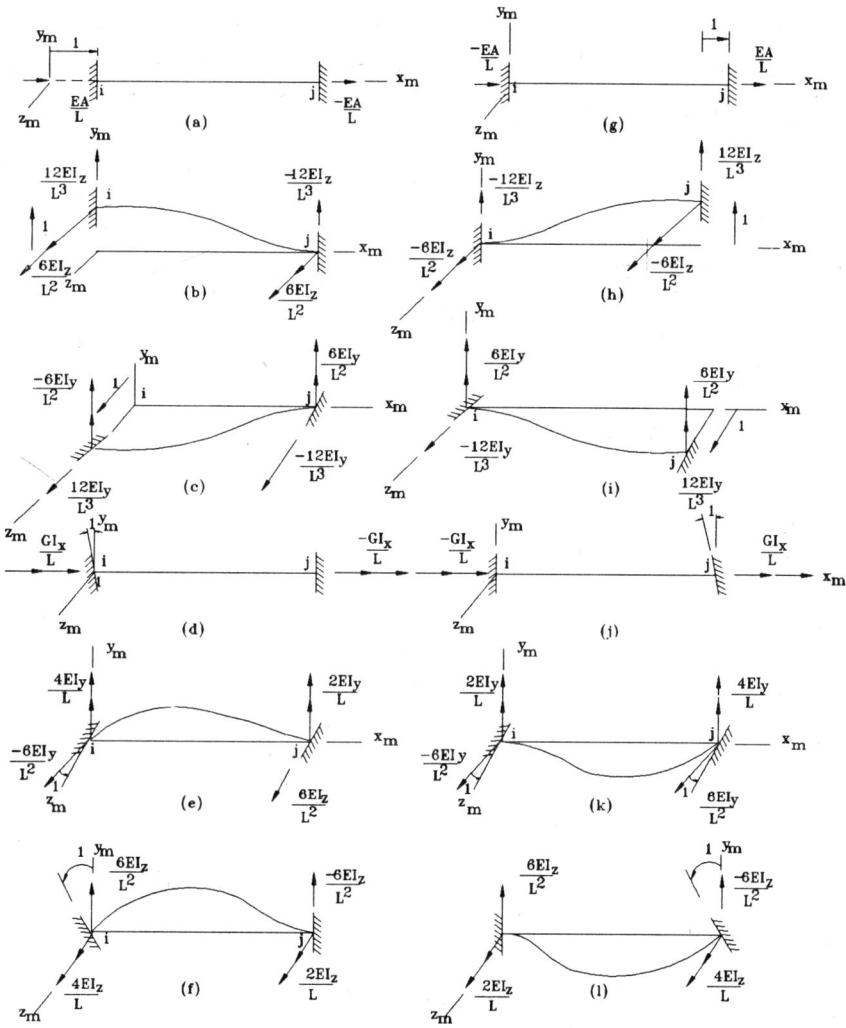

Fig. 7.16 Three-dimensional beam element-evaluation of stiffness coefficients

Figure 7.16 shows the evaluation of the stiffness coefficients of a three-dimensional beam by imposing unit displacements along the twelve possible degrees of freedom at the ends i and j. Figures 7.16 (a), (b), (f), (g), (h) and (l) show the unit values of end displacements along the degrees of freedom d_1, d_2, d_6, d_7, d_8 and d_{12} directions and it can be observed that they are of the same nature as those of the two-dimensional beam whose stiffness coefficients are shown in Fig.7.8.

The displacements along d_3, d_5, d_9 and d_{11} as shown in Figs. 7.16(c), (e), (i) and (k) correspond to bending in $x_m - z_m$ plane, and the stiffness coefficients can be written down as they are similar to two-dimensional beam. The stiffness coefficients due to unit rotation about the x_m axis at i and j ends, i.e., along d_4 and d_{10} directions, are calculated using Eq.7.69.

Now the stiffness coefficients shown in Fig.7.16 can be arranged in the matrix form given in Eq.7.70.

7.4.3   Transformation Matrix

The stiffness matrix of a three-dimensional beam element $[k_m]$ with reference to the member axes x_m, y_m and z_m has been derived in the earlier section. In general the member may be arbitrarily oriented in space as shown in Fig.7.17. Hence, it is required to transform the element stiffness matrix to correspond to the global degrees of freedom as shown in Fig.7.18.

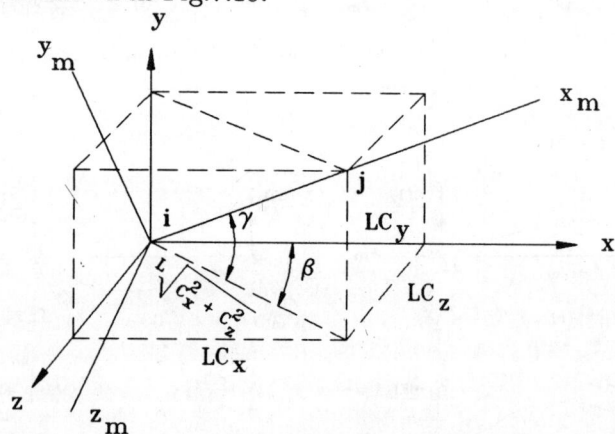

Fig. 7.17 Orientation of beam element with respect to global axes

For a truss element the position of the two end nodes fully defines the orientation of the element. In the case of a three-dimensional beam element the line joining i and j define the x_m axis, but the y_m and z_m axes should be along the principal planes of bending and hence their orientation has to be specified. A third node k is used to define these principal planes of bending.

$$(7.70)$$

$$[k_m] = \begin{array}{c} \\ 1 \\ 2 \\ 3 \\ 4 \\ 5 \\ 6 \\ 7 \\ 8 \\ 9 \\ 10 \\ 11 \\ 12 \end{array} \begin{bmatrix} \frac{EA}{L} & 0 & 0 & 0 & 0 & 0 & -\frac{EA}{L} & 0 & 0 & 0 & 0 & 0 \\ 0 & \frac{12EI_z}{L^3} & 0 & 0 & 0 & \frac{6EI_z}{L^2} & 0 & -\frac{12EI_z}{L^3} & 0 & 0 & 0 & \frac{6EI_z}{L^2} \\ 0 & 0 & \frac{12EI_y}{L^3} & 0 & -\frac{6EI_y}{L^2} & 0 & 0 & 0 & -\frac{12EI_y}{L^3} & 0 & -\frac{6EI_y}{L^2} & 0 \\ 0 & 0 & 0 & \frac{GI_x}{L} & 0 & 0 & 0 & 0 & 0 & -\frac{GI_x}{L} & 0 & 0 \\ 0 & 0 & -\frac{6EI_y}{L^2} & 0 & \frac{4EI_y}{L} & 0 & 0 & 0 & \frac{6EI_y}{L^2} & 0 & \frac{2EI_y}{L} & 0 \\ 0 & \frac{6EI_z}{L^2} & 0 & 0 & 0 & \frac{4EI_z}{L} & 0 & -\frac{6EI_z}{L^2} & 0 & 0 & 0 & \frac{2EI_z}{L} \\ -\frac{EA}{L} & 0 & 0 & 0 & 0 & 0 & \frac{EA}{L} & 0 & 0 & 0 & 0 & 0 \\ 0 & -\frac{12EI_z}{L^3} & 0 & 0 & 0 & -\frac{6EI_z}{L^2} & 0 & \frac{12EI_z}{L^3} & 0 & 0 & 0 & -\frac{6EI_z}{L^2} \\ 0 & 0 & -\frac{12EI_y}{L^3} & 0 & \frac{6EI_y}{L^2} & 0 & 0 & 0 & \frac{12EI_y}{L^3} & 0 & \frac{6EI_y}{L^2} & 0 \\ 0 & 0 & 0 & -\frac{GI_x}{L} & 0 & 0 & 0 & 0 & 0 & \frac{GI_x}{L} & 0 & 0 \\ 0 & 0 & -\frac{6EI_y}{L^2} & 0 & \frac{2EI_y}{L} & 0 & 0 & 0 & \frac{6EI_y}{L^2} & 0 & \frac{4EI_y}{L} & 0 \\ 0 & \frac{6EI_z}{L^2} & 0 & 0 & 0 & \frac{2EI_z}{L} & 0 & -\frac{6EI_z}{L^2} & 0 & 0 & 0 & \frac{4EI_z}{L} \end{bmatrix}$$

$$\begin{array}{cccccccccccc} 1 & 2 & 3 & 4 & 5 & 6 & 7 & 8 & 9 & 10 & 11 & 12 \end{array}$$

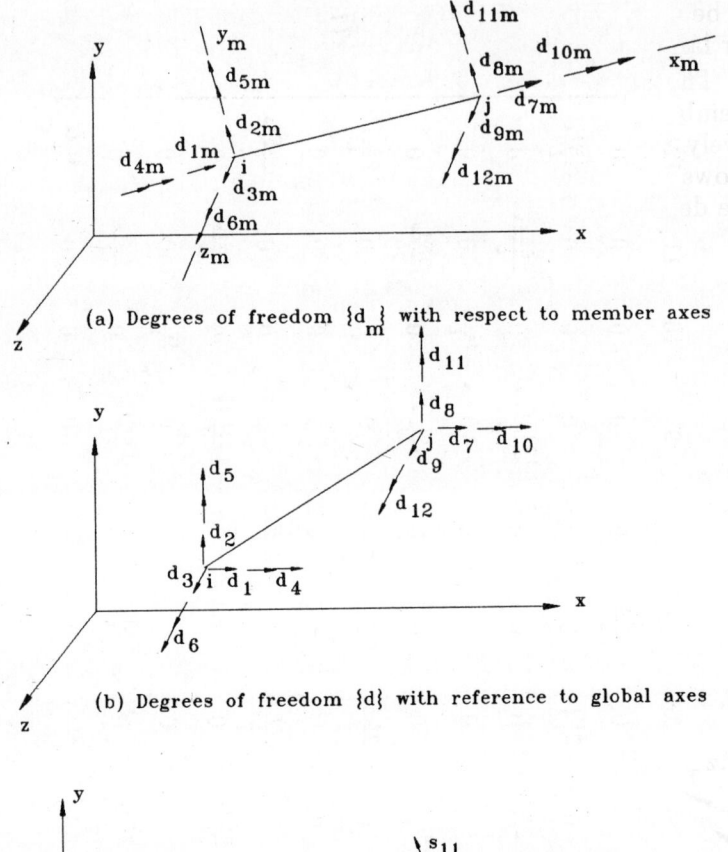

(a) Degrees of freedom $\{d_m\}$ with respect to member axes

(b) Degrees of freedom $\{d\}$ with reference to global axes

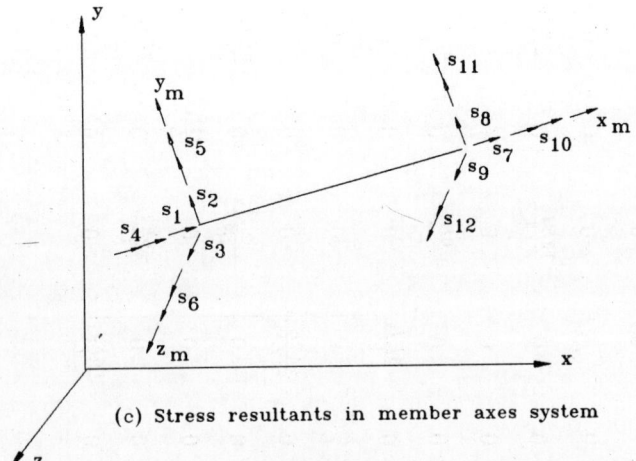

(c) Stress resultants in member axes system

Fig. 7.18 Three-dimensional beam element

The rotation transformation matrix can be worked out by either vector algebra principles which is explained in Chapter 11 for a similar situation or by successive rotation of axes. The second method is explained in this section and the concepts involved will also be found to be useful in the transformation of stiffness matrix of substructures for large scale analysis of structures (Chapter 12).

The global (structure) axes are brought to coincide with the local member axes by sequence of rotation about y, z and x axes respectively. This is refered to as $y - z - x$ transformation [5]. Figure 7.19 shows the three rotations of the global axes. The angles of rotation are denoted as β, γ and α respectively.

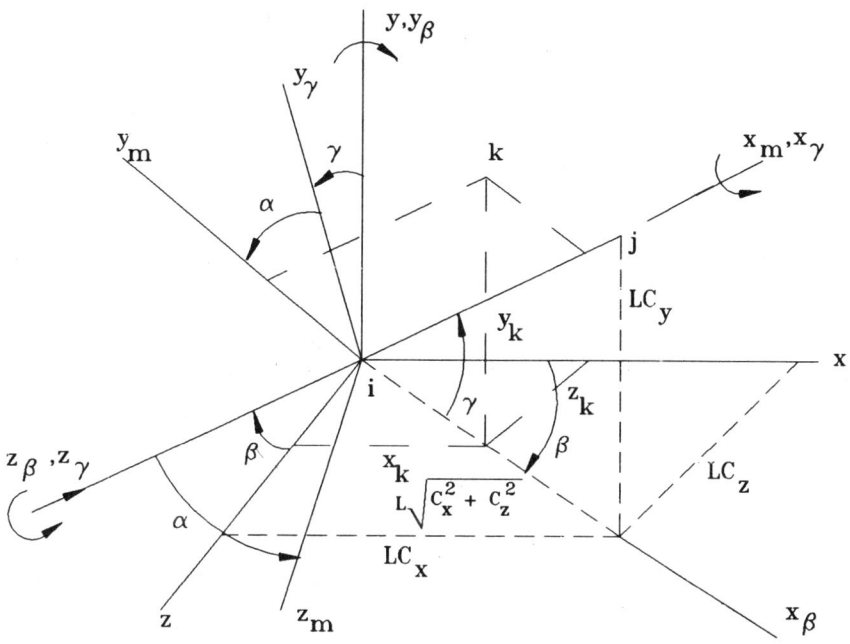

Fig. 7.19 Rotation transformation of axes for a 3-D beam element

The first rotation about the y-axis by an angle β, places x-axis in position denoted as x_β which is the projection of the axis of the member on $x - z$ plane i.e., x_β axis will coincide with the line of intersection of $x - z$ and $y - x_m$ planes. The z-axis is placed in the position z_β and y-axis remains unchanged. The vector components Q_1, Q_2, Q_3 in the $x - y - z$ axes can be resolved into vector components in the new $x_\beta, y_\beta, z_\beta$ axes and can be expressed as

$$\left\{ \begin{array}{c} Q_{1\beta} \\ Q_{2\beta} \\ Q_{3\beta} \end{array} \right\} = \left[\begin{array}{ccc} \cos\beta & 0 & \sin\beta \\ 0 & 1 & 0 \\ -\sin\beta & 0 & \cos\beta \end{array} \right] \left\{ \begin{array}{c} Q_1 \\ Q_2 \\ Q_3 \end{array} \right\} \tag{7.71}$$

where $cos\beta$ and $sin\beta$ can be expressed in terms of the direction cosines of the member which is aligned in the direction of x_m axis (Fig.7.19).

$$\cos\beta = \frac{C_x}{\sqrt{C_x^2 + C_z^2}} \tag{7.72a}$$

$$\sin\beta = \frac{C_z}{\sqrt{C_x^2 + C_z^2}} \tag{7.72b}$$

and $\qquad C_x = \dfrac{x_j - x_i}{L} \qquad C_y = \dfrac{y_j - y_i}{L}$

$$C_z = \frac{z_j - z_i}{L} \tag{7.73}$$

$$L = \sqrt{(x_j - x_i)^2 + (y_j - y_i)^2 + (z_j - z_i)^2}$$

Thus the transformation matrix $[T_\beta]$ for rotation about y-axis consists of direction cosines of $x_\beta, y_\beta, z_\beta$ axes with respect to the global axes x, y, z and is given by

$$[T_\beta] = \begin{bmatrix} \dfrac{C_x}{\sqrt{C_x^2 + C_z^2}} & 0 & \dfrac{C_z}{\sqrt{C_x^2 + C_z^2}} \\ 0 & 1 & 0 \\ \dfrac{-C_z}{\sqrt{C_x^2 + C_z^2}} & 0 & \dfrac{C_x}{\sqrt{C_x^2 + C_z^2}} \end{bmatrix} \tag{7.74}$$

Now rotate $x_\beta, y_\beta, z_\beta$ axes about the z_β axis through the angle γ so that new axes $x_\gamma, y_\gamma, z_\gamma$ will be such that x_γ will coincide with the longitudinal member axis x_m, the z_γ axis will be the same as z_β and the $y_\gamma - z_\gamma$ plane will contain the y_m and z_m axes of the member. The vector components $Q_{1\beta}, Q_{2\beta}, Q_{3\beta}$ with respect to $x_\beta, y_\beta, z_\beta$ axes can now be resolved into components $Q_{1\gamma}, Q_{2\gamma}, Q_{3\gamma}$ with respect to $x_\gamma, y_\gamma, z_\gamma$ axes system as

$$\begin{Bmatrix} Q_{1\gamma} \\ Q_{2\gamma} \\ Q_{3\gamma} \end{Bmatrix} = \begin{bmatrix} cos\gamma & sin\gamma & 0 \\ -sin\gamma & cos\gamma & 0 \\ 0 & 0 & 1 \end{bmatrix} \begin{Bmatrix} Q_{1\beta} \\ Q_{2\beta} \\ Q_{3\beta} \end{Bmatrix} \tag{7.75}$$

Thus the transformation matrix corresponding to this rotation is

$$[T_\gamma] = \begin{bmatrix} cos\gamma & sin\gamma & 0 \\ -sin\gamma & cos\gamma & 0 \\ 0 & 0 & 1 \end{bmatrix} \tag{7.76}$$

as observed earlier the terms in $[T_\gamma]$ are the direction cosines of $x_\gamma, y_\gamma, z_\gamma$ with respect to $x_\beta, y_\beta, z_\beta$ axes. Also it can be shown that (Fig.7.19)

$$cos\gamma = \sqrt{C_x^2 + C_z^2} \qquad (7.77)$$

$$sin\gamma = C_y$$

Thus we get

$$[T_\gamma] = \begin{bmatrix} \sqrt{C_x^2 + C_z^2} & C_y & 0 \\ -C_y & \sqrt{C_x^2 + C_z^2} & 0 \\ 0 & 0 & 1 \end{bmatrix} \qquad (7.78)$$

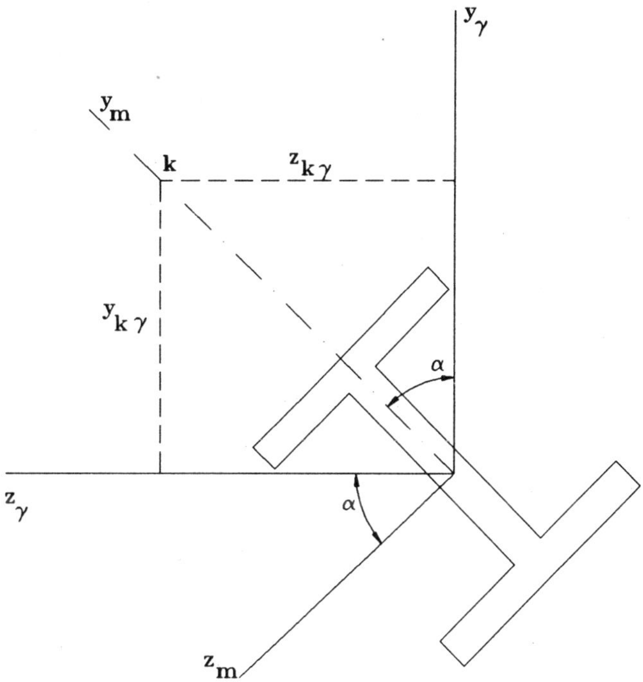

Fig. 7.20 Rotation of 3-D beam member about x_m axis

Finally the axes $x_\gamma, y_\gamma, z_\gamma$ are rotated about the x_m axis so that x_γ and y_γ axes are brought to coincide with the y_m and z_m axes of the member. Figure 7.20 shows the last rotation with the cross-sectional view of the member looking in the negative x_m direction. The $x_m - y_\gamma$ plane is a vertical plane through the axis of the member and defining the angle between y_γ and y_m axes or between the z_γ and z_m axes as α (i.e., the angle α is measured from that plane to one of the principal axes of the cross-section), the vector components Q_{1m}, Q_{2m}, Q_{3m} with respect to x_m, y_m, z_m axes can now be expressed in terms of $Q_{1\gamma}, Q_{2\gamma}, Q_{3\gamma}$, as

$$\begin{Bmatrix} Q_{1m} \\ Q_{2m} \\ Q_{3m} \end{Bmatrix} = \begin{bmatrix} 1 & 0 & 0 \\ 0 & cos\ \alpha & sin\ \alpha \\ 0 & -sin\ \alpha & cos\ \alpha \end{bmatrix} \begin{Bmatrix} Q_{1\gamma} \\ Q_{2\gamma} \\ Q_{3\gamma} \end{Bmatrix} \qquad (7.79)$$

Thus the transformation matrix corresponding to this rotation is

$$[T_\alpha] = \begin{bmatrix} 1 & 0 & 0 \\ 0 & cos\ \alpha & sin\ \alpha \\ 0 & -sin\ \alpha & cos\ \alpha \end{bmatrix} \qquad (7.80)$$

It may be noted that the elements in $[T_\alpha]$ are the direction cosines of the member axes x_m, y_m, z_m with respect to the $x_\gamma, y_\gamma, z_\gamma$ axes.

Combining Eqs. 7.71, 7.75 and 7.79, the components of vector with reference to the global axes can be resolved into the components with reference to the local member axes. Thus,

$$\{Q'_m\} = [T_\alpha]\,[T_\gamma]\,[T_\beta]\,\{Q'\} \qquad (7.81)$$

$$\text{i.e.,} \quad \{Q'_m\} = [T']\,\{Q'\} \qquad (7.82a)$$

$$\text{where} \quad [T'] = [T_\alpha][T_\gamma][T_\beta] \qquad (7.82b)$$

Substituting for $[T_\beta], [T_\gamma]$ and $[T_\alpha]$ from Eqs.7.74, 7.78 and 7.80 we get

$$[T'] =$$

$$\begin{bmatrix} C_x & C_y & C_z \\ \dfrac{-C_x\,C_y\,cos\ \alpha\ -\ C_z\,sin\ \alpha}{\sqrt{C_x^2 + C_z^2}} & \sqrt{C_x^2 + C_z^2}\,cos\ \alpha & \dfrac{-C_y\,C_z\,cos\ \alpha\ +\ C_x\,sin\alpha}{\sqrt{C_x^2 + C_z^2}} \\ \dfrac{C_x C_y sin\alpha - C_z cos\alpha}{\sqrt{C_x^2+C_z^2}} & -\sqrt{C_x^2 + C_z^2}\,sin\alpha & \dfrac{C_y C_z sin\alpha + C_x\ cos\alpha}{\sqrt{C_x^2+C_z^2}} \end{bmatrix}$$

$$(7.83)$$

As there are twelve degrees of freedom for a three-dimensional beam element the rotation transformation matrix can now be expressed as

$$\{Q_m\} = [T]\,\{Q\}$$

$$\text{where} \quad [T] = \begin{bmatrix} [T'] & [0] & [0] & [0] \\ [0] & [T'] & [0] & [0] \\ [0] & [0] & [T'] & [0] \\ [0] & [0] & [0] & [T'] \end{bmatrix} \qquad (7.84)$$

Proceeding along the same lines of derivation for two-dimensional beam element in subsection 7.3.4 it can be shown that

$$[k] = [T]^T [k_m] [T] \tag{7.85}$$

where $[k]$ is the stiffness matrix of the element with reference to global system of axes and $[k_m]$ is the stiffness matrix with reference to the local member axes (Eq.7.70).

Also the nodal loads computed in member axes system can be transformed to the nodal loads in global system.

The matrix $[T']$ in Eq.7.83 is fully described except for the angle α. For this an additional node 'k' is chosen lying in the $y_m - x_m$ principal plane but not along the x_m axis. Let the coordinates of node k with respect to the global axes be x_k, y_k, z_k. Then the coordinates of the point k with respect to the node i in the global system will be

$$
\begin{aligned}
x_{ki} &= x_k - x_i \\
y_{ki} &= y_k - y_i \\
z_{ki} &= z_k - z_i
\end{aligned}
\tag{7.86}
$$

As shown in Fig.7.20 let the coordinates of point k with respect to $x_\gamma, y_\gamma, z_\gamma$ axes be $x_{k\gamma}, y_{k\gamma}$ and $z_{k\gamma}$. The coordinates with respect to γ system of axes can be evaluated by applying the first two transformations i.e., using Eq.7.74 and 7.76, thus

$$
\begin{Bmatrix} x_{k\gamma} \\ y_{k\gamma} \\ z_{k\gamma} \end{Bmatrix} = [T_\gamma][T_\beta] \begin{Bmatrix} x_{ki} \\ y_{ki} \\ z_{ki} \end{Bmatrix}
\tag{7.87a}
$$

$$
= \begin{bmatrix} \sqrt{C_x^2 + C_z^2} & C_y & 0 \\ -C_y & \sqrt{C_x^2 + C_z^2} & 0 \\ 0 & 0 & 1 \end{bmatrix} \begin{bmatrix} \dfrac{C_x}{\sqrt{C_x^2+C_z^2}} & 0 & \dfrac{C_z}{\sqrt{C_x^2+C_z^2}} \\ 0 & 1 & 0 \\ \dfrac{-C_z}{\sqrt{C_x^2+C_z^2}} & 0 & \dfrac{C_x}{\sqrt{C_x^2+C_z^2}} \end{bmatrix} \begin{Bmatrix} x_{ki} \\ y_{ki} \\ z_{ki} \end{Bmatrix}
\tag{7.87b}
$$

Simplifying we get,

$$x_{k\gamma} = C_x\, x_{ki} + C_y\, y_{ki} + C_z\, z_{ki}$$

$$y_{k\gamma} = \frac{-C_x\, C_y}{\sqrt{C_x^2 + C_z^2}}\, x_{ki} + \sqrt{C_x^2 + C_z^2}\, y_{ki} - \frac{C_y\, C_z}{\sqrt{C_x^2 + C_z^2}}\, z_{ki}$$

$$z_{k\gamma} = \frac{-C_z}{\sqrt{C_x^2 + C_z^2}} x_{ki} + \frac{C_x}{\sqrt{C_x^2 + C_z^2}} z_{ki} \qquad (7.88)$$

Thus the coordinates $y_{k\gamma}$ and $z_{k\gamma}$ locate point k in the $y_m - z_m$ plane. The angle α can now be calculated as (Fig. 7.20).

$$sin\alpha = \frac{z_{k\gamma}}{\sqrt{y_{k\gamma}^2 + z_{k\gamma}^2}}$$

$$cos\alpha = \frac{y_{k\gamma}}{\sqrt{y_{k\gamma}^2 + z_{k\gamma}^2}} \qquad (7.89)$$

Using Eq.7.88 the values of sin α and cos α can be computed. When these values are substituted in Eq.7.83 all the elements in the matrix $[T']$ are defined.

The above transformation is not generally applicable to a vertical member, i.e., a member which is parallel to the global y axis. In this case C_x and C_z have zero values and the transformation does not reduce to the correct form which will be described below.

In a general case of a vertical number, the principal axes of bending will be rotated about the x_m axis so that they form an angle α with the directions of the structure axes.

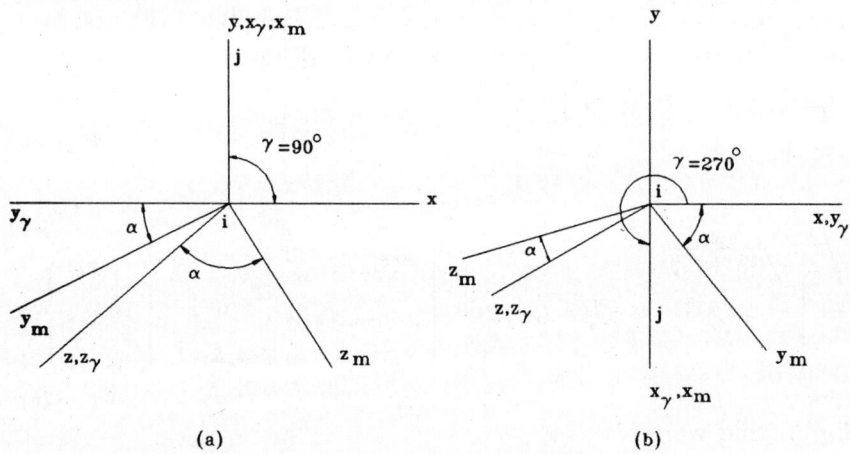

(a)   (b)

Fig. 7.21 Rotation of axes for a vertical 3-D beam member

Referring to Fig. 7.21 it can be seen that for a vertical member there are only two rotations. The first rotation is through an angle which is either 90° or 270° (Fig.7.21a and 7.21b) depending on whether C_y is +1 or −1. The second rotation is through the angle α about the x_m

axis. The rotation matrix for either of the cases shown in Fig.7.21 can be given by,

$$[T']_{vert} = \begin{bmatrix} 0 & C_y & 0 \\ -C_y \cos\alpha & 0 & \sin\alpha \\ C_y \sin\alpha & 0 & \cos\alpha \end{bmatrix} \qquad (7.90)$$

The angle α necessary to use Eq.7.90 can be calculated by locating a point k lying in a principal plane of bending. Figure 7.22(a) shows a vertical member with the lower end designated as node i and the other end designated as node j. The coordinates of the point k shown in the Fig.7.22 are positive and the angle α lies between $90°$ and $180°$. It can be observed that

$$\sin\alpha = \frac{z_k}{\sqrt{x_k^2 + z_k^2}} \quad \text{and} \quad \cos\alpha = \frac{-x_k}{\sqrt{x_k^2 + z_k^2}} \qquad (7.91)$$

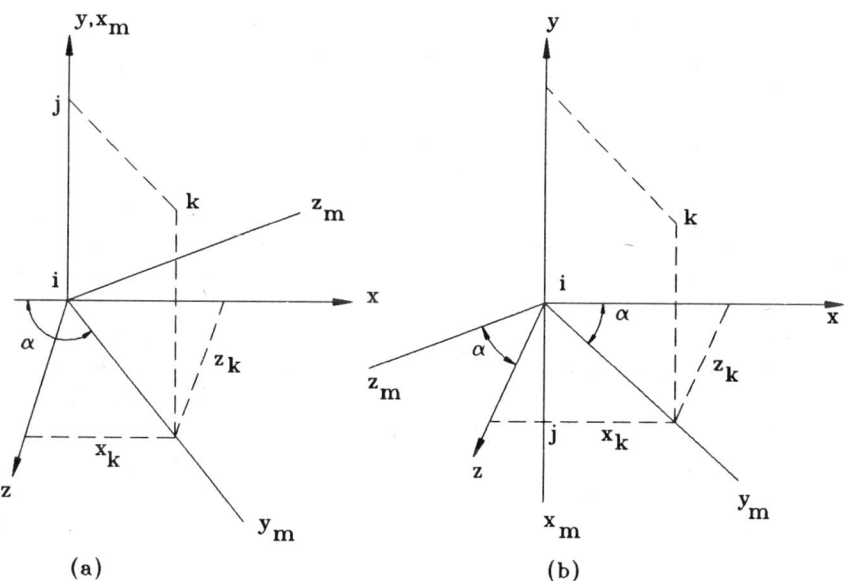

(a)　　　　　　　　　　　(b)

Fig. 7.22 Vertical 3-D beam element-Point k
and calculation of angle α

In Fig.7.22(b) the other alternative of the member with the upper end designated as i is also shown. In this case the angle α lies between $0°$ and $90°$ for the positive coordinates of the point k. Then,

$$\sin\alpha = \frac{z_k}{\sqrt{x_k^2 + z_k^2}} \quad \text{and} \quad \cos\alpha = \frac{x_k}{\sqrt{x_k^2 + z_k^2}} \qquad (7.92)$$

The Eqs. 7.91 and 7.92 can be combined as

$$sin\ \alpha\ =\ \frac{z_k}{\sqrt{x_k^2\ +\ z_k^2}}\ and\ cos\ \alpha\ =\ \frac{-x_k}{\sqrt{x_k^2\ +\ z_k^2}}\ C_y \qquad (7.93)$$

where C_y is the direction cosine of the member which has values $+1$ and -1 for cases shown in Fig.7.22a and Fig.7.22b respectively.

Thus for a vertical member the values of sin α and cos α evaluated using Eq.7.93 can be substituted in Eq.7.90 to compute the matrix $[T']$ and subsequently construct the rotation matrix $[T]$ of Eq.7.84.

7.5 Shear Deformation in Beams

In the simple theory of bending of beams it is assumed that plane sections before bending remain plane even after bending. When shear forces are present shear stresses are induced which in turn cause angular distortion of the element and as a consequence the plane sections no longer remain plane after deformation. Also, warping of the adjacent sections will be different in the problems of varying shear force and this invariably introduces longitudinal stresses and strains. The limitations of the simple theory of bending are now apparent. Not only does the shear force introduce longitudinal stresses as well as shear stresses, but our assumption that the longitudinal stresses vary linearly is no longer applicable and the parabolic distribution of the shear stresses will be affected by shear deformation. The designer is to be careful in situations where if shear deformation is neglected may lead to serious errors in the magnitude of the stresses calculated and the distribution across the depth. Structures which require consideration of the shear deformation include deep beams, flanged sections, thick plates and shells, and laminated plate constructions. In this section, the shear deformation in beams is briefly explained and a finite element formulation for a plane beam element is given. This concept is extended in Chapter 10 for application to plate bending problems.

Figure 7.23c shows an elementary slice of a rectangular beam (Fig. 7.23a) before deformation. The shear force causes this element to deform (or warp) to the shape shown in Fig.7.23d. It may be observed that axial strains no longer remain proportional to the distance from the neutral axis. Vertical strains are very small and variation of vertical strains between sections CC_1 and DD_1 are negligible. Hence, we may assume that layers parallel to the beam axis $M'N$ before deformation will still be parallel to the axis $M'N$ after deformation (Fig.7.23d). The deformation considered here is due to shear alone and we first compute the shape of the warped cross-section DND_1. Line ENE_1 is drawn perpendicular to the axis NM' and is also perpendicular to the top and bottom surfaces. The shear strain at a distance z from the neutral axis is given by [7].

$$\phi_{(z)} = \frac{q}{Gb} = \frac{6Q}{Gbh}\{\tfrac{1}{4} - (\tfrac{z}{h})^2\}$$

$$= \frac{6Q}{GA}\{\tfrac{1}{4} - (\tfrac{z}{h})^2\} \tag{7.94}$$

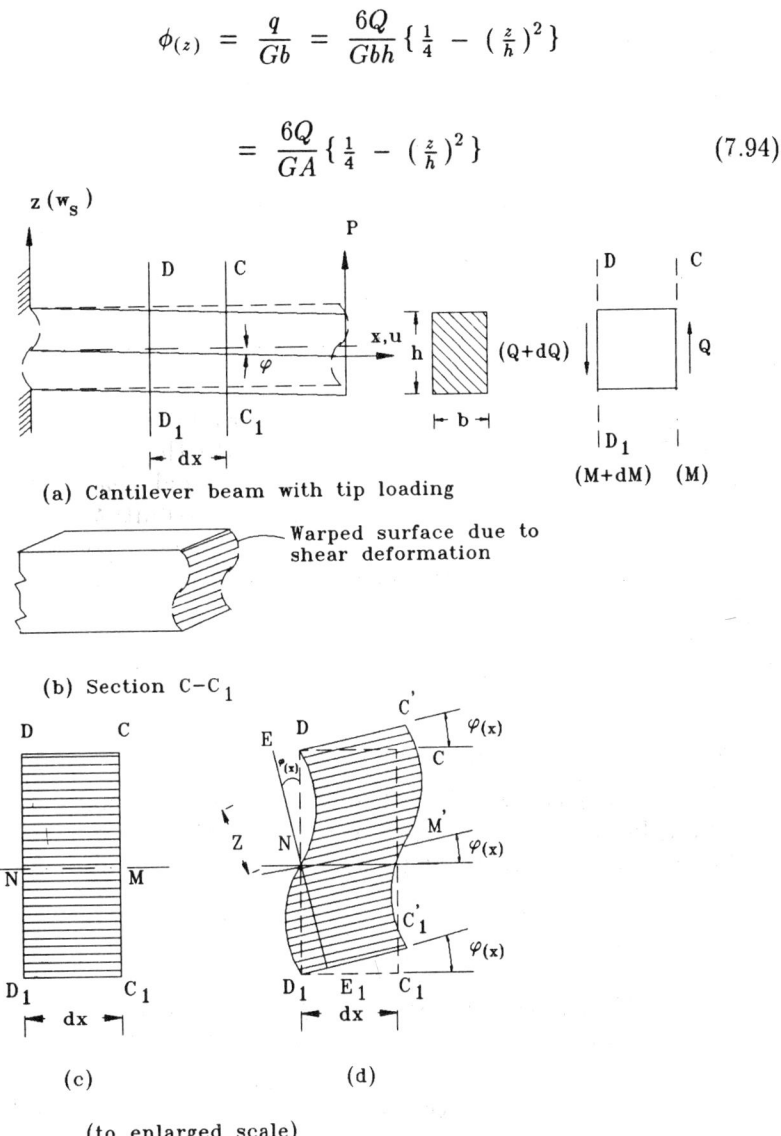

(a) Cantilever beam with tip loading

(b) Section $C-C_1$

Warped surface due to shear deformation

(c)

(d)

(to enlarged scale)

Fig. 7.23 Shear deformation in beams

As shown in Fig. 7.23d,

$$g = \int_o^z \phi_{(z)} dz = \frac{6Q}{GA} \left\{ \frac{z}{4} - \frac{z^3}{3h^2} \right\} \tag{7.95}$$

$$g \cdot = g_{max} \quad \text{at} \quad z = \frac{h}{2}$$

$$g_{max} = DE = \frac{Qh}{2GA}$$

$$\phi_{(x)} = \frac{DE}{h/2} = \frac{Q}{GA} \tag{7.96}$$

where $\phi_{(x)}$ is the change in angle between the cross-section and the centroidal axis, and may be regarded as an overall shear strain at the given cross-section.

The shear stress τ_{xz}, in general, can be expressed as,

$$\tau_{xz} = G \phi (x, z) \tag{7.97a}$$

where $\phi(x, z)$ is the shear strain at a distance x and depth z. However the use of general Eq.7.97a would complicate the problem and a simplified approach is possible in one dimensional beam. In order to account for the non-uniform shear stress distribution at a cross-section while still retaining the one dimensional approach we can modify the Eq.7.97a by introducing a shear correction factor α as follows,

$$\tau_{xz} = \alpha[G \phi_{(x)}] \tag{7.97b}$$

$$\text{and} \quad Q = \tau_{xz} A = \alpha AG \phi_{(x)} \tag{7.98}$$

The shear correction factor, α, is a function of the cross-sectional shape and Poisson's ratio μ. The term (αA) is the 'shear area' of the section associated with shear parallel to the z axis and denoting it by A_s we have

$$A_s = \alpha A, \quad \alpha < 1.0 \tag{7.99}$$

Values of α for various cross-sectional shapes are given in references [8,9,10]. The values of α for a rectangular section is 5/6.

$$\text{Hence,} \quad \phi_{(x)} = \frac{Q}{GA_s} \tag{7.100}$$

For short stubby beams the effect of shear deformation can not be ignored. Timoshenko beam theory [10] provides a means of accounting for the effects of shear deformation in a simple manner. Figure

7.24(a) shows a cantilever with an end force P. Any shear deformation will depend entirely on the boundary conditions assumed at the support. The fixity at the support may be interpreted in the following two ways.

(a) The axis of the beam is prevented from rotation at $x = 0, z = 0$, i.e., as shown in Fig.7.24b,

$$\left(\frac{\partial w}{\partial x}\right) = 0 \qquad (7.101)$$

where w = total deflection due to combined bending and shear.

(b) A vertical line element GH (Fig.7.24a) at $x = 0, z = 0$, is restricted against rotation in which case the beam axis rotates by an angle equal to the shear strain, $\phi_{(x)}$, i.e., as shown in Fig.7.24c.

$$\frac{\partial u}{\partial z} = 0 \qquad (7.102)$$

The above two possibilities represent extreme situations. The actual condition depends on the nature of the support i.e., of the adjoining member since, in general, members are joined so that their end faces match, it is the inclination of the end face rather than the rotation of the end tangent of the member axis is of interest (Fig.7.24d). This situation is further complicated by the fact that the strain distribution in the vicinity of the support cannot be computed from elementary beam theory, as the deformed shape of the cross-section at the support is not well defined. So, a simplified assumption is required to overcome this difficulty. The axis of the beam is assumed to undergo a rotation which is equal to the shear deformation $\phi_{(x)}$. That is, line elements GH normal to the center line (Fig.7.24a) remain vertical but the axis of the beam itself rotates through an angle equal to $\phi_{(x)}$. Accordingly $\phi_{(x)}$ gives the shear angle (or shear deformation) at points along the center line considering the effect of shear alone and referring to Fig.7.23(a).

$$\phi_{(x)} = \frac{dw_s}{dx} \qquad (7.103)$$

where w_s is the deflection due to shear and it means that the shear angle is a measure of the change in slope of the beam axis due to shear. Substituting from Eq.7.101, we get,

$$\frac{Q}{GA_s} = \frac{dw_s}{dx}$$

$$\text{or} \quad \frac{dQ}{dx} = q_x = GA_s = \frac{d^2w_s}{dx^2} \qquad (7.104)$$

where q_x is the intensity of loading.

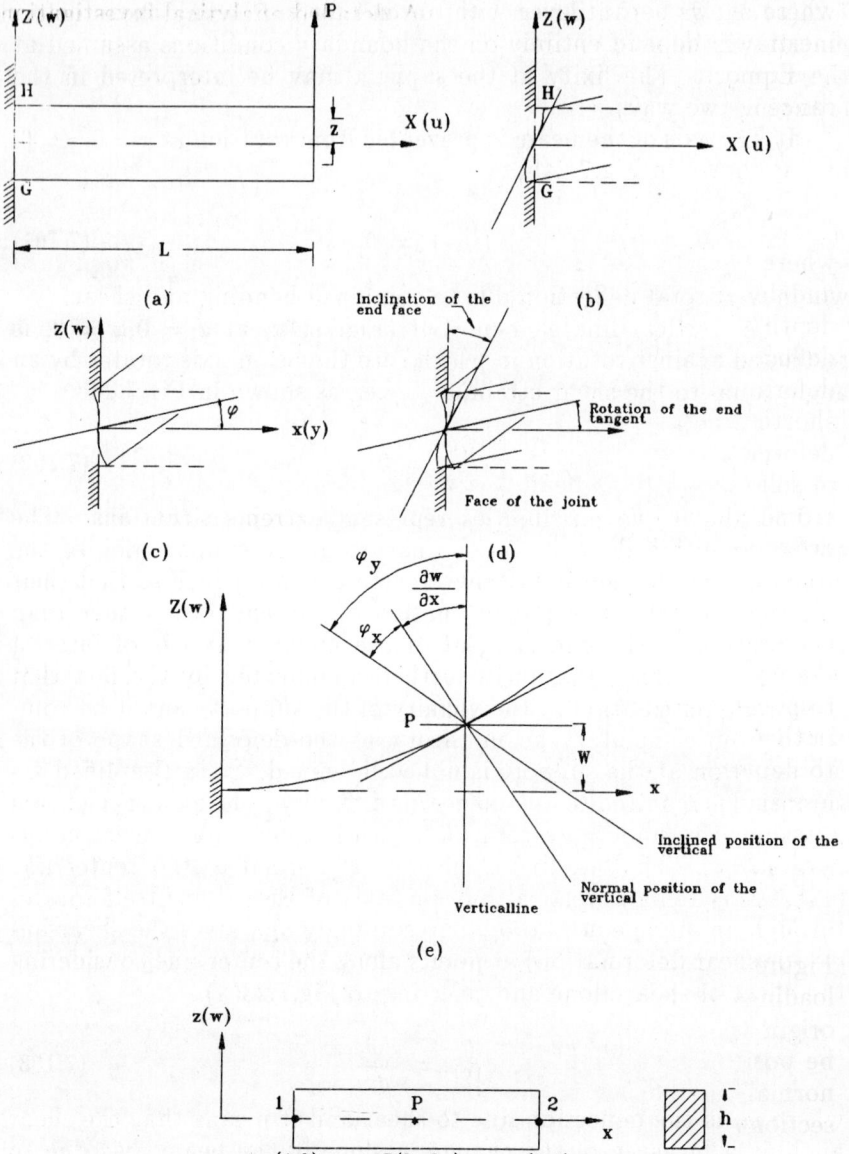

Fig. 7.24 Timoshenko beam theory

It is also known from simple theory of bending that

$$M_x = EI_y \frac{d^2 w_b}{dx^2} \tag{105}$$

where w_b is the deflection due to bending and I_y is the moment of inertia of the section about the y-axis.

Equations 7.104 and 7.105 can be utilised to evaluate the significance of shear deformation.

It has been shown in reference [11], that

$$\frac{w_s}{w_b} = C \left(\frac{h}{L}\right)^2 \tag{106}$$

where C is a constant depending on E, G, I, A_s, and type of beam and loading. The value of w_s/w_b is proportional to the square of depth/span ratio. For long slender beams (say $h/L = 1/15$), the deflection due to shear will be negligibly small. Therefore, the shear deformation becomes increasingly significant as the beam becomes shorter and deeper. It has also been shown that the effect of shear deformation in flanged sections is more pronounced than in the case of solid rectangular beam [7]. It should also be noted here that the cross-sectional details of beam are of importance in assessing the deflections due to shear.

When $L/h = 15$ (a common value), the shear deflection is less than 2% of the bending deflection [11]. In the case of simply supported deeper beams with $L/h = 8$, the corresponding values are approximately 3% and 6% for the distributed and central concentrated loads respectively. Thus the effect of shear deformation can be neglected in the slender beam range without any serious error. But as the span to depth ratio of a beam decreases, the simple theory of bending is less accurate and a more rigorous analysis is necessary.

7.5.1 Modified Strain Energy to Include Shear Deformation

Figure 7.24(e) shows the deformed shape of a beam under arbitrary loading. Let w be the deflection at any point P distant x from the origin O. Assuming anti-clockwise rotation of the cross-section to be positive, and denoting the additional rotation of the midsurface normal due to shear alone as $\phi_{(x)}$, the total rotation, θ_y, of the cross-section can be expressed as

$$\theta_y = \phi_{(x)} + \frac{dw}{dx} \tag{107}$$

$$\phi_{(x)} = \theta_y - \frac{dw}{dx} \tag{108}$$

Since $Q = GA_s \phi$, substituting for $\phi_{(x)}$ from Eq.7.108 we get

$$Q = GA_s \left(\theta_y - \frac{dw}{dx}\right)$$

Denoting strain energy due to shear deformation as U_s,

$$U_s = \int_o^L \frac{Q^2}{2GA_s} \, dx$$

$$= \frac{1}{2}\alpha \, GA \int_o^L \left(\theta_y - \frac{dw}{dx}\right)^2 \, dx \qquad (7.109)$$

The curvature of the deflected beam is $d\theta_y/dx$ and the strain energy due to bending is given by

$$U_b = \frac{1}{2}EI \int_o^L \left(\frac{d\theta_y}{dx}\right)^2 \, dx \qquad (7.110)$$

Let U be the total strain energy due to bending and shear

$$U = U_b + U_s \qquad (7.111)$$

Substituting from Eqs. 7.109 and 7.110, we get

$$U = \frac{1}{2}EI \int_o^L \left(\frac{d\theta_y}{dx}\right)^2 \, dx + \frac{1}{2}\,\alpha\, GA \int_o^L \left(\theta_y - \frac{dw}{dx}\right)^2 \, dx \qquad (7.112)$$

7.5.2 A Finite Element Formulation of Beam Element to Include Shear Deformation

The strain energy expression (Eq.7.112) for the beam has w and θ_y as deformation parameters. Hence, in the finite element formulation the variation of the transverse displacement w and the rotation θ_y of the cross-section within the element are expressed by independent functions. In the first order linear displacement model, it is assumed that both w and θ_y vary linearly within the element of length l (Fig. 7.24f). This leads to a four degree of freedom beam element, i.e., the displacements w_1, w_2, and rotations θ_1 and θ_2 at the nodes 1 and 2 are the nodal degrees of freedom.

$$\begin{Bmatrix} w \\ \theta_y \end{Bmatrix} = \begin{bmatrix} L_1 & L_2 & 0 & 0 \\ 0 & 0 & L_1 & L_2 \end{bmatrix} \begin{Bmatrix} w_1 \\ w_2 \\ \theta_1 \\ \theta_2 \end{Bmatrix} \qquad (7.113)$$

where L_1 and L_2 are natural co-ordinates given by Eq.3.12 as

$$L_1 = 1 - \frac{x}{l} \quad \text{and} \quad L_2 = \frac{x}{l}$$

Using isoparametric element concept, we have the following relations

$$x = \sum_{i=1}^{2} L_i \, x_i \tag{7.114}$$

$$\frac{d\theta_y}{dx} = \sum_{i=1}^{2} \frac{\partial L_i}{\partial x} \, \theta_i \tag{7.115}$$

$$\text{and } \phi = \sum_{i=1}^{2} L_i \, \theta_i - \sum_{i=1}^{2} \left(\frac{\partial L_i}{\partial x} \, w_i \right) \tag{7.116}$$

(for simplicity $\phi_{(x)}$ has been denoted as ϕ in the above equation)

Now, the strain-displacement relation can be expressed as

$$\{\epsilon\}_b = [B] \, \{d\} \tag{7.117}$$

$$\text{where } \{\epsilon\}_b^T = [k_x \;\; \phi] \tag{7.118}$$

$$\{d\}^T = [\, w_1 \;\;\; \theta_1 \;\;\; w_2 \;\;\; \theta_2 \,] \tag{7.119}$$

$$k_x = \frac{d\theta_y}{dx} \tag{7.120a}$$

$$\phi = \theta_y - \frac{dw}{dx} \tag{7.120b}$$

and

$$[B] = \begin{bmatrix} 0 & \frac{\partial L_1}{\partial x} & 0 & \frac{\partial L_2}{\partial x} \\ \frac{-\partial L_1}{\partial x} & L_1 & \frac{-\partial L_2}{\partial x} & L_2 \end{bmatrix} \tag{7.121}$$

The stress-resultants M and Q are related to $\{\epsilon\}_b$ as,

$$\{\sigma\}_b = \left\{ \begin{array}{c} M \\ Q \end{array} \right\} = [C]_b \, \{\epsilon\}_b \tag{7.122}$$

The terms in matrix $[C]_b$ account for the integration of the stresses σ_x and τ_{xz} over the cross-section to define the moment M and the shear force Q in the above equation. For a rectangular section $[C]_b$ is given by, (μ is assumed to be zero).

$$[C]_b = \frac{Ebh}{12} \begin{bmatrix} h^2 & 0 \\ 0 & 6\alpha \end{bmatrix} \tag{7.123}$$

The element stiffness matrix can be evaluated as

$$[k] = \int_o^l [B]^T [C]_b [B] \, dx \qquad (7.124)$$

substituting Eq.7.121 and 7.123 into the above Eq.7.124, we get,

$$[k] = \frac{Ebh}{12l^2} \times$$

$$\int_o^l \begin{bmatrix} 6\alpha & 6\alpha(l-x) & -6\alpha & 6\alpha x \\ 6\alpha(l-x) & h^2+6\alpha(l-x)^2 & -6\alpha(l-x) & -h^2+6\alpha x(l-x) \\ -6\alpha & -6\alpha(l-x) & 6\alpha & -6\alpha x \\ 6\alpha x & -h^2+6\alpha x(l-x) & -6\alpha x & h^2+6\alpha x^2 \end{bmatrix} dx$$

$$(7.125)$$

Separating the bending and shear terms we get

$$[k] = \frac{Ebh}{12l^2} \int_o^l \left[\begin{bmatrix} 0 & 0 & 0 & 0 \\ 0 & h^2 & 0 & -h^2 \\ 0 & 0 & 0 & 0 \\ 0 & -h^2 & 0 & h^2 \end{bmatrix} \right.$$

$$+6\alpha \left. \begin{bmatrix} 1 & (l-x) & -1 & x \\ (l-x) & (l-x)^2 & -(l-x) & x(l-x) \\ -1 & -(l-x) & 1 & -x \\ x & x(l-x) & -x & x^2 \end{bmatrix} \right] dx \qquad (7.126)$$

Thus

$$[k] = [k]_b + [k]_s \qquad (7.127)$$

where $[k]_b$ contributes to the bending stiffness and $[k]_s$ to the shear stiffness.

The integration of Eq.7.126 can be performed exactly and for a beam with unit thickness (b=1) the coefficients of the $[k]_b$ and $[k]_s$ matrices are given below.

$$[k]_b = \frac{Eh^3}{12l} \begin{bmatrix} 0 & 0 & 0 & 0 \\ 0 & 1 & 0 & -1 \\ 0 & 0 & 0 & 0 \\ 0 & -1 & 0 & 1 \end{bmatrix} \qquad (7.128)$$

$$[k]_s = \frac{Gh\alpha}{l} \begin{bmatrix} 1 & \frac{l}{2} & -1 & \frac{l}{2} \\ \frac{l}{2} & \frac{l^2}{3} & \frac{-l}{2} & \frac{l^2}{6} \\ -1 & \frac{-l}{2} & 1 & \frac{-l}{2} \\ \frac{l}{2} & \frac{l^2}{6} & \frac{-l}{2} & \frac{l^2}{3} \end{bmatrix} \qquad (7.129)$$

Since $d\theta_y/dx$ is constant within the element the bending stiffness $[k]_b$ (bending strain energy) can be exactly evaluated by one point Gauss quadrature. On the other hand, the shear stiffness matrix $[k]_s$ (shear strain energy) contains second order terms and two point Gauss quadrature is required to exactly integrate it [12]. Denoting shear stiffness evaluated by one point and two point Gauss quadrature as $[k^{(1)}]_s$ and $[k^{(2)}]_s$ respectively.

$$[k^{(1)}]_s = \frac{\alpha G h}{l} \begin{bmatrix} 1 & \frac{l}{2} & -1 & \frac{l}{2} \\ \frac{l}{2} & \frac{l^2}{4} & \frac{-l}{2} & \frac{l^2}{4} \\ -1 & \frac{-l}{2} & 1 & \frac{-l}{2} \\ \frac{l}{2} & \frac{l^2}{4} & \frac{-l}{2} & \frac{l^2}{4} \end{bmatrix} \tag{7.130}$$

and $$[k^{(2)}]_s = [k]_s = \frac{\alpha G h}{l} \begin{bmatrix} 1 & \frac{l}{2} & -1 & \frac{l}{2} \\ \frac{l}{2} & \frac{l^2}{3} & \frac{-l}{2} & \frac{l^2}{6} \\ -1 & \frac{-l}{2} & 1 & \frac{-l}{2} \\ \frac{l}{2} & \frac{l^2}{6} & \frac{-l}{2} & \frac{l^2}{3} \end{bmatrix} \tag{7.131}$$

It is clear from Eqs. 7.130 and 7.131 that one-point quadrature 'under integrates' the shear stiffness (energy) and two point quadrature integrates exactly the shear stiffness (energy) causing an overly stiff element [12].

These aspects are explained in the following.

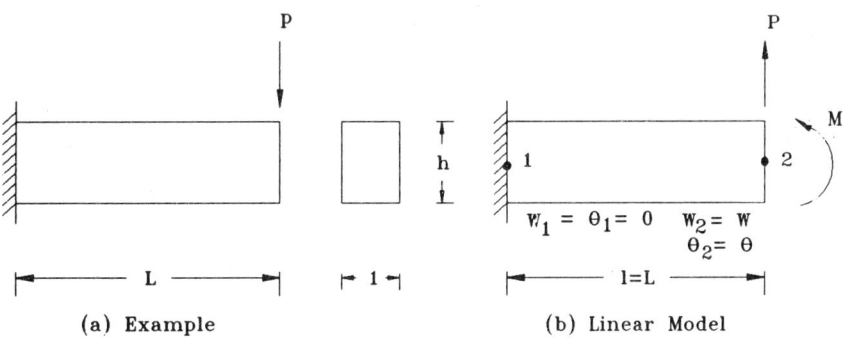

(a) Example   (b) Linear Model

Fig. 7.25 Cantilever beam

A detailed analytical investigation was carried out by Hughes et al. [12] to study the behaviour of the linear finite element model and the effect of shear stiffness evaluated using one and two point Gauss rule. The example chosen was a cantilever beam subjected to an end load of P, shown in Fig.7.25.

As a first case the beam has been analysed for various discretizations using the finite element model described here. The following data were used:

$$E = 1000; \quad G = 375; \quad h = 1; \quad L = 4; \quad b = 1$$

The ratio of the tip displacement obtained from finite element analysis to the value obtained using simple theory of bending is given in Table 7.2 [12].

Table 7.2 Normalised Tip Displacement Based on Finite Element Analysis of Cantilever Beam

| No. of Elements | Tip Displacement FEA | |
| | Tip Displacement simple theory of bending | |
	One-Point Quadrature	Two-Point Quadrature
1	0.762	0.416×10^{-1}
2	0.940	0.445
4	0.985	0.762
8	0.996	0.927
16	0.999	0.981

It can be observed from Table 7.2 that the linear finite element using one point quadrature gives much better results compared to using two point quadrature. The severe test to which the element must be subjected to is the satisfaction of Bernoulli-Euler Theory and accordingly the shear strains are to be equal to zero. This case can be tested by making the beam depth very small compared to the element length l or by assuming a very large fictitious value of G. The later approach has been used in reference [12] by assigning $G = 375 \times 10^5$. The tip displacements obtained for different discretizations are given in Table 7.3

It can be seen that the one-point quadrature results are quite accurate, whereas the two-point quadrature results are grossly in error. This is because use of two-point quadrature leads to overly stiff element. The reason for this situation is further explained below.

Table 7.3 Normalised Tip Displacement of Cantilever
Beam for large value of G

| No. of Elements | Tip Displacement FEA | |
| | Tip Displacement simple theory of bending | |
	One-Point Quadrature	Two-Point Quadrature
1	0.750	0.200×10^{-4}
2	0.938	0.800×10^{-4}
4	0.984	0.320×10^{-3}
8	0.996	0.128×10^{-3}
16	0.999	0.512×10^{-3}

Consider the case of a cantilever beam shown in Fig.7.25b discretized by one linear element. The shear stiffness contribution $[k]_s$ can be computed using Eqs. 7.130 and 7.131, one-point and two-point quadratures. The results of the analysis are described below.

Case (i) One Point Quadrature $[k^{(1)}]_s$

The stiffness matrix for the beam is given by,

$$[k] = [k]_b + [k^{(1)}]_s$$

where $[k]_b$ and $[k^{(1)}]_s$ are evaluated using Eqs.7.128 and 7.130. Applying the boundary conditions and defining $\psi = Eh^3/12l$ and $\beta = \alpha Gh/l$, we get the equilibrium equation for the beam as,

$$\begin{bmatrix} \beta & \frac{1}{2}\beta & -\beta & \frac{1}{2}\beta \\ \frac{1}{2}\beta & \psi + \beta\frac{l^2}{4} & \frac{-\beta l}{2} & -\psi + \frac{\beta l^2}{4} \\ -\beta & -\frac{1}{2}\beta & \beta & -\frac{1}{2}\beta \\ \frac{1}{2}\beta & -\psi + \frac{l^2}{4}\beta & -\beta\frac{l}{2} & \psi + \beta\frac{l^2}{4} \end{bmatrix} \begin{Bmatrix} 0 \\ 0 \\ w \\ \theta \end{Bmatrix} = \begin{Bmatrix} 0 \\ 0 \\ P \\ M \end{Bmatrix} \qquad (7.132)$$

Solving Eq.7.132 we get,

$$w = (\frac{l^2}{4\psi} + \frac{1}{\beta})P + \frac{l}{2\psi}M \qquad (7.133a)$$

$$\text{and} \quad \theta = \frac{(\frac{l}{2}P + M)}{\psi} \qquad (7.133b)$$

In the limiting condition i.e., tending to thin beam limit $\beta \gg \psi$, the equation (7.133a) becomes,

$$w = l \frac{\left(\frac{lP}{2} + M\right)}{2\psi} \tag{7.134}$$

and expression for θ (equation 7.133b) remains unchanged. Thus the beam deformation (w and θ) is solely due to bending as given by Eq.7.134 and 7.133b and this is correct.

Case (ii) Two-Point Quadrature - $[k^{(2)}]_s$

The shear stiffness $[k^{(2)}]_s$ contribution is evaluated using two-point Gauss rule as given in Eq.7.131. Thus the stiffness matrix can be calculated as,

$$[k] = [k]_b + [k^{(2)}]_s$$

The equilibrium equation for the beam after imposing boundary conditions is given by,

$$\begin{bmatrix} \beta & -\beta\frac{l}{2} \\ -\frac{\beta l}{2} & \psi + \frac{\beta l^3}{3} \end{bmatrix} \begin{Bmatrix} w \\ \theta \end{Bmatrix} = \begin{Bmatrix} P \\ M \end{Bmatrix} \tag{7.135}$$

solving for w and θ we get

$$w = \left[\frac{\psi + \frac{\beta l^2}{3}}{\beta\left(\psi + \frac{\beta l^2}{12}\right)}\right] P + \frac{lM}{2\left(\psi + \frac{\beta l^2}{12}\right)} \tag{7.136a}$$

$$\text{and} \quad \theta = \frac{M + \frac{l}{2}P}{\left(\psi + \frac{\beta l^2}{12}\right)} \tag{7.136b}$$

In the thin beam limit, Eqs. 7.136a and 7.136b become

$$w = \frac{\left(4P + \frac{6M}{l}\right)}{\beta} \tag{7.137a}$$

$$\theta = \frac{6(lP + 2M)}{l^2 \beta} \tag{7.137b}$$

It can be observed that Eqs. 7.137a and 7.137b lead to erroneous results as they contain only the coefficient β corresponding to shear deformation.

The above example illustrates the influence of shear stiffness matrix and also the use of one-point and two-point quadrature approach in evaluating $[k]_s$. Mathematically, the rank of the stiffness matrix $[k^{(1)}]_s$ is one and that of $[k^{(2)}]_s$ is two, and the rank of the exactly integrated $[k]_s$ is also two. Since the rank of $[k^{(1)}]_s$ is one less than that of the matrix evaluated exactly, one point integration leads to a more flexible mesh as in this case only one constraint is imposed on the element. In the case of two point integration, the stiffness $[k]$ is completely dominated by the shear stiffness and the mesh gets 'locked' as two constraints (i.e., one constraint at each Gauss point) are imposed on the element.

7.6 Offset Beam Element (BEAM 2)

This element is useful to model the connection between beam element with other elements when the shear and flexural centre of the beam do not coincide with the common element node and thus there is eccentricity of the beam flexural centre with respect to the element node to which it is connected.

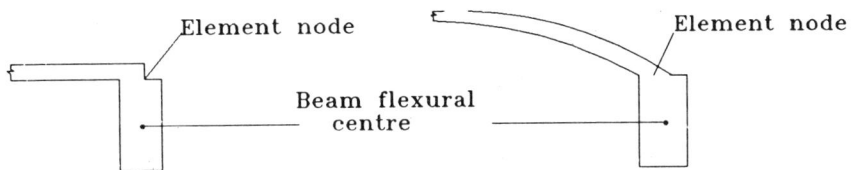

Fig. 7.26a Beam centre and element node at eccentricity

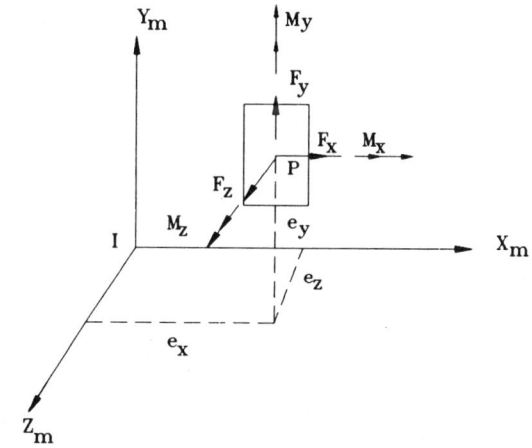

Fig. 7.26b Eccentricity of beam and element nodes

Such problems occur in the case of plate and shell structures which may be stiffened at the end or other places by a beam as shown in Fig.7.26a.

Figure 7.26b shows a general case where the shear and flexural centre P of the beam element has eccentricity e_x, e_y, e_z with respect to element node I measured along the local beam member axes.

The force vector at flexural centre P of the beam, $\{Q'_m\}_p^T = [F_x \ F_y \ F_z \ M_x \ M_z \ M_z]_p$ referred to the local axes of the beam can be transferred to the element node I. Assuming rigid connection between P and I, the transformation is a parallel axes transformation which can be worked out as given below [2].

Consider the transfer of force F_x alone. It is seen that when F_x is transferred to 'I' it produces F_x at I together with moments about the axes Iy_m and Iz_m equal to F_xe_z and $-F_xe_y$ respectively. Similarly, the transfer of forces F_y and F_z produce couples about the axes Ix_m, Iz_m and Ix_m, Iy_m respectively. The moments M_x, M_y and M_z produce effects that are independent of their points of application. So the forces at I due to the forces applied at P are as follows:

$$
\begin{aligned}
F_{xi} &= F_{xp} \\
F_{yi} &= F_{yp} \\
F_{zi} &= F_{zp} \\
M_{xi} &= M_{xp} - F_{yp}e_z + F_{zp}e_y \\
M_{yi} &= M_{yp} + F_{xp}e_z - F_{zp}e_x \\
M_{zi} &= M_{zp} - F_{xp}e_y + F_{yp}e_x
\end{aligned}
$$

$$(7.138)$$

These equations may be written in a matrix form as

$$
\left\{
\begin{array}{c}
F_x \\ F_y \\ F_z \\ M_x \\ M_y \\ M_z
\end{array}
\right\}_i
=
\begin{bmatrix}
1 & 0 & 0 & 0 & 0 & 0 \\
0 & 1 & 0 & 0 & 0 & 0 \\
0 & 0 & 1 & 0 & 0 & 0 \\
0 & -e_z & e_y & 1 & 0 & 0 \\
e_z & 0 & -e_x & 0 & 1 & 0 \\
-e_y & e_x & 0 & 0 & 0 & 1
\end{bmatrix}
\left\{
\begin{array}{c}
F_x \\ F_y \\ F_z \\ M_x \\ M_y \\ M_z
\end{array}
\right\}_p
$$

$$(7.139a)$$

i.e.,

$$\{Q'_m\}_i = [T'_p]^T \{Q'_m\}_p \qquad (7.139b)$$

Again, the displacement vector at I, $\{d_m\}_i^T = [u \ v \ w \ \theta_x \ \theta_y \ \theta_z]$ can be transformed to the point P as described below.

At point P the translational displacement components will be produced by the combination of the two effects:

(a) the displacement u, v, w at I and

(b) the displacement components produced by $\theta_x, \theta_y, \theta_z$.

The rotations are assumed to be small so that the resulting displacements are linearised. Thus, for example (Fig.7.26c) if the point

P is located in $x_m y_m$ plane and if the rotation is about z_m axis, θ_z, the displacement components produced at P are,

$$u = -l\,\theta_z\,\sin\phi = -\theta_z e_y$$

$$v = l\,\theta_z\,\cos\phi = \theta_z e_x \qquad (7.140)$$

For a general point P with eccentricity e_x, e_y, e_z the displacement components due to $\theta_x, \theta_y, \theta_z$ can be calculated in a similar manner by considering each rotation in turn.

Thus,

$$
\begin{aligned}
u_p &= u_i + \theta_{yi}\,e_z - \theta_{zi}\,e_y \\
v_p &= v_i - \theta_{xi}\,e_y + \theta_{zi}\,e_x \\
w_p &= w_i + \theta_{xi}\,e_y - \theta_{yi}\,e_x \\
\theta_{xp} &= \theta_{xi} \\
\theta_{yp} &= \theta_{yi} \\
\theta_{zp} &= \theta_{zi}
\end{aligned}
$$

$$(7.141a)$$

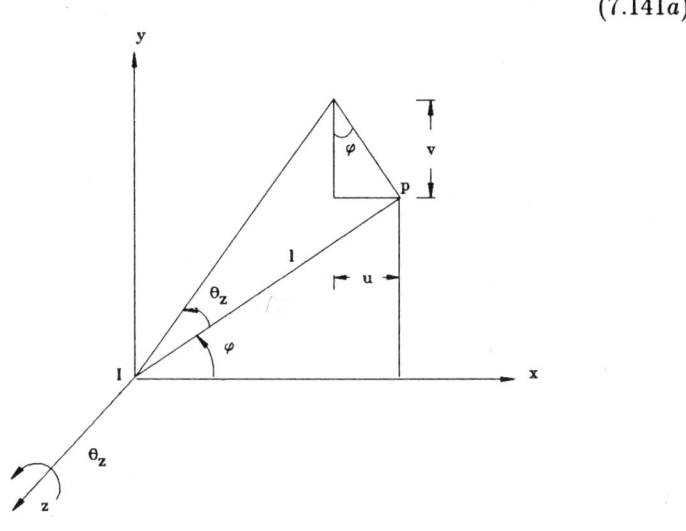

Fig. 7.26c Displacement at P due to θ_z

The above equations may be written in a matrix form as

$$
\begin{Bmatrix} u \\ v \\ w \\ \theta_x \\ \theta_y \\ \theta_z \end{Bmatrix}_p = \begin{bmatrix} 1 & 0 & 0 & 0 & e_z & -e_y \\ 0 & 1 & 0 & -e_z & 0 & e_x \\ 0 & 0 & 1 & e_y & -e_x & 0 \\ 0 & 0 & 0 & 1 & 0 & 0 \\ 0 & 0 & 0 & 0 & 1 & 0 \\ 0 & 0 & 0 & 0 & 0 & 1 \end{bmatrix} \begin{Bmatrix} u \\ v \\ w \\ \theta_x \\ \theta_y \\ \theta_z \end{Bmatrix}_i \qquad (7.141b)
$$

It can be seen that the matrix in the above equation is the transpose of the matrix in Eq.(7.139b), i.e.,

$$
\{d'_m\}_p = [T'_p]\{d'_m\}_i \qquad (7.141c)
$$

Assuming the beam to be prismatic and the eccentricities of the other end of the beam Q with respect to the node J are the same, the transformation matrices for the displacement and force vectors from the other centre Q to the node J will be the same as given in Eqs. 7.139b and 7.141c. Thus the transformations of the displacement and force vectors for both the ends of the beam from the element nodes I and J to the beam centres P and Q are given by

$$
\{d_m\}_{pq} = [T_p]\{d_m\} \qquad (7.142a)
$$

$$
\{Q_m\}_{pq} = [T_p^T]^{-1}\{Q_m\} \qquad (7.142b)
$$

where

$$
[T_p] = \begin{bmatrix} [T'_p] & [0] \\ [0] & [T''_p] \end{bmatrix} \qquad (7.142c)
$$

It may be noted that the displacement and force vectors referred to in Eqs. 7.142a and 7.142b are with respect to the beam local axes. These quantities can now be transformed to the global axes using the rotation transformation matrix [T] given in Eq.7.84. Thus,

$$
\{d_m\} = [T]\{d\} \qquad (7.143a)
$$

$$
\{Q_m\} = [T]\{Q\} \qquad (7.143b)
$$

Now the equation of equilibrium for the beam element in the local member axes system can be written as

$$
[k_m]_{pq}\{d_m\}_{pq} = \{Q_m\}_{pq} \qquad (7.144)
$$

where $[k_m]_{pq}$ is the stiffness matrix of the beam corresponding to the flexural centres P and Q in the local axes systems and is given by

Eq.7.70. Substituting from Eq.7.142a and 7.142b into Eq.7.144, we get

$$[k_m]_{pq}[T_p]\{d_m\} = [T_p^T]^{-1}\{Q_m\} \qquad (7.145)$$

Substituting from Eqs.7.143a and 7.143b into Eq.7.145, we get

$$[k_m]_{pq} [T_p] [T] \{d\} = [T_p^T]^{-1} [T] \{Q\} \qquad (7.146)$$

Premultiplying Eq.7.145 by $[T_p]^T$ and $[T]^{-1}$ successively, we get,

$$[T]^{-1}[T_p]^T[k_m]_{pq}[T_p][T]\{d\} = \{Q\} \qquad (7.147)$$

Thus, noting $[T]^{-1} = [T]^T$, we get

$$[T]^T [k_m] [T] \{d\} = \{Q\} \qquad (7.148a)$$

where

$$[k_m] = [T_p]^T [k_m]_{pq} [T_p] \qquad (7.148b)$$

and it is the stiffness matrix of the beam corresponding to the element nodes I and J with respect to local member axes system. And, $[k] = [T]^T [k_m] [T]$ in Eqs.7.148a is the stiffness matrix of the beam in the global axes system with respect to the element nodes I and J.

7.7 Subroutine For 3D Truss Element

The control routine for the truss elements is THREDT and it is called by the FELIB routine. As indicated in the discussion on the element library, provision is made for three shapes of truss elements and the general arrangement is given in Fig.7.27. However, only one shape for truss element is incorporated in the program presented here and the routine called TRUS1 is organised following the guide lines given in Chapter 6. The stiffness matrix computation is based on the theory described in section 7.1. The listing of the program is given in Appendix 2.

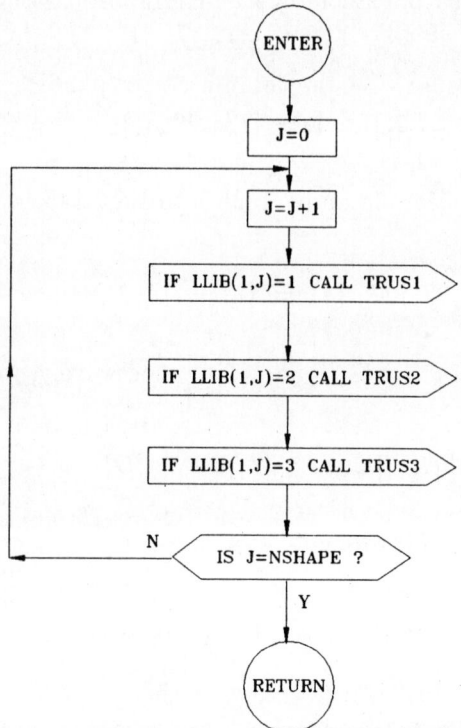

Fig. 7.27 Subroutine THREDT

7.7.1   Input Details for Truss Element

The input details for truss element are given below. This data set pertains to the data sets 5 and 6 required for PASSFEM described in section 6.11.

5. ELEMENT TYPE DATA (5I5)

Columns	1-5	Element type number or row number of the element library (LTYPE). (For truss elements, LTYPE = 1)
	6-10	Number of shapes (NSHAPE) belonging to the element type with a minimum of 1 to a maximum of 3.
	11-15	Element shape number or column number of the element library. (Element shape number is 1 for TRUS1).

16-20   Element shape number. If NSHAPE is more than one, the second shape number is keyed in this column. Otherwise, the column is left blank.

21-25   Element shape number. If NSHAPE = 3, third shape number is keyed in this column.

6. ELEMENT DATA SET

A. *Control Data (2I5)*

Columns   1 - 5   Number of space truss elements in the structure
6 - 10   Number of geometric property groups.

B. *Geometric Property Data (I5, F 10.3)*

Columns   1 - 5   Geometric property group number.
6 - 15   Area of cross-section.

This set of data is repeated for each different geometric property group.

C. *Element Data (6I5)*

Columns   1 - 5   Element Number
6 - 10   The material property group number for the element
11 - 15   Node i of the element
16 - 20   Node j of the element
21 - 25   Geometric property group number for the element.
26 - 30   Automatic generation parameter.

D. *Self-weight Factors (3F 5.2)*

Columns   1 - 5   Factor for self-weight in the global +x direction
6 - 10   Factor for self-weight in the global +y direction
11 - 15   Factor for self-weight in the global +z direction.

The element data must be given in sequence. Generation can be used only if all the elements to be generated have the same material group and cross sectional area. The generation parameter should be given only on the second line of the sequence.

7.7.2   Output Details for Truss Element

The output results are the force and stress for each element.

7.8   Subroutine for 3D Beam Element

The control routine for the beam element is THREDB and the general flow chart is given in Fig.7.28. Provision is made for three shapes of beam element. In the present version of the program two shapes are included; one for a three-dimensional prismatic beam element (BEAM1) and the other for an off-set beam (BEAM2). The listing of the program is given in Appendix 3.

The element type data of data set 5 required for PASSFEM described in section 6.11 follows.

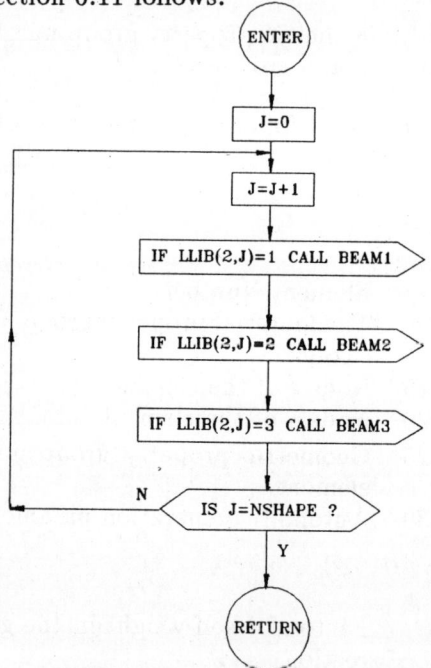

Fig. 7.28 Subroutine THRED8

5. ELEMENT TYPE DATA (5I5)

Columns	1 - 5	Element type number or row number of the element library (LTYPE). For beam elements LTYPE = 2.
	6 - 10	Number of shapes (NSHAPE) belonging to the element type with a minimum of 1 to a maximum of 3.

11 - 15   Element shape number or column number of the element library. (Element shape number is 1 for BEAM1 (prismatic beam element).

16 - 20   Element shape number. If NSHAPE is more than one, the second shape number is keyed in this column. (Element shape number is 2 for BEAM2 (offset beam element)). Otherwise, the column is left blank.

21 - 25   Element shape number. If NSHAPE = 3, third shape number is keyed in this column.

7.8.1 Three-dimensional Prismatic Beam Element (BEAM1)

The routine corresponding to three-dimensional prismatic beam element is named as BEAM1 and the routine is organised following the guide lines given in Chapter 6. The element properties required in addition to the general data given in the flow chart of Fig.6.13 are the cross-sectional area, torsional moment of inertia and moment of inertia about minor and major principal planes respectively. These quantities are specified under geometric properties and the geometric property group number is input for each element. In addition the element loads are also read as input.

For computing the element stiffness matrix the procedure given in section 7.4 is used. Using the 'k' node to specify the principal planes of bending, the rotation transformation matrix described in subsection 7.4.3. is computed and the member stiffness matrix is transformed to global system of axes.

Next the fixed end forces are calculated. These are used to compute the global load vector and the element end forces in the third stage.

INPUT Details for BEAM1

The input details for prismatic beam element are given below. This data set pertain to the data set 6 required for PASSFEM as described in section 6.11 of Chapter 6.

6. ELEMENT DATA SET

A. Control Data (3I5)

Columns   1 - 5   Number of beam elements in the structure.

6 - 10   Number of geometric property groups (NGP).

11 - 15   Number of element load sets (NLS)

B. *Geometric Property Data (I 10, 4F 10.2)*

Columns   1 - 10   Geometric property group number
· 11 - 20   Area of cross-section
21 - 30   Torsional moment of inertia
31 - 40   Moment of inertia about the principal y_m axis
41 - 50   Moment of inertia about the principal z_m axis.

This set of data is repeated for each different geometric property group.

C. *Element Load Data (I10, 7F 10.3)*

This data set is required only if $NLS > 0$; otherwise go to D-Element Data

Columns   1 - 10   Element load set number
11 - 20   Concentrated load along the member y_m axis
21 - 30   Distance of the load from node I
31 - 40   Concentrated load along the member z_m axis.
41 - 50   Distance of the above load from node I.
51 - 60   Intensity of uniformly distributed load along the member y_m axis.
61 - 70   Intensity of uniformly distributed load along the member z_m axis.
71 - 80   Axial force in the member.

The above data must be given for each element load set.

D. *Element Data (8I5)*

Columns   1 - 5   Element Number
6 - 10   Material group
11 - 15   Node I
16 - 20   Node J
21 - 25   Node K
26 - 30   Geometric property group number
31 - 35   Element load set number
36 - 40   Node generation parameter

For the generated elements, the material group number, the 'k' node, the geometric property group number, and load set number should be the same. The parameter for generation should be specified on the second line. The value of this parameter is equal to the difference between the node numbers of successive elements to be generated.

E. *Self-weight Factors (3F 5.2)*

Columns   1 - 5   Factor for self-weight in the global $+x$ direction.

            6 - 10   Factor for self-weight in the global $+y$ direction.

           11 - 15   Factor for self-weight in the global $+z$ direction.

Output

The output for the beam element are the axial force and shear forces, and bending and torsional moments at both ends of the member. These stress resultants are given with respect to the member axes system (Fig. 7.18c).

Input Details for BEAM2

The subroutine BEAM2 developed for off-set beam uses many of the routines for BEAM1 which are common for both the elements. The input details for BEAM2 are the same for BEAM1 except for the element data and is given below:

E. *Element Data (8I5, 2F 10.4)*

Columns   1 - 5   Element number

            6 - 10   Material group

           11 - 15   Node I

           16 - 20   Node J

           21 - 25   Node K

           26 - 30   Geometric property group number

           31 - 35   Element load set number

           36 - 40   Node generation parameter

           41 - 50   Eccentricity e_y, measured from element node I or J to the node P or Q along y_m.

           51 - 60   Eccentricity e_z, measured from element node I or J to the node P or Q along z_m.

For the generated elements besides the material group number, 'k' node, the geometric property group number and load set number, the eccentricities e_y and e_z should be the same.

Output

The output results are the stress resultants similar to BEAM1 except they are calculated at the beam centres P and Q in the local members axes system.

7.9   Subroutine For Boundary Element

The controlling routine BOUND calls three shapes of boundary elements as shown in Fig. 7.29.

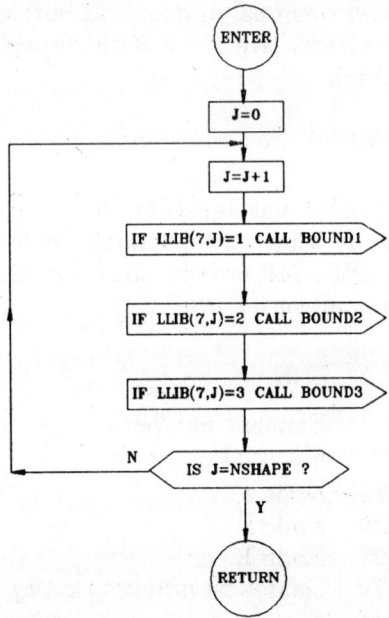

Fig. 7.29 Subroutine BOUND

In the present version of the program only one shape is developed and the listing of the routine BOUND is given in Appendix 4.

    The main purpose of the boundary element is to apply displacement boundary conditions and to compute the values of the support reactions. The boundary element is a spring with axial and torsional stiffness (Fig. 7.30). To get the value of a reaction in the direction of a particular degree of freedom a very high stiffness coefficient is added to the corresponding digonal coefficient. This yields a very small but finite displacement along that degree of freedom which when multiplied by the imposed stiffness k gives the desired reaction.

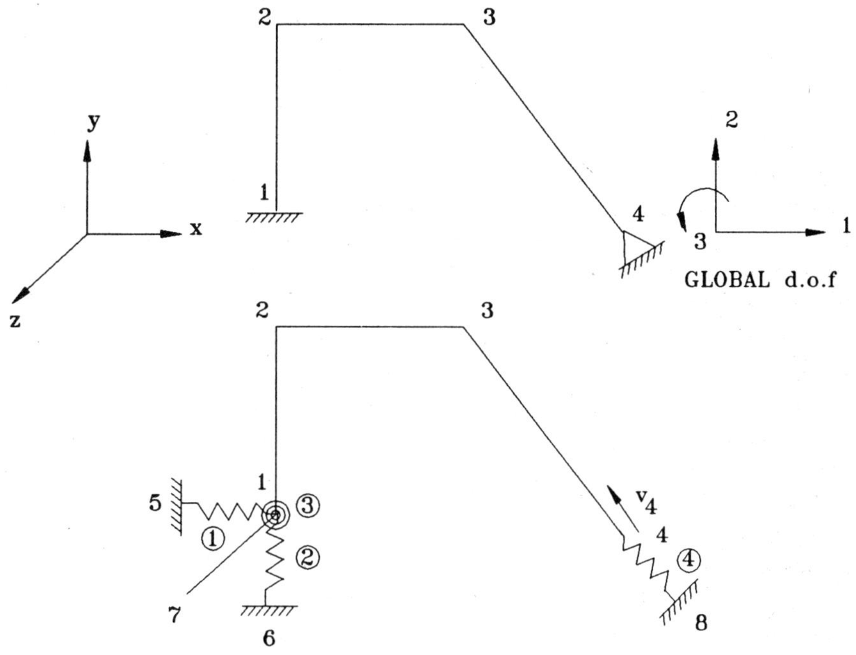

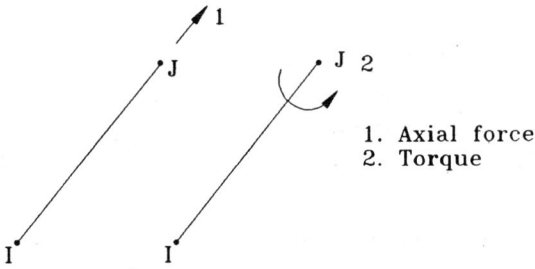

$$k_1 \;>>\; k_{11}$$
$$k_2 \;>>\; k_{22}$$
$$k_3 \;>>\; k_{33} \text{ along} \left.\begin{array}{c} \\ \\ \\ \\ \end{array}\right\} \Rightarrow r_1 = r_2 = r_3 \cong 0$$
$$k_4 \;>>\; k_{44}$$

Fig. 7.30a Boundary elements

1. Axial force
2. Torque

Fig. 7.30b DOF of the boundary elements

For applying a non-zero displacement, b, at a degree of freedom

r_i the load vector is modified by the product of the stiffness and the specified displacement as,

$$kr_i = kb \qquad (7.149)$$

where $k \gg k_{ij}$. This Eq. 7.149 when added to the overall equilibrium equation $[K]\{r\} = \{P\}$ and solved will give $r_i = b$. The physical interpretation of this is that a spring of large stiffness k is added to the structure at the degree of freedom i and because of relatively flexible element assemblage the load is specified to produce the desired displacement [6].

Like other element subroutines for various shapes the subroutine BOUND1 is also divided into three parts.

In the first part the number of boundary elements is read. For each element, the connected nodes, the code for translation and rotational degrees of freedom, specified displacement and rotation, and stiffness of the spring are read. If the stiffness of the spring is input as zero the default value is assumed. The coordinates of the connected nodes and other data are stored on the NDARAY file.

In the second stage of the processing, the direction cosines of the longitudinal direction of the boundary element are calculated. The stiffness contribution towards different degrees of freedom at the attached node are calculated and added to the global stiffness matrix in the routine PASSEM. Depending on the code for displacement and rotation the axial or torsional stiffness is imposed. In case of specified displacements the load vector is modified as indicated in Eq.7.149.

During the third stage of processing the displacements of the nodes at which the boundary elements are attached, are identified and the forces in the boundary elements are computed and printed out.

7.9.1   Input Details For Boundary Element

The input details for boundary are given below. This data set pertains to the data sets 5 and 6 required for PASSFEM described in Chapter 6.

5. ELEMENT TYPE DATA (5I5)

> Columns   1 - 5   Element type number or row number of the
> element library (LTYPE). For boundary elements, LTYPE = 7.

6 - 10	Number of shapes (NSHAPE) belonging to the element type with a minimum of 1 to a maximum of 3.
11 - 15	Element shape number or column number of the element library. (Element shape number is 1 for BOUND1).
16 - 20	Element shape number of the second shape used. This is keyed in only if NSHAPE > 1 .
21 - 25	Element shape number of the third shape used. This is keyed in only if NSHAPE = 3.

6. ELEMENT DATA SET

A. *Control Data (I5)*

Columns   1 - 5   Number of Boundary Elements in the structure.

B. *Element Data (6I5, 3F 10.3)*

Columns		
	1 - 5	Element Number
	6 - 10	Node I - the node to which the boundary element is attached
	11 - 15	Node J - this decides the direction of the boundary element
	16 - 20	Code for displacement 1 for attaching the spring at node I and 0 for not attaching it
	21 - 25	Code for rotation – 1 for attaching the spring at node I and 0 for not attaching it
	26 - 30	Node generation parameter
	31 - 40	Specified displacement at node I
	41 - 50	Specified rotation at node I
	51 - 60	Spring stiffness. If the input is zero it is assumed equal to 10^{10}.

7.9.2   Output Details for Boundary Element

The axial force in case of translational spring, along the direction IJ, and torque about IJ axis in case of rotational spring are printed out for boundary elements.

7.10   Examples

Number of examples have been solved using the routines described above. The salient features of the examples along with the results are presented in this section. In order to establish the validity of PASSFEM, the results of examples 1 to 7 are compared with SAP IV [13] and in all the cases they are found to be in agreement.

7.10.1   Plane Truss

A 47-member plane truss is shown in Fig.7.31.

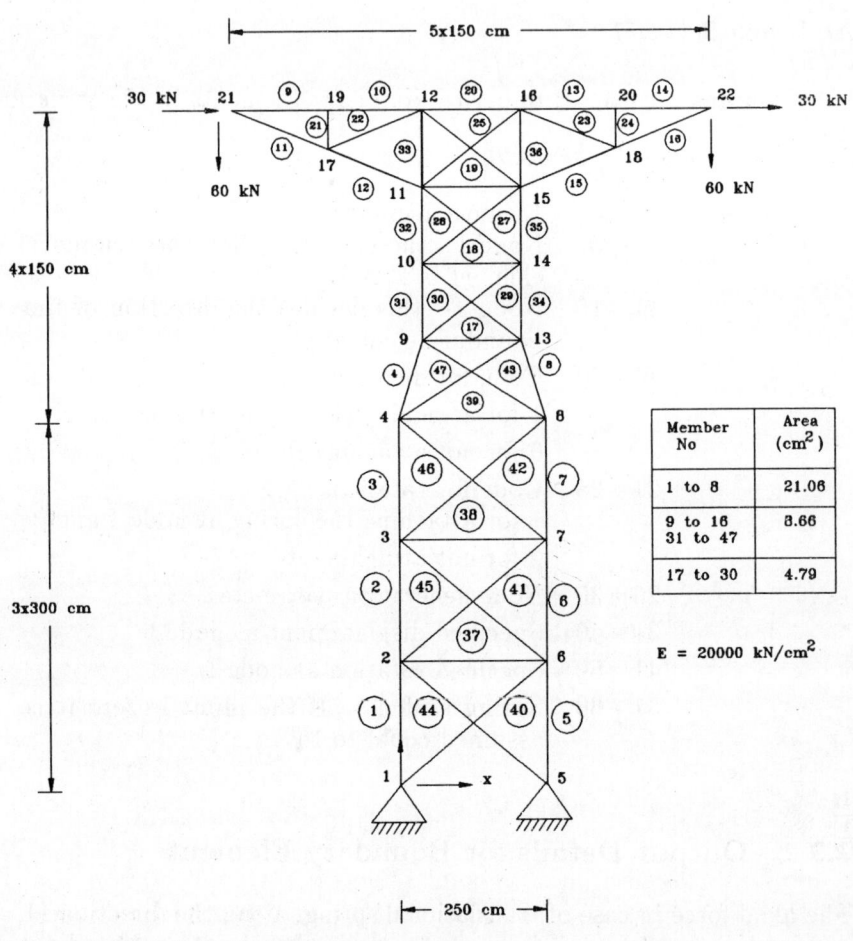

Fig. 7.31 Plane truss

The area of cross-section of the members and the loading are also shown therein. The forces in some of the members calculated by the method of joints are compared in Table 7.4 with the values obtained by PASSFEM.

Table 7.4 Comparison of Axial Forces in Members of Plane Truss

Member No.	13	14	15	16	23	24
Force (kN) by Method of joints	150.0	150.0	−134.16	−134.16	0	0
Force (kN) by PASSFEM	150.0	150.0	−134.16	−134.16	0	0

7.10.2 Space Truss

A 25 member space truss is shown in Fig.7.32. The member properties and load data are given in the Tables below. $E = 2.1 \times 10^4$ kN/cm^2.

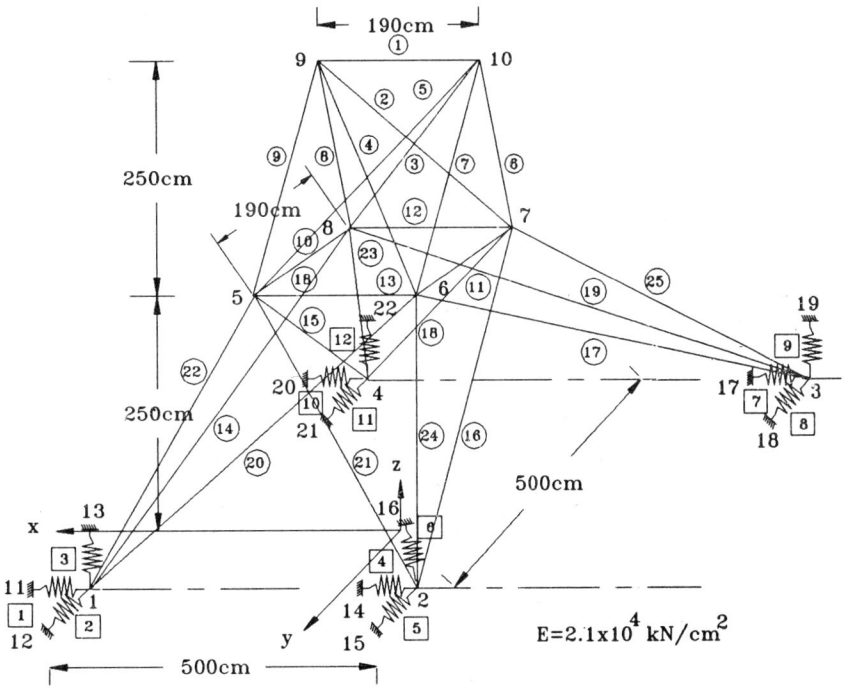

Fig. 7.32 Space Truss

Table 7.5(a) Member Properties

Member No.	Area (cm^2)
1 to 5	10.47
6 to 9	19.03
10 to 21	21.06
22 to 25	42.12

Table 7.5(b) Load Data

Node	Load (kN)		
	x-direction	y-direction	z-direction
5	−2.30	−	−
8	−2.30	−	−
9	−4.50	−45.0	−23.0
10	−	−45.0	−23.0

The forces obtained by PASSFEM are compared with SAP IV results. The values for typical members are presented in Table 7.5c. Boundary elements are used to compute the reactions at the supports and the values are compared in Table 7.5d.

Table 7.5(c) Comparison of Member Forces (kN)

Leg Members	PASSFEM	SAP IV
22	49.707	49.707
23	−63.973	−63.973
24	42.562	42.562
25	−71.117	−71.117
Bracings		
2	−23.664	−23.664
3	−19.779	−19.779
4	12.996	12.996
5	16.851	16.851

Table 7.5(d) Comparison of Support Reactions - Boundary Element Forces (kN)

Element	PASSFEM	SAP IV
1	−28.716	−28.716
2	−16.679	−16.679
3	36.900	36.900
4	24.166	24.166
5	−11.712	−11.712
6	30.100	30.100

(contd.)

Element	PASSFEM	SAP IV
7	−45.872	−45.872
8	−33.288	−33.288
9	−59.900	−59.900
10	41.322	41.322
11	−28.321	−28.321
12	−53.100	−53.100

7.10.3 Two-Span Continuous Beam

A two-span continuous beam with an overhang on one side is shown in Fig.7.33.

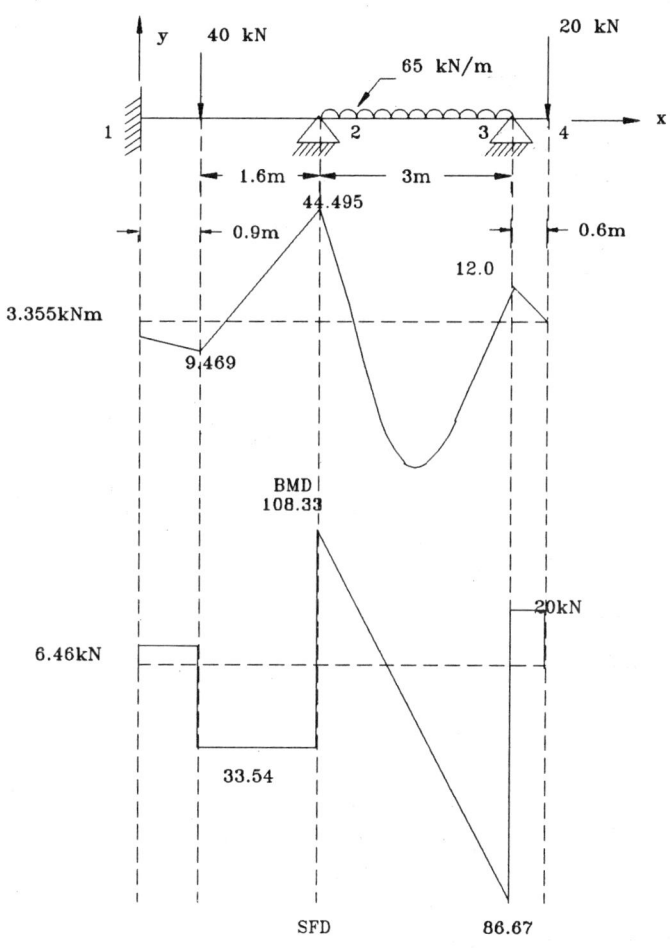

Fig. 7.33 Continuous Beam

The breath and depth of the section are 30 cm and 45 cm respectively, and $E = 2 \times 10^3$ kN/cm^2. As the principal axes of the beam members required definition of 'k' node it has been assumed to be located at x = 650 cm, y = 50.0 from the origin, for the two beam members. The bending moment and shear force diagrams are plotted in Fig. 7.33 and the values exactly coincide with the values obtained by moment distribution method.

7.10.4   Plane Frame

Figure 7.34 shows a two bay, six storey frame subjected to uniformly distributed load on the beams and lateral loads at each storey level.

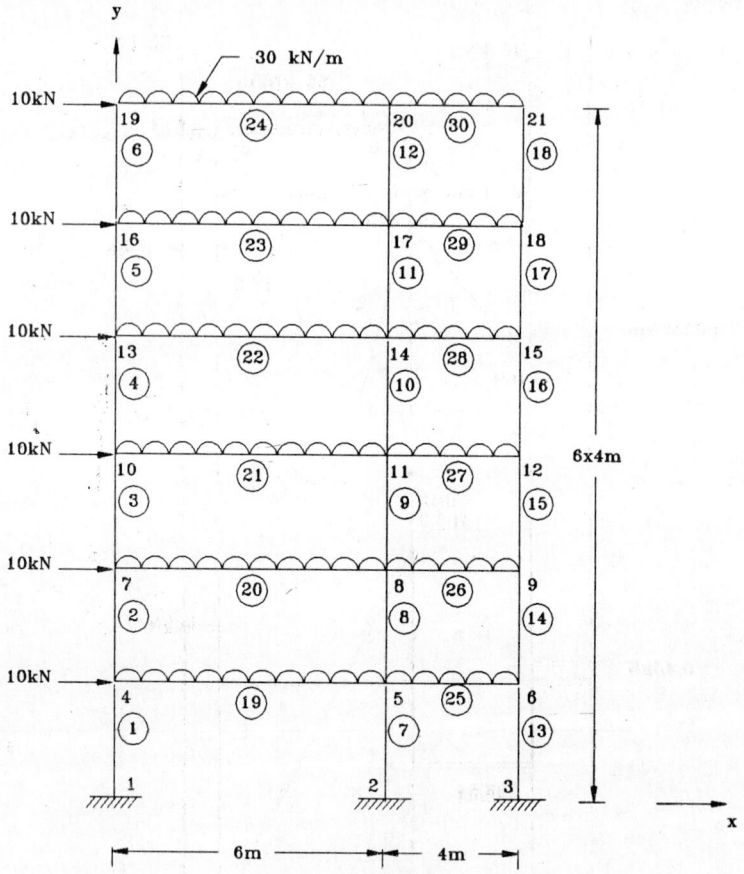

Fig. 7.34 Plane frame

The sizes of members are given below.

All Columns: 30 × 50*cm*
All beams : 30 × 60*cm*

$$E = 2 \times 10^3 \text{ kN/cm}^2$$

For all the members 'k' node has been assumed to be located at $x = -100.0$ cm, $y = 2500.0$ cm.

The results of analysis for maximum end moments for some of the members are presented in Table 7.6.

Table 7.6 Comparison of End Moments (kN.cm)

Analysis	Member No.				
	1	7	13	19	25
PASSFEM	2921	5949	5263	−12340	−9256
SAP IV	2921	5949	5263	−12340	−9256

7.10.5 Grid Frame

A grid frame supported on columns and subjected to a vertical load is shown in Fig.7.35. The section properties are also given therein. The three dimensional beam element has been used to analyse the grid. The coordinates of "k" node for all the members have been assumed to be at $x = -100$ cm and $y = 800$ cm. Typical values of twisting moments (TM) and bending moments (BM) for some of the members are compared in Table 7.7.

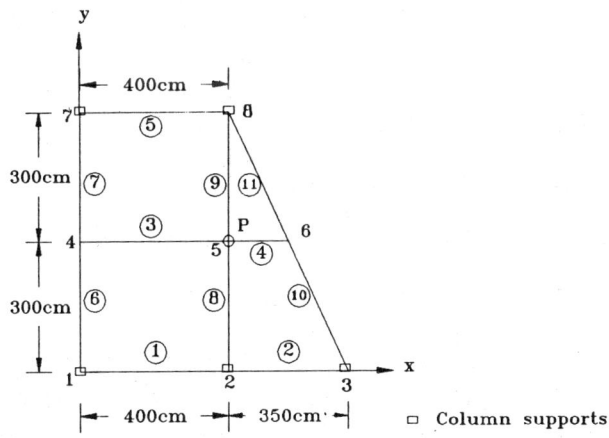

Fig. 7.35 Grid Frame

Section Properties Area =   110.74 cm²   P =   60 kN($\downarrow$)
Moment of Inertia about Member Axes

$I_x =$ 82.5 cm⁴
$I_y =$ 1369.8 cm⁴ E =   21000 kN/cm²
$I_z =$ 45218.3 cm⁴ $\mu =$ 0.3

Table 7.7 Comparison of End Moments (kN.cm)

Member No.	PASSFEM		SAPIV
3	TM	± 0.8405	± 0.8405
	BM	−2606.6	−2607.0
4	TM	± 14.400	± 14.400
	BM	43.94	43.94
8	TM	± 2.279	±2.280
	BM	−5402.4	−5402.4
9	TM	±1.223	±1.223
	BM	606.3	606.4

7.10.6   Space Frame 1

Figure 7.36 shows a space frame analysed by PASSFEM using the three dimensional beam element.

The coordinates of "k" node for various members are given in Table 7.8(a).

Table 7.8(a) "k" Node Details

Member No.	coordinates of "k" Node(cm)		
	x	y	z
1 to 6	0.0	0.0	6.0
7,8	−30.5	−52.83	2.0
9, 10	−30.5	52.83	2.0
11, 12	61.0	0.0	2.0

The results of analysis by PASSFEM are compared in Table 7.8b with the results of SAP IV for two members.

Table 7.8(b) Comparison of Forces and Moments

Analysis	Member			
	1		7	
	Axial force (kN)	B.M. (kNcm)	Axial Force (kN)	B.M. (kN cm)
PASSFEM	± 489.5	260.0	± 487.7	± 86.8
SAP IV	± 489.5	260.0	± 487.7	± 86.8

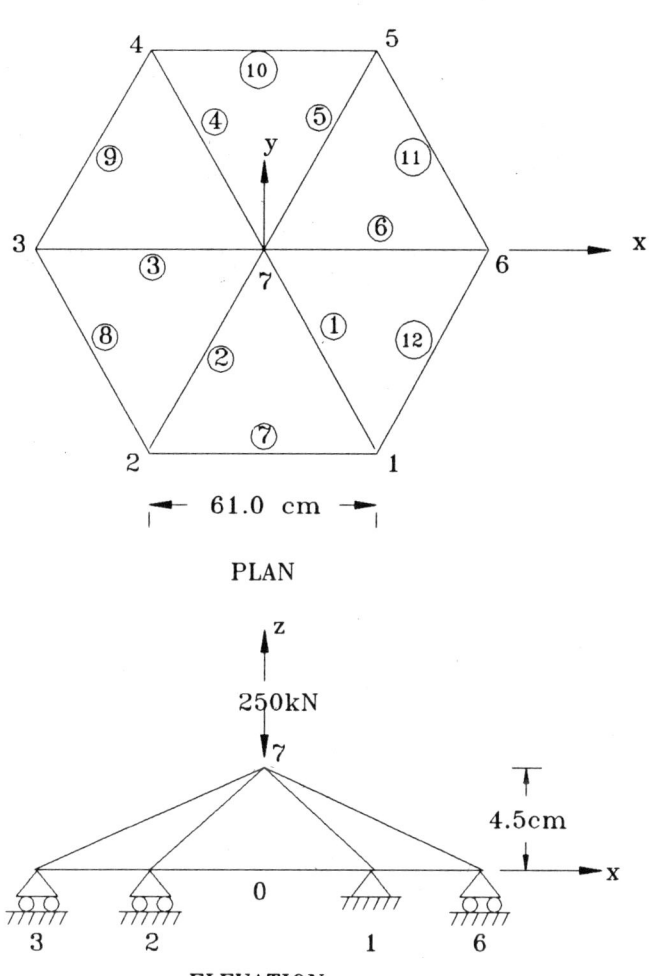

PLAN

ELEVATION

$E = 21000 \text{kN/cm}^2$ For all members

Area = 4 cm^2, $I_x = 2.667$ cm^4, $I_y = 1.333$ cm^4, $I_z = 1.333$ cm^4

Fig. 7.36 Space frame 1

7.10.7   Space Frame 2

A two storey space frame subject to lateral loads is shown in Fig.7.37.
The section properties of beams and columns are given therein. The

'k' nodes for various members are referred to the node numbers and are given below.

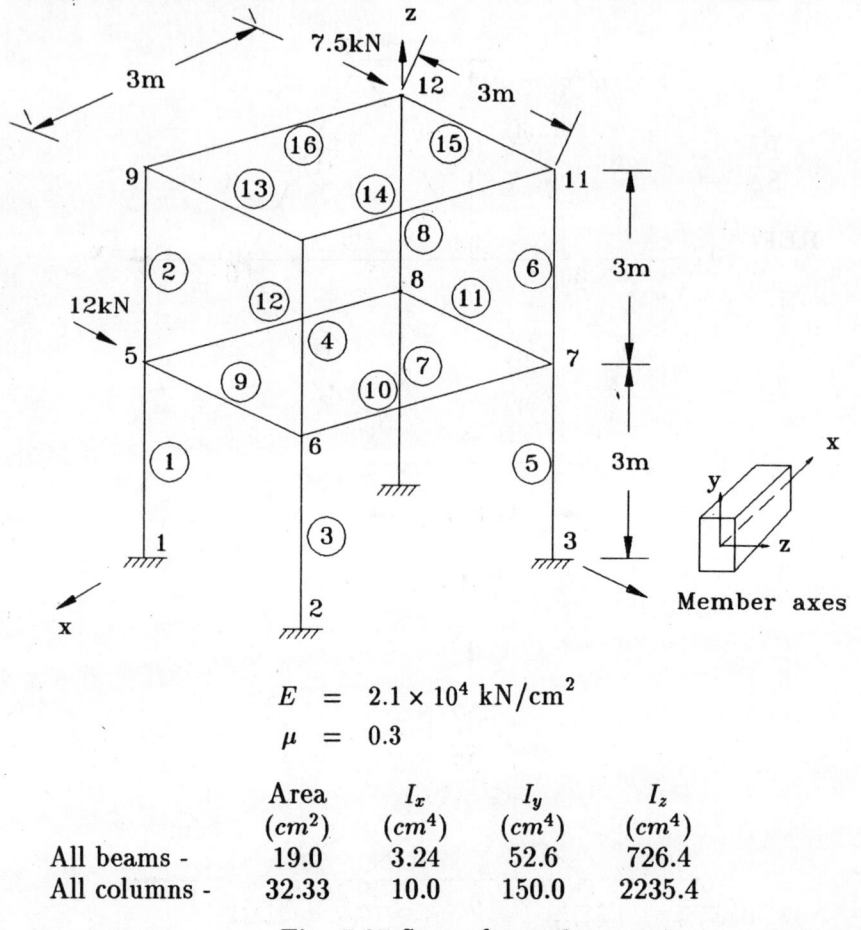

$$E \;=\; 2.1 \times 10^4 \text{ kN/cm}^2$$
$$\mu \;=\; 0.3$$

	Area (cm^2)	I_x (cm^4)	I_y (cm^4)	I_z (cm^4)
All beams -	19.0	3.24	52.6	726.4
All columns -	32.33	10.0	150.0	2235.4

Fig. 7.37 Space frame 2

Member No.	Node No. (for 'k' node)
1,2	2
3,4	1
5,6	4
7,8	3
9,10	10
11,12	12
13,14	6
15,16	8

The results of analysis are compared in Table 7.9 for maximum values for two typical members.

Table 7.9 Comparison of Moments (kNcm)

Analysis	Member No.			
	1		11	
	M_y	M_z	M_y	M_z
PASSFEM	7.35	−1170.0	3.27	−697.95
SAP IV	7.35	−1170.0	3.27	−697.95

REFERENCES

1. Gere, J.M. and W.Weaver, *Analysis of Framed Structures*. Van Nostrand Company Inc., Princeton, N.J., 1965.

2. Meek, J.L., *Matrix Structural Analysis*, McGraw-Hill, Inc., N.Y., 1971.

3. Rubinstein, M.R., *Matrix Computer Analysis of Structures*, Prentice-Hall, Inc., Englewood Cliffs, N.J., 1966.

4. Desai, C.S. and J.F., Abel, *Introduction to the Finite Element Method*, Affiliated East-West Press Pvt. Ltd., New Delhi, East-West Student Edition, 1977.

5. Beaufait, F.W., W.H. Rowan, P.G. Hoadley, and R.M.M.Hackett. *Computer Methods of Structural Analysis*, Prentice-Hall Inc., Englewood Cliffs, N.J., 1970.

6. Bathe, K.J., and E.L.Wilson, *Numerical Methods in Finite Element Analysis*, Prentice-Hall of India Private Limited, New Delhi, 1978.

7. Hall, A.S., *An introduction to the Mechanics of Solids*, John Wiley and Sons, N.Y., 1969.

8. Cowper, G.R., *The Shear Coefficients in Timoshenko's Beam Theory*, ASME Transactions, Journal of Applied Mechanics, Vol.33, June 1966.

9. Dym, C.L. and I.H. Shames, *Solid Mechanics*, International Student Edition, McGraw-Hill Kogakusha Ltd., 1973.

10. Ugural, A.G. and S.K.Fenster, *Advanced Strength and Applied Elasticity*, American Elsevier Publishing Co., N.Y., 1975.

11. Coull,A and A.R.Dykes, *Fundamentals of Structural Theory*, McGraw-Hill Book Co., U.K., 1972.

12. Hughes, T.J.R., Taylor, R.L. and Kanoknukulchai, W., *A Simple and Efficient Finite Element for Plate Bending*, International Journal for Numerical Methods in Engineering, Vol.11, 1977.

EXERCISES

7.1 Derive the stiffness matrix for a truss element with varying cross-section shown in Fig. E7.1 by using three nodes. Use static condensation procedure given in Chapter 3 to condense out the internal degrees of freedom. Also use the Gauss elimination procedure directly on the internal degree of freedom.

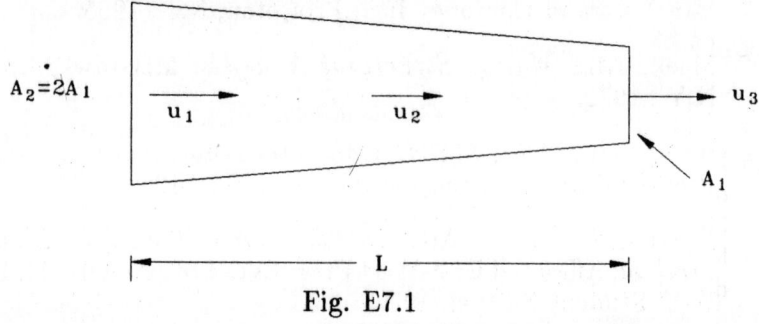

Fig. E7.1

7.2 An axial bar with non-uniform section is shown in Fig. E7.2. Treating this as consisting of two bars, obtain the stiffness matrix by direct stiffness method. Condense out the internal degree of freedom by Gauss elimination procedure.

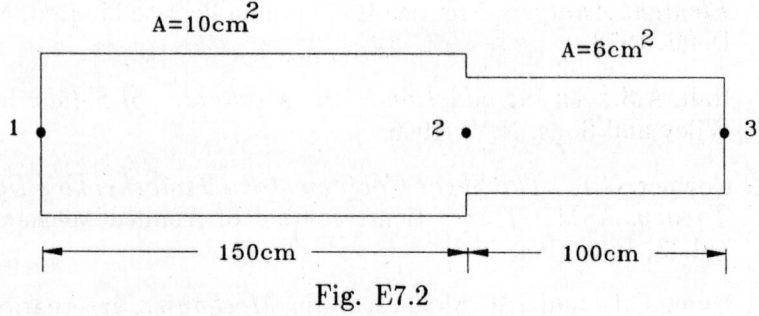

Fig. E7.2

7.3 Assume the bar shown in Fig. E7.2 has both the nodes 1 and 3 fixed. Calculate the axial force in each segment of the bar when it is subjected to a temperature rise of $10°$ C.

$$E \; = \; 20 \times 10^6 \text{ N/cm}^2 \; \text{ and } \; \alpha \; = \; 11 \times 10^{-6}/°C$$

7.4 Analyse the truss shown in Fig. E7.4. Compute the forces in each bar. The area of cross section is marked against each member.

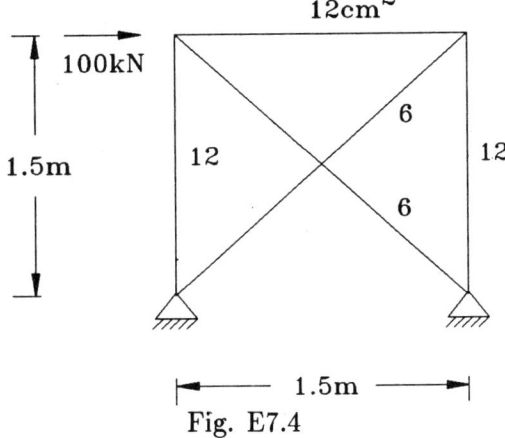

Fig. E7.4

7.5 In the guyed truss shown in Fig. E7.5, member 1 is a 7.5 mm dia steel wire and members 2,3 and 4 are aluminium alloy tubes, 30 mm in external dia with 2.0 mm wall thickness. A horizontal force of 10.0 kN is applied at the top of the structure in the X direction as shown. Calculate the force in the steel wire.

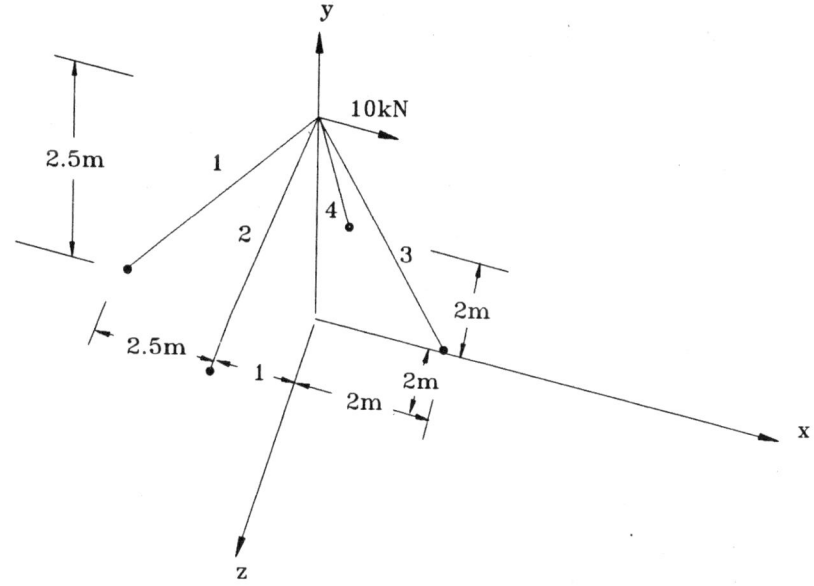

Fig. E7.5

7.6 A cantilever beam is shown in Fig. E7.6. Analyse the beam and compute the deflection and slope at the tip for the following cases of modeling:

Case 1. one non-uniform beam element
Case 2. two non-uniform beam elements
Case 3. one uniform beam element
Case 4. two uniform beam elements

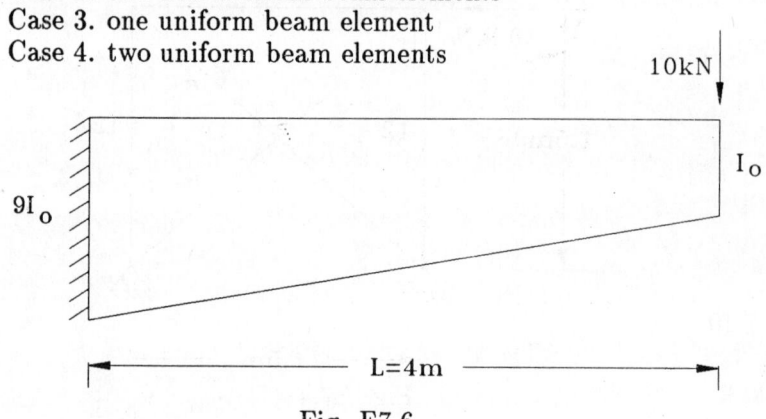

Fig. E7.6

The variation of moment of inertia is given by

$$I(x) = I_o\left[1 + r\left(\frac{x}{L}\right)^\alpha\right]$$

where $r = 8$ and $\alpha = 1$. (Hint: Refer Gallagher, R.H. and Lee, C.H., Matrix Dynamic and Instability Analysis with Non-uniform Elements, International Journal for Numerical Methods in Engineering, Vol.2, No.2, pp.265-276, 1970).

7.7 Derive the stiffness matrix of a two noded 2-D beam element with two d.o.f at each node by using a natural coordinate system with the origin at the centre of the element. Also use Gauss quadrature for integrations.

7.8 For the beam of problem 7.7, calculate the nodal load vector for the following cases: a) U.D.L. of intensity p, (b) noncentral concentrated load P, (c) uniformly varying load as shown in Fig. E7.13. Use numerical integration procedure.

7.9 Derive the stiffness matrix of 2D beam element shown in Fig. E7.9 using the static condensation procedure (to condense out the internal degrees of freedom).

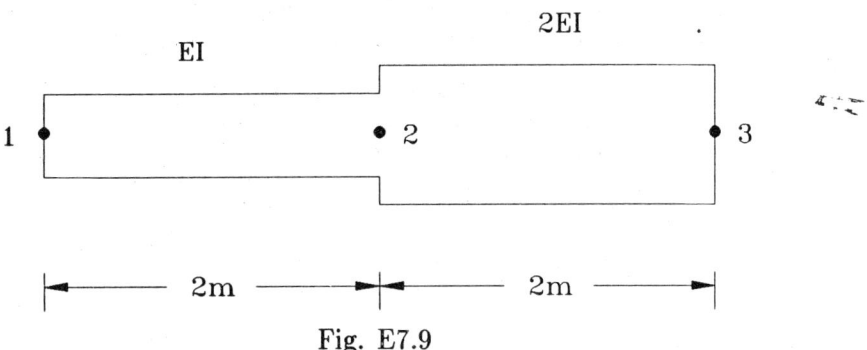

Fig. E7.9

7.10 Use the stiffness matrix of problem 9 to calculate the end moments and shear forces of a beam shown in Fig. E7.10 subjected to a load of 5 kN/m.

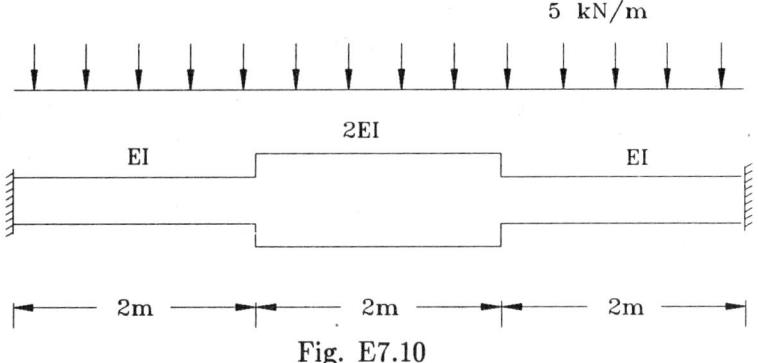

Fig. E7.10

7 11 A beam connected to truss bars is shown in Fig. E7.11.

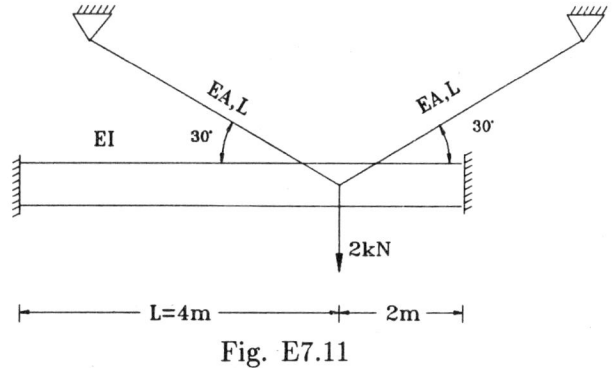

Fig. E7.11

The axial deformation in beam can be neglected. The axial

rigidity of the truss bars can be assumed to be equal to $24EI/L^2$.
Draw the bending moment and shear force diagrams for the
beam. Sketch the deflected shape.

7.12 Analyse the frame shown in Fig. E7.12. Draw the bending
moment and shear force diagrams. Sketch the deflected shape
of the structure. $E = 2 \times 10^3 \text{ kN/cm}^2$

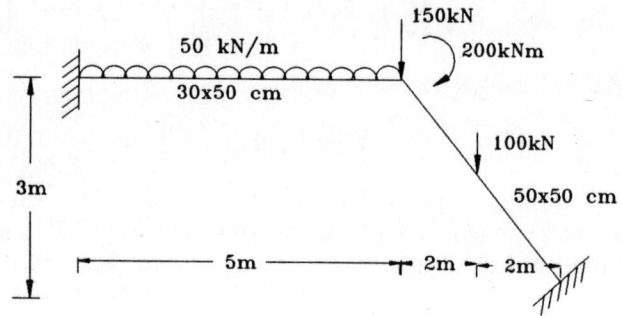

Fig. E7.12

7.13 Compute the nodal load vector for a 2D beam member subjected to an uniformly varying load as shown in Fig. E7.13.

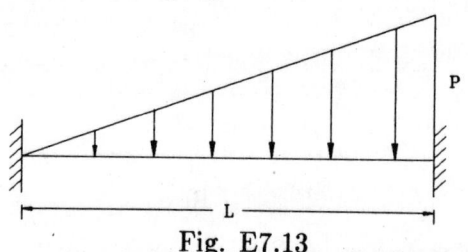

Fig. E7.13

7.14 Compute the support reaction at the node 3 of a continuous
beam shown in Fig. E7.14.

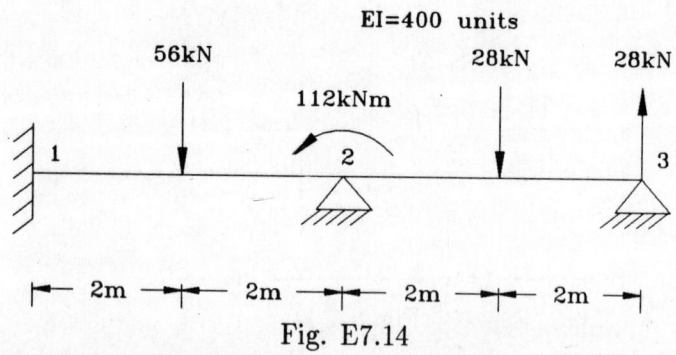

Fig. E7.14

Use the concept of boundary element. EI = 400 units.

7.15 For the beam shown in Fig. E7.14, compute the moment at the support 1 using the boundary element.

7.16 The support 2 of a continuous beam shown in Fig. E7.16 sinks vertically by 12 mm. $I_{12} = 18000 \text{cm}^4$ $I_{23} = 12000 \text{cm}^4$ $E = 20 \times 10^6 \text{ N/cm}^2$.

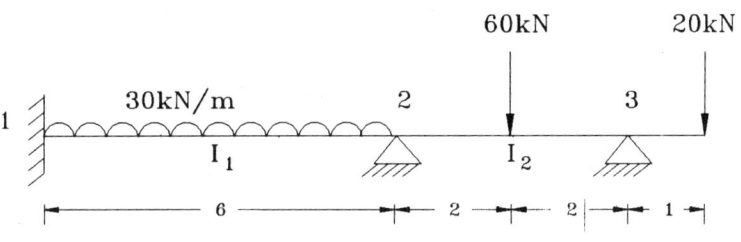

Fig. E7.16

Using the boundary element, impose the boundary condition and analyse the structure. Draw the bending moment and shear force diagrams. Sketch the deflected shape of the structure.

7.17 A truss is shown in Fig. E7.17. Analyse the truss using the PASSFEM program. Assume $E = 210 \text{ kN/mm}^2$. Area of cross section of members is given in mm^2.

Member	A
1-2 and 5-6	375
3-2	875
3-4 and 4-6	1150
2-4 and 4-5	750
2-5	500

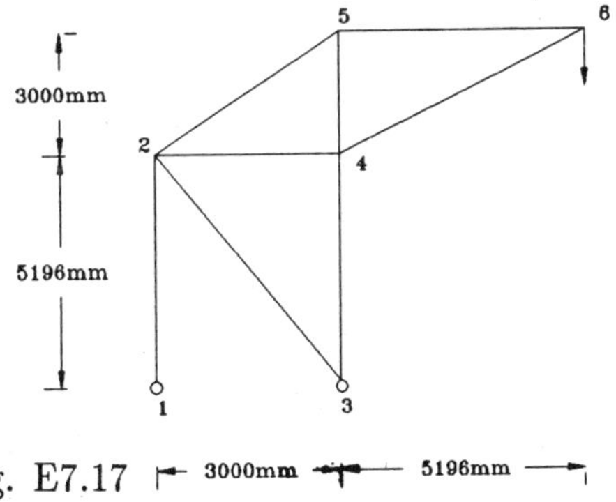

Fig. E7.17

7.18 The frame shown in Fig. E7.18 is subjected to a normal wind pressure. The cross-sectional properties can be assumed to be uniform. Analyse the frame using the PASSFEM program. Draw the bending moment and shear force diagrams. Sketch the deflected shape of the structure.

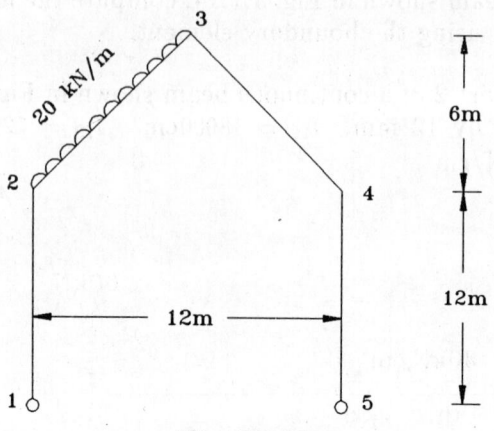

Fig. E7.18

7.19 Analyse the plane grid shown in Fig. E7.19. Assume $E = 14 \times 10^6$ kN/m^2. $G = 6.36 \times 10^6$ kN/m^2. The properties of the members are given below:

Member	L	$I \times 10^6$	$I_x \times 10^6$	
1-2	2	66.67	45.67	All the properties
3-4				are in metre units
2-3	3	27.31	19.12	

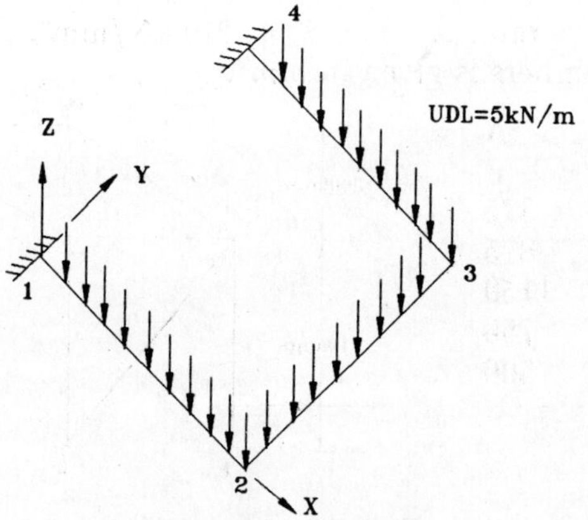

Fig. E7.19

7.20 Analyse the structure shown in Fig. E7.20. Assume

$$I_z = I_{major} = 361.6 \times 10^{-6} \ m^4$$
$$I_y = I_{minor} = 10.93 \times 10^{-6} \ m^4$$

$$I_x = 0.93 \times 10^{-6} \ m^4, \quad A = 10440 \times 10^{-6} \ m^2$$
$$E = 210 \times 10^6 \ kN/m^2 \ \ G = 80.77 \times 10^6 \ kN/m^2$$
$$L = 4m \quad udl = 10 \ kN/m$$

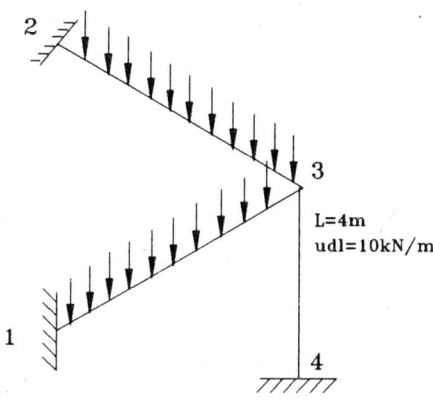

Fig. E7.20

Chapter 8

Plane Stress and Plane Strain Analysis

There are many problems in engineering which can be treated as plane stress or plane strain cases and the examples of which are presented in Chapter 2. The constant strain triangular element and four noded quadrilateral element used for such analyses are described in Chapters 3 and 4 and these elements are briefly reviewed here. The other triangular, quadrilateral and isoparametric elements which are used for plane stress or strain analysis are explained here.

The formulation can easily be extended to the analysis of axisymmetric solids subject to axisymmetric loading and this is illustrated through the derivation of element properties for a four noded isoparametric element.

For the plane stress analysis of problems involving bending the inadequacy of the lower order quadrilateral element PSQ4 is discussed. The use of incompatible modes to improve the bending properties of the element is described along with the derivation of the element matrices.

A reinforced concrete element is presented for the plane stress analysis of concrete structures, with reinforcement considered as an integral part of the element.

The subroutine PLANE has two element shapes CST and PSQ4. It provides option for the analysis of plane stress/strain and axisymmetric solids and for inclusion of incompatible mode to PSQ4. The input and output details are described. Examples are presented to illustrate the use of the routine.

8.1 Triangular Elements

Triangular elements are relatively simple in their formulation and can easily be programmed. They can be used along with the other element shapes to discretize regions of steep strain gradients and irregular boundaries. Two types of triangular elements are described in the following.

8.1.1 Constant Strain Triangle (CST)

The element shown in Fig. 8.1 has three nodes and six unknown nodal displacements. The shape function for the element in natural coordinates is given in Eq.3.39. The properties of the element are given in Eqs. 3.85, 3.99 and 3.105. This is one of the simple elements and can easily be programmed using the above equations.

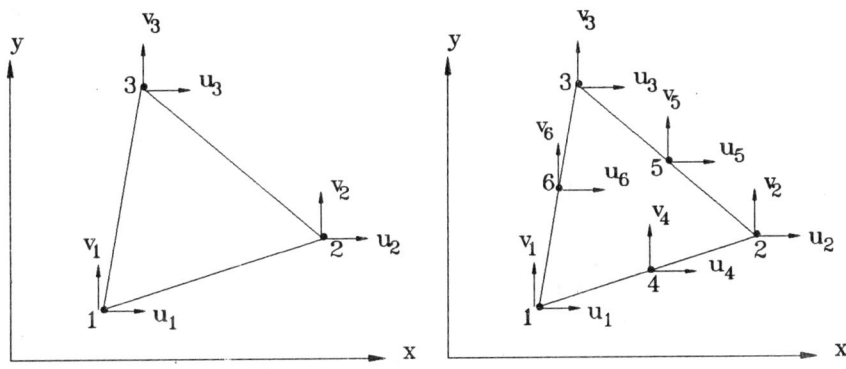

Fig. 8.1 CST Element   Fig. 8.2 LST Element

8.1.2 Linear Strain Triangle LST

The constant strain triangular element provides in general a reasonable displacement pattern but the interpretation of element stresses is often not adequate in regions of high strain gradients. The linear strain triangle, explained in reference [1,2], has been extensively tried and found to represent a significant improvement over CST meshes having similar or larger number of degrees of freedom.

The LST element shown in Fig.8.2 has three primary nodes and three secondary nodes at the mid-points of the sides. The shape functions used are quadratic and are given by Eqs. 3.46 and 3.47.

$$\{u\} = \begin{Bmatrix} u \\ v \end{Bmatrix} = \begin{bmatrix} \{N_2\}^T & \{0\}^T \\ \{0\}^T & \{N_2\}^T \end{bmatrix} \{d\} \qquad (8.1)$$

where $\{N_2\}^T = [L_1(2L_1-1) \quad L_2(2L_2-1) \quad L_3(2L_3-1) \quad 4L_1L_2 4L_2L_3 \quad 4L_3L_1]$
and $\{d\}^T = [u_1 \quad u_2 \quad u_3 \quad u_4 \quad u_5 \quad u_6 \quad v_1 \quad v_2 \quad v_3 \quad v_4 \quad v_5 \quad v_6]$

In practice, the derivation of stiffness matrix [k] using the Eq.3.97a is unnecessarily cumbersome, especially for refined (higher-order) elements. A more elegant procedure has been used by Felippa[1] and is briefly described below.

In this approach the stresses and strains are expressed in terms of the nodal stresses and strains respectively. Because the order of the polynomials representing the variation of stress and strain is less than the order of the displacement functions by atleast one, a smaller number of nodal values of the stresses and strains is usually sufficient to define the element stresses and strains. For example, to specify the displacements u and v for the LST element six nodal values of each of the two displacements are needed. However, the strain variation in this case is linear, so only three nodal values of each strain, say at corner nodes, will suffice to define the element strains.

The stresses and strains may be expressed in terms of their nodal values by interpolation functions.

$$\{\epsilon\} = [N_\epsilon] \{\epsilon_n\} \tag{8.2a}$$

and

$$\{\sigma\} = [N_\sigma] \{\sigma_n\} \tag{8.2b}$$

where the subscript n indicates a vector of nodal values. If the material properties are uniform for the element, the interpolation functions $[N_\epsilon]$ and $[N_\sigma]$ are identical. Furthermore the nodal values of the strains can be expressed in terms of the nodal displacements by evaluating the matrix [B] of the Eq. 3.91 at the appropriate nodes. Hence,

$$\{\epsilon_n\} = [B_n] \{d\} \tag{8.3}$$

The stresses at the nodes are related to the strains at the nodes by the constitutive matrix which has the appropriate nodal values of the material properties.

$$\{\sigma_n\} = [C_n] \{\epsilon_n\} \tag{8.4}$$

As a result of the above expressions, the strain energy of the element can be expressed as follows.

$$U = \frac{1}{2} \int\int\int_V \{\epsilon\}^T \{\sigma\} \, dV = \frac{1}{2} \{\epsilon_n\}^T \int\int\int_V [N_\epsilon]^T [N_\sigma] \{\sigma_n\} \, dV \tag{8.5a}$$

Substituting from Eqs. 8.2 and 8.3, we get

$$U = \frac{1}{2} \{d\}^T [B_n]^T \int\int\int_V [N_\epsilon]^T [N_\sigma] [C_n] [B_n] \{d\} \, dV \qquad (8.5b)$$

Therefore, the stiffness matrix can be given by

$$[k] = [B_n]^T [D] [C_n] [B_n] \qquad (8.6a)$$

where

$$[D] = \int\int\int_V [N_\epsilon]^T [N_\sigma] \, dV \qquad (8.6b)$$

The integration of Eq. 8.6b is much easier than that of Eq.3.97a, because advantage can be had of repetitive patterns of the products in it.

The strain displacement relation for the LST can be worked out following the procedure indicated for the CST element in section 3.6. It can be expressed as

$$\{\epsilon\} = \begin{bmatrix} \{B_1\}^T & \{0\}^T \\ \{0\}^T & \{B_2\}^T \\ \{B_2\}^T & \{B_1\}^T \end{bmatrix} \left\{ \begin{matrix} \{d_u\} \\ \{d_v\} \end{matrix} \right\} = [B] \{d\} \qquad (8.7)$$

where

$$\{\epsilon\} = \left\{ \begin{matrix} \epsilon_x \\ \epsilon_y \\ \gamma_{xy} \end{matrix} \right\}$$

$$\{B_1\}^T = \frac{1}{2A} [(4L_1 - 1)b_1 \ (4L_2 - 1)b_2 \ (4L_3 - 1)b_3$$
$$4(L_2 b_1 + L_1 b_2) \ 4(L_3 b_2 + L_2 b_3) \ 4(L_1 b_3 + L_3 b_1)] \qquad (8.8a)$$

and

$$\{B_2\}^T = \frac{1}{2A} [(4L_1 - 1)a_1 \ (4L_2 - 1)a_2 \ (4L_3 - 1)a_3$$
$$4(L_2 a_1 + L_1 a_2) \ 4(L_3 a_2 + L_2 a_3) \ 4(L_1 a_3 + L_3 a_1)] \qquad (8.8b)$$

The nodal strain vector is

$$\{\epsilon_n\}^T = [\epsilon_{x1} \ \ \epsilon_{x2} \ \ \epsilon_{x3} \ \ \epsilon_{y1} \ \ \epsilon_{y2} \ \ \epsilon_{y3} \ \ \gamma_{xy1} \ \ \gamma_{xy2} \ \ \gamma_{xy3}] \qquad (8.9)$$

$$\{\epsilon_n\} = \begin{bmatrix} [B_{n1}] & [0] \\ [0] & [B_{n2}] \\ [B_{n2}] & [B_{n1}] \end{bmatrix} \{d\} \qquad (8.10)$$

where

$$[B_{n1}] = \frac{1}{2A} \begin{bmatrix} 3b_1 & -b_2 & -b_3 & 4b_2 & 0 & 4b_3 \\ -b_1 & 3b_2 & -b_3 & 4b_1 & 4b_3 & 0 \\ -b_1 & -b_2 & +3b_3 & 0 & 4b_2 & 4b_1 \end{bmatrix} \qquad (8.11a)$$

and

$$[B_{n2}] = \frac{1}{2A} \begin{bmatrix} 3a_1 & -a_2 & -a_3 & 4a_2 & 0 & 4a_3 \\ -a_1 & 3a_2 & -a_3 & 4a_1 & 4a_3 & 0 \\ -a_1 & -a_2 & +3a_3 & 0 & 4a_2 & 4a_1 \end{bmatrix} \qquad (8.11b)$$

The nodal stress vector is

$$\{\sigma_n\}^T = [\sigma_{x1} \quad \sigma_{x2} \quad \sigma_{x3} \quad \sigma_{y1} \quad \sigma_{y2} \quad \sigma_{y3} \quad \tau_{xy1} \quad \tau_{xy2} \quad \tau_{xy3}] \qquad (8.12)$$

Assuming the material properties are constant over the element the strains and stresses vary linearly and can be expressed by the following interpolation functions,

$$[N_\epsilon] = [N_\sigma] = \begin{bmatrix} \{N_1\}^T & \{0\}^T & \{0\}^T \\ \{0\}^T & \{N_1\}^T & \{0\}^T \\ \{0\}^T & \{0\}^T & \{N_1\}^T \end{bmatrix} \qquad (8.13a)$$

where,

$$\{N_1\}^T = [L_1 \quad L_2 \quad L_3] \qquad (8.13b)$$

For two-dimensional problems the constitutive matrix [C] is of the form

$$[C] = \begin{bmatrix} C_{11} & C_{12} & C_{13} \\ C_{21} & C_{22} & C_{23} \\ C_{31} & C_{32} & C_{33} \end{bmatrix}$$

The nodal values of the constitutive matrix $[C_n]$ can now be expressed as,

$$[C_n] = \begin{bmatrix} C_{11}[I] & C_{12}[I] & C_{13}[I] \\ C_{21}[I] & C_{22}[I] & C_{23}[I] \\ C_{31}[I] & C_{32}[I] & C_{33}[I] \end{bmatrix} \qquad (8.14)$$

where [I] is the 3 × 3 identity matrix.

Assuming that the thickness of the element is constant and of value h, the matrix $[D]$ in Eq. 8.6b can be evaluated as follows.

$$[D] = \int \int \int_V [N_\epsilon]^T \, [N_\sigma] \, dV = h \int \int_A [N_\epsilon]^T \, [N_\sigma] \, dA$$

$$= \begin{bmatrix} [D_1] & [0] & [0] \\ [0] & [D_1] & [0] \\ [0] & [0] & [D_1] \end{bmatrix} \qquad (8.15a)$$

where

$$[D_1] = h \int \int_A \{N_1\} \, \{N_1\}^T \, dA \qquad (8.15b)$$

And using the formula given in Eq. (3.24) the above integral can be evaluated as

$$[D_1] = \frac{Ah}{12} \begin{bmatrix} 2 & 1 & 1 \\ 1 & 2 & 1 \\ 1 & 1 & 2 \end{bmatrix} \qquad (8.16)$$

For the evaluation of the stiffness matrix $[k]$ given in Eq. (8.6a) as

$$[k] = [B_n]^T \, [D] \, [C_n] \, [B_n]$$

The matrix $[B_n]$ is evaluated by substituting the values of $[B_{n1}]$ and $[B_{n2}]$ from Eq.8.11(a) and 8.11(b) into the Eq.8.10 and $[C_n]$ is calculated by using Eq.8.14. The matrix $[D]$ is computed by evaluating $[D_1]$ using Eq. 8.16 and then substituting into Eq. 8.15(a). Then the matrix multiplication can be performed to compute the matrix $[k]$.

The properties of isoparametric curved six noded element can be evaluated following the procedure described in Chapter 4. Numerical integration for triangular domain is adopted for computing the stiffness matrix $[k]$.

8.1.3 Computation of Nodal Load Vector

The nodal load due to body forces and surface traction for a CST element has been worked out in subsection 3.7.1. In the following description the nodal load computation is illustrated for a LST element.

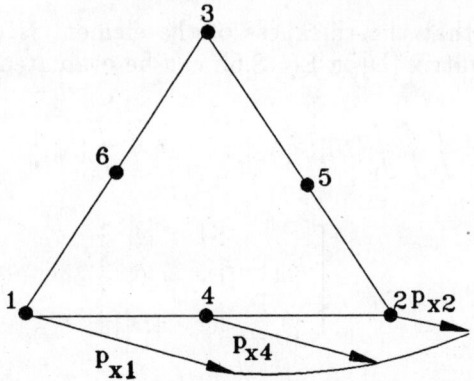

Fig. 8.3 Nodal Load Computation-LST Element

Consider a LST element subjected to quadratically varying surface force acting in the x-direction on one of the sides say 1-4-2 (Fig.8.3). The variation of the load on the side can be represented by the following interpolation function.

$$p_x = [L_1\,(2L_1 - 1) \quad L_2\,(2L_2 - 1) \quad 4L_1L_2] \begin{Bmatrix} p_{x1} \\ p_{x2} \\ p_{x4} \end{Bmatrix} \qquad (8.17)$$

The nodal load vector given in Eq. 3.92 can be split up into two parts as,

$$\{Q\} = \begin{Bmatrix} \{Q_x\} \\ \{Q_y\} \end{Bmatrix} = \int\int [N^s]^T \begin{Bmatrix} p_x \\ p_y \end{Bmatrix} dS \qquad (8.18)$$

Since in this illustration the surface traction is applied parallel to the x-axis only, $\{Q_y\} = \{0\}$ and assuming p_x is the intensity per unit length,

$$\{Q_x\} = \int [N_2^s]\, p_x\, dS \qquad (8.19)$$

which integral is to be evaluated as a line integral along the side 3, i.e., 1-4-2. Thus

$$\{Q_x\} = \int \begin{bmatrix} L_1(2L_1 - 1) \\ L_2(2L_2 - 1) \\ L_3(2L_3 - 1) \\ 4L_1L_2 \\ 4L_2L_3 \\ 4L_3L_1 \end{bmatrix} [L_1(2L_1 - 1) \; L_2(2L_2 - 1) \; 4L_1L_2] \begin{Bmatrix} p_{x1} \\ p_{x2} \\ p_{x4} \end{Bmatrix} ds_3$$

$$(8.20)$$

It may be observed that along the side $L_3 = 0; L_2 = 1 - L_1$ and $ds_3 = -l_3dL_1$. Substituting in Eq.8.20 leads to $\{Q_x\} =$

$$= l_3 \int_0^1 \begin{bmatrix} L_1(2L_1 - 1) \\ (1 - L_1)(1 - 2L_1) \\ 0 \\ 4L_1(1 - L_1) \\ 0 \\ 0 \end{bmatrix} [L_1(2L_1 - 1)(1 - L_1)(1 - 2L_1) \; 4L_1(1 - L_1)] \begin{Bmatrix} p_{x1} \\ p_{x2} \\ p_{x4} \end{Bmatrix} dL_1$$

$$(8.21)$$

The above integration can easily be evaluated and results in,

$$\begin{Bmatrix} Q_{x1} \\ Q_{x2} \\ Q_{x4} \end{Bmatrix} = \frac{l_3}{15} \begin{bmatrix} 2 & -1/2 & 1 \\ -1/2 & 2 & 1 \\ 1 & 1 & 8 \end{bmatrix} \begin{Bmatrix} p_{x1} \\ p_{x2} \\ p_{x4} \end{Bmatrix} \qquad (8.22)$$

The other components of $\{Q_x\}$i.e., Q_{x3}, Q_{x5} and Q_{x6} are zero as there are no surface tractions on the other two sides.

8.2   Rectangular Elements

In the case of stress analysis problems with simple straight boundaries like beams or plates subject to inplane forces, rectangular elements can be used. The main advantage of these elements is the simplicity in formulation and the students may also find it easy to develop the relevant routines giving them good exercise to deal with more complex elements later. Rectangular elements with 4 and 8 nodes are described in the following pages.

8.2.1   Four Noded Rectangle (PSR4)

The rectanglular element shown in Fig.8.4 has four nodes and this element used for plane stress and strai analysis will be referred to as PSR4. The natural coordinates of the element are given by Eq.3.48. And the shape functions for the displacement variation are given by Eqs. 3.52 and 3.53.

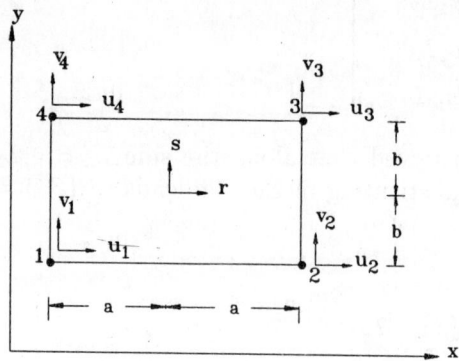

Fig. 8.4 Four Noded Rectangle-PSR4

Following the reasons given in section 4.2, these equations can be rearranged as,

$$
\begin{Bmatrix} u \\ v \end{Bmatrix} =
\begin{bmatrix}
N_1 & 0 & N_2 & 0 & N_3 & 0 & N_4 & 0 \\
0 & N_1 & 0 & N_2 & 0 & N_3 & 0 & N_4
\end{bmatrix}
\begin{Bmatrix} u_1 \\ v_1 \\ u_2 \\ v_2 \\ u_3 \\ v_3 \\ u_4 \\ v_4 \end{Bmatrix}
\tag{8.23a}
$$

where

$$
N_1 = \frac{(1-r)(1-s)}{4} \qquad N_3 = \frac{(1+r)(1+s)}{4}
$$

$$
N_2 = \frac{(1+r)(1-s)}{4} \qquad N_4 = \frac{(1-r)(1+s)}{4}
\tag{8.23b}
$$

For the chosen shape functions, the displacement variation will be linear along the edges and bi-linear inside the element. For the computation of element properties, the derivatives with respect to cartesian coordinates are given by

$$
\frac{\partial f}{\partial x} = \frac{1}{a}\frac{\partial f}{\partial r} \qquad \text{and} \qquad \frac{\partial f}{\partial y} = \frac{1}{b}\frac{\partial f}{\partial s}
\tag{8.24}
$$

The strain-displacement matrix [B] can be computed using the above equation and thus,

$$\{\epsilon\} = [B]\{d\}$$

where $[B] =$

$$\frac{1}{4}\begin{bmatrix} \frac{-(1-s)}{a} & 0 & \frac{(1-s)}{a} & 0 & \frac{(1+s)}{a} & 0 & \frac{-(1+s)}{a} & 0 \\ 0 & \frac{-(1-r)}{b} & 0 & \frac{-(1+r)}{b} & 0 & \frac{(1+r)}{b} & 0 & \frac{(1-r)}{b} \\ \frac{-(1-r)}{b} & \frac{-(1-s)}{a} & \frac{-(1+r)}{b} & \frac{(1-s)}{a} & \frac{(1+r)}{b} & \frac{(1+s)}{a} & \frac{(1-r)}{b} & \frac{-(1+s)}{a} \end{bmatrix}$$

$$(8.25a)$$

and

$$\{d\}^T = [u_1 \quad v_1 \quad u_2 \quad v_2 \quad u_3 \quad v_3 \quad u_4 \quad v_4] \qquad (8.25b)$$

Assuming constant thickness the stiffness matrix for the element is given by,

$$[k] = h \int \int [B]^T [C] [B] \, dA$$

It may be noted that $dA = ab \, dr \, ds$ and hence

$$[k] = abh \int_{-1}^{+1} \int_{-1}^{+1} [B]^T [C] [B] \, dr \, ds \qquad (8.26)$$

8.2.2 Computation of Nodal Load Vector

The nodal load vector due to surface traction can be calculated by following the procedure indicated in section 8.1 for LST element. Figure 8.5 shows a rectangular element subject to surface force only on the side 2-3 with intensities per unit length as p_{x2} and p_{x3} at nodes 2 and 3 respectively.

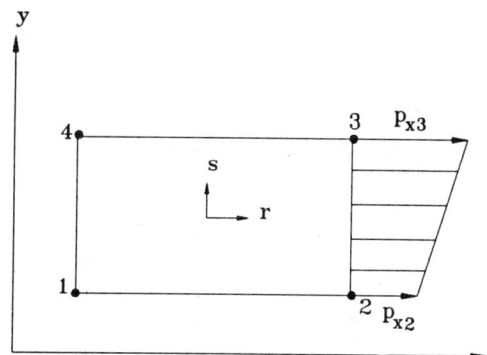

Fig. 8.5 PSR4 Element Subjected to Surface Force

At any point on the side 2-3 the intensity of the surface traction can be expressed by an interpolation function as,

$$p_x = \frac{1}{2}\left[(1-s)(1+s)\right]\begin{Bmatrix} p_{x2} \\ p_{x3} \end{Bmatrix} \tag{8.27}$$

The nodal load vector is evaluated as,

$$\{Q_x\} = \int \{N_2^s\}\, p_x\, ds_1 \tag{8.28}$$

Along the side 2-3, $r = 1$, Hence,

$$\{N_2^s\} = \frac{1}{2}\begin{Bmatrix} 0 \\ (1-s) \\ (1+s) \\ 0 \end{Bmatrix} \tag{8.29}$$

Also $\int_{-1}^{+1} ds_1 = 2b$ and therefore, $ds_1 = bds$. Substituting Eq. 8.27 in Eq.8.28,

$$\{Q_x\} = \frac{b}{4}\int_{-1}^{+1}\begin{Bmatrix} 0 \\ (1-s) \\ (1+s) \\ 0 \end{Bmatrix}\left[(1-s)(1+s)\right]\begin{Bmatrix} p_{x2} \\ p_{x3} \end{Bmatrix} ds = \frac{b}{3}\begin{Bmatrix} 0 \\ 2p_{x2}+p_{x3} \\ p_{x2}+2p_{x3} \\ 0 \end{Bmatrix} \tag{8.30}$$

8.2.3　Eight Noded Rectangle (PSR8)

To improve on the first order rectangular element PSR4, explained above, the displacement along the sides are made to vary quadratically. This requires one additional node along each side as shown in Fig.8.6.

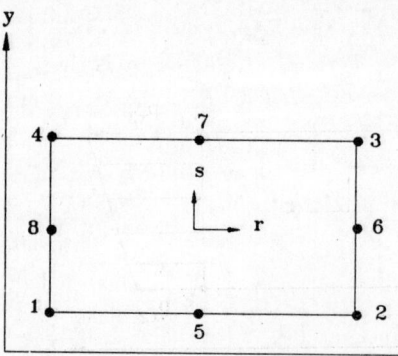

Fig. 8.6 Eight Noded Rectangle-PSR4

The shape function for displacement for this element is given by Eqs. 3.59 and 3.60. As observed earlier the terms can be rearranged to match the displacements $u_i, v_i (i = 1, ...8)$ taken successively at each node as

$$
\left\{ \begin{array}{c} u \\ v \end{array} \right\} = \left[\begin{array}{cccccccccccc} N_1 & 0 & N_2 & 0 & N_3 & 0 & N_4 & 0 & N_5 & 0 & N_6 & 0 \\ 0 & N_1 & 0 & N_2 & 0 & N_3 & 0 & N_4 & 0 & N_5 & 0 & N_6 \end{array} \right.
$$

$$
\left. \begin{array}{cccc} N_7 & 0 & N_8 & 0 \\ 0 & N_7 & 0 & N_8 \end{array} \right]
\left\{ \begin{array}{c} u_1 \\ v_1 \\ u_2 \\ v_2 \\ u_3 \\ v_3 \\ u_4 \\ v_4 \\ \vdots \\ \vdots \\ \vdots \\ u_8 \\ v_8 \end{array} \right\}
$$

i.e., $$\{u\} = [N]\{d\} \tag{8.30a}$$

where the elements $N_i (i = 1, ...8)$ are defined by Eq. 3.59a.

The strain displacement matrix [B] can now be evaluated as

$$\{\epsilon\} = [B]\{d\}$$

where

$$
[B] = \left[\begin{array}{cccccccc} \frac{\partial N_1}{\partial x} & 0 & \frac{\partial N_2}{\partial x} & 0 & \frac{\partial N_3}{\partial x} & 0 & \frac{\partial N_4}{\partial x} & 0 \\ 0 & \frac{\partial N_1}{\partial y} & 0 & \frac{\partial N_2}{\partial y} & 0 & \frac{\partial N_3}{\partial y} & 0 & \frac{\partial N_4}{\partial y} \\ \frac{\partial N_1}{\partial y} & \frac{\partial N_1}{\partial x} & \frac{\partial N_2}{\partial y} & \frac{\partial N_2}{\partial x} & \frac{\partial N_3}{\partial y} & \frac{\partial N_3}{\partial x} & \frac{\partial N_4}{\partial y} & \frac{\partial N_4}{\partial x} \end{array} \right.
$$

$$
\left. \begin{array}{cccccccc} \frac{\partial N_5}{\partial x} & 0 & \frac{\partial N_6}{\partial x} & 0 & \frac{\partial N_7}{\partial x} & 0 & \frac{\partial N_8}{\partial x} & 0 \\ 0 & \frac{\partial N_5}{\partial y} & 0 & \frac{\partial N_6}{\partial y} & 0 & \frac{\partial N_7}{\partial y} & 0 & \frac{\partial N_8}{\partial y} \\ \frac{\partial N_5}{\partial y} & \frac{\partial N_5}{\partial x} & \frac{\partial N_6}{\partial y} & \frac{\partial N_6}{\partial x} & \frac{\partial N_7}{\partial y} & \frac{\partial N_7}{\partial x} & \frac{\partial N_8}{\partial y} & \frac{\partial N_8}{\partial x} \end{array} \right] \tag{8.31}
$$

The derivatives of the shape functions required in the above equation are given below.

$$\frac{\partial N_1}{\partial x} = (1-s)(2r+s)/4a \qquad \frac{\partial N_2}{\partial x} = (1-s)(2r-s)/4a$$

$$\frac{\partial N_3}{\partial x} = (1+s)(2r+s)/4a \qquad \frac{\partial N_4}{\partial x} = (1+s)(2r-s)/4a$$

$$\frac{\partial N_5}{\partial x} = -(1-s)r/a \qquad \frac{\partial N_6}{\partial x} = -(1-s^2)/2a$$

$$\frac{\partial N_7}{\partial x} = -(1+s)r/a \qquad \frac{\partial N_8}{\partial x} = -(1-s^2)r/a$$

$$\frac{\partial N_1}{\partial y} = (1-r)(2s+r)/4b \qquad \frac{\partial N_2}{\partial y} = (1+r)(2s-r)/4b$$

$$\frac{\partial N_3}{\partial y} = (1+r)(2s+r)/4b \qquad \frac{\partial N_4}{\partial y} = (1-r)(2s-r)/4b$$

$$\frac{\partial N_5}{\partial y} = -(1-r^2)/2b \qquad \frac{\partial N_6}{\partial y} = -(1+r)s/b$$

$$\frac{\partial N_7}{\partial y} = (1-r^2)/2b \qquad \frac{\partial N_8}{\partial y} = -(1-r)s/b$$

$$(8.32)$$

Assuming constant thickness h, the element stiffness matrix is given by

$$[k] = abh \int_{-1}^{+1} \int_{-1}^{+1} [B]^T [C] [B] \, dr \, ds \qquad (8.33)$$

The above integration can easily be evaluated using Gauss quadrature explained in Chapter 4. And a 2 × 2 scheme has been found to give good results.

8.3   Isoparametric Elements

It has been observed in Chapter 4 that isoparametric elements are widely used in practice due to the generality of approach suitable for computer programming and their adaptability for the idealisation of complex shapes. Two elements, viz., four noded quadrilateral (PSQ4) and eight noded element (PSQ8) are described here.

8.3.1   Four Noded Quadrilateral (PSQ4)

The shape functions for the element and the evaluation of the element stiffness matrix [k] using the numerical integration procedure have been explained in Chapter 4. And the equations given therein are adequate for the development of the routines to compute the element stiffness matrix [k]. However, it is necessary to illustrate the computation of nodal load vector $\{Q\}$ due to surface traction for this isoparametric element [3].

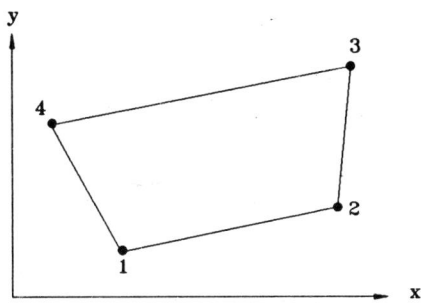

Fig. 8.7 Isoparametric Element - PSQ4

Let the element shown in Fig.8.8 be subjected to uniformly varying surface pressure of intensities p_x and p_y per unit area of the surface along the side 1-2.

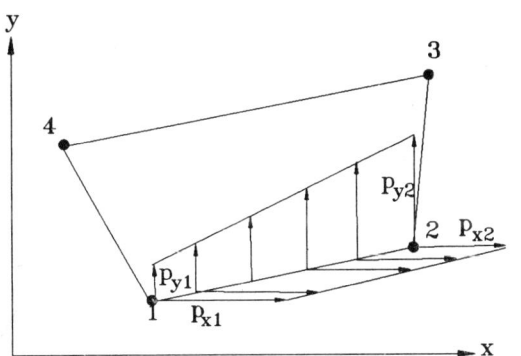

Fig. 8.8 PSQ 4 Element Subjected to Surface Pressure

The pressure p_x and p_y at any point along the side 1-2 can be expressed by the following interpolation functions

$$\left\{\begin{matrix} p_x \\ p_y \end{matrix}\right\} = \{p\} = \begin{bmatrix} \frac{(1-r)}{2} & 0 & \frac{(1+r)}{2} & 0 \\ 0 & \frac{(1-r)}{2} & 0 & \frac{(1+r)}{2} \end{bmatrix} \left\{\begin{matrix} p_{x1} \\ p_{y1} \\ p_{x2} \\ p_{y2} \end{matrix}\right\} \qquad (8.34)$$

Since the pressure is applied on the side 1-2, the shape function for displacement variation along the edges can be got by setting $s = -1$ in Eqs. 8.23a and 8.23b. Thus,

$$[N^s] = \begin{bmatrix} \frac{(1-r)}{2} & 0 & \frac{(1+r)}{2} & 0 & 0 & 0 & 0 & 0 \\ 0 & \frac{(1-r)}{2} & 0 & \frac{(1+r)}{2} & 0 & 0 & 0 & 0 \end{bmatrix} \tag{8.35}$$

We know that the nodal load vector due to surface traction is calculated as

$$\{Q\} = \int\int_s [N^s]^T \{p\} \, dS \tag{8.18}$$

where in this case $[N^s]$ and $\{p\}$ are given by Eqs. 8.35 and 8.34 respectively. If the thickness of the element at any point r along this edge is h_r, $dS = h_r \, dl$ where dl is the differential length. It can be shown that

$$dl = \{ (\frac{\partial x}{\partial r})^2 + (\frac{\partial y}{\partial r})^2 \}^{1/2} dr \tag{8.36}$$

The derivatives $\frac{\partial x}{\partial r}$ and $\frac{\partial y}{\partial r}$ are the elements of the Jacobian given by Eq.4.6. And substituting for $s = -1$, we get

$$\frac{\partial x}{\partial r} = \frac{x_2 - x_1}{2} \qquad \frac{\partial y}{\partial r} = \frac{y_2 - y_1}{2} \tag{8.37}$$

Substituting in Eqn (8.36),

$$dl = \frac{l}{2} \, dr \tag{8.38}$$

where l is the length of the side 1-2.
Hence, the integral of Eq.(3.18) reduces to

$$\{Q\} = \int_{-1}^{+1} [N^s]^T \{p\} \, h_r \, dl \tag{8.39}$$

The above integral is a line integral and in this case after substituting for all the relevant quantities from Eqs.8.32, 8.35 and 8.38, it can be evaluated explicitly. But in general it becomes necessary to perform the integration numerically. In order to illustrate this procedure the integral of Eq.8.39 is evaluated as described below.

The numerical integration, using Gauss Quadrature is explained in section 4.2.1. Thus

$$\{Q\} = \frac{l}{2} \int_{-1}^{+1} [N^s]^T \{p\} \, h_r \, dr = \sum_{i=1}^{n} W_i \, f(r_i) \tag{8.40}$$

where

$$f(r) = \frac{l}{2} [N^s]^T \{p\} h_r$$

If the thickness is assumed constant,

$$f(r) = \frac{lh}{2} [N^s]^T \{p\} \tag{8.41}$$

Selecting two sampling points, the integral is evaluated as,

$$\{Q\} = W_1 f(r_1) + W_2 f(r_2)$$

where

$$r_1 = \frac{1}{\sqrt{3}}$$

$$r_2 = \frac{-1}{\sqrt{3}}$$

and

$$W_1 = W_2 = 1$$

The above values for r_1 and r_2 are substituted in Eqs 8.34 and 8.35 to evaluate $\{Q\}$ using the Eqs. 8.40 and 8.41. The process is indicated in Eq.8.42a. Simplifying Eq. 8.42a, we get

$$\{Q\} = \frac{lh}{6} \begin{bmatrix} 2p_{x1} + p_{x2} \\ 2p_{y1} + p_{y2} \\ p_{x1} + 2p_{x2} \\ p_{y1} + 2p_{y2} \\ 0 \\ 0 \\ 0 \\ 0 \end{bmatrix} \tag{8.42b}$$

$$
\{Q\} =
\overbrace{
\begin{Bmatrix}
Q_{x1} \\ Q_{y1} \\ Q_{x2} \\ Q_{y2} \\ \vdots \\ Q_{x3} \\ Q_{y3} \\ Q_{x4} \\ Q_{y4}
\end{Bmatrix}
}
= \frac{lh}{8}
\begin{bmatrix}
\left(1-\tfrac{1}{\sqrt{3}}\right) & 0 & \left(1-\tfrac{1}{\sqrt{3}}\right) & 0 \\
0 & \left(1-\tfrac{1}{\sqrt{3}}\right) & 0 & \left(1-\tfrac{1}{\sqrt{3}}\right) \\
\left(1+\tfrac{1}{\sqrt{3}}\right) & 0 & \left(1+\tfrac{1}{\sqrt{3}}\right) & 0 \\
0 & \left(1+\tfrac{1}{\sqrt{3}}\right) & 0 & \left(1+\tfrac{1}{\sqrt{3}}\right) \\
0 & 0 & 0 & 0 \\
0 & 0 & 0 & 0 \\
0 & 0 & 0 & 0 \\
0 & 0 & 0 & 0
\end{bmatrix}
\overbrace{
\begin{Bmatrix}
p_{x1} \\ p_{y1} \\ p_{x2} \\ p_{y2}
\end{Bmatrix}
}
$$

$$
+\frac{8}{lh}
\begin{bmatrix}
\left(1+\tfrac{1}{\sqrt{3}}\right) & 0 \\
0 & \left(1+\tfrac{1}{\sqrt{3}}\right) \\
\left(1-\tfrac{1}{\sqrt{3}}\right) & 0 \\
0 & \left(1-\tfrac{1}{\sqrt{3}}\right) \\
0 & 0 \\
0 & 0 \\
0 & 0 \\
0 & 0
\end{bmatrix}
\begin{bmatrix}
\left(1+\tfrac{1}{\sqrt{3}}\right) & 0 & \left(1-\tfrac{1}{\sqrt{3}}\right) & 0 \\
0 & \left(1+\tfrac{1}{\sqrt{3}}\right) & 0 & \left(1-\tfrac{1}{\sqrt{3}}\right)
\end{bmatrix}
\begin{Bmatrix}
p_{x1} \\ p_{y1} \\ p_{x2} \\ p_{y2}
\end{Bmatrix}
\tag{8.42a}
$$

8.3.2 Eight Noded Isoparametric Element (PSQ8)

For plane stress or plane strain analysis of problems with complex shapes or curved boundaries we may have to use curved elements for better idealisation. The eight noded isoparametric element shown in Fig.8.9 is a higher order element with quadratic variation of displacement along its boundaries.

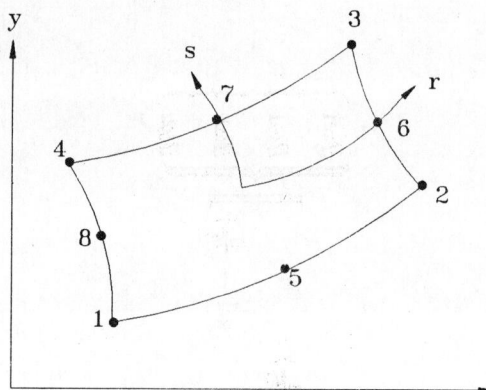

Fig. 8.9 Eight Noded Isoparametric Element - PSQ8

The concept of isoparametric elements is discussed in Chapter 4. Thus it may be observed that the element can be derived from the parent eight noded rectangle mapped into a square in natural coordinates $r - s$. Hence the shape functions given in Eq.3.59 would be used to describe the geometry of this element. Thus

$$
\begin{Bmatrix} x \\ y \end{Bmatrix} = \begin{bmatrix} N_1 & 0 & N_2 & 0 & N_3 & 0 & N_4 & 0 & N_5 & 0 & N_6 \\ 0 & N_1 & 0 & N_2 & 0 & N_3 & 0 & N_4 & 0 & N_5 & 0 \end{bmatrix}
$$

$$
\left. \begin{matrix} 0 & N_7 & 0 & N_8 & 0 \\ N_6 & 0 & N_7 & 0 & N_8 \end{matrix} \right] \{x_n\} \qquad (8.43a)
$$

i.e.,

$$
\{x\} = [N]\{x_n\}
$$

where

$$
\{x_n\}^T = \begin{bmatrix} x_1 & y_1 & x_2 & y_2 & x_3 & y_3 & x_4 & y_4 & x_5 & y_5 & x_6 & y_6 & x_7 \end{bmatrix}
$$

$$
y_7 \quad x_8 \quad y_8 \end{bmatrix}
$$

and the terms $N_i(i, 1, ..., 8)$ are defined by Eq.3.59a.

Similarly the displacement variation can be expressed by the same shape functions as,

$$
\left\{ \begin{matrix} u \\ v \end{matrix} \right\} = \begin{bmatrix} N_1 & 0 & N_2 & 0 & N_3 & 0 & N_4 & 0 & N_5 & 0 & N_6 \\ 0 & N_1 & 0 & N_2 & 0 & N_3 & 0 & N_4 & 0 & N_5 & 0 \end{bmatrix}
$$

$$
\left. \begin{matrix} 0 & N_7 & 0 & N_8 & 0 \\ N_6 & 0 & N_7 & 0 & N_8 \end{matrix} \right] \{d\} \tag{8.43b}
$$

where

$$
\{d\}^T = [\, u_1 \quad v_1 \quad u_2 \quad v_2 \quad u_3 \quad v_3 \quad u_4 \quad v_4 \quad u_5 \quad v_5
$$

$$
u_6 \quad v_6 \quad u_7 \quad v_7 \quad u_8 \quad v_8 \,]
$$

and the terms $N_i(i = 1, 2, ..., 8)$ are defined by the Eqn.3.59(a).

The stiffness matrix for this element can be evaluated using the numerical integration procedure explained in Chapter 4.

8.4 Incompatible Displacement Models

In finite element analysis the accuracy can be improved by using a fine mesh size or by using refined (higher order) elements. Some-times the requirement of a fine mesh over a region of rapidly varying stress leads to bad numerical condition so that larger or more compli-cated elements are mandatory. With the introduction of isoparamet-ric elements many refined elements with higher order displacement fields have been developed. Eventhough these elements give good re-sults they require more computer time and storage. Other attempts have been made by Wilson et.al.[4] to improve the basic accuracy of the simpler (first order) elements, without very much increasing the computer time and storage, by introducing the incompatible dis-placement modes. This formulation is explained here with respect to the quadrilateral element PSQ4 explained in section 8.2.

8.4.1 Sources of Error

One of the main causes of the inaccuracies in the linear strain quadri-lateral element PSQ4 is its inability to represent certain simple stress gradients. The displacement function given by Eqs. 3.52 is incom-plete in that it does not contain all the quadratic terms in shape functions. Therefore, this element is not adequate in representing flexural response. This is clearly illustrated by subjecting a sim-ple rectangular element to pure bending stress as shown in Fig.8.10. Taking the origin of the coordinates at the centroid of the element, the stress components given by the elementary bending theory are:

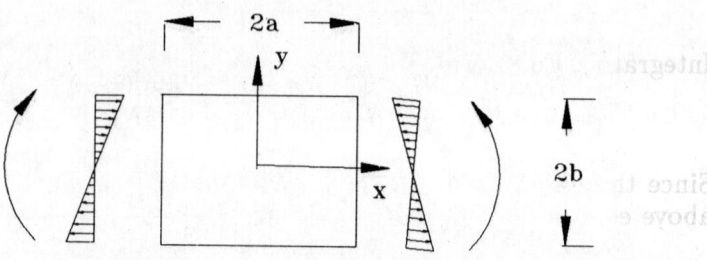

(a) Simple bending stress

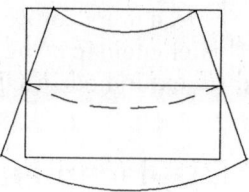

(b) Exact displacements

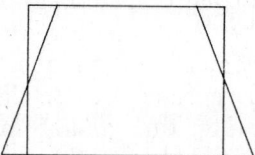

(c) Finite element displacements

Fig. 8.10 Rectangular element

$$\sigma_x = \frac{E}{R} y \quad \sigma_y = \tau_{xy} = 0 \tag{8.44}$$

where   E   is the modulus of elasticity
    y   is the distance of the point from the neutral axis
    R   is the radius of curvature.

If the results in Eq.8.44. are substituted in the Hooke's law, the strain components are given by,

$$\epsilon_x = \frac{\partial u}{\partial x} = \frac{y}{R} \tag{8.45a}$$

$$\epsilon_y = \frac{\partial v}{\partial y} = \frac{-\mu y}{R} \tag{8.45b}$$

$$\gamma_{xy} = \frac{\partial u}{\partial y} + \frac{\partial v}{\partial x} = 0 \qquad (8.45c)$$

Integrating Eq.8.45(a)

$$u = \frac{xy}{R} + g_1(y)$$

Since the displacement in the x -direction at the origin is zero, the above equation becomes,

$$u = \frac{xy}{R} \qquad (8.46)$$

From Eq.8.45b.

$$v = -\frac{\mu y^2}{2R} + f_1(x) \qquad (8.47a)$$

Eq.8.45c and 8.46 give,

$$\frac{\partial v}{\partial x} = -\frac{\partial u}{\partial y} = \frac{-x}{R}$$

Integrating the above equation

$$v = -\frac{x^2}{2R} + g_2(y) \qquad (8.47b)$$

From Eqs. 8.47a and 8.47b, the displacement v can be expressed as,

$$v = -\frac{x^2}{2R} - \frac{\mu y^2}{2R} + C \qquad (8.47c)$$

where C is a constant. The constant C is evaluated by using the condition that the v displacements at the four corners of the element are zero.
i.e.,

$$C - \frac{a^2}{2R} - \frac{\mu b^2}{2R} = 0$$

or

$$C = \frac{a^2}{2R} + \frac{\mu b^2}{2R}$$

Substituting for C in Eq.8.47c

$$v = \left(1 - \frac{x^2}{a^2}\right)\frac{a^2}{2R} + \left(1 - \frac{y^2}{b^2}\right)\frac{\mu b^2}{2R} \qquad (8.48)$$

From the Eqs. 8.46 and 8.48, the true displacements under pure bending can be expressed in the form

$$u = \alpha_1\, xy \qquad (8.49a)$$

$$v = \alpha_2\left(1 - \frac{x^2}{a^2}\right) + \alpha_3\left(1 - \frac{y^2}{b^2}\right) \tag{8.49b}$$

It is seen that the general four node quadrilateral element includes only the first part of the Eq.8.49 in its displacement assumption. The second part of Eq.8.49 is not adequately represented. This is the primary source of error in the solution when the element is under flexural load.

To simulate the flexural response more precisely Wilson et.al.[4] introduced a quadrilateral element with additional displacement modes which has the same form as the errors in the simple displacement assumptions.

For a general quadrilateral element the displacement approximation may be of the following form.

$$u = N_1 u_1 + N_2 u_2 + N_3 u_3 + N_4 u_4 + P_1 \alpha_1 + P_2 \alpha_2$$
$$v = N_1 v_1 + N_2 v_2 + N_3 v_3 + N_4 v_4 + P_1 \alpha_3 + P_2 \alpha_4$$

$$\tag{8.50}$$

where $P_1 = (1 - r^2)$ and $P_2 = (1 - s^2)$

It may be noted that the functions P_1 and P_2 are chosen such that they are zero at the four nodes in order to maintain the displacement compatibility at the nodes and of the same form as the errors in the bending deformation. The displacement amplitudes α_i are the additional degrees of freedom; therefore, the resulting stiffness matrix will be 12×12. However, if the strain energy within the element is minimised with respect to α_i four additional equations can be generated and the additional displacements, can be eliminated, and a reduced 8×8 stiffness matrix can be obtained. This is identical to the standard static condensation procedure.

The incompatible quadrilateral element discussed above was an improvement over linear strain quadrilateral PSQ4. However, its accuracy is found to be dependent upon the discretization of the structure. It yields excellent results for flexural problems if the elements are rectangular, but if they are not, then the stresses calculated are found to be erroneous [4]. Also under constant deformation state even the rectangular elements give erroneous results.

8.4.2 Improved Incompatible Quadrilateral Element (PSQI6)

The inconsistencies in the incompatible elements have been investigated and an improvement has been suggested by Taylor et.al.[5].

When incompatible modes are added to the four noded quadrilateral element, the displacements u and v are given by Eq.8.50 and the global coordinates are expressed as,

$$x = N_1 x_1 + N_2 x_2 + N_3 x_3 + N_4 x_4$$

$$y = N_1 y_1 + N_2 y_2 + N_3 y_3 + N_4 y_4 \qquad (8.51)$$

Rewriting Eq.8.50

$$\{u\} = [N] \{d_a\} + [P] \{\alpha\} \qquad (8.52)$$

where

$$[P] = \begin{bmatrix} P_1 & P_2 & 0 & 0 \\ 0 & 0 & P_1 & P_2 \end{bmatrix}$$

and

$$\{\alpha\}^T = [\alpha_1 \quad \alpha_2 \quad \alpha_3 \quad \alpha_4]$$

Then the strain vector is given by,

$$\{\epsilon\} = [B] \{d_a\} + [P'] \{\alpha\} \qquad (8.53)$$

where $[P']$ represents differentiation with respect to the global coordinates x and y.

We know that the strain energy of the element is given by

$$U = \frac{1}{2} \int \int \int_V \{\epsilon\}^T [C] \{\epsilon\} \, dV$$

Substituting Eq.8.53 into the above expression, we get

$$U = \frac{1}{2} \int \int \int_V [\{d_a\}^T [B]^T + \{\alpha\}^T [P']^T][C] \times [[B]\{d\} + [P']\{\alpha\}] dV$$

i.e.,

$$U = \frac{1}{2} \int \int \int_V [\{d_a\}^T \ \{\alpha\}^T] \begin{Bmatrix} [B]^T \\ [P']^T \end{Bmatrix} [C] \times [[B][P']] \begin{Bmatrix} \{d_a\} \\ \{\alpha\} \end{Bmatrix} dV \qquad (8.54)$$

The element equilibrium equation is given by

$$[k] \{d\} = \{Q\}$$

Since there are no external forces associated with α's the above equation can be expressed as,

$$\begin{bmatrix} [k_{aa}] & [k_{a\alpha}] \\ [k_{\alpha a}] & [k_{\alpha\alpha}] \end{bmatrix} \begin{Bmatrix} \{d_a\} \\ \{\alpha\} \end{Bmatrix} = \begin{Bmatrix} \{Q_a\} \\ \{0\} \end{Bmatrix} \qquad (8.55)$$

where $\{Q_a\}$ represents nodal forces, $\{d_a\}$ represents the nodal displacements, $[k_{aa}]$ is the stiffness matrix associated with displacements $\{d_a\}$, $[k_{\alpha\alpha}]$ is the stiffness matrix associated with the nonconforming generalised displacements $\{\alpha\}$ and $[k_{a\alpha}] = [k_{\alpha a}]^T$.

The three criteria to be satisfied for convergence are explained in Chapter 3 under section 3.3. It may be noted that the first two conditions are already satisfied for the element considered. Hence, the following condition is derived to satisfy the constant strain state of convergence [3].

Let $\{d_a^c\}$ be the displacement vector for a constant deformation state. Under this condition, generalised displacement vector $\{\alpha\}$ is to be zero in order that the incompatible modes are not activated. The second equation of 8.55 can be expressed as

$$[k_{\alpha a}] \{d_a^c\} + [k_{\alpha\alpha}] \{\alpha\} = \{0\} \tag{8.56}$$

For the above equation to be valid when $\{\alpha\} = \{0\}$ the first term $[k_{\alpha a}] \{d_a^c\} = \{0\}$. From Eq. 8.54

$$[k_{\alpha a}] = \int \int \int_V [P']^T [C] \{B\} \, dV \tag{8.57}$$

Substituting for $[k_{\alpha a}]$ it can be written as,

$$\int \int \int_V [P']^T [C] [B] \{d_a^c\} \, dV = \{0\} \tag{8.58}$$

Now $[C]$ is constant and $[B] \{d_a^c\} = \{\epsilon_a^c\}$ is also constant (since constant deformation state is assumed). Thus Eq. 8.58 reduces to

$$\int \int \int_V [P']^T \, dV = [O] \qquad \text{i.e.,} \qquad \int \int \int_V [P'] \, dV = [O]^T \tag{8.59}$$

For two-dimensional problem the Eq.8.59 becomes

$$h \int \int_A [P'] \, dA = [O]^T \tag{8.60}$$

In isoparametric formulation Eq.8.60 becomes

$$h \int_{-1}^{+1} \int_{-1}^{+1} [P'] |J| \, dr \, ds = [O]^T \tag{8.61}$$

From Eqs. 4.3 and 4.4

$$[P'] = \frac{1}{|J|} \begin{bmatrix} \frac{\partial y}{\partial s} & -\frac{\partial y}{\partial r} \\ -\frac{\partial x}{\partial s} & \frac{\partial x}{\partial r} \end{bmatrix} \begin{bmatrix} -2r & 0 & 0 & 0 \\ 0 & 0 & -2s & 0 \end{bmatrix} \tag{8.62}$$

Substituting for $[P']$ in Eq. 8.61

$$\int_{-1}^{+1} \int_{-1}^{+1} \begin{bmatrix} -2r\frac{\partial y}{\partial s} & 0 & +2s\frac{\partial y}{\partial r} & 0 \\ 2r\frac{\partial x}{\partial s} & 0 & -2s\frac{\partial x}{\partial r} & 0 \end{bmatrix} dr\ ds = [O] \tag{8.63}$$

In order to ensure that Eq.8.63 is identically zero, the variation in the derivatives $\frac{\partial x}{\partial r}, \frac{\partial x}{\partial s}, \frac{\partial y}{\partial r}, \frac{\partial y}{\partial s}$ at the quadrature points is eliminated by using constant values i.e., those at the centroid of the element, for all integration points. Thus, since $\int_{-1}^{+1} rdr = \int_{-1}^{+1} sds = 0$, the Eq.8.63 will be identically zero.

The generalised displacement vector $\{\alpha\}$ is thus constrained to be zero under constant deformation conditions, thereby ensuring convergence under all conditions.

The Jacobian matrix $[J]$ evaluated at the centroid of the element is

$$[J] = \begin{bmatrix} J_{11} & J_{12} \\ J_{21} & J_{22} \end{bmatrix} = \frac{1}{4} \begin{bmatrix} (-x_1 + x_2 + x_3 - x_4) & (-y_1 + y_2 + y_3 - y_4) \\ (-x_1 - x_2 + x_3 + x_4) & (-y_1 - y_2 + y_3 + y_4) \end{bmatrix}$$

and

$$[J]^{-1} = \begin{bmatrix} J_{11}^{**} & J_{12}^{**} \\ J_{21}^{**} & J_{22}^{**} \end{bmatrix} \tag{8.64}$$

where

$$J_{11}^{**} = \frac{(-y_1 - y_2 + y_3 + y_4)}{4|J|^{**}}$$

$$J_{12}^{**} = \frac{(y_1 - y_2 - y_3 + y_4)}{4|J|^{**}}$$

$$J_{21}^{**} = \frac{(x_1 + x_2 - x_3 - x_4)}{4|J|^{**}}$$

$$J_{22}^{**} = \frac{(-x_1 + x_2 + x_3 - x_4)}{4|J|^{**}}$$

and

$$|J|^{**} = J_{11}\ J_{22} - J_{12}\ J_{21} \tag{8.64a}$$

For evaluating the strain vector $\{\epsilon\}$ given by Eq.(4.10) the derivatives of u and v with respect to r and s can be computed as follows:

$$
\begin{Bmatrix} \dfrac{\partial u}{\partial r} \\[2mm] \dfrac{\partial u}{\partial s} \\[2mm] \dfrac{\partial v}{\partial r} \\[2mm] \dfrac{\partial v}{\partial s} \end{Bmatrix} = \frac{1}{4}
\begin{bmatrix}
-(1-s) & 0 & (1-s) & 0 & (1+s) & 0 \\
-(1-r) & 0 & -(1+r) & 0 & (1+r) & 0 \\
0 & -(1-s) & 0 & (1-s) & 0 & (1+s) \\
0 & -(1-r) & 0 & -(1-r) & 0 & (1+r)
\end{bmatrix}
$$

$$
\begin{bmatrix}
-(1+s) & 0 & -8r & 0 & 0 & 0 \\
(1-r) & 0 & 0 & -8s & 0 & 0 \\
0 & -(1+s) & 0 & 0 & -8r & 0 \\
0 & (1-s) & 0 & 0 & 0 & -8s
\end{bmatrix} \{d\} \qquad (8.65)
$$

and $\{d\}^T = \begin{bmatrix} u_1 & v_1 & u_2 & v_2 & u_3 & v_3 & u_4 & v_4 & \alpha_1 & \alpha_2 & \alpha_3 & \alpha_4 \end{bmatrix}$

Now, while substituting the values from Eq.8.65 into Eq.4.10 to get the [B] matrix it may be noted that the inverse Jacobian coefficients J^*_{ij} for the terms corresponding to the nodal displacements are calculated using Eq.4.72 at any point (i.e., at a Gauss point for numerical integration) and the coefficients J^{**}_{ij} pertaining to the terms for incompatible modes are calculated at the centroid of the element as given by Eq.8.64a. The [B] matrix is given in Eq. 8.66.

8.5   The Patch Test

The patch test was originally proposed by Irons [6,7] as a simple means of testing the validity of an element formulation and its implementation in a computer program. This test can be used to test whether an assemblage of non conforming elements is complete. In this test, a 'patch' of elements is subjected to boundary nodal forces that in an exact analysis correspond to constant strain conditions. If the element strains do actually represent the constant strain conditions, the patch test is passed. We assume that the element is stable, i.e. the element admits no zero energy deformation states when adequately supported against rigid body motions.

     If the element passes the patch test, we have assurance that when this element is used to model any structure, mesh refinement will produce a sequence of approximate solutions that converges to the exact solution. In other words, the patch test serves as a necessary and sufficient condition for correct convergence of a finite element formulation.

$$[B] = \frac{1}{4}\begin{bmatrix}
-J_{11}^*(1-s)-J_{12}^*(1-r) & 0 & J_{11}^*(1-s)-J_{12}^*(1+r) & 0 & J_{11}^*(1+s)+J_{12}^*(1+r) & 0 & -J_{11}^*(1+s)+J_{12}^*(1-r) & 0 & -8J_{11}^{**}r & 0 & -8J_{12}^{**}s & 0 \\[4pt]
0 & -J_{21}^*(1-s)-J_{22}^*(1-r) & 0 & J_{21}^*(1-s)-J_{22}^*(1+r) & 0 & J_{21}^*(1+s)+J_{22}^*(1+r) & 0 & -J_{21}^*(1+s)+J_{22}^*(1-r) & 0 & -8J_{21}^{**}r & 0 & -8J_{22}^{**}s \\[4pt]
-J_{21}^*(1-s)-J_{22}^*(1-r) & -J_{11}^*(1-s)-J_{12}^*(1-r) & J_{21}^*(1-s)-J_{22}^*(1+r) & J_{11}^*(1-s)-J_{12}^*(1+r) & J_{21}^*(1+s)+J_{22}^*(1+r) & J_{11}^*(1+s)+J_{12}^*(1+r) & -J_{21}^*(1+s)+J_{22}^*(1-r) & -J_{11}^*(1+s)+J_{12}^*(1-r) & -8J_{21}^{***}r & -8J_{11}^{***}r & -8J_{22}^{***}s & -8J_{12}^{***}s
\end{bmatrix}$$

$$(8.66)$$

The stiffness matrix of the element is computed from Eqn. 4.15.

Procedure

A small number of elements are assembled into a patch. There should be at least one node within that patch, so that the node is shared by two or more elements, and one or more inter element boundaries must exist, Fig. E8.24 shows an example for conducting a patch test for two-dimensional plane stress problem. Boundary nodes of the patch are to be loaded by consistently derived nodal loads appropriate to a state of constant stress. The patch is provided with enough supports to prevent rigid body motions. Next a standard finite element solution for the problem is obtained and stresses are computed. If throughout the element the computed stresses agree with the exact stresses for the problem, the patch test is passed. For the exercise problem 8.24 the student should carryout the patch test for the constant stress condition σ_x. Similarly, we must test the patch for constant σ_y and again for constant τ_{xy}.

Similarly the patch test can be conducted for other types of elements viz., plate bending, shell and three-dimensional solid.

Though the original concept of Irons was intuitive and physical, mathematical rigor was given by later investigations [8,9,10]. The role of patch test can be summarised in the following.

(a) It is a necessary and sufficient condition for convergence.

(b) It provides an assessment of the convergence rate of the element.

(c) It checks the robustness of the algorithm.

(d) It provides as a check whether correct coding was achieved in a finite element computer program.

(e) It is also a tool to check the accuracy of the finite elements which incorporate incompatible modes to improve the performance of the element in some situations.

8.6 Reinforced Concrete Element

The finite element discretization of reinforced concrete structures essentially depends on the presence of reinforcements. In most of the cases the mesh size is decided by the spacing of reinforcements when the reinforcing bar is idealised as one-dimensional bar element and used along with the rectangular or quadrilateral element to idealise concrete. This, many times, results in finer mesh due to close spacing of lateral reinforcements. Computationally, such a fine discretization may be uneconomical in many cases, since results with the same degree of accuracy can be obtained by adopting a coarser discretization at regions of minimum strain gradients. This is possible only when the discretization is independent of reinforcement spacing. Such a discretization as indicated in references [11,12] is computationally more economical, especially in the case of non-linear analysis because of less number of nodal points and elements. The formulation of reinforced concrete element which included the reinforcement in any orientation is described in the following. It has been shown [11,12] that the use of such elements permits a finite element discretization independent of spacing and orientation of reinforcements, thus allowing a coarser mesh and resulting in increased computational efficiency.

8.6.1 Formulation of Reinforced Concrete Element

Figure 8.11 shows a reinforced concrete element with an inclined reinforcing bar.

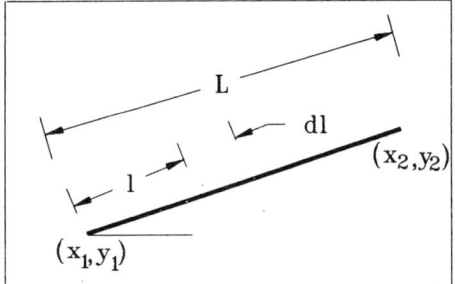

Fig. 8.11 Reinforced Concrete Element

The total potential energy for the element is written as,

$$\Pi = U_c + U_s + W \qquad (8.67)$$

where U_c is the strain energy of elastic uncracked concrete

           U_s is the strain energy of steel

and   W is the potential energy due to external loads.

For finite element displacement formulation the total energy of an element is given by

$$\Pi = \frac{1}{2}\{d\}^T \left(\int\int\int_{V_c} [B]^T\,[C_c]\,[B]\,dV_c + \int\int\int_{V_s} [B]^T\,[C_s]\,[B]\,dV_s \right) \{d\}$$

$$-\{d\}^T \int\int\int_V [N]^T\,\{X\}\,dV$$

$$-\{d\}^T \int \int_{S_1} [N]^T \{p\} \, dS_1 \qquad (8.68)$$

where
    $[C_c]$ is the material constitutive matrix for elastic
           uncracked concrete.
    $[C_s]$ is the material constitutive matrix for un-
           yielded steel.

From the above equation the internal strain energy of elastic un-cracked element including reinforcement is given by

$$U = \frac{1}{2}[d]^T \left(\int \int \int_{V_c} [B]^T [C_c] [B] \, dV_c + \int \int \int_{V_s} [B]^T [C_s] [B] \, dV_s \right) \{d\}$$

$$\frac{1}{2}\{d\}^T [[k_c] + [k_s]] \{d\} = \frac{1}{2} \{d\}^T [k_{cs}] \{d\} \qquad (8.69)$$

where $[k_c]$ is the stiffness matrix of concrete
           $[k_s]$ is the stiffness matrix of steel
and $[k_{cs}]$ is the stiffness matrix of reinforced concrete
              element.

For concrete the stiffness matrix $[k_c]$ is computed from Eq.(4.15) which is expressed as

$$[k_c] = [k] = h \int_{-1}^{+1} \int_{-1}^{+1} [B]^T [C_c] [B] \, |J| \, dr \, ds \qquad (8.70)$$

For steel the stiffness matrix $[k_s]$ is computed as the sum of the stiffness of number of reinforcing bars n in the element. Thus,

$$[k_s] = \sum_{i=1}^{n} [k_s]_i \qquad (8.71a)$$

and

$$[k_s]_i = \int \int \int_{V_s} [B]^T [C_s]_i [B] \, dV_s \qquad (8.71b)$$

where $[C_s]_i$ is the material constitutive matrix of the steel bar i and the value of the elastic modulus can be modified to account for the replacement of concrete by steel. The volume integral for steel is computed by converting the integral into a line integral of the rod. Referring to Fig. 8.11

$$[k_s]_i = \int \int \int_{V_s} [B]^T [C_s]_i [B] \, dV_s$$

$$= \int [B]^T \ [C_s] \ [B] \ [A_i] \ dl$$

$$= A_i \int [B]^T \ [C_s]_i \ [B] \ dl \tag{8.72}$$

where A_i is the area of the reinforcement of uniform diameter.

8.7 Axisymmetric Solid Element

Analysis of solids of revolution under axisymmetric loading is of considerable practical interest.

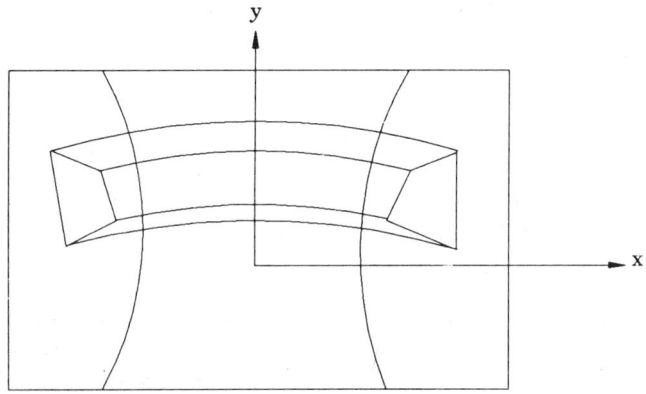

Fig. 8.12 Solid of revolution - an axisymmetric element

A common example may be the case of a thick walled cylinder subjected to internal or external radial pressure. In such cases, due to the symmetry about the circumferential (θ) direction, displacements are confined to the radial (x) and axial (y) directions only, and the problem can simply be treated as two-dimensional. Thus with little modifications, any of the plane elements described in the earlier sections of this chapter, can be successfully employed in solving the axisymmetric problems. However, it may be noted that the axisymmetric problem, unlike the plane problem, involves an additional non-zero hoop normal strain ϵ_θ due to the presence of non-zero radial displacement.

8.7.1 Formulation of Axisymmetric Solid Element for Axisymmetric Loading Case

Consider a typical solid of revolution, discretized by four noded axisymmetric solid elements belonging to isoparametric family, as shown in Fig.8.12. The element can be easily identified as a ring of constant cross section each of its nodes denoting a concentric circle with centre on the vertical y-axis.

Let u and v be the displacements in x and y directions, respectively, the circumferential displacement being zero everywhere due to axisymmetric loading. Assuming the element as a linear isoparametric PSQ4 element, the displacements are denoted by Eqs.8.23.

The stress-strain relation for axisymmetric load case is given by Eq.2.25 and can be expressed as

$$\{\sigma\} = [C]\{\epsilon\} \tag{8.73}$$

where

$$\{\sigma_x\} = [\sigma_x \quad \sigma_y \quad \sigma_\theta \quad \gamma_{xy}]^T \tag{8.74a}$$

$$\{\epsilon\} = [\epsilon_x \quad \epsilon_y \quad \epsilon_\theta \quad \gamma_{xy}]^T \tag{8.74b}$$

and

$$[C] = \frac{E}{(1+\mu)(1-2\mu)} \begin{bmatrix} 1-\mu & \mu & \mu & 0 \\ \mu & 1-\mu & \mu & 0 \\ \mu & \mu & 1-\mu & 0 \\ 0 & 0 & 0 & \frac{1-2\mu}{2} \end{bmatrix} \tag{8.74c}$$

in which E and μ denote Young's modulus and Poisson's ratio of the material under consideration, as described earlier. Likewise, from the theory of elasticity the strain-displacement relations for the same case are given by 2.24. Thus,

$$\begin{Bmatrix} \epsilon_x \\ \epsilon_y \\ \epsilon_\theta \\ \gamma_{xy} \end{Bmatrix} = \begin{bmatrix} 1 & 0 & 0 & 0 & 0 \\ 0 & 0 & 0 & 1 & 0 \\ 0 & 0 & 0 & 0 & \frac{1}{x} \\ 0 & 1 & 1 & 0 & 0 \end{bmatrix} \begin{Bmatrix} \frac{\partial u}{\partial x} \\ \frac{\partial u}{\partial y} \\ \frac{\partial v}{\partial x} \\ \frac{\partial v}{\partial y} \\ u \end{Bmatrix} \tag{8.75}$$

In deriving the stiffness matrix of the isoparametric element, the relations between displacements quantities in $x - y$ and $r - s$ are needed and may be deduced easily from Eq.4.8 as

$$\begin{Bmatrix} \frac{\partial u}{\partial x} \\ \frac{\partial u}{\partial y} \\ \frac{\partial v}{\partial x} \\ \frac{\partial v}{\partial y} \\ u \end{Bmatrix} = \begin{bmatrix} J_{11}^* & J_{12}^* & 0 & 0 & 0 \\ J_{21}^* & J_{22}^* & 0 & 0 & 0 \\ 0 & 0 & J_{11}^* & J_{12}^* & 0 \\ 0 & 0 & J_{21}^* & J_{22}^* & 0 \\ 0 & 0 & 0 & 0 & 1 \end{bmatrix} \begin{Bmatrix} \frac{\partial u}{\partial r} \\ \frac{\partial u}{\partial s} \\ \frac{\partial v}{\partial r} \\ \frac{\partial v}{\partial s} \\ u \end{Bmatrix} \tag{8.76}$$

where J_{ij}^* is the inverse of the Jacobian matrix $[J]$ and is described by Eqs. 4.6 and 4.7. Substituting Eqs. 8.76 and 4.12 in Eq.8.75, we get the following general form.

$$\{\epsilon\} = [B]\{d\} \tag{8.77}$$

where

$$[B] = \begin{bmatrix} J_{11}^* & J_{12}^* & 0 & 0 & 0 \\ 0 & 0 & J_{21}^* & J_{22}^* & 0 \\ 0 & 0 & 0 & 0 & \frac{1}{x} \\ J_{21}^* & J_{22}^* & J_{11}^* & J_{12}^* & 0 \end{bmatrix} \times$$

$$\begin{bmatrix} \frac{(s-1)}{4} & 0 & \frac{(1-s)}{4} & 0 & -\frac{(1+s)}{4} & 0 & -\frac{(1+s)}{4} & 0 \\ \frac{(r-1)}{4} & 0 & -\frac{(1+r)}{4} & 0 & \frac{(1+r)}{4} & 0 & \frac{(1-r)}{4} & 0 \\ 0 & \frac{(s-1)}{4} & 0 & \frac{(1-s)}{4} & 0 & \frac{(1+s)}{4} & 0 & -\frac{(1+s)}{4} \\ 0 & \frac{(r-1)}{4} & 0 & -\frac{(1+r)}{4} & 0 & \frac{(1+r)}{4} & 0 & \frac{(1-r)}{4} \\ N_1 & 0 & N_2 & 0 & N_3 & 0 & N_4 & 0 \end{bmatrix}$$

$$\tag{8.78}$$

Furthermore, it should be noted that $\{d\}$ describes the vector of nodal displacements, Eq.8.25b, and N_1 ... N_4 are the shape functions defined by Eq.8.23b.

The element stiffness matrix can now be derived as

$$[k] = \int_{-1}^{1}\int_{-1}^{1} [B]^T [C] [B]\, 2\pi x\, det[J]\, dr\, ds \tag{8.79}$$

wherein x is taken corresponding to the Gauss point when a quadrature rule is employed in evaluating the above integral.

The corresponding nodal load vector due to live loads can be obtained as described in section 8.3. Live load intensity at node i is obtained by multiplying the corresponding pressure intensity by a factor $2\pi x_i$ where x_i denotes the x coordinate of the node i.

Asymmetric Loading

There are situations in practical applications when an axisymmetric structure may be subjected to a nonaxisymmetric loading. In that case, strictly it becomes a three-dimensional analysis problem since the displacements and stresses are not axially symmetric. However, it is possible to take advantage of the axisymmetric geometry of the structure, by expanding the externally applied load into Fourier series consiting of symmetric and antisymmetric load contributions.

A two-dimensional analysis is performed for each load component (Fourier harmonic) and the solutions are superposed [7,13].

8.8   Subroutine PLANE

This subroutine controls all the subroutines of element shapes used for plane stress/strain and axisymmetric analysis (Fig.8.13). There is a provision for three element shapes and in the present version two shapes, a constant strain triangle and a four noded isoparametric quadrilateral (with and without incompatible modes), are included. The routines for the two element shapes are organised following the guidelines given in Chapter 6.

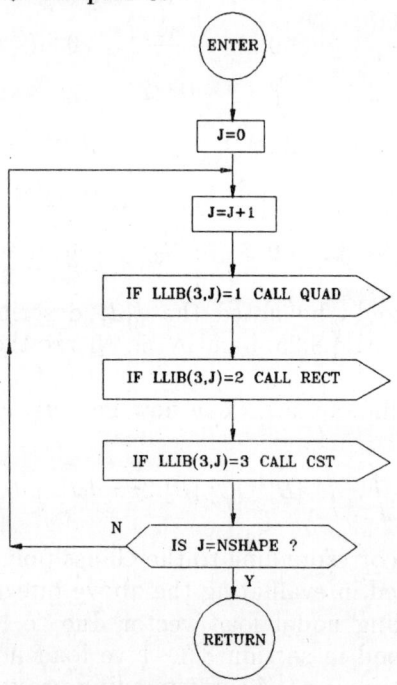

Fig. 8.13 Subroutine PLANE

The CST element is the simplest of all the plane stress elements. The strain-displacement matrix and the stiffness matrix are directly computed using the Eqs. 3.75 and 3.87.

The other element shape is a four noded quadrilateral. A parameter 'KOPT' is to be read as data and if KOPT equals one then incompatible modes are included and if KOPT is 3, axisymmetic analysis is invoked using the quadrilateral element. The incompatible modes are included for the analysis of problems involving bending and the use of such element is explained in the earlier section.

The stress-displacement matrix is calculated at the centroid of the element and stored on the scratch file ISTRES. This is used for the calculation of element stresses later. The stiffness matrix is calculated using the numerical integration with 2 ×2 Gauss points as described in Chapter 4.

In the case of the element with incompatible modes, the stiffness matrix will be of size 12 × 12, as explained in the previous section. This is reduced to 8 × 8 corresponding to the nodal displacements by employing the static condensation procedure.

After the evaluation of stiffness matrix, the element loads are read. For every loaded element the number of loaded faces are read and for every loaded face the value of pressure intensity in the two directions at each node of the face is also specified through input data. A linear variation of pressure intensity is assumed between the nodes. To simplify calculations, the element load vector for the element with incompatible modes is approximated to the corresponding load vector for the four noded quadrilateral element.

8.8.1 Input Details

The input details for plane stress/strain or axisymmetric element are given below. This data set pertains to the data sets 5 and 6 required for PASSFEM described in section 6.11

5. ELEMENT TYPE DATA (5I5)

columns 1-5   Element type number or row number of the element library (LTYPE). (For plane stress/ strain/ axisymmetric elements LTYPE=3)

6-10   Number of shapes (NSHAPE) belonging to the element type with a minimum of 1 to a maximum of 3.

11-15   Element shape number or column number of the element library. (Element shape number is 1 for Quadrilateral Element)

16-20   Element shape number. If NSHAPE is more than one, the second shape number is typed in this column, otherwise, it is left blank.

21-25   Element shape number. If NSHAPE=3, third shape number is typed. Element shape number is 3 for CST element.

6. ELEMENT DATA SET

Analysis Type Data (I5)

Columns   1-5   NOPT-Option for plane stress/strain analysis. Specify 2 for plane stress and any other value for plane strain.

SHAPE NO.1 FOUR NODED QUADRILATERAL

A. Control Data (3I5)

Columns   1-5   Number of quadrilateral elements in the structure.

6-10   Number of loaded quadrilateral elements (NLE)

11-15   KOPT=0 for quadrilateral element; =3 for analysis of axisymmetric solid using the quadrilateral element.

B. Element Data (6I5, F 10.2, 2I5)

Columns   1-5   Element Number

6-10   Material group number

11-15   Node 1 of the element

16-20   Node 2 of the element

21-25   Node 3 of the element

26-30   Node 4 of the element

31-40   Thickness of the element. Specify 1.0 for plane strain case.

41-45   Option for incompatible modes. 0 without incompatible modes, 1 for inclusion of incompatible modes.

46-50   Automatic element data generation parameter (specify KI value for generation in the second line of the series)

C. Self-weight Factor (2F 5.2)

Columns   1-5   Factor for self-weight in the global +x direction

6-10   Factor for self-weight in the global +y direction.

D. Element Load Data

This data is required only if $NLE > 0$, otherwise this data set is not required.

1. Control Data (6I5)

Columns   1-5   Element Number of the loaded element
               6-10   Number of loaded faces
              11-15   Number of the first loaded face
              16-20   Number of the second loaded face
              21-25   Number of the third loaded face
              26-30   Number of the fourth loaded face

The face number 1 means between node 1 and 2 of the element, 2 between node 2 and 3 and so on.

2. Pressure Intensity Data (4F10.3)

One line per loaded face of the element is required.

Columns   1-10   Pressure intensity at the first node in x-direction
              11-20   Pressure intensity at the second node in x-direction
              21-30   Pressure intensity at the first node in y-direction
              31-40   Pressure intensity at the second node in y-direction

Note : The Control Data (1) and Pressure Intensity Data (2) are to be given for every loaded element.

The pressure intensity between the first and second node of any face is assumed to vary linearly. For axisymmetric solids the pressure intensity is to be multiplied by $2\pi r$.

SHAPE NO.3 CONSTANT STRAIN TRIANGLE (CST)

A. Control Data (I5)

Columns   1-5   Number of CST elements in the structure

B. Element Data (5I5, F10.3, I5)

    Columns   1-5   Element number

                      6-10   Material Group Number

                    11-15   Node 1 of the element

                    16-20   Node 2 of the element

                    21-25   Node 3 of the element

                    26-35   Thickness of the element.   Specify 1.0 for plane strain case

                    36-40   Automatic element data generation parameter

C. Self-Weight Factor (2F 5.2)

    Columns   1-5   Factor for self-weight in the global+x direction

                      6-10   Factor for self-weight in the, global+y direction

D. Element Load Data

1. Control Data (I5)

    Columns   1-5   Number of element load data sets (i.e., the number of pressure intensity data set, to be specified for each loaded face.

Note : If no element load is applied, type '0' and skip the following Pressure Intensity Data.

2. Pressure Intensity Data (3I5, 4F10.3)

    Columns   1-5   Element Number

                      6-10   Node number I

                    11-15   Node number J

                    16-25   Pressure intensity at I in x-direction

                    26-35   Pressure intensity at J in x-direction

                    35-45   Pressure intensity at I in y-direction

                    46-55   Pressure intensity at J in y-direction

8.8.2   Output Details

Stresses $\sigma_x, \sigma_y, \tau_{xy}$ and the principal stresses are printed out at the centroid of each element; and in the case of axisymmetric analysis $\sigma_x, \sigma_y, \sigma_\theta$ and τ_{xy}. In addition, stresses at the nodes after smoothening are also printed out. The stress smoothening technique is explained in Chapter 10.

8.9   Examples

Five examples are presented here to illustrate the use of the program PASSFEM for plane stress/strain and axisymmetric analyses. Two discretizations are tried and the results are compared with the analytical solutions.

8.9.1   Cantilever Beam

Figure 8.14 shows a cantilever beam with rectangular cross section subjected to an end moment which is replaced by a couple of forces. The results of analysis for deflection by using quadrilateral element with and without incompatible modes are given in Table 8.1 for two discretizations. It can be observed that the element with incompatible modes gives excellent results for this bending type of problem even for a coarse discretization.

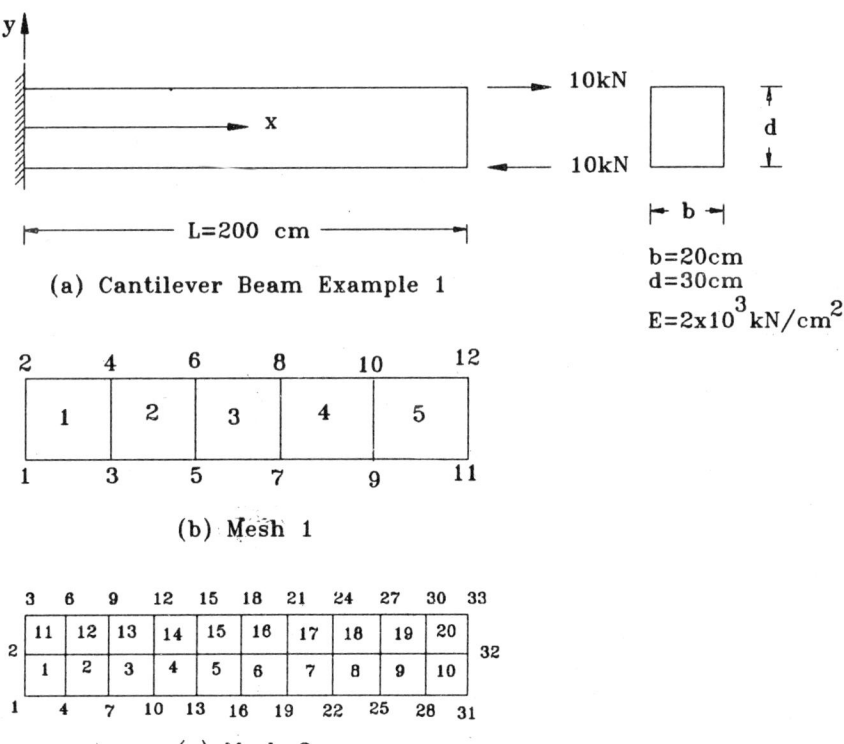

(a) Cantilever Beam Example 1

b=20cm
d=30cm
$E=2\times10^{3}\,kN/cm^{2}$

(b) Mesh 1

(c) Mesh 2

Fig. 8.14

Table 8.1 Comparison of Deflection for Cantilever Beam

Distance from fixed end (cm)	Theoretical value of deflection	Quadrilateral Element with incompatible modes		Quadrilateral Element	
		Mesh 1	Mesh 2	Mesh 1	Mesh 2
40	0.002667	0.002667	0.002667	0.001418	0.002182
80	0.010667	0.010667	0.010667	0.005647	0.008727
120	0.023990	0.024000	0.024000	0.012706	0.019636
160	0.042666	0.042666	0.042666	0.022588	0.034909
200	0.066667	0.066667	0.066667	0.035294	0.054346

8.9.2 Tapered Cantilever Beam

A tapered cantilever beam subjected to a concentrated load at the free end is shown in Fig.8.15.

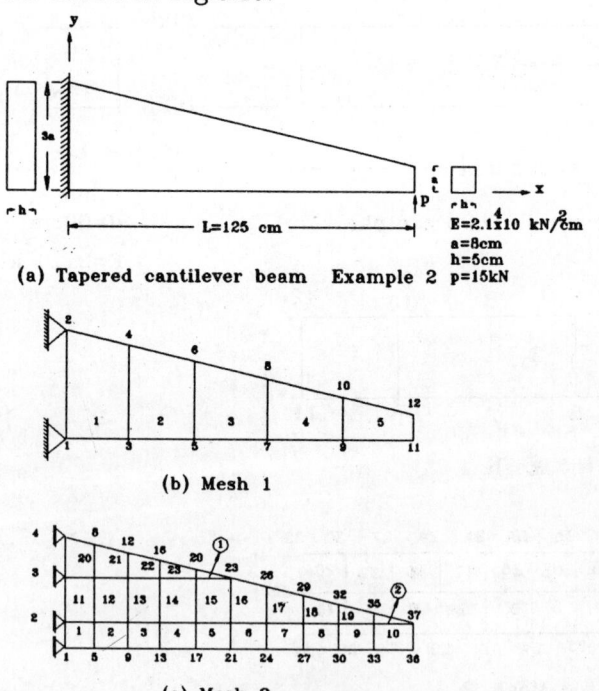

(a) Tapered cantilever beam   Example 2

$E=2.1 \times 10^4$ kN/cm^2
a=8cm
h=5cm
p=15kN

(b) Mesh 1

(c) Mesh 2

Fig. 8.15

Table 8.2 Comparison of Deflection for
Tapered Cantilever Beam

Distance from fixed end (cm)	Theoretical value of deflection	Quadrilateral Element with incompatible modes		Quadrilateral Element	
		Mesh 1	Mesh 2	Mesh 1	Mesh 2
25	0.005189	0.005349	0.005479	0.003379	0.004798
50	0.022340	0.022680	0.023050	0.013620	0.019800
75	0.054260	0.054760	0.055390	0.031080	0.046580
100	0.103990	0.104430	0.105530	0.055490	0.086230
125	0.171430	0.170830	0.173303	0.084790	0.136860

Two types of discretizations are used and in mesh 2 the quadrilateral and triangular elements are combined. Table 8.2 gives the comparison of deflections for the two meshes. As the structure is subjected to predominantly bending, the quadrilateral element with incompatible modes gives excellent results for coarser discretization.

8.9.3 Deep Beam

When the height of the beam is comparable to the span, the assumptions made in the simple theory of bending are not valid. Li Chow, Conway and Winter [14] used finite difference technique to analyse deep beams of various height-span ratios. Fig.8.16 shows an example of deep beam subjected to uniformly distributed load. Due to symmetry only one half of the structure is considered for analysis and the finite element discretizations are shown in Fig.8.16 b and c. The results of finite element analysis using the program PASSFEM are compared with the finite difference technique and are shown in Fig.8.17.

8.9.4 Plate with a Circular Hole

The effect of circular holes on the stress distribution in plates is well treated in text books such as given in reference 15. Figure 8.18 shows an example of a rectangular plate with a circular hole subjected to uniform tension on both sides. The results of finite element analysis using quadrilateral elements are given in Table 8.3 and good agreement has been observed with the analytical solution for σ_θ [15].

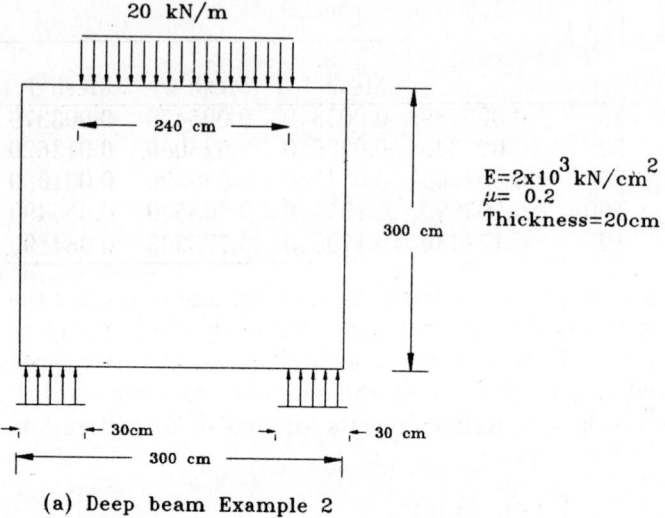

(a) Deep beam Example 2

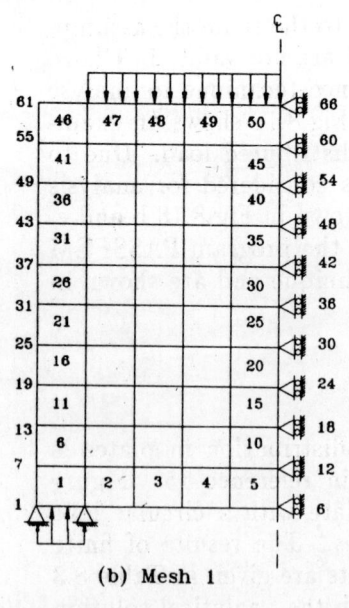

(b) Mesh 1

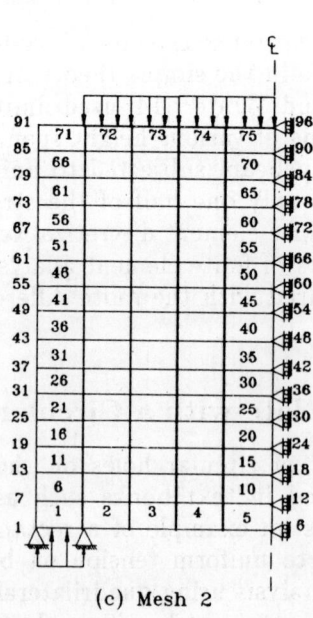

(c) Mesh 2

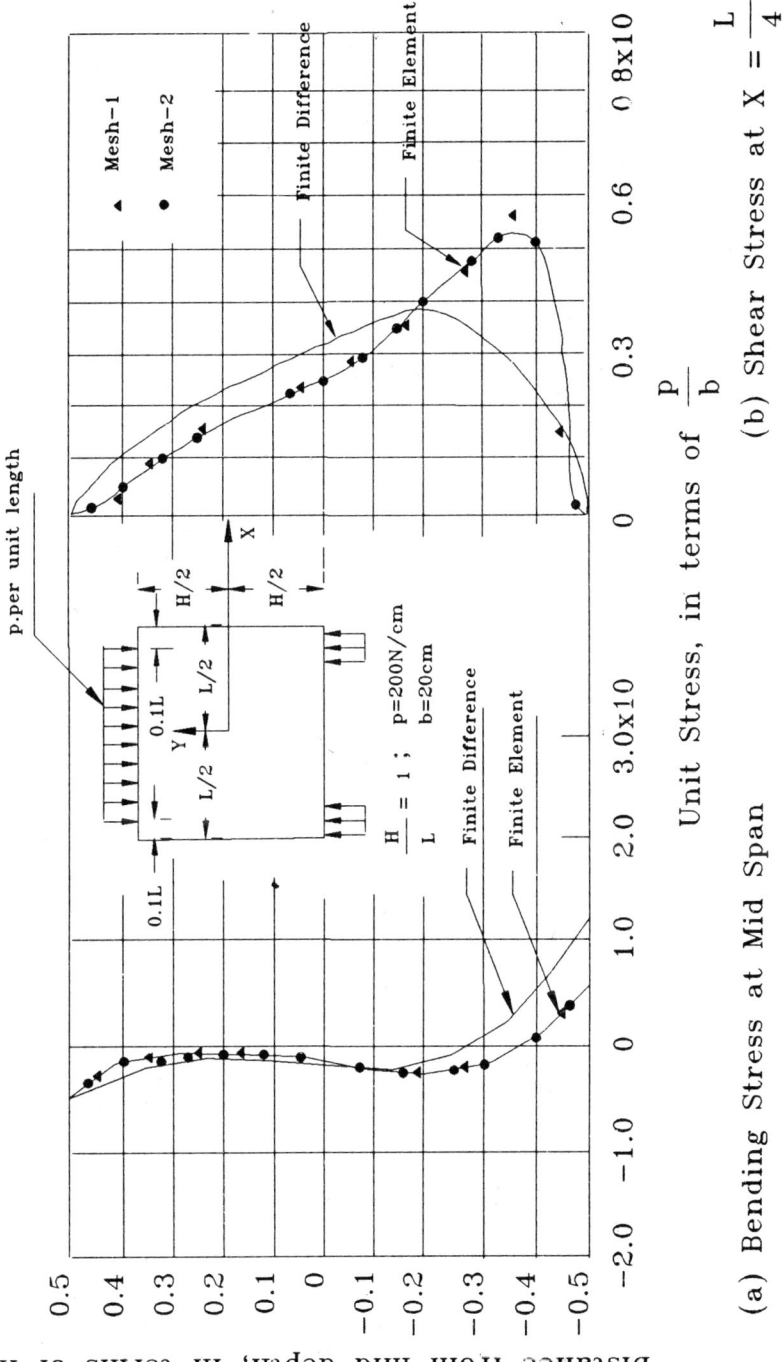

(a) Bending Stress at Mid Span

(b) Shear Stress at $X = \dfrac{L}{4}$

Fig.8.17 Comparison of results—deep beam

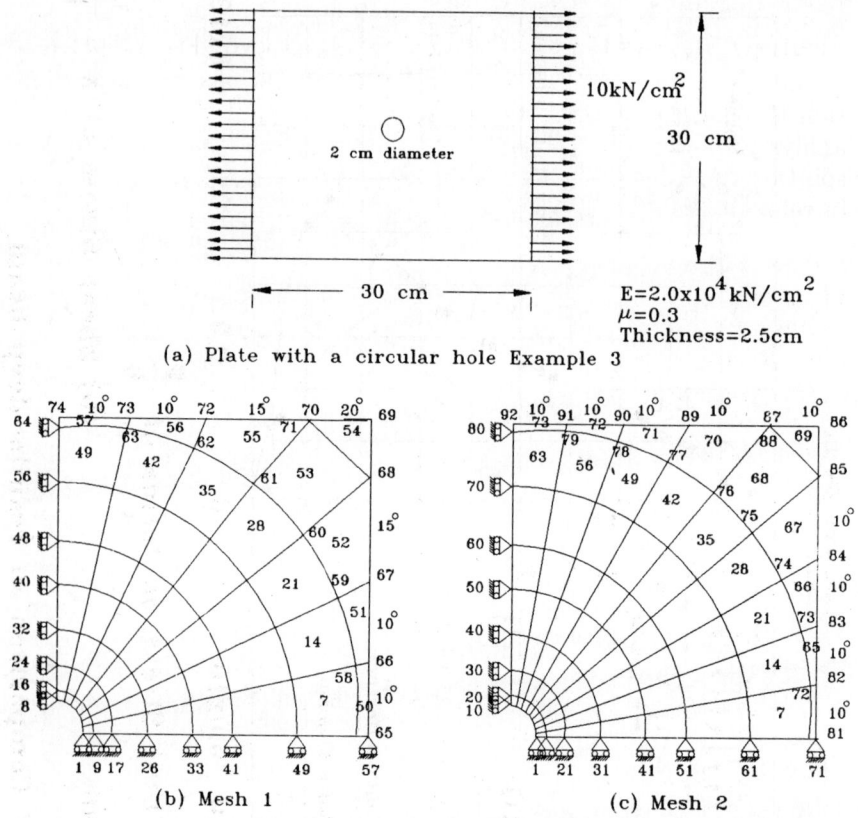

(a) Plate with a circular hole Example 3

(b) Mesh 1

(c) Mesh 2

Fig. 8.18

Table 8.3 Comparison of Stresses for Plate with Circular Hole

r (cms)	Analy. soln. σ_θ $\frac{N}{mm^2}$	Elem. No.	FE soln. Mesh 1 σ_θ $\frac{N}{mm^2}$	r (cms)	Analy. soln. σ_θ $\frac{N}{mm^2}$	Elem. No.	FE soln. Mesh 2 σ_θ $\frac{N}{mm^2}$
1.245	192.98	43	200.28	1.245	192.98	57	199.78
1.992	121.22	44	124.08	1.941	122.85	58	125.48
3.238	105.35	45	106.24	3.087	106.11	59	107.22
4.981	101.49	46	102.09	4.831	101.65	60	102.38
6.973	100.33	47	100.76	6.923	100.35	61	100.86
9.464	99.82	48	99.89	9.364	99.83	62	99.99
12.950	99.54	49	98.75	12.850	99.55	63	98.76

8.9.5 Thick Cylinder

Figure 8.19 shows the cross section of a thick cylinder subjected to an internal pressure of 1.5 kN/cm^2. The finite element discretization is also shown in the above Figure. The results of axisymmetric analysis are given in Table 8.4 and they coincide with the analytical solution obtained by using the expression for $\sigma_r(\sigma_x)$ and σ_θ are given in reference [15].

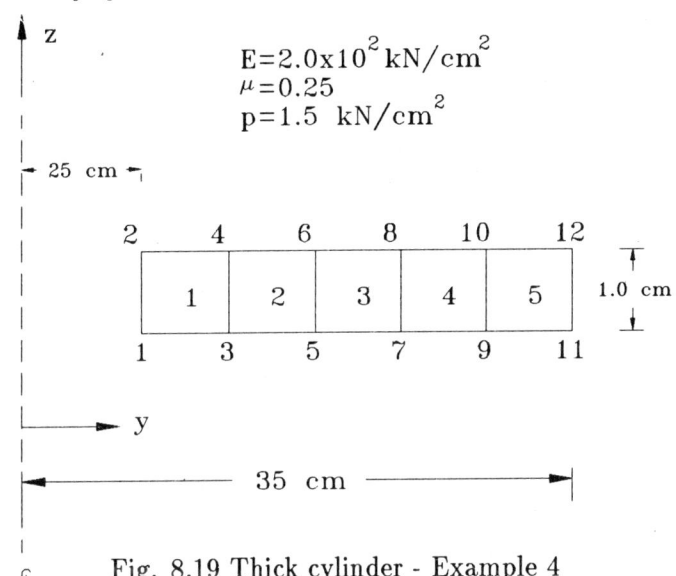

E=2.0x10^2 kN/cm^2
μ=0.25
p=1.5 kN/cm^2

Fig. 8.19 Thick cylinder - Example 4

Table 8.4 Comparison of Results for Thick Cylinder

	Analytical Soln.			FE Analysis	
r(cm)	σ_x (N/mm^2)	σ_θ (N/mm^2)	Element	σ_x (N/mm^2)	σ_θ (N/mm^2)
26	-12.689	43.939	1	-12.69	43.95
28	-8.789	40.039	2	-8.788	40.04
30	-5.642	36.893	3	-5.641	36.89
32	-3.067	34.317	4	-3.066	34.31
34	-0.933	32.183	5	-0.932	32.18

REFERENCES

1. Felippa, C.A. *Refined Finite Element Analysis of Linear and Non-linear Two Dimensional Structures*, Ph.D. Thesis, Structural Engineering Division, University of California, Berkeley, 1966.

2. Desai, C.S. and J.F. Abel *Introduction to the Finite Element Method*, Van Nostrand, N.Y. 1972.

3. Bathe, K.J. and E.L. Wilson, *Numerical Method in Finite Element Analysis*, Prentice-Hall of India Private Limited, New Delhi, 1978.

4. Wilson, E.L., R.L.Taylor, W.P. Doherty and T.Ghabussi, *Incompatible Displacement Models, Numerical and Computer Methods in Structural Mechanics*, (Ed.Fenves, S.J. et.al), Academic Press pp.43-57, 1973.

5. Taylor, R.L., P.J.Beresford and E.L.Wilson, *A Non conforming Element for Stress Analysis*, International Journal for Numerical Methods in Engineering, vol.10 pp.1121-20. 1976.

6. Irons, B.M. and A.Razzaque, *Experience with the Patch Test for Convergence of Finite Element Methods, Math Foundations of the Finite Element Method*, (Ed. A.K.Aziz), Academic Press, 1972, pp.557-587.

7. Irons, B.M. and S.Ahmad, *Techniques of Finite Elements*, Ellis Horwood, Chichester, 1980.

8. Zienkiewicz, O.C. and R.L.Taylor, *The Finite Element Method*, Vol.1, Basic Formulation and Linear Problems, McGraw Hill International Edition, 1989.

9. Strang, G. and G.J.Fix, *An Analysis of the Finite Element Method*, Prentice Hall, Englewood Cliffs, N.J., 1973.

10. Taylor, R.L., O.C.Zienkiewicz, J.C.Sims and A.H.C.Chan, *The Patch Test - A Condition for Assessing FEM Convergence*, International Journal for Numerical Methods in Engineering, Vol.22, pp.39-62, 1986.

11. Krishnamoorthy, C.S., and A.Panneerselvam A. *Finite Element Model for Non-linear Analysis of Reinforced Concrete Framed Structures*, Structural Engineer Journal (London), Vol.55, pp.331-338, 1977.

12. Krishnamoorthy, C.S., and A.Paneerselvam, FEPASI, *A Finite Element Program for Non-linear Analysis of Reinforced Concrete Framed Structures*, Computers and Structures, vol.9, pp.451-461, 1978.

13. Cook, R.D., D.S.Malkus and M.E.Plesha, *Concepts and Applications of Finite Element Analysis*, Third Edition, John Wiley & Sons, New York, 1989.

14. Li Chow, H.D. Conway and G.Winter, *Stresses in Deep Beams,* *Transactions ASCE*, Paper No.2557, vol.118, pp.686-702, 1953.

15. Timoshenko, S. and J.N.Goodier, *Theory of Elasticity*, McGraw-Hill Co., New York, 1970.

16. Fenner, D.N., *Engineering Stress Analysis*, Ellis Horwood, Chichester, 1987.

17. Boresi, A.P., O.M.Sidebottom, *Advanced Mechanics of Materials*, John Wiley & Company, New York, Fourth Edition, 1985.

EXERCISES

8.1 A six noded triangular element is subjected to an uniformly varying surface traction along x direction on the side 1-4-2. Compute the nodal load vector.

8.2 For the six noded triangular element, compute the nodal load vector due to self-weight.

8.3 For the six noded triangular element shown in Fig. E8.3. compute the stiffness matrix using isoparametric formulation and numerical integration scheme described in Chapter 4. Compare the results using the procedure explained in this Chapter.

$$E = 2.1 \times 10^{11} \text{ N/m}^2 \quad \mu = 0.3 \quad t = 2 \text{ cm}$$

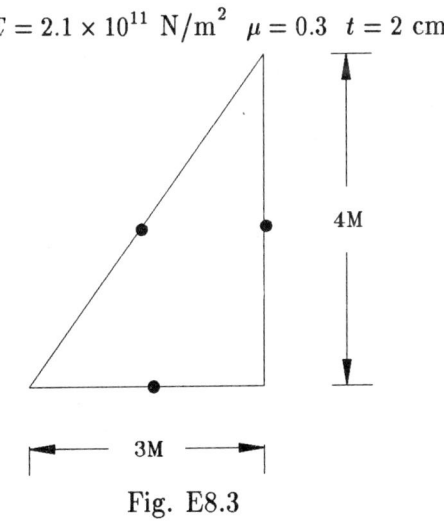

Fig. E8.3

8.4 For the central 8-noded element of exercise 6.15 in Chapter 6, compute the nodal load vector. Use numerical integration procedure.

8.5 Develop the necessary routines for adding eight noded isoparametric element to the PASSFEM library for plane stress/strain analysis.

8.6 Using eight noded isoparametric element analyse the cantilever beam shown in Fig. 8.14. Use atleast two discretizations and compare the results with analytical solution.

8.7 Using eight noded isoparametric element analyse the tappered cantilever beam shown in Fig. 8.15. Use atleast two discretizations and compare the results with analytical solution.

8.8 Analyse the deep beam shown in Fig. 8.16 using eight noded element. Use atleast two discretizations and compare the results with the analytical results given in Fig. 8.17.

8.9 Develop necessary subroutines to add a variable noded element, nine to four noded element to the PASSFEM library for plane stress/strain analysis.

8.10 Using the routines of problem 9, analyse the plate with a circular hole shown in Fig. 8.18. Discretize the region in such a way to illustrate the use of higher order elements in the regions of steep strain gradients and lower order elements in other areas.

8.11 Analyse the problem given in exercise 6.15 in Chapter 6, using the routines developed in exercise 8.5.

8.12 A circular plate is clamped along its outer edge and carries a concentrated load of 4 kN/cm along the edge of a central hole shown in Fig. E8.12. Analyse the plate and compare the result with the plate theory solution which is given by, $\delta = 0.081 \frac{pa^2}{Eh^3}$ $E = 2 \times 10^5$ N/mm^2 $\mu = 0.3$

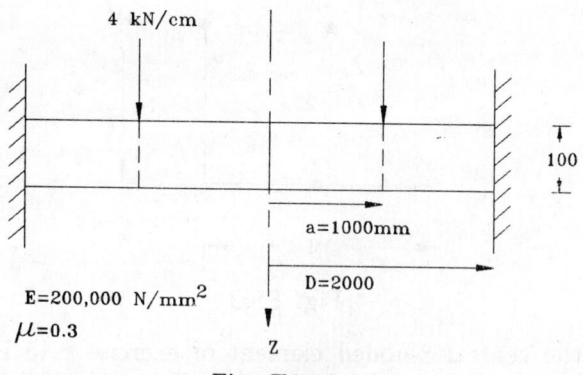

Fig. E8.12

8.13 Analyse the nonprismatic simply supported beam shown in Fig. E8.13. Use (a) triangular element and (b) four noded quadrilateral elements with incompatible modes. Compare the results. width of beam = 50cm $E = 2 \times 10^3$ kN/cm^2 $\mu = 0.0$

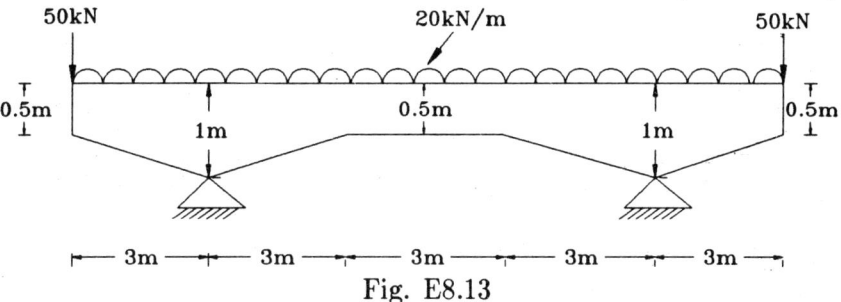

Fig. E8.13

8.14 A member with a notch is shown in Fig. E8.14. Analyse the problem for stress concentration using PASSFEM. Thickness = 1.6 cm. $E = 2.1 \times 10^4$ kN/cm^2. [17]

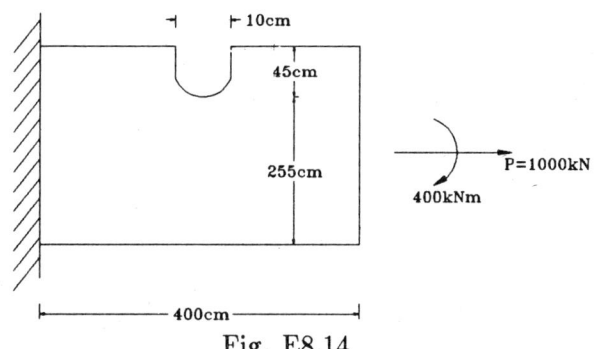

Fig. E8.14

8.15 The square plate shown in Fig. E8.15 has an elliptical hole and is subjected to tensile stress as shown therein. $t = 2.5$ cms $E = 2 \times 10^4$ kN/cm^2 $\mu = 0.3$.

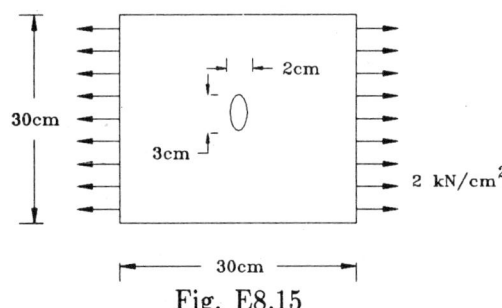

Fig. E8.15

Analyse the problem for stress concentration and compare the results with analytical solution [17].

8.16 Analyse a simply supported beam with a square opening in the centre as shown in Fig. E8.16. Discuss the stress concentration around the opening. $E = 2 \times 10^3$ kN/cm^2 $b = 20$ cm $d = 30$ cm.

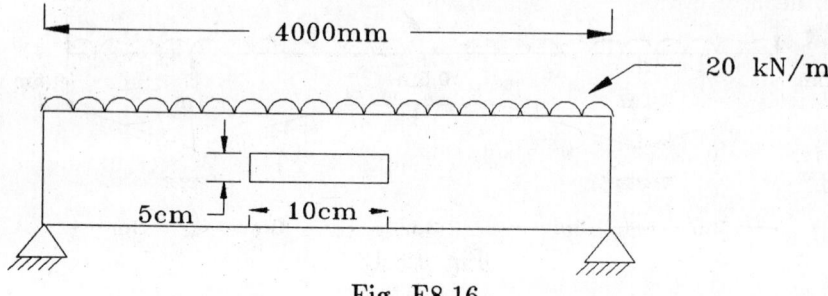

Fig. E8.16

8.17 A circular disk of unit thickness shown in Fig. E8.17 is loaded by a concentrated load of 100 kN. Analyse the problem using 2D elements [17]. $E = 2 \times 10^4$ kN/cm^2 $\mu = 0.3$.

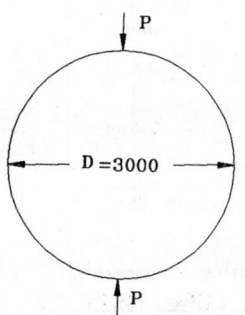

Fig. E8.17

8.18 Fig. E8.18 shows a concrete pipe with a thin steel cylinder.

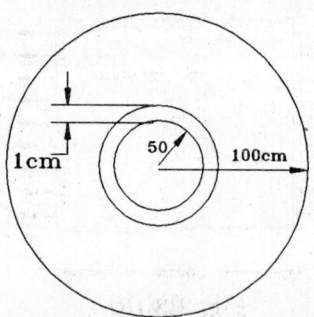

Fig. E8.18

The conduit is subjected to an internal pressure of 1 N/mm^2

$E_c = 2 \times 10^3$ kN/cm^2 $\mu_c = 0.15$ $E_s = 2 \times 10^4$ kN/cm^2 $\mu_s = 0.3$

8.19 Analyse the plate shown in Fig. E8.19. Comment on the results near the tapered portion of the plate.

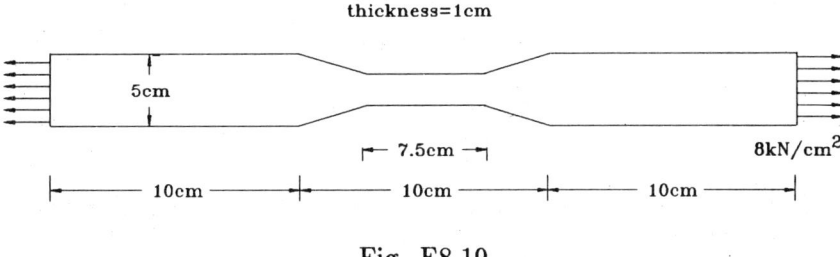

thickness=1cm

5cm

7.5cm

8kN/cm^2

10cm 10cm 10cm

Fig. E8.19

8.20 Analyse the frame shown in Fig. E8.20. It has a square cross-section of 50 mm with load p located 100 mm from the centre of curvature of the curved portion of the frame. Compare the maximum vertical deflection with the analytical solution of 0.1757161 mm [17]. $E = 20$ kN/mm^2 $\mu = 0.3$

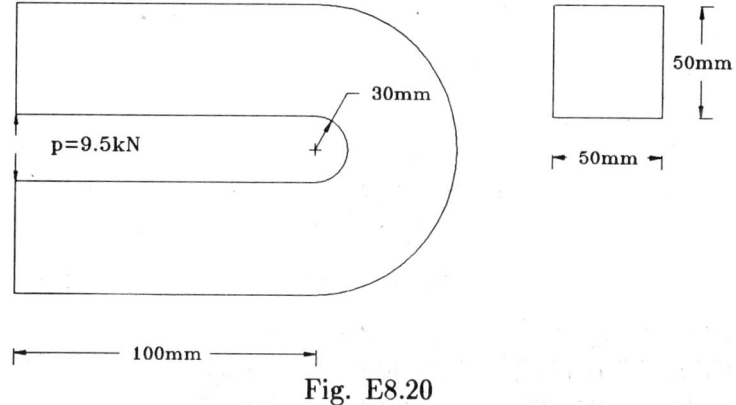

30mm

p=9.5kN

50mm

50mm

100mm

Fig. E8.20

8.21 Analyse the plate in tension with a central crack which is shown in Fig. E8.21. The stress variation near the tip of the stationary crack is given in reference [17].

$$\sigma_y = \frac{\sigma x}{\sqrt{x^2 - a^2}} \qquad \sigma_x = \sigma\left(\frac{x}{\sqrt{x^2 - a^2}} - 1\right)$$

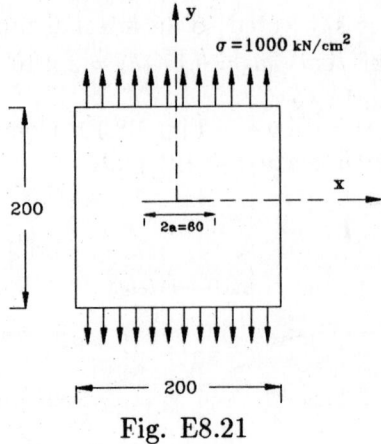

Fig. E8.21

8.22 A long thick walled cylindrical pressure vessel of circular cross section is subjected to an internal pressure p. Analyse the problem assuming plane strain condition. Compare the results with the analytical solution given in Table E8.22 [16]. $E = 2.1 \times 10^4$ kN/cm^2 $\mu = 0.3$ $p = 10$ kN/cm^2.

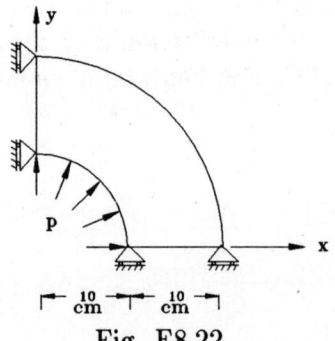

Fig. E8.22

Table E8.22

r	u	σ_r	σ_θ
10.00	0.09533	-10.000	16.6666
13.33	0.07655	-4.1667	10.8333
16.66	0.06644	-1.4666	8.13590
20.00	0.06066	0.0000	6.66660

8.23 Analyse the thick walled spherical pressure vessel shown in Fig. E8.23, using axisymmetric condition. Analytical results by strength of materials solution [16] are given in Table E8.23. $p = 10$ kN/cm^2 $E = 2.1 \times 10^4$ kN/cm^2 $\mu = 0.3$.

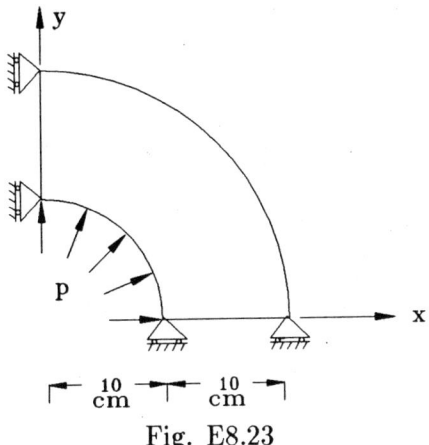

Fig. E8.23

Table E8.23

r	u	σ_r	σ_θ
1000	4.000	-10.000	7.143
1250	2.736	-4.422	4.354
1500	2.079	-1.957	3.122
1750	1.713	-0.7038	2.495
2000	1.500	0.000	2.143

8.24 Fig. E8.24 shows a plate subjected to inplane stress. **Perform the 'Patch Test' for the quadrilateral element.** A discretization is suggested in the Fig. E8.24. Discuss the results with respect to quadrilateral element with incompatible modes.

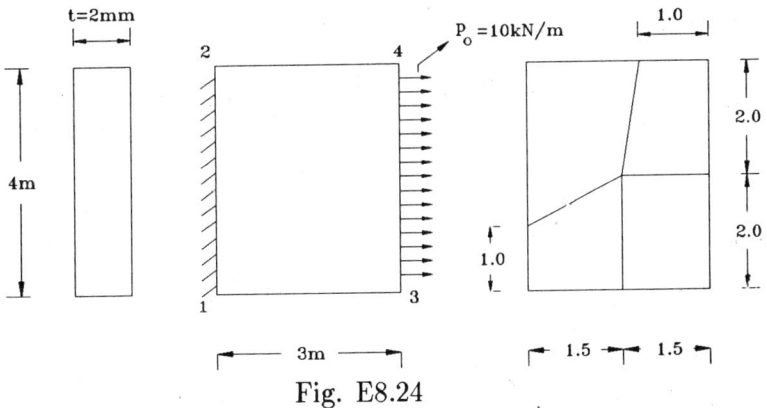

Fig. E8.24

Chapter 9

Three-Dimensional Stress Analysis

Solid or three-dimensional elements enable the solution of problems for a general three-dimensional stress analysis. There are many problems such as concrete dams, stress distribution in soils and rocks, ring beams, pressure vessels, pipe intersections, stresses around openings, machine components, etc. where three-dimensional stress analysis is required. For such problems finite element analysis provides a powerful tool for getting numerical solution.

Figure 9.1 shows typical elements belonging to tetrahedron, triangular prism and hexahedron family of elements. A brief description of these elements will be given in this chapter.

The eight-noded isoparametric element is one of the simplest and its performance in situations where bending is involved can be improved by the addition of incompatible modes similar to the quadrilateral elements discussed in Chapter 8. A detailed derivation of the element properties is presented here. In case of 3D analysis of curved solids, twenty noded isoparametric element has been found to be useful and the element properties are described in this chapter.

Two subroutines THRD08 and THRD20 have been developed using the eight and twenty noded isoparametric elements. The program has been applied to solve typical problems of 3D analysis and the salient features of these examples are given in this chapter.

9.1 Three-Dimensional Solid Elements

Three-dimensional solid elements can be broadly grouped under tetrahedral, triangular prism and hexahedral family of elements [1, 2, 3, 4].

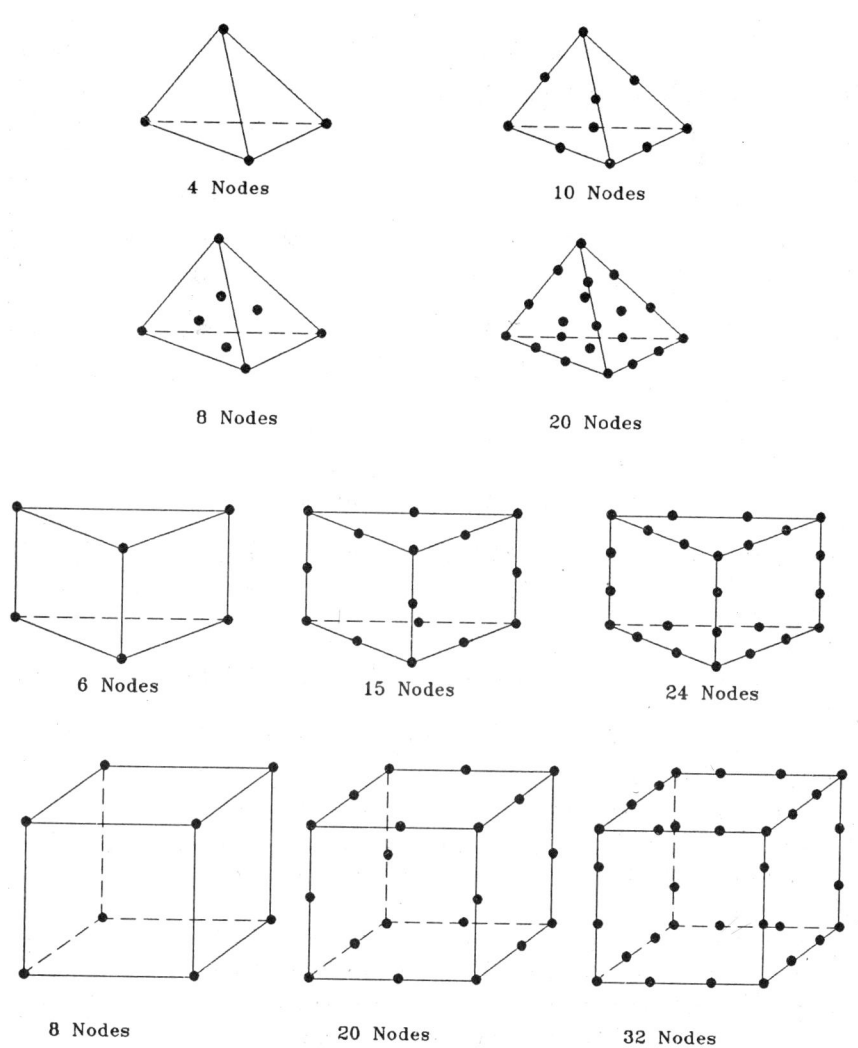

4 Nodes   10 Nodes

8 Nodes   20 Nodes

6 Nodes   15 Nodes   24 Nodes

8 Nodes   20 Nodes   32 Nodes

Fig. 9.1 Three-dimensional solid elements

9.1.1 Tetrahedral Elements

The simplest element of the tetrahedral family is a four noded tetrahedron shown in Fig. 9.2.

The linear shape function for this element can be expressed as,

$$\{N_3\}^T = [L_1 \quad L_2 \quad L_3 \quad L_4] \tag{9.1}$$

where L_i is defined by Eq.3.25.

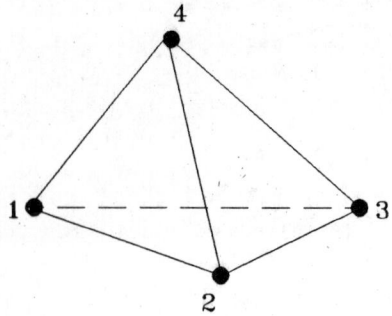

Fig. 9.2 Four noded   Fig. 9.3 Ten noded
    tetrahedron   tetrahedron

Figure 9.3 shows a ten noded tetrahedran. The shape function for this element is given by,

$$\{N_3\}^T = [L_1(2L-1)\ L_2(2L_2-1)\ L_3(2L_3-1)\ L_4(2L_4-1)$$

$$4L_1L_2\ 4L_2L_3\ 4L_3L_1\ 4L_1L_4\ 4L_2L_4\ 4L_3L_4] \tag{9.2}$$

where L_i is defined by Eq.3.25.

The strain variation is linear within the element.

Two other types of eight and twenty noded tetrahedral elements are shown in Fig.9.1. The main disadvantages of tetrahedral family of elements are:

(i) it requires small and costly subdivision and (ii) the division of a space volume into individual tetrahedron sometimes presents difficulties of visualisation and could lead to errors in nodal numbering and element connectivity in data preparation.

9.1.2 Triangular Prism Isoparametric Elements

The simplest triangular prism element is a six noded element shown in Fig.9.4. The polynomial function in natural curvilinear coordinates r, s, and t describing the geometry and the variation of displacement over the element is

$$\{\phi_3\}^T = [1 \quad r \quad s \quad t \quad rs \quad st] \tag{9.3}$$

The six nodes form a solid bounded by two triangular and three quadrilateral faces and the sides are non-intersecting. This element is compatible with eight noded isoparametric hexahedral element.

When quadratic variation is required, fifteen noded triangular prism can be used. The fifteen nodes form a solid bounded by two curved or straight sided triangular faces and three curved or straight sided quadrilateral faces. Neither the faces nor the edges should intersect each other. The polynomial function defining the geometry and displacement variation is $\{\phi_3\}^T =$

$$= [1 \quad r \quad s \quad t \quad rs \quad st \quad rt \quad rst \quad r^2 \quad s^2 \quad t^2 \quad r^2t \quad s^2t \quad rt^2 \quad st^2]$$
$$\tag{9.4}$$

This element is compatible with twenty noded isoparametric hexahedral element.

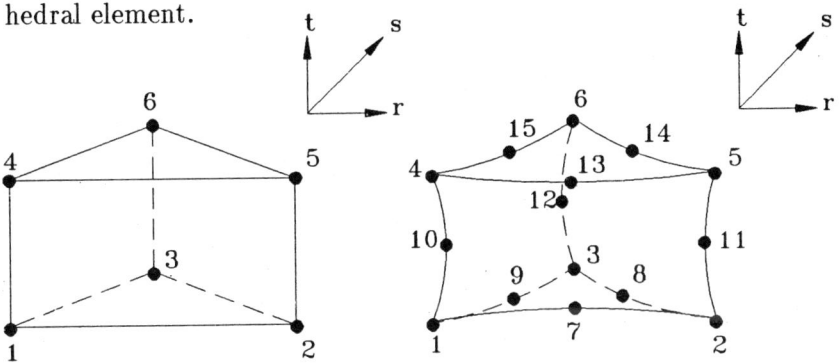

Fig. 9.4 Six noded solid   Fig. 9.5 Fifteen noded solid
triangular element   triangular element

Figure 9.6 shows a twelve noded triangular prism element similar to the fifteen noded element but with no mid-side nodes along the thickness direction t. The twelve nodes form a solid bounded by two curved or straight sided triangular faces and three quadrilateral faces two edges of which must be straight. The polynomial function defining the geometry and displacement variation is the same as six-noded solid triangular element but contains additional terms in r and s directions, viz.,

$$\{\phi_3\}^T = [1 \quad r \quad s \quad t \quad rs \quad st \quad rt \quad r^2 \quad s^2 \quad rst \quad r^2t \quad s^2t] \tag{9.5}$$

This element is compatible with the sixteen noded isoparametric element.

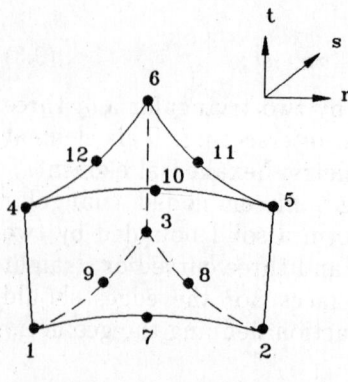

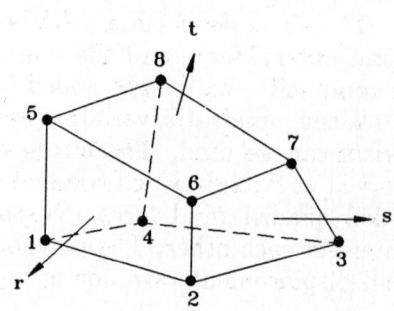

Fig. 9.6 Twelve noded solid
triangular element

Fig. 9.7 Eight noded hexahedral
element

9.1.3   Hexahedral Isoparametric Elements

The eight noded element shown in Fig. 9.7 is the simplest hexahedral element. The polynomial function defining the geometry and displacement variation is

$$\{\phi_3\}^T = [1 \quad r \quad s \quad t \quad rs \quad st \quad rt \quad rst] \qquad (9.6)$$

The twenty noded element shown in Fig.9.8 has eight corner nodes and twelve nodes located at the midpoints of the edges thus capable of accommodating curved boundaries.

The polynomial function defining the geometry and variation of displacements is given by,

$$\{\phi_3\}^T = [1 \quad r \quad s \quad t \quad rs \quad st \quad rt \quad rst \quad r^2 \quad s^2 \quad t^2 \quad r^2s \quad s^2t \quad t^2r \quad r^2t$$

$$rs^2 \quad st^2 \quad r^2st \quad rs^2t \quad rst^2] \qquad (9.7)$$

Figure 9.9 shows a sixteen noded hexahedral element which is similar to the twenty noded element but with no mid-side nodes along the thickness direction t. The sixteen nodes form a solid bounded by six quadrilateral faces, two of each can be curved.

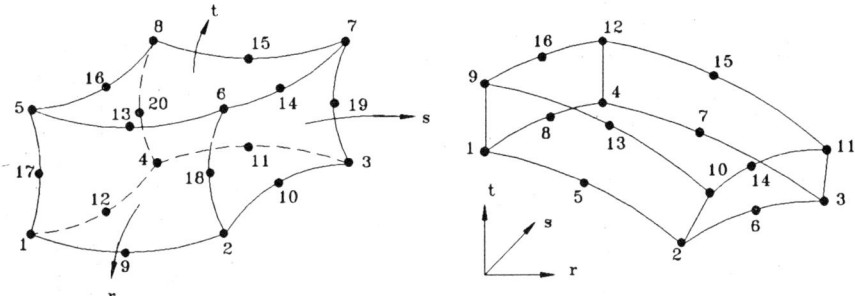

Fig. 9.8 Twenty noded curved solid   Fig. 9.9 Sixteen noded
element   hexahedral element

The polynomial function defining the geometry and displacement variation is the same as for the eight noded hexahedral element but contains additional terms in the r and s direction and is given by

$$\{\phi_3\}^T = [\,1 \quad r \quad s \quad t \quad rs \quad st \quad rt \quad rst \quad r^2 \quad s^2 \quad r^2 s \quad s^2 t$$

$$r^2 t \quad rs^2 \quad r^2 st \quad rs^2 t\,] \tag{9.8}$$

Apart from the above mentioned elements, there are certain other isoparametric elements that have different number of mid surface nodes in the curvilinear directions. These elements may be useful for problems where the variation in displacement function in one direction is of a higher order than in other directions. In the subsequent sections the formulation of two elements, eight noded and twenty noded isoparametric elements are presented.

9.2   Eight Noded Isoparametric Solid Element

The three-dimensional eight noded isoparametric element shown in Fig. 9.10 has eight nodes located at the corners and has three translational degrees of freedom at each node. The shape function defining the geometry and variation of displacement is given by

$$N_i = \frac{1}{8}\,(1 + rr_i)\,(1 + ss_i)\,(1 + tt_i) \qquad i = 1,\,2,\,\ldots,\,8 \tag{9.9}$$

where r, s, t are natural coordinates and, r_i, s_i, t_i are the values of natural coordinates for a node i.

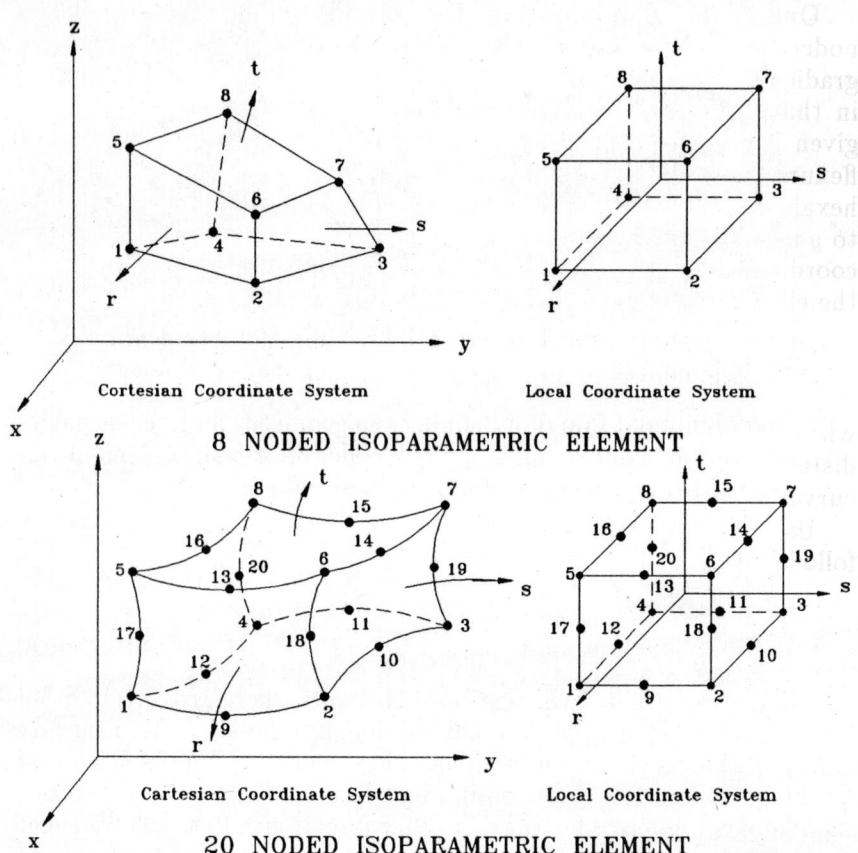

8 NODED ISOPARAMETRIC ELEMENT

20 NODED ISOPARAMETRIC ELEMENT

Fig. 9.10 Isoparametric solid elements

Thus, the geometry of the element is described as,

$$\begin{Bmatrix} x \\ y \\ z \end{Bmatrix} = \sum_{i=1}^{8} N_i \begin{Bmatrix} x_i \\ y_i \\ z_i \end{Bmatrix} \tag{9.10}$$

where x_i, y_i, and z_i are the global coordinates of node i.
The variation of displacement inside the element can be expressed using the same function as,

$$\begin{Bmatrix} u \\ v \\ w \end{Bmatrix} = \sum_{i=1}^{8} N_i \begin{Bmatrix} u_i \\ v_i \\ w_i \end{Bmatrix} \tag{9.11}$$

where u_i, v_i, and w_i are nodal displacements of node i in the cartesian (global) coordinate system.

One of the main causes of the inaccuracies in general of eight noded solid element is its inability to represent some simple stress gradients. The displacement function given by Eq.9.11 is incomplete, in that it does not contain all quadratic terms in shape functions given by Eq.9.9. Hence, this element is not adequate in simulating flexural response. This can be illustrated by subjecting a regular hexahedral element to pure bending moment acting on faces normal to y-axis and in yz plane as shown in Fig.9.11. Taking the origin of coordinates at the centroid of the element, the stress components by the elementary bending theory, are

$$\sigma_y = \frac{E}{R}z, \ \sigma_x = \sigma_z = \tau_{xy} = \tau_{yz} = \tau_{zx} = 0 \qquad (9.12)$$

where E is the modulus of elasticity of the material and z is the distance of the layer from the neutral axis and R is the radius of curvature.

Using the Hooke's law, the strain components are computed as follows:

$$\epsilon_y = \frac{\partial v}{\partial y} = \frac{z}{R} \qquad (9.12a)$$

$$\epsilon_x = \frac{\partial u}{\partial x} = -\frac{\mu z}{R} \qquad (9.12b)$$

$$\epsilon_z = \frac{\partial w}{\partial z} = -\frac{\mu z}{R} \qquad (9.12c)$$

$$\gamma_{xy} = \frac{\partial u}{\partial y} + \frac{\partial v}{\partial x} = 0 \qquad (9.12d)$$

$$\gamma_{yz} = \frac{\partial v}{\partial z} + \frac{\partial w}{\partial y} = 0 \qquad (9.12e)$$

$$\gamma_{zx} = \frac{\partial w}{\partial x} + \frac{\partial u}{\partial z} = 0 \qquad (9.12f)$$

where μ is the Poisson's ratio.

Integrating Eqs. 9.12a, 9.12b and 9.12c we get

$$v = \frac{zy}{R} + f_1(x, z) \qquad (9.13a)$$

$$u = \frac{-\mu zx}{R} + g_1(y, z) \qquad (9.13b)$$

$$w = \frac{-\mu z^2}{2R} + h_1(x, y) \qquad (9.13c)$$

where f_1, g_1 and h_1 are constants of integration.

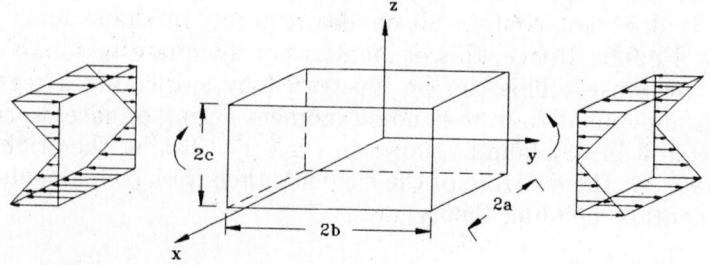

SIMPLE STRESS DISTRIBUTION IN ISOMETRIC VIEW

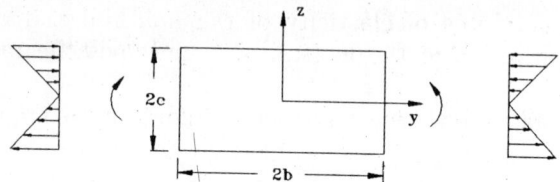

SIMPLE STRESS DISTRIBUTION IN ELEVATION

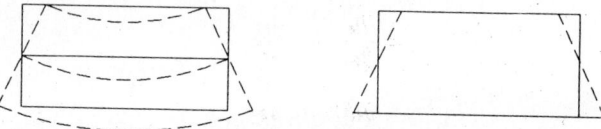

EXACT DISPLACEMENT   FINITE ELEMENT DISPLACEMENT

Fig. 9.11 Stress distribution in eight-noded element

Since the displacements u, v are zero at the origin, Eqs. 9.13a and 9.13b become

$$v = \frac{zy}{R} \qquad (9.14a)$$

$$u = \frac{-\mu\, zx}{R} \qquad (9.14b)$$

From Eqs. 9.12c and 9.14a, we get,

$$\frac{\partial w}{\partial y} = -\frac{\partial v}{\partial z} = -\frac{y}{R}$$

or

$$w = -\frac{y^2}{2R} + h_2(x, z) \qquad (9.15)$$

where h_2 is a constant of integration
From Eqs. 9.12f and 9.14b we get,

$$\frac{\partial w}{\partial x} = -\frac{\partial u}{\partial z} = \frac{\mu x}{R}$$

or

$$w = \frac{\mu x^2}{2R} + h_3\,(y,z) \tag{9.16}$$

From Eqs. 9.13c, 9.15 and 9.16 we get,

$$w = \frac{-y^2}{2R} - \frac{\mu z^2}{2R} + \frac{\mu x^2}{2R} + c_1 \tag{9.17}$$

The constant c_1 in Eq.(9.17) is evaluated by using the condition w
displacements at the eight corners of the element are zero.

$$0 = -\frac{b^2}{2R} - \frac{\mu c^2}{2R} + \frac{\mu a^2}{2R} + c_1$$

$$c_1 = \frac{b^2}{2R} + \frac{\mu c^2}{2R} - \frac{\mu a^2}{2R}$$

Substituting the above value of c_1 into Eq.9.17,

$$w = \frac{b^2}{2R}\left(1 - \frac{y^2}{b^2}\right) + \frac{\mu c^2}{2R}\left(1 - \frac{z^2}{c^2}\right) - \frac{\mu a^2}{2R}\left(1 - \frac{x^2}{a^2}\right) \tag{9.18}$$

Generalising, Eq.9.14a, 9.14b and 9.18 can be expressed as,

$$u = \alpha_1\,xz \tag{9.19a}$$

$$v = \alpha_2\,yz \tag{9.19b}$$

$$w = \alpha_3\left(1 - \frac{y^2}{b^2}\right) + \alpha_4\left(1 - \frac{z^2}{c^2}\right) + \alpha_5\left(1 - \frac{x^2}{a^2}\right) \tag{9.19c}$$

where $\alpha_1, ..., \alpha_5$ are constants which are functions of μ, R, a, b and c.
Similarly, if the element is subject to pure bending moment on
faces normal to the x axis, and in the xy plane, the displacement
variation can be expressed as,

$$u = \beta_1\,xy \tag{9.19d}$$

$$v = \beta_2\left(1 - \frac{y^2}{b^2}\right) + \beta_3\left(1 - \frac{z^2}{c^2}\right) + \beta_4\left(1 - \frac{x^2}{a^2}\right) \tag{9.19e}$$

$$w = \beta_5\,yz \tag{9.19f}$$

where $\beta_1, ..., \beta_5$ are constants.

When the element is subjected to pure bending moment on faces normal to z-axis and in xz plane, the displacement variation can be expressed as

$$u = \gamma_1 \left(1 - \frac{y^2}{b^2}\right) + \gamma_2 \left(1 - \frac{z^2}{c^2}\right) + \gamma_3 \left(1 - \frac{x^2}{a^2}\right) \qquad (9.19g)$$

$$v = \gamma_4\, xy \qquad (9.19h)$$

$$w = \gamma_5\, xz \qquad (9.19i)$$

where $\gamma_1, ..., \gamma_5$ are constants.

When we refer Eqs.9.19c, 9.19e and 9.19g these displacement variations are not represented by the shape functions used for the eight noded solid element. This element is able to represent only displacements given by Eqs. 9.19a, 9.19b, 9.19d, 9.19f, 9.19h and 9.19i. Absence of quadratic terms,

$\left(1 - \frac{x^2}{a^2}\right), \left(1 - \frac{y^2}{b^2}\right), \left(1 - \frac{z^2}{c^2}\right)$ which can be expressed as $(1 - r^2), (1 - s^2), (1 - t^2)$ by putting $r = \frac{x}{a}, s = \frac{y}{b}$ and $t = \frac{z}{c}$ for the element shown in Fig.9.11, is the primary source of error in the solution when the element is under flexural action.

To simulate adequately the flexural response, Wilson et.al.[5,6] introduced additional displacement modes to the general eight noded solid element and these additional modes have the same form and order as the terms in Eqs.9.19c, 9.19e and 9.19g. These additional modes are represented by functions of the type,

$$P_1 = (1 - r^2), \quad P_2 = (1 - s^2), \quad P_3 = (1 - t^2) \qquad (9.20)$$

These additional modes are called incompatible modes and they are not activated at the nodes of the element. So the displacement variation for an eight noded solid element with incompatible modes can be expressed as,

$$\begin{Bmatrix} u \\ v \\ w \end{Bmatrix} = \sum_{i=1}^{8} N_i \begin{Bmatrix} u_i \\ v_i \\ w_i \end{Bmatrix} + [P]\{\alpha\} \qquad (9.21)$$

where

$$[P] = \begin{bmatrix} P_1 & P_2 & P_3 & 0 & 0 & 0 & 0 & 0 & 0 \\ 0 & 0 & 0 & P_1 & P_2 & P_3 & 0 & 0 & 0 \\ 0 & 0 & 0 & 0 & 0 & 0 & P_1 & P_2 & P_3 \end{bmatrix}$$

and

$$\{\alpha\}^T = [\alpha_1 \quad \alpha_2 \quad \alpha_3 \quad \alpha_4 \quad \alpha_5 \quad \alpha_6 \quad \alpha_7 \quad \alpha_8 \quad \alpha_9]$$

9.2.1 Strain-Displacement Matrix [B]

As the formulation is done in terms of natural coordinates, we have to establish the relationship between the derivatives of functions in the natural coordinates and the derivates in cartesian coordinates. This is derived by evaluating the Jacobian given by Eq.4.16 as,

$$\left\{ \begin{array}{c} \frac{\partial}{\partial r} \\ \frac{\partial}{\partial s} \\ \frac{\partial}{\partial t} \end{array} \right\} = [J] \left\{ \begin{array}{c} \frac{\partial}{\partial x} \\ \frac{\partial}{\partial y} \\ \frac{\partial}{\partial z} \end{array} \right\} \tag{9.22}$$

where the Jacobian [J] is given by,

$$[J] = \begin{bmatrix} \frac{\partial x}{\partial r} & \frac{\partial y}{\partial r} & \frac{\partial z}{\partial r} \\ \frac{\partial x}{\partial s} & \frac{\partial y}{\partial s} & \frac{\partial z}{\partial s} \\ \frac{\partial x}{\partial t} & \frac{\partial y}{\partial t} & \frac{\partial z}{\partial t} \end{bmatrix} \tag{9.23}$$

Substituting Eq.9.10 for x, y and z into Eq.9.23 we get,

$$[J] = \sum_{i=1}^{8} \begin{bmatrix} x_i \frac{\partial N_i}{\partial r} & y_i \frac{\partial N_i}{\partial r} & z_i \frac{\partial N_i}{\partial r} \\ x_i \frac{\partial N_i}{\partial s} & y_i \frac{\partial N_i}{\partial s} & z_i \frac{\partial N_i}{\partial s} \\ x_i \frac{\partial N_i}{\partial t} & y_i \frac{\partial N_i}{\partial t} & z_i \frac{\partial N_i}{\partial t} \end{bmatrix} \tag{9.24}$$

The inverse of the above matrix can be symbolically written as.

$$[J]^{-1} = \begin{bmatrix} J_{11}^* & J_{12}^* & J_{13}^* \\ J_{21}^* & J_{22}^* & J_{23}^* \\ J_{31}^* & J_{32}^* & J_{33}^* \end{bmatrix} \tag{9.25}$$

The strain-displacement relation is given by Eq.3.69 as,

$$\{\epsilon\} = [B]\{d\} \tag{9.26a}$$

where

$$\{d\}^T = [\, u_1 \quad v_1 \quad w_1 \quad . \quad . \quad . \quad u_8 \quad v_8 \quad w_8$$

$$\alpha_1 \quad \alpha_4 \quad \alpha_7 \quad \alpha_2 \quad \alpha_5 \quad \alpha_8 \quad \alpha_3 \quad \alpha_6 \quad \alpha_9 \,]$$

Equation 9.26a can be written as

$$\{\epsilon\} = \sum_{i=1}^{8} [B_i]\{d_i\} + [P']\{\alpha'\} \tag{9.26b}$$

$$= [B_a]\{d_a\} + [P']\{\alpha'\} \tag{9.26c}$$

where $\{d_i\}^T = [\,u_i \quad v_i \quad w_i\,]$

$$\{d_a\}^T = [\,u_1 \quad v_1 \quad w_1 \quad . \quad . \quad . \quad u_8 \quad v_8 \quad w_8\,]$$

$$\{\alpha'\}^T = [\,\alpha_1 \quad \alpha_4 \quad \alpha_7 \quad \alpha_2 \quad \alpha_5 \quad \alpha_8 \quad \alpha_3 \quad \alpha_6 \quad \alpha_9\,]$$

and $[P']$ represents the differentiation of $[P]$ with respect to the global coordinates x, y and z.

From Eq.2.10 the components of strains can be expressed as,

$$\{\epsilon\} = \begin{Bmatrix} \epsilon_x \\ \epsilon_y \\ \epsilon_z \\ \gamma_{xy} \\ \gamma_{yz} \\ \gamma_{zx} \end{Bmatrix} = \begin{Bmatrix} \dfrac{\partial u}{\partial x} \\[6pt] \dfrac{\partial v}{\partial y} \\[6pt] \dfrac{\partial w}{\partial z} \\[6pt] \dfrac{\partial u}{\partial y} + \dfrac{\partial v}{\partial x} \\[6pt] \dfrac{\partial v}{\partial z} + \dfrac{\partial w}{\partial y} \\[6pt] \dfrac{\partial u}{\partial z} + \dfrac{\partial w}{\partial x} \end{Bmatrix} \tag{9.27}$$

In order to find the derivatives of the displacements in global coordinates, we have to first obtain the derivatives of the displacements in natural coordinates. Differentiating Eq.9.21 with respect to r, s, t we get,

$$\begin{bmatrix} \dfrac{\partial u}{\partial r} & \dfrac{\partial v}{\partial r} & \dfrac{\partial w}{\partial r} \\[6pt] \dfrac{\partial u}{\partial s} & \dfrac{\partial v}{\partial s} & \dfrac{\partial w}{\partial s} \\[6pt] \dfrac{\partial u}{\partial t} & \dfrac{\partial v}{\partial t} & \dfrac{\partial w}{\partial t} \end{bmatrix} = \sum_{i=1}^{8} \begin{bmatrix} \dfrac{\partial N_i}{\partial r} u_i & \dfrac{\partial N_i}{\partial r} v_i & \dfrac{\partial N_i}{\partial r} w_i \\[6pt] \dfrac{\partial N_i}{\partial s} u_i & \dfrac{\partial N_i}{\partial s} v_i & \dfrac{\partial N_i}{\partial s} w_i \\[6pt] \dfrac{\partial N_i}{\partial t} u_i & \dfrac{\partial N_i}{\partial t} v_i & \dfrac{\partial N_i}{\partial t} w_i \end{bmatrix}$$

$$+ \begin{bmatrix} -2r\alpha_1 & -2r\alpha_4 & -2r\alpha_7 \\ -2s\alpha_2 & -2s\alpha_5 & -2s\alpha_8 \\ -2t\alpha_3 & -2t\alpha_6 & -2t\alpha_9 \end{bmatrix} \tag{9.28}$$

Now using the Jacobian inverse we get,

$$\begin{bmatrix} \dfrac{\partial u}{\partial x} & \dfrac{\partial v}{\partial x} & \dfrac{\partial w}{\partial x} \\[6pt] \dfrac{\partial u}{\partial y} & \dfrac{\partial v}{\partial y} & \dfrac{\partial w}{\partial y} \\[6pt] \dfrac{\partial u}{\partial z} & \dfrac{\partial v}{\partial z} & \dfrac{\partial w}{\partial z} \end{bmatrix} = \begin{bmatrix} J^*_{11} & J^*_{12} & J^*_{13} \\ J^*_{21} & J^*_{22} & J^*_{23} \\ J^*_{31} & J^*_{32} & J^*_{33} \end{bmatrix} \begin{bmatrix} \dfrac{\partial u}{\partial r} & \dfrac{\partial v}{\partial r} & \dfrac{\partial w}{\partial r} \\[6pt] \dfrac{\partial u}{\partial s} & \dfrac{\partial v}{\partial s} & \dfrac{\partial w}{\partial s} \\[6pt] \dfrac{\partial u}{\partial t} & \dfrac{\partial v}{\partial t} & \dfrac{\partial w}{\partial t} \end{bmatrix} \tag{9.29}$$

Expanding the above equation, we get,

$$\frac{\partial u}{\partial x} = J_{11}^* \frac{\partial u}{\partial r} + J_{12}^* \frac{\partial u}{\partial s} + J_{13}^* \frac{\partial u}{\partial t} \tag{9.30a}$$

Substituting from Eq.9.21 for u in the above equation,

$$\frac{\partial u}{\partial x} = \sum_{i=1}^{8} (J_{11}^* \frac{\partial N_i}{\partial r} + J_{12}^* \frac{\partial N_i}{\partial s} + J_{13}^* \frac{\partial N_i}{\partial t}) u_i \tag{9.30b}$$

$$-2r J_{11}^* \, \alpha_1 - 2s J_{12}^* \, \alpha_2 - 2t J_{13}^* \, \alpha_3$$

Simplifying Eq.9.30b we get,

$$\frac{\partial u}{\partial x} = \sum_{i=1}^{8} \frac{\partial N_i}{\partial x} u_i - 2r J_{11}^* \, \alpha_1 - 2s J_{12}^* \, \alpha_2 - 2t J_{13}^* \, \alpha_3 \tag{9.31a}$$

Similarly we can derive expressions for the remaining derivatives of displacement with respect to the cartesian system and they are as follows:

$$\frac{\partial v}{\partial x} = \sum_{i=1}^{8} \frac{\partial N_i}{\partial x} v_i - 2r J_{11}^* \, \alpha_4 - 2s J_{12}^* \, \alpha_5 - 2t J_{13}^* \, \alpha_6 \tag{9.31b}$$

$$\frac{\partial w}{\partial x} = \sum_{i=1}^{8} \frac{\partial N_i}{\partial x} w_i - 2r J_{11}^* \, \alpha_7 - 2s J_{12}^* \, \alpha_8 - 2t J_{13}^* \, \alpha_9 \tag{9.31c}$$

$$\frac{\partial u}{\partial y} = \sum_{i=1}^{8} \frac{\partial N_i}{\partial y} u_i - 2r J_{21}^* \, \alpha_1 - 2s J_{22}^* \, \alpha_2 - 2t J_{23}^* \, \alpha_3 \tag{9.31d}$$

$$\frac{\partial w}{\partial y} = \sum_{i=1}^{8} \frac{\partial N_i}{\partial y} v_i - 2r J_{21}^* \, \alpha_4 - 2s J_{22}^* \, \alpha_5 - 2t J_{23}^* \, \alpha_6 \tag{9.31e}$$

$$\frac{\partial w}{\partial y} = \sum_{i=1}^{8} \frac{\partial N_i}{\partial y} w_i - 2r J_{21}^* \, \alpha_7 - 2s J_{22}^* \, \alpha_8 - 2t J_{23}^* \, \alpha_9 \tag{9.31f}$$

$$\frac{\partial u}{\partial z} = \sum_{i=1}^{8} \frac{\partial N_i}{\partial z} u_i - 2r J_{31}^* \, \alpha_1 - 2s J_{32}^* \, \alpha_2 - 2t J_{33}^* \, \alpha_3 \tag{9.31g}$$

$$\frac{\partial v}{\partial z} = \sum_{i=1}^{8} \frac{\partial N_i}{\partial z} v_i - 2r J_{31}^* \ \alpha_4 - 2s J_{32}^* \ \alpha_5 - 2t J_{33}^* \ \alpha_6 \qquad (9.31h)$$

$$\frac{\partial w}{\partial z} = \sum_{i=1}^{8} \frac{\partial N_i}{\partial z} w_i - 2r J_{31}^* \ \alpha_7 - 2s J_{32}^* \ \alpha_8 - 2t J_{33}^* \ \alpha_9 \qquad (9.31i)$$

Using the equations (9.31). we can obtain the values of the matrices $[B_i]$ and $[P']$ of Eq.9.26b as,

$$[B_i] = \begin{bmatrix} \frac{\partial N_i}{\partial x} & 0 & 0 \\ 0 & \frac{\partial N_i}{\partial y} & 0 \\ 0 & 0 & \frac{\partial N_i}{\partial z} \\ \frac{\partial N_i}{\partial y} & \frac{\partial N_i}{\partial x} & 0 \\ 0 & \frac{\partial N_i}{\partial z} & \frac{\partial N_i}{\partial y} \\ \frac{\partial N_i}{\partial z} & 0 & \frac{\partial N_i}{\partial x} \end{bmatrix} \qquad (9.32a)$$

and $[P'] =$

$$= \left[\begin{array}{ccc|ccc} -2r J_{11}^* & 0 & 0 & -2s J_{12}^* & 0 & 0 \\ 0 & -2r J_{21}^* & 0 & 0 & -2s J_{22}^* & 0 \\ 0 & 0 & -2r J_{31}^* & 0 & 0 & -2s J_{32}^* \\ -2r J_{21}^* & -2r J_{31}^* & 0 & -2s J_{22}^* & -2s J_{12}^* & 0 \\ 0 & -2r J_{31}^* & -2r J_{21}^* & 0 & -2s J_{32}^* & -2s J_{22}^* \\ -2r J_{31}^* & 0 & -2r J_{11}^* & -2s J_{32}^* & 0 & -2s J_{12}^* \end{array} \right.$$

$$\left. \begin{array}{ccc} -2t J_{13}^* & 0 & 0 \\ 0 & -2t J_{23}^* & 0 \\ 0 & 0 & -2t J_{33}^* \\ -2t J_{23}^* & -2t J_{13}^* & 0 \\ 0 & -2t J_{33}^* & -2t J_{23}^* \\ -2t J_{33}^* & 0 & -2t J_{13}^* \end{array} \right] \qquad (9.32t)$$

$$= [[B_9] \quad [B_{10}] \quad [B_{11}]]$$

where $[B_9]$ $[B_{10}]$ and $[B_{11}]$ are submatrices of size 6×3 shown in partition in Eq.9.32b. Hence we can write the $[B]$ matrix of Eq.9.26a as,

$$[B] =$$

$$[[B_1] \quad [B_2] \quad [B_3] \quad [B_4] \quad [B_5] \quad [B_6] \quad [B_7] \quad [B_8] \quad [B_9] \quad [B_{10}] \quad [B_{11}]]$$

6 × 3   6 × 3   6 × 3   6 × 3   6 × 3   6 × 3   6 × 3   6 × 3   6 × 3   6 × 3   6 × 3

$$(9.32c)$$

9.2.2   Convergence Criteria

The eight noded solid element with incompatible modes was an improvement over the one without incompatible modes in representing the bending behaviour of the element. However, its accuracy is found to be dependent upon the discretization of the structure. It yields excellent results for flexure problems if the elements are regular hexahedrons, but if they are not, then the stresses calculated are found to be erroneous [5, 6]. Also under constant deformation state even the regular hexahedron elements give erroneous results. The inconsistencies have been investigated by Taylor et.al.[6] and the condition to be satisfied for constant strain state has been derived. The procedure is exactly similar to the derivation presented in Chapter 8 for isoparametric four noded quadrilateral element with incompatible modes. Hence, a full derivation is left to be done by the students as an exercise and the final condition to be satisfied is stated below.

$$\int_{-1}^{+1} \int_{-1}^{+1} \int_{-1}^{+1} [P'] \, |J| \, dr \, ds \, dt = [0] \qquad (9.33)$$

where $|J|$ is the determinant of the Jacobian matrix $[J]$. For eight noded solid element with incompatible modes $[P']$ is given in Eq.9.32b.

Equation 9.33 is automatically satisfied if the element is parallelopiped in which $|J|$ is constant. For arbitrary shaped elements, the Jacobian is not constant and since the terms in $[P']$ are first powers of r, s, t, Eq.9.33 will be satisfied if $|J|$ is made constant. This can be done by evaluating $|J|$ from a Jacobian matrix $[J]$ computed at $r = s = t = 0$ regardless of actual coordinates of the integration points. The Jacobian matrix $[J]$ used to evaluate $[B_a]$ in Eq.9.26c is not modified. In essence the Jacobian matrix involved in the formation of $[B_a]$ in Eq.9.26c is evaluated at integration points, while the Jacobian involved in $[P']$ in Eq.9.26c is evaluated at $r = s = t = 0$.

9.2.3   Element Stiffness Matrix

The constitutive matrix $[C]$ for isotropic three-dimensional stress analysis is given in Eq.2.16. The element stiffness matrix is evaluated using the Eq.4.19 as,

$$[K] = \int_{-1}^{+1} \int_{-1}^{+1} \int_{-1}^{+1} [B]^T \quad [C] \quad [B] \, |J| dr \, ds \, dt \qquad (9.34)$$

33 × 33   33 × 6   6 × 6   6 × 33

Numerical integration procedure explained in Chapter 4 (Eq.4.24) wi'l be used for evaluating the stiffness matrix. Using Gauss quadrature, the 2 × 2 × 2 scheme has been found to be adequate.

The stiffness matrix as given by Eq.9.34 is of size 33 × 33 and includes coefficients pertaining to incompatible modes. However, these terms can be eliminated by using the static condensation procedure and the condensed stiffness matrix will be of the order 24 × 24 pertaining to the nodal degrees of freedom.

9.3   Twenty   Noded   Isoparametric   Solid Element

The three-dimensional twenty noded isoparametric element shown in Fig.9.10 has eight corner nodes and other nodes at mid points of the lines joining the successive corner nodes. The shape functions are given below:

$$N_i = \frac{1}{8} \left(1 + r_i r\right) \left(1 + s_i s\right) \left(1 + t_i t\right) \left(r_i r + s_i s + t_i t - 2\right)$$

for nodes $i = 1$ to 8 $\qquad (9.35a)$

$$N_i = \frac{1}{4} \left(1 - r^2\right) \left(1 + s_i s\right) \left(1 + t_i t\right)$$

for nodes $i = 10, 12, 14, 16$ $\qquad (9.35b)$

$$N_i = \frac{1}{4} \left(1 - s^2\right) \left(1 + r_i r\right) \left(1 + t_i t\right)$$

for nodes $i = 9, 11, 13, 15,$ $\qquad (9.35\text{-})$

$$N_i = \frac{1}{4} \left(1 - t^2\right) \left(1 + r_i r\right) \left(1 + s_i s\right)$$

for nodes $i = 17, 18, 19, 20$ $\qquad (9.35d)$

The geometry of the element is described as,

$$\begin{Bmatrix} x \\ y \\ z \end{Bmatrix} = \sum_{i=1}^{20} N_i \begin{Bmatrix} x_i \\ y_i \\ z_i \end{Bmatrix} \qquad (9.36)$$

The variation of displacement over the element is expressed by the same shape function as,

$$\left\{\begin{array}{c} u \\ v \\ w \end{array}\right\} = \sum_{i=1}^{20} N_i \left\{\begin{array}{c} u_i \\ v_i \\ w_i \end{array}\right\} \tag{9.37}$$

The element properties can be derived similar to the eight noded element except that no incompatible modes are assumed in this case. The Jacobian matrix [J] and the strain-displacement matrix [B] are given below.

$$[J] = \sum_{i=1}^{20} \begin{bmatrix} x_i \frac{\partial N_i}{\partial r} & y_i \frac{\partial N_i}{\partial r} & z_i \frac{\partial N_i}{\partial r} \\ x_i \frac{\partial N_i}{\partial s} & y_i \frac{\partial N_i}{\partial s} & z_i \frac{\partial N_i}{\partial s} \\ x_i \frac{\partial N_i}{\partial t} & y_i \frac{\partial N_i}{\partial t} & z_i \frac{\partial N_i}{\partial t} \end{bmatrix} \tag{9.38}$$

$$[B] = [[B_1] \quad [B_2] \quad . \quad . \quad . \quad [B_{20}]] \tag{9.39}$$

$$6 \times 3 \quad 6 \times 3 \quad\quad 6 \times 3$$

where $[B_i], i = 1, ...20$ is given by Eq.(9.32a)

The element stiffness matrix is given by the following expression,

$$[k] = \int_{-1}^{+1} \int_{-1}^{+1} \int_{-1}^{+1} [B]^T \quad [C] \quad [B] \quad |J| \ dr \ ds \ dt \tag{9.40}$$

$$60 \times 6 \quad 6 \times 6 \quad 6 \times 60$$

The stiffness matrix is evaluated using $3 \times 3 \times 3$ Gauss points.

Instead of using $3 \times 3 \times 3$ integration scheme, a reduced 14 point integration rule [7, 8, 9] can also be used, since this rule gives the same accuracy with less computational effort. Of these 14 points, six correspond to three pairs of points situated symmetrically along each axis of symmetry and the remaining eight correspond to those situated symmetrically about each plane of symmetry just similar to that $2 \times 2 \times 2$ Gauss rule in the case of eight noded solid element. The values of weight functions and the coordinates of the sampling points are given below in Eq.9.41. If $f(r,s,t)$ is the function to be integrated numerically, its value is given by,

$$\int \int \int f(r,s,t) \ dr \ ds \ dt = B_6 \ [f(-b,o,o) + f(b,0,0) + f(o,-b,o)$$

$$+ \ . \ . \ . \ (6 \ \text{terms})] \ +$$

$$+ C_8 \ [f(-c,-c,-c) + f(-c,-c,+c) + f(-c,c,-c)$$

$$+ \ . \ . \ . \ (8 \ \text{terms})] \tag{9.41}$$

where B_6 and C_8 are the weights given by

$$B_6 = 0.8864265927977839, \quad C_8 = 0.3351800554016621$$

where b and c are sampling point locations from centre of element given by,

$$b = 0.798224257542215, \quad c = 0.7587869106393281$$

9.4 Properties of Element Faces

For calculating element load vector and stresses, element faces are to be defined and the orientation with respect to global coordinates are to be worked out. These details are presented below.

The element faces are numbered as follows:

Face 1   corresponds to $+r$ direction
      2   corresponds to $-r$ direction
      3   corresponds to $+s$ direction
      4   corresponds to $-s$ direction
      5   corresponds to $+t$ direction
      6   corresponds to $-t$ direction
      0   corresponds to the centre of the element.

Thus, faces 1, 3, 5 are positive, faces 2, 4, 6 are negative faces.

In addition to the above, a set of local axes (x', y', z') are considered for element faces. These local axes are defined for each face. Let I, J, K and L be the four corners of the element face as shown in Fig.9.12.

Then x' is specified by LI-JK where LI and JK are midpoints of sides LI and JK.

z' is normal to x', and to the line joining midpoints IJ and KL

y' is normal to x' and z' and form the right handed system. The corresponding nodal points I, J, K and L for each face are given in the following Table 9.1.

Table 9.1 Element Face and Nodal Points

Face	Nodal Points			
	I	J	K	L
1	1	2	6	5
2	4	3	7	8
3	3	7	6	2
4	4	8	5	1
5	8	5	6	7
6	4	1	2	3

Let i, j, k be unit vectors along x, y and z directions.,

e_1, e_2, e_3 vectors along r, s, and t directions

e_1', e_2', e_3' vectors along x', y' and z' directions.

The vectors along the global and natural coordinate directions are related by the Jacobian matrix [J] as given below.

$$\begin{Bmatrix} e_1 \\ e_2 \\ e_3 \end{Bmatrix} = \begin{bmatrix} J_{11} & J_{12} & J_{13} \\ J_{21} & J_{22} & J_{23} \\ J_{31} & J_{32} & J_{33} \end{bmatrix} \begin{Bmatrix} i \\ j \\ k \end{Bmatrix} = [J] \begin{Bmatrix} i \\ j \\ k \end{Bmatrix} \qquad (9.42)$$

Depending on the face number, the vector e_i' will be identical to one of the vectors along the natural coordinate axes. For example, consider the face number 1. For this face e_1' is identical to e_2 and e_3' is perpendicular to the element face.

$$e_1' = e_2' = J_{21}i + J_{22}j + J_{23}k$$
$$= J_{11}'i + J_{12}'j + J_{13}'k \qquad (9.43a)$$

$$e_3' = e_2 \times e_3 = (J_{22} J_{33} - J_{23} J_{32})i + (J_{23} J_{31} - J_{21} J_{33})j +$$

$$(J_{21} J_{32} - J_{22} J_{31})k$$

$$= (J_{11}^* i + J_{21}^* j + J_{31}^*)k$$

where J_{11}^*, J_{21}^* and J_{31}^* are the values of inverse Jacobian as given in Eq.9.25. Hence, e_3' can be written as

$$e_3' = J_{31}'i + J_{32}'j + J_{33}'k \qquad (9.43b)$$

Now e_2' can be computed such that x', y' and z' form an orthogonal system.

$$e_2' = e_3' \times e_1' = (J_{21}^* J_{23} - J_{31}^* J_{22})i + (J_{31}^* J_{21} - J_{11}^* J_{23})j$$

$$+ (J_{11}^* J_{22} - J_{21}^* J_{22})k$$

$$= (J_{21}'i + J_{22}'j + J_{23}'k) \qquad (9.43c)$$

Thus we get,

$$\begin{Bmatrix} e_1' \\ e_2' \\ e_3' \end{Bmatrix} = \begin{bmatrix} J_{11}' & J_{12}' & J_{13}' \\ J_{21}' & J_{22}' & J_{23}' \\ J_{31}' & J_{32}' & J_{33}' \end{bmatrix} \begin{Bmatrix} i \\ j \\ k \end{Bmatrix} \qquad (9.44)$$

The direction cosines $[l_1 \quad m_1 \quad n_1]$ for the x' axis can be got by normalising the vector $[J'_{11} \quad J'_{12} \quad J'_{13}]$ as,

$$\begin{Bmatrix} l_1 \\ m_1 \\ n_1 \end{Bmatrix} = \frac{1}{\sqrt{J'^2_{11} + J'^2_{12} + J'^2_{13}}} \begin{Bmatrix} J'_{11} \\ J'_{12} \\ J'_{13} \end{Bmatrix} \tag{9.45}$$

Similarly after normalising the vectors $[J'_{21} \quad J'_{22} \quad J'_{23}]$ and $[J'_{31} \quad J'_{32} \quad J'_{33}]$, we obtain the $[D]$ matrix representing direction cosines of the new axes x'; y', and z' as given in Eq.(9.46)

$$[D] = \begin{bmatrix} l_1 & l_2 & l_3 \\ m_1 & m_2 & m_3 \\ n_1 & n_2 & n_3 \end{bmatrix} \tag{9.46}$$

Similarly for other faces, we can find the direction cosines of the local axes.

9.5 Element Load Vector

The following three element load cases are considered in the program:
- (i) Gravity load in the global negative z direction
- (ii) Uniform pressure normal to the element surface
- (iii) Hydrostatic pressure normal to the element surface.

9.5.1 Gravity Loads

The load vector due to body forces is given by the Eq.3.82 as,

$$\{Q\} = \int_V [N]^T \{X\} \, dV \tag{9.47}$$

where $\{X\}$ is the vector of body force components per unit volume. Let ρ be the weight density of the material. The nodal load vector at any node i is given by,

$$\{Q_i\} = \int_V [N_i]\{X\} \, dV$$

where

$$[N_i] = \begin{bmatrix} N_i & 0 & 0 \\ 0 & N_i & 0 \\ 0 & 0 & N_i \end{bmatrix} \quad \text{and} \quad \{X\} = \begin{Bmatrix} 0 \\ 0 \\ -\rho \end{Bmatrix}$$

Thus,

$$\{Q_i\} = \int_V \left\{ \begin{array}{c} 0 \\ 0 \\ -N_i\rho \end{array} \right\} dV$$

For isoparametric element with functions expressed in natural coordinates, the above equation becomes,

$$\{Q_i\} = \int_{-1}^{+1} \int_{-1}^{+1} \int_{-1}^{+1} \left\{ \begin{array}{c} 0 \\ 0 \\ -N_i\rho \end{array} \right\} |J| \ dr \ ds \ dt \qquad (9.48a)$$

Using Gauss quadrature the integral in Eq.9.48a is evaluated as,

$$\{Q_i\} = \sum_{i=1}^{n} \sum_{j=1}^{n} \sum_{k=1}^{n} w_i \ w_j \ w_k |J|(r_i, \ s_j, \ t_k) \left\{ \begin{array}{c} 0 \\ 0 \\ -N_i\rho \end{array} \right\}_{(r_i,s_j,t_k)}$$
$$(9.48b)$$

For eight noded element the integration order suggested is $2 \times 2 \times 2$ and for twenty noded element it is $3 \times 3 \times 3$.

9.5.2   Uniform Pressure Normal to Element Face

In this case a uniform pressure of intensity q is considered normal to the element face. A positive surface load acts in the direction of the outward normal of a positive element face and along the inward normal of a negative element face as shown in Fig.9.12. The load vector due to surface pressure is given by,

$$\{Q\} = \int \int [N^s]^T \ \{p\} \ dA \qquad (9.49a)$$

Thus, the nodal load vector at any node i is given by,

$$\{Q_i\} = \int \int [N_i^s] \ \{p\} \ dA \qquad (9.49b)$$

The value of $[N_i^s]$ in the above equation is

$$[N_i^s] = \left[\begin{array}{ccc} N_i^s & 0 & 0 \\ 0 & N_i^s & 0 \\ 0 & 0 & N_i^s \end{array} \right] \qquad (9.49c)$$

where N_i^s is the shape function for the node i. For face number 1, the value of N_i^s is got by substituting r=1 in N_i. The surface pressure is expressed as,

$$\{p\} = \begin{Bmatrix} q\ l_3 \\ q\ m_3 \\ q\ n_3 \end{Bmatrix} \qquad (9.50a)$$

where l_3, m_3, n_3 are the direction cosines of the z' axis. The value of dA is calculated by considering the cross product of vectors along the natural coordinates parallel to the loaded face of the element, i.e.,

$$dA = |e_2 \times e_3| ds\ \ dt \qquad (9.50b)$$

Hence, Eq.9.49b can be written as

$$\{Q_i\} = \int_{-1}^{+1} \int_{-1}^{+1} \begin{Bmatrix} N_i^s\ q\ l_3 \\ N_i^s\ q\ m_3 \\ N_i^s\ q\ n_3 \end{Bmatrix} dA \qquad (9.50c)$$

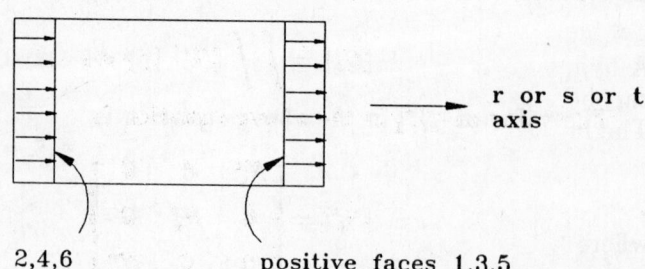

negative faces 2,4,6   positive faces 1,3,5

Positive Surface Loading P

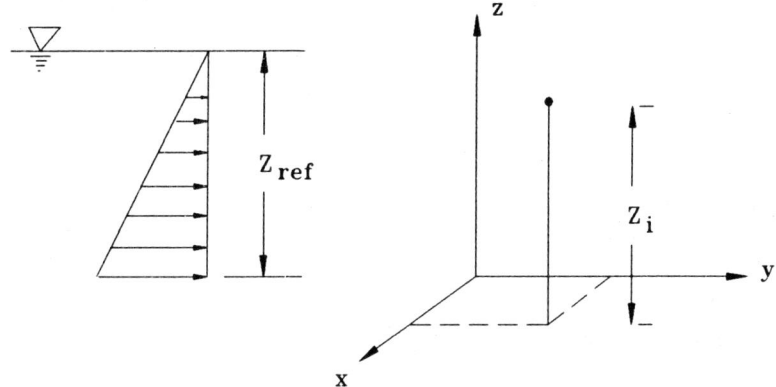

Hydrostatic Pressure Loading

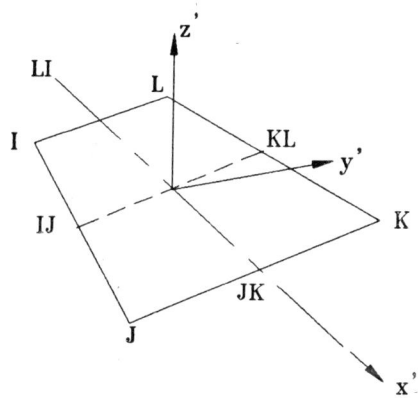

Local Axes at Element Faces
Fig. 9.12 Element loading and local axes at faces

9.5.3 Hydrostatic Pressure Loading Normal to the Surface

A hydrostatically varying surface pressure on element face 2, 4 or 6 can be specified by a reference fluid surface and a fluid density ρ_L. The intensity of pressure q normal to the surface is calculated as

$$q = \rho_L(z - z_{ref}) \tag{9.51}$$

where z is the global z coordinate of the point at which q is calculated and z_{ref} specifies the fluid surface assuming gravity acts along the negative z axis as shown in Fig.9.12.

The value of q from Eq.9.51 is substituted in Eq.(9.50c) to find the nodal load vector at any node i.

Using numerical integration procedure Eq.(9.50c) is evaluated as,

$$\{Q_i\} = \sum_{i=1}^{n} \sum_{i=1}^{n} \left\{ \begin{array}{c} N_i^s q l_3 \\ N_i^s q m_3 \\ N_i^s q n_3 \end{array} \right\}_{(r_i, s_j)} dA(r_i, s_j) \qquad (9.52)$$

Numerical integration of order 2×2 is used for eight noded element and 4×4 for twenty noded element.

9.6   Evaluation of Stresses

The stresses are calculated at the centre of the element and at the centres of the element faces. The stresses at the centre of the element are expressed in the global axes and the stresses on the faces in the local axes directions.

9.6.1   Stresses at the Centre

The [B] matrix for eight noded element is given by Eq.9.32c and for twenty noded element by Eq.9.39. For eight noded element, the additional displacement amplitudes due to incompatible modes are eliminated using static condensation so that the final size of the matrix will be 6×24.

The stress-displacement matrix is given by,

$$[CB] = [C]\,[B] \qquad (9.53)$$

where [C] is the constitutive matrix given by Eq.2.16. Then the stress components at the centroid of the element are given by,

$$\{\sigma\} = [CB]\,\{d_a\} \qquad (9.54)$$

where

$$\{\sigma\}^T = [\sigma_x \quad \sigma_y \quad \sigma_z \quad \tau_{xy} \quad \tau_{yz} \quad \tau_{zx}]$$

and

$$\{d_a\}^T = [u_1 \quad v_1 \quad w_1 \quad . \quad . \quad . \quad u_i \quad v_i \quad w_i]$$

$i = 8$ for eight noded element.

$i = 20$ for twenty noded element.

9.6.2 Stresses on the Element Faces

The strain vector with respect to the local axes is given by

$$
\{\epsilon'\} =
\begin{Bmatrix}
\epsilon_{x'} \\
\epsilon_{y'} \\
\epsilon_{z'} \\
\gamma_{x'y'} \\
\gamma_{y'z'} \\
\gamma_{z'x'}
\end{Bmatrix}
=
\begin{Bmatrix}
\dfrac{\partial u'}{\partial x'} \\[6pt]
\dfrac{\partial v'}{\partial y'} \\[6pt]
\dfrac{\partial w'}{\partial z'} \\[6pt]
\dfrac{\partial u'}{\partial y'} + \dfrac{\partial v'}{\partial x'} \\[6pt]
\dfrac{\partial v'}{\partial z'} + \dfrac{\partial w'}{\partial y'} \\[6pt]
\dfrac{\partial u'}{\partial z'} + \dfrac{\partial w'}{\partial x'}
\end{Bmatrix}
\tag{9.55}
$$

where u', v' and w' are the displacement components along the local axes.

The strain components along the local axes can be obtained by using the following relation [9],

$$
\begin{bmatrix}
\dfrac{\partial u'}{\partial x'} & \dfrac{\partial v'}{\partial x'} & \dfrac{\partial w'}{\partial x'} \\[6pt]
\dfrac{\partial u'}{\partial y'} & \dfrac{\partial v'}{\partial y'} & \dfrac{\partial w'}{\partial y'} \\[6pt]
\dfrac{\partial u'}{\partial z'} & \dfrac{\partial v'}{\partial z'} & \dfrac{\partial w'}{\partial z'}
\end{bmatrix}
= [D]^{T}
\begin{bmatrix}
\dfrac{\partial u}{\partial x} & \dfrac{\partial v}{\partial x} & \dfrac{\partial w}{\partial x} \\[6pt]
\dfrac{\partial u}{\partial y} & \dfrac{\partial v}{\partial y} & \dfrac{\partial w}{\partial y} \\[6pt]
\dfrac{\partial u}{\partial z} & \dfrac{\partial v}{\partial z} & \dfrac{\partial w}{\partial z}
\end{bmatrix}
[D]
\tag{9.56}
$$

Substituting the value of [D] from Eq.9.46 into Eq.9.56 we get

$$
\frac{\partial u'}{\partial x'} = l_1\left(l_1\frac{\partial u}{\partial x} + m_1\frac{\partial v}{\partial x} + n_1\frac{\partial w}{\partial x}\right) + m_1\left(l_1\frac{\partial u}{\partial y} + m_1\frac{\partial v}{\partial y} + n_1\frac{\partial w}{\partial y}\right)
$$

$$
+ n_1\left(l_1\frac{\partial u}{\partial z} + m_1\frac{\partial v}{\partial z} + n_1\frac{\partial w}{\partial z}\right)
\tag{9.57a}
$$

$$
\frac{\partial v'}{\partial x'} = l_1\left(l_2\frac{\partial u}{\partial x} + m_2\frac{\partial v}{\partial x} + n_2\frac{\partial w}{\partial x}\right) + m_1\left(l_2\frac{\partial u}{\partial y} + m_2\frac{\partial v}{\partial y} + n_2\frac{\partial w}{\partial y}\right)
$$

$$
+ n_1\left(l_2\frac{\partial u}{\partial z} + m_2\frac{\partial v}{\partial z} + n_2\frac{\partial w}{\partial z}\right)
\tag{9.57b}
$$

$$
\frac{\partial w'}{\partial x'} = l_1\left(l_3\frac{\partial u}{\partial x} + m_3\frac{\partial v}{\partial x} + n_3\frac{\partial w}{\partial x}\right) + m_1\left(l_3\frac{\partial u}{\partial y} + m_3\frac{\partial v}{\partial y} + n_3\frac{\partial w}{\partial y}\right)
$$

$$+n_1\left(l_3\frac{\partial u}{\partial z} + m_3\frac{\partial v}{\partial z} + n_3\frac{\partial w}{\partial z}\right) \qquad (9.57c,$$

$$\frac{\partial u'}{\partial y'} = l_2\left(l_1\frac{\partial u}{\partial x} + m_1\frac{\partial v}{\partial x} + n_1\frac{\partial w}{\partial x}\right) + m_2\left(l_1\frac{\partial u}{\partial y} + m_1\frac{\partial v}{\partial y} + n_1\frac{\partial w}{\partial y}\right)$$

$$+n_2\left(l_1\frac{\partial u}{\partial z} + m_1\frac{\partial v}{\partial z} + n_1\frac{\partial w}{\partial z}\right) \qquad (9.57d)$$

$$\frac{\partial v'}{\partial y'} = l_2\left(l_2\frac{\partial u}{\partial x} + m_2\frac{\partial v}{\partial x} + n_2\frac{\partial w}{\partial x}\right) + m_2\left(l_2\frac{\partial u}{\partial y} + m_2\frac{\partial v}{\partial y} + n_2\frac{\partial w}{\partial y}\right)$$

$$+n_2\left(l_2\frac{\partial u}{\partial z} + m_2\frac{\partial v}{\partial z} + n_2\frac{\partial w}{\partial z}\right) \qquad (9.57e)$$

$$\frac{\partial w'}{\partial y'} = l_2\left(l_3\frac{\partial u}{\partial x} + m_3\frac{\partial v}{\partial x} + n_3\frac{\partial w}{\partial x}\right) + m_2\left(l_3\frac{\partial u}{\partial y} + m_3\frac{\partial v}{\partial y} + n_3\frac{\partial w}{\partial y}\right)$$

$$+n_2\left(l_3\frac{\partial u}{\partial z} + m_3\frac{\partial v}{\partial z} + n_3\frac{\partial w}{\partial z}\right) \qquad (9.57f)$$

$$\frac{\partial u'}{\partial z'} = l_3\left(l_1\frac{\partial u}{\partial x} + m_1\frac{\partial v}{\partial x} + n_1\frac{\partial w}{\partial x}\right) + m_3\left(l_1\frac{\partial u}{\partial y} + m_1\frac{\partial v}{\partial y} + n_1\frac{\partial w}{\partial y}\right)$$

$$+n_3\left(l_1\frac{\partial u}{\partial z} + m_1\frac{\partial v}{\partial z} + n_1\frac{\partial w}{\partial z}\right) \qquad (9.57g)$$

$$\frac{\partial v'}{\partial z'} = l_3\left(l_2\frac{\partial u}{\partial x} + m_2\frac{\partial v}{\partial x} + n_2\frac{\partial w}{\partial x}\right) + m_3\left(l_2\frac{\partial u}{\partial y} + m_2\frac{\partial v}{\partial y} + n_2\frac{\partial w}{\partial y}\right)$$

$$+n_3\left(l_2\frac{\partial u}{\partial z} + m_2\frac{\partial v}{\partial z} + n_2\frac{\partial w}{\partial z}\right) \qquad (9.57h)$$

$$\frac{\partial w'}{\partial z'} = l_3\left(l_3\frac{\partial u}{\partial x} + m_3\frac{\partial v}{\partial x} + n_3\frac{\partial w}{\partial x}\right) + m_3\left(l_3\frac{\partial u}{\partial y} + m_3\frac{\partial v}{\partial y} + n_3\frac{\partial w}{\partial y}\right)$$

$$+n_3\left(l_3\frac{\partial u}{\partial z} + m_3\frac{\partial v}{\partial z} + n_3\frac{\partial w}{\partial z}\right) \qquad (9.57i)$$

Arranging the above equations we get,

$$\{\epsilon'\} = [T_\epsilon]\,\{\epsilon\}$$

where the strain transformation matrix $[T_\epsilon]$ is given by

$$[T_\epsilon] = \begin{bmatrix} l_1^2 & m_1^2 & n_1^2 & l_1m_1 & m_1n_1 & n_1l_1 \\ l_2^2 & m_2^2 & n_2^2 & l_2m_2 & m_2n_2 & n_2l_2 \\ l_3^2 & m_3^2 & n_3^2 & l_3m_3 & m_3n_3 & n_3l_3 \\ 2l_1l_2 & 2m_1m_2 & 2n_1n_2 & l_1m_2+ & m_1n_2+ & n_1l_2+ \\ & & & l_2m_1 & m_2n_1 & n_2l_1 \\ 2l_2l_3 & 2m_2m_3 & 2n_2n_3 & l_3m_2+ & m_3n_2+ & n_3l_2+ \\ & & & l_2m_3 & m_2n_3 & n_2l_3 \\ 2l_3l_1 & 2m_3m_1 & 2n_3n_1 & l_3m_1+ & m_3n_1+ & n_3l_1+ \\ & & & l_1m_3 & m_1n_3 & n_1l_3 \end{bmatrix} \qquad (9.58)$$

But we have,

$$\{\epsilon\} = [B]\{d_a\}$$

where [B] is the strain-displacement matrix (after condensation for eight noded element). Hence,

$$[\epsilon'] = [T_\epsilon][B]\{d_a\}$$

i.e.,

$$\{\epsilon'\} = [B']\{d_a\}$$

where

$$[B'] = [T_\epsilon][B] \qquad (9.59)$$

Then, the stresses on the faces are given by,

$$\{\sigma'\} = [CB']\{d_a\} \qquad (9.60)$$

where $\{\sigma'\}$ is the stress vector referred to local axes.

9.7 Subroutine THREDS

This control routine calls different three-dimensional element routines as shown in Fig.9.13. Though provision for three shapes of element is made in the element library, only the following two shapes are incorporated in the program.

1. Shape 1 - Eight noded isoparametric solid element with incompatible modes.

2. Shape 2 - Twenty noded isoparametric solid element.

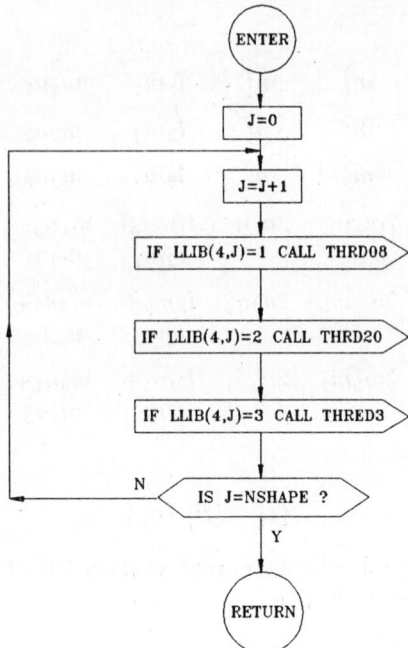

Fig. 9.13 Subroutine THREDS

9.7.1 Subroutine THRD08 and THRD20

The listing of the subroutine THRD08 and THRD20 is given in Appendix 6.

Each element shape routine is called thrice by the main program and is divided into two parts. The three segments are executed depending on the value of the flag IND.

When IND equals 1 the first segment is invoked. The nodal point data and material information are obtained through the argument list. Number of elements (eight or twenty noded depending on the routine called) and the number of load sets acting on the element are read. Then for each element its nodal connectivity (LNC), material group (MG), order of integration (NINT), load set identification number (LSN), number of faces at which stresses are to be calculated are read. The nodal degree of freedom array, (ND), is generated for each element and is stored in the NDARAY file in a sequential manner. The data read and generated is printed out if the option IPR is zero. The above information will only be read and not printed out if IPR is greater than zero. Finally the subroutine COLUMH is called to calculate the column heights of the global stiffness matrix.

When IND=2 the control transfers to the second segment of the element routine. At this stage in addition to the nodal data and material properties, the addresses of the stiffness matrix, and load vector are also obtained from the argument list. For each element ND array is read from the scratch file along with the other information stored there. The stress displacement matrix [CB] is calculated and is written on the ISTRES file. Next the element stiffness matrix is calculated and is passed on to the subroutine PASSEM for assembling element stiffness matrix. The element load vector is calculated and added to the global load vector.

After the global displacements have been obtained, IND is set equal to 3 and control is transferred to the third segment of the element routine. At this stage the ND array and stress-displacement matrix are read from the NDARAY and ISTRES files respectively. Then for each element the stress at the specified points are computed and printed out.

9.7.2   Input Details for Solid Element

The input details for solid element type are given below. This data set would pertain to the data set 5 and 6 required for PASSFEM described in section 6.11 of Chapter 6.

5. ELEMENT TYPE DATA (5I5)

Columns   1-5   Element type number or row number of the element library (LTYPE). (For three-dimensional elements, LTYPE=4)

6-10   Number of shapes (NSHAPE) belonging to the element type with a minimum of 1 to a maximum of 3.

11-15   Element shape number or column number of the element library (Element shape number is 1 for THRD08 and 2 for THRD20).

16-20   Element shape number. If NSHAPE is more than one, the second shape number is typed in this column. Otherwise the column is left blank.

21-25   Element shape number. If NSHAPE=3 third shape number is typed in this column.

6. ELEMENT DATA SET

This data set is element dependent and the input details for THRD08 and THRD20 are given below:

1. *Eight Noded Isoparametric Element with Incompatible Modes (THRD08)*

A. *Control Data (2I5)*

This data specifies the number of eight noded elements in the structure and the number of element distributed load sets.

Columns 1-5   Number of eight noded elements (NEL)

             6-10   Number of element distributed load sets (NDL).

B. *Distributed Surface Loads (2I5, 2F10.3, I5)*

If $NDL > 0$, one line is required for each unique set of uniformly distributed surface loads and for each reference fluid level for hydrostatically varying pressure loads (refer Fig.9.12 for sign convention). (The data set is not required if NDL=0).

Columns 1-5   Load set identification number (N)

             6-10   LT (Load Type)

                        LT=1 specifies a uniformly distributed load LT=2 specifies a hydrostatically varying pressure.

             11-20 PN

                        If LT=1, PN is the magnitude of the uniformly distributed load. If LT=2, PN is the weight density of the fluid causing the hydrostatic pressure.

             21-30 ZREF

                        If LT=1, leave blank. If LT=2, ZREF is the global z coordinate of the surface of the fluid causing hydrostatic pressure loading.

             31-35 Element face number on which surface load acts (NF). Face numbers are from 1 to 6 as described in section 9.4 for uniformly distributed loads and it can be only on faces 2, 4 or 6 for hydrostatically varying pressure.

C. *Element Data* (10I5, 13I2)

One line per element with the following information must be given:

Columns 1-5   Element number (LNUM)

             6-10   Material group of the element (MG)

             11-50 Node numbers of the element in the order shown in Fig.9.10.

51-52	Order of integration (NINT)
53-54	Element Data generator (KI)
55 – 56 57 – 58	Load set identification number of the distributed load acting on this element LSN(1) and LSN(2)
59-60	Number of faces at which stresses are to be calculated (NSS). Stresses can be calculated at centres of the six surfaces and at the centroid of the element.
61-74	Face numbers for stress output
75-76	Option for incompatible modes (IOPT=1 for inclusion) (IOPT=0 for suppression)

2. *Twenty Noded Isoparametric Element THRD20*

A. *Control Data (2I5)*

Columns 1-5 Number of twenty noded elements (NEL)

6-10 Number of element distributed load sets (NDL)

B. *Distributed Surface Loads*: (2I5, 2F10.3, I5)

If $NDL > 0$, one line of data is required for each load set

Columns 1-5 Load set identification number (N)

6-10 LT (load type)

11-20 PN

21-30 ZREF

31-35 ELEMENT face numbers on which surface load acts (NF)

For details of LT, PN, ZREF, NF see THRD08 input data preparation.

C. *Element Data* (23I3, 1I1)

Columns 1-3 Element Number (LNUM)

4-6 Material group of the element (MG)

7-66 Node numbers of the element in the order shown in Fig.9.10

67-69 Element data generator (KI)

70   Order of integration NINT

$\left.\begin{array}{c}71\\72\end{array}\right\}$ Load set identification number of the distributed load acting on this element (LSN(1) LSN(2))

73   Number of faces at which stresses are calculated (NSS)

74-80   Face numbers for stress output.

NOTES

1. *Element Data Generation*

Element data lines must be in element number sequence. A series of elements occurs in which each element number $(LNUM)_i$ is one greater than the previous element number $(LNUM)_{i-1}$, only the first and last element cards in that series need be given and the program automatically generates the intermediate omitted elements of the series as follows:

The increment for element is *one* i.e.,

$$(LNUM)_i = (LNUM)_{i-1} + 1$$

The corresponding increment for nodal number is KI. i.e.,

$$[LNC(j)]_i = [LNC(j)]_{i-1} + KI; \quad j = 1 \text{ to } 8 \text{ or } 20$$

The nodal number of the omitted elements will be generated with regard to the information on the first line in the series.

The generated elements will have the same material properties, order of integration, load set numbers and face numbers for stress output as that of the last element in the generated series.

2. *Integration Order*

For THRD08, NINT = 2 is used.

For this $2 \times 2 \times 2$ sampling points are used for integration of the stiffness matrix. For THRD20, when NINT=2, 14 point integration rule is used and when NINT=3, $3 \times 3 \times 3$ points are used for integration of the stiffness matrix. When NINT = 0, the value of stiffness matrix is taken as the same as the preceding element value.

9.7.3   Output

The program prints out the global displacements u, v and w at each node. The element stresses are output as follows:

1. At the centroid of the element, stresses are referred to the global axes. Three principal stresses are also printed.
2. At the centre of an element face, stresses are referred to local axes as given in section 9.4. Two surface principal stresses and the angle between the algebraically largest principal stress and the local x-axis are printed with the output.

9.8 Examples

Six numerical examples have been solved using three-dimensional solid elements. The type of examples varies from a simple straight cantilever beam to complex arch dam which is curved both in plan and elevation. The six examples are:

1. Straight cantilever beam
2. Curved cantilever beam
3. Framed structure
4. Thick plate
5. Arch dam type 1
6. Arch dam type 2.

The discretizations used for analysis and the results are described below. In all examples, except 4, eight noded element with incompatible modes have been used.

9.8.1 Straight Cantilever Beam

The cantilever beam shown in Fig.9.14 has been analysed for the following loading conditions:

(i) Moment of 300 kN cm at the free end.

(ii) Point load of 10kN at the free end in the vertical direction.

$E = 2 \times 10^3 \text{ kN/cm}^2$; $\mu = 0.0$

The discretization adopted for analysis using eight noded solid element with incompatible modes and twenty noded elements are shown in Fig.9.14. The results of analysis are presented in Tables 9.2, 9.3, 9.4 and 9.5. The finite element analysis results are compared with the theoretical values based on elementary strength of materials approach.

It is observed that the results compare very well with the theoretical values, even for a coarse discretization of three elements. The eight noded element with incompatible modes has been found to give exact results for this problem.

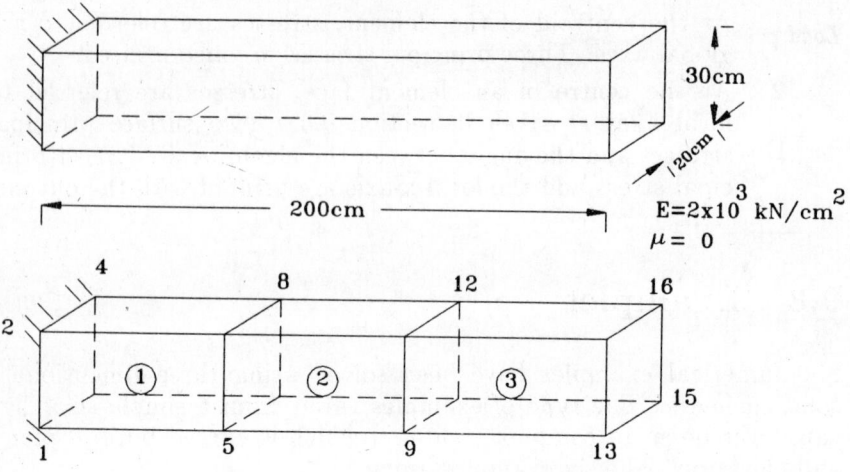

THRD08-3 ELEMENTS

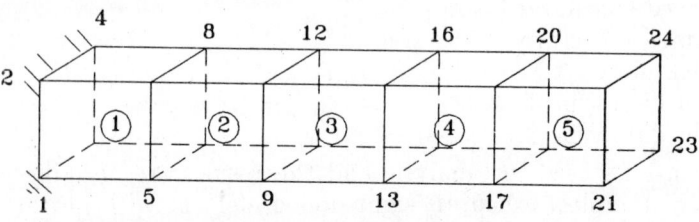

THRD08-5 ELEMENTS

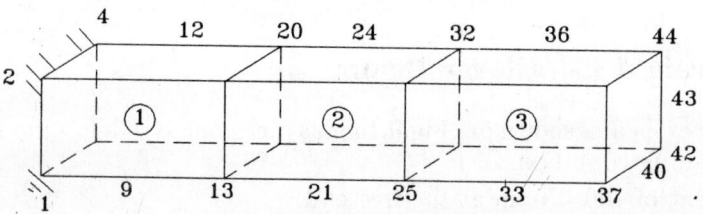

THRD20-3 ELEMENTS

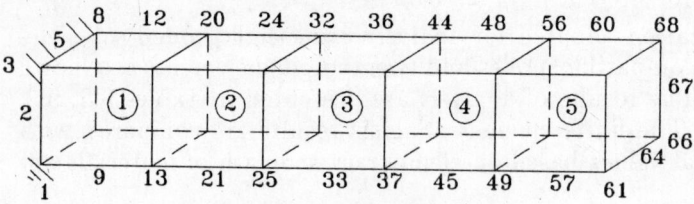

THRD20-5 ELEMENTS

Fig. 9.14 Cantilever beam and its discretizations

Load case 1-Moment at the free end

Table 9.2 Comparison of vertical deflections (cm)

Distance from free end	Theory	THRD08 No. of elements		THRD20 No. of Elements	
		3	5	3	5
00.00	0.06667	0.06667	0.06667	0.06667	0.06667
40.00	0.04267	-	0.04267	-	0.04267
66.67	0.02963	0.02963	-	0.02963	-
80.00	0.02400	-	0.02400	-	0.02400
120.00	0.01067	-	0.01067	-	0.01067
133.33	0.00741	0.00741	-	0.00741	-
160.00	0.00267	-	0.00267	-	0.00267
200.00	0.00000	0.00000	0.00000	0.00000	0.00000

Table 9.3 Comparison of bending stresses (in N/mm^2)

Distance from free end	Theory	THRD08 No. of elements		THRD20 No. of Elements	
		3	5	3	5
20.00	1.00	-	1.00	-	0.92
33.33	1.00	1.00	-	1.00	-
60.00	1.00	-	1.00	-	1.01
100.00	1.00	1.00	1.00	1.00	1.00
140.00	1.00	-	1.00		1.00
167.00	1.00	1.00	-	1.00	-
180.00	1.00	-	1.00	-	1.00

Load Case 2 - Point Load at the free end

Table 9.4 Comparison of vertical deflections (in cm)

Distance from free end	Theory	THRD08 No. of elements		THRD20 No. of Elements	
		3	5	3	5
0.00	0.29629	0.29150	0.29630	0.29732	0.29983
40.00	0.20859	-	0.20884	-	0.21084
66.67	0.15364	0.15040	-	0.15042	-
80.00	0.12800	-	0.12821	-	0.12973
120.00	0.06163	-	0.06177	-	0.06278
133.33	0.04380	0.04227	-	0.04409	-
160.00	0.01659	-	0.01667	-	0.01717
200.00	0.00000	0.00000	0.00000	0.00000	0.00000

Table 9.5 Comparison of bending stresses (in N/mm^2)

Distance from free end	Theory	THRD08 No. of elements		THRD20 No. of Elements	
		3	5	3	5
20.00	0.67	-	0.67	-	0.67
33.33	1.11	1.11	-	1.11	-
60.00	2.00	-	2.00	-	2.00
100.00	3.33	3.33	3.33	3.33	3.33
140.00	4.67	-	4.67	-	4.67
166.67	5.56	5.56	-	5.56	-
180.00	6.00	-	6.00	-	6.00

9.8.2　Curved Cantilever Beam

A curved cantilever beam shown in Fig.9.15 has been analysed for the following two loading conditions:

(i) Point load of 10 kN at the free end in the negative x direction so that the beam acts as a cantilever beam curved in plan.

(ii) Point load of 10 kN at the free end in the negative z direction so that the beam acts as a cantilever beam curved in elevation. $E = 2 \times 10^3$ kN/cm^2　$\mu = 0.15$

Two discretizations are used for each element type and these are shown in Fig. 9.15. Results of finite element analysis are compared with the values obtained by using the formulae given in reference [11] for beam curved in plan and in reference [10] for beam curved in elevation. The results are tabulated in Tables 9.6 to 9.9.

The maximum percentage error for a discretization with nine eight noded elements is 1.7% for deflection and is 0.5% for stress. The maximum percentage error for a discretization with five twenty noded elements is 2.6% for deflection and is 4% for stress.

Load Case 1 - Cantilever beam curved in plan

Table 9.6 Comparison of x displacement at the free end (in cm)

value from ref.11	THRD08 No. of Elements		THRD20 No. of Elements	
	5	9	3	5
-0.26263	-0.24496	-0.25829	-0.22534	-0.25577

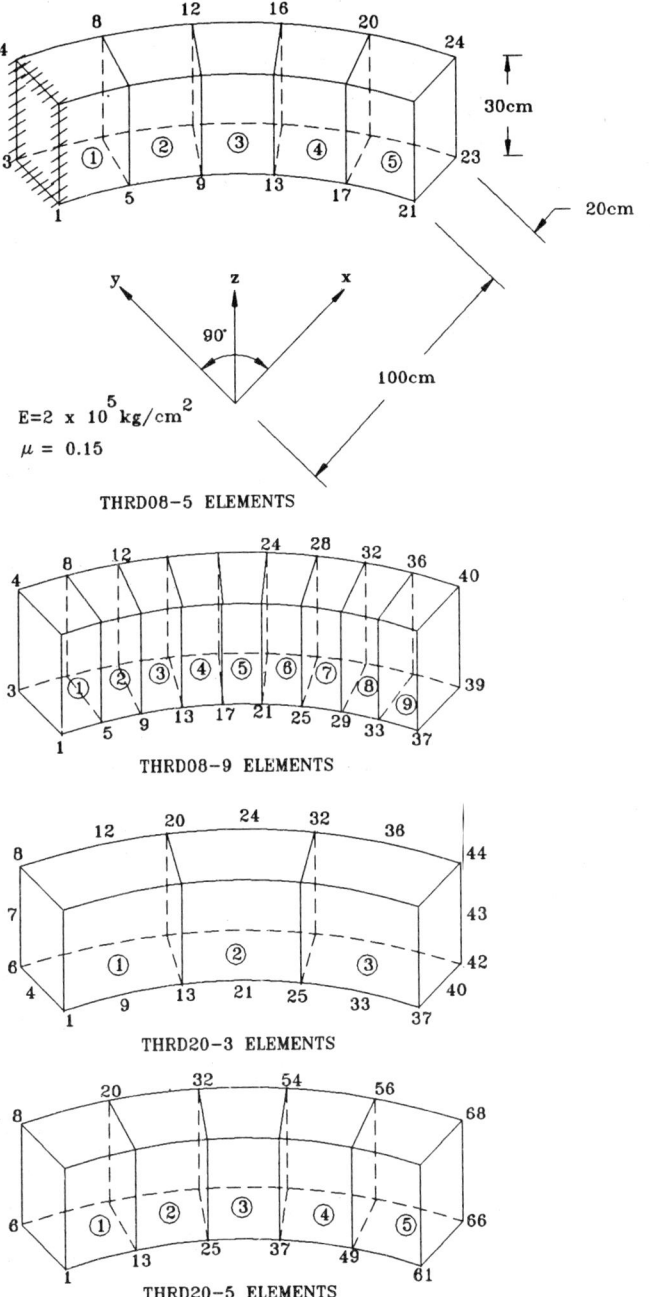

Fig. 9.15 Curved cantilever beam and its discretizations

Table 9.7 Comparison of bending stress
(σ_θ) along the inner radius (inN/mm^2)

Angle θ^o	Value from ref.11	THRD08 No. of Elements		THRD20 No. of Elements	
		5	9	3	5
5	-0.53	-	-0.53	-	-
9	-0.94	-0.92	-	-	-0.91
15	-1.56	-	-1.56	-1.59	-
25	-2.55	-	-2.55	-	-
27	-2.74	-2.71	-	-	-2.93
35	-3.46	-	-3.46	-	-
45	-4.26	-4.10	-4.26	-4.58	-4.54
55	-4.94	-	-4.94	-	-
63	-5.37	-5.39	-	-	-
65	-5.47	-	-5.46	-	-5.77
75	-5.83	-	-5.83	-6.17	-
81	-5.96	-5.49	-	-	-6.31
85	-6.01	-	-5.97	-	-

Table 9.8 Comparison of bending stress along the
outer radius (inN/mm^2)

Angle θ^o	Value from ref.11	THRD08 No. of Elements		THRD20 No. of Elements	
		5	9	3	5
5	0.44	-	0.44	-	-
9	0.79	0.76	-	-	0.78
15	1.30	-	1.29	0.86	-
25	2.12	-	2.11	-	-
27	2.28	2.24	-	-	1.94
35	2.88	-	2.86	-	-
45	3.55	3.46	3.52	2.13	3.08
55	4.12	-	4.08	-	-
63	4.48	4.46	-	-	3.83
65	4.55	-	4.52	-	-
75	4.85	-	4.82	2.99	-
81	4.96	4.53	-	-	4.31
85	5.01	-	4.94	-	-

Load Case 2: Cantilever beam curved in elevation

Table 9.9 Comparison of bending stress
(σ_θ) on the horizontal faces (inN/mm^2)

Angle θ°	Value from ref.10	THRD08 No. of Elements		THRD20 No. of Elements	
		5	9	3	5
5	0.32	-	0.32	-	-
9	0.57	0.57	-	-	0.57
15	0.95	-	0.94	0.96	-
25	1.55	-	1.54	-	-
27	1.66	1.65	-	-	1.68
35	2.10	-	2.09	-	-
45	2.59	2.55	2.57	2.65	2.61
55	3.00	-	2.98	-	-
63	3.27	3.25	-	-	3.29
65	3.32	-	3.30	-	-
75	3.54	-	3.53	3.59	-
81	3.62	3.62	-	-	3.62
85	3.65	-	3.70	-	-

9.8.3 Framed Structure

Figure 9.16 shows a framed structure discretized using eight and twenty noded elements. The frame is subjected to a concentrated load of 100 kN. $E = 2 \times 10^3$ kN/cm^2 $\mu = 0.0$.

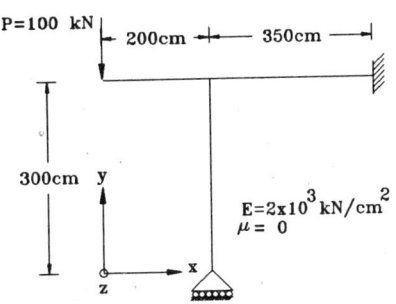

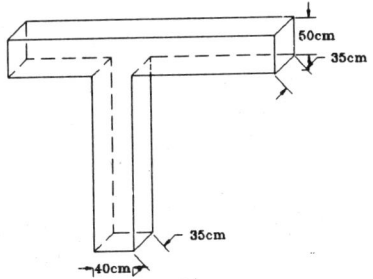

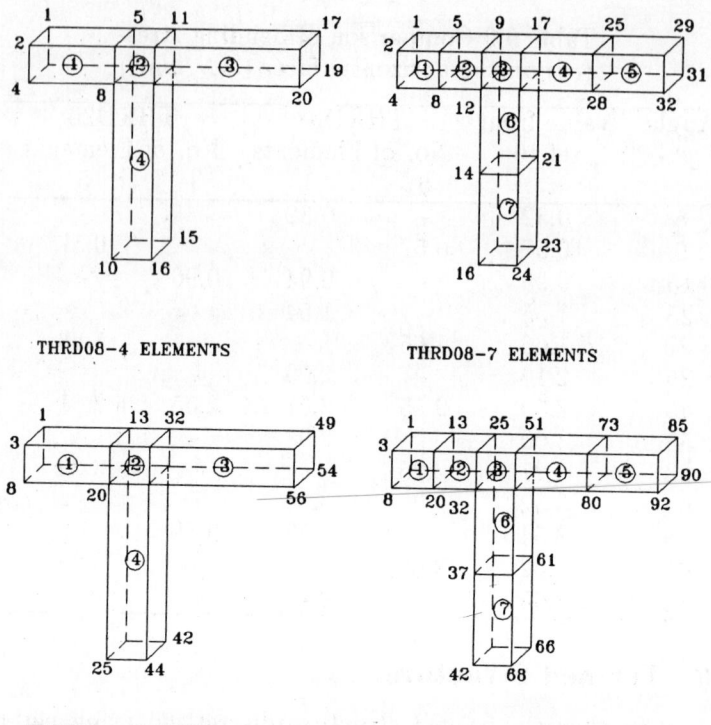

THRD08-4 ELEMENTS

THRD08-7 ELEMENTS

THRD20-4 ELEMENTS

THRD20-7 ELEMENTS

Fig. 9.16 Framed structure and its discretizations

Tables 9.10 to 9.12 give the comparison of theoretical and finite element analysis results. The theoretical values are calculated using elementary structural analysis principles considering flexural effects only. In this case it has been noted that the finer discretization gives results which are close to the analytical solution.

Table 9.10 Comparison of vertical deflections
at the free end (cm)

Analytical	THRD08		THRD20	
	No. of elements		No. of elements	
	4	7	4	7
0.3086	0.1833	0.2933	0.1905	0.3055

Table 9.11 Comparison of vertical stress
in vertical member (N/mm^2)

Analytical	THRD08		THRD20	
	No. of elements		No. of elements	
	4	7	4	7
-1.52	-1.58	-1.43	-1.48	-1.40

Table 9.12 Comparison of bending stresses
in the horizontal member (N/mm^2)

Dist. from free end	Analytical	THRD08 No. of elements		THRD20 No. of elements	
		4	7	4	7
32.50	2.23		2.23	-	2.23
65.00	4.46	4.46	-	4.46	-
97.50	6.69	-	6.69	-	6.69
215.00	5.29	-	5.05	-	5.27
260.00	1.80	0.81	-	1.50	-
305.00	1.67	-	1.12	-	0.77

9.8.4 Thick Plate

Thick plate shown in Fig.9.17 has been analysed using twenty noded element for varying h/L (thickness/span) ratios of $0.05, 0.10, 0.15, 0.20$ and 0.25. The plate is simply supported along the boundaries and is subjected to a central concentrated load P of 16 kN. The maximum deflection may be expressed as $w_{max} = \beta_1 \frac{PL^2}{D}$ where D is the flexural rigidity of the plate $\left(\frac{Eh^3}{12(1-\mu^2)}\right)$. The values of β_1 are calculated for varying h/L ratios and compared in Table 9.13 with the theoretical values given in reference [12]. The results are also plotted in Fig.9.17. $E = 2 \times 10^3$ kN/cm^2 $\mu = 0.15$.

Table 9.13 Comparison of β_1 values

h/L	Classical thin plate theory	Ref.12	THRD20 4 Elms	9 Elms
0.05	0.01160	0.01219	0.01105	0.01178
0.10	0.01160	0.01353	0.01273	0.01336
0.15	0.01160	0.01551	0.01483	0.01558
0.20	0.01160	0.01801	0.01753	0.01871
0.25	0.01160	0.02101	0.02099	0.02297

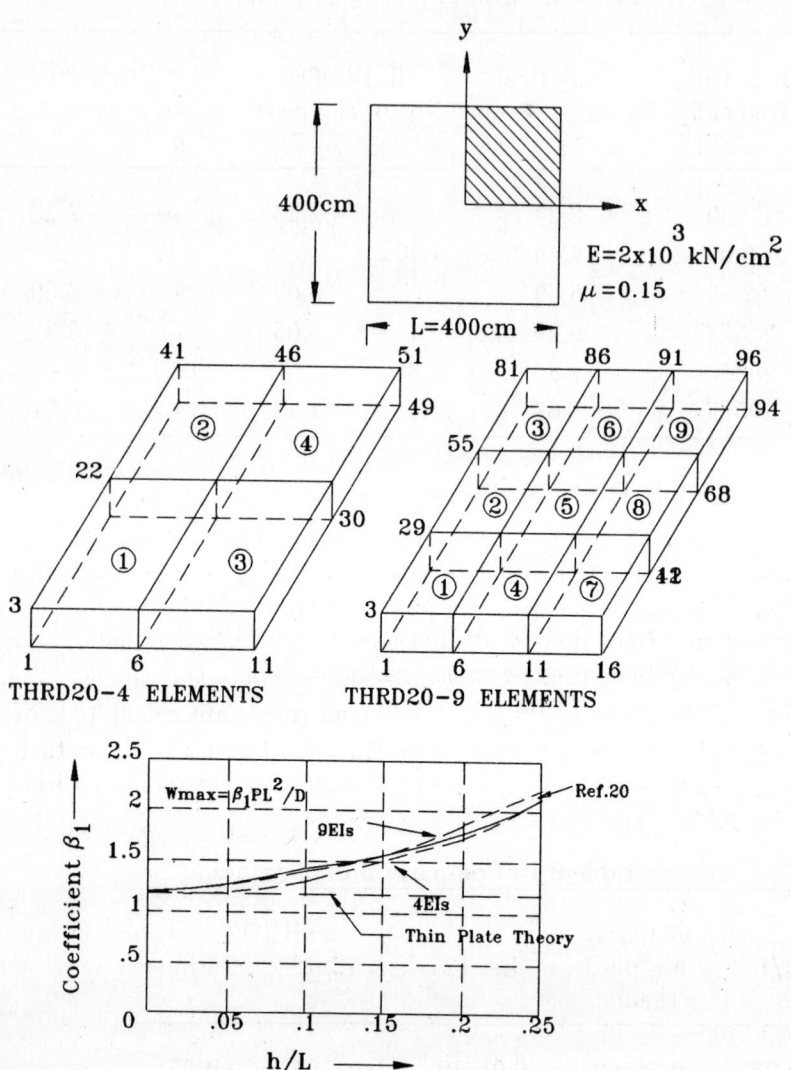

Fig. 9.17 Simply supported square plate

9.8.5 Arch Dam Type 1

Arch Dam type 1 (according to USBR classification) shown in Fig.
9.18 has constant thickness and radius. It is subjected to water
pressure on the upstream side. Two discretizations have been tried

for eight noded element.

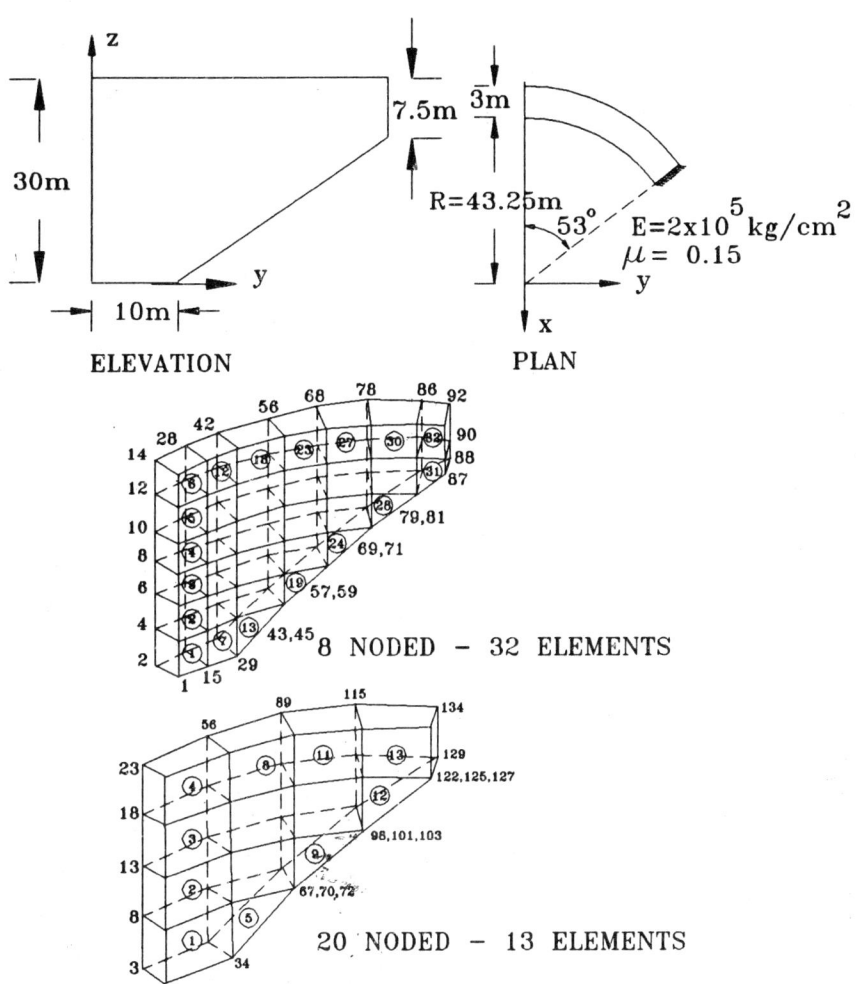

Fig. 9.18 Arch dam type 1 and its discretizations

In the first discretization thirty elements and in the second, thirty two elements have been used. For discretization using twenty noded solid element, thirteen elements have been used as shown in Fig.9.18.

Finite Element Analysis results for hoop and vertical stresses along the centre line section, and radial displacement along the crown section at the downstream side are compared with the values reported in references (13, 14). The results are shown in Fig. 9.19. The results agree well with the values reported in references (13,14). In this case it has been observed, that more number of eight noded elements are required for good accuracy. However, the execution time for 32 eight noded element was found to be nearly half of that of 13 twenty noded element.

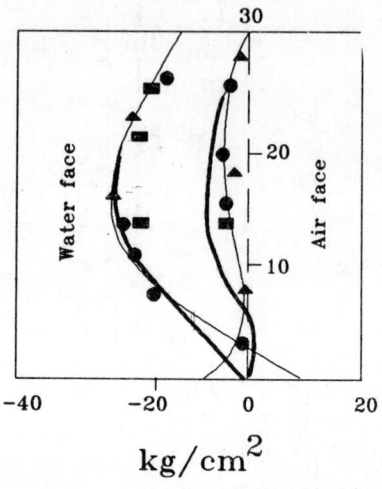

Hoop stress on centre
line section

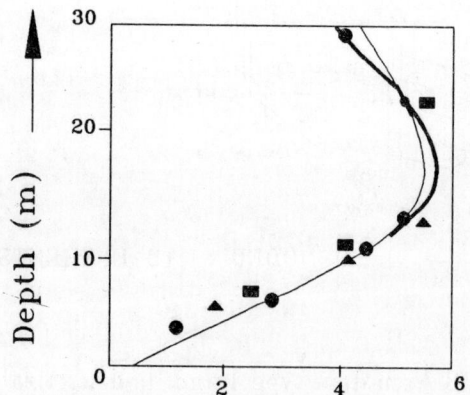

Radial displacement (mm)

Crown section downstream

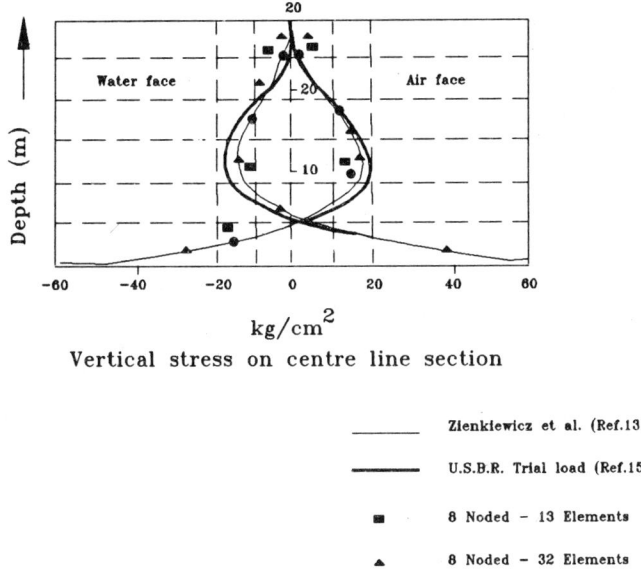

Vertical stress on centre line section

	Zienkiewicz et al. (Ref.13)
	U.S.B.R. Trial load (Ref.15)
■	8 Noded - 13 Elements
▲	8 Noded - 32 Elements
●	20 NOded - 13 Elements

Fig. 9.19 Arch dam type 1 - Results of Analysis

9.8.6 Arch Dam Type 5

An example of arch dam type 5 (according to USBR classification) is shown in Fig.9.20.

Due to symmetry only one half of the structure is shown in the Figure. The arch dam is having curvature in both plan and elevation. The discretizations used for eight and twenty noded solid elements are also shown in Fig.9.20. Water pressure on the upstream side has been considered for analysis.

Finite element analysis results for hoop and vertical stresses and radial deflection at the crown section are compared in Fig.9.21 with the results of reference [13]. It has been observed that nine twenty noded elements give results in good agreement with the published results whereas in the case of eight noded elements thirty two elements are required. The execution time for both the cases was almost the same.

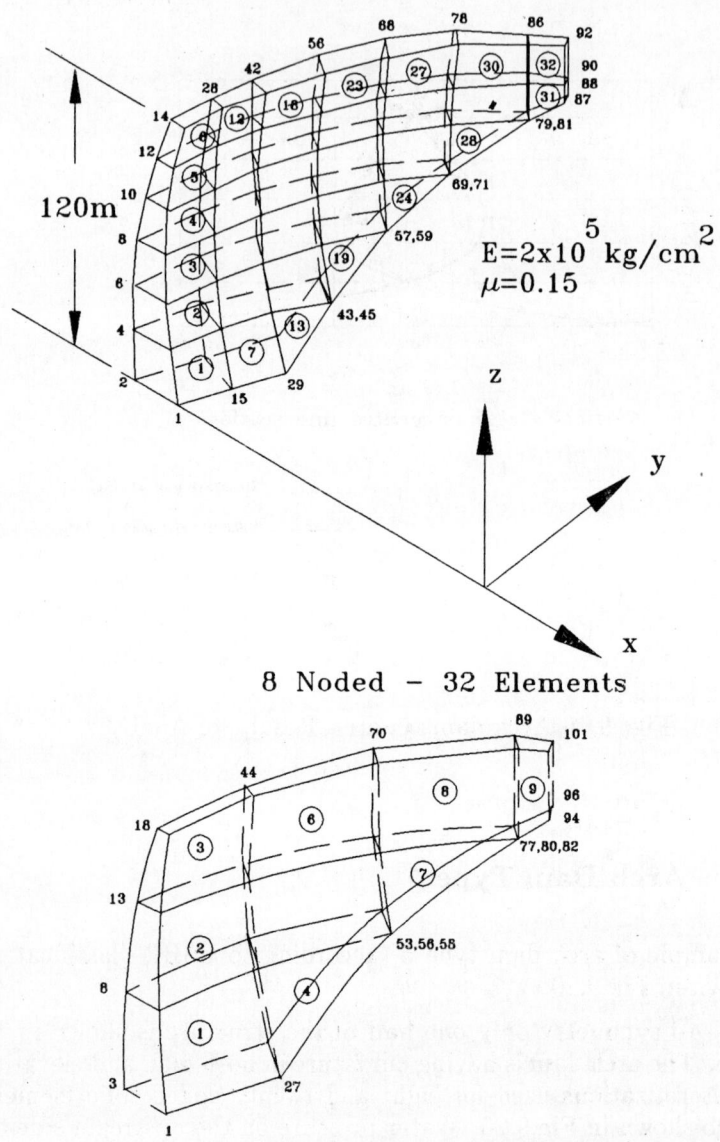

8 Noded − 32 Elements

20 Noded − 9 Elements

Fig. 9.20 Arch dam type 5 and its discretizations

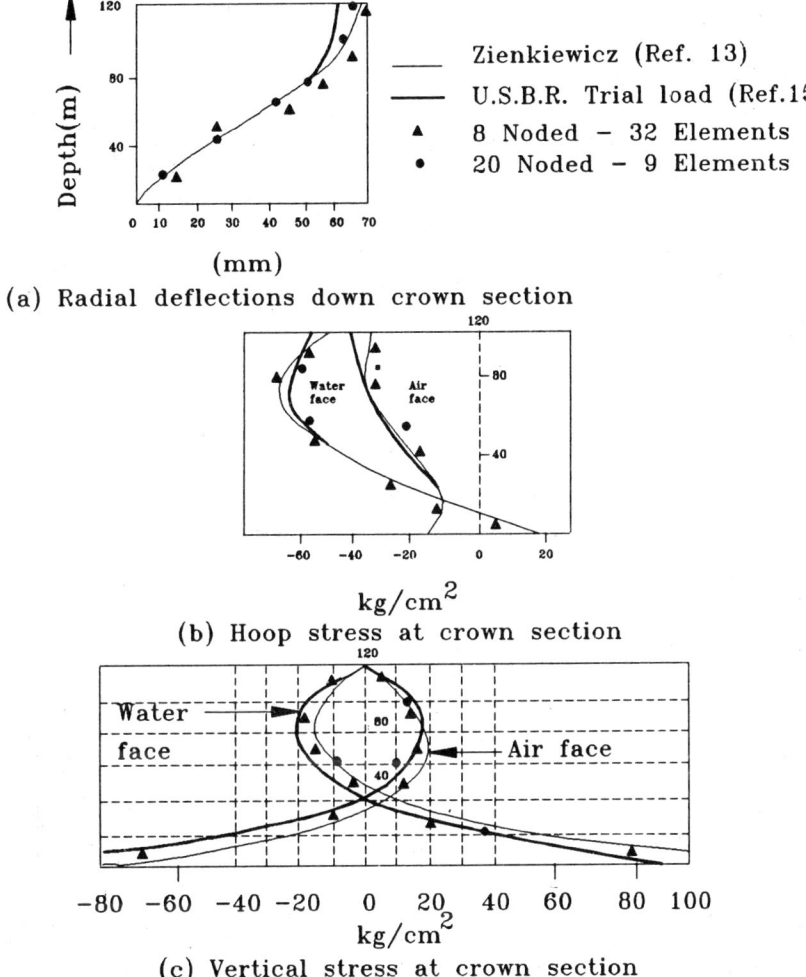

Zienkiewicz (Ref. 13)

U.S.B.R. Trial load (Ref.15)

▲   8 Noded – 32 Elements

●   20 Noded – 9 Elements

(a) Radial deflections down crown section

(b) Hoop stress at crown section

(c) Vertical stress at crown section

Fig. 9.21 Arch dam type 5 - Results of Analysis

It has been generally observed that eight noded element with incompatible modes is computationally efficient for elastic analysis of solids when they are subject to bending also. Twenty noded elements are more efficient when the structure to be analysed has considerable thickness in all the three directions and is also curved.

REFERENCES

1. Argyris, J.H., *Matrix Analysis of Three-Dimensional Elastic Media Small and Large Displacements*, Journal A.I.A.A., Vol.3., pp.45-51, Jan.1965.

2. Argyris, J.H. and D.W.Scharpf, *The Curved Tetrahedral Element TEO and TRIC for the Matrix Displacement Method.* Royal Aeronautical Society, Vol.73, pp. 55-65, 1969.

3. Zienkiewicz, O.C. et. al. *Isoparametric and Associated Element Families for Two and Three-Dimensional Analysis*, Chapter 13, Finite Element Methods in Stress Analysis, edited by Holand. I and Bell, K., Technical University of Norway, Tapir Press, Norway, 1969.

4. Clough, R.W. *Comparison of Three-Dimensional Finite Elements*, Proceedings of Symposium on Application of Finite Elements in Civil Engineering, Vanderbilt University, Nashville, Ten., pp.1-26, 1969.

5. Wilson, E.L., R.L.Taylor, W.P.Doherty and T.Ghabussi, *Incompatible Displacement Models, Numerical and Computer Methods in Structural Mechanics*, Edited by Fenves,S.J. et. al. Academic Press, pp.43-57, 1973.

6. Taylor, R.L., P.J.Beresford and E.L.Wilson, *A Nonconforming Element for Stress Analysis*, International Journal for Numerical Methods in Engineering, Vol.10, pp.1211-20, 1976.

7. Irons, B.M., *Quadrature Rules for Brick Based Finite Finite Elements*, International Journal for Numerical Methods in Engineering, Vol.3. pp.293-294, 1971.

8. Hellen, T.K., *Effective Quadrature Rules for Quadratic Solid Isoparametric Finite Element*, International Journal for Numerical Methods in Engineering, Vol.4, pp.597-600, 1972.

9. Cook, R.D., D.S.Malkus and Michael E Plesha, *Concepts and Applications of Finite Element Analysis*, John Wiley, Third Edition, New York, 1989.

10. Timoshenko, S., *Elements of Strength of Materials*, Van Nostrand, New Jersy 1956.

11. Timoshenko, S. and J.N.Goodier, *Theory of Elasticity*, Third edition, McGraw-Hill, New York, 1970.

12. Pragor, C.W., R.M.Barker, and D.Frederick, *Finite Element Bending Analysis of Reissner Plate*, Journal of Engineering Mechanics Division, A.S.C.E., Vol.96, 1970.

13. Ergatoudis, J., B.M. Irons, and O.C.Zienkiewicz, *Three- Dimensional Analysis of Arch Dams and their Foundations*, Proceedings of Symposium on Arch Dams, Inst. Civil Engineers, London, 1968.

14. Zienkiewicz, O.C., and Y.K.Cheung, *Finite Element Method of Analysis for Arch Dam Shells and Comparison with Finite Difference Procedures*, Paper presented at the Int. Symposium held at Santhampton University, April 1964.

15. Arch Dams: *A Review of British Research and Development*, Institution of Civil Engineers, London, 1968.

16. Cook, R.D., and W.C.Wang, *Advanced Mechanics of Materials*, Macmillon Publishing Company, 1985.

EXERCISES

9.1 For the eight noded solid element 5 of the beam shown in Fig. 9.15 derive the [D] matrix for faces 2 and 3.

9.2 For the above element compute the nodal load vector $\{Q\}$ assuming density of 24000 N/m^3.

9.3 Compute the nodal load vector $\{Q\}$ for the element of problem 9.1 due to a normal pressure of 10 kN/cm^2 on face 2.

9.4 Show that a tetrahedral element can be created from an eight noded element by collapsing the sides as given in Fig. E9.4 and the strain in the element is constant.

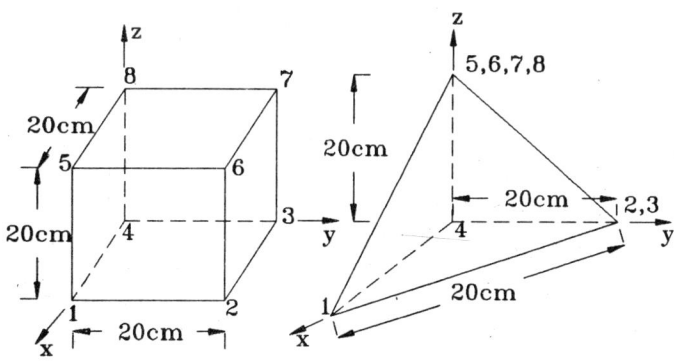

Fig. E9.4

9.5 Develop the necessary expressions for fast stiffness computation for an eight noded solid element without incompatible modes. Draw a flow chart for developing a subroutine for this computation. Discuss the reduction in arithmetic operations by this procedure.

9.6 Derive the shape functions for a variable noded 3-D isoparametric solid element, 8 nodes to 27 nodes. Draw a flow chart for computer implementation of this shape function.

9.7 For the twenty noded solid element shown in Fig. 9.15 compute the nodal load vector $\{Q\}$ due to self-weight assuming density of 24000 N/m^3. Use numerical integration procedure.

9.8 Analyse a machine frame with T cross section as shown in Fig. E9.8 to find out radial and circumferential stresses at the following locations along the dashed line: r = 50 mm, r = 100 mm and r = 200 mm. $E = 1 \times 10^5$ kN/mm^2 $\mu = 0.3$.

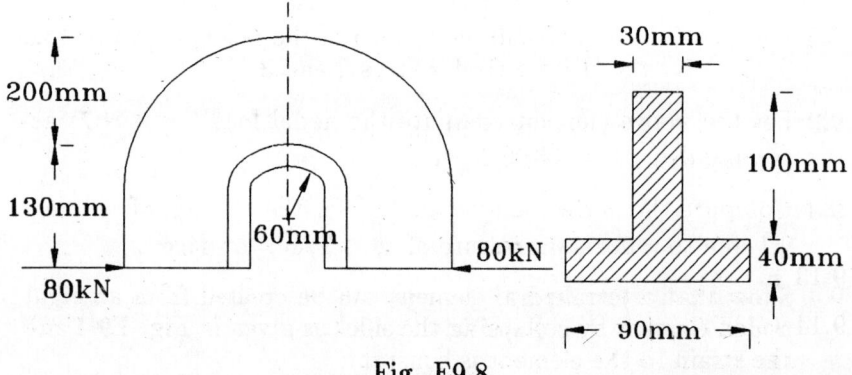

Fig. E9.8

Use 8 noded solid elements. Compare the results with analytical solution [16]

9.9 Solve the problem 9.8 using 20 noded solid element.

9.10 Find the vertical deflection and horizontal normal stresses at point B in a ring shown in Fig. E9.10. Use 8 noded solid elements. Compare the results with analytical solution. (Hint: exploit symmetry about two axes). Q = 50 kN, a = 350 mm, b = 100 mm, E = 10^5 kN/mm^2, $\mu = 0.3$, thickness = 10 mm.

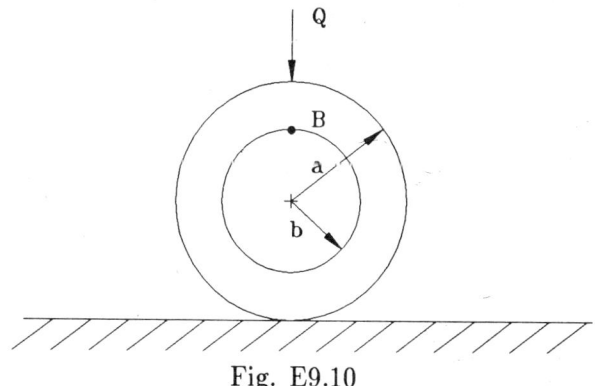

Fig. E9.10

9.11 Solve the problem 9.10 using 20 noded solid element.

9.12 Find out the maximum radial and circumferential stresses for the frame shown in Fig. E9.12. Use 8 noded solid element. $E = 10^5$ kN/mm^2 $\mu = 0.3$.

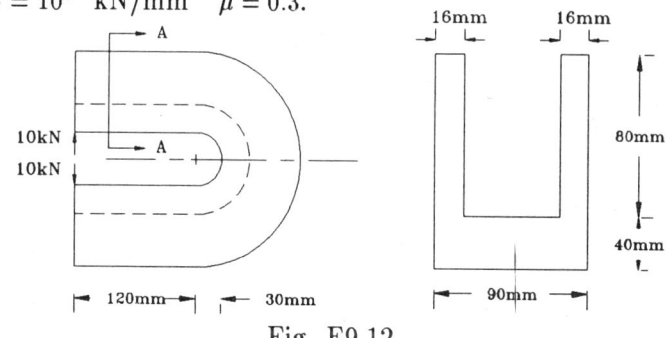

Fig. E9.12

9.13 Solve problem 9.12 by using 20 noded solid element.

9.14 Check the maximum stresses in the footing shown in Fig. E9.14. $E = 2 \times 10^3$ kN/cm^2, $\mu = 0.15$.

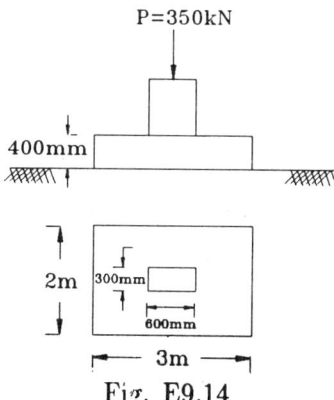

Fig. E9.14

Chapter 10

Analysis of Plate Bending

In many areas of structural design we require analysis of plates subjected to lateral loads. It is well known from the elastic theory of plates [1] that the classical solution involves tedious calculations especially when the plates are arbitrarily shaped and are anisotropic. As a numerical solution to the problem a number of finite element models have been developed and some of the models are described in this chapter. In order to help the student to understand the finite element formulation a brief review is made of the basic Kirchhoff's theory of plates. The shear deformation in plates is also described using the Mindlin's theory.

Though a number of finite element models are available, a simple and efficient four noded isoparametric element (PLATE4) is explained in detail and this element gives good results for problem of moderately thick and thin plates. A higher order 8 noded isoparametric element (PLATE8) is also described. The subroutines (PLATE 4 and PLATE 8) which form the element library of PASSFEM are presented here. A number of examples are given at the end of the chapter to illustrate the performance of the elements described here.

10.1 Basic Theory of Plate Bending

According to the nature of stress states the plates are classified as follows [2]:

1. Thick plate, in which triaxial state of stress is developed, is defined by a complete set of differential equations of three-dimensional theory of elasticity. Plates for which the ratio of thickness to least dimension on plan exceeds 1/10 may be taken as belonging to this class.

2. Thin plates with small deflection in which the membrane stresses are very small compared to flexual stresses under deformation due to transverse loading. This class may be taken to comprise plates for which the ratio of thickness to span does not exceed $1/10$ and the maximum deflection w is less than $h/10$ to $h/5$.

3. Thin plates with large deflection are characterised by the fact that the flexural stresses are accompanied by relatively large tensile or compressive stresses in the middle plane. These membrane stresses significantly affect the bending moment.

10.1.1 Thin Plate Theory

Classical thin plate theory (also referred to as Kirchhoff's theory) is based upon the assumptions initiated for beams by Bernoulli but first applied to plates and shells by Love and Kirchhoff. The assumptions are as follows:

1. A lineal element of the plate normal to the midsurface, on application of the load (a) undergoes at most a translation and rotation and (b) remains normal to the deformed mid-surface.

2. The stresses normal to the plate can be neglected.

From assumption 1(a) it can be noted that a lineal element through the thickness does not elongate or contract. And following the assumption 1(b) the shear strains γ_{xz} and γ_{yz} becomes zero, i.e., $\gamma_{xz} = \gamma_{yz} = 0$. Assumption (2) implies that $\sigma_z = 0$. Thus the three- dimensional problem has been reduced to a two-dimensional problem. It has been shown that the errors introduced due to these assumptions are negligible for thin plates of homogeneous materials under most of the loading conditions.

10.1.2 Basic Relationships

The theory of plates is well treated in the texts [1,2] and is only outlined here to furnish the necessary equations for subsequent illustrations of element formulations. A differential element of thin plate of thickness h is shown in Fig.10.1. With the usual assumptions of thin plate bending the stresses σ_x, σ_y and τ_{xy} vary linearly across the thickness and hence the integration of these stresses across the thickness yields the stress resultants M_x, M_y and M_{xy}.

$$M_x = \int_{-h/2}^{h/2} \sigma_x \, z dz \quad M_y = \int_{-h/2}^{h/2} \sigma_y \, z dz \quad \text{and} \quad M_{xy} = M_{yx} = \int_{-h/2}^{h/2} \tau_{xy} \, z dz$$

$$(10.1a)$$

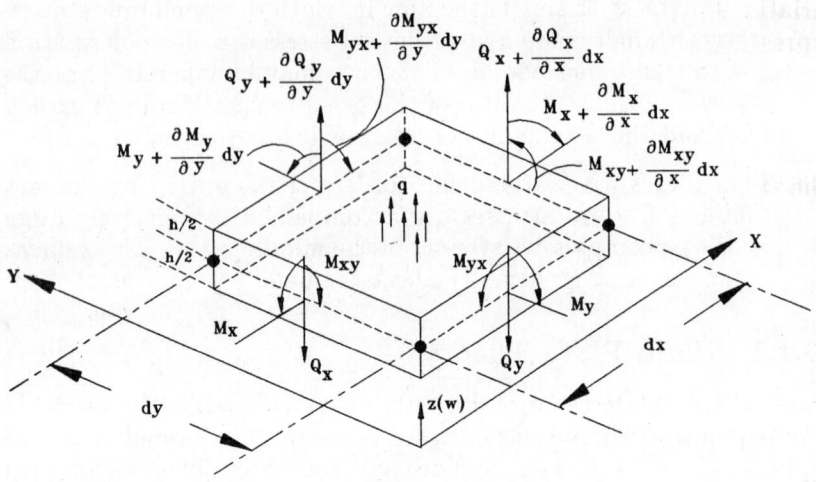

(a) Detailed illustration

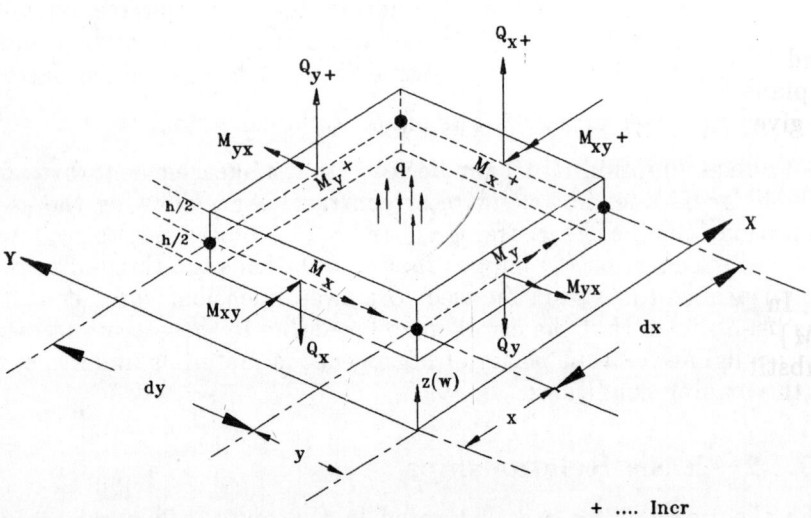

(b) Vectorial notation

Fig. 10.1 External and internal forces on the element
of the middle surface

Similarly, the stress resultants Q_x and Q_y are given by

$$Q_x = \int_{-h/2}^{h/2} \tau_{xz} \, dz \qquad Q_y = \int_{-h/2}^{h/2} \tau_{yz} \, dz \qquad (10.1b)$$

Let u, v and w be the displacements at any point (x, y, z) and the variation of the displacement u and v across the thickness can be expressed in terms of the displacement w as

$$u = -z\frac{\partial w}{\partial x} \qquad v = -z\frac{\partial w}{\partial y} \qquad (10.2)$$

The strain distribution corresponding to Eq.10.2 given by

$$\epsilon_x = \frac{\partial u}{\partial x} = -z\frac{\partial^2 w}{\partial x^2} = zk_x$$

$$\epsilon_y = \frac{\partial v}{\partial y} = -z\frac{\partial^2 w}{\partial x^2} = zk_y$$

$$\gamma_{xy} = \frac{\partial u}{\partial y} + \frac{\partial v}{\partial x} = -2z\frac{\partial^2 w}{\partial x \partial y} = zk_{xy} \qquad (10.3)$$

where

$$k_x = -\frac{\partial^2 w}{\partial x^2}, \quad k_y = -\frac{\partial^2 w}{\partial y^2} \text{ and } k_{xy} = -\frac{2\partial^2 w}{\partial x \partial y} \qquad (10.4)$$

and the shear strains $\gamma_{xz} = \gamma_{yz} = 0$. Thus the problem is reduced to a plane stress problem. The general constitutive law for plane stress is given by

$$\left\{ \begin{array}{c} \sigma_x \\ \sigma_y \\ \tau_{xy} \end{array} \right\} = \left[\begin{array}{ccc} C_{11} & C_{12} & C_{13} \\ C_{21} & C_{22} & C_{23} \\ C_{31} & C_{32} & C_{33} \end{array} \right] \left\{ \begin{array}{c} \epsilon_x \\ \epsilon_y \\ \gamma_{xy} \end{array} \right\} \qquad (10.5)$$

In the case of plates, it is convenient to regard the stress resultants $\{M\}^T = \{ M_x \ \ M_y \ \ M_{xy} \}$ instead of $\{\sigma\}^T = \{ \sigma_x \ \ \sigma_y \ \ \tau_{xy} \}$. Thus substituting the Eqs. 10.5 and 10.3 in Eq.10.1a, we get,

$$\left\{ \begin{array}{c} M_x \\ M_y \\ M_z \end{array} \right\} = \frac{h^3}{12} \left[\begin{array}{ccc} C_{11} & C_{12} & C_{13} \\ C_{21} & C_{22} & C_{23} \\ C_{31} & C_{32} & C_{33} \end{array} \right] \left\{ \begin{array}{c} k_x \\ k_y \\ k_{xy} \end{array} \right\}$$

i.e $\{M\} = [C_f] \{k_c\}$
where

$$\{k_c\}^T = [k_x \ \ k_y \ \ k_{xy}] \qquad (10.6)$$

This equation is exactly similar to $\{\sigma\} = [C] \{\epsilon\}$. In the case of isotropic plates, the constitutive matrix is given by,

$$[C_f] = \frac{Eh^3}{12(1 - \mu^2)} \left[\begin{array}{ccc} 1 & \mu & 0 \\ \mu & 1 & 0 \\ 0 & 0 & \frac{1-\mu}{2} \end{array} \right] \qquad (10.7)$$

Thus in the case of isotropic plates we get the familiar relationships as,

$$M_x = -D_p\left(\frac{\partial^2 w}{\partial x^2} + \mu\frac{\partial^2 w}{\partial y^2}\right)$$

$$M_y = -D_p\left(\frac{\partial^2 w}{\partial y^2} + \mu\frac{\partial^2 w}{\partial x^2}\right)$$

$$M_{xy} = -D_p(1-\mu)\frac{\partial^2 w}{\partial x \partial y} \tag{10.8}$$

where

$$D_p = \frac{Eh^3}{12(1-\mu^2)}$$

We observe that the solution of thin plate flexure problem depends entirely on the selection of a single component w, the transverse displacement.

10.2   Displacement Functions

The convergence and compatibility requirements of displacement function have been discussed in section 3.2 and 3.3. It may be noted that in the case of plates the constant strain state amounts to constant curvature condition and the displacement function must satisfy this requirement. For satisfaction of compatibility conditions, $w(x,y)$ and the normal slope $\partial w/\partial n$ must be uniquely specified along any element interface.

In general, there are three classes of displacement functions which can be chosen for plate element and they are as follows:

1. Class,C^2:   The assumed function $w(x,y)$ has continuous second derivatives (and hence curvatures) at element corners and inside the element.
2. Class,C^1:   The assumed function for $w(x,y)$ has continuous first derivatives but may have discontinuous corner curvatures.
3. Class,C^o: The assumed function $w(x,y)$ only is continuous and independent functions are assumed to represent the variation of w and slopes.

In the early stages of development, a large amount of work has been carried out using shape functions for w belonging to Classes C^2 and C^1. To construct a compatible element with C^2 type displacement function, a minimum of six degrees of freedom (w, $\frac{\partial w}{\partial x}$, $\frac{\partial w}{\partial y}$, $\frac{\partial^2 w}{\partial x^2}$, $\frac{\partial^2 w}{\partial x \partial y}$

and $\frac{\partial^2 w}{\partial y^2}$) are required at each non-right angled corner. And four degrees of freedom ($w, \frac{\partial w}{\partial y}, \frac{\partial w}{\partial x}$ and $\frac{\partial^2 w}{\partial x \partial y}$) are required at each right angled corner. It follows, therefore, that atleast eighteen degrees of freedom are required for a general triangle and twenty-four for an arbitrary quadrilateral and sixteen for a rectangle. A number of researchers have contributed to the development of these types of elements with second derivatives also as degrees of freedom. The problem with these types of higher order elements is that it becomes difficult to match the boundary conditions with the other types of elements.

It has been shown in reference [3] (also illustrated in the next section) that if we take $w, \frac{\partial w}{\partial x}$ and $\frac{\partial w}{\partial y}$ as nodal degrees of freedom then there will be discontinuity in the curvatures at the corners. However, a number of elements have been developed with three degrees of freedom at each node and in some cases full compatibility has been ensured by properly constraing the slopes along the edges and by adopting subdomain technique. A brief description of some of the elements based on models of C^1 and C^2 types is given in section 10.3.

It has been observed that the above approaches to the plate bending elements based on Kirchhoff's theory of thin plates led to either incompatible elements or involved complicated formulation and programming. In the recent past several attempts have been made to develop simple and efficient plate bending element using displacement models satisfying C^o continuity requirement. These models are based on Mindlin's theory which considers shear deformation in plates. The shear deformation and the details of finite element formulation are explained in subsequent sections.

10.3 Plate Bending Elements

A variety of plate elements have been proposed since the early days of the finite element method. A survey of these elements has been presented in references [3] and [4]. Some of the elements which are commonly used are briefly described below.

10.3.1 Rectangular Plate Element with 12 Degrees of Freedom

A rectangular plate bending element with 12 degrees of freedom due to Melosh [5] is one of the oldest and best known element for the analysis of plates.

The displacement field is given by

$$w = \alpha_1 + \alpha_2 x + \alpha_3 y + \alpha_4 x^2 + \alpha_5 xy + \alpha_6 y^2 + \alpha_7 x^3 + \alpha_8 x^2 y + \alpha_9 xy^2$$
$$+\alpha_{10} y^3 + \alpha_{11} x^3 y + \alpha_{12} xy^3 \qquad (10.9)$$

The three variables at each corner node are

$$w_i, \qquad \left(\frac{\partial w}{\partial x}\right)_i \qquad \text{and} \qquad \left(\frac{\partial w}{\partial y}\right)_i$$

i.e.,

$$\{d_i\}^T = [w_i \ \left(\frac{\partial w}{\partial x}\right)_i \ \left(\frac{\partial w}{\partial y}\right)_i] \qquad (10.10)$$

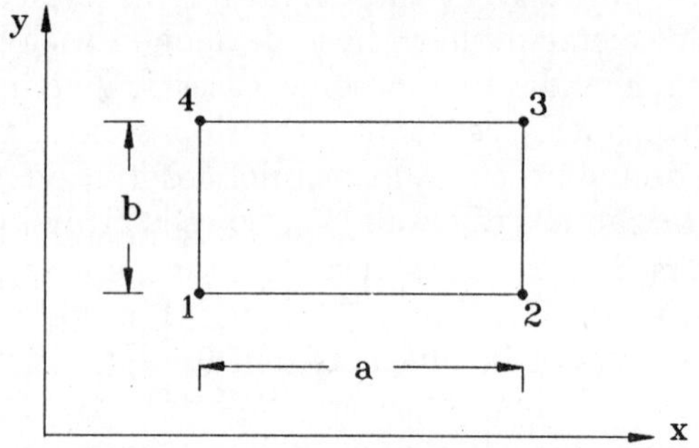

Fig. 10.2 Rectangular Element with 12 degrees of freedom

We would like to study the compatibility of displacement and slopes along the interface edges. Consider the edge 1-2 which is assumed to be the interelement boundary between two elements A and B joining along this edge (Fig.10.3). The displacement w along this edge is given by

$$w = \alpha_1 + \alpha_2 x + \alpha_4 x^2 + \alpha_7 x^3 \qquad (10.11)$$

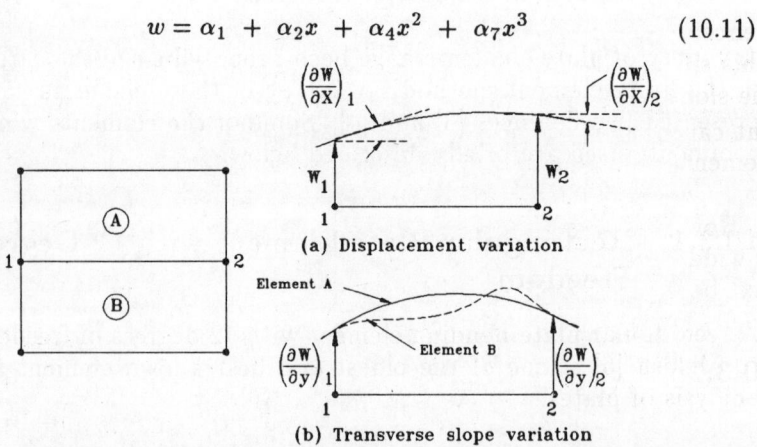

(a) Displacement variation

(b) Transverse slope variation

Fig. 10.3 Compatibility conditions along the edge 1-2

The slope with respect to x, $\frac{\partial w}{\partial x}$ is given by

$$\frac{\partial w}{\partial x} = \alpha_2 + 2\alpha_4 x + 3\alpha_7 x^2 \tag{10.12}$$

For the nodes 1 and 2, we will be specifying w_1, $\left(\frac{\partial w}{\partial x}\right)_1$, w_2 and $\left(\frac{\partial w}{\partial x}\right)_2$. And these four nodal displacements will be the same for the two elements joining along the edge 1-2. By substituting these quantities for $w = w_1, \frac{\partial w}{\partial x} = \left(\frac{\partial w}{\partial x}\right)_1$ at x=0, and $w = w_2, \frac{\partial w}{\partial x} = \left(\frac{\partial w}{\partial x}\right)_2$ at $x = a$, in Eq.10.11 and 10.12, the coefficients $\alpha_1, \alpha_2, \alpha_4$ and α_7 can be uniquely determined. Therefore w and $\frac{\partial w}{\partial x}$ are fully defined along the edge 1-2 and continuity of displacement and slope $\frac{\partial w}{\partial x}$ along the edge is ensured (Fig. 10.3a).

The above conclusion does not apply to the normal slope $\frac{\partial w}{\partial y}$ along the interelement edge 1-2. This can be examined as follows.

$$\left(\frac{\partial w}{\partial y}\right) = \alpha_3 + \alpha_5 x + \alpha_8 x^2 + \alpha_{11} x^3 \tag{10.13}$$

It can be seen from Eq.10.13, that to define $\frac{\partial w}{\partial y}$ along the edge 1-2 we need to calculate four coefficients α_3, α_5, α_8 and α_{11}. But only two equations can be obtained by substituting in equation (10.13) the conditions

$$\left(\frac{\partial w}{\partial y}\right) = \left(\frac{\partial w}{\partial y}\right)_1 \quad \text{for} \quad x = 0 \quad \text{and}$$

$$\left(\frac{\partial w}{\partial y}\right) = \left(\frac{\partial w}{\partial y}\right)_2 \quad \text{for} \quad x = a$$

Hence, the slope normal to the boundary 1-2 is not fully defined. This situation is illustrated in Fig. 10.3b. It can be observed that the slope $\frac{\partial w}{\partial y}$ is the same for both the elements at the nodes 1 and 2 but can be different at other points along the edge 1-2. Hence this element is not fully compatible.

However the performance of the element is reasonably good and is widely used. The properties of the element are available in the text books by Zienkiewicz[3] and Przemieniecki[6].

10.3.2 Rectangular Plate Element with 16 Degrees of Freedom

Bogner, Fox and Schmit[7] developed a sixteen degrees of freedom element with displacement field assumed as

$$w = \alpha_1 + \alpha_2 x + \alpha_3 y + \alpha_4 x^2 + \alpha_5 xy + \alpha_6 y^2 + \alpha_7 x^3 + \alpha_8 x^2 y$$
$$+ \alpha_9 xy^2 + \alpha_{10} y^3 + \alpha_{11} x^3 y + \alpha_{12} xy^3 + \alpha_{13} x^2 y^2$$
$$+ \alpha_{14} x^3 y^2 + \alpha_{15} x^2 y^3 + \alpha_{16} x^3 y^3 \qquad (10.14)$$

It may be noted that Eq.(10.14) is a 'complete polynomial' for the terms of the expression correspond to the product $(1 + x + x^2 + x^3)(1 + y + y^2 + y^3)$. This displacement function results in a cubic polynomial for the displacements and the slopes along the edges of the elements. Following the procedure described in section 10.3.1 for the rectangular element with 12 degrees of freedom, it can be proved that the normal slope along any edge is completely defined, because of the choice of $\frac{\partial^2 w}{\partial x \partial y}$ as a degree of freedom at each node. Thus the element is a fully compatible element. The nodal degrees of freedom at each node i is given by

$$\{d_i\}^T = [w_i \ (\frac{\partial w}{\partial x})_i \ (\frac{\partial w}{\partial y})_i \ (\frac{\partial^2 w}{\partial x \partial y})_i] \qquad (10.15)$$

For more details of the element the reader may refer to Bogner et.al.[7].

10.3.3   A Refined Quadrilateral Element

This element is due to Clough and Felippa [8] and it has 12 degrees of freedom, three at each node. The formulation of this element is different from other types of element in that a sub-domain approach is used. The quadrilateral is divided into four triangles. Now to obtain a compatible triangular element, using a cubic polynomial for defining the displacement pattern, the triangle is subdivided into three subtriangles shown in Fig. 10.4.

The nodal degrees of freedom which are to be considered for the computation of the stiffness matrix are shown in Fig.10.4. They are as follows:

1. the transverse displacement at each corner node w_i

2. the rotation at each corner node θ_{xi} and θ_{yi}

3. the rotation at each midside node about an axis parallel to each side θ_4, θ_5 and θ_6.

In order to develop the displacement model for the complete triangular element, cubic interpolation functions are assumed independently for each subtriangle. The nodal displacement vector for each subelement, for example subtriangle 1, is

$$\{d^{(1)}\}^T = [\,w_2 \quad \theta_{x2} \quad \theta_{y2} \quad w_3 \quad \theta_{x3} \quad \theta_{y3} \quad w_o \quad \theta_{xo} \quad \theta_{yo} \quad \theta_5\,] \quad (10.16)$$

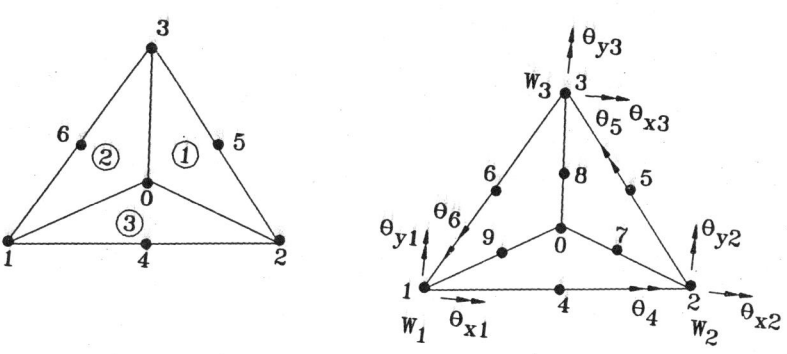

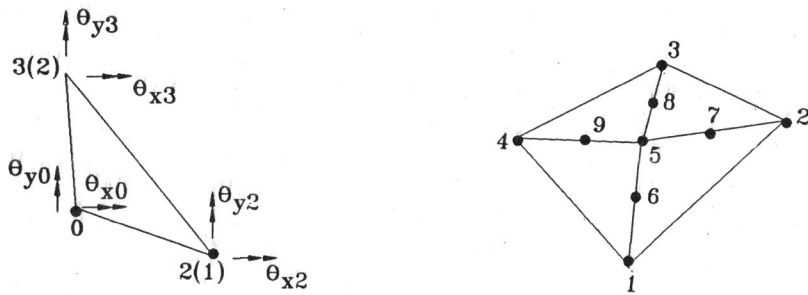

Fig. 10.4 Plate element due to Clough and Felippa

Hence, we can write the cubic interpolation to express the displacement variation within a subelement in terms of the ten nodal degrees of freedom. Now consider two adjacent elements with common interface.

For example, along the interface 02 of subtriangles 1 and 3 transverse displacements are uniquely specified by nodal displacement values at 0 and 2. The transverse displacements are compatible along the interface 02. But their normal slopes differ between nodes. To establish slope compatibility along the interface edges of the subelements, additional nodes 7, 8 and 9 are introduced at the midpoints of these edges. To maintain internal slope compatibility, it is necessary to match the normal slopes of the mid-nodes of adjacent subelements, as for example

$$\theta_7^{(1)} = -\theta_7^{(3)}$$

Thus we would get three such compatibility requirements and the displacements of the node 0, $\{w_o, \theta_{xo} \text{ and } \theta_{yo}\}$ can be computed to satisfy these conditions. Then the compatibile displacement field for the complete triangle can be expressed in terms of the displacements of the nodes 1, 2, 3 and 4, 5, 6. Once the displacement field has been specified the properties of the element can be worked out and the details of this element referred to as LCCT-12 Linear Curvature Compatible Triangle with 12 degrees of freedom, can be found in reference [8].

Although the LCCT-12 element employs an optimum compatible cubic displacement field, its midside nodes (4, 5, 6) and their rotational degrees of freedom (θ_4, θ_5, θ_6) complicate the analysis. To overcome this difficulty, a special version of the element has been developed which constraints the normal slope to vary linearly along the side, thus eliminating one of the midside nodes. The resulting partially constrained element is designated as LCCT-11. Four such LCCT-11 elements may then be assembled into the Q-19 quadrilateral having no midside nodes on the exterior edges as shown in Fig. 10.4d. Although the element has 19 degrees of freedom, the internal degrees of freedom of the assemblage are eliminated by the static condensation procedure. The final quadrilateral element has only 12 degrees of freedom, as w, θ_x and θ_y displacements at each of the four nodes. It is a fully compatible element, having linear variations of normal slopes along all the exterior edges. This element has been incorporated in the general purpose program, SAP IV and has been found to give good results for the analysis of plate bending problems.

10.4   Shear Deformation in Plates

In thin plate theory, the transverse shear deformation is neglected and deformation is completely described by a function $w(x, y)$ where w is the lateral displacement perpendicular to the plate. If the effects of transverse shear deformations are to be considered, it is important to include these deformations in the rotational degrees of freedom of the nodes. Let θ_x and θ_y be the rotations of a line that was normal to the midsurface of the undeformed plate. Unless transverse shear deformation is taken as zero, θ_x and θ_y will not be equal in magnitude to the slopes $\frac{\partial w}{\partial x}$ and $\frac{\partial w}{\partial y}$ of the tangent to the deformed surfaces.

Accordingly, in finite element formulation it is the rotations θ_x and θ_y of the midsurface normal that must be made compatible between elements and be included in the nodal degrees of freedom.

Transverse shear deformation must be considered in the case of thick plates and sandwich plates of laminated construction. Reissner [9] considered the effect of transverse shear deformation by relaxation of assumptions made in Krichhoff's theory and later Mindlin [10] considered it in a slightly different manner. Mindlin's approximation has been found to be quite useful to consider shear deformation in plates and is widely adopted in finite element formulation [9]. These modified postulates facilitate to extend the theory of thin plates to thick plates and it is referred to in literature as Reissner-Mindlin theory [3] or Mindlin's theory

10.4.1 Mindlin's Theory to Include Shear Deformation in Plates

Mindlin's approximation is that straight lines originally normal to the midsurface, before deformation, remain straight but not normal to the deformed midsurface, i.e., the average rotation of the section may be taken as the rotation in which normals remain perpendicular to the midsurface plus an additional rotation due to transverse shear (Fig. 10.5). Thus the actual shear deformation is assumed to be equivalent to a straight line rotation representing a uniform shear strain through the thickness and this assumption is similar to the one made in the case of shear deformation in beams described in section 7.5 of Chapter 7.

Now, the three assumptions made in Mindlin's Theory of plates are given below:

1. The deflections of the plate, w, are small.

2. Normals to the plate midsurface before deformation remain straight but are not necessarily normal to it after deformation.

3. Stresses normal to the midsurface are negligible.

The second assumption to account for shear deformation is different from Kirchhoff's theory while the first and third assumptions are the same for both the theories.

Referring to Fig. 10.5b, in which ϕ_x denotes an average transverse shear strain for a section $x = $ constant, the total rotation θ_y can be expressed as,

$$\theta_y = -\frac{\partial w}{\partial x} + \phi_x \qquad (10.17)$$

and similarly for section $y = $ constant (Fig. 10.5a),

$$-\theta_x = -\frac{\partial w}{\partial y} + \phi_y \qquad (10.18)$$

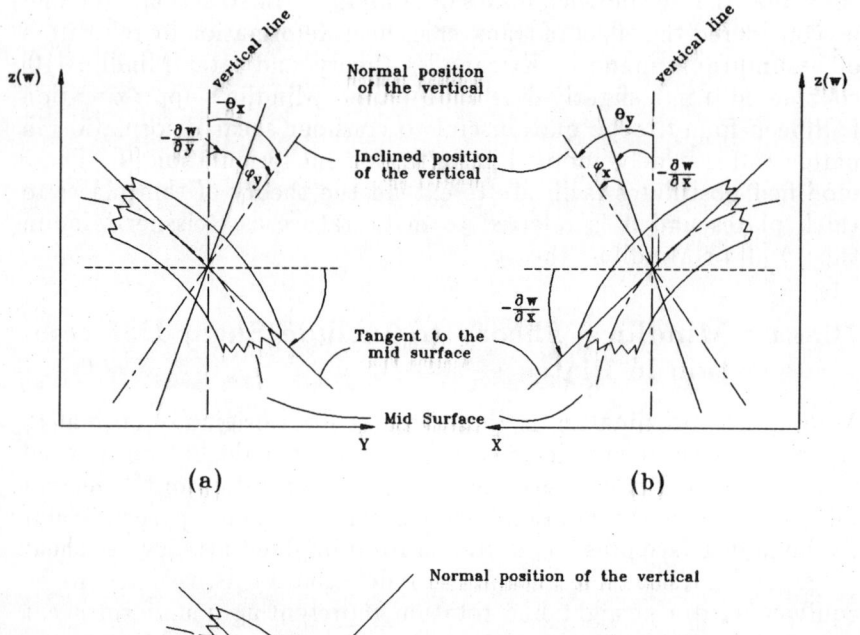

(a)   (b)

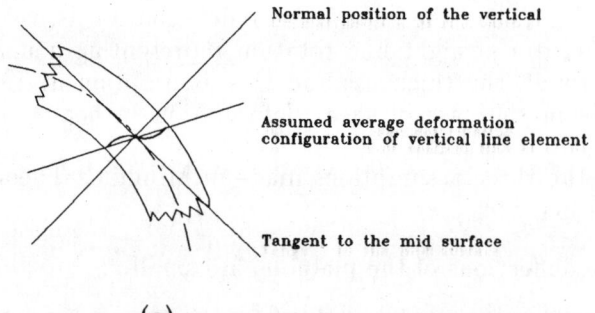

(c)

Fig. 10.5 Rotation of the normals about x and y axes
considering average shear deformation

Hence, the average shear deformations, ϕ_x and ϕ_y are given by

$$\phi_x = \theta_y + \frac{\partial w}{\partial x}$$

$$\phi_y = -\theta_x + \frac{\partial w}{\partial y} \qquad (10.19)$$

Equations 10.19 are based on the assumption that the total rotations θ_x and θ_y are small and the transverse strain ϵ_z is negligible. In the limit tending to thin plate range the shear strains tend to become zero and hence $\theta_y \rightarrow -\frac{\partial w}{\partial x}$ and $\theta_x \rightarrow \frac{\partial w}{\partial y}$.

Following the discussion in section 7.5 of Chapter 7 the expression for strain energy to include the contribution due to shear can be worked out. The shear strain energy is given by,

$$U_s = \frac{1}{2} \, \alpha \, GA \, \int \int_A [(\phi_x)^2 + (\phi_y)^2] dx \, dy \tag{10.20}$$

Substituting for ϕ_x and ϕ_y from Eqs. 10.19, we get, for isotropic plate,

$$U_s = \frac{Eh^3}{24(1-\mu^2)} [\frac{24(1-\mu^2)}{Eh^3} \times \frac{1}{2} \, \alpha \, GA] \int \int_A [(\frac{\partial w}{\partial x} + \theta_y)^2$$

$$+ (\frac{\partial w}{\partial y} - \theta_x)^2] dx \, dy \tag{10.21}$$

Since $G = \frac{E}{2(1+\mu)}$, Eq(10.21) can be written as,

$$U_s = \frac{Eh^3}{24(1-\mu^2)} \times \frac{6\alpha(1-\mu)}{h^2} \int \int_A [(\frac{\partial w}{\partial x} + \theta_y)^2 + (\frac{\partial w}{\partial y} - \theta_x)^2] dx \, dy \tag{10.22}$$

The expression for strain energy due to bending of isotropic plates can be obtained following the usual theory of plates [1],

$$U_b = \frac{Eh^3}{24(1-\mu^2)} \int \int_A \{(\frac{\partial \theta_x}{\partial x})^2 + 2\mu \, (\frac{\partial \theta_x}{\partial x}) \, (\frac{\partial \theta_y}{\partial y}) + (\frac{\partial \theta_y}{\partial y})^2$$

$$+ \frac{(1-\mu)}{2} (\frac{\partial \theta_x}{\partial y} + \frac{\partial \theta_y}{\partial x})^2\} \, dx \, dy \tag{10.23}$$

Therefore, the total strain energy is given by,

$$U = U_b + U_s \tag{10.24}$$

10.4.2 Basic Relationships for Finite Element Formulation

In section 10.1.2 the basic relationships for the curvatures and stress resultants are described for thin plates based on Kirchhoff's theory. Similar relationships will be derived here for plates considering shear deformation using Mindlin's approximation.

The average shear deformation ϕ_x and ϕ_y are expressed in Eqs. 10.19. The shear stresses τ_{xz} and τ_{yz} and the shear deformation are related as follows.

$$\left\{ \begin{matrix} \tau_{xz} \\ \tau_{yz} \end{matrix} \right\} = \left[\begin{matrix} C_{44} & C_{45} \\ C_{45} & C_{55} \end{matrix} \right] \left\{ \begin{matrix} \phi_x \\ \phi_y \end{matrix} \right\} \tag{10.25}$$

For an isotropic material the above relation can be written as

$$\left\{ \begin{matrix} \tau_{xz} \\ \tau_{yz} \end{matrix} \right\} = \frac{E}{2(1+\mu)} \left[\begin{matrix} 1 & 0 \\ 0 & 1 \end{matrix} \right] \left\{ \begin{matrix} \phi_x \\ \phi_y \end{matrix} \right\} \tag{10.26}$$

The average shear deformation ϕ_x and ϕ_y are constant over the thickness, and allowing for warping of the cross-section the stress resultants Q_x and Q_y can be computed as

$$\left\{ \begin{matrix} Q_x \\ Q_y \end{matrix} \right\} = \frac{Eh\alpha}{2(1+\mu)} \left[\begin{matrix} 1 & 0 \\ 0 & 1 \end{matrix} \right] \left\{ \begin{matrix} \phi_x \\ \phi_y \end{matrix} \right\} \tag{10.27}$$

or in general,

$$\{Q\} = [C_s] \{\phi\} \tag{10.28}$$

where α is a numerical correction factor used to represent the restraint of the cross-section against warping. The value of α commonly used is 5/6 but may be assumed between 2/3 for section having no restraint against warping and 1.0 for sections having complete restraint against warping [11].

Now the curvatures k_x and k_y and the twist k_{xy} can be expressed in terms of the rotations θ_x and θ_y as,

$$k_x = \frac{\partial \theta_y}{\partial x} \qquad k_y = -\frac{\partial \theta_x}{\partial y} \qquad k_{xy} = \left(\frac{\partial \theta_y}{\partial y} - \frac{\partial \theta_x}{\partial x} \right) \tag{10.29}$$

The stress resultants $\{M\}$ and $\{Q\}$ given by Eqs. 10.6 and 10.28 can be combined and for homogeneous plates, it can be expressed as,

$$\left\{ \begin{matrix} M_x \\ M_y \\ M_{xy} \\ \text{--} \\ Q_x \\ Q_y \end{matrix} \right\} = \left[\begin{array}{c|c} \frac{h^3}{12} \left[\begin{matrix} C_{11} & C_{12} & C_{13} \\ Sym. & C_{22} & C_{23} \\ & & C_{33} \end{matrix} \right] & \left[\begin{matrix} 0 & 0 \\ 0 & 0 \\ 0 & 0 \end{matrix} \right] \\ \hline \left[\begin{matrix} 0 & 0 & 0 \\ 0 & 0 & 0 \end{matrix} \right] & h \left[\begin{matrix} C_{44} & C_{45} \\ C_{54} & C_{55} \end{matrix} \right] \end{array} \right] \left\{ \begin{matrix} k_x \\ k_y \\ k_{xy} \\ \text{--} \\ \phi_x \\ \phi_y \end{matrix} \right\} \tag{10.30}$$

For isotropic plates, using the specific values of coefficient for the constitutive matrix, we get

$$\begin{Bmatrix} M_x \\ M_y \\ M_{xy} \\ - \\ Q_x \\ Q_y \end{Bmatrix} = \left[\begin{array}{ccc|cc} & \dfrac{Eh^3}{12(1-\mu^2)} \begin{bmatrix} 1 & \mu & 0 \\ \mu & 1 & 0 \\ 0 & 0 & \frac{1-\mu}{2} \end{bmatrix} & & \begin{bmatrix} 0 & 0 \\ 0 & 0 \\ 0 & 0 \end{bmatrix} & \\ \hline & \begin{bmatrix} 0 & 0 & 0 \\ 0 & 0 & 0 \end{bmatrix} & & \dfrac{Eh}{2(1+\mu)} \begin{bmatrix} \alpha & 0 \\ 0 & \alpha \end{bmatrix} & \end{array} \right] \begin{Bmatrix} k_x \\ k_y \\ k_{xy} \\ - \\ \phi_x \\ \phi_y \end{Bmatrix}$$

$$(10.30a)$$

These relations can be expressed in compact form as

$$\begin{Bmatrix} \{M\} \\ --- \\ \{Q\} \end{Bmatrix} = \begin{bmatrix} [C_f] & | & [0] \\ -- & - & -- \\ [0]^T & | & [C_s] \end{bmatrix} \begin{Bmatrix} \{k_c\} \\ -- \\ \{\phi\} \end{Bmatrix} \qquad (10.30b)$$

The above relation can be compared with $\{\sigma\} = [C]\{\epsilon\}$ for the usual stress-strain relation. Thus in the case of plate bending, the stress resultants and the corresponding curvatures and shear deformations can be considered analogous to stresses and strains. Hence, for uniformity and convenience we may express Eq.10.30b as

$$\{\sigma\}_p = [C]_p \{\epsilon\}_p \qquad (10.31)$$

where

$$\{\sigma\}_p = \begin{Bmatrix} M_x \\ M_y \\ M_{xy} \\ Q_x \\ Q_y \end{Bmatrix} \qquad [C]_p = \begin{bmatrix} [C_f] & | & [0] \\ -- & - & -- \\ [0]^T & | & [C]_s \end{bmatrix}$$

and

$$\{\epsilon\}_p = \begin{Bmatrix} k_x \\ k_y \\ k_{xy} \\ \phi_x \\ \phi_y \end{Bmatrix} \qquad (10.31a)$$

Using Eqs. 10.19 and 10.29 the curvature and shear deformation vector $\{\epsilon\}_p$ can be expressed in terms of the three displacements w, θ_x and θ_y as,

$$\{\epsilon\}_p = \begin{Bmatrix} \dfrac{\partial \theta_y}{\partial x} \\ -\dfrac{\partial \theta_x}{\partial y} \\ \dfrac{\partial \theta_y}{\partial y} - \dfrac{\partial \theta_x}{\partial x} \\ \theta_y + \dfrac{\partial w}{\partial x} \\ -\theta_x + \dfrac{\partial w}{\partial y} \end{Bmatrix} \qquad (10.31b)$$

We can express the strain energy of the plate in terms of stress resultants as,

$$U = \frac{1}{2} \int\int_A (\{M\}^T \{k_c\} + \{Q\}^T \{\phi\}) \, dA \qquad (10.32a)$$

Substituting from Eqs. 10.6 and 10.28, the above expression may also be written as,

$$U = \frac{1}{2} \int\int_A (\{k_c\}^T [C_f] \{k_c\} + \{\phi\}^T [C_s] \{\phi\}) \, dA \qquad (10.32b)$$

In order to get an insight into the understanding of the problems due to inclusion of shear deformation and later application to thin plate situations, the strain energy expressions (Eqs. 10.22 and 10.23) are expressed in terms of non-dimensional variables as given below [12].

$$\overline{w} = \frac{w}{a}, \quad \overline{x} = \frac{x}{a}, \quad \overline{y} = \frac{y}{a}, \quad \overline{D} = \frac{D}{Eh^3}$$

$$\overline{q} = \frac{qa^2}{Eh^2}, \quad \overline{h} = \frac{h}{a}, \quad \lambda = \frac{G}{E} \left(\frac{a}{h}\right)^2 \qquad (10.33)$$

where a is the length along the x-axis of the plate and q is the lateral load. The total potential energy Π can now be expressed as,

$$\Pi = Eh^3 \left\{ \frac{\overline{D}}{2} \int\int_A \left[\left(\frac{\partial \theta_x}{\partial \overline{x}}\right)^2 + 2\mu \left(\frac{\partial \theta_x}{\partial \overline{x}}\right) \left(\frac{\partial \theta_y}{\partial \overline{y}}\right) + \left(\frac{\partial \theta_y}{\partial \overline{y}}\right)^2 \right. \right.$$

$$\left. \frac{1-\mu}{2} \left(\frac{\partial \theta_x}{\partial \overline{y}} + \frac{\partial \theta_y}{\partial \overline{x}}\right)^2 \right] d\overline{x} \, d\overline{y} +$$

$$\frac{\lambda}{2} \int_A \int_A \left[\left(\frac{\partial \overline{w}}{\partial \overline{x}} + \theta_y\right)^2 + \left(\frac{\partial \overline{w}}{\partial \overline{y}} - \theta_x\right)^2 \right] d\overline{x} d\overline{y} -$$

$$\left. \int\int_A \frac{\overline{q}\,\overline{w}}{\overline{h}} \, d\overline{x} \, d\overline{y} \right\} \qquad (10.34)$$

In the above equation, λ tends to infinity as the plate thickness is decreased and it is termed as the penalty parameter. As $\frac{a}{h} \to \infty$, (i.e., in thin plate limit), the shear deformation tends to become zero,

$$\text{i.e.,} \quad \frac{\partial w}{\partial x} + \theta_y = 0 \quad \text{and} \quad \frac{\partial w}{\partial y} - \theta_x = 0 \qquad (10.35)$$

The penalty parameter enforces the above conditions for thin plate and the conditions stated in Eqs. 10.35 are referred to as *Kirchhoff/Constraints*. The potential energy given in Eq. 10.34 reduces to that of classical thin plate theory.

Following the derivation of finite element equilibrium equation explained in Chapter 3, by minimizing the potential energy Π of Eq. 10.34 we get,

$$[[k]_b + \lambda[k]_s]] \{d\} = \{Q\} \tag{10.36}$$

where $[k]_b$ is the bending stiffness matrix, $[k]_s$ is the shear stiffness matrix, $\{d\}$ is the element displacement vector and $\{Q\}$ is the element model load vector. In the case of thin plates, when $\lambda \to \infty$, $[k]_s \{d\}$ must become zero for the potential energy Π to remain bounded. Since $[k]_s \{d\}$ is obtained by minimizing U_s, it tends to zero as the Kirchhoff constraints of Eq.10.35 are enforced. This implies that the shear strain in each element should tend to zero as the thickness is decreased. The problem of plate elements based on Mindlin theory is precisely due to the inability of these elements to model a state of zero transverse shear strain and that it leads to an overstiff element and the phenomenon is known as *shear locking*.

10.4.3 Plate Bending Elements

In this section, a brief review will be made of the various types of elements developed using the Mindlin's theory of plates. It may be noted that there are three independent displacement quantities w, θ_x, and θ_y to be considered for inclusion of shear deformation. Hence for finite element formulation three shape functions are chosen to represent the variation of w, θ_x and θ_y and they belong to the class C^o. Since the satisfaction of C^o continuity requirement can be easily met with, plate elements based on Mindlin's theory have been successfully formulated [13, 14, 15]. Such formulations showed promise for application to thick or thin plates, curved boundaries and laminated materials. However, the main difficulty experienced in the use of such elements was that they exhibited overstiff locking behaviour in thin plate situations as explained above. This led to an extensive investigation reported in references [12, 15]. An eight noded isoparametric element has been proposed by Zienkiewicz et.al.[17] and a reduced integration has been used for application to thin plate situations. This element will be described in detail in section 10.6 as a higher order element for PASSFEM library.

Another approach to the development of elements for thin plates involves the use of Discrete Kirchhoff Theory [18, 19]. In this approach, the independent displacement quantities assumed for finite element formulation are w, θ_x and θ_y and only C^o continuity requirements need to be satisfied. The transverse shear energy is neglected altogether and Kirchhoff hypothesis is introduced in a discrete way along the edges of the element to relate the rotations to the transverse displacements. That is, the constraint of zero shear strain ($\gamma_{xz} = \gamma_{yz} = 0$) is imposed at discrete number of points along the edges of the element to represent the behaviour of thin plates. Each constraint removes one degree of freedom and thus yields a flexible mesh. This property makes it possible to avoid element locking associated with lower order elements applied to very thin plates. The practical application of DKT is difficult and the implementation is complicated. Furthermore, the element predicts stresses relatively poorly. The DKT element has not received wide adoption and has not been implemented in any major computer package. As the implementation of reduced or selective integration to avoid mesh locking is easier compared to DKT, the former is given more importance in plate bending application.

In summarizing the review of all the above developments one finds that for linear problems of plate bending many accurate elements exist. Many users of the finite element programs find a basic four noded quadrilateral element more appealing due to its simplicity. This preference will become greater in non-linear applications where frequent updating of stiffness matrix require computationally efficient formulation. A successful attempt towards developing a simpler and efficient plate bending element has been made by Hughes et.al.[14]. This element is a four noded quadrilateral with the three degrees of freedom per node. The element shape functions are bilinear for transverse displacement and rotations. Mindlin's plate theory is used and C^o continuity for the displacement model is ensured. The shear locking associated with such low order displacement functions for application to thin plates is alleviated by separating the shear and bending energy terms and using selective integration procedure. The simplicity of the element lends itself to concise and efficient computer implementation. However, in some situations the element may cause problem due to spurious deformation mode and it is discussed later. This element it described in detail in the next section and a subroutine PLATE4 using this element is included in PASSFEM.

10.5   Four Noded Isoparametric Element (Plate 4)

The finite element formulation of a four noded quadrilateral element. (bilinear plate bending element) due to Hughes et al [14] is explained in this section (Fig. 10.6). The element properties required for programming are also presented here.

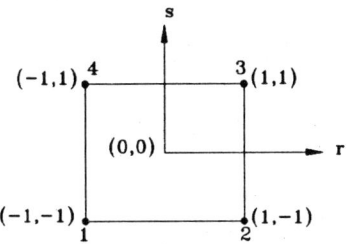

(a) Bilinear element with
normalised co-ordinates

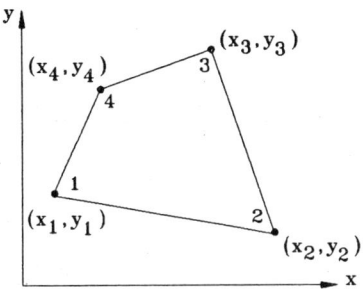

(b) Four noded isoparametric
element with straight boundaries

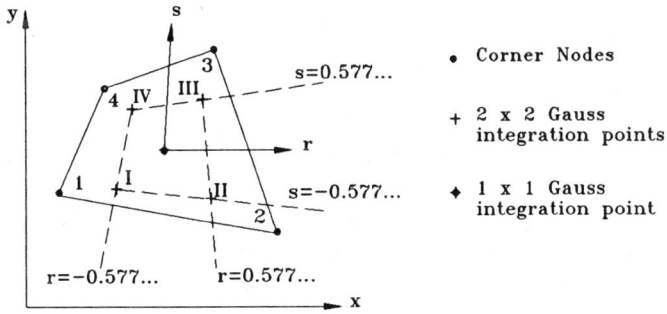

(c) Gauss quadrature and stress evaluation points

Fig. 10.6 Details of bilinear plate element

10.5.1   Displacement Method

Isoparametric element concept is adopted in the element formulation. The geometry of the element, and the variation of the displacement w and the rotations θ_x and θ_y are exposed by the shape functions described in section 4.1.1 of Chapter 4. Thus, the geometry of the element is given by

$$x = \sum_{i=1}^{4} N_i\, x_i$$

$$y = \sum_{i=1}^{4} N_i\, y_i \tag{10.37}$$

The variation of displacements within an element is expressed in terms of the nodal values as

$$w = \sum_{i=1}^{4} N_i\, w_i$$

$$\theta_x = \sum_{i=1}^{4} N_i\, \theta_{xi}$$

$$\theta_y = \sum_{i=1}^{4} N_i\, \theta_{yi} \tag{10.38}$$

where the shape functions N_i in Eqs. 10.37 and 10.38 are given in Eq.4.26b as

$$N_i = \frac{1}{4}\,(1 + rr_i)\,(1 + ss_i)$$

where r_i and s_i are local coordinates r and s of the node i and $w_i, \theta_{xi}, \theta_{yi}$ are the values of $w, \theta_x,$ and θ_y at the node i.

The nodal displacement vector is given by

$$\{d\}^T = [\,w_1 \quad \theta_{x1} \quad \theta_{y1} \quad w_2 \quad \theta_{x2} \quad \theta_{y2} \quad w_3 \quad \theta_{x3} \quad \theta_{y3} \quad w_4 \quad \theta_{x4} \quad \theta_{y4}\,] \tag{10.39}$$

The Jacobian matrix $[J]$ and the relation between the derivatives of N_i with respect to r and s $\left(\frac{\partial N_i}{\partial r}, \frac{\partial N_i}{\partial s}\right)$ with regard to x and y, $\left(\frac{\partial N_i}{\partial x}, \frac{\partial N_i}{\partial y}\right)$ are derived in section 4.1.1 (Eqs. 4.3, 4.4, 4.5, 4.6 and 4.7). Thus the following relations can be obtained.

$$[J] = \begin{bmatrix} \dfrac{\partial N_1}{\partial r} & \dfrac{\partial N_2}{\partial r} & \dfrac{\partial N_3}{\partial r} & \dfrac{\partial N_4}{\partial r} \\ \dfrac{\partial N_1}{\partial s} & \dfrac{\partial N_2}{\partial s} & \dfrac{\partial N_3}{\partial s} & \dfrac{\partial N_4}{\partial s} \end{bmatrix} \begin{bmatrix} x_1 & y_1 \\ x_2 & y_2 \\ x_3 & y_3 \\ x_4 & y_4 \end{bmatrix} \qquad (10.40)$$

$$\begin{bmatrix} \dfrac{\partial N_1}{\partial x} & \dfrac{\partial N_2}{\partial x} & \dfrac{\partial N_3}{\partial x} & \dfrac{\partial N_4}{\partial x} \\ \dfrac{\partial N_1}{\partial y} & \dfrac{\partial N_2}{\partial y} & \dfrac{\partial N_3}{\partial y} & \dfrac{\partial N_4}{\partial y} \end{bmatrix} = [J]^{-1} \begin{bmatrix} \dfrac{\partial N_1}{\partial r} & \dfrac{\partial N_2}{\partial r} & \dfrac{\partial N_3}{\partial r} & \dfrac{\partial N_4}{\partial r} \\ \dfrac{\partial N_1}{\partial s} & \dfrac{\partial N_2}{\partial s} & \dfrac{\partial N_3}{\partial s} & \dfrac{\partial N_4}{\partial s} \end{bmatrix}$$
$$(10.41)$$

10.5.2   Strain-Displacement Matrix [B]

The element curvatures and shear deformation $\{\epsilon\}_p$ given in Eq. 10.31b and the nodal displacements $\{d\}$ given in Eq.10.39 are related by,

$$\{\epsilon\}_p = [B]\{d\} \qquad (10.42)$$

In order to compute the matrix $[B]$, we shall first differentiate Eq.10.38 with respect to x and y to get each of the terms of $\{\epsilon\}_p$ of Eq.10.40. Thus we get,

$$k_x = \sum_{i=1}^{4} \theta_{yi} \frac{\partial N_i}{\partial x}$$

$$k_y = \sum_{i=1}^{4} -\theta_{xi} \frac{\partial N_i}{\partial y}$$

$$k_{xy} = \sum_{i=1}^{4} \theta_{yi} \frac{\partial N_i}{\partial y} - \sum_{i=1}^{4} \theta_{xi} \frac{\partial N_i}{\partial x}$$

$$\phi_x = \sum_{i=1}^{4} w_i \frac{\partial N_i}{\partial x} + \sum_{i=1}^{4} \theta_{yi} N_i$$

$$\phi_y = \sum_{i=1}^{4} w_i \frac{\partial N_i}{\partial y} - \sum_{i=1}^{4} \theta_{xi} N_i$$

$$(10.43)$$

where $\frac{\partial N_i}{\partial x}$ and $\frac{\partial N_i}{\partial y}$ are to be computed using Eq.10.41. We observe from Eq.10.43 that the elements of the vector $\{\epsilon\}_p$ for curvatures and shear deformations are expressed in terms of the nodal displacements, w, θ_{xi} and θ_{yi}. Thus,

$$\{\epsilon\}_p = \left\{ \begin{array}{c} k_x \\ k_y \\ k_{xy} \\ \phi_x \\ \phi_y \end{array} \right\} = [B] \left\{ \begin{array}{c} w_1 \\ \theta_{x1} \\ \theta_{y1} \\ \cdots \\ \vdots \\ \cdots \\ w_4 \\ \theta_{x4} \\ \theta_{y4} \end{array} \right\} \tag{10.44}$$

and Eq.10.44 can be expressed as

$$\{\epsilon\}_p = \sum_{i=1}^{4} [B_i]\{d_i\} \tag{10.45}$$

where

$$[B_i] = \begin{bmatrix} 0 & 0 & \frac{\partial N_i}{\partial x} \\ 0 & -\frac{\partial N_i}{\partial y} & 0 \\ 0 & -\frac{\partial N_i}{\partial x} & \frac{\partial N_i}{\partial y} \\ \frac{\partial N_i}{\partial x} & 0 & N_i \\ \frac{\partial N_i}{\partial y} & -N_i & 0 \end{bmatrix}_{i=1,\ 2,\ 3,\ 4} \tag{10.46}$$

and

$$\{d_i\} = \left\{ \begin{array}{c} w_i \\ \theta_{xi} \\ \theta_{yi} \end{array} \right\} \tag{10.47}$$

The stress resultants, $\{\sigma\}_p$ of Eq.10.31 can be expressed in terms of nodal displacements after substituting for $\{\epsilon\}_p$ from Eq. 10.45. Thus,

$$\{\sigma\}_p = \left\{ \begin{array}{c} M_x \\ M_y \\ M_{xy} \\ Q_x \\ Q_y \end{array} \right\} = [C]_p \{\epsilon\}_p$$

$$= [C]_p \sum_{i=1}^{4} [B_i]\{d_i\} \tag{10.48}$$

or

$$\{\sigma\}_p = [C_p][B]\{d\} \tag{10.49}$$

where

$$[B] = \sum_{i=1}^{4} [B_i] \tag{10.50a}$$

$$[[B_1] \mid [B_2] \mid [B_3] \mid [B_4]] \qquad (10.50b)$$

$$(5 \times 3) \qquad (5 \times 3) \qquad (5 \times 3) \qquad (5 \times 3)$$

$[B] =$

$$
\begin{bmatrix}
0 & 0 & \frac{\partial N_1}{\partial x} & \Big| & 0 & 0 & \frac{\partial N_2}{\partial x} & \Big| & 0 & 0 & \frac{\partial N_3}{\partial x} & \Big| \\
0 & -\frac{\partial N_1}{\partial y} & 0 & \Big| & 0 & -\frac{\partial N_2}{\partial y} & 0 & \Big| & 0 & -\frac{\partial N_3}{\partial y} & 0 & \Big| \\
0 & -\frac{\partial N_1}{\partial x} & \frac{\partial N_1}{\partial y} & \Big| & 0 & -\frac{\partial N_2}{\partial x} & \frac{\partial N_2}{\partial y} & \Big| & 0 & -\frac{\partial N_3}{\partial x} & \frac{\partial N_3}{\partial y} & \Big| \\
\frac{\partial N_1}{\partial x} & 0 & N_1 & \Big| & \frac{\partial N_2}{\partial x} & 0 & N_2 & \Big| & \frac{\partial N_3}{\partial x} & 0 & N_3 & \Big| \\
\frac{\partial N_1}{\partial y} & -N_1 & 0 & \Big| & \frac{\partial N_2}{\partial y} & -N_2 & 0 & \Big| & \frac{\partial N_3}{\partial y} & -N_3 & 0 & \Big|
\end{bmatrix}
$$

$$
\begin{bmatrix}
0 & 0 & \frac{\partial N_4}{\partial x} \\
0 & -\frac{\partial N_4}{\partial y} & 0 \\
0 & -\frac{\partial N_4}{\partial x} & \frac{\partial N_4}{\partial y} \\
\frac{\partial N_4}{\partial x} & 0 & N_4 \\
\frac{\partial N_4}{\partial y} & -N_4 & 0
\end{bmatrix} \qquad (10.51)
$$

Using the expression for $[C]_p$ from Eq. 10.30a and for $[B_i]$ from Eq. 10.46 the product of $[C]_p \, [B]$ can be computed as,

$$[C]_p \, [B] = [C]_p \sum_{i=1}^{4} [B_i]$$

$$= [[CB_1] \quad [CB_2] \quad [CB_3] \quad [CB_4]] \qquad (10.52)$$

$$(5 \times 3) \quad (5 \times 3) \quad (5 \times 3) \quad (5 \times 3)$$

$$(5 \times 12)$$

where the submatrices $[CB_i], i = 1, 2, 3, 4$ are given by

$$
[CB_i] = \frac{Eh}{12(1+\mu)}
\begin{bmatrix}
0 & \frac{-\mu h^2}{1-\mu}\left(\frac{\partial N_i}{\partial y}\right) & \frac{h^2}{1-\mu}\left(\frac{\partial N_i}{\partial x}\right) \\
0 & \frac{-h^2}{1-\mu}\left(\frac{\partial N_i}{\partial y}\right) & \frac{\mu h^2}{1-\mu}\left(\frac{\partial N_i}{\partial y}\right) \\
0 & \frac{-h^2}{2}\left(\frac{\partial N_i}{\partial x}\right) & \frac{h^2}{2}\left(\frac{\partial N_i}{\partial y}\right) \\
6\alpha\left(\frac{\partial N_i}{\partial x}\right) & 0 & 6\alpha N_i \\
6\alpha\left(\frac{\partial N_i}{\partial y}\right) & -6\alpha N_i & 0
\end{bmatrix} \qquad (10.53)
$$

$$5 \times 3$$

Separating the bending and shear terms, we get,

$$[CB_i] = \frac{Eh}{12(1+\mu)}$$

$$\times \left[\begin{bmatrix} 0 & \frac{-\mu h^2}{1-\mu}\left(\frac{\partial N_i}{\partial y}\right) & \frac{h^2}{1-\mu}\left(\frac{\partial N_i}{\partial x}\right) \\ 0 & \frac{-h^2}{1-\mu}\left(\frac{\partial N_i}{\partial y}\right) & \frac{\mu h^2}{1-\mu}\left(\frac{\partial N_i}{\partial y}\right) \\ 0 & \frac{-h^2}{2}\left(\frac{\partial N_i}{\partial x}\right) & \frac{h^2}{2}\left(\frac{\partial N_i}{\partial y}\right) \\ 0 & 0 & 0 \\ 0 & 0 & 0 \end{bmatrix} + 6\alpha \begin{bmatrix} 0 & 0 & 0 \\ 0 & 0 & 0 \\ 0 & 0 & 0 \\ \frac{\partial N_i}{\partial x} & 0 & N_i \\ \frac{\partial N_i}{\partial y} & -N_i & 0 \end{bmatrix} \right] \quad (10.54)$$

$$(5 \times 3) \qquad\qquad (5 \times 3)$$

or in compact form the above equation can be expressed as

$$[CB_i] = [CB_i]_b + [CB_i]_s \quad (10.55)$$

where $[CB_i]_b$ and $[CB_i]_s$ are the bending and shear contributions to the stress displacement matrix $[C]_p\,[B]$. The bending contribution $[CB]_b$ of Eq. 10.55 is evaluated at 2×2 Gauss points and the shear contribution $[CB]_s$ is evaluated at 1×1 Gauss Point.

10.5.3   Element Stiffness Matrix

The element stiffness matrix $[k]$ is given by

$$[k] = \int\int_A [B]^T \qquad [C]_p \qquad [B]\; dx\; dy \quad (10.56)$$

$$(12 \times 12) \qquad (12 \times 5) \quad (5 \times 5) \quad (5 \times 12)$$

The above expression in local coordinates is written as

$$[k] = \int\int_A [B]^T\,[C]_p\,[B]\,|J|\; dr\; ds \quad (10.57)$$

Using the expression for $[C_p]$ from Eq.10.30 and for $[B_i]$ from Eq.10.46, the product $[[B]^T\,[C]_p\,[B]]$ can be written as,

$$[\bar{k}] = [B]^T\,[C]_p\,[B] = \begin{bmatrix} [\bar{k}_{11}] & [\bar{k}_{12}] & [\bar{k}_{13}] & [\bar{k}_{14}] \\ [\bar{k}_{21}] & [\bar{k}_{22}] & [\bar{k}_{23}] & [\bar{k}_{24}] \\ [\bar{k}_{33}] & [\bar{k}_{32}] & [\bar{k}_{33}] & [\bar{k}_{34}] \\ [\bar{k}_{41}] & [\bar{k}_{42}] & [\bar{k}_{43}] & [\bar{k}_{44}] \end{bmatrix} \quad (10.58)$$

where

$$[\bar{k}_{ij}] = \frac{Eh}{12(1+\mu)}$$

$$\times \begin{bmatrix} 6\alpha\left[\frac{\partial N_i}{\partial x}\frac{\partial N_j}{\partial x} + \frac{\partial N_i}{\partial y}\frac{\partial N_j}{\partial y}\right] & -6\alpha\frac{\partial N_i}{\partial y}N_j & 6\alpha\frac{\partial N_i}{\partial y}N_j \\ & \cdots & \cdots \\ & \frac{h^2}{1-\mu}\left[\frac{\partial N_i}{\partial y}\frac{\partial N_j}{\partial y}\right] & -\frac{\mu h^2}{1-\mu}\left(\frac{\partial N_i}{\partial y}\frac{\partial N_j}{\partial x}\right) \\ -6\alpha N_i\frac{\partial N_j}{\partial y} & +\frac{h^2}{2}\left[\frac{\partial N_i}{\partial x}\frac{\partial N_j}{\partial x}\right] & -\frac{h^2}{2}\left(\frac{\partial N_i}{\partial x}\frac{\partial N_i}{\partial y}\right) \\ & +6\alpha N_i N_j & \\ \cdots & \cdots & \cdots \\ & -\frac{\mu h^2}{1-\mu}\left(\frac{\partial N_i}{\partial x}\frac{\partial N_j}{\partial y}\right) & \frac{h^2}{1-\mu}\left(\frac{\partial N_i}{\partial x}\frac{\partial N_j}{\partial x}\right) \\ 6\alpha N_i\frac{\partial N_j}{\partial x} & -\frac{h^2}{2}\left(\frac{\partial N_i}{\partial y}\frac{\partial N_j}{\partial x}\right) & +\frac{h^2}{2}\left(\frac{\partial N_i}{\partial y}\frac{\partial N_j}{\partial y}\right) \\ & & +6\alpha N_i N_j \end{bmatrix}$$

(10.59)

The size of each submatrix in Eq. 10.58 is 3 × 3. The submatrix $[\bar{k}_{11}]$ is computed by substituting $i = j = 1$ in Eq. 10.59. Similarly, $[\bar{k}_{12}], [\bar{k}_{13}]$, and $[\bar{k}_{14}]$ are generated by substituting $i = 1,\ j = 2, 3, 4$ and $j = 1, i = 4$ respectively. Thus the first three rows of the matrix $[\bar{k}]$ is generated by assigning $i = 1,\ j = 1, 2, 3, 4$ to the equation 10.59. Similarly, the fourth, fifth and sixth rows of $[\bar{k}]$ are generated by substituting $i = 2,\ j = 1, 2, 3, 4$. Likewise substituting $i = 3,\ j = 1, 2, 3, 4$ and $i = 4,\ j = 1, 2, 3, 4$ in Eq.10.59, the seventh to ninth rows and tenth to twelvth rows of $[\bar{k}]$ can be obtained.

At this stage it is required to separate out bending and shear energy terms in Eq.10.59 and it can easily be done as the shear energy terms are all associated with the shear correction factor α. This is an important step in stiffness computation as the numerical integration scheme to be adopted to evaluate bending and shear energy contributions to the total element stiffness is not the same. Thus, separating bending and shear contribution in Eq.10.59, we get Eq. 10.60 and 10.60a.

The matrix $[\bar{k}]$ can now be written as the sum of bending and shear contributions.

$$[\bar{k}] = [\bar{k}]_b + [\bar{k}]_s \tag{10.61}$$

$$[\bar{k}]_b = \begin{bmatrix} [\bar{k}_{11}]_b & [\bar{k}_{12}]_b & [\bar{k}_{13}]_b & [\bar{k}_{14}]_b \\ [\bar{k}_{21}]_b & [\bar{k}_{22}]_b & [\bar{k}_{23}]_b & [\bar{k}_{24}]_b \\ [\bar{k}_{33}]_b & [\bar{k}_{32}]_b & [\bar{k}_{33}]_b & [\bar{k}_{34}]_b \\ [\bar{k}_{41}]_b & [\bar{k}_{42}]_b & [\bar{k}_{43}]_b & [\bar{k}_{44}]_b \end{bmatrix} \tag{10.62}$$

$$[\bar{k}_{ij}] = \frac{Eh}{12(1+\mu)} \begin{bmatrix} 0 & 0 & 0 \\[2mm] \dfrac{h^2}{1-\mu}\left(\dfrac{\partial N_i}{\partial y}\dfrac{\partial N_i}{\partial y}\right) & +\dfrac{h^2}{2}\left(\dfrac{\partial N_i}{\partial x}\dfrac{\partial N_i}{\partial x}\right) & -\dfrac{\mu h^2}{1-\mu}\left(\dfrac{\partial N_i}{\partial y}\dfrac{\partial N_i}{\partial x}\right) & \dfrac{h^2}{2}\left(\dfrac{\partial N_i}{\partial y}\dfrac{\partial N_i}{\partial y}\right) & 0 \\[2mm] -\dfrac{\mu h^2}{1-\mu}\left(\dfrac{\partial N_i}{\partial x}\dfrac{\partial N_i}{\partial y}\right) & -\dfrac{h^2}{2}\left(\dfrac{\partial N_i}{\partial y}\dfrac{\partial N_i}{\partial x}\right) & \dfrac{h^2}{1-\mu}\left(\dfrac{\partial N_i}{\partial x}\dfrac{\partial N_i}{\partial x}\right) & +\dfrac{h^2}{2}\left(\dfrac{\partial N_i}{\partial y}\dfrac{\partial N_i}{\partial y}\right) \end{bmatrix}$$

$$+\,6\alpha \begin{bmatrix} \left(\dfrac{\partial N_i}{\partial x}\dfrac{\partial N_i}{\partial x} + \dfrac{\partial N_i}{\partial y}\dfrac{\partial N_i}{\partial y}\right) & -N_i\dfrac{\partial N_i}{\partial y} & N_i\dfrac{\partial N_i}{\partial x} \\[2mm] -\dfrac{\partial N_i}{\partial y}N_j & N_i\,N_j & 0 \\[2mm] \dfrac{\partial N_i}{\partial x}N_j & 0 & N_i\,N_j \end{bmatrix} \qquad (10.60)$$

$$= [\bar{k}_{ij}]_b + [\bar{k}_{ij}]_s \qquad i = 1 \text{ to } 4, \quad j = 1 \text{ to } 4 \qquad (10.60a)$$

$$(3 \times 3) \quad (3 \times 3)$$

The matrix $[\bar{k}]_s$ can be expressed in a similar manner. The size of each submatrix $[\bar{k}_{ij}]_b$ or $[\bar{k}_{ij}]_s$ is 3 × 3 and they are computed exactly following the procedure indicated for the Eq.10.59.

Substituting $[\bar{k}]$ for $[B]^T [C]_p [B]$ in Eq.10.57 the stiffness matrix $[k]$ is given by

$$[k] = \int_{-1}^{+1} \int_{-1}^{+1} [\bar{k}]|J| \ dr \ ds$$

Substituting from Eq.10.61,

$$[k] = \int_{-1}^{+1} \int_{-1}^{+1} [[\bar{k}]_b + [\bar{k}]_s] \ |J| \ dr \ ds \qquad (10.63)$$

Gauss quadrature is used to compute the stiffness matrix [k] and the procedure is well explained in Chapter 4. The mesh 'locking' due to shear stiffness is described in section 7.5. Following the explanation given therein the bending stiffness contribution $[[\bar{k}]_b]$ is evaluated using 2×2 Gauss quadrature and the shear stiffness contribution $[[\bar{k}]_s]$ is calculated using 1 × 1 rule. These are indicated in Fig.10.6.

10.5.4 Element Nodal Load Vector

The nodal load $\{Q_i\}$ at the node i for a uniformly distributed load q is given by

$$\{Q_i\} = \left\{ \begin{matrix} F_z \\ M_x \\ M_y \end{matrix} \right\} = \int \int_A N_i \left\{ \begin{matrix} q \\ 0 \\ 0 \end{matrix} \right\} |J| \ dr \ ds \qquad (10.64)$$

Combining the nodal load vectors $\{Q_i\}$, the element load vector $\{Q\}$ can be numerically evaluated as,

$$\{Q\} = \begin{Bmatrix} Q_1 \\ Q_2 \\ Q_3 \\ \dots \\ Q_4 \\ Q_5 \\ Q_6 \\ \dots \\ Q_7 \\ Q_8 \\ Q_9 \\ \dots \\ Q_{10} \\ Q_{11} \\ Q_{12} \end{Bmatrix} = q \sum_{i=1}^{n} \sum_{j=1}^{n} W_i W_j |J| \begin{Bmatrix} N_1 \\ 0 \\ 0 \\ \dots \\ N_2 \\ 0 \\ 0 \\ \dots \\ N_3 \\ 0 \\ 0 \\ \dots \\ N_4 \\ 0 \\ 0 \end{Bmatrix} \qquad (10.65)$$

Gauss quadrature with 2×2 sampling points is used to evaluate Eq.10.65.

10.6 Eight Noded Isoparametric Element (Plate8)

In the previous section a simple four noded isoparametric element has been described. Following a similar approach, a higher order eight noded element has been presented by Hinton and Owen [20]. The element is briefly described here and is included in PASSFEM library as PLATE8 element. The element can be used for moderately thick and thin plates, and also for modelling curved boundaries.

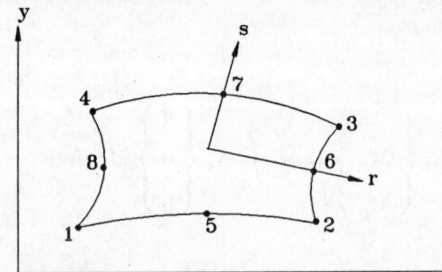

Fig. 10.7 Eight noded isoparametric element

The element shown in Fig.10.7 is an eight noded isoparametric element. The element has eight nodes and is exactly of the same type of element used for plane stress/strain analysis described in subsection 8.3.2. Hence the shape functions given in Eq.3.59 can be used to describe the geometry and displacement variation.

As in the case of PLATE4 element, Mindlin's theory is used in the element formulation to include shear deformation. Hence, the displacements at any point in an element are the lateral displacement w and rotations about x and y axes. θ_x and θ_y. Hence the nodal displacement vector is given by

$$\{d\}^T = [w_1 \quad \theta_{x1} \quad \theta_{y1} \quad w_2 \quad \theta_{x2} \quad \theta_{y2} \ldots w_8 \quad \theta_{x8} \quad \theta_{y8}] \tag{10.66}$$

Following the isoparametric element concept the geometry of the element is described by

$$x = \sum_{i=1}^{8} N_i x_i$$

$$y = \sum_{i=1}^{8} N_i y_i \tag{10.67}$$

where the shape functions $N_i (i = 1, ..., 8)$ are given by Eq. 3.59.

The same shape functions are used to express the variation of displacement in an element as,

$$w = \sum_{i=1}^{8} N_i w_i$$

$$\theta_x = \sum_{i=1}^{8} N_i \theta_{xi}$$

$$\theta_y = \sum_{i=1}^{8} N_i \theta_{yi} \tag{10.68}$$

As explained in the section 10.5., the Jacobian matrix needed for computing the derivatives of shape functions with respect to local and global axes can be shown to be,

$$[J] = \begin{bmatrix} \dfrac{\partial N_1}{\partial r} & \dfrac{\partial N_2}{\partial r} & \cdots & \dfrac{\partial N_8}{\partial r} \\[2ex] \dfrac{\partial N_1}{\partial s} & \dfrac{\partial N_2}{\partial s} & \cdots & \dfrac{\partial N_8}{\partial s} \end{bmatrix} \begin{bmatrix} x_1 & y_1 \\ x_2 & y_2 \\ \cdot & \cdot \\ \cdot & \cdot \\ \cdot & \cdot \\ x_8 & y_8 \end{bmatrix} \tag{10.69}$$

Now the derivatives with respect to global axis are found by inverting the Jacobian [J] and is given by

$$
\begin{bmatrix}
\dfrac{\partial N_1}{\partial x} & \dfrac{\partial N_2}{\partial x} & \cdots\cdots & \dfrac{\partial N_8}{\partial x} \\[2ex]
\dfrac{\partial N_1}{\partial y} & \dfrac{\partial N_2}{\partial y} & \cdots\cdots & \dfrac{\partial N_8}{\partial y}
\end{bmatrix}
= [J]^{-1}
\begin{bmatrix}
\dfrac{\partial N_1}{\partial r} & \dfrac{\partial N_2}{\partial r} & \cdots\cdots & \dfrac{\partial N_8}{\partial r} \\[2ex]
\dfrac{\partial N_1}{\partial s} & \dfrac{\partial N_2}{\partial s} & \cdots\cdots & \dfrac{\partial N_8}{\partial s}
\end{bmatrix}
\tag{10.70}
$$

10.6.1   Computation of Stiffness Matrix and Nodal Load Vector

The relation between the element curvatures and shear deformation, $\{\epsilon\}_p$ and the nodal displacements $\{d\}$ are given in Eq.10.42.

Differentiating Eq. 10.68 with respect to x and y, the curvatures and shear deformations can be expressed in terms of nodal displacements through the shape functions. Thus, we get

$$
k_x = \sum_{i=1}^{8} \theta_{yi}\frac{\partial N_i}{\partial x}
$$

$$
k_y = -\sum_{i=1}^{8} \theta_{xi}\frac{\partial N_i}{\partial y}
$$

$$
k_{xy} = \sum_{i=1}^{8} \theta_{yi}\frac{\partial N_i}{\partial y} - \sum_{i=1}^{8} \theta_{xi}\frac{\partial N_i}{\partial x}
$$

$$
\phi_x = \sum_{i=1}^{8} w_i\frac{\partial N_i}{\partial x} + \sum_{i=1}^{8} \theta_{yi} N_i
$$

$$
\phi_y = \sum_{i=1}^{8} w_i\frac{\partial N_i}{\partial y} - \sum_{i=1}^{8} \theta_{xi} N_i
$$

$$
\tag{10.71}
$$

where $\frac{\partial N_i}{\partial x}$ and $\frac{\partial N_i}{\partial y}$ are given by Eq. 10.70.

Substituting for $\{\epsilon\}_p$ from Eq. 10.71 into Eq. 10.42.

$$\{\epsilon\}_p = \left\{ \begin{array}{c} k_x \\ k_y \\ k_{xy} \\ \phi_x \\ \phi_y \end{array} \right\} = [B] \left\{ \begin{array}{c} w_1 \\ \theta_{x1} \\ \theta_{y1} \\ --- \\ \cdot \\ \cdot \\ --- \\ w_8 \\ \theta_{x8} \\ \theta_{y8} \end{array} \right\} \tag{10.72}$$

i.e

$$\{\epsilon\}_p = \sum_{i=1}^{8} [B_i] \{d_i\} \tag{10.73}$$

where $[B_i]$ and $\{d_i\}$ are given by Equation 10.46 and 10.47 respectively.

$$[B_i] = \begin{bmatrix} 0 & 0 & \frac{\partial N_i}{\partial x} \\ 0 & -\frac{\partial N_i}{\partial y} & 0 \\ 0 & -\frac{\partial N_i}{\partial x} & \frac{\partial N_i}{\partial y} \\ \frac{\partial N_i}{\partial x} & 0 & N_i \\ \frac{\partial N_i}{\partial y} & -N_i & 0 \end{bmatrix} \tag{10.73a}$$

and

$$\{d_i\} = \left\{ \begin{array}{c} w_i \\ \theta_{xi} \\ \theta_{yi} \end{array} \right\} \tag{10.73b}$$

The stress resultants $\{\sigma\}_p$ are given by,

$$\{\sigma\}_p = \left\{ \begin{array}{c} M_x \\ M_y \\ M_{xy} \\ Q_x \\ Q_y \end{array} \right\} = [C]_p \{\epsilon\}_p = [C]_p \sum_{i=1}^{8} [B_i]\{d_i\} \tag{10.74}$$

where $[C]_p$ is computed from Eq. 10.30 and $[B]_i$ and $\{d_i\}$ are given by Eq. 10.73. The derivatives of N_i with respect to x and y are determined using Eq. 10.70.

The stiffness **matrix** of the element is given by,

$$[k] = \int_{-1}^{+1} \int_{-1}^{+1} [B]^T \, [C]_p \, [B] \, |J| \, dr \, ds \qquad (10.75)$$

Gauss quadrature is used to evaluate the stiffness matrix. It has been found that a reduced integration technique of 2×2 Gauss rule gives satisfactory results for thin plate problems [17].

For an uniformly distributed load q the nodal load $\{Q_i\}$ at a node i is given by Eq. 10.64.

$$\{Q_i\} = \left\{ \begin{array}{c} F_z \\ M_x \\ M_y \end{array} \right\} = \int \int N_i \left\{ \begin{array}{c} q \\ 0 \\ 0 \end{array} \right\} |J| dr \, ds \qquad (10.64)$$

Now the element load vector for the eight noded element can be computed by numerical integration as given by the following expression.

$$\left\{ \begin{array}{c} Q_1 \\ Q_2 \\ Q_3 \\ --- \\ \cdot \\ \cdot \\ --- \\ Q_{22} \\ Q_{23} \\ Q_{24} \end{array} \right\} = q \sum_{i=1}^{n} \sum_{i=1}^{n} W_i W_j |J| \left\{ \begin{array}{c} N_1 \\ 0 \\ 0 \\ --- \\ \cdot \\ \cdot \\ --- \\ N_8 \\ 0 \\ 0 \end{array} \right\} \qquad (10.76)$$

10.6.2   Selective/Reduced Integration and Behaviour of Plate Elements

In the earlier section, it was shown that as $\frac{a}{h} \to \infty$, the energy due to shear must vanish and the Kirchhoff constraints in Eq.10.35 are to be enforced for the analysis of thin plates. In the finite element models based on Mindlin's theory, the Kirchhoff constraints are approximated within each element according to the assumed form of the shape functions. As it has been pointed out in the earlier section 10.4.3, intensive research has been conducted to alleviate the problems associated with shear locking and methods have been proposed on the use of selective and reduced integration techniques. While such integration techniques have been shown to yield excellent results in many cases, the analysis using exact integration orders gave erroneous results. However, such inexact integration results in rank deficient element stiffness matrix. Averill and Reddy [21] have conducted a detailed investigation on the behaviour of plate elements and presented a new analytical technique to determine a priori whether a given element will lock when used to model thin structures.

The selective or reduced integration technique introduces spurious modes in an element referred to as *mechanism, hourglass mode* or *zero-energy mode*. The eigenvalue test helps, to detect zero-energy modes, lack of invariance and absence of rigid body modes in an element. The test is briefly described below.

The equilibrium equation of an element is given by Eq.3.

$$[k]\{d\} = \{Q\}$$

Let us assume the element nodal load $\{Q\}$ to be proportional to element nodal displacements $\{d\}$ as $\lambda\{d\}$. Thus,

$$[k]\{d\} = \{Q\} = \lambda\{d\} \quad \text{or}$$

$$[[k] - \lambda[I]]\{d\} = \{0\} \tag{10.77}$$

The above equation represents an eigen problem and the values of λ_i are called eigenvalues of $[k]$. The number of eigenvalues $\lambda_i s$ is equal to the number of element degrees of freedom ($\{d\}$). For each value of λ_i there is a corresponding $\{d\}_i$, called eigen vector. If each $\{d\}_i$ is normalised such that $\{d\}_i^T\{d_i\} = 1$, premultiplication of Eq.10.77 by $\{d\}_i^T$ yields,

$$\{d_i\}^T[k]\{d_i\} = \lambda_i \tag{10.78}$$

As observed in Chapter 3, the strain energy U of an element can be expressed as $U = \frac{1}{2}\{d\}^T[k]\{d\}$. Then the Eq.10.78 reduces to,

$$2U_i = \lambda_i \tag{10.79}$$

Equation 10.79 shows that when the eigenvalue $\lambda_i = 0$ (of the stiffness matrix [k]), the strain energy is zero. The situation may correspond to nodal displacement $\{d_i\}$ due to rigid body motion. Also the above equation suggests that mechanisms may also yield zero eigen values. Since the stiffness matrix is computed numerically by evaluating at each Gauss point the value of $[B]^T[C][B]^T|J|$, it indicates that the deformation mode causes zero strain at the Gauss points.

Thus, in testing an element, we can compute the eigen values of $[k]$. The stiffness matrix $[k]$ should have as many $\lambda_i = 0$ values as they are needed to represent the required rigid body modes for the element. If they are less, the element lacks the desired capability for

representing rigid body motion without strain. If they are more than
the required number, it suggests that the element has one or more
mechanisms or zero-energy modes.

Averill and Reddy [21] have analysed various plate elements, i.e.,
four noded bilinear, nine noded biquadrature, eight noded serendipity and Heterosis elements. It was shown that the number of shear
constraints imposed on an element is equal to the rank of the shear
stiffness matrix $[k]_s$. The rank is found by subtracting the number of zero eigen values from the order of the matrix. As the plate
element has three rigid body modes, the element stiffness matrix
$[k]$ $([k]_b + [k]_s)$ should contain three zero eigenvalues corresponding
to these rigid body modes. Any additional zero eigenvalues indicate
presence of zero energy modes or mechanisms. For example, for four
noded plate element, the number of zero eigen values of $[k]_b + [k]_s$
under selective integration is 5. Thus, it indicates two mechanisms
and they are inplane twist and w - hourglass mode as shown in Fig.
10.8.

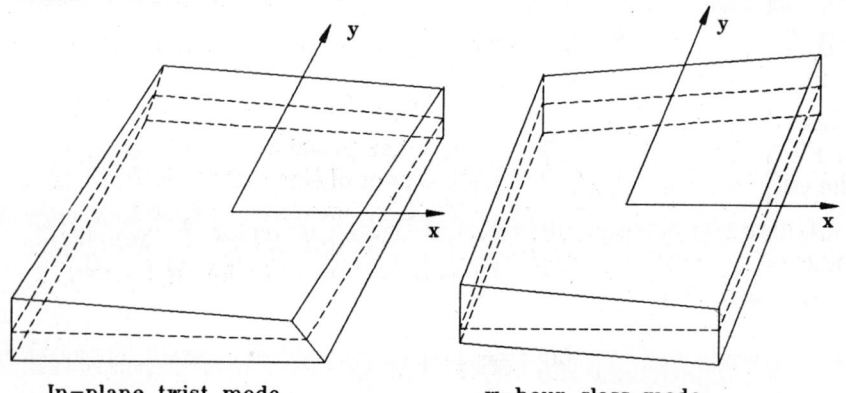

In-plane twist mode   w-hour glass mode

Fig. 10.8 Two mechanisms of four noded element

The inplane twist mode seldom causes difficulties since it is ap-
parently eliminated upon assembly of elements and imposition of
support restraints. However, the hourglass mode pertaining to w
displacements affects the performance of the element under lightly
constrained boundaries, for example, plate supported at corners only.
These problems are investigated by various authors and methods are
also suggested to control mechanisms. [22,23]

The eight noded serendipity element exhibits one mechanism under reduced 2×2 integration but this mechanism does not propogate
in an assembled mesh. It has been observed in clamped boundary
conditions, this element produces over-stiff solutions when coarse
meshes are used. However, when integrated selectively using 3×3
scheme for $[k]_b$ and 2×2 for $[k]_s$ it exhibits no mechanism. It gives

acceptable results for this plate situations for values of $h/l = 0.01$.

The 'heterosis' element is reported to be one of the best elements [24]. It is a nine noded element similar to Lagrange type but it has only 26 degrees of freedom - three degrees of freedom at the boundary nodes but only θ_x and θ_y at the central node. The selective integration scheme of 3×3 for $[k]_b$ and 2×2 for $[k]_s$ is adopted. It does not have mechanism, locking is avoided and in general, the element gives good results. However, the element is relatively uneconomical due to additional computing time needed for static condensation of the central node and for the mixed integration scheme.

Readers can refer to the recent publications [25,26,27,28] dealing with various aspects of shear deformation problems, integration schemes, patch tests and development of *robust* elements for thick and thin plates based on Reissner-Mindlin Theory.

10.7 Subroutine PLATE

This control routine calls different plate element routines as shown in Fig. 10.9. The following two shapes of the elements, described in the earlier sections 10.5 and 10.6, are included in the program.

1. Shape 1-8 noded isoparametric element - PLATE8
2. Shape 2-4 noded (bilinear) plate bending element PLATE4

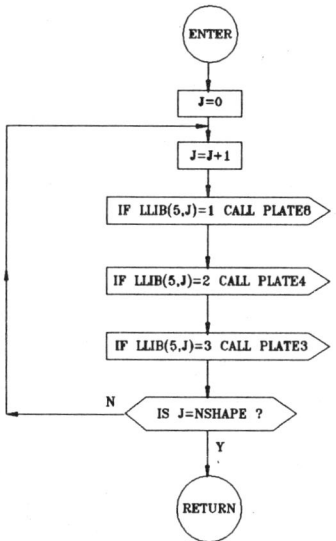

Fig. 10.9 Subroutine PLATE

10.7.1   Organisation of the Routines PLATE8   and PLATE4

The general guidelines for adding an element routine is explained in Chapter 6. Each element shape is called thrice by the main program depending on the value of the flag IND.

When IND equals one, the first segment of the element routine (PLATE8 or PLATE4) is evoked. The nodal point data and material property information are obtained through the argument list. The number of 8-noded or 4-noded elements is to be read as input. Then for each element its nodal connectivity (LNC), material group (MG), the thickness (TH) and the uniformly distributed load (UDL) are read. The nodal degree of freedom array (ND), is generated for each element and is stored on NDARAY file in a sequential manner. The subroutine COLUMH is called to calculate the column heights of the global stiffness matrix.

When IND=2 the control transfers to the second segment of the element routine. At this stage in addition to the nodal data and material properties the addresses of the stiffness matrix, and the load vector are also obtained from the argument list. The ND array for each element is read from the scratch file, NDARAY. For each Gauss point the stress-displacement matrix $[CB]$ is calculated and is written on the ISTRES file. Next, the element stiffness matrix is calculated and is passed on to the subroutine PASSEM for assembling element stiffness matrix. The element load vector is calculated and added to the global load vector.

After obtaining the global displacements, IND is set equal to 3 and control is transferred to the third segment of the element routine. At this stage the ND array and the stress-displacement matrix are read from the NDARAY and ISTRES files respectively. Then for each element the stress resultants at the 2×2 Gauss points are computed. Then by using the bilinear extrapolation explained in the next section, the stress resultants at the nodes are calculated and printed out.

The listing of the program is given in Appendix 7.

10.7.2   Stress Smoothing Technique

In the finite element analysis using displacement method, the stresses are discontinuous between elements, because of the nature of the assumed displacement variation. A typical stress distribution is shown in Fig. 10.10a.

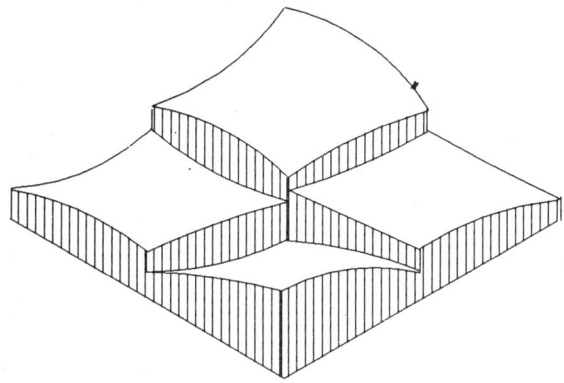

(a)   Unsmoothed stress distribution

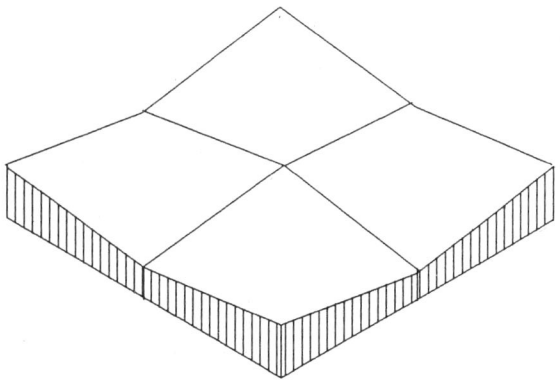

(b) Stress distribution after smoothing

Fig. 10.10 Local Stress Smoothing

Experience has shown that in the case of isoparametric elements the integration (Gauss) points are the best stress sampling points. The nodes, which are the most useful points for output and interpretation of stresses, appear to be the worst sampling points [3]. However, it has been observed that shape function derivatives (and hence stresses) evaluated at interior of the element is more accurate than those calculated on the element boundary.

Many analysts have taken the nodal average of stresses, i.e., average of the nodal stresses of all the elements meeting at a common node. This economic and simple solution works very well on the whole but does not consider the effect of the size of the adjacent elements.

Barlow [29] has shown that for two-dimensional isoparametric elements the 2×2 Gauss integration (sampling) points are the optimal sampling points. Further Hinton and Campbell [30] have demonstrated that a technique known as 'local stress smoothing' is the natural method for sampling stresses in finite element using reduced integration. In this context, local smoothing is simply a bilinear extrapolation of the stress values computed at the Gauss points. By using the least squares fit the following expression has been obtained to get the smoothed corner node stresses [30].

$$
\begin{Bmatrix} \overline{\sigma}_1 \\ \overline{\sigma}_2 \\ \overline{\sigma}_3 \\ \overline{\sigma}_4 \end{Bmatrix} = \begin{bmatrix} \left(1+\frac{\sqrt{3}}{2}\right) & -\frac{1}{2} & \left(1-\frac{\sqrt{3}}{2}\right) & -\frac{1}{2} \\ -\frac{1}{2} & \left(1+\frac{\sqrt{3}}{2}\right) & -\frac{1}{2} & \left(1-\frac{\sqrt{3}}{2}\right) \\ \left(1-\frac{\sqrt{3}}{2}\right) & -\frac{1}{2} & \left(1+\frac{\sqrt{3}}{2}\right) & -\frac{1}{2} \\ -\frac{1}{2} & \left(1-\frac{\sqrt{3}}{2}\right) & -\frac{1}{2} & \left(1+\frac{\sqrt{3}}{2}\right) \end{bmatrix} \begin{Bmatrix} \sigma_I \\ \sigma_{II} \\ \sigma_{III} \\ \sigma_{IV} \end{Bmatrix}
$$

$$(10.80)$$

where $\overline{\sigma}_1$ to $\overline{\sigma}_4$ are the smoothed nodal values and $\overline{\sigma}_I$ to $\overline{\sigma}_{IV}$ are the stresses at the 2×2 Gauss points shown in Fig. 10.6c.

In the case of selective integration, the transverse shear stresses may be greatly in error except at the appropriate Gauss points. Hence, for the four noded element, the shear stresses τ_{xz} and τ_{yz} are calculated at the centre of the element and it is assumed that they are constant throughout the element. In the case of eight noded element they are calculated at the 2×2 Gauss points and extrapolated using the above equation (10.80).

In the computer program, the smoothed stress resultants are then modified by finding the average of the nodal stress resultants of all the elements meeting at a common node.

10.7.3   Input Details for Plate Element

The input details for plate element are given below. This data set pertains to the data sets 5 and 6 required for PASSFEM described under section 6.11.

5. ELEMENT TYPE DATA (5I5)

      Columns   1-5   Element type number or row number of the element library (LTYPE). For plate elements LTYPE = 5.

6-10   Number of shapes (NSHAPE) belonging to the element type with a minimum of 1 to a maximum of 3.

11-15   Element shape number or column number of the element library (element shape number is 1 for PLATE8 and 2 for PLATE4).

16-20   Element shape number. If NSHAPE is more than one, the second shape number is typed in this column. Otherwise this column is left blank.

21-25   Element shape number. If NSHAPE=3, the third shape number is typed in this column.

6. ELEMENT DATA SET

This data set is element shape dependent and the following two sets of data must be supplied for each shape.

A. *Control Data (I5)*

Columns   1-5   Number of elements (NEL)

B. *Element Data (2I5, 4I5, 2F10.4, 3X, I2)*

One line per element with the following information is required.

Columns   1-5   Element Number (LNUM)

6-10   Material Group of the Element (MG)

11-30   Node numbers of the element in the order shown in Fig.10.6 for PLATE4 (LNC) and Fig. 10.7 for PLATE8.

31-40   Thickness of the element (TH)

41-50   Intensity of the uniformly distributed load on the element (UDL)

54-55   Element data generator (KI) for PLATE4.

Element data set must be in element number sequence. In the case of PLATE4 automatic generation is possible, i.e., if the node numbers of a series of elements increase by certain specified number, KI, the omitted elements can be generated. This generation is possible only if all the elements in a particular series have the same material identification number, thickness and UDL given in the first line of the series. The element data generator KI should be given on the second line which gives the data for the last element of the series.

10.8   Examples

Several numerical examples which have become more or less standard
ones for evaluating the plate bending elements are presented in this
section. The examples are solved using the program described in the
earlier section and are also compared with the results obtained from
SAP IV package. The examples are categorized under (i) thin plates,
and (ii) thick plates.

10.8.1   Examples of Thin Plates

Table 10.1 gives the details of the examples selected to study the
behaviour of the plate elements described in the earlier sections 10.5
and 10.6.

The details of discretization of square and circular plates for analysis using PLATE4 and PLATE8 are given in Figs. 10.11, 10.12,
10.13 and 10.14. It may be noted that because of symmetry, only
one quadrant of the plate is considered for analysis. All the above
examples are analysed for the following two load cases:

(i) uniformly distributed lateral load of 5.88 kN/m^2 (the live load
on the plate is $3.0 kN/m^2$ and self weight is 2.88 kN/m^2.

(ii) A central concentrated load of 16 kN along with the dead load
of 2.88 kN/m^2.

Table 10.1 Examples of Thin Plates

Example No.	Shape of the plate	Dimension of the plate	Boundary conditions
1(a)	Square	$L \times L \times h$ ($400 \times 400 \times 12$cm)	Simply supported
1(b)	Square	,, ,,	Clamped
2(a)	Circular	Radius=a=150 cm thickness=h=12 cm	Simply supported
2(b)	Circular	,, ,,	Clamped

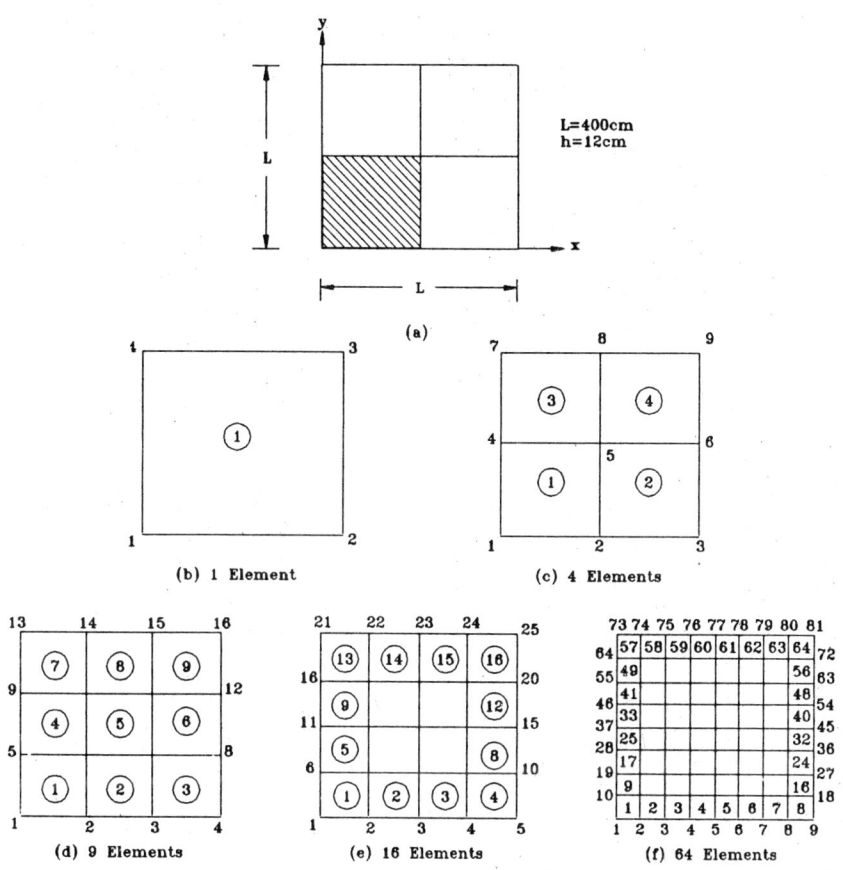

Fig. 10.11 Square Plate and its discretizations of
one quadrant - PLATE4

Elastic constants used in the analysis are $E = 2 \times 10^3$ kN/cm^2 and $\mu = 0.15$

In order to compare the values of finite element analysis with the theoretical value of a particular quantity obtained from classical thin plate theory [1, 2] , the following ratio is defined.

$$R = \frac{\text{Finite Element Analysis result for the variable}}{\text{Theoretical Value of the variable}}$$

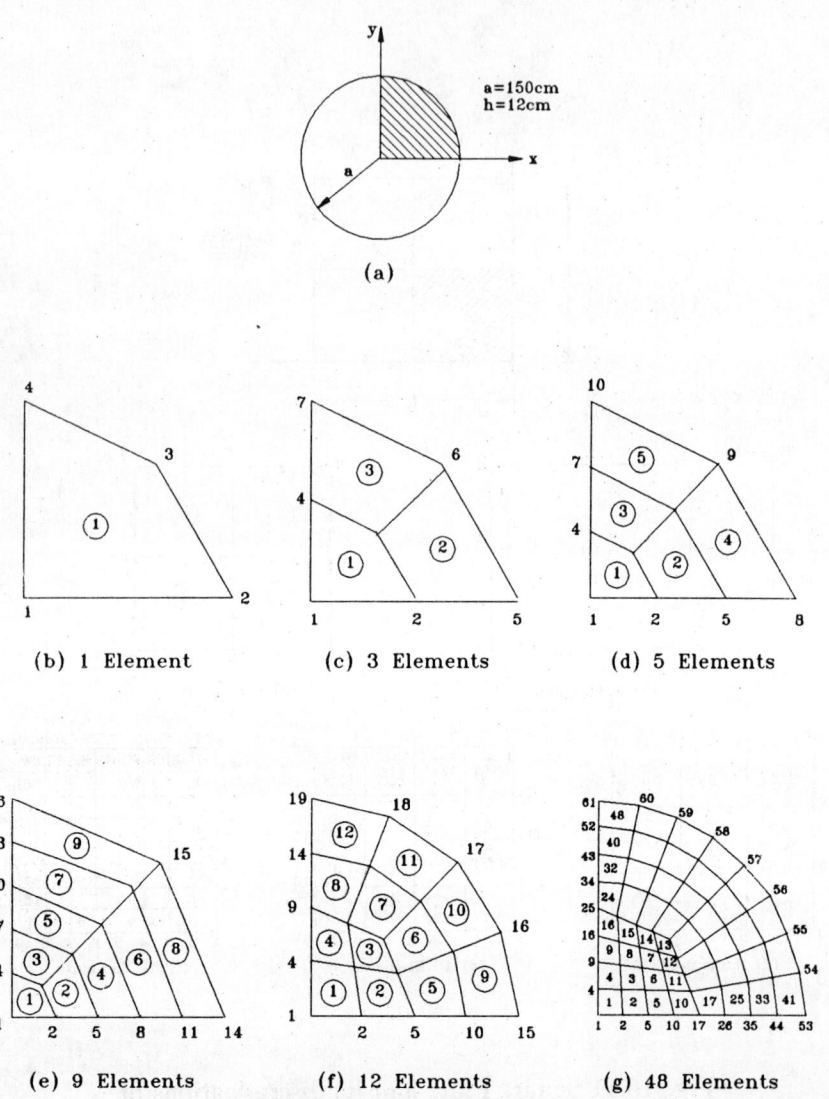

Fig. 10.12 Circular plate and its discretizations of
one quadrant - PLATE4

The variables chosen for comparison are the deflection and the
stress-resultants.

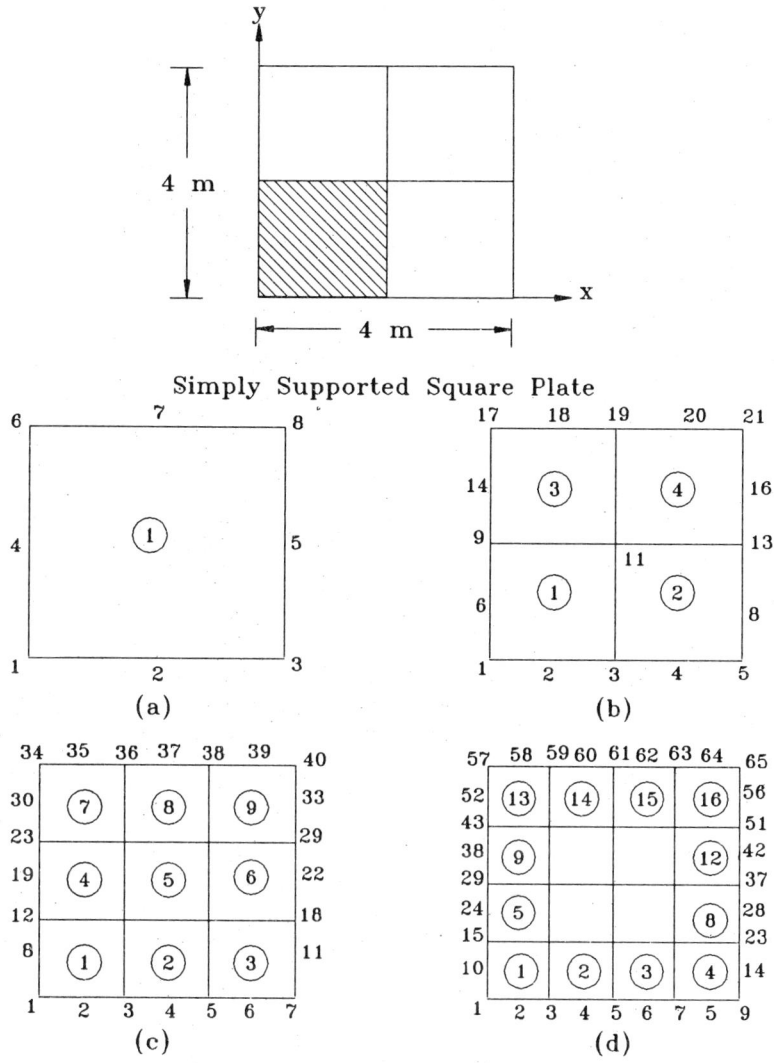

Fig. 10.13 Square plate and its discretizations of
one quadrant-PLATE8

To study the performance of plate elements, we have chosen PLATE
4 and PLATE 8 which are available in PASSFEM library and the element due to Clough and Felippa [8] used in SAP IV package. Figure
10.15 shows the variation of M_x and w for various discretizations (N
= Number of elements) of simply supported plate subjected to self
weight and central concentrated load.

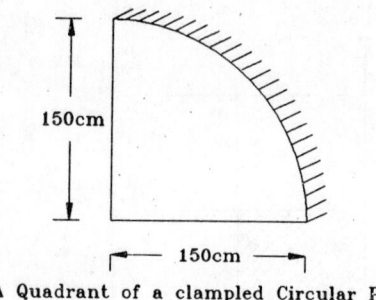

A Quadrant of a clampled Circular Plate

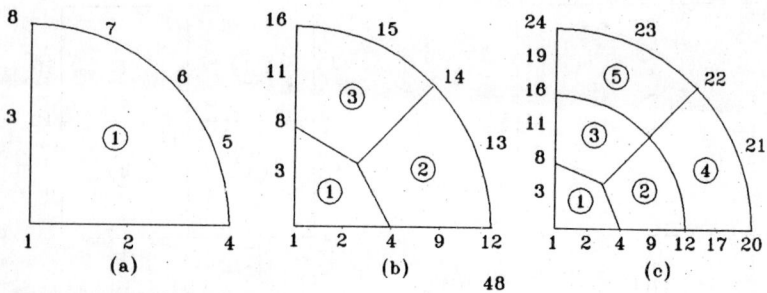

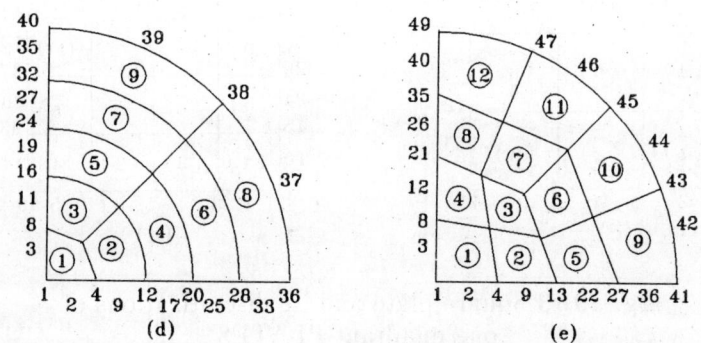

Fig. 10.14 Circular plate and its discretization of
one quadrant-PLATE8

The ratio R for displacement w and various stress resultants M_x, M_y, M_{xy} and Q_x have been calculated using the results obtained by PLATE4, PLATE8 and plate bending element in SAP IV. These values are given in Tables 2 to 6. It may be mentioned here that in the case of SAP IV the stress resultants M_x, M_y and M_{xy} are calculated at the centroid of the elements and hence these values are extrapolated by drawing the variation along the section for computing the value of R.

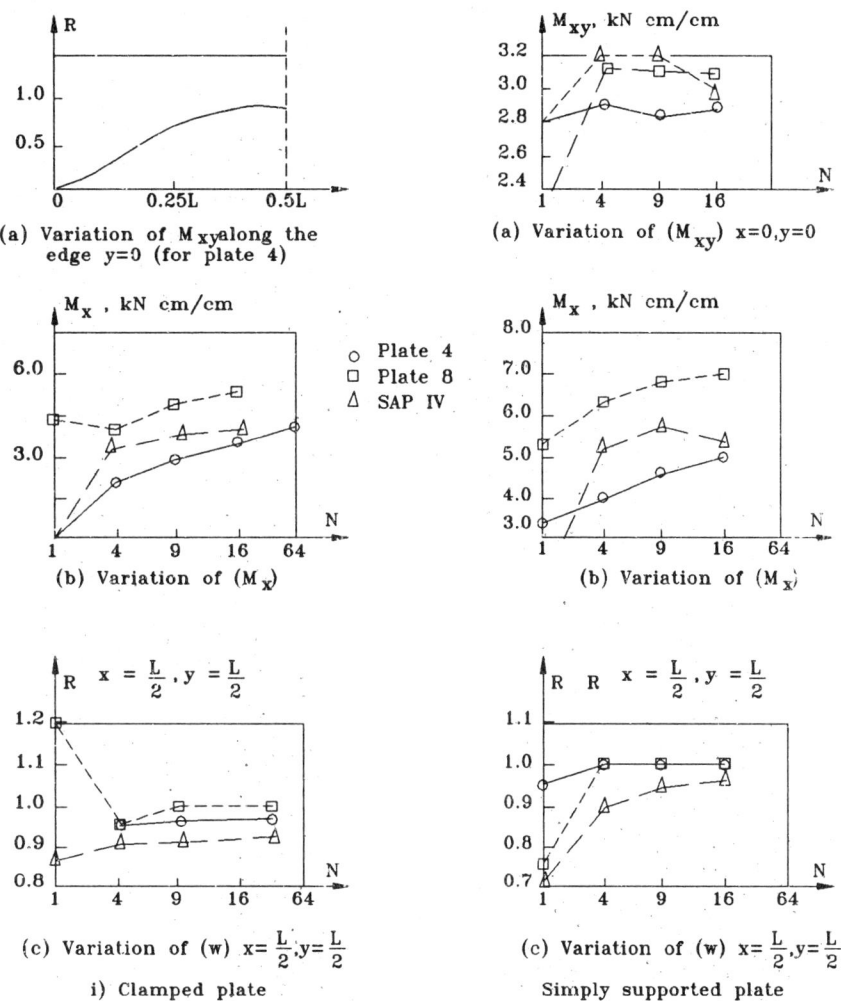

(a) Variation of M_{xy} along the edge y=0 (for plate 4)

(a) Variation of (M_{xy}) x=0,y=0

o Plate 4
□ Plate 8
△ SAP IV

(b) Variation of (M_x)

(b) Variation of (M_x)

(c) Variation of (w) $x=\frac{L}{2}, y=\frac{L}{2}$

(c) Variation of (w) $x=\frac{L}{2}, y=\frac{L}{2}$

i) Clamped plate

Simply supported plate

Fig. 10.15 Square plate subjected to self weight and central concentrated load

The same procedure is adopted in the case of PLATE4 element for arriving at the value of R for Q_x. However, SAP IV does not print out shear stress resultants.

Table 10.2a Comparison of Deflection and Bending Moment Values For Example 1(a) Under the Load Case (i)

No. of	(w) at x=y=L/2			(M_x or M_y) at x=y=L/2		
	R					
elem. (N)	SAP IV	PLATE		SAP IV	PLATE	
		8	4		8	4
1	0.5992	0.7148	0.7919	0.3798	1.2395	0.6873
4	0.8799	1.0044	0.9860	1.1698	1.0810	0.9967
9	0.9469	1.0061	1.0000	0.9685	1.0354	1.0025
16	0.9710	1.0282	1.0046	0.9811	0.9992	1.0025

Table 10.2b Comparison of Twisting Moment and Shear Values for Example 1(a) under the Load Case (i)

No. of	(M_{xy}) at x=y=0			(Q_x) at x=0,y=L/2		
	R					
elem. (N)	SAP IV	PLATE		SAP IV	PLATE	
		8	4		8	4
1	0.6579	0.7089	0.5472	-	2.9133	
4	1.2194	1.0703	0.8285	-	1.0797	1.0566
9	1.1381	1.0484	0.9069	-	0.9966	1.0000
16	0.9348	1.1063	0.9417	-	0.9856	0.9812

Table 10.3 Comparison of Displacement and Moment Values for Example 1(b) Under the Load Case (i)

No. of	(w) at x=y=L/2			(M_x or M_y) at x=y=L/2		
	R					
elem. (N)	SAP IV	PLATE		SAP IV	PLATE	
		8	4		8	4
1	0.9490	1.3242	0.0444	–	2.1196	–
4	0.9749	0.9798	1.0455	1.4627	1.0980	1.0595
9	1.0164	1.0729	1.0557	1.0709	1.0654	1.0344
16	1.0353	1.0774	1.0657	1.0187	1.0354	1.0089

Table 10.4 Comparison of Displacement Values for
Examples 1(a) and 1(b) Under the Load Case (ii)

No. of elem.	R					
	(w) at x=y=L/2 for example 1(a)			$(M_x$ or $M_y)$ at x=y=L/2 for example 1(b)		
(N)	SAP IV	PLATE		SAP IV	PLATE	
		8	4		8	4
1	0.7175	0.7382	0.9405	0.8608	1.2055	0.0201
4	0.9106	1.0011	0.9913	0.9225	0.9464	0.9413
9	0.9583	1.0063	1.0004	9.9692	1.0378	1.0009
16	0.9761	1.0078	1.0047	0.9908	1.0449	1.0239

Table 10.5 Comparison of Displacement and Moment Values
for Example 2(a) Under the Load Case (i)

No. of elem. (N)	R					
	(w) at x=y=0			(M_x) at x=y=0		
	SAP IV	PLATE		SAP IV	PLATE	
		8	4		8	4
3	0.8522	0.4228	0.7738	0.9982	0.7794	0.8100
12	0.7523	0.7096	0.9442	0.8638	0.8696	0.9841

Table 10.6 Comparison of Central Displacement for
Circular Plate Under the Load Case (ii)

No. of elem.	R					
	(w) at x=y=0 for s.s. plate			(w) at x=y=0 for clamped plate		
(N)	SAP IV	PLATE		SAP IV	PLATE	
		8	4		8	4
3	0.9558	0.5286	0.9448	0.7777	0.8888	0.9367
12	0.8160	0.7728	1.0062	0.8723	1.0073	1.0549

While the performance of the three types of elements is evident
from the Tables and Figures presented above it has been observed in
general that PLATE4 takes less computing time compared to SAPIV
and PLATE8 elements. And the time taken by PLATE8 elements is
the highest in all the cases.

10.8.2 Examples of Thick Plates

Following two examples are considered to study the influence of transverse shear on deflection.

(i) Simply supported square plate with varying thickness/span (h/L) ratios.

(ii) Clamped circular plate with varying thickness/radius (h/a) ratios.

Reissner's theory is a widely accepted theory for analysing thick plates and closed form solution is available in reference [9]. The theory considers the effect of transverse shear deformation and normal pressure and is characterised by a sixth-order system of linear partial differential equation in terms of the transverse deflection, w, and the two shear stress resultants Q_x and Q_y. Finite element analysis results are obtained using PLATE4 and PLATE8 elements.

(i) Simply Supported Square Plate (Thick)

The simply supported square plate shown in Fig. 10.11 is analysed for varying h/L (thickness/span) ratios of 0.05, 0.10, 0.15, 0.20 and 0.25. The material properties of the plate are the same as described in section 10.8.1. The plate is analysed for an uniformly distributed load of intensity, q.

The maximum deflection may be expressed as

$$w_{max} = q_1 \frac{\alpha L^4}{E h^3}$$

For comparison with Reissner's theory, the non-dimensional deflection parameter α_1 is calculated for various plate thickness to lateral dimension ratios. The results are given in Table 10.7 for a discretization of 16 elements for one quarter of the plate. Figure 10.16 shows the variation of the values for PLATE4 and PLATE8 elements.

10.7 Coefficients for Central Deflection of Simply Supported Square Plate (Thick)

h/L	$\alpha_1 = w_{max} \frac{Eh^3}{qL^4}$;		$q = $ UDL	
	Classical thin plate theory	Reissner's theory	FEM solution	
			PLATE4	PLATE8
0.01	0.04437	0.04768	-	-
0.05	0.04437	0.04819	0.04769	0.04816
0.10	0.04437	0.04976	0.04930	0.04968
0.15	0.04437	0.05238	0.05197	0.05222
0.20	0.04437	0.05604	0.05571	0.05578
0.25	0.04437	0.06076	0.06051	0.06036

The results of PLATE4 and PLATE8 differ but on an average closely agree with those of Reissner's theory. The increase in deflection due to transverse shear compared to the classical thin plate theory is found to be 11% and 11.96% in case of analysis using PLATE4 and PLATE8 elements, for $h/L = 0.1$. For $h/L = 0.2$, the increase in deflection due to shear is found to be 25.5% for analysis using PLATE4 and 25.72% for PLATE8 elements.

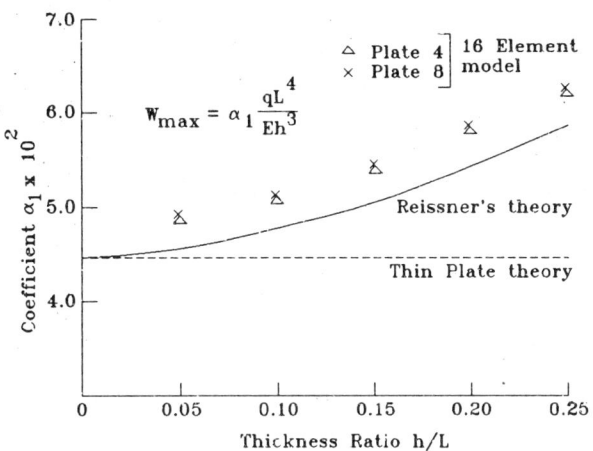

Fig. 10.16 Influence of transverse shear on maximum deflection of a simply supported square plate subjected to UDL

(ii) Clamped Circular Plate (Thick)

The clamped circular plate shown in Fig. 10.12 is analysed for varying thickness/radius (h/a) ratios. The material properties are the same as given in section 10.8.1 and the plate is subjected to a uniformly distributed load of q.

The deflection of the plate can be expressed as,

$$w = \alpha_2 \frac{qa^4}{64D}$$

Figure 10.17 shows the graphical plot of the coefficient α_2 for two different values of h/a ratio. For $h/a = 0.08$, the results of PLATE4 and PLATE8 agree well with the solution obtained by Reissner's theory. For $h/a = 0.40$, the level of accuracy of PLATE4 and PLATE8 is reasonably good compared with the Reissner's theory.

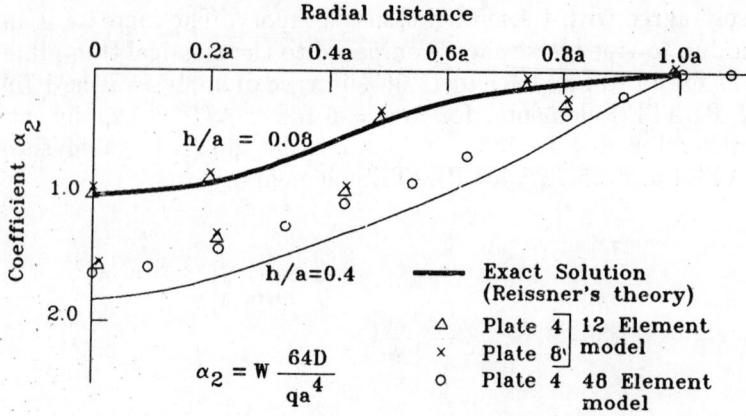

Fig. 10.17 Influence of transverse shear on the deflection
of a clamped circular plate subjected to UDL

10.8.3   Additional Examples

Two more examples are described below to illustrate the finite element analysis of thin plates supported on edge beams of continuous plates.

Thin Plate Supported On Edge Beams

Figure 10.18 shows a plate supported on beams along the edges. The following two cases are considered for analysis:

Case (i)   the mid surface of the plate and the centre of the beam are at the same level i.e., there is no eccentricity of the beam with respect to the plate.

Case (ii)   The centre of the beam is at eccentricity to the mid surface of the plate.

The following data is used for the analysis of the two cases.

Dimensions of the plate:	$300 \times 300 \times 12$ cm
Breadth of edge beam	$b = 25$ cm
Depth of edge beam	$D = 40$ cm
Eccentricity of the beam,	$e = 14.0$ cm
Uniformly distributed load on the plate	$= 5$ kN/m^2
	$\mu = 0.25$
	$E = 2.6 \times 10^3$ kN/cm^2

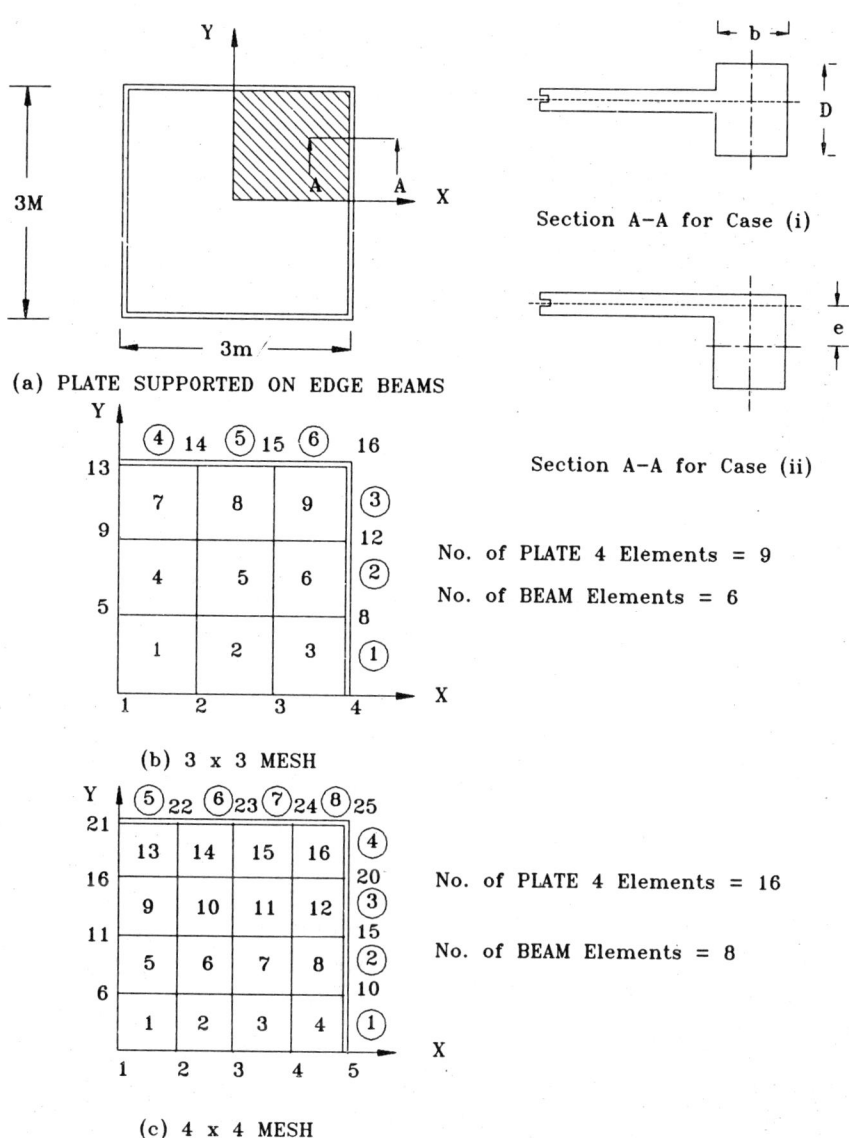

(a) PLATE SUPPORTED ON EDGE BEAMS

Section A–A for Case (i)

Section A–A for Case (ii)

No. of PLATE 4 Elements = 9

No. of BEAM Elements = 6

(b) 3 x 3 MESH

No. of PLATE 4 Elements = 16

No. of BEAM Elements = 8

(c) 4 x 4 MESH

Fig. 10.18 Discretization of one quadrant

The details of discretization are given in Fig. 10.18. Due to symmetry only one quadrant of the plate is considered for analysis. The analysis is done using PLATE4, BEAM1 and BEAM2 (offset beam element) elements. The results of the analysis are compared in Tables 10.8 and 10.9.

Table 10.8 Comparison of Deflection and Bending Moment Values for Case (i)

No. of elem.	(w) at x=y=0 (cm)			(M_x or M_y) at x=y=0 (kN cm/cm)		
	Theory	PASS-FEM	R	Theory	PASS-FEM	R
9	0.06046	0.05919	0.9790	2.328	2.291	0.9829
16	0.06046	0.05979	0.9889	2.328	2.299	0.9876

Table 10.9 Comparison of Deflection and Bending Moment Values for Case (ii)

No. of elem.	(w) at x=y=0			(M_x or M_y) at x=y=0 (kN cm/cm)		
	Theory	PASS-FEM	R	Theory	PASS-FEM	R
9	0.06046	0.04894	0.8095	2.328	2.171	0.9323
16	0.06046	0.04925	0.8146	2.328	2.166	0.9304

Note: R Denotes the ratio as defined in section 10.8.1.

The theoretical values [1] correspond to the solution in which the eccentricity of the edge beam and its torsional resistance are not taken into account. To compare with these theoretical values for case (i), a low value of torsional moment of inertia is used in finite element analysis. However, for case (ii) actual eccentricity and low value of torsional resistance of the edge beam are considered and the values are compared with the theoretical solution (Table 10.9)

It has been reported that there exits possibility of an error due to incompatibility in the axial displacement field in the beam element, when its reference axis is eccentric to the middle plane of the plate [31,32]. The error caused may be significant depending on the section properties of the beam and the plate it is attached to, the distribution of shear force and moment, and discretization. This error can be eliminated through the addition of one more degree of freedom for axial displacement, i.e., the mid-side node in beam element, which allows quadratic variation of axial displacement [33,34].

Continuous Plate

A rectangular plate continuous over three spans, and simply supported alround and along two intermediate supports is shown in Fig. 10.19. The geometry and the loading details are as follows:

$$
\begin{aligned}
\text{Length of each span} &= 4 \text{ m}\\
\text{Width of the plate} &= 4 \text{ m}\\
\text{Thickness of the plate} &= 12 \text{ cm}\\
\text{Uniformly distributed load on the plate} &= 5 \text{ kN/m}^2\\
\mu &= 0.20\\
E &= 2.6 \times 10^3 \text{ kN/cm}^2
\end{aligned}
$$

The details of discretizations are shown in Fig. 10.19.

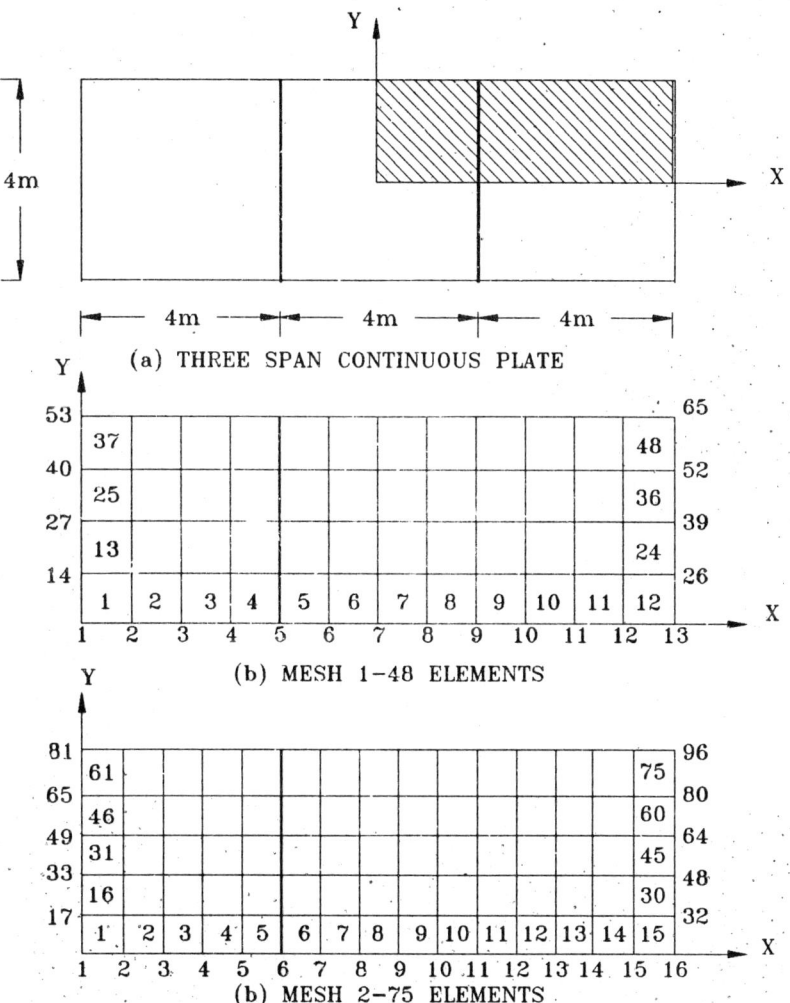

Fig. 10.19 Discretizations of one quadrant

Due to symmetry only one quadrant as shown in Fig. (10.9) (hatched portion) is considered for analysis. The analysis is done using

PLATE4 element. The values of bending moments at salient points are compared with the theoretical values [1] and are presented in the Table 10.10.

The bending moment in the vicinity of the intermediate supports has a very steep variation and hence the stress smoothing technique does not give correct value of moment at the support point. So, the value of moment at this point has been obtained by graphical interpolation.

Table 10.10 Comparison of Bending Moment Values
for Continuous plate

	At x=0 y=0		At x=400 cm y=0		At x = 200 cm y=0
	M_x	M_y	M_x	M_y	M_x
	(All moments in kN cm)				
Theory	2.781	1.901	3.230	2.701	-6.767
Mesh 1 (48 elems.)	2.494	1.679	3.089	2.682	−5.800*
R	0.8968	0.08832	0.9565	0.9930	0.8571
Mesh 2 (75 elems.)	2.685	1.734	3.288	2.812	−6.350*
R	0.9655	0.9122	1.0180	1.0411	0.9384

* Values are obtained by graphical interpolation.

REFERENCES

1. Timoshenko, S and S.W.Krieger, *Theory of Plates and Shells*, Second edition, McGraw-Hill Book Company Inc., N.Y., 1959.

2. Szilard, R., *Theory and Analysis of Plates (Classical and Numerical Methods)*, Prentice Hall Inc., Englewood Cliffs, N.J., 1974.

3. Zienkiewicz, O.C, and R.L.Taylor, *The Finite Element Method, Vol.2, Solid and Fluid Mechanics, Dynamics and Non-linearity*, McGraw-Hill Book Co. U.K. 1991.

4. Gallagher, R.H., *Analysis of Plate and Shell Structures, Applications of Finite Element Method in Engineering*, Vanderbilt University, ASCE, 1969.

5 Melosh, R.H., *Basis for Derivation of Matrices by the Direct Stiffness Method*, AIAA Journal, Vol. 1, pp. 1631-1637, 1963.

6. Przemieniecki, J.S., *Theory of Matrix Structural Analysis*, McGraw-Hill, New York, 1968.

7. Bogner, F.K., R.L.Fox and L.A.Schmidt., *The Generation of Interelement Compatible Stiffenss and Mass Matrices by the Use of Interpolation Formulae*, Proc. of the Conference on Matrix Methods in Structural Mechanics, Wright Patterson Air Force Base, Ohio, Oct. 1965.

8. Clough, R.W. and C.A., Felippa, *A Refined Quadrilateral Element for the Analysis of Plate Bending* Proc. of the Second Conf. on Matrix Methods in Structural Mechanics, Wright Patterson Air Force Base, Ohio, Oct. 1968.

9. Reissner, E., *The Effect of Transverse Shear Deformation on the Bending of Elastic Plates*, Journal of Applied Mechanics, Vol.12, pp.69-77, 1945.

10. Mindlin, R.D., *Influence of Rotary Inertia and Shear on flexural Motions of Isotropic Elastic Plates*, Journal of Applied Mechanics, Vol. 18, pp.31-38, 1951.

11. Progor, C.W., R.M.Barker and D.Frederick, *Finite Element Bending Analysis of Reissner Plate*, Journal of Engineering Mechanics Division, ASCE, Vol.96. 1970.

12. Pugh, E.D.L., E.Hinton and O.C.Zienkiewicz, *A Study of Quadrilateral Plate Bending Elements with Reduced Integration*, International Journal for Numerical Methods in Engineering, Vol. 12, pp.1059-1078, 1978.

13. Hughes, T.J.R., Cohen, M and M.Haroun, *Reduced and Selective Integration Techniques in the Finite Element Analysis of Plates*, Nuclear Engineering Design, Vol. 46, pp.203-222, 1978.

14. Hughes, T.J.R., R.L.Taylor and W.Kanoknukulchai, *A Simple and Efficient Finite Element for Plate Bending*, International Journal for Numerical Methods in Engineering, Vol.11, pp.1529-1543, 1977.

15. Wood, R.D., and E.Hinton, *Finite Element Analysis of Geometrically Nonlinear Plate Behaviour using Mindlin Formulation*, Computers and Structures, Vol.II, pp.203-215, 1980.

16. Hinton, E, and N.Bicanic, *A Comparison of Lagrangian and Serendipity Mindlin Plate Elements for Free Vibration*, Computers and Structures, Vol.10, pp.483-493, 1979.

17. Zienkiewicz, O.C., R.L.Taylor and J.M.Too. *Reduced Integration Technique in General Analysis of Plates and Shells*, International Journal for Numerical Methods in Engineering, Vol.3, pp.275-290, 1971.

18. Batoz, J.L., K.J.Bathe and L.W.Ho, *A Study of Three Node Triangular Plate Bending Elements*, International Journal for Numerical Methods in Engineering, Vol.15, pp.1771-1812, 1980.

19. Bathe, K.J., *Finite Element Procedures in Engineering Analysis*, Prentice-Hall Inc. Englewood Cliffs, N.J., 1982.

20. Hinton, E. and D.R.J. Owen., *Finite Element Programming*, Academic Press. Inc. (London) Ltd., 1977.

21. Averill, R.C. and J.N.Reddy, *Behaviour of Plate Elements Based on the First-Order Shear Deformation Theory*, Engineering Computations, Vol.7, pp.57-74, March 1990.

22. Belytschko, T. and C.S.Tsay, *A Stability Procedure for the Quadrilateral Plate with One Point Quadrature*, International Journal for Numerical Methods in Engineering, Vol.19, pp.405-420, 1983.

23. Belytschko, T. and C.S.Tsay, and W.K.Liu, *A Stabilisation Matrix for the Bilinear Mindlin Plate Element*, Computer Methods in Applied Mechanics and Engineering, Vol29, pp.313-327, 1981.

24. Hughes, T.J.R. and M.Cohen, *The 'Heterosis' Finite Element for Plate Bending*, Computers and Structures, Vol.9, No.5, pp.445-450, 1978.

25. Hughes, T.J.R., M.Cohen and M.Haroun, *Reduced and Selective Integration Techniques in the Finite Element Analysis of Plates*, Nuclear Engineering Design, Vol.46, No.1, pp.203-222, 1978.

26. Hughes, T.J.R. and T.E.Tezduyar, *Finite Elements Based on Mindlin Plate Theory with Particular Reference to the Four-Node Bilinear Isoparametric Element*, Journal of Applied Mechanics, Vol.48, No.3, pp.587-596, 1981.

27. Hrabok, M.M. and T.M.Hrudey, *A Review and Catalog of Plate Bending Finite Elements*, Computers and Structures, Vol.19, No.3, pp.479-495, 1984.

28. Hinton,E. and H.C.Huang, *A Family of Quadrilateral Mindlin Plate Elements with Substitute Shear Strain Fields*, Computers and Structures, Vol.23, No.3, pp.409-431, 1986.

29. Barlow, J., *Optimal Stress Locations in Finite Element Method*, International Journal for Numerical Methods in Engineering, Vol 10, pp.243-51, 1976.

30. Hinton, E. and J.S.Campbell, *Local and Global Smoothing of Discontinuous Finite Element Function using a Least Square Method*, International Journal for Numerical Methods in Engineering, Vol.8, pp.461-480, 1974.

31. Gupta, A.K. and P.S.Ma, *Error in Eccentric Beam Formulation*, International Journal for Numerical Methods in Engineering, Vol.11, pp.1473-1477, 1977.

32. Balmer, H.A., *Another Aspect of the Error in Eccentric Beam Formulation*, International Journal for Numerical Methods in Engineering, Vol.12, pp.1761-1763, 1978.

33. Miller, R.E., *Reduction of the Error in Eccentric Beam Modelling*, International Journal for Numerical Methods in Engineering, Vol.15, pp.575-582, 1980.

34. Crisfield, M.A., *The Eccentricity Issue in the Design of Plate and Shell Elements*, Communications in Applied Numerical Methods, Vol.7, pp.47-56, 1991.

EXERCISES

10.1 For the element shown in Fig. 10.11c, derive the [B] matrix using 4 noded plate bending element based on Mindlin's theory.

10.2 Compute the product matrix [CB] = [C][B] for the element of problem 10.1 assuming $E = 2 \times 10^3$ kN/cm^2 and $\mu = 0.15$.

10.3 Following the selective integration scheme, compute $[k]_b$ and $[k]_s$ for the element 4 using the expressions obtained from problems 10.1 and 10.2.

10.4 Calculate the element nodal load vector $\{Q\}$ for the element 4 of problem 10.1 due to uniformly distributed lateral load of 5.88 kN/m^2.

10.5 Compute the [B] matrix at the Gauss point (0.57735, 0.57735) for the four noded plate bending element 3 shown in Fig. 10.12d.

10.6 Calculate the element nodal load vector $\{Q\}$ for the element 3 of problem 10.5 due to uniformly distributed lateral load of 5.88 kN/m².

10.7 For the eight noded element 3 shown in Fig.10.14c, calculate the [B] matrix at the Gauss point (0.57735, 0.57735).

10.8 Calculate the element nodal load vector $\{Q\}$ for the element 3 of problem 10.7 due to uniformly distributed lateral load of 5.88 kN/m².

10.9 Derive an expression to compute the nodal load vector $\{Q\}$ for a four noded element when it is subjected to varying pressure load. Indicate the numerical integration procedure that can be used for computation of $\{Q\}$. (Refer Problem 10.20)

10.10 Using PASSFEM analyse the plate shown in Fig. 10.11 for the discretization given in (d). For a typical element 5, compare the nodal values of stress resultants obtained (a) without stress smoothening and (b) with stress smoothening.

10.11 Using PASSFEM analyse the plate of problem 10.10 using eight noded element and discretization shown in Fig.10.13c. For a typical element 5, compare the nodal values of stress resultants obtained (a) without stress smoothening and (b) with stress smoothening.

10.12 Analyse the simply supported square plate subjected to uniformly distributed lateral load shown in Fig.10.13b, adopting 4-element discretization. Use 8 noded element described in the text and compare the results by adopting selective integration scheme of 3×3 for computing $[k]_b$ and 2×2 for $[k]_s$.

10.13 Analyse the plate shown in Fig. E10.13.

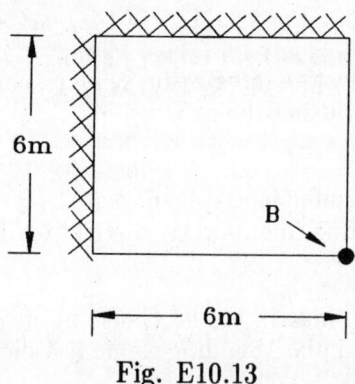

Fig. E10.13

The two adjacent edges are fixed against rotation and translation, and the other two edges are free. Consider the following two load cases: (a) uniform load of 12 kN/m ^{2}, (b) a concentrated load of 10 kN at B. Compare the results with analytical solution. $E = 2 \times 10^5$ N/mm^2, $\mu = 0.3$, $h = 25$ mm.

10.14 Analyse the rectangular plate subjected to uniform load of 4 kN/m^2 acting over a central rectangular area. All edges are simply supported. Compare the results with the analytical solution. $h = 200$ mm, $E = 2 \times 10^4$ N/mm^2, $\mu = 0.15$.

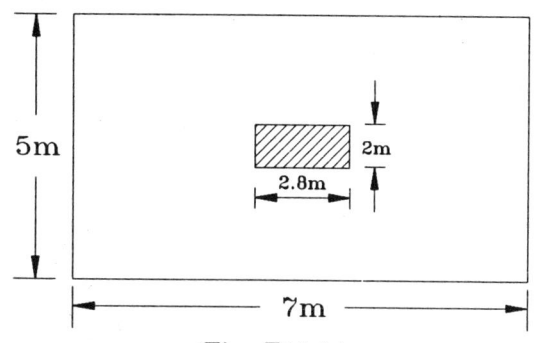

5m

2m

2.8m

7m

Fig. E10.14

10.15 Analyse a simply supported circular plate subjected to an uniformly distributed load of 5 kN/m^2 over the outer annular ring as shown in Fig. E10.15. Compare the results with the analytical solution. $h = 15$ mm, $E = 2 \times 10^5$ N/mm^2, $\mu = 0.3$.

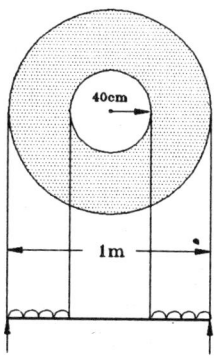

40cm

1m

Fig. E10.15

10.16 Fig. E10.16 shows a simply supported skew plate and it is subjected to uniformly distributed load of 4 kN/m^2. Analyse the plate and compare the results with theoretical solution. $h =$

200 mm, $E = 2 \times 10^4$ N/mm^2, $\mu = 0.15$, $\theta = 30°$

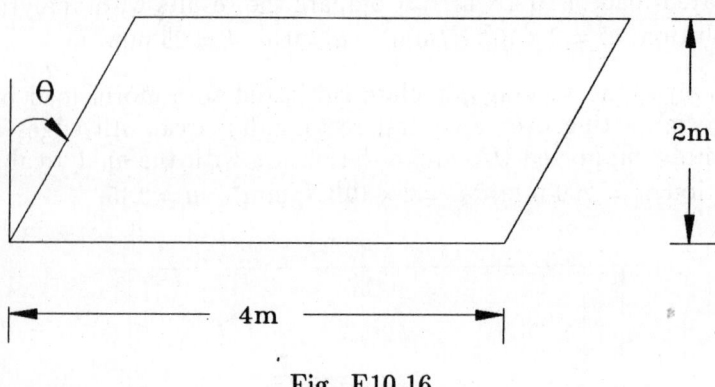

Fig. E10.16

10.17 Analyse the skew plate of problem 10.14 with all edges fixed.

10.18 Analyse a simply supported equilateral triangular plate shown in Fig. E10.18. It is subjected to an uniformly distributed load of 5 kN/m^2. $h = 25$ mm, $E = 2 \times 10^5$ N/mm^2, $\mu = 0.3$.

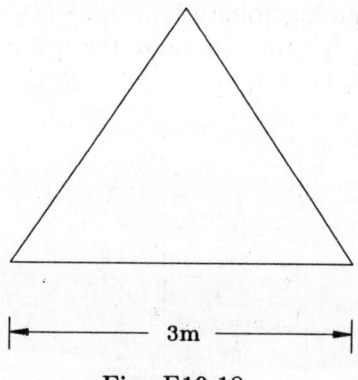

Fig. E10.18

10.19 Fig. E10.19 shows a simply supported plate subjected to a lateral load of 2.5 kN/m^2 and uniform tension of 1.0 kN/m applied along the edges. Calculate the maximum stress and compare the result with analytical solution. $h = 40$ mm, $E = 2 \times 10^5$ N/mm^2, $\mu = 0.3$. (Hint: Use plate as well as plane stress element).

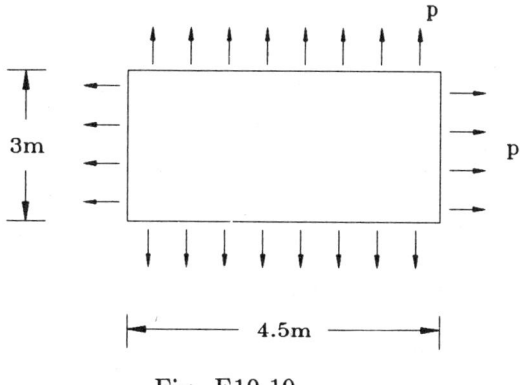

Fig. E10.19

10.20 Using the element nodal load vector computed in exercise 10.9, analyse the sidewall of a water tank for the loading shown in Fig. E10.20. Compare the results with the analytical solution. $h = 400$ mm, $E = 2 \times 10^4$ N/mm^2, $\mu = 0.2$

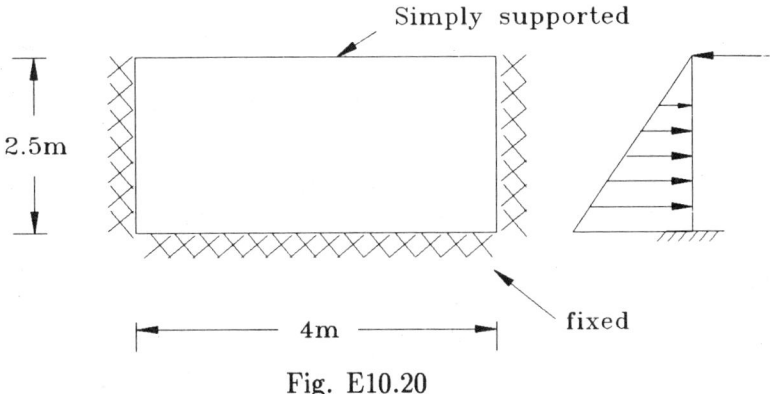

Fig. E10.20

10.21 Analyse the problem given in Exercise 9.14 under Chapter 9, using plate elements and compare the results.

10.22 For an eight noded element, develop the necessary expressions for the computation of stiffness matrix using selective integration scheme as suggested in the text. $[3 \times 3$ for $[k]_b$ and 2×2 for $[k]_s]$.

10.23 Develop the program segments to incorporate selective integration scheme for eight noded element using the expressions derived for problem 10.22.

10.24 Compare the performance of eight noded element with selective and reduced integration schemes through the example given in the text (Fig. 10.13).

10.25 Develop the necessary expressions for calculating the element properties of 'heterosis' element with 26 degrees of freedom. Adopt selective integration scheme of 3×3 for evaluating $[k]_b$ and 2×2 for $[k]_s$.

10.26 Develop the routines to include 'heterosis' element in the PASS-FEM library.

10.27 Investigate the performance of four noded, eight noded and 'heterosis' elements using selective integration schemes for various values of h/L. Choose any one of the examples given in this Chapter.

10.28 Plot the variation of stress resultants M_x, M_y, M_{xy}, Q_x and Q_y for the plate example 1(b) given in the text (square plate with clamped edges) using the four-noded eight noded and 'heterosis' elements and discuss the results in comparison to analytical solution given in reference [1].

10.29 Taking any one of the typical four noded element, determine the number of zero eigenvalues for $[k]_b + [k]_s$ using full integration scheme of 2×2 for $[k]_b$ and $[k]_s$. Plot the rigid body modes for the element.

10.30 For the problem 10.29 determine the number of zero eigenvalues for $[k]_b + [k]_s$ using selective integration scheme of 2×2 for $[k]_b$ and 1×1 for $[k]_s$.

10.31 Using the routines developed for **problem** 10.22 determine the number of zero eigenvalues for a typical eight noded element under selective and reduced integration schemes.

10.32 Show that heterosis element does not have any spurious zero energy modes under selective integration scheme. Use the routines developed for problem 10.26.

Chapter 11

Analysis of Shells

In many areas of structural design, we require analysis of shells subjected to different types of loads. It is well known from theory of shells [1,2,3] that classical solution involves tedious calculations and is extremely difficult especially for shells of arbitrary shape. The finite element method is very much suited for the analysis of shells of general shape because of its flexibility in accounting for arbitrary geometry, loadings and variation in material properties. A number of elements have been developed and as reported in the reference [4], considerable amount of effort has been made to develop elements that perform satisfactorily in many situations in practice and are also computationally economical. As larger number of publications have appeared and continue to appear on this subject, only a brief review is presented here highlighting the important strategies used in the development of shell elements.

The concept of treating a shell element as a special case of three-dimensional analysis was used by Ahmad, Irons and Zienkiewicz[5] and Pawsey[6] and it seems to provide a simple and efficient strategy for development of isoparametric shell elements. Two such elements, a four noded bilinear degenerated shell element due to Kanok-Nukulchai[7] and an eight noded (eight external and one internal node) curved shell element due to Pawsey[6] are described in this Chapter. Two subroutines SHELL4 and SHELL8 have been developed and are included in PASSFEM. The application of these elements is illustrated through several practical examples.

11.1   Thin Shell Theory

If the thickness of the shell is small compared to the radii of curvature of the midsurface, the shell is referred to as geometrically thin shell. If the transverse shear force per unit length is small enough so that transverse shear deflections are insignificant, then it is referred to as structurally thin shell. There are classifications of shells based on the closed form methods of analysis and hence, depend on the shape of the midsurface, for example cylindrical shells, shells of revolution, shells of translation, etc.

As in the case of plates, thin shell theories are based on Love–Kirchhoff assumption that as the shell deforms and the midsurface stretches and bends, the fibres of the shell initially straight and normal to the mid-surface, remain straight and normal to the midsurface. Also it is usually assumed that the normal stress is zero. These assumptions permit to define the displacement of every point in the shell and hence the stresses and strains of everypoint in terms of the displacement of the midsurface of the shell. This in effect represents the reduction of the problem from a three-dimensional to a two-dimensional case.

Figure 11.1 shows the stress resultants (forces per unit length) acting on a shell element.

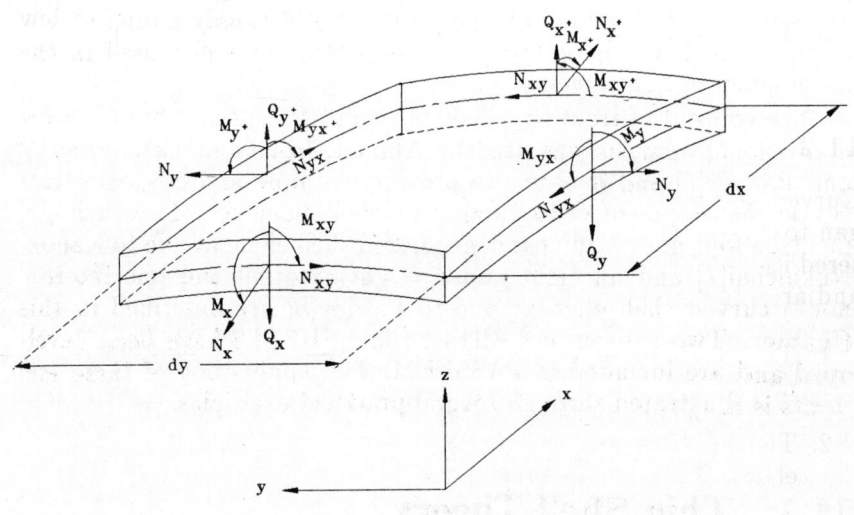

Fig. 11.1 Stress resultants

11.2 Review of Shell Elements

11.2.1 Flat Plate Elements

The earliest attempts at constructing suitable shell elements were based on combining membrane elements with plate bending elements that enforced Kirchhoff's hypothesis. Assembly of these elements gives a geometry which only approximates the actual shell surface. This class of shell elements [8] has proven to possess good convergence characteristics and one of the widely used SAPIV package has such an element [9] for the analysis of thin shells. The development of shell elements of this type closely paralleled the development of good plate bending elements.

The difficulties and the shortcomings of the flat plate element used for the analysis of shells have been pointed out by Gallagher [10]. They are briefly stated below.

1. The behaviour of the shells as represented by the differential equations is not approached in the limit of refinement of flat plate approximation.

2. The discontinuities of slope between adjacent plate elements may produce bending moments in the regions of shells where they do not exist.

3. The coupling of membrane and bending effects due to curvature of the shell is absent in the interior of the individual elements.

11.2.2 Curved Shell Elements

Curved elements based on exact or approximate shapes of shells began to appear in the late 1960s. Four general difficulties are encountered in the development of curved elements for the analysis of shells and are summarised below [10].

1. The choice of an appropriate shell theory as there are quite a number of theories being available.

2. The description of the geometry of the elements using the given element data.

3. The satisfaction of rigid body modes of behaviour is acute in curved shell analysis.

4. The inter element compatibility condition is difficult to achieve as pointed out in the case of plate elements.

Bogner, Fox and Schmit[11] described a cylindrical shell element, which used interpolation functions defined in shell coordinates. Nodal values involved the three displacements and derivatives of these elements with respect to the local coordinates. This results in an element with twelve degrees of freedom for each of the four nodes. Utku [12] had represented a shallow curved triangular shell element. Using three displacements and two rotations at each node, by use of thin shallow shell equations, he defined the internal strains from an assumption of linear variation of the displacement quantities within the triangular element. Reference [4] contains exhaustive information on the curved elements developed for the analysis of shells.

11.2.3   Three-Dimensional Solid Element for Shell Analysis

We noted earlier the complexities involved in developing a curved element using the general shell theory. Three-dimensional solid element like 8 noded solid element can be used but in this case more than one layer of elements may be needed across the thickness to simulate the bending behaviour of shell. As an alternative, higher order elements like 20 noded isoparametric solid element can be used. But these approaches to treat the analysis of shell as three- dimensional stress analysis are very costly.

Degenerated Shell Elements Shell elements have been developed by degenerating the solid element formulations [5,6,7] by introduction of assumptions that normals to the midsurface remain straight and that the normal stress is zero. By this approach the displacements and rotations of the shell midsurface are taken as degrees of freedom. This formulation makes it possible to develop elements for the analysis of moderately thick shells but in the case of thin shells selective and reduced integration technique have to be used due to shear locking effect explained in Chapter 10 in connection with the formulation of plate element based on Mindlin's theory. Detailed formulations of four noded and eight noded shell elements are presented in the following sections.

11.3   Bilinear Degenerated Shell Element

A simple and efficient four noded shell element was presented by Kanok-Nukulchai[7] and its application to a number of practical cases was illustrated. The following two assumptions are made:

1. Normals to the midsurface remain straight after deformation. Thus, the formulation includes transverse shear deformation

Kirchhoff-Love hypothesis is not assumed.

2. Stresses normal to the midsurface are zero.

11.3.1 Shape Functions for Geometry and Displacement

The four noded element is evolved from an eight noded solid element. The midsurface enclosed by four straight sides forms a hyperbolic paraboloid. The shell element is shown in Fig. 11.2. The shape function to describe the midsurface in terms of natural coordinates is the same as given by Eq. 4.2(b) for two-dimensional isoparametric element. Thus,

$$N_i = \frac{1}{4}(1 + r_i r)(1 + s_i s) \quad i = 1, ..., 4 \tag{11.1}$$

where r_i and s_i are the natural coordinates of node i.

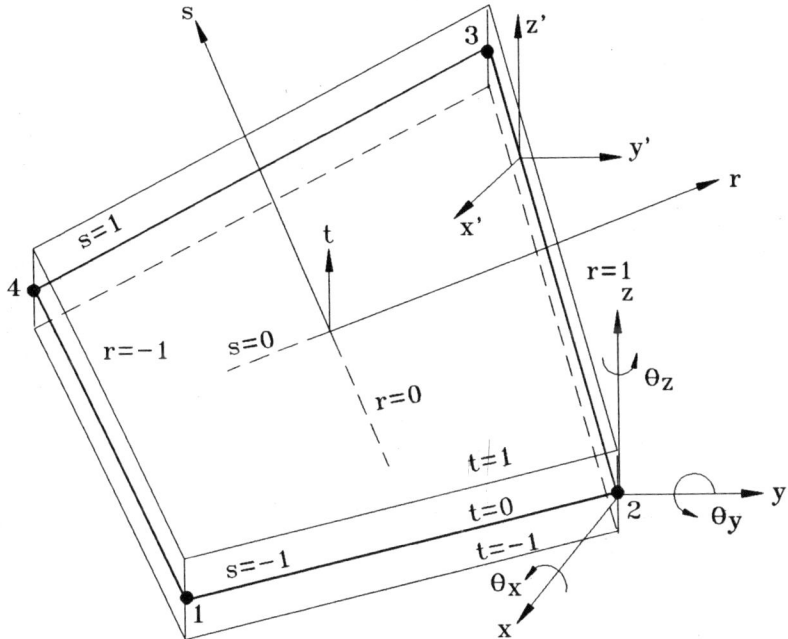

Fig. 11.2 Four noded shell element

The thickness of the shell element in the direction normal to the midsurface at each node is required and is specified as input. Using the shape functions, the coordinates of any point in the element can be uniquely given in terms of nodal coordinates and thicknesses as,

$$\left\{ \begin{array}{c} x \\ y \\ z \end{array} \right\} = \sum_{i=1}^{4} N_i \left\{ \left\{ \begin{array}{c} x_i \\ y_i \\ z_i \end{array} \right\} + \frac{1}{2} th_i \left\{ \begin{array}{c} l_{3i} \\ m_{3i} \\ n_{3i} \end{array} \right\} \right\} \qquad (11.2)$$

where x_i, y_i, z_i are the global coordinates of the mid-surface node i.

h_i is the thickness at node i and

l_{3i}, m_{3i} and n_{3i} are the normal unit vectors at node i.

At any point (r, s) on the midsurface $(t = 0)$ an orthogonal set of local coordinate axes x', y', z' are constructed, e_3' is the normal unit vector and e_1' and e_2' are tangent to the midsurface. It is well known from vector algebra that the cross product of two vectors gives a vector oriented normally to the plane given by the two vectors and unit vector is obtained by dividing it by its scalar length. For details of vector products reference [13] may be consulted. Thus,

$$\{e_3'\} = \left\{ \begin{array}{c} l_3 \\ m_3 \\ n_3 \end{array} \right\}_{(r,s)} = \dfrac{\left\{ \begin{array}{c} \frac{\partial x}{\partial r} \\ \frac{\partial y}{\partial r} \\ \frac{\partial z}{\partial r} \end{array} \right\}_{(r,s)} \times \left\{ \begin{array}{c} \frac{\partial x}{\partial s} \\ \frac{\partial y}{\partial s} \\ \frac{\partial z}{\partial s} \end{array} \right\}_{(r,s)}}{\left| \left\{ \begin{array}{c} \frac{\partial x}{\partial r} \\ \frac{\partial y}{\partial r} \\ \frac{\partial z}{\partial r} \end{array} \right\}_{(r,s)} \times \left\{ \begin{array}{c} \frac{\partial x}{\partial s} \\ \frac{\partial y}{\partial s} \\ \frac{\partial z}{\partial s} \end{array} \right\}_{(r,s)} \right|} \qquad (11.3a)$$

$$\{e_2'\} = \left\{ \begin{array}{c} l_2 \\ m_2 \\ n_2 \end{array} \right\}_{(r,s)} = \dfrac{\left\{ \begin{array}{c} l_3 \\ m_3 \\ n_3 \end{array} \right\}_{(r,s)} \times \left\{ \begin{array}{c} \frac{\partial x}{\partial r} \\ \frac{\partial y}{\partial r} \\ \frac{\partial z}{\partial r} \end{array} \right\}_{(0,0)}}{\left| \left\{ \begin{array}{c} l_3 \\ m_3 \\ n_3 \end{array} \right\}_{(r,s)} \times \left\{ \begin{array}{c} \frac{\partial x}{\partial r} \\ \frac{\partial y}{\partial r} \\ \frac{\partial z}{\partial r} \end{array} \right\}_{(0,0)} \right|} \qquad (11.3b)$$

$$\{e_1'\} = \begin{Bmatrix} l_1 \\ m_1 \\ n_1 \end{Bmatrix}_{(r,s)} = \begin{Bmatrix} l_2 \\ m_2 \\ n_2 \end{Bmatrix}_{(r,s)} \times \begin{Bmatrix} l_3 \\ m_3 \\ n_3 \end{Bmatrix}_{(r,s)} \quad (11.3c)$$

The partial derivatives such as $\frac{\partial x}{\partial r}, \frac{\partial y}{\partial r}$ etc. can be obtained from Eq.(11.1). Now the direction cosines of the new axes x', y', z' with respect to x, y, z are defined by [D] matrix, Eq. (9.46), as

$$[D] = \begin{bmatrix} l_1 & l_2 & l_3 \\ m_1 & m_2 & m_3 \\ n_1 & n_2 & n_3 \end{bmatrix} \quad (11.4)$$

The displacement variation in the element can be expressed as

$$\begin{Bmatrix} u \\ v \\ w \end{Bmatrix} = \sum_{i=1}^{4} N_i \left\{ \begin{Bmatrix} u_i \\ v_i \\ w_i \end{Bmatrix} + \begin{Bmatrix} u_i^* \\ v_i^* \\ w_i^* \end{Bmatrix} \right\} \quad (11.5)$$

where u_i, v_i, w_i are the displacements of the node i on the midsurface along the global x, y, z directions; u_i^*, v_i^*, w_i^* are the relative nodal displacements along x, y, z directions produced by the rotation of the normal at the node i.

The displacements u_i^*, v_i^*, w_i^* are to be expressed explicitly in terms of the rotations $\theta_{xi}, \theta_{yi}, \theta_{zi}$ at each node i about the global axes. Using the shell assumption that straight normals to the midsurface remain straight after deformation, the displacements produced by the normal rotations α_{1i}' and α_{2i}' can be calculated as (Fig. 11.3)

$$\begin{Bmatrix} u_i' \\ v_i' \\ w_i' \end{Bmatrix} = \frac{1}{2} t h_i \begin{Bmatrix} \alpha_{2i}' \\ -\alpha_{1i}' \\ 0 \end{Bmatrix} \quad (11.6)$$

where u_i', v_i', w_i' are displacement components along x', y', z' at node i and α_{1i}' and α_{2i}' are rotations about x' and y' respectively.

The components of these displacements along the global directions, u_i^*, v_i^*, w_i^*, can now be got by knowing the direction cosines of x', y', z' with respect to x, y, z (Eq.11.4)

$$u_i^* = l_{1i} u_i' + l_{2i} v_i'$$

$$v_i^* = m_{1i} u_i' + m_{2i} v_i'$$

$$w_i^* = n_{1i} u_i' + n_{2i} v_i' \quad (11.7)$$

Substituting from Eq.(11.6) into Eq.(11.7) and arranging the terms in matrix form we get,

$$\left\{\begin{array}{c} u_i^* \\ v_i^* \\ w_i^* \end{array}\right\} = \frac{1}{2}th_i \begin{bmatrix} l_{1i} & -l_{2i} \\ m_{1i} & -m_{2i} \\ n_{1i} & -n_{2i} \end{bmatrix} \left\{\begin{array}{c} \alpha_{2i}' \\ \alpha_{1i}' \end{array}\right\} \tag{11.8}$$

We can now express α_{1i}' and α_{2i}' in terms of global rotations θ_{xi}, θ_{yi} and θ_{zi} as,

$$\alpha_{1i}' = l_{1i}\theta_{xi} + m_{1i}\theta_{yi} + n_{1i}\theta_{zi}$$

$$\alpha_{2i}' = l_{2i}\theta_{xi} + m_{2i}\theta_{yi} + n_{2i}\theta_{zi} \tag{11.9}$$

Arranging the terms in matrix form we get

$$\left\{\begin{array}{c} \alpha_{2i}' \\ \alpha_{1i}' \end{array}\right\} = \begin{bmatrix} l_{2i} & m_{2i} & n_{2i} \\ l_{1i} & m_{1i} & n_{1i} \end{bmatrix} \left\{\begin{array}{c} \theta_{xi} \\ \theta_{yi} \\ \theta_{zi} \end{array}\right\} \tag{11.10}$$

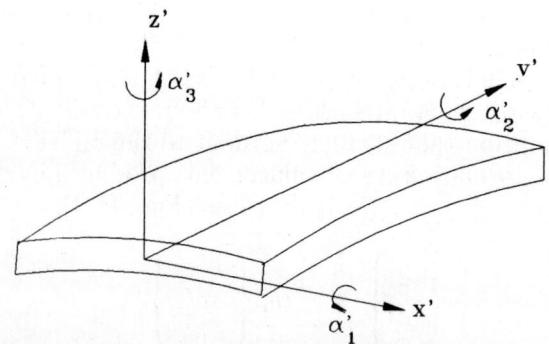

Fig. 11.3a Rotation about local axis

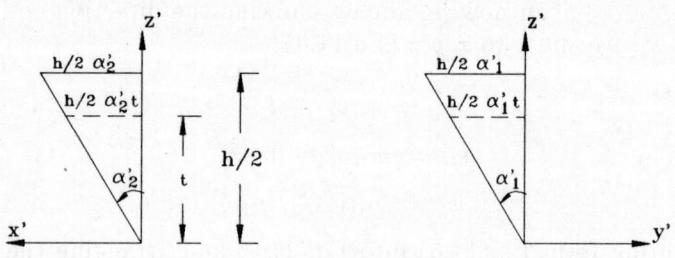

Fig. 11.3b Rotation of normal due to α_2' and α_1'

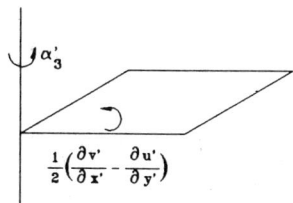

Fig. 11.3c Torsional stiffness

Substituting Eq.(11.10) into Eq.(11.8) we get,

$$\left\{ \begin{array}{c} u_i^* \\ v_i^* \\ w_i^* \end{array} \right\} = \frac{1}{2} th_i \ [\overline{D}_i] \left\{ \begin{array}{c} \theta_{xi} \\ \theta_{yi} \\ \theta_{zi} \end{array} \right\} \tag{11.11}$$

where

$$[\overline{D}_i] = \begin{bmatrix} l_{1i} & -l_{2i} \\ m_{1i} & -m_{2i} \\ n_{1i} & -n_{2i} \end{bmatrix} \begin{bmatrix} l_{2i} & m_{2i} & n_{2i} \\ l_{1i} & m_{1i} & n_{1i} \end{bmatrix}$$

which results in

$$[\overline{D}_i] = \begin{bmatrix} 0 & n_{3i} & -m_{3i} \\ -n_{3i} & 0 & l_{3i} \\ m_{3i} & -l_{3i} & 0 \end{bmatrix} \tag{11.12}$$

The terms l_{3i}, m_{3i}, and n_{3i} are direction cosines of unit vector e_3' as defined in equation (11.3a) and are to be evaluated at node i.

Substituting Eq.(11.12) into Eq.(11.5) the displacement variation is expressed in terms of nodal values, as,

$$\left\{ \begin{array}{c} u \\ v \\ w \end{array} \right\} = \sum_{i=1}^{4} N_i \left\{ \left\{ \begin{array}{c} u_i \\ v_i \\ w_i \end{array} \right\} + \frac{1}{2} th_i \ [\overline{D}_i] \left\{ \begin{array}{c} \theta_{xi} \\ \theta_{yi} \\ \theta_{zi} \end{array} \right\} \right\} \tag{11.13}$$

Substituting from Eq.(11.12) we get,

$$\left\{ \begin{array}{c} u \\ v \\ w \end{array} \right\} = \sum_{i=1}^{4} N_i \left\{ \left\{ \begin{array}{c} u_i \\ v_i \\ w_i \end{array} \right\} + \frac{t}{2} th_i \left\{ \begin{array}{cccc} n_{3i} & \theta_{yi} & -m_{3i} & \theta_{zi} \\ l_{3i} & \theta_{zi} & -n_{3i} & \theta_{xi} \\ m_{3i} & \theta_{xi} & -l_{3i} & \theta_{yi} \end{array} \right\} \right\} \tag{11.13a}$$

11.3.2   Strain-Displacement Matrix

Assuming $\epsilon_z = 0$, the strain components along the local axes of the shell element are given by,

$$\{\epsilon'\} = \begin{Bmatrix} \epsilon_{x'} \\ \epsilon_{y'} \\ \gamma_{x'y'} \\ \gamma_{x'z'} \\ \gamma_{y'z'} \end{Bmatrix} = \begin{Bmatrix} \dfrac{\partial u'}{\partial x'} \\[2mm] \dfrac{\partial v'}{\partial y'} \\[2mm] \dfrac{\partial u'}{\partial y'} + \dfrac{\partial v'}{\partial x'} \\[2mm] \dfrac{\partial u'}{\partial z'} + \dfrac{\partial w'}{\partial x'} \\[2mm] \dfrac{\partial v'}{\partial z'} + \dfrac{\partial w'}{\partial y'} \end{Bmatrix} \tag{11.14}$$

The strain components in the local axes system can be obtained through the Eq.(9.56) and the [D] matrix is given by Eq.(11.4). The derivative of u, v,w with respect to x,y,z to be used in Eq.(9.56) are computed from Eq.(9.29). For this purpose the derivatives of u,v,w with respect to r,s,t are required and they are obtained by differentiating Eq.(11.13). Thus,

$$\begin{bmatrix} \dfrac{\partial u}{\partial r} & \dfrac{\partial v}{\partial r} & \dfrac{\partial w}{\partial r} \\[2mm] \dfrac{\partial u}{\partial s} & \dfrac{\partial v}{\partial s} & \dfrac{\partial w}{\partial s} \\[2mm] \dfrac{\partial u}{\partial t} & \dfrac{\partial v}{\partial t} & \dfrac{\partial w}{\partial t} \end{bmatrix} = \sum_{i=1}^{4} \begin{bmatrix} \dfrac{\partial N_i}{\partial r}u_i & \dfrac{\partial N_i}{\partial r}v_i & \dfrac{\partial N_i}{\partial r}w_i \\[2mm] \dfrac{\partial N_i}{\partial s}u_i & \dfrac{\partial N_i}{\partial s}v_i & \dfrac{\partial N_i}{\partial s}w_i \\[2mm] 0 & 0 & 0 \end{bmatrix}$$

$$+ \sum_{i=1}^{4} \frac{h_i}{2} \begin{bmatrix} t\frac{\partial N_i}{\partial r}(n_{3i}\theta_{yi} - m_{3i}\theta_{zi}) & t\frac{\partial N_i}{\partial r}(l_{3i}\theta_{zi} - n_{3i}\theta_{xi}) \\[2mm] t\frac{\partial N_i}{\partial s}(n_{3i}\theta_{yi} - m_{3i}\theta_{zi}) & t\frac{\partial N_i}{\partial s}(l_{3i}\theta_{zi} - n_{3i}\theta_{xi}) \\[2mm] N_i(n_{3i}\theta_{yi} - m_{3i}\theta_{zi}) & N_i(l_{3i}\theta_{zi} - n_{3i}\theta_{xi}) \end{bmatrix}$$

$$\begin{matrix} t\frac{\partial N_i}{\partial r}(m_{3i}\theta_{xi} - l_{3i}\theta_{yi}) \\[2mm] t\frac{\partial N_i}{\partial s}(m_{3i}\theta_{xi} - l_{3i}\theta_{yi}) \\[2mm] N_i(m_{3i}\theta_{xi} - l_{3i}\theta_{yi}) \end{matrix} \tag{11.15}$$

Making use of the procedure and equations indicated above, all the derivatives necessary to compute $\{\epsilon'\}$ of Eq.11.14 can be obtained. The strain displacement matrix [B] may be split up conveniently into two matrices $[B_m]$ and $[B_s]$ such that

$$\{\epsilon'_m\} = \begin{Bmatrix} \epsilon_{x'} \\ \epsilon_{y'} \\ \gamma_{x'y'} \end{Bmatrix} = \sum_{i=1}^{4}[B_{mi}]\{d_i\} \tag{11.16a}$$

$$\{\epsilon'_s\} = \left\{ \begin{array}{c} \gamma_{x'z'} \\ \gamma_{y'z'} \end{array} \right\} = \sum_{i=1}^{4} [B_{st}] \{d_i\} \qquad (11.16b)$$

where $\{d_i\}$ represents the global displacements and rotations at each node.

The strain-displacement matrices $[B_{mi}]$ is further split as

(i) $[B_{1mi}]$, (ii) $[B_{2mi}]$, and (iii) $[B_{3mi}]$.

$[B_{1mi}]$ is formed considering only in plane displacements u_i, v_i, and w_i.

$[B_{2mi}]$ and $[B_{3mi}]$ are formed considering rotations $\theta_{xi}, \theta_{yi}, \theta_{zi}$.

Similarly the strain-displacements matrix $[B_{si}]$ is split into (i) $[B_{1si}]$, (ii) $[B_{2si}]$, and (iii) $[B_{3si}]$. $[B_{1si}]$ is formed considering only in plane displacements and $[B_{2si}]$ and $[B_{3si}]$ are formed considering rotations only.

(a) Formulation of $[B_{1mi}]$: Making use of Eq.11.15, the derivatives of u' and v' with respect to x' and y' are computed. Arranging the terms with respect to inplane displacements u_i, v_i, and $w_i, [B_{1mi}]$ matrix is constructed. A typical term is given below:

$$\frac{\partial u'}{\partial x'} = \sum_{i=1}^{4} l_i \left[\left(\frac{\partial N_i}{\partial x} l_1 u_i + \frac{\partial N_i}{\partial x} m_1 v_i + \frac{\partial N_i}{\partial x} n_1 w_i \right) \right.$$

$$+ m_i \left(\frac{\partial N_i}{\partial y} l_1 u_i + \frac{\partial N_i}{\partial y} m_1 v_i + \frac{\partial N_i}{\partial y} n_1 w_i \right)$$

$$\left. + n_i \left(\frac{\partial N_i}{\partial z} l_1 u_i + \frac{\partial N_i}{\partial z} m_1 v_i + \frac{\partial N_i}{\partial z} n_1 w_i \right) \right] \qquad (11.17)$$

Similarly $\frac{\partial v'}{\partial x'}$ and $\left(\frac{\partial u'}{\partial y'} + \frac{\partial v'}{\partial x'} \right)$ can be computed. Now the inplane strains due to u_i, v_i and w_i can be expressed as,

$$\left\{ \begin{array}{c} \epsilon_{x'} \\ \epsilon_{y'} \\ \gamma_{x'y'} \end{array} \right\} = \left\{ \begin{array}{c} \dfrac{\partial u'}{\partial x'} \\[2mm] \dfrac{\partial v'}{\partial y'} \\[2mm] \dfrac{\partial u'}{\partial y'} + \dfrac{\partial v'}{\partial x'} \end{array} \right\} = \sum_{i=1}^{4} [B_{1mi}] \left\{ \begin{array}{c} u_i \\ v_i \\ w_i \end{array} \right\} \qquad (11.18)$$

where

$$[B_{1mi}] = \begin{bmatrix} l_1 B'(1,i) & | & m_1 B'(1,i) & | & n_1 B'(1,i) \\ \hline l_2 B'(2,i) & | & m_2 B'(2,i) & | & n_2 B'(2,i) \\ \hline l_1 B'(2,i)+ & | & m_1 B'(2,i)+ & | & n_1 B'(2,i)+ \\ l_2 B'(1,i) & | & m_2 B'(1,i) & | & n_2 B'(1,i) \end{bmatrix} \quad (11.19)$$

and

$$B'(1,i) = \frac{\partial N_i}{\partial x} l_1 + \frac{\partial N_i}{\partial y} m_1 + \frac{\partial N_i}{\partial z} n_1$$

$$B'(2,i) = \frac{\partial N_i}{\partial x} l_2 + \frac{\partial N_i}{\partial y} m_2 + \frac{\partial N_i}{\partial z} n_2 \quad (11.20)$$

(b) *Formulation of* $[B_{2mi}]$ and $[B_{3mi}]$. The same procedure as in the case of $[B_{1mi}]$ is followed and here the contributions due to rotations are considered. A typical term is given below

$$\frac{\partial u'}{\partial x'} = \sum_{i=1}^{4} \frac{h_i}{2} \left[l_1 \left(t \frac{\partial N_i}{\partial x} + J_{13}^* N_i \right) + m_1 \left(t \frac{\partial N_i}{\partial y} + J_{23}^* N_i \right) \right.$$

$$\left. + n_1 \left(t \frac{\partial N_i}{\partial z} + J_{33}^* N_i \right) \right] \times \left[(n_{3i} \theta_{yi} - m_{3i} \theta_{zi}) l_1 \right.$$

$$\left. + (l_{3i} \theta_{zi} - n_{3i} \theta_{xi}) m_1 + (m_{3i} \theta_{xi} - l_{3i} \theta_{yi}) n_1 \right] \quad (11.21)$$

Arranging the relation between the inplane strains and rotations in matrix form we get,

$$\begin{Bmatrix} \epsilon_{x'} \\ \epsilon_{y'} \\ \gamma_{x'y'} \end{Bmatrix} = \begin{Bmatrix} \dfrac{\partial u'}{\partial x'} \\[2mm] \dfrac{\partial v'}{\partial y'} \\[2mm] \left(\dfrac{\partial u'}{\partial y'} + \dfrac{\partial v'}{\partial x'} \right) \end{Bmatrix} = \sum_{i=1}^{4} [[B_{2mi}] + t[B_{3mi}]] \begin{Bmatrix} \theta_{xi} \\ \theta_{yi} \\ \theta_{zi} \end{Bmatrix}$$

$$(11.22)$$

In $[B_{2mi}]$ we have terms such as $l_1 J_{13}^* + m_1 J_{23}^* + n_1 J_{33}^*$ and $l_2 J_{13}^* + m_2 J_{23}^* + n_2 J_{33}^*$

It can be shown that $J_{13}^* = l_3; J_{23}^* = m_3$ and $J_{33}^* = n_3$.

And by orthogonality condition we have,

$$l_1 l_3 + m_1 m_3 + n_1 n_3 = 0$$

$$l_2 l_3 + m_2 m_3 + n_2 n_3 = 0 \tag{11.23}$$

Hence, it can be observed that

$$[B_{2mi}] = [0] \tag{11.24}$$

Therefore, Eq. 11.22 reduces to,

$$\begin{Bmatrix} \epsilon_{x'} \\ \epsilon_{y'} \\ \gamma_{x'y'} \end{Bmatrix} = \sum_{i=1}^{4} t[B_{3mi}] \begin{Bmatrix} \theta_{xi} \\ \theta_{yi} \\ \theta_{zi} \end{Bmatrix} \tag{11.25}$$

where

$$[B_{3mi}] = \frac{h_i}{2} \times \begin{bmatrix} B'(1,i)(m_{3i}n_i - n_{3i}m_1) & | & B'(1,i)(n_{3i}l_1 - l_{3i}n_1) & | \\ \hline B'(2,i)(m_{3i}n_2 - n_{3i}m_2) & | & B'(2,i)(n_{3i}l_2 - l_{3i}n_2) & | \\ \hline B'(2,i)(m_{3i}n_i - n_{3i}m_1) & | & B'(2,i)(n_{3i}l_1 - l_{3i}n_1) & | \\ + & & + & \\ B'(1,i)(m_{3i}n_2 - n_{3i}m_2) & | & B'(1,i)(n_{3i}l_2 - l_{3i}n_2) & | \end{bmatrix}$$

$$\begin{matrix} B'(1,i)(l_{3i}m_1 - m_{3i}l_1) \\ \hline B'(2,i)(l_{3i}m_2 - m_{3i}l_2) \\ \hline B'(2,i)(l_{3i}m_1 - m_{3i}l_1) \\ + \\ B'(1,i)(l_{3i}m_2 - m_{3i}l_2) \end{matrix} \tag{11.26}$$

(c) *Formulation of* $[B_{1si}]$: Similarly the derivatives are worked out for computing $\gamma_{x'z'}$ and $\gamma_{y'z'}$ due to displacements u', v', w'. Thus,

$$\begin{Bmatrix} \gamma_{x'z'} \\ \gamma_{y'z'} \end{Bmatrix} = \begin{Bmatrix} \frac{\partial u'}{\partial z'} + \frac{\partial w'}{\partial x'} \\ \frac{\partial v'}{\partial z'} + \frac{\partial w'}{\partial y'} \end{Bmatrix} = \sum_{i=1}^{4} [B_{1si}] \begin{Bmatrix} u_i \\ v_i \\ w_i \end{Bmatrix} \tag{11.27}$$

where

$$[B_{1si}] = \begin{bmatrix} l_1 B'(3,i)+ & | & m_1 B'(3,i)+ & | & n_1 B'(3,i)+ \\ l_3 B'(1,i) & | & m_3 B'(1,i), & | & n_3 B'(1,i) \\ \hline l_2 B'(3\ i)+ & | & m_2 B'(3,i)+ & | & n_2 B'(3,i)+ \\ l_3 B'(2,i) & | & m_3 B'(2,i) & | & n_3 B'(2,i) \end{bmatrix} \tag{11.28}$$

and

$$B'(3,i) = \frac{\partial N_i}{\partial x} l_3 + \frac{\partial N_i}{\partial y} m_3 + \frac{\partial N_i}{\partial z} n_3 \qquad (11.29)$$

(d) *Formulation of* $[B_{2si}]$ *and* $[B_{3si}]$. The strains $\gamma_{x'z'}$ and $\gamma_{y'z'}$ due to rotations θ_{xi}, θ_{yi} and θ_{zi} are expressed as,

$$\left\{ \begin{array}{c} \gamma_{x'z'} \\ \gamma_{y'z'} \end{array} \right\} = \left\{ \begin{array}{c} \frac{\partial u'}{\partial z'} + \frac{\partial w'}{\partial x'} \\ \frac{\partial v'}{\partial z'} + \frac{\partial w'}{\partial y'} \end{array} \right\} = \sum_{i=1}^{4} [[B_{2si}] + t[B_{3si}]] \left\{ \begin{array}{c} \theta_{xi} \\ \theta_{yi} \\ \theta_{zi} \end{array} \right\} \qquad (11.30)$$

The matrices $[B_{2si}]$ and $[B_{3si}]$ are constructed in a similar manner as before and are given as follows:

$[B_{2si}] = \frac{h_i}{2} N_i B'' \times$

$$\begin{bmatrix} (m_{3i}n_1 - n_{3i}m_1) & | & (n_{3i}l_1 - l_{3i}n_1) & | & (l_{3i}m_1 - m_{3i}l_1) \\ \text{-----} & + & \text{-----} & + & \text{-----} \\ (m_{3i}n_2 - n_{3i}m_2) & | & (n_{3i}l_2 - l_{3i}n_2) & | & (l_{3i}m_2 - m_{3i}l_2) \end{bmatrix} \qquad (11.31)$$

where

$$B'' = l_3 J_{13}^* + m_3 J_{23}^* + n_3 J_{33}^* \qquad (11.32)$$

and

$$[B_{3si}] = \frac{h_i}{2} \begin{bmatrix} \begin{array}{c} B'(3,i)(m_{3i}n_1 - n_{3i}m_1) \\ + \\ B'(1,i)(m_{3i}n_3 - n_{3i}m_3) \\ \text{-------} \\ B'(3,i)(m_{3i}n_2 - n_{3i}m_2) \\ + \\ B'(2,i)(m_{3i}n_3 - n_{3i}m_3) \end{array} & \bigg| & \begin{array}{c} B'(3,i)(n_{3i}l_1 - l_{3i}n_1) \\ + \\ B'(1,i)(n_{3i}l_3 - l_{3i}n_3) \\ \text{-------} \\ B'(3,i)(n_{3i}l_2 - l_{3i}n_2) \\ + \\ B'(2,i)(n_{3i}l_3 - l_{3i}n_3) \end{array} & \bigg| \end{bmatrix}$$

$$\begin{bmatrix} B'(3,i)(l_{3i}m_1 - m_{3i}l_1) \\ + \\ B'(1,i)(l_{3i}m_3 - m_{3i}l_3) \\ \text{------} \\ B'(3,i)(l_{3i}m_2 - m_{3i}l_2) \\ + \\ B'(2,i)(l_{3i}m_3 - m_{3i}l_3) \end{bmatrix} \qquad (11.33)$$

We can now arrange the strain-displacement matrix in the following form:

$$\left\{ \begin{array}{c} \{\epsilon_m\} \\ \text{---} \\ \{\epsilon_s\} \end{array} \right\} = \sum_{i=1}^{4} \begin{bmatrix} [B_{1mi}] & | & t[B_{3mi}] \\ \text{---} & + & \text{-------} \\ [B_{1si}] & | & [B_{2si}] + t[B_{3si}] \end{bmatrix} \left\{ \begin{array}{c} \{d_i^u\} \\ \text{---} \\ \{d_i^\theta\} \end{array} \right\}$$

where

$$\{d_i^u\} = [u_i \ v_i \ w_i]^T \qquad (11.34)$$

$$\{d_i^\theta\} = [\theta_{xi} \ \theta_{yi} \ \theta_{zi}]^T$$

11.3.3 Stress-Displacement Matrix

The element stresses and nodal displacements are related as (equation 3.70)

$$\{\sigma\} = [C] \ [B] \ \{d\}$$

$$= [CB] \ \{d\}$$

For the case of isotropic material; the consitutive relation is given by Eq.2.16. Now imposing the condition that $\sigma_{z'} = 0$, the following relation is obtained for the stress-strain relation in x', y', z' coordinates;

$$\{\sigma'\} = [C] \ \{\epsilon'\}$$

i.e.,

$$
\begin{Bmatrix} \sigma_{x'} \\ \sigma_{y'} \\ \tau_{x'y'} \\ \tau_{x'z'} \\ \tau_{y'z'} \end{Bmatrix} = \frac{E}{1-\mu^2}
\begin{bmatrix}
1 & \mu & 0 & 0 & 0 \\
\mu & 1 & 0 & 0 & 0 \\
0 & 0 & \frac{(1-\mu)}{2} & 0 & 0 \\
0 & 0 & 0 & \frac{\alpha(1-\mu)}{2} & 0 \\
0 & 0 & 0 & 0 & \frac{\alpha(1-\mu)}{2}
\end{bmatrix}
\begin{Bmatrix} \epsilon_{x'} \\ \epsilon_{y'} \\ \gamma_{x'y'} \\ \gamma_{x'z'} \\ \gamma_{y'z'} \end{Bmatrix}
$$

$$(11.35)$$

where α is a numerical correction factor used to account for a better representation of shear deformation and is explained in Chapters 7 and 10. A value of 5/6 has been suggested.

To facilitate adoption of different numerical integration schemes for bending and shear contributions to the stiffness matrix, the constitutive matrix is split into $[C_m]$ and $[C_s^1]$ as,

$$[C] = \left[\begin{array}{c|c} [C_m] & [0] \\ \hline [0] & [C_s^1] \end{array} \right] \qquad (11.36)$$

$$[C_m] = \frac{E}{1-\mu^2} \begin{bmatrix} 1 & \mu & 0 \\ \mu & 1 & 0 \\ 0 & 0 & \frac{1-\mu}{2} \end{bmatrix} \qquad (11.37)$$

and

$$[C_s^1] = \frac{E\alpha}{2(1+\mu)} \begin{bmatrix} 1 & 0 \\ 0 & 1 \end{bmatrix} \qquad (11.38)$$

It may be noted that Eq.11.37 is the same as for the plane stress conditions given by Eq.2.17 and Eq.11.38 is similar to Eq.10.28 except that the terms are multiplied by h to get the corresponding shear forces in the case of plate bending problems.

11.3.4   Element Stiffness Matrix

It is convenient to split the stiffness matrix into two parts; the bending and membrane effects and transverse shear effects. This will allow the use of appropriate order of numerical integration of each part.
Thus,

$$[k] = [k]_m + [k]_s \qquad (11.39a)$$

i.e.,

$$[k] = \sum_{i=1}^{4} \sum_{j=1}^{4} [[k_{ij}]_m + [k_{ij}]_s] \qquad (11.39b)$$

$$[k_{ij}]_m = \int_V [B_{mi}]^T [C_m] [B_{mj}] \, dV \qquad (11.40a)$$

$$[k_{ij}]_s = \int_V [B_{si}]^T [C_s] [B_{sj}] \, dV \qquad (11.40b)$$

Where Eq.11.40a gives the contribution due to bending and membrane effects and Eq.11.40b gives transverse shear contribution to stiffness matrix.

Substituting from Eq.11.34 for bending and membrane contribution into Eq.11.40a we get,

$$[k_{ij}]_m = \int_v \begin{bmatrix} [B_{1mi}]^T \\ ----- \\ t[B_{3mi}]^T \end{bmatrix} [C_m] \Big[[B_{1mj}] \;\vdots\; t[B_{3mj}] \Big] \, dV \qquad (11.41)$$

Simplifying the above equation and expressing it in natural coordinates we get, $[k_{ij}]_m =$

$$\int_{-1}^{+1} \int_{-1}^{+1} \int_{-1}^{+1} \begin{bmatrix} [B_{1mi}]^T [C_m] [B_{1mj}] & | & t[B_{1mi}]^T [C_m] [B_{3mj}] \\ ----------- & + & ----------- \\ t[B_{3mi}]^T [C_m] [B_{1mi}] & | & t^2[B_{3mi}]^T [C_m] [B_{3mj}] \end{bmatrix}$$

$$|J_{(r,s,t)}| \; dr \; ds \; dt \qquad (11.42)$$

where $|J|$ is the determinant of the Jacobian matrix defined in Eq.9.23. To be consistent with the shell assumption $|J_{(r,s,t)}|$ can be approximated by $|J_{(r,s,o)}|$.

Since $[B_{1mi}]$ and $[B_{3mj}]$ are functions of r and s only, the integral of Eq.11.42 can be analytically integrated with respect to r and thus we get,

$$[k_{ij}]_m = \int_{-1}^{+1} \int_{-1}^{+1} \left[\begin{array}{c|c} 2[B_{1mi}]^T[C_m][B_{1mj}] & [0] \\ \hline [0] & \frac{2}{3}[B_{3mi}]^T[C_m][B_{3mj}] \end{array} \right]$$

$$\times |J_{r,s,o}| \, dr \, ds \qquad (11.43)$$

Similarly substituting from Eq.11.34 for shear contribution into Eq.11.40b we get,

$$[k_{ij}] = \int_V \left[\begin{array}{c} [B_{1si}]^T \\ \hline [B_{2si}]^T + t[B_{3si}]^T \end{array} \right] [C_s] \left[[B_{1sj}] \; \vdots \; [B_{2sj}] + t[B_{3sj}] \right] \, dV$$

$$(11.44)$$

Integrating across the thickness as before, we get

$$[k_{ij}]_s = \int_{-1}^{+1} \int_{-1}^{+1} \left[\begin{array}{c|c} 2[B_{1si}]^T [C_s] [B_{1sj}] & 2[B_{1si}]^T [C_s] [B_{2sj}] \\ \hline 2[B_{2si}] [C_s] [B_{1sj}] & 2[B_{2si}] [C_s] [B_{2si}] \\ & +\frac{2}{3}[B_{3si}]^T [C_s] [B_{3sj}] \end{array} \right] \times$$

$$|J_{(r,s,o)}| \, dr \, ds \qquad (11.45)$$

The size of each sub matrix in Eqs. 11.43 and 11.45 is 6×6. Thus the bending and membrane, and shear stiffness contribution to the element stiffness matrix can be computed as,

$$[k]_m \text{ or } [k]_s = \begin{bmatrix} [k_{11}] & [k_{12}] & [k_{13}] & [k_{14}] \\ [k_{21}] & [k_{22}] & [k_{23}] & [k_{24}] \\ [k_{31}] & [k_{32}] & [k_{33}] & [k_{34}] \\ [k_{41}] & [k_{42}] & [k_{43}] & [k_{44}] \end{bmatrix} \qquad (11.46)$$

The Eq.11.46 is similar to Eq.10.58 for the computation of plate bending stiffness. The sub-matrix $[k_{11}, [k_{12}], [k_{13}], [k_{14}]$ are evaluated by letting $i = 1, j = 1, 2, 3, 4$. The other sub-matrices are evaluated in a similar manner.

Numerical integration procedure is used to evaluate the stiffness matrix of Eqs.11.43 and 11.45. A 2 × 2 Gauss quadrature is used to evaluate the integral 11.43, bending and membrane contribution. To avoid shear locking effect as explained in Chapter 10, one point Gauss quadrature is used to evaluate the integral 11.45, shear contribution to the stiffness matrix.

11.3.5 Torsional Stiffness

The four noded degenerated shell element described earlier employs six degrees of freedom at each node. However, no stiffness corresponding to the torsional rotation degree of freedom exists in the local axes system in the formulation. But when non-planar elements join at a node i, the resistance to this rotation is got due to transformation of stiffness coefficients parallel to global direction, i.e., along θ_{zi}. When the finite element mesh is refined, the angles between elements may become close to 2π and this will result in a very small amount of stiffness for the torsional rotation. Therefore, any slight disturbance in the load corresponding to this degree of freedom will amplify the torsional rotation degree of freedom to an unrealistic value and thus will affect the global solution.

This problem is common to all shell elements that use six global degrees of freedom at each node. But the difficulty is got over by providing a fictitious torsional spring along the local normal direction at each node of the element. This technique increases the amount of work in data preparation and is also unsatisfactory since for a flexible system it produces an unrealistic amount of strain energy in a spring by a rigid body motion.

For the four noded shell element, the rotation of the normal and the mid-surface field are independent. As shown in Fig.11.3c the rotation of the midsurface is $\frac{1}{2}\left(\frac{\partial v'}{\partial x'} - \frac{\partial u'}{\partial y'}\right)$.

The derivation of the torsional rotation of the normal from that of the midsurface is assumed to have the governing strain energy,

$$U_t = \alpha_t Gh \int_A \int \left[\alpha_3' - \frac{1}{2}\left(\frac{\partial v'}{\partial x'} - \frac{\partial u'}{\partial y'}\right)\right]^2_{(r,s,o)} dA \qquad (11.47)$$

where α_t is known as torsional coefficient.

If $\alpha_t Gh$ is chosen to be large relative to the factor Eh^3 used in bending energy calculations, Eq.11.47 will play the role of penalty function and results in the desired constraint at the Gauss points [7] as

$$\alpha_3' \approx \frac{1}{2}\left(\frac{\partial v'}{\partial x'} - \frac{\partial u'}{\partial y'}\right) \qquad (11.48)$$

The use of penalty functions in the potential energy functional and illustration of constraints for beam and plate problems are described in reference[13].

Now the torsional stiffness coefficient is derived from Eq.11.47. Following the procedure indicated in the earlier section we get,

$$
\begin{aligned}
\left(\frac{\partial v'}{\partial x'} - \frac{\partial u'}{\partial y'}\right) =\ & \sum_{i=1}^{4} [l_1\left(\frac{\partial N_i}{\partial x} l_2 u_i + \frac{\partial N_i}{\partial x} m_2 v_i + \frac{\partial N_i}{\partial x} n_2 w_i\right) \\
& + m_1\left(\frac{\partial N_i}{\partial y} l_2 u_i + \frac{\partial N_i}{\partial y} m_2 v_i + \frac{\partial N_i}{\partial y} n_2 w_i\right) \\
& + n_1\left(\frac{\partial N_i}{\partial z} l_2 u_i + \frac{\partial N_i}{\partial z} m_2 v_i + \frac{\partial N_i}{\partial z} n_2 w_i\right) \\
& - l_2\left(\frac{\partial N_i}{\partial x} l_1 u_i + \frac{\partial N_i}{\partial x} m_1 v_i + \frac{\partial N_i}{\partial x} n_1 w_i\right) \\
& + m_2\left(\frac{\partial N_i}{\partial y} l_1 u_i + \frac{\partial N_i}{\partial y} m_1 v_i + \frac{\partial N_i}{\partial y} n_1 w_i\right) \\
& + n_2\left(\frac{\partial N_i}{\partial z} l_1 u_i + \frac{\partial N_i}{\partial z} m_1 v_i + \frac{\partial N_i}{\partial z} n_1 w_i\right)]
\end{aligned}
$$

$$(11.49)$$

If α_{3i} is the rotation about the local z' axis at node i, then it can be expressed in terms of global rotations $\theta_{xi}, \theta_{yi}, \theta_{zi}$ as,

$$\alpha'_{3i} = l_{3i}\, \theta_{xi} + m_{3i}\, \theta_{yi} + n_{3i}\, \theta_{zi} \qquad (11.50)$$

If α'_3 is the rotation at any point (r, s) on the midsurface, then the variation can be expressed through the same function N_i, as

$$\alpha'_3 = N_i\, (l_3\, \theta_{xi} + m_3\, \theta_{yi} + n_3\, \theta_{zi}) \qquad (11.51)$$

Arranging Eq.11.49 and 11.50 in matrix form and substituting in Eq.11.47 we can express U_t in terms of the torsional stiffness matrix as,

$$U_t = \{d\}^T\, [k]_t\, \{d\} \qquad (11.52)$$

where, the submatrix $[k_{ij}]_t$ for torsional stiffness, is given by

$$
[k_{ij}]_t = \alpha_t\, G\, h \int_{-1}^{+1}\int_{-1}^{+1}
\begin{bmatrix}
[R_{mi}]^T\, [R_{mj}] & [R_{mi}]^T\, [R_{nj}] \\
[R_{ni}]^T\, [R_{mj}] & [R_{ni}]^T\, [R_{nj}]
\end{bmatrix}
|a|\, dr\, ds
$$

$$(11.53)$$

where $\alpha_t =$ torsional coefficient
   $G =$ shear modulus
   $h =$ thickness

$$[R_{mi}] = \frac{1}{2} \ [B'(2,i) \ - \ B'(1,i) \ 0] \ [D]^T \qquad (11.54a)$$

$$[R_{ni}] \ = \ N_i[l_3 \ m_3 \ n_3] \qquad (11.54b)$$

$$dA \ = \ |e_1' \times e_2'| \ = \ |a| \ dr \ ds$$

$$|a| \ =$$

$$\sqrt{(\frac{\partial x}{\partial r} \frac{\partial y}{\partial s} \ - \ \frac{\partial x}{\partial s} \frac{\partial y}{\partial r})^2 \ + \ (\frac{\partial x}{\partial s} \frac{\partial z}{\partial r} \ - \ \frac{\partial x}{\partial r} \frac{\partial z}{\partial s})^2 \ + \ (\frac{\partial y}{\partial r} \frac{\partial z}{\partial s} \ - \ \frac{\partial y}{\partial s} \frac{\partial z}{\partial r})^2}$$

$$\qquad (11.54c)$$

It can also be shown that

$$|a| \ = \ |J| \times \sqrt{(J_{13}^*)^2 \ + \ (J_{23}^*)^2 + (J_{33}^*)^2} \qquad (11.54d)$$

Now the torsional stiffness matrix $[k]_t$ is evaluated from $[k_{ij}]_t$ following the procedure described for evaluating bending and membrane contribution, and shear contribution by letting $i = 1, 2, ...4$ and $j = 1, ...4$ (Refer description under Eq.11.46.)

A 1×1 Gauss quadrature is used in evaluating $[k]_t$ at the centre of the element to avoid an overconstrained situation similar to shear locking behaviour explained earlier.

It was demonstrated in reference [7] that the converged solution in insensitive to α_t as long as α_t is large enough (> 0.1) to sufficiently restrain the torsional modes. This experiment also indicates that the addition of torsional stiffness will not degrade the behaviour of the system.

11.3.6   Element Load Vector

Computation of element load vector due to gravity load, uniform normal surface pressure, uniform vertical loading and also water pressure, is described in the following. For all the cases 2×2 Gauss points are used for numerical integration of the nodal load vector.

(a) *Gravity Loads*   Gravity loads are calculated from uniform weight density (ρ) throughout the element. From Eq.11.13 the vertical displacement is given by,

$$w \ = \ \sum_{i=1}^{4} N_i \ w_i \ + \ \sum_{i=1}^{4} \frac{1}{2} \ th_i \ N_i \ (m_{3i} \ \theta_{xi} \ - \ l_{3i} \ \theta_{yi}) \qquad (11.55)$$

Hence, the load vector at node i is given by,

$$\{Q_i\} = \int_{-1}^{+1} \int_{-1}^{+1} \int_{-1}^{+1} \begin{Bmatrix} 0 \\ 0 \\ N_i \\ \frac{1}{2} th_i N_i m_{3i} \\ -\frac{1}{2} th_i N_i l_{3i} \\ 0 \end{Bmatrix} |J| \, dr \, ds \, dt \qquad (11.56)$$

If we integrate Eq.11.56 analytically with respect to t we get,

$$\{Q_i\} = \int_{-1}^{+1} \int_{-1}^{+1} \rho \begin{Bmatrix} 0 \\ 0 \\ N_i \\ 0 \\ 0 \\ 0 \end{Bmatrix} |J| \, dr \, ds \qquad (11.56a)$$

(b) *Uniform Normal Surface Pressure* To evaluate the nodal loads for normal pressure, the displacement normal to the surface is required and it is given by,

$$u_n = [u \quad v \quad w] \begin{Bmatrix} l_3 \\ m_3 \\ n_3 \end{Bmatrix} \qquad (11.57)$$

Let p_n be the normal pressure applied at the top surface, $t = 1$. Substituting from equation (11.13a) for u_n we get

$$\{Q_i\} = \int \int p_n \begin{Bmatrix} l_3 N_i \\ m_3 N_i \\ n_3 N_i \\ \frac{1}{2} h_i (n_3 m_{3i} - m_3 n_{3i}) \\ \frac{1}{2} h_i (l_3 n_{3i} - n_3 l_{3i}) \\ \frac{1}{2} h_i (m_3 l_{3i} - l_3 m_{3i}) \end{Bmatrix} dA \qquad (11.58a)$$

where dA is defined in Eq.11.54c.

(c) *Uniform Vertical Loading* Let p_v is the vertical pressure applied at the top surface, $t = 1$. Making use of Eq.11.55 for vertical displacement w, the nodal load vector is given by,

$$\{Q_i\} = \int\int p_v \left\{ \begin{array}{c} 0 \\ 0 \\ N_i \\ \frac{1}{2} h_i N_i m_{3i} \\ -\frac{1}{2} h_i N_i l_{3i} \\ 0 \end{array} \right\} dA \qquad (11.58b)$$

where dA is given by Eq.11.54c.

(d) *Water Pressure* The expression derived for normal pressure is applicable for this case. Let the unit weight of water be p_w and H is height of water level. Then,

$$\{Q_i\} = \int\int p_w H \left\{ \begin{array}{c} l_3 N_i \\ m_3 N_i \\ n_3 N_i \\ \frac{1}{2} h_i (n_3 m_{3i} - m_3 n_{3i}) \\ \frac{1}{2} h_i (l_3 n_{3i} - n_3 l_{3i}) \\ \frac{1}{2} h_i (m_3 l_{3i} - l_3 m_{3i}) \end{array} \right\} dA \qquad (11.58c)$$

where dA is defined by Eq.11.54c.

11.4 Eight Noded Shell Element

The eight noded curved shell element due to Pawsey[6] is also based on the assumptions that normals to the midsurface remain straight after deformation and the normal stress is zero. This shell element is based on the degeneration concept and numerical integration technique is adopted for evaluation of the integral across the thickness. Hence, the element properties can be derived using the procedure and expressions described in the case of three-dimensional solid elements in Chapter 9. The basic expressions to define the shell characteristics and necessary equations leading to the formulation of the stiffness matrix and element load vector are explained in this section.

11.4.1 Shape Functions for Geometry

A typical curved shell element, with eight nodes on the mid-surface and an additional node at the centre of the midsurface is shown in Fig.11.4. along with the global coordinate system (x, y, z). The shape functions treating it as a two-dimensional isoparametric element in (r, s) coordinates, are given below.

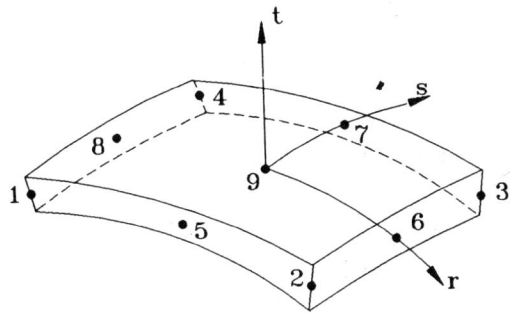

Fig. 11.4 Curved shell element

For nodes 1,2,3 and 4

$$N_i = \frac{1}{4}\left[(1 + rr_i)(1 + ss_i)(rr_i + ss_i - 1)\right]$$

For nodes 5 and 7

$$N_i = \frac{1}{2}(1 - r^2)(1 + ss_i)$$

For nodes 6 and 8

$$N_i = \frac{1}{2}(1 - s^2)(1 + rr_i)$$

For node 9,

$$N_i = (1 - r^2)(1 - s^2) \tag{11.59}$$

where r_i and s_i are the coordinates r and s of the node i of the mid-surface.

It may be noted that the shape functions for all the nodes except the central node are the same as those used for eight noded plates.

Assuming the lines joining the top and bottom nodes to be straight, the shape of the element is defined by the eight nodal values, as

$$\left\{\begin{array}{c} x \\ y \\ z \end{array}\right\} = \sum_{i=1}^{8} N_i \left\{\left\{\begin{array}{c} x_i \\ y_i \\ z_i \end{array}\right\} + t \left\{\begin{array}{c} x_i^* \\ y_i^* \\ z_i^* \end{array}\right\}\right\} \tag{11.60}$$

where x_i, y_i, z_i are the global coordinates of the midsurface node i which are computed by taking the average of the top and bottom node coordinates and x_i^*, y_i^*, z_i^* are the global coordinates of the point $(r_i, s_i, 1)$ with respect to the nodes $(r_i, s_i, 0)$ (Fig.11.5) and are obtained by dividing the difference between the top and bottom node coordinates by 2.

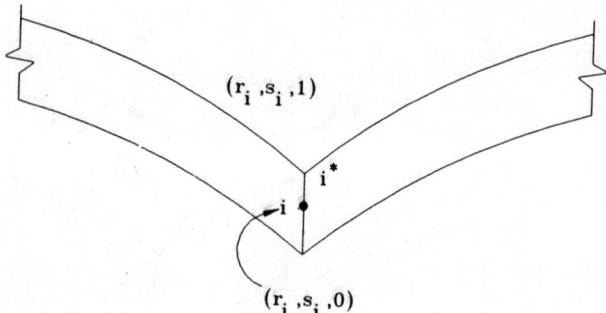

Fig. 11.5 Local coordinates of top and mid surface

11.4.2   Displacement Field

Analytical investigations have been reported that adding an internal node improved the performance of the element [6,18]. Thus, using the shape functions, Eq.11.59, the displacement field within the element is given by [6],

$$\left\{\begin{matrix} u \\ v \\ w \end{matrix}\right\} = \sum_{i=1}^{9} N_i \left\{ \left\{\begin{matrix} u_i \\ v_i \\ w_i \end{matrix}\right\} + t \left\{\begin{matrix} u_i^* \\ v_i^* \\ w_i^* \end{matrix}\right\} \right\} \qquad (11.61)$$

where u_i, v_i, w_i are the displacements of the node i in the x, y, z directions respectively, and u_i^*, v_i^*, w_i^* are the relative global displacements at point i caused by the rotations of the normal. As N_9 vanishes at $r, s = +1$, inter element compatibility is unaffected.

Now the relative displacements u_i^*, v_i^*, and w_i^* should be expressed in terms of the rotations α_i and β_i at the node i, in order to specify the displacements in terms of the nodal displacements and rotations. Let α_i and β_i be defined as the rotations of the normal about the axes $\bar{a}$ and $\bar{b}$ which lie in the mid- surface. The two vectors $\bar{a}$ and $\bar{b}$ are obtained as follows:

Let $\bar{i}, \bar{j}, \bar{k}$ be the unit vectors in the x, y and z directions respectively (Fig.11.6). Vector $\bar{n}$ is defined by input data so that it is normal to the mid-surface, i.e., the top and bottom nodes, whose coordinates are given as input should lie on the normal to the mid-surface. Now x^*, y^*, z^* are the coordinates of a vector along $\bar{n}$ i.e.,

$$\bar{n} = x^*\bar{i} + y^*\bar{j} + z^*\bar{k} \qquad (11.62)$$

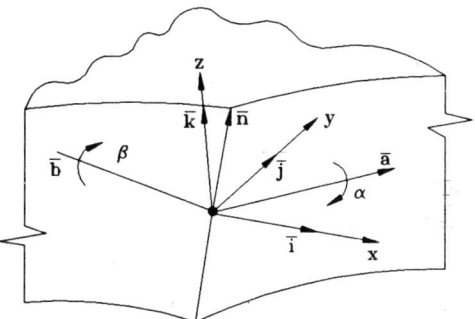

Fig. 11.6 Global and rotational axis

Now $\bar{a}$ is defined as a vector perpendicular to $\bar{n}$ and also to vertical $\bar{k}$
i.e.,

$$\bar{a} = \bar{k} \times \bar{n}$$
$$= \bar{k} \times (x^*\bar{i} + y^*\bar{j} + z^*\bar{k})$$
$$-y^*\bar{i} + x^*\bar{j} \qquad (11.63)$$

We now define $\bar{b}$ as a vector perpendicular to both $\bar{n}$ and $\bar{a}$
i.e.,

$$\bar{b} = \bar{n} \times \bar{a}$$
$$= (x^*\bar{i} + y^*\bar{j} + z^*\bar{k}) \times (-y^*\bar{i} + x^*\bar{j})$$
$$= -x^*z^*\bar{i} - y^*x^*\bar{j} + (y^*y^* + x^*x^*)\bar{k} \qquad (11.64)$$

It is seen that when $\bar{k}$ and $\bar{n}$ coincides, the vectors $\bar{a}$ and $\bar{b}$ are not specific and in that case they are defined as

$$\bar{a} = \bar{j} \text{ and } \bar{b} = \bar{i}$$

Now by normalising the vectors $\bar{a}$ and $\bar{b}$ we can get the direction cosines of these vectors with respect to x, y and z directions.

Let a_1, a_2 and a_3 be the direction cosines of $\bar{a}$ with respect to x, y and z axis respectively, and b_1, b_2 and b_3 be the direction cosines of $\bar{b}$ with respect to x, y and z axis respectively.

As α and β are the rotations with respect to $\bar{a}$ and $\bar{b}$ respectively, the displacement at $t = 1$ is $\frac{h}{2}(-\bar{b}\,\alpha + \bar{a}\,\beta)$ where h is the thickness at the point. Hence, the relative displacements at i in the global directions are given by,

$$[u_i^* \ v_i^* \ w_i^*] = \frac{-1}{2}h_i\,\alpha_i[b_1 \ b_2 \ b_3]_i + \frac{1}{2}\,h_i\,\beta_i[a_1 \ a_2 \ a_3]_i \qquad (11.65)$$

Substituting in equation (11.61), we get,

$$\begin{Bmatrix} u \\ v \\ w \end{Bmatrix} = \sum_{i=1}^{9} N_i \left\{ \begin{Bmatrix} u_i \\ v_i \\ w_i \end{Bmatrix} - \frac{th_i\alpha_i}{2} \begin{Bmatrix} b_1 \\ b_2 \\ b_3 \end{Bmatrix} + \frac{th_i\beta_i}{2} \begin{Bmatrix} a_1 \\ a_2 \\ a_3 \end{Bmatrix} \right\} \quad (11.66)$$

11.4.3   Jacobian Matrix

Noting that the shape functions are functions of r and s alone, the Jacobian matrix may be computed as shown below.

$[J] =$

$$\begin{bmatrix} \sum_{i=1}^{8} (x_i + tx_i^*) \frac{\partial N_i}{\partial r} & \sum_{i=1}^{8} (y_i + ty_i^*) \frac{\partial N_i}{\partial r} & \sum_{i=1}^{8} (z_i + tz_i^*) \frac{\partial N_i}{\partial r} \\ \sum_{i=1}^{8} (x_i + tx_i^*) \frac{\partial N_i}{\partial s} & \sum_{i=1}^{8} (y_i + ty_i^*) \frac{\partial N_i}{\partial s} & \sum_{i=1}^{8} (z_i + tz_i^*) \frac{\partial N_i}{\partial s} \\ \sum_{i=1}^{8} N_i \, x_i^* & \sum_{i=1}^{8} N_i \, y_i^* & \sum_{i=1}^{8} N_i \, z_i^* \end{bmatrix}$$
$$(11.67)$$

By inverting the matrix [J] we can get the derivatives with respect to global coordinates.

11.4.4   Strain Displacement Matrix

The element strains and the nodal displacements are related as (Eq. 3.69)

$$\{\epsilon\} = [B]\{d\} \quad (11.68a)$$

where

$$\{d\}^T = [u_1 v_1 w_1 \; \alpha_1 \beta_1 \; u_2 v_2 w_2 \; \alpha_2 \beta_2 \; \dots \; u_9 v_9 w_9 \; \alpha_9 \beta_9] \quad (11.68b)$$

and $\{\epsilon\}$ is given by Eqs. 2.10 and 9.27

In order to establish the strain displacement relation we have to first obtain the derivatives of the displacements with respect to local coordinates and they are obtained by differentiating Eq.11.66 with respect to r, s and t. Thus,

$$\begin{bmatrix} \frac{\partial u}{\partial r} & \frac{\partial v}{\partial r} & \frac{\partial w}{\partial r} \\ \frac{\partial u}{\partial s} & \frac{\partial v}{\partial s} & \frac{\partial w}{\partial s} \\ \frac{\partial u}{\partial t} & \frac{\partial v}{\partial t} & \frac{\partial w}{\partial t} \end{bmatrix} = \sum_{i=1}^{9} \begin{Bmatrix} \frac{\partial N_i}{\partial r} \\ \frac{\partial N_i}{\partial s} \\ 0 \end{Bmatrix} [u_i \; v_i \; w_i] - \frac{1}{2} \sum_{i=1}^{9} \begin{Bmatrix} th_i \frac{\partial N_i}{\partial r} \\ th_i \frac{\partial N_i}{\partial s} \\ h_i \, N_i \end{Bmatrix} \times$$

$$\alpha_i [b_1 \quad b_2 \quad b_3]_i + \frac{1}{2} \sum_{i=1}^{9} \begin{Bmatrix} th_i \frac{\partial N_i}{\partial r} \\ th_i \frac{\partial N_i}{\partial s} \\ h_i \ N_i \end{Bmatrix} \beta_i \ [a_1 \quad a_2 \quad a_3]_i \qquad (11.69)$$

Now using the Jacobian inverse and Eq.9.29 we can find the derivatives with respect to global coordinates and obtain the strain displacement matrix [B].

11.4.5 Strain Transformation Matrix

As explained earlier, the stresses normal to the midsurface is negligible and is taken as zero. Since the expressions for stresses obtained are in the global directions, we must rotate the axes from x, y, z system to a set x', y', z' as shown in Fig. 11.7. This set of mutually perpendicular axes at the point being considered, is determined from the local axes, r, s, t as follows.

x' is identical to r

z' is perpendicular to r and s (r,s,z' forming a right handed set)

y' is perpendicular to x' and z' (x', y', z' forming a right handed set)

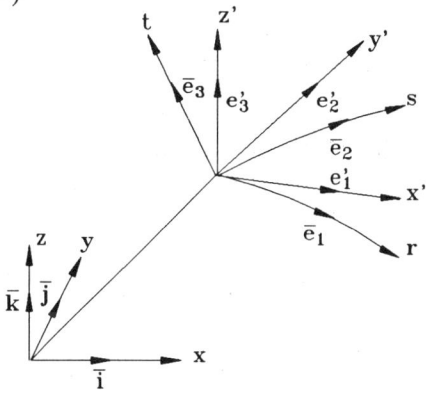

Fig. 11.7 Global and local axes for strain

Let $\bar{e}_1, \bar{e}_2, \bar{e}_3$ be unit vectors along r, s and t directions respectively and $\bar{i}, \bar{j}, \bar{k}$ be unit vectors along x, y and z directions. These vectors are related by the Jacobian matrix [J] as given below [6].

$$\begin{Bmatrix} \bar{e}_1 \\ \bar{e}_2 \\ \bar{e}_3 \end{Bmatrix} = [J] \begin{Bmatrix} \bar{i} \\ \bar{j} \\ \bar{k} \end{Bmatrix} \qquad (11.70)$$

We have to define three vectors e_1', e_2', e_3' along x', y' and z' respectively. From the definition of x', y', z' system, we have

$e_1' = \bar{e}_1$

e_3' perpendicular to $\bar{e}_1$ and $\bar{e}_2$

e_2' perpendicular to e_1' and e_3'

where

$$\begin{Bmatrix} \bar{e}_1 \\ \bar{e}_2 \\ \bar{e}_3 \end{Bmatrix} = \begin{bmatrix} J_{11} & J_{12} & J_{13} \\ J_{21} & J_{22} & J_{23} \\ J_{31} & J_{32} & J_{33} \end{bmatrix} \begin{Bmatrix} \bar{i} \\ \bar{j} \\ \bar{k} \end{Bmatrix} \tag{11.71}$$

$$e_1' = \bar{e}_1 = J_{11}\bar{i} + J_{12}\bar{j} + J_{13}\bar{k} = J_{11}'\bar{i} + J_{12}'\bar{j} + J_{13}'\bar{k}$$

$$e_3' = \bar{e}_1 \times \bar{e}_2 = (J_{12}J_{23} - J_{13}J_{22})\bar{i} + (J_{13}J_{21} - J_{11}J_{23})\bar{j} + (J_{11}J_{22} - J_{12}J_{21})\bar{k}$$

$$= J_{31}'\bar{i} + J_{32}'\bar{j} + J_{33}'\bar{k}$$

$$e_2' = e_3' \times e_1' = (J_{32}'J_{13}' - J_{33}'J_{12}')\bar{i} + (J_{11}'J_{33}' - J_{13}'J_{31}')\bar{j} + (J_{31}'J_{12}' - J_{32}'J_{11}')\bar{k}$$

$$= J_{21}'\bar{i} + J_{22}'\bar{j} + J_{23}'\bar{k} \tag{11.72}$$

$$\begin{Bmatrix} e_1' \\ e_2' \\ e_3' \end{Bmatrix} = \begin{bmatrix} J_{11}' & J_{12}' & J_{13}' \\ J_{21}' & J_{22}' & J_{23}' \\ J_{31}' & J_{32}' & J_{33}' \end{bmatrix} \begin{Bmatrix} \bar{i} \\ \bar{j} \\ \bar{k} \end{Bmatrix} \tag{11.73}$$

If we normalise the vectors $[J_{11}' \quad J_{12}' \quad J_{13}']$, $[J_{21}' \quad J_{22}' \quad J_{23}']$ and $[J_{31}' \quad J_{32}' \quad J_{33}']$ we get the $[D]$ matrix (Eq.9.46) representing direction cosines of the new axes directions e_1', e_2' and e_3'

The strains $\{\epsilon'\}$ with respect to x', y', z' system (Eq.9.55) are related to $\{\epsilon\}$ in the global x, y, z system by the matrix $[T_\epsilon]$ given by Eq.9.58 as,

$$\{\epsilon'\} = [T_\epsilon]\{\epsilon\}$$

Also from Eq.9.59 we have $\{\epsilon'\} = [T_\epsilon][B]\{d\} = [B']\{d\}$ where $[B'] = [T_\epsilon][B]$

11.4.6   Stress Displacement Matrix

The element stresses and nodal displacements are related as given by Eq.9.60,

$$\{\sigma\} = [C][B]\{d\}$$

$$= [CB]\,\{d\}$$

Considering the case of isotropic material, the constitutive relation is given in Eq.2.16. Now imposing the conditions that $\sigma_{z'} = 0$, the following relation is obtained for stress-strain relation in x', y', z' coordinates,

$$\{\sigma'\} = [C]\,\{\epsilon'\} \tag{11.74a}$$

i.e.,
$$
\begin{Bmatrix} \sigma_{x'} \\ \sigma_{y'} \\ \sigma_{z'} \\ \tau_{x'y'} \\ \tau_{y'z'} \\ \tau_{z'x'} \end{Bmatrix}
= \frac{E}{1-\mu^2}
\begin{bmatrix}
1 & \mu & 0 & 0 & 0 & 0 \\
\mu & 1 & 0 & 0 & 0 & 0 \\
0 & 0 & 0 & 0 & 0 & 0 \\
0 & 0 & 0 & \frac{1-\mu}{2} & 0 & 0 \\
0 & 0 & 0 & 0 & \frac{\alpha(1-\mu)}{2} & 0 \\
0 & 0 & 0 & 0 & 0 & \frac{\alpha(1-\mu)}{2}
\end{bmatrix}
\begin{Bmatrix} \epsilon_{x'} \\ \epsilon_{y'} \\ \epsilon_{z'} \\ \gamma_{x'y'} \\ \gamma_{y'z'} \\ \gamma_{z'x'} \end{Bmatrix}
\tag{11.74b}
$$

where α is a factor used to account for a better representation of shear deformation when a constant strain is assumed across the thickness, rather than the correct quadratic and a value of $\alpha = 5/6$ has been suggested in reference [6].

It may be noted that the third row of $[C]$ is zero to make the stress normal to the midsurface as zero. As C_{3j} is taken as zero, the terms multiplied by C_{3j} in $[B']$ matrix (i.e., the third row) need not be calculated. This is achieved by excluding the third row of $[T_\epsilon]$ matrix. Thus the dimensions of the matrices become $[B]_{6\times 45}$, $[T_\epsilon]_{5\times 6}$, $[B']_{5\times 45}$, $[C]_{5\times 5}$, $[CB]_{5\times 45}$ and $[k]_{45\times 45}$.

Substituting for $\{\epsilon'\} = [B']\,\{d\}$ in Eq.11.74a we get

$$\{\sigma'\} = [C]\,[B']\,\{d\}$$
$$= [CB']\,\{d\} \tag{11.74c}$$

The stresses thus obtained will be in the local coordinate system x', y', z' which is convenient in the case of curved surfaces.

11.4.7 Element Stiffness Matrix

As we have obtained the [B] matrix and [CB] matrix in local coordinates x', y', z' coordinates we should use the same matrices for the computation of element stiffness matrix. Here the transformation is derived based on the concept that during any virtual displacement, the resulting increment in strain energy density must be the same regardless of the coordinate system in which it is computed [6]. Hence the matrix is given by

$$[k] = \int\int\int [B]^T [C] [B] \, dV$$

$$= \int\int\int [B']^T [C] [B'] \, dV$$

$$= \int\int\int [B']^T [CB'] \, dV \qquad (11.75)$$

where $[B']$ and $[CB']$ matrices are computed as explained in sections 11.4.4 and 11.4.5.

The integration is carried out numerically resulting in an element stiffness matrix of size 45×45. Since the node 9 is internal to the element, the degrees of freedom associated with it are removed by static condensation procedure resulting in a 40×40 element stiffness matrix.

The integration scheme, i.e., choice of Gauss points to calculate the stiffness matrix requires careful study. The reference [6] contains detailed description about the selective integration to account for the appropriate bending and shear energy terms. It has been noted that a $3 \times 3 \times 2$ integration gives good results for thick shells, whereas a reduced integration scheme of $2 \times 2 \times 2$ is economical and reasonably accurate for thin shells. For moderately thick to thin shells a selective integration scheme has been found to give good results[6]. In the program presented here any of the three schemes can be adopted by the user depending on the type of the shell to be analysed.

11.4.8 Element Load Vector

Element Load vector resulting from gravity, uniform surface pressure and uniform vertical load are considered here.

(a) *Gravity Loads* Gravtity loads are calculated from a uniform density throughout each element. From Eq.11.66, the vertical displacement is given by

$$w = \sum_{i=1}^{9} N_i \, w_i - \frac{1}{2}t \sum_{i=1}^{9} N_i \, h_i \, b_{3i} \, \alpha_i + \frac{1}{2}t \sum_{i=1}^{9} N_i \, h_i \, a_{3i} \, \beta_i \qquad (11.76)$$

Hence the load vector at node i is given by,

$$\{Q_i\} = \int\int\int \rho \left\{ \begin{array}{c} 0 \\ 0 \\ N_i \\ -\frac{1}{2}t \, N_i \, h_i \, b_{3i} \\ +\frac{1}{2}t \, N_i \, h_i \, a_{3i} \end{array} \right\} |J| \, dr \, ds \, dt \qquad (11.77)$$

The integration is carried out numerically using the same integration points used for stiffness computation, to avoid using more integration points than necessary.

The nodal loads due to uniform surface pressure and vertical loading are calculated by adopting the same procedure followed for four noded shell element explained in section 11.3. The integration in these cases is carried out using a 2 × 2 Gauss quadrature.

11.5   Subroutine SHELL

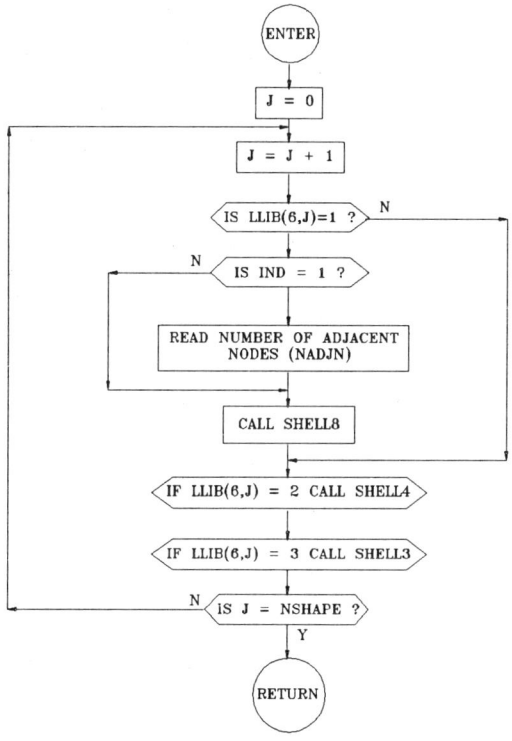

Fig. 11.8 Subroutine SHELL

This control routine calls three different routines as shown in Fig. 11.8. The following two shapes of elements described in the earlier sections are included in the program.

1.   Shape 1: Eight Noded Curved Shell Element, SHELL8
2.   Shape 2: Four Noded Shell Element, SHELL4.

11.5.1   Organisation of Routines SHELL4 and SHELL8

The general organisation of the routines follows the procedure given
in Chapter 6. Each element shape is called thrice by the main program depending on the value of the flag IND.

When IND equals one the first segment is evoked. For each element, the nodal connectivity, material group, loading type and element loads are read. For the eight noded element the connectivity
of bottom nodes (NADJ) are also read to compute x^*, y^* and z^*
in Eq.11.60. The ND array is generated for each element and is
stored on the NDARAY file in a sequential manner. The subroutine COLUMH is called to calculate the column heights of the global
stiffness matrix.

When IND equals two, in the next segment of the program the
[CB] matrix is computed and stored on ISTRES file. Next the element stiffness matrix and load vector are computed and are added
to global values.

In the third stage the [ND] array and [CB] matrices are retrieved
from the NDARAY and ISTRES files. The stresses or stress resultants are computed for each element depending on the option exer-
cised by the user. The listing of the routines for shell elements is
given in Appendix 8.

11.5.2   Input Details for Shell Elements

The input details for shell element type are given below. This data
set would pertain to the data sets 5 and 6 required for PASSFEM
described in Chapter 6.

   5.ELEMENT TYPE DATA (5I5)

   Columns   1 - 5   Element type number or row number of the
                       element library (LTYPE). For shell elements
                       LTYPE=6.

             6-10   Number of shapes (NSHAPE) belonging to
                       the element type with a minimum of 1 to a
                       maximum of 3.

             11-15   Element shape number or column number of
                       the element library. (Element shape number
                       is 1 for SHELL 8 and 2 for SHELL4).

             16-20   Element shape number. If NSHAPE is more
                       than one, the second shape number is typed
                       in this column; otherwise it is left blank.

             21-25   Element shape number. If NSHAPE=3, third
                       shape number is typed in this column.

6. ELEMENT DATA SET

This data set is element dependent and the input details for SHELL4 and SHELL8 are given below.

(i) FOUR NODED SHELL ELEMENT (SHELL 4)

A. Control Data (I5)

Columns 1-5 Number of four noded shell elements

B. Element Data (6I5, 4F10.4/3I2, 5F10.4, I5, I2, I3, 2I2)

Two lines per element with the following information must be given:

First Line

Columns		
	1-5	Element Number (LNUM)
	6-10	Material Group Number (MG)
	11-30	Node numbers of the element in the order shown in Fig.11.2 (LNC)
	31-70	Thickness of the element at each node (T(I))

Second Line

Columns		
	1-2	Loading Type 1 (LOTYPE)
	3-4	Loading Type 2(LOTYPE)
	5-6	Loading Type 3(LOTYPE)
	7-16	Pressure corresponding to Loading Type 1
	17-26	Pressure corresponding to Loading Type 2
	27-36	Pressure corresponding to Loading Type 3
	37-46	Water table
	47-56	Torsional coefficient
	57-61	Integration scheme
	62-63	Membrane element Option
		EQ 0 Membrane element
		GE 1 Bending cum membrane element
	64-66	Element Data Generator (to be specified on the second line of the series)
	67-68	Output option - stress resultants at the centre of elements (GE.1)
	69-70	Principal stress at the centre of the element (GE. 1)

Four types of loading are specified in the program.

LOTYPE   = 1   Self weight (Gravity load) of the element in the global z direction.

              = 2   Vertical pressure in the global z direction

              = 3   Normal pressure in the local z' direction

              = 4   Water pressure in the local z' direction

The integration scheme in columns 57-61 is generally 4 41 ie., 4 integration points for bending, 4 integration points for in plane shear and 1 integration point for transverse shear. However, if bending behaviour is predominant as in the case of webs of box girder, then the integration scheme is 421, i.e., 4 points for bending, 2 for inplane shear and 1 for transverse shear.

The torsional coefficient specified in Columns 47-56 may be taken as 10.0 (> 0.1).

If the element is a membrane element as in the case of rigid diaphragms, then columns 62-64 are left blank; otherwise any number greater than or equal to 1 is typed.

(ii) EIGHT NODED SHELL ELEMENT

A. Control Data for Bottom Nodes (I5)

     Columns   1-5   The number of bottom nodes NADJN for the 8 noded shell element.

B. Nodal Data - Bottom Nodes (2I5, 3 F10.3)

The following information is required for each bottom node.

     Columns   1-5   Node Number

                    6-10   Nodal increment for automatic generation. Generation will be carried out along a straight line as explained under Nodal Data in Chapter 6

                  11-20 x - Coordinate of the bottom node

                  21-30 y -Coordinate of the bottom node

                  31-40 z -Coordinate of the bottom node

C. Integration Scheme Data (2I5)

     Columns   1-5   Number of 8 noded shell element in the structure

                    6-10   Code for integration

                            1-for 3 × 3 × 2 integration

                            2-for selective integration

                            3-for 2 × 2 × 2 integration

D. Element Data (2I3, 16I3)

The following information has to be supplied per element.

Columns   1-3   Element Number

                  4-6   Material Group

                  7-30   Nodal connectivity of top nodes

                  31-54   Nodal connectivity of bottom nodes

E. Element Load Data (4 F10.3)

Columns   1-10   Uniform surface pressure on bottom face.

                  11-20   Uniform surface pressure on top face

                  21-30   Uniform vertical load on bottom face

                  31-40   Uniform vertical load on top face

If the load acts towards the surface, it is positive and if it acts away from the surface it is negative.

11.5.3 Output Details

In the case of four noded shell elements the following three options are available.

(i)   Stress resultants, $M_x, M_y, M_{xy}, N_x, N_y$ and N_{xy} at the centre of the element are calculated as the average of the values at the Gauss points. The shear forces Q_x and Q_y are calculated directly at the centre of the element.

(ii)   Principal stresses and directions are calculated at the centre of the element.

(iii)   Stress resultants, $M_x, M_y, M_{xy}, N_x, N_y, N_{xy}$ at the nodes are calculated by adopting the stress smoothing technique explained in Chapter 10 and finally average values at the nodes are computed.

For the eight noded curved shell element $\sigma_x, \sigma_y, \tau_{xy}, \tau_{yz}$ and τ_{zx} are computed at the Gauss points of the top and bottom faces of each element. Extrapolation to the values at the nodes and averaging the values at each node of top and bottom are also carried out.

11.6   Examples

Five practical shell structures have been analysed using the SHELL4 and SHELL8 routines presented in this chapter. The results of analysis are compared with theoretical values and are described in the following.

11.6.1   Cylindrical Shell

Fig. 11.9 shows a thin cylindrical shell supported on rigid diaphragms.

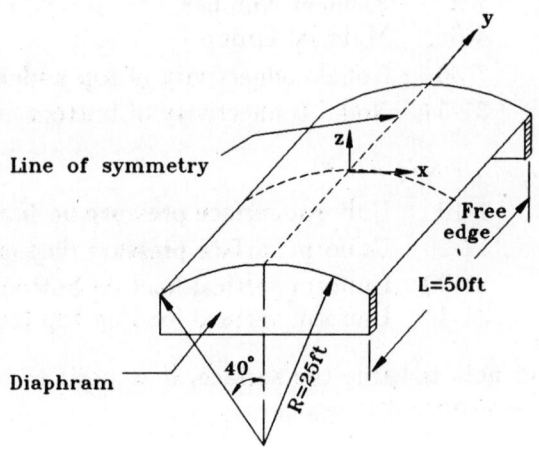

Fig. 11.9 Thin cylindrical shell

This particular example was originally solved by Scordelis and
Lo[14] and was extensively used for checking the performance of various
types of shell elements.

The shell was analysed for self weight of 90 lbs/sq.ft. The thickness of the shell is 3″ and the radius of the midsurface is 25 ft.
Because of symmetry only a quarter of the shell was actually analysed. The diaphragm does not allow displacement in its own plane
but offers no resistance to displacements perpendicular to it. Hence,
the boundary conditions along the nodes on the diaphragm were assumed as, $u = 0, w = 0, \theta_y = 0, \theta_z = 0$. Two discretizations used for
four and eight noded shell elements are shown in Fig.11.10.

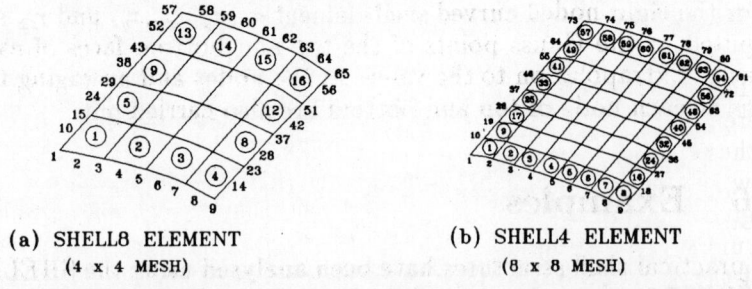

(a) SHELL8 ELEMENT   (b) SHELL4 ELEMENT
   (4 x 4 MESH)   (8 x 8 MESH)

Fig. 11.10 Discretizations of one quadrant of the shell

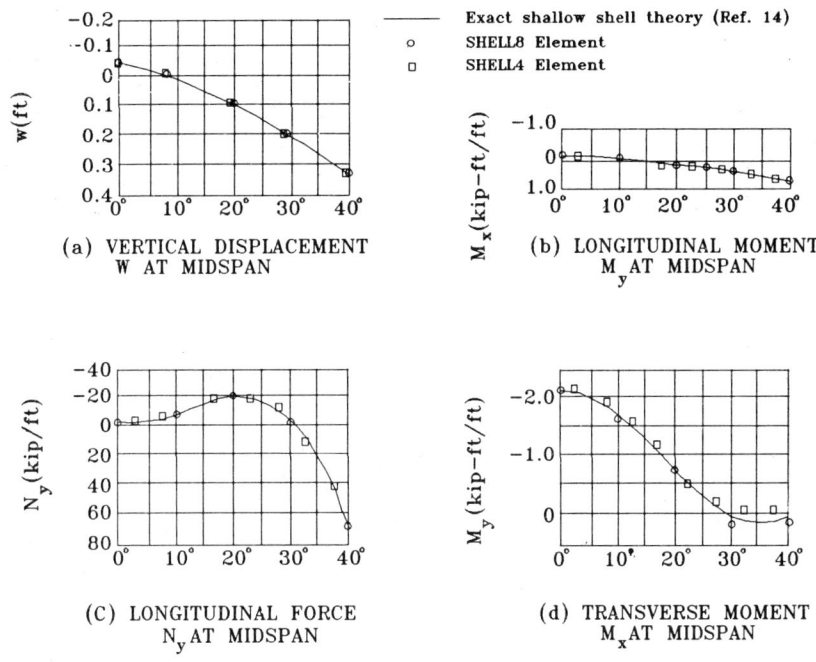

Fig. 11.11 Comparison of results for cylindrical shell

The results of analysis are compared with the values reported in reference [14]. The displacements of the shell in the vertical direction at the midspan are shown in Fig.11.11a. Figures, 11.11b and 11.11c show the variation of longitudinal moment per unit length and the longitudinal force per unit length at midspan. Transverse moment per unit length at midspan is shown in Fig.11.11d. It has been found that the results of both the elements are in good agreement with the published results.

11.6.2   Clamped Hyper Shell

Figure 11.12 shows a hyperbolic paraboloid shell clamped all along the boundary. The shell is subjected to a normal pressure loading of 1 lb/sq.in.

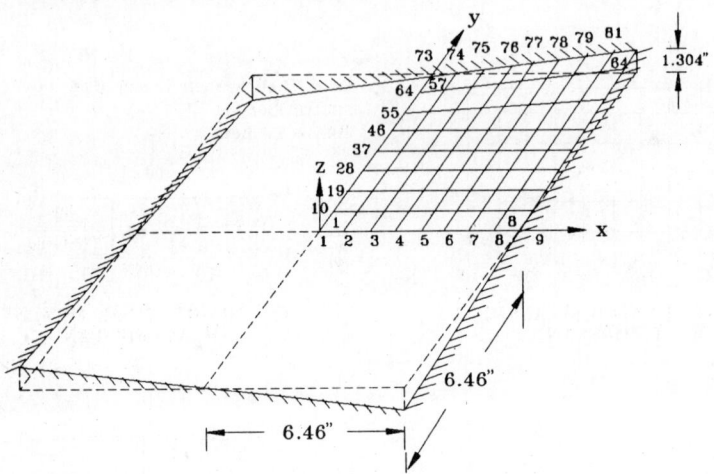

Thickness = 0.25 in   E = 5×10^5 lb/sq.in

$\mu = 0.39$ Loading (vertical) = 1.0 lb/sq.in

Fig. 11.12 Clamped hyperbolic paraboloid

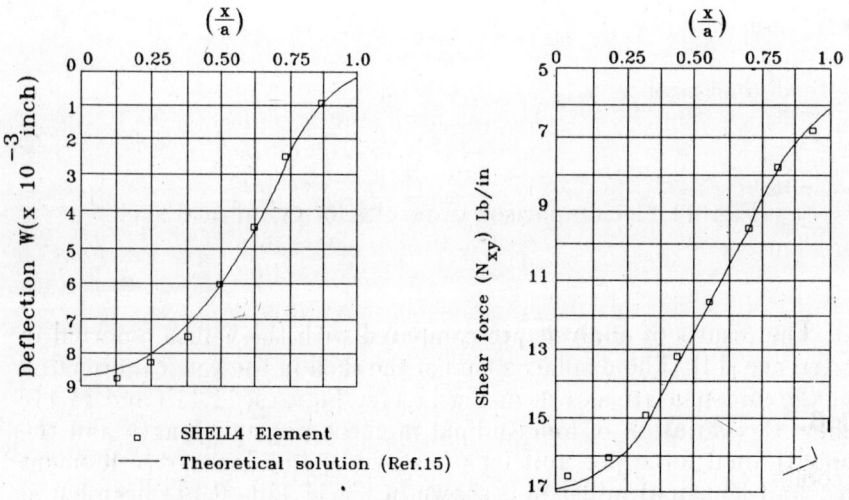

Fig. 11.13a Comparison of results for clamped hyper shell

One quadrant of the shell was analysed using four noded degenerated shell element. This problem was solved analytically by Chetty and Tottenham[15]. The results are plotted in Fig.11.13a and b and compared with the theoretical values. The stress resultants have been obtained at the centre of the elements. A good agreement between finite element analysis and theoretical solution has been observed.

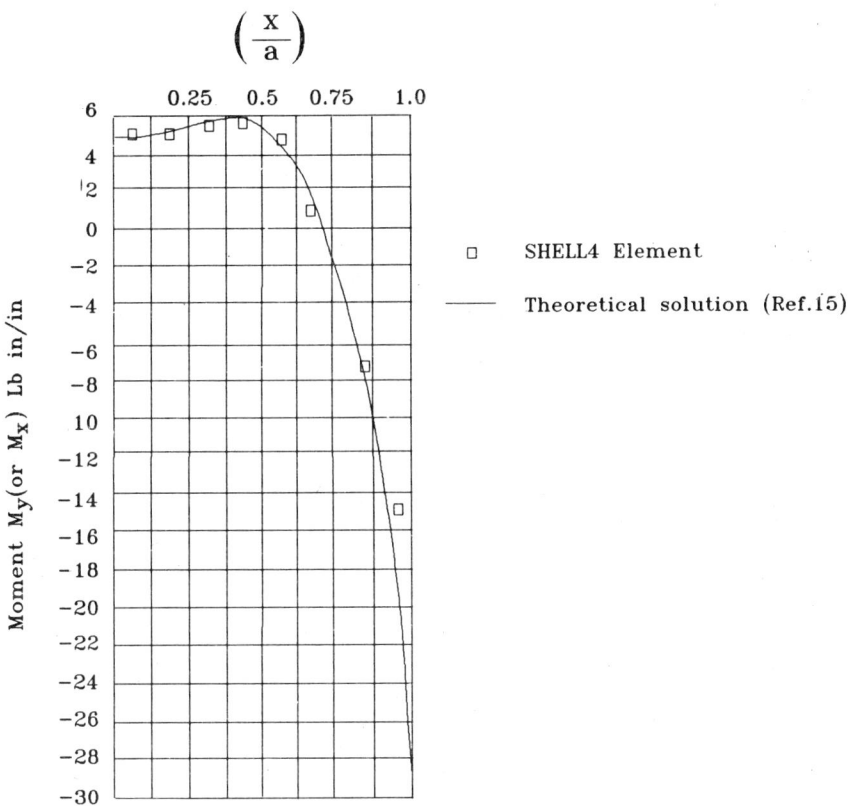

Fig. 11.13b Comparison of results for clamped hyper shell

11.6.3 Cooling Tower

A cooling tower of height 330 ft is shown in Fig.11.14. The tower was subject to wind loading and the variation of wind pressure across the height as reported in reference [16] is given in Fig.11.15. The problem was analysed by using four noded element and the discretization is shown in Fig.11.14. The results of analysis are compared in Fig.11.16 with the solution by finite difference method given in reference [16]. Top diameter = 174.22 ft, Bottom diameter = 274.0 ft, Neck diameter = 168.0 ft, Shell thickness = 5 in, $E = 0.432 \times 10^6$ kips/sq.ft, $\mu = 0.15$.

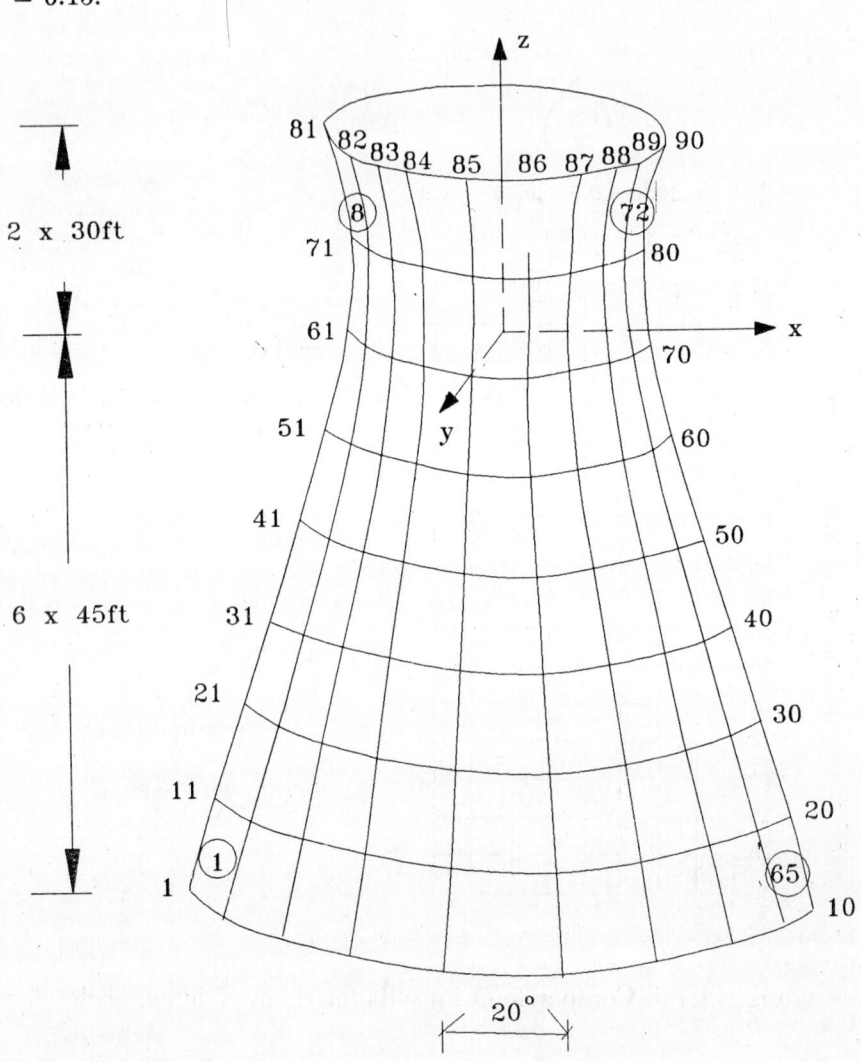

Fig. 11.14 Cooling Tower Mesh Divisions

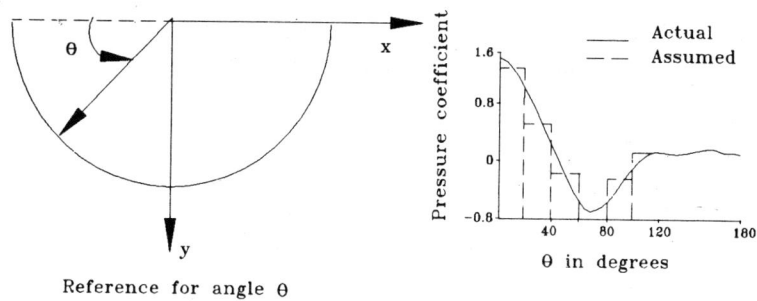

Reference for angle θ

Fig. 11.15 Pressure load variation about circumference

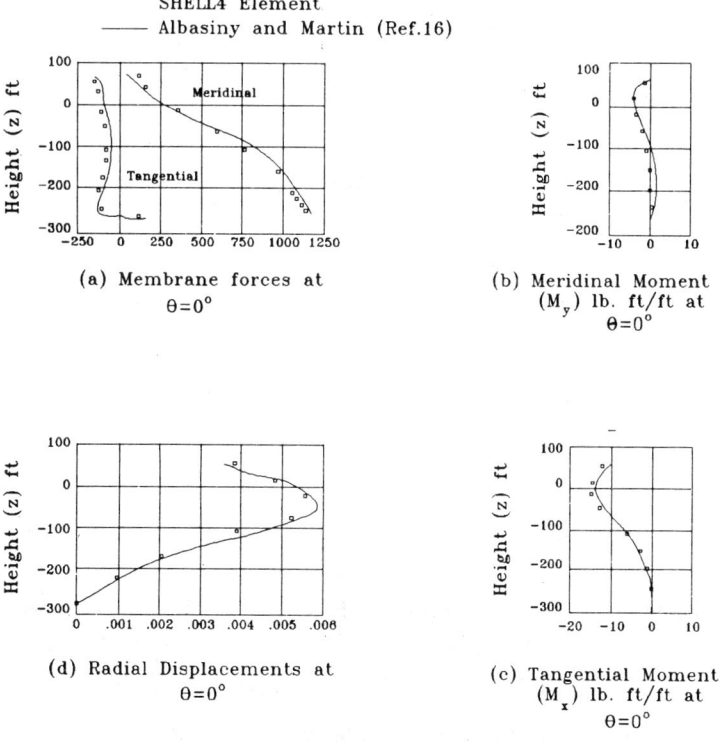

Fig. 11.16 Comparison of results-cooling tower

11.6.4   Box Girder

A horizontally curved box girder model tested by Fa and Turkstra [17] is shown in Fig.11.16. The finite element idealisation using four noded shell element is shown in Fig.11.17.

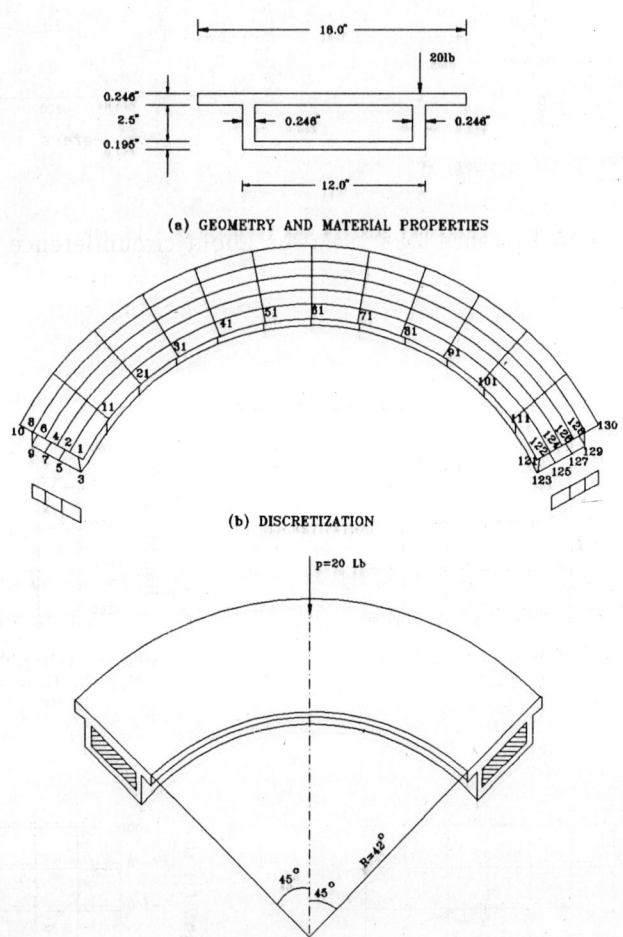

(a) GEOMETRY AND MATERIAL PROPERTIES

(b) DISCRETIZATION

Model   :   Plexiglass,   E : 4×10^5 psi μ : 0.36
Inner radius   :   42 inch   Outer radius : 60 inch

Fig. 11.17 A horizontally curved box girder

The integration scheme for this problem was adopted from the values used by Kanok-Nukulchai[7]. Accordingly for top and bottom flange elements, 2×2 Gauss quadrature was used to consider membrane and bending effects. For web elements, 2×2 scheme for bending, and 2×1 for membrane effects were used. For transverse shear effects 1x1 scheme was used for flange and web elements. The end diaphragms were analysed considering only membrane effects, using 2×2 scheme.

The results of finite element analysis for vertical deflection along half span are compared with the experimental values [17] in Fig.11.18a. The results for tangential stresses at the top and bottom flanges at mid span are compared in Figs.11.18b and (c) with the results of using higher order elements as reported in reference [17].

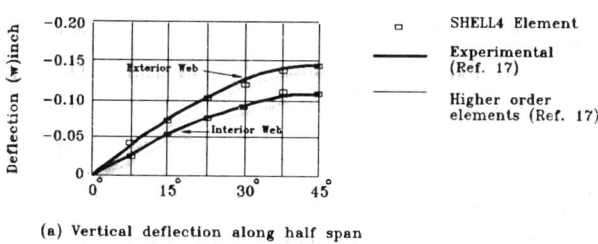

(a) Vertical deflection along half span

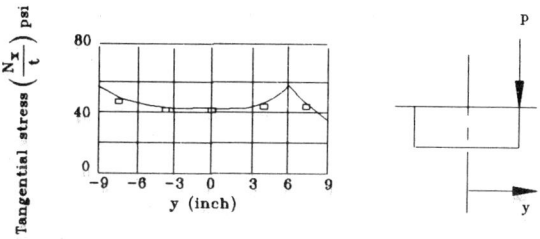

(b) Tangential stresses in top flange at mid span

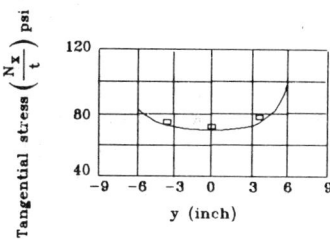

(c) Tangential stresses in bottom flange at mid span

Fig. 11.18 Comparison of results for box girder

11.6.5   Folded Plate

A simply supported folded plate is shown in Fig.11.19. The geometry and loading details are shown therein. ⋅Because of symmetry only one quadrant of the shell was analysed using four noded shell element and the discretization used is shown in Fig.11.20. The results of analysis due to combined dead and live loads are compared in Table 11.1 with the theoretical values given in reference[3].

Geometry

Thickness of top & bottom plates = 4 in (i.e. plates (1),(3), etc.)
Thickness of inclined plates (2), (4) = 3.5in
Span length (L) = 60 ft

Material Properties

Modulus of elasticity (E) = 0.432 × 10^6 kips/sq.ft
Poisson's Ratio (μ) = 0.15

Loading :

Dead load = 12.5 psf of plate/inch thickness
Live load = 15.0 psf of surface area

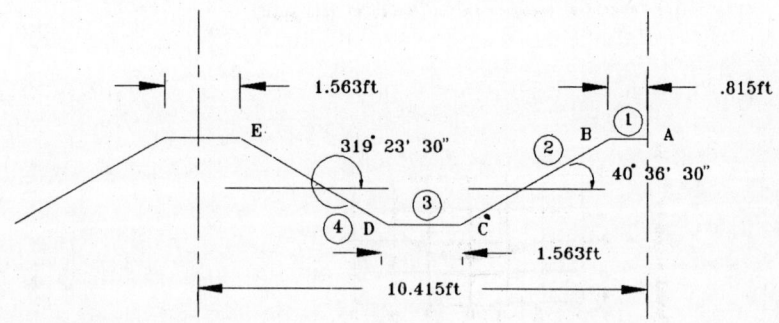

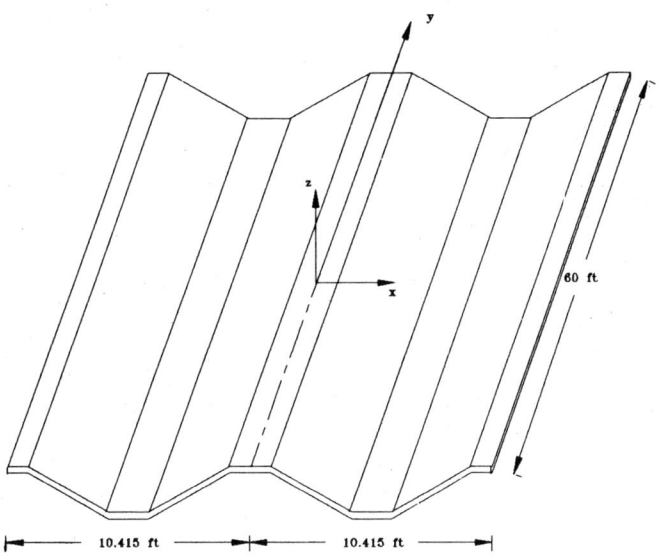

Fig. 11.19 Simply supported folded plate

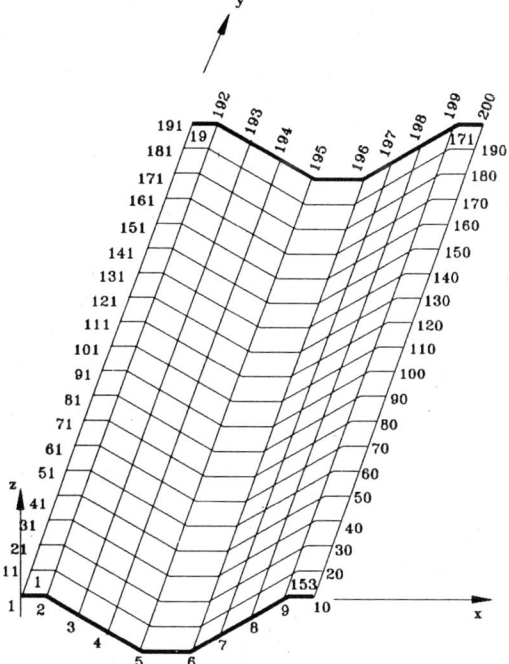

Fig. 11.20 Discretization of one quadrant of folded plate

Table 11.1 Comparison of Results for Folded Plate Structure

Stress Resultant	Units	A*	B	C	D	E
1. Longitudinal Stress (N_y/t) At Midspan	Kip/Sq.Ft	-145.30	-104.20	106.80	113.30	-108.30
		(-155.00)	(-105)	(106)	(114)	(-109)
2. Transverse Moment (M_y) At Midspan	Ft-Lb	0	+30	+980.0	960.0	-210
		(0)	(20)	(986)	(953)	(-270)
3. Longitudinal Stress (N_y/t) At Quarter Span	Kip.Sq/Ft.	-52	-95	81.1	80	-78
		(-41)	(-99)	(89)	(82)	(-81)
4. Transverse Moment (M_y) At Quarter Span	Ft-Lb	0	+29	718.2	682.3	-165.50
		(0)	(20)	(+725)	(-695)	(-169)

Theoretical Values in Brackets

* Ref. Fig. 11.19

REFERENCES

1. Flügge, W., *Stresses in Shells*, Springer-Verlag, New York, 1969.

2. Donnell, L.H., *Beams, Plates and Shells*, McGraw-Hill, N.Y., 1976.

3. Ramaswamy, G.S., *Design and Construction of Concrete Shell Roofs*, Tata McGraw Hill, New Delhi, 1971.

4. Ashwell, D. and R.H. Gallagher (Ed), *Finite Element for Thin shells Analysis, Members*, Wiley, New York, 1976.

5. Ahmad, S., B.M.Irons and O.C.Zienkiewicz, *Analysis of Thick and Thin Shell Structures by Curved Finite Elements*, International Journal for Numerical Methods in Engineering, Vol.2, pp.419-451, 1970.

6. Pawsey, S.F., *The Analysis of Moderately Thick and Thin Shells*, Ph.D. thesis, Department of Civil Eng. University of California, Berkeley, 1970.

7. Worsak Kanok-Nukulchai, *A Simple and Efficient Finite Element for General Shell Analysis*, International Journal for Numerical Methods in Engineering, Vol.14, pp.179-200, 1979.

8. Clough, R.W and C.P.Johnson, *A Finite Element Approximation for the Analysis of Thin Shells*, International Journal for Solids and Structures, Vol.4, 1968.

9. Clough, R.W. and C.A.Fellipa, *A Refined Quadrilateral Element for Analysis of Plate Bending*, Proc. of Second Conference on Matrix Methods in Structural Mechanics, Wright-Patersen, A.F.B., Ohio, 1968.

10. Gallagher, R.H., *Analysis of Plate and Shell Structures, Application of Finite Element Method in Engineering*, Vanderbilt University, ASCE Publication, 1969.

11. Bogner, F.K., R.L.Fox and L.A.Schmit, *A Cylindrical Shell Discrete Element*, A.I.A.A.Journal, Vol.5, No.4, 1967.

12. Utku, S., *Stiffness Matrices for Thin Triangular Elements of Non-zero Gaussian Curvature*, A.I.A.A. Journal, Vol.5, No.9, 1967.

13. Zienkiewicz, O.C, and R.L.Taylor, *The Finite Element Method, Vol.2, Solid and Fluid Mechanics, Dynamics and Nonlinearity*, McGraw-Hill Book Co., U.K. 1991.

14. Scordelis, A.C and K.S.Lo, *Computer Analysis of Cylindrical Shells*, ACI Journal, Vol.61, pp.539-561, 1964.

15. Chetty, S.M.K and H.Tottenham., *An Investigation into the Bending Analysis of Hyperbolic Paraboloid Shells*, Indian Concrete Journal, Vol.38, pp.248-258, 1964.

16. Albasiny, E.L., and D.W.Martin, *Bending and Membrane Equilibrium in Cooling Towers*, Proc. ASCE, Jl. of Engineering Mechanics Division, Vol.93, EM3, pp.1-17, 1967.

17. Fa, A.R.M., and C.Turkstra, *Model Study of Horizontally Curved Box Girder*, Proc. ASCE, Jl. of Structures Division, Vol.102, ST5, pp.1097-2008, 1976.

18. Cook, R.D., D.S.Malkus, and Michael E.Plesha, *Concepts and Applications of Finite Element Analysis*, Third Edition, John Wiley, New York, 1989.

19. Pica, A, and R.D.Wood, *Postbuckling Behaviour of Plates and Shells Using a Mindlin Shallow Shell Formulation*, Computers and Structures, Vol.12, pp.759-768, 1980.

20. Chan, H.C., and W.C.Chung, *Geometrically Nonlinear Analysis of Shallow Shells Using Higher Order Finite Elements*, Computers and Structures, Vol.31, pp.329-338, 1989.

EXERCISES

11.1 For the four noded shell element 11 in Fig. 11.10a derive the shape function and express the relation between u, v, w displacements and nodal degrees of freedom as in Eq. 11.13a.

11.2 Derive the relation between u, v, w displacements and the nodal degrees of freedom for an eight noded shell element 6 (Eq. 11.66) shown in Fig. 11.10a.

11.3 Compute the nodal load vector $\{Q\}$ for the four noded shell element 11 shown in Fig. 11.12 due to vertical loading.

11.4 Compute the nodal load vector $\{Q\}$ due to self weight for eight noded shell element 6 shown in Fig. 11.10a.

11.5 Derive the transformation matrix for computing the global stiffness matrix of a four noded plate/shell element. The four noded plate/shell element is formed by combining the four noded plate bending element and plane stress element described in Chapters 10 and 8 respectively.

11.6 Develop the necessary subroutines to include the four noded plate/shell element in PASSFEM library. The transformation matrix developed in problem 11.5 can also be made use of.

11.7 Fig. E11.7 shows a spherical shell resting on a square platform subjected to a single central concentrated load.

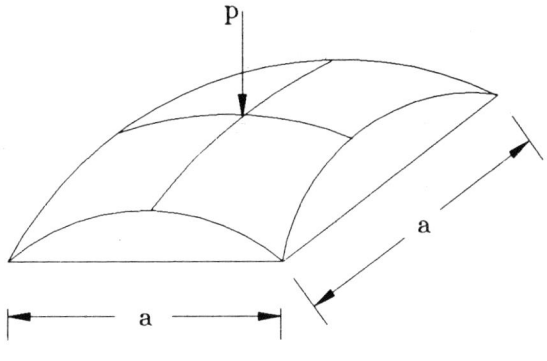

Fig. E11.7

Analyse the shell using four and eight noded shell elements. Compare the results with the solution given in reference [19]. $P = 10.0$ kN $E = 68.95$ kN/mm^2 $t = 99.45$ mm $\mu = 0.30$ $R = 2540$ mm $a = 1569.8$ mm.

11.8 Compute the torsional stiffness matrix for the element in exercise 11.1.

11.9 The straight edges of a cylindrical shell shown in Fig. E11.9 are hinged and the curved edges are free. Analyse the shell using four and eight noded shell elements and compare the results with the solution given in reference [19]. $E = 3.10275$ kN/mm^2 $t = 12.70$ mm $\mu = 0.30$ $R = 2540$ mm $L = 508$ mm $P = 0.8859$ kN.

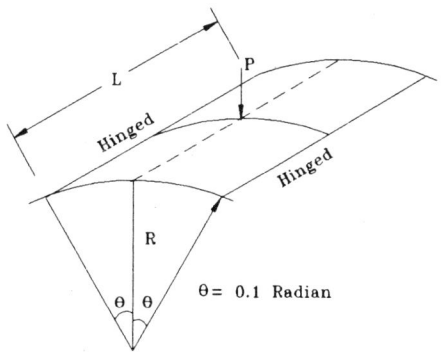

Fig. E11.9

$E = 3.10275$ kN/mm^2 $t = 12.70$ mm $\mu = 0.30$ $R = 2540$ mm $L = 508$ mm $P = 0.8859$ kN $\theta = 0.1$ radian.

11.10 The geometry of a concave shell is shown in Fig. E11.10.

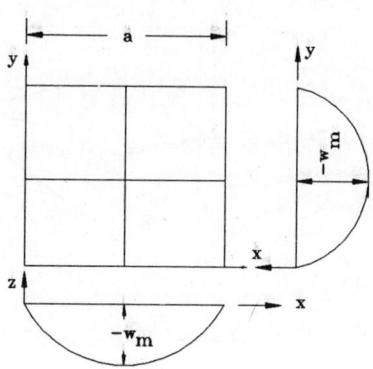

Fig. E11.10

The drop of the shell is given by the equation, $w_o(x,y) = -w_m \sin\left(\frac{\pi x}{a}\right) \sin\left(\frac{\pi y}{a}\right)$ where w_m represents the maximum vertical deflection at the centre of the shell and a refers to the side of its vertical projection as shown in Fig. E11.10. The boundary edges of the shell are clamped. Analyse the shell using four noded shell element and compare the results with the solution given in reference [20]. $E = 3.0 \times 10^7$ lb/in^2 $\mu = 0.316$ $t = 3.0''$ $a = 300.0''$ Analyse for $w_m/t = -5.0$ $q = -120.0 lb/in^2$.

Chapter 12

Conduction Heat Transfer

The majority of the finite element practioners address problems in design oriented structural mechanics. Conversely, others tend to be mathematically formal in content, without addressing the design issue. For engineers and scientists whose expertise lie outside mathematics and structural mechanics, and in heat transfer in particular, the approaches followed by mathematicians and structural analysts are probably confusing if not incomprehensible and frustrating. The engineer or scientist, with a basic interest in heat transfer, has probably sensed that the finite element methodology may possess considerable power and versatility for the problem class. However, with no comprehensive and easily readable literature, these individuals will most likely resort directly to finite difference method rather than spending time to gain the required theoretical proficiencies. In the present Chapter, it is attempted to derive, develop and apply finite element solution methodology to the governing differential equation for practical problems in conduction heat transfer.

Conduction heat transfer takes place in a body by virtue of temperature difference. Conduction is the only mode of heat transfer that takes place in a solid body which is influenced by other modes of heat transfer at the boundary surfaces. Temperature distribution in a body is important in many problems in engineering practice. If a heated body is not permitted to expand or contract freely in all the directions, some stresses are developed within the body. The magnitude of these thermal stresses will influence the design of various equipments like boilers, steam and gas turbines, nuclear reactors, heat exchangers, jet engines and rocket motors. The first step in the determination of thermal stresses is to determine the temperature distribution within the body. The heat added or removed from a body is also important in many situations like buildings, castings, crystal growth, cooling of electronic systems and super conducting magnets. For satisfactory operation of some of the equipments like steam and gas turbine rotors, high speed machine tools, it is essential to restrict the thermal deformations. This also needs temperature distribution within the body.

12.1   Basic Equations of Heat Transfer

The basic equations of heat transfer are essentially the rate equations and the conservation of energy equation.

12.1.1   Rate Equations

The rate equations describe the rate of energy flow within a body as in conduction or between bodies as in convection or radiation.
(i) Conduction is the mode of heat transfer within a body without any net motion of the mass of the material. Fourier's law of heat conduction gives

$$Q = -kA\frac{\partial T}{\partial \eta} \tag{12.1}$$

where Q = rate of heat flow in, W
   k = thermal conductivity of the material, $W/m^\circ C$
   A = area normal to the direction of heat flow, m^2
   T = temperature ,$^\circ C$
   η = length parameter, m(may be x, y or z)

In many situations, the rate of heat transfer is required per unit area which is termed heat flux denoted by q.

$$q = -k\frac{\partial T}{\partial \eta} \tag{12.2}$$

(ii) Convection is the mode of heat transfer between a solid and a fluid surrounding it. The rate of heat transfer by convection is given as

$$Q = hA\,(T_w - T_\infty) \tag{12.3}$$

where h = surface heat transfer coefficient, $W/m^2 K$
   A = surface area of the body from which heat flows, m^2
   T_w = surface (or wall) temperature, $^\circ C$.
   T_∞ = temperature of the surrounding medium, $^\circ C$

$$q = h(T_w - T_\infty) \tag{12.4}$$

(iii) Radiation heat transfer is the mode of heat transfer between two surfaces or bodies obeying laws of electromagnetics. This is the only mode of heat transfer which takes place in vacuum and also when two bodies are not in direct contact with each other.

The rate of heat flow by radiation between two surfaces is given by,

$$Q = \sigma \, \epsilon \, A \, (T_1^4 - T_2^4) \tag{12.5}$$

where σ = Stefan-Boltzmann constant $5.669 \times 10^{-8} \ W/m^2 K^4$
  ϵ = emissivity of the surface
  A = surface area of the body from which heat flows, m^2
  T_1 = absolute temperature of the body 1, K
  T_2 = absolute temperature of the body 2, K

The Eqn. 12.5 implies that body 2 is very large compared to the body 1 and the body 1 is completely enclosed in body 2, in which case $A = A_1$ and $\epsilon = \epsilon_1$. In case the above restriction is not met, one has to use the radiation configuration factor (fraction of energy leaving body 1 reaching body 2 denoted by F_{12}) and also bring the effect of the different emissivities of the two surfaces (ϵ_1 and ϵ_2). For further details, the reader can refer to any text on heat transfer [1].

12.1.2 Energy Balance Equation

The equation of conservation of energy is one of the key equations in the heat transfer analysis which states that

$$E_{in} + E_g = E_o + E_{int} \tag{12.6}$$

where E_{in} = energy inflow into the system, W
  E_g = energy generated in the system, W
  E_o = energy leaving the system, W
  E_{int} = change in internal energy of the system, W

12.2 Governing Differential Equation for Heat Conduction

The governing differential equation will be derived for a three-dimensional stationary system in cartesian coordinates. Consider an elemental volume of a solid body as shown in Fig. 12.1. The energy balance for the elemental volume can be written as

$$\begin{bmatrix} \text{Heat in flow} \\ \text{in time } dt \end{bmatrix} + \begin{bmatrix} \text{Heat generated within} \\ \text{the body in time } dt \end{bmatrix}$$

$$= \begin{bmatrix} \text{Heat leaving the} \\ \text{body in time } dt \end{bmatrix} + \begin{bmatrix} \text{change in internal} \\ \text{energy during } dt \end{bmatrix}$$

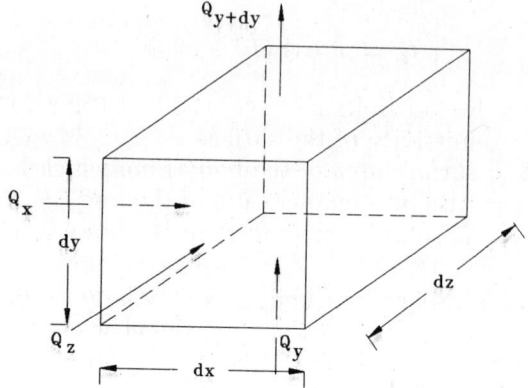

Fig. 12.1 Elemental volume

$$(Q_x+Q_y+Q_z)dt+G\, dxdydz\, dt = (Q_{x+dx}+Q_{y+dy}+Q_{z+dz})dt+\rho c\, dxdydz\, dT \tag{12.7}$$

where G = rate of heat generated per unit volume, W/m^3

dT = rise in temperature during dt $°C$

ρ = density of the material, kg/m^3

C = specific heat of the material

$Q_x = -k_x\, A_x\, \frac{\partial T}{\partial x} = -k_x\, dydz\, \frac{\partial T}{\partial x}$

where Q_x is the heat flow into the face $dydz$ located at x. Similarly,

$$\begin{aligned} Q_{x+dx} &= \text{Heat flow from the face } dydz \text{ located at } x + dx \\ &= Q_x + \frac{\partial Q_x}{\partial x}\, dx \\ &= -k_x\, dydz\, \frac{\partial T}{\partial x} - \frac{\partial}{\partial x}\left(k_x\, A_x\, \frac{\partial T}{\partial x}\right) dx \\ &= -k_x\, dydz\, \frac{\partial T}{\partial x} - \frac{\partial}{\partial x}\left(k_x\, \frac{\partial T}{\partial x}\right) dxdydz \end{aligned}$$

where k_x is the thermal conductivity of the material in the x - direction. By writing down similar expression in the y and z directions and substituting in the energy balance equation yields (after dividing throughout by $dx\, dy\, dz\, dt$)

$$\frac{\partial}{\partial x}\left(k_x\frac{\partial T}{\partial x}\right) + \frac{\partial}{\partial y}\left(k_y\frac{\partial T}{\partial y}\right) + \frac{\partial}{\partial z}\left(k_z\frac{\partial T}{\partial z}\right) + G = \rho c\frac{\partial T}{\partial t} \tag{12.8}$$

Eq.12.8 is the differential equation governing heat conduction in a solid body in which k_x, k_y and k_z are different. If the thermal conductivities in x, y and z directions are assumed to be same $k_x = k_y = k_z = k =$ constant as in an isotropic material, the Eq. 12.8 reduces to

$$\frac{\partial^2 T}{\partial x^2} + \frac{\partial^2 T}{\partial y^2} + \frac{\partial^2 T}{\partial z^2} + \frac{G}{k} = \frac{\rho c}{k} \frac{\partial T}{\partial t} \qquad (12.9)$$

where $\frac{k}{\rho c}$ is denoted by α, thermal diffusivity whose units are m^2/s.

If there are no heat sources or sinks in a body, Eq.12.9 reduces to

$$\frac{\partial^2 T}{\partial x^2} + \frac{\partial^2 T}{\partial y^2} + \frac{\partial^2 T}{\partial z^2} = \frac{1}{\alpha} \frac{\partial T}{\partial t} \qquad (12.10)$$

If the body is in a steady state, the temperature is independent of time in which case $\frac{\partial T}{\partial t}$ is zero. Hence, Eq. 12.9 reduces to Poissons equation,

$$\frac{\partial^2 T}{\partial x^2} + \frac{\partial^2 T}{\partial y^2} + \frac{\partial^2 T}{\partial z^2} + \frac{G}{k} = 0 \qquad (12.11)$$

If the body is in a steady state without heat sources or sinks, Eq. 12.9 reduces to Laplace equation,

$$\frac{\partial^2 T}{\partial x^2} + \frac{\partial^2 T}{\partial y^2} + \frac{\partial^2 T}{\partial z^2} = 0 \qquad (12.12)$$

If the cylindrical coordinate system (with r, θ, z coordinates) is used instead of the Cartesian x, y, z system, Eq. 12.8 will have the form

$$\frac{1}{r} \frac{\partial}{\partial r} \left(k_r \, r \, \frac{\partial T}{\partial r} \right) + \frac{1}{r^2} \frac{\partial}{\partial \theta} \left(k_\theta \frac{\partial T}{\partial \theta} \right) + \frac{\partial}{\partial z} \left(k_z \frac{\partial T}{\partial z} \right) + G = \rho c \frac{\partial T}{\partial t} \quad (12.13)$$

For an isotropic material this reduces to

$$\frac{1}{r} \frac{\partial}{\partial r} \left(r \frac{\partial T}{\partial r} \right) + \frac{1}{r^2} \frac{\partial^2 T}{\partial \theta^2} + \frac{\partial^2 T}{\partial z^2} + \frac{G}{k} = \frac{1}{\alpha} \frac{\partial T}{\partial t} \qquad (12.14)$$

If a spherical coordinate system (r, θ, ϕ) is used in place of x, y, z system, Eq. 12.9 reduces (for an isotropic material) to

$$\frac{1}{r^2} \frac{\partial}{\partial r} \left(r^2 \frac{\partial T}{\partial r} \right) + \frac{1}{r^2 sin\theta} \frac{\partial}{\partial \theta} \left(sin\theta \frac{\partial T}{\partial \theta} \right) +$$

$$\frac{1}{r^2 sin^2\theta} \frac{\partial^2 T}{\partial \phi^2} + G = \frac{1}{\alpha} \frac{\partial T}{\partial t} \qquad (12.15)$$

Boundary and initial conditions

The differential equation, Eq. 12.8 or Eq. 12.9 is second order since the highest derivative with respect to x, y and z is two, and two boundary conditions, in each of the three directions, x, y and z are to be specified.

$$\text{Let} \quad T(x, y, z, t) = T_o \quad \text{for} \quad t > 0 \quad \text{on the surface} \quad S_1 \qquad (12.16)$$

$$k_x \frac{\partial T}{\partial x} l + k_y \frac{\partial T}{\partial y} m + \frac{\partial T}{\partial z} n + q = 0 \text{ on } S_2 \text{ for } t > 0 \qquad (12.17)$$

$$k_x \frac{\partial T}{\partial x} l + k_y \frac{\partial T}{\partial y} m + k_z \frac{\partial T}{\partial z} n + h(T - T_\infty) = 0 \text{ on } S_3 \text{ for } t > 0 \quad (12.18)$$

where q = surface heat flux, W/m^2
        h = convection heat transfer coefficient, $W/m^2 K$
        T_∞ = surrounding temperature ,$^\circ C$
        l, m, n = direction cosines of outward normal to the boundary
        S_1 = boundary on which temperature is specified
        S_2 = boundary on which heat flux is specified
        S_3 = boundary on which convective heat loss $h(T - T_\infty)$ is specified

The differential equation 12.8 or 12.9 is first order in time t. Hence it requires one initial condition

$$T(x, y, z, 0) = T_i(x, y, z) \text{ in } V \qquad (12.19)$$

In many industrial applications like wire drawing, crystal growth, continuous casting etc., the material will have motion in space and this motion may be restricted to one direction as in examples cited above. The general energy equation for heat conduction taking into account the motion of the body in space is given by

$$\frac{\partial}{\partial x}(k_x \frac{\partial T}{\partial x}) + \frac{\partial}{\partial y}(k_y \frac{\partial T}{\partial y}) + \frac{\partial}{\partial z}(k_z \frac{\partial T}{\partial z}) + G$$

$$= \rho c (\frac{\partial T}{\partial t} + u_x \frac{\partial T}{\partial x} + u_y \frac{\partial T}{\partial y} + u_z \frac{\partial T}{\partial z}) \qquad (12.20)$$

where u_x, u_y, and u_z are the components of the velocity in the three directions x, y and z respectively.

12.3   Discrete System

In order to illustrate the nature of the finite element equations to be solved to obtain temperature distribution in a heated body, a discrete sytem of two composite slabs, as used in insulation of refrigerators, roof of air conditioned rooms etc is considered here. Fig. 12.2 shows a composite slab consisting of two homogeneous materials. The steady state conditions are assumed. The thermal conductivity is assumed to be independent of temperature. Heat sources and sinks are absent in the system.

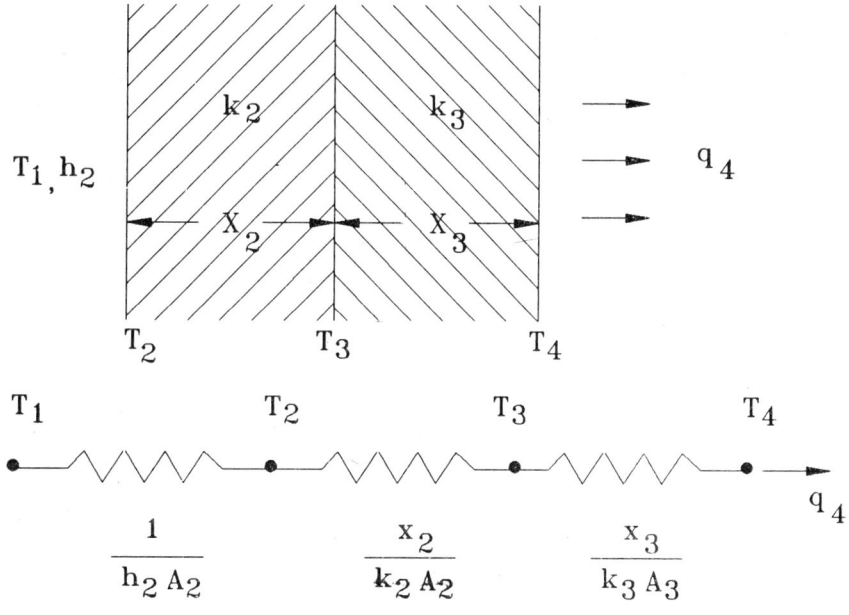

Fig. 12.2 Composite slab

The following equations are written by inspecting the Fig.12.2.

$$h_2 A_2 (T_1 - T_2) = Q_1$$

$$\frac{k_2 A_2}{X_2} (T_2 - T_3) = Q_2$$

$$\frac{k_3 A_3}{X_3} (T_3 - T_4) = Q_3$$

$$q_4 A_4 = Q_4$$

Under steady state conditions $Q_1 = Q_2 = Q_3 = Q_4$
Hence,

$$h_2 A_2 (T_1 - T_2) = \frac{k_2 A_2}{X_2} (T_2 - T_3)$$

$$\frac{k_2 A_2}{X_2} (T_2 - T_3) = \frac{k_3 A_3}{X_3} (T_3 - T_4) \text{ and}$$

$$\frac{k_3 A_3}{X_3} (T_3 - T_4) = q_4 A_4$$

Rearranging we get,

$$\left(h_2 A_2 + \frac{k_2 A_2}{X_2}\right) T_2 - \frac{k_2 A_2}{X_2} T_3 = h_2 A_2 T_1$$

$$-\frac{k_2 A_2}{X_2} T_2 + \left(\frac{k_2 A_2}{X_2} + \frac{k_3 A_3}{X_3}\right) T_3 - \frac{k_3 A_3}{X_3} T_4 = 0$$

$$-\frac{k_3 A_3}{x_3} T_3 + \frac{k_3 A_3}{X_3} T_4 = -q_4 A_4$$

The equations can be written in matrix form in terms of unknown temperatures T_2, T_3, T_4

$$\begin{bmatrix} \left(h_2 A_2 + \frac{k_2 A_2}{X_2}\right) & -\frac{k_2 A_2}{X_2} & 0 \\ -\frac{k_2 A_2}{X_2} & \left(\frac{k_2 A_2}{X_2} + \frac{k_3 A_3}{X_3}\right) & -\frac{k_3 A_3}{X_3} \\ 0 & -\frac{k_3 A_3}{X_3} & \frac{k_3 A_3}{X_3} \end{bmatrix} \begin{Bmatrix} T_2 \\ T_3 \\ T_4 \end{Bmatrix} = \begin{Bmatrix} h_2 A_2 T_1 \\ 0 \\ -q_4 A_4 \end{Bmatrix}$$

$$(12.21)$$

which can be written as

$$[k] [T] = \{f\} \tag{12.22}$$

where $[k]$ = conduction stiffness matrix
              $[T]$ = vector of unknown temperatures
              $\{f\}$ = thermal load vector

Compare this with $[k] \{d\} = \{Q\}$ for the structural analysis case. In the case of heat conduction there is only one degree of freedom at each node as temperature is a scalar. Thus the conduction heat transfer problems are simple when compared to structural mechanics problems. Several important features can be observed from Eq. 12.21.

The conduction stiffness matrix is symmetric and positive definite

The characteristics of each layer of the composite slab for heat conduction can be written as

$$\frac{k_2 A_2}{x_2} \begin{bmatrix} +1 & -1 \\ -1 & +1 \end{bmatrix} \begin{Bmatrix} T_i \\ T_j \end{Bmatrix} = \begin{Bmatrix} Q_i \\ Q_j \end{Bmatrix} \qquad (12.23)$$

The global stiffness matrix can be obtained by assemblage of the stiffness matrix of each sublayer.

The convection effect appears both in the stiffness matrix and the loading vector and adds to the conduction portion in the stiffness matrix.

The effect of heat flux boundary condition appears only in the loading term.

The thermal force vector consists of known values.

The method of assemblage can be extended to more than two layers of insulation.

The effect of natural boundary conditions enters at the formulation stage itself.

Thus the main objective in the following pages is to show how to evaluate $[k]$ and $\{f\}$ for various situations one comes across in practice and to solve the system of assembled equations to obtain the temperature distribution in the body.

12.4 Formulation of the Finite Element Method for Heat Conduction

In finite element method we consider a given region is composed of a number of subregions called elements connected at the nodes at which the values have to be evaluated. In finite element method, we assume a variation within an element and the continuous variable is replaced by piece-wise continuous functions defined in the element in terms of the nodal values. Whereas in the finite difference method, we replace the differential coefficient by a difference equation. By formulation of finite element method, we attempt to find a relation of the various quantities in the element in the form

$$[k] \{T\} = \{f\} \qquad (12.24)$$

There are two ways the formulation can be made. (i) Variational or (ii) Weighted residual method. Weighted residual method is more versatile in nature and the same is used for the formulation of conduction heat transfer.

12.4.1   Weighted Residual Method

To solve the differential equation

$$L(T) - f = 0 \tag{12.25}$$

where L is an operator.

Assume

$$T = \overline{T} = \sum_{i=1}^{n} N_i \, T_i$$

(N_i - assumed function, T_i - Unknown parameter)

$$L(\overline{T}) - f \neq 0 \text{ but } L(\overline{T}) - f = R \tag{12.26}$$

where R is the residue.

$$\int_D R \, W_i \, dD = 0, \quad i = 1, 2, \dots n \tag{12.27}$$

The method of weighted residuals seeks to determine the n unknown T_i in such a way that the error R over the entire solution domain is small. This is accomplished by forming a weighted average of the error and specifying that this weighted average vanish over the solution domain. Hence we choose n linearly independent weighting functions W_i. Once we specify the weighting functions, the Eq. 12.27 represents a set of equations, either algebraic or ordinary differential equations to be solved for T_i to obtain an approximate representation of the unknown field variables.

12.4.2   Various Weighted Residual Techniques

Depending on the choice of the weighting function, various names are given to particular cases.

(i) Point Collocation Method

$$W_i = \delta(x - x_i) \tag{12.28a}$$

$$\int_D W_i \, R \, dD = (R)_{x_i} = 0$$

(ii) Sub domain collocation method

$$W_i = \begin{cases} 1 & \text{in } D_j \\ = 0 & \text{elsewhere} \end{cases} \tag{12.28b}$$

(iii) Galerkin's method

$$W_i = N_i, \quad \text{the shape function} \qquad (12.28c)$$

(iv) Least squares method

$$W_i = R, \quad \text{the residue itself} \qquad (12.28d)$$

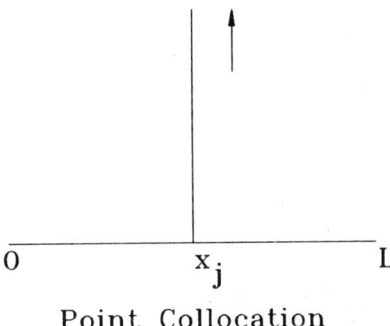

Point Collocation

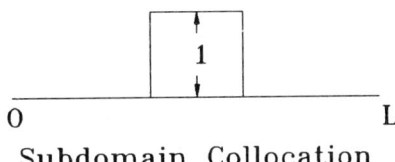

Subdomain Collocation

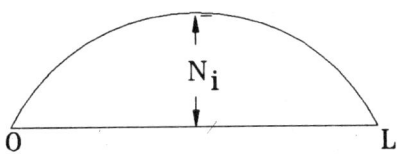

Galerkin

Fig. 12.3 Weighted residual methods

12.5 Galerkin's Method for Quasi Harmonic Equation - Heat Conduction Equation

$$\frac{\partial}{\partial x}\left(k_x \frac{\partial T}{\partial x}\right) + \frac{\partial}{\partial y}\left(k_y \frac{\partial T}{\partial y}\right) + \frac{\partial}{\partial z}\left(k_z \frac{\partial T}{\partial z}\right) + G = 0 \qquad (12.29)$$

$$T = T_B \quad \text{on} \quad S_1$$

$$k_x \frac{\partial T}{\partial x} l + k_y \frac{\partial T}{\partial y} m + k_z \frac{\partial T}{\partial z} n + q + h(T - T_\infty) = 0$$

on S_2 where q is specified
S_3 where h, T_∞ are specified.

$$T \sim \overline{T} = \sum_{i=1}^{r} N_i T_i = [N] \{T\} \tag{12.30}$$

Applying Galerkin criterion,

$$\int_V N_i \left[\frac{\partial}{\partial x} \left(k_x \frac{\partial \overline{T}}{\partial x} \right) + \frac{\partial}{\partial y} \left(k_y \frac{\partial \overline{T}}{\partial y} \right) + \frac{\partial}{\partial z} \left(k_z \frac{\partial \overline{T}}{\partial z} \right) + G \right] dx \, dy \, dz = 0 \tag{12.31}$$

To introduce the influence of natural boundary conditions, integration is carried out by parts. Consider a typical term for integration,

$$\int_v N_i \frac{\partial}{\partial x} \left(k_x \frac{\partial \overline{T}}{\partial x} \right) dx \, dy \, dz$$

$$= \int_s N_i k_x \frac{\partial \overline{T}}{\partial x} \, dy \, dz - \int_v k_x \frac{\partial \overline{T}}{\partial x} \frac{\partial N_i}{\partial x} \, dx \, dy \, dz$$

$$= \int_s N_i k_x \frac{\partial \overline{T}}{\partial x} \, lds - \int_v k_x \frac{\partial [N]}{\partial x} \{T\} \frac{\partial N_i}{\partial x} \, dx \, dy \, dz$$

Final equation takes the form,

$$\int_v G N_i \, dx \, dy \, dz + \int_s N_i \left(k_x \frac{\partial \overline{T}}{\partial x} l + k_y \frac{\partial \overline{T}}{\partial y} m + k_z \frac{\partial \overline{T}}{\partial z} n \right) ds$$

$$- \int_v \left[k_x \frac{\partial [N]}{\partial x} \frac{\partial N_i}{\partial x} + k_y \frac{\partial [N]}{\partial y} \frac{\partial N_i}{\partial y} + k_z \frac{\partial [N]}{\partial z} \frac{\partial N_i}{\partial z} \right] \{\overline{T}\} \, dx \, dy \, dz = 0$$

But, from boundary condition equation, we have,

$$k_x \frac{\partial T}{\partial x} l + k_y \frac{\partial T}{\partial y} m + k_z \frac{\partial T}{\partial z} n = -q - h(\overline{T} - T_\infty)$$

Hence

$$\int_v G N_i \, dx \, dy \, dz - \int_s q N_i \, dS + \int_s h T_\infty N_i \, dS - \int_s h [N] N_i \{\overline{T}\} \, ds$$

$$-\int_v \left(k_x \frac{\partial[N]}{\partial x}\frac{\partial N_i}{\partial x} + k_y \frac{\partial[N]}{\partial y}\frac{\partial N_i}{\partial y} + k_z \frac{\partial[N]}{\partial z}\frac{\partial N_i}{\partial z}\ dx\ dy\ dz\right)\{\overline{T}\} = 0$$

$$(12.32)$$

which can be written, in final form as,

$$[k]\ \{T\} = \{f\}$$

where

$$k_{ij} = \int_V \left(k_x \frac{\partial N_i}{\partial x}\frac{\partial N_j}{\partial x} + k_y \frac{\partial N_i}{\partial y}\frac{\partial N_i}{\partial y} + k_z \frac{\partial N_i}{\partial z}\frac{\partial N_j}{\partial z}\right)dx\ dy\ dz$$

$$+ \int_S h\ N_i\ N_j\ dS \quad i = 1, 2, \dots n \quad j = 1, 2, \dots n \qquad (12.33)$$

$$f_i = \int_v G\ N_i\ dV - \int_s q\ N_i\ dS + \int_s h\ T_\infty\ N_i\ dS \qquad (12.34)$$

$$\{T\}^T = [T_1 \quad T_2 \quad . \qquad . \quad T_n\,] \qquad (12.35)$$

The above equation can be recast in matrix form as

$$[k] = \int_V [B]^T\ [D]\ [B]\ dV + \int_S h\ [N]^T\ [N]\ dS \qquad (12.36)$$

$$\{f\} = \int_v G\ [N]^T\ dV - \int_s q\ [N]^T\ dS + \int_s h\ T_\infty\ [N]^T\ dS \quad (12.37)$$

where B = gradient matrix

$$[B] = \begin{bmatrix} \dfrac{\partial N_i}{\partial x} & \dfrac{\partial N_j}{\partial x} & \cdots\cdots & \dfrac{\partial N_n}{\partial x} \\[2mm] \dfrac{\partial N_i}{\partial y} & \dfrac{\partial N_j}{\partial y} & \cdots\cdots & \dfrac{\partial N_n}{\partial y} \\[2mm] \dfrac{\partial N_i}{\partial z} & \dfrac{\partial N_j}{\partial z} & \cdots\cdots & \dfrac{\partial N_n}{\partial z} \end{bmatrix} \qquad (12.38a)$$

$$\text{and,} \quad [D] = \begin{bmatrix} k_x & 0 & 0 \\ 0 & k_y & 0 \\ 0 & 0 & k_x \end{bmatrix} \qquad (12.38b)$$

12.6　Field　of　Application　of　Quasi-Harmonic Equation

Quasi-harmonic equation is one of the important differential equations dealt in mathematical physics as it has many applications in various fields as indicated below.

(i) Torsion of Noncircular Sections

$$k_x = k_y = 1 \quad \text{and} \quad G = 2G\,\theta$$

$$\frac{\partial^2 \phi}{\partial x^2} + \frac{\partial^2 \phi}{\partial y^2} + 2G\,\theta = 0 \qquad (12.39a)$$

where ϕ = stress function

(ii) Irrotational Flow of Fluids

$$k_x = k_y = 1, \quad G = 0$$

$$\frac{\partial^2 \psi}{\partial x^2} + \frac{\partial^2 \psi}{\partial y^2} = 0 \qquad (12.39b)$$

where ψ = potential or stream function

(iii) Flow Through Porous Media

$$k_x \frac{\partial^2 \phi}{\partial x^2} + k_y \frac{\partial^2 \phi}{\partial y^2} + G = 0 \qquad (12.39c)$$

ϕ　=　piezometric head
G　=　source or sink
q　=　seepage of water into or out of aquifer (comes as a boundary condition)

$$k_x \frac{\partial \phi}{\partial x}\, l + k_y \frac{\partial \phi}{\partial y}\, m = q$$

(iv) Film Lubrication (Reynolds Equation)

$$\frac{\partial}{\partial x}\Big(H^3\, \frac{\partial P}{\partial x}\Big) + \frac{\partial}{\partial y}\Big(H^3\, \frac{\partial P}{\partial y}\Big) = \lambda\, \frac{\partial H}{\partial x} \qquad (12.40)$$

where H = film thickness and P = pressure in the film. This can also be used for pneumatic conveyance of packages.

(v) Helmholtz Equations: (Steady state field equations)

Propagation problems involving wave motion of type

$$\frac{\partial}{\partial x}\Big(k_x \frac{\partial \phi}{\partial x}\Big) + \frac{\partial}{\partial y}\Big(k_y \frac{\partial \phi}{\partial y}\Big) + \frac{\partial}{\partial z}\Big(k_z \frac{\partial \psi}{\partial z}\Big) + \phi = 0 \qquad (12.41)$$

Particular Cases

(a) Seiche Motion

Standing waves on a bounded shallow body of water

$$\frac{\partial}{\partial x}\left(h\frac{\partial w}{\partial x}\right) + \frac{\partial}{\partial y}\left(h\frac{\partial w}{\partial y}\right) + \frac{4\pi^2}{gt_p^2}w = 0 \qquad (12.42)$$

h   =   water depth of quiescent state
w   =   elevation of free surface above quiescent level
t_p =   period of oscillation

$$\frac{\partial w}{\partial x}l + \frac{\partial w}{\partial y}m = 0 \text{ on all solid boundaries}$$

(b) Electromagnetic Waves

The propagation of electromagnetic waves in a wave guide filled
with a dielectric material obeys the equation

$$\frac{\partial}{\partial x}\left(\frac{1}{\epsilon_d}\frac{\partial \psi}{\partial x}\right) + \frac{\partial}{\partial y}\left(\frac{1}{\epsilon_d}\frac{\partial \phi}{\partial y}\right) + \frac{\partial}{\partial z}\left(\frac{1}{\epsilon_d}\frac{\partial \phi}{\partial z}\right) + \omega^2 \mu_o \epsilon_o \phi = 0 \quad (12.43)$$

ψ =   A component of the magnetic field strength
vector H or a component of the electric field
vector E.

ω =   wave frequency
μ =   permeability of free space
ϵ =   permitivity of free space
ϵ_d =   permitivity of dielectric

Eq. 12.43 can be derived from Farady's law and Maxwell's equations. If ψ represents an H wave component, say $\phi = H_x$, then ψ
must satisfy the Neumann boundary conditions at solid boundaries.
But, if ψ represents an E wave component, Eq. 12.43 normally not
solved for both the E and H waves because one vector field can be
obtained from the other via the equation,

$$\nabla \times E = -\sqrt{-1}\omega \mu_o \overline{H}$$

(c) Acoustic Vibrations

Fluid vibrations, in a closed volume represents a sonic field of
spherical waves governed by

$$\frac{\partial^2 P}{\partial x^2} + \frac{\partial^2 P}{\partial y^2} + \frac{\partial^2 P}{\partial z^2} + \frac{\omega^2}{C^2}P = 0 \qquad (12.44)$$

P = pressure excess above ambient pressure
ω = frequency
C = wave velocity in the medium

12.7   One-Dimensional Heat Conduction

12.7.1   Element Characteristics

A fin dissipating heat to the ambient is an example of one-dimensional heat conduction. Assume for simplicity, a linear variation of temperature in the element,

$$T = C_o + mx \tag{12.45}$$

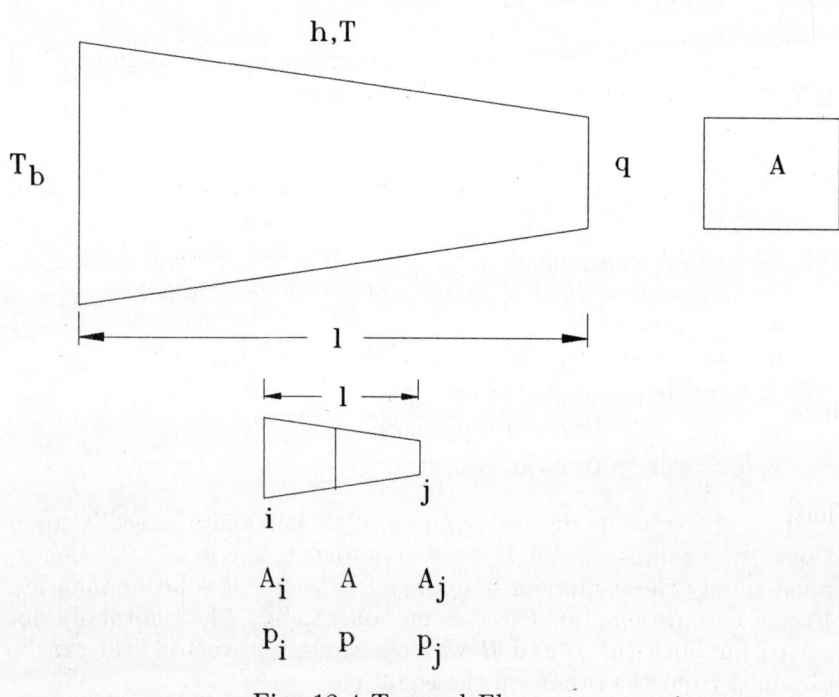

Fig. 12.4 Tapered Element

$$T_i = C_o$$
$$T_j = C_o + ml$$

Hence $C_o = T_i$ and $m = \dfrac{T_j - T_i}{l}$

Therefore $T = T_i + (T_j - T_i)\dfrac{x}{l}$.

$$T = T_i\left(1 - \frac{x}{l}\right) + T_j\,\frac{x}{l}$$

$$T = T_i\,N_i + T_j\,N_j = [N_i \quad N_j]\begin{Bmatrix} T_i \\ T_j \end{Bmatrix} = [N]\,\{T\}$$

$$(12.46)$$

where $N_i = \left(1 - \frac{x}{l}\right)$ and $N_j = \frac{x}{l}$

N_i and N_j are called shape functions or interpolation function or basis functions and are the same as defined by Eq.3.12. Let us consider the property of the above shape functions.

x	N_i	N_j	$N_i + N_j$
0 i.e., i	1	0	1
1 i.e. j	0	1	1
Any x	N_i	N_j	1

12.7.2 Element Characteristics for 1-D Problem to Derive the Expression for [k] and {f}

$$T = (N_i)\,T_i + (N_j)\,T_j = [N_i \quad N_j]\begin{Bmatrix} T_i \\ T_j \end{Bmatrix} \qquad (12.47)$$

$$\frac{dT}{dx} = T_i\left(\frac{dN_i}{dx}\right) + T_j\left(\frac{dN_j}{dx}\right) = [\frac{-1}{l} \quad \frac{1}{l}]\begin{Bmatrix} T_i \\ T_j \end{Bmatrix} \qquad (12.48)$$

Since,

$$\frac{d(N_i)}{dx} = -\frac{1}{l} \quad \text{and} \quad \frac{d(N_j)}{dx} = \frac{1}{l}$$

Thus,

$$[B] = [-1/l \quad 1/l]$$

$$[k] = \int_V [B]^T\,[D]\,[B]\,dV + \int_S h\,[N]^T\,[N]\,dS \qquad (12.49)$$

$$= \int_l \begin{bmatrix} -1/l \\ 1/l \end{bmatrix}[k_x]\,[-1/l \quad 1/l]\,A dx + \int_l h \begin{bmatrix} N_i \\ N_j \end{bmatrix}[N_i \quad N_j]\,P dx$$

where A = area of cross section and P = perimeter of fin.

$$[k] = \int_l \frac{Ak_x}{l^2}\begin{bmatrix} 1 & -1 \\ -1 & 1 \end{bmatrix} dx + \int_l hP \begin{bmatrix} N_i\,N_i & N_i\,N_j \\ N_i\,N_j & N_j\,N_j \end{bmatrix} dx$$

In case of linear problem, N_i and N_j are also local coordinates, i.e., L_i, L_j. As indicated in Eq.3.16, explicit integration can be carried out as,

$$\int_l L_i^a \, L_j^b \, dl = \frac{a! \, b!}{(a+b+1)!} \, l \tag{12.50}$$

Using the Eq.12.50,

$$\int_l N_i \cdot N_j \, dl = \frac{1! \, 1!}{(1+1+1)!} \, l = l/6$$

$$\int_l N_i \cdot N_i \, dl = \frac{2! \, 0!}{(2+0+1)!} \, l = l/3$$

$$[k] = \frac{Ak_x}{l} \begin{bmatrix} 1 & -1 \\ -1 & 1 \end{bmatrix} + \frac{hPl}{6} \begin{bmatrix} 2 & 1 \\ 1 & 2 \end{bmatrix} \tag{12.51}$$

$$\{f\} = \int_v G[N]^T \, dV - \int_s q[N]^T \, ds + \int_s hT_\infty \, \{N\}^T \, ds \tag{12.52}$$

$$= \int_l G \begin{Bmatrix} N_i \\ N_j \end{Bmatrix} A dx - \int_A q \begin{Bmatrix} N_i \\ N_j \end{Bmatrix} dA + \int_l hT_\infty \begin{Bmatrix} N_i \\ N_j \end{Bmatrix} Pdx$$

Either we can substitute the value of N_i, N_j and integrate or use the relation 12.50.

$$\int_l N_i \, dx = \frac{1! \, 0!}{(1+0+1)!} \, l = l/2$$

Therefore,

$$\{f\} = \frac{GAL}{2} \begin{Bmatrix} 1 \\ 1 \end{Bmatrix} - q \begin{Bmatrix} 0 \\ 1 \end{Bmatrix} A + \frac{hT_\infty \, P}{2} \begin{Bmatrix} 1 \\ 1 \end{Bmatrix} \tag{12.53}$$

Finally,

$$[k] \begin{Bmatrix} T_i \\ T_j \end{Bmatrix} = \{f\}$$

$$2 \times 2 \quad 2 \times 1 \qquad 2 \times 1$$

Example

Calculate the temperature distribution in stainless steel fin shown in Fig. 12.5. The region is discretized into 5 elements and 6 nodes. The properties of an element are computed as follows.

$$
\begin{aligned}
[k] &= \frac{k\,A}{l}\begin{bmatrix} +1 & -1 \\ -1 & +1 \end{bmatrix} + \frac{hPl}{6}\begin{bmatrix} 2 & 1 \\ 1 & 2 \end{bmatrix} \\
&= \frac{0.17(\pi)}{2}\begin{bmatrix} +1 & -1 \\ -1 & +1 \end{bmatrix} + \frac{0.0025(2\pi)2}{6}\begin{bmatrix} 2 & 1 \\ 1 & 2 \end{bmatrix} \\
&= \pi \begin{bmatrix} 0.0866 & -0.0834 \\ -0.0834 & 0.0866 \end{bmatrix}
\end{aligned}
$$

$$
\begin{aligned}
\{f_1\} &= \frac{h\,T_\infty Pl}{2}\begin{Bmatrix} 1 \\ 1 \end{Bmatrix} = \frac{0.0025(25)(2\pi)(2)}{2}\begin{Bmatrix} 1 \\ 1 \end{Bmatrix} = \pi \begin{Bmatrix} 0.125 \\ 0.125 \end{Bmatrix} \\
&= \{f_2\} = \{f_3\} = \{f_4\}
\end{aligned}
$$

$$
\begin{aligned}
\{f_5\} &= \{f_1\} - qA\begin{Bmatrix} 0 \\ 1 \end{Bmatrix} = \pi \begin{Bmatrix} 0.125 \\ 0.125 \end{Bmatrix} - 0.0625(\pi)\begin{Bmatrix} 0 \\ 1 \end{Bmatrix} \\
&= \pi \begin{Bmatrix} 0.125 \\ 0.0625 \end{Bmatrix}
\end{aligned}
$$

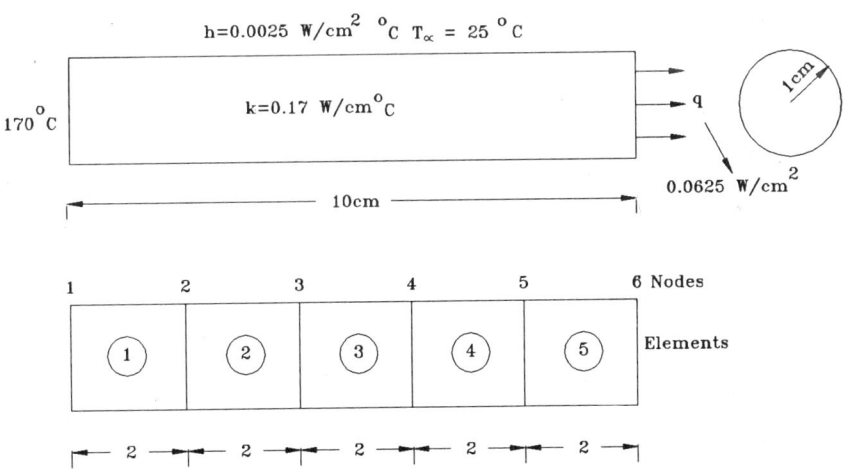

Fig. 12.5 One-dimensional example with linear elements

Assembling the elements similar to stress analysis problem, we get,

$$\begin{array}{cccccc} 1 & 2 & 3 & 4 & 5 & 6 \end{array}$$

$$
\begin{array}{c} 1 \\ 2 \\ 3 \\ 4 \\ 5 \\ 6 \end{array}
\begin{bmatrix}
0.0866 & -0.0834 & & & & \\
-0.0834 & 0.0866 & -0.0834 & & & \\
 & 0.0866 & & & & \\
 & -0.0834 & 0.0866 & & & \\
 & & 0.0866 & -0.0834 & & \\
 & & -0.0834 & 0.0866 & -0.834 & \\
 & & & 0.0866 & 0.0866 & \\
 & & & & 0.0866 & -0.0834 \\
 & & & -0.0834 & & \\
 & & & & -0.0834 & \\
 & & & & & 0.0866
\end{bmatrix}
\begin{Bmatrix}
T_1 \\ T_2 \\ T_3 \\ T_4 \\ T_5 \\ T_6
\end{Bmatrix}
$$

$$
= \begin{Bmatrix}
0.125 \\
0.125 + 0.125 \\
+0.0834(170) \\
0.125 + 0.125 \\
0.125 + 0.125 \\
0.125 + 0.125 \\
0.065
\end{Bmatrix}
\qquad (12.54)
$$

Solving the above system of equations, we get, the values for $\{T\}$. The Finite Element solution of $\{T\}$ is compared with the exact solution and the results are given below:

	T_1	T_2	T_3	T_4	T_5	T_6
FEM°C	170	132.30	106.88	90.68	81.76	79.06
EXACT°C	170	131.39	105.43	89.02	80.22	77.98

The exact solution for this problem is given by [1]

$$
\frac{T - T_\infty}{T - T_1} = \frac{\cosh m(l - x) + \left(\frac{h}{mk}\right) \sinh m(l - x)}{\cosh ml + \left(\frac{h}{mk}\right) \sinh ml}
$$

where $m = \sqrt{\frac{hp}{kA}}$

l = length of the fin.

T_∞ = ambient temperature

Heat dissipated from the fin = $-k \times A \frac{dT}{dx}$

$$
= -0.17(\pi) \left(\frac{-T_1}{l} + \frac{T_2}{l}\right)
$$

$$= \frac{-0.17(\pi)}{2} (-170 + 132.3) \; ; \; Q_d = \frac{37.7 \times 0.17 \times \pi}{2} = 10.06 \text{ watts}$$

12.7.3 Variable Cross Section Along the Length

Consider linear variation in cross section and perimeter as in trapezoidal or rectangular cross section.

$$
\begin{aligned}
A &= A_i + \frac{(A_j - A_i)}{l}x \\
&= A_i \left(1 - \frac{x}{l}\right) + A_j\left(\frac{x}{l}\right) \\
A &= A_i (N_i) + A_j (N_j)
\end{aligned}
$$

Similarly $P = N_i(P_i) + N_j(P_j)$

$$
\begin{aligned}
k &= \int_l \begin{bmatrix} -1/l \\ 1/l \end{bmatrix} k_x \begin{bmatrix} -1/l & 1/l \end{bmatrix} (N_i(A_i) + N_j(A_i))dx \\
&+ \int_l h \begin{bmatrix} N_i \\ N_j \end{bmatrix} \begin{bmatrix} N_i & N_j \end{bmatrix} (N_i(P_i) + N_j(P_i))dx \\
k &= \frac{k_x}{l} \begin{bmatrix} 1 & -1 \\ -1 & 1 \end{bmatrix} \frac{(A_i + A_j)}{2} + \frac{hl}{12} \begin{bmatrix} 3P_i + P_j & P_i + P_j \\ P_i + P_j & P_i + 3P_j \end{bmatrix}
\end{aligned}
$$

$$\text{(12.55)}$$

$$\{f\} = \int_l G \begin{Bmatrix} N_i \\ N_j \end{Bmatrix} (A_i(N_i) + A_j(N_j))dx - \int_A q \begin{Bmatrix} N_i \\ N_j \end{Bmatrix} dA + \int_l hT_\infty \begin{Bmatrix} N_i \\ N_j \end{Bmatrix}$$

$$(N_i(P_i) + N_j(P_j))dx$$

$$\{f\} = \frac{Gl}{6} \begin{Bmatrix} 2A_i + A_j \\ A_i + 2A_j \end{Bmatrix} - qA_j \begin{Bmatrix} 0 \\ 1 \end{Bmatrix} + \frac{hT_\infty l}{6} \begin{Bmatrix} 2P_i + P_j \\ P_i + 2P_j \end{Bmatrix} \quad \text{(12.56)}$$

Stiffness matrix and loading terms indicate unequal contribution to the nodes. Variation in properties can be dealt with in a similar way.

12.8 Two-Dimensional Conduction Heat Transfer

12.8.1 Element Characteristics

We shall use linear triangular elements for discretization as shown in Fig. 12.6. The temperature in an element is represented by

$$T = N_i T_i + N_j T_j + N_k T_k \tag{12.57}$$

where
$$N_i = \frac{1}{2A}(c_i + b_i x + a_i y)$$
$$N_j = \frac{1}{2A}(c_j + b_j x + a_j y)$$
$$N_k = \frac{1}{2A}(c_k + b_k x + a_k y)$$

in which
$$c_i = x_j y_k - x_k y_j \ ; \ b_i = y_i - y_k \ ; \ a_i = x_k - x_j$$
$$c_j = x_k y_i - x_i y_k \ ; \ b_j = y_k - y_i \ ; \ a_j = x_i - x_k$$
$$c_k = x_i y_j - x_j y_i \ ; \ b_k = y_i - y_j \ ; \ a_k = x_j - x_i$$

$$N_i = L_i \ , \quad N_j = L_j \ , \quad N_k = L_k$$

The following relations can be derived.

$$\left\{ \begin{array}{c} \frac{\partial \overline{T}}{\partial x} \\ \frac{\partial \overline{T}}{\partial y} \end{array} \right\} = \frac{1}{2A} \begin{bmatrix} b_i & b_j & b_k \\ a_i & a_j & a_k \end{bmatrix} \left\{ \begin{array}{c} T_i \\ T_j \\ T_k \end{array} \right\}$$

$$\int_A L_i^a L_j^b L_k^c \, dA = \frac{a! \, b! \, c!}{(a+b+c+2)!} \, 2A \tag{12.58}$$

$$[B] = \frac{1}{2A} \begin{bmatrix} b_i & b_j & b_k \\ a_i & a_j & a_k \end{bmatrix}$$

$$[D] = \begin{bmatrix} k_x & 0 \\ 0 & k_y \end{bmatrix}$$

$$2A = \begin{vmatrix} 1 & x_i & y_i \\ 1 & x_j & y_j \\ 1 & x_k & y_k \end{vmatrix}$$

$$[k] = \int_A \begin{bmatrix} b_i & a_i \\ b_j & a_j \\ b_k & a_k \end{bmatrix} \begin{bmatrix} k_x & 0 \\ 0 & k_y \end{bmatrix} \begin{bmatrix} b_i & b_j & b_k \\ a_i & a_j & a_k \end{bmatrix} t \, dA$$

$$+ \int_{l_{ik}} h \left\{ \begin{array}{c} N_i \\ N_j \\ N_k \end{array} \right\} [N_i \ N_j \ N_k] \, t \, dl$$

where t is the thickness of the lamina considered.

$$[k] = \frac{t}{4A} \left[k_x \begin{bmatrix} b_i b_i & b_i b_j & b_i b_k \\ & b_j b_j & b_j b_k \\ \text{sym.} & & b_k b_k \end{bmatrix} + k_y \begin{bmatrix} a_i a_i & a_j a_j & a_i a_k \\ & a_j a_j & a_j a_k \\ \text{sym.} & & a_k a_k \end{bmatrix} \right]$$

$$+ \frac{htl_{ik}}{6} \begin{bmatrix} 2 & 0 & 1 \\ 0 & 0 & 0 \\ 1 & 0 & 2 \end{bmatrix} \qquad (12.59)$$

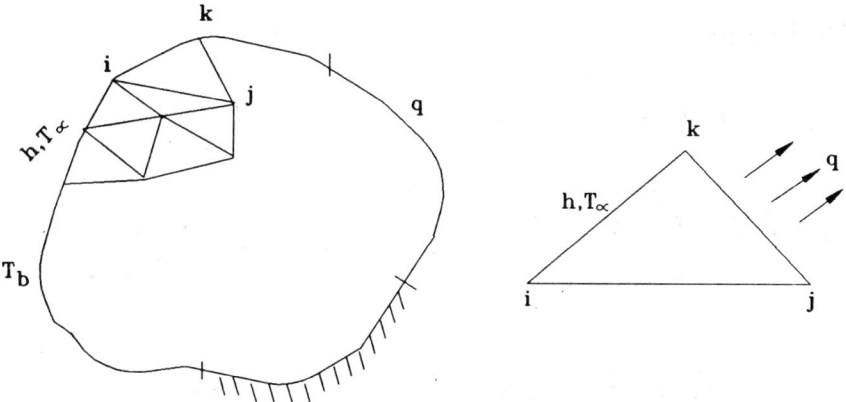

Fig. 12.6 Two-dimensional problem

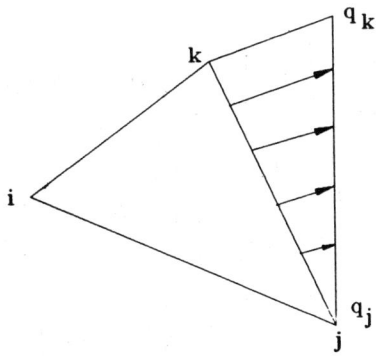

Fig. 12.7 Linear variation in q on side jk

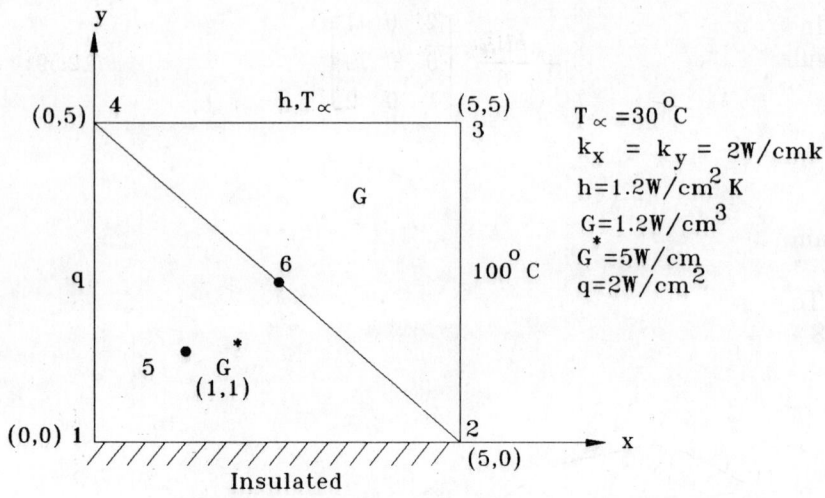

Fig. 12.8 Example in Two-dimensional Conduction

$$\{f\} = \int_A G \begin{Bmatrix} N_i \\ N_j \\ N_k \end{Bmatrix} t dA - \int_{l_{jk}} q \begin{Bmatrix} N_i \\ N_j \\ N_k \end{Bmatrix} t dl + \int_{l_{ik}} hT_\infty \begin{Bmatrix} N_i \\ N_j \\ N_k \end{Bmatrix} t dl_{ik}$$

$$= \frac{GA\,t}{3} \begin{Bmatrix} 1 \\ 1 \\ 1 \end{Bmatrix} - \frac{qtl_{jk}}{2} \begin{Bmatrix} 0 \\ 1 \\ 1 \end{Bmatrix} + \frac{hT_\infty\,tl_{ik}}{2} \begin{Bmatrix} 1 \\ 0 \\ 1 \end{Bmatrix} \qquad (12.60)$$

It can be observed that the heat generation is uniformly distributed among 3 nodes and the loads are shared equally by participating nodes. In case the boundary conditions vary it can be treated as follows: (q varies linearly from j to k as shown in Fig. 12.7)

$$\int_{l_{ik}} a \begin{Bmatrix} N_i \\ N_j \\ N_k \end{Bmatrix} (N_j(q_j)) + (N_k(q_k))\,t dl_{jk}$$

$$= \int_{l_{ik}} q \begin{bmatrix} 0 \\ N_j\,N_j\,q_j + N_k\,N_j\,q_k \\ N_j\,N_k\,q_k + N_k\,N_k\,q_k \end{bmatrix} t dl_{jk}$$

$$= \frac{t\,l_{jk}}{6} \begin{Bmatrix} 0 \\ 2q_j + q_k \\ a_i + 2a_k \end{Bmatrix} \qquad (12.61)$$

In case there is a line source G^* at (x_o, y_o), then the loads are calculated as,

$$= G^* \left\{ \begin{array}{c} N_i \\ N_j \\ N_k \end{array} \right\}_{x_o, y_o} \tag{12.62}$$

Example

To find the temperature distribution in the body shown in Fig. 12.8

$$[k]^e = \frac{t \, k_x}{4A} \begin{bmatrix} b_i b_i & b_i b_j & b_i b_k \\ b_i b_j & b_j b_j & b_j b_k \\ b_i b_k & b_j b_k & b_k b_k \end{bmatrix} + \frac{t \, k_y}{4A} \begin{bmatrix} a_i a_i & a_i a_j & a_i a_k \\ a_i a_j & a_j a_j & a_j a_k \\ a_i a_k & a_j a_k & a_k a_k \end{bmatrix}$$

$$+ \frac{th \, l_{ik}}{6} \begin{bmatrix} 2 & 0 & 1 \\ 0 & 0 & 0 \\ 1 & 0 & 2 \end{bmatrix}$$

$$\{f\} = \frac{GV}{3} \left\{ \begin{array}{c} 1 \\ 1 \\ 1 \end{array} \right\} - \frac{qt \, l_{jk}}{2} \left\{ \begin{array}{c} 0 \\ 1 \\ 1 \end{array} \right\} + \frac{hT \, tl_{ik}}{2} \left\{ \begin{array}{c} 1 \\ 0 \\ 1 \end{array} \right\} + G^* \left\{ \begin{array}{c} N_i \\ N_j \\ N_k \end{array} \right\}_{x_o, y_o}$$

Element Properties

Element Triangle 124

$$\begin{aligned}
c_1 &= x_2 \, y_4 - x_4 \, y_2 = 5*5 - 0*0 = 25 \\
b_1 &= y_2 - y_4 = 0 - 5 = -5 \\
a_1 &= x_4 - x_2 = 0 - 5 - -5
\end{aligned}$$

Similarly,

$$\begin{aligned}
c_2 &= x_4 \, y_1 - y_4 \, x_1 = 0 \\
b_2 &= y_4 - y_1 = 5 \\
a_2 &= x_1 - x_4 = 0 \\
c_4 &= x_1 \, y_2 - x_2 \, y_1 = 0 \\
h_4 &= y_1 - y_2 = 0 \\
a_4 &= x_2 - x_1 = 5
\end{aligned}$$

$$2 \times \text{area of element} = \begin{vmatrix} 1 & 0 & 0 \\ 1 & 5 & 0 \\ 1 & 0 & 5 \end{vmatrix} = 25$$

Therefore, $A_1 = 12.5$

$$[k_1] = \frac{2}{4*12.5} \begin{bmatrix} 25 & -25 & 0 \\ -25 & 25 & 0 \\ 0 & 0 & 0 \end{bmatrix} + \frac{2}{4*12.5} \begin{bmatrix} 25 & 0 & -25 \\ 0 & 0 & 0 \\ -25 & 0 & 25 \end{bmatrix}$$

$$= \begin{bmatrix} 2 & -1 & -1 \\ -1 & 1 & 0 \\ -1 & 0 & 1 \end{bmatrix}$$

$$l_{ik} = \sqrt{(x_i - x_k)^2 + (y_i - y_k)^2} = 5$$

$$\{f_1\} = 0 - \frac{2 \times 5}{2} \begin{Bmatrix} 1 \\ 0 \\ 1 \end{Bmatrix} + 0 + 5 \begin{Bmatrix} 3/5 \\ 1/5 \\ 1/5 \end{Bmatrix} = \begin{Bmatrix} -2 \\ 1 \\ -4 \end{Bmatrix}$$

where $N_i = \frac{1}{2 \times 12.5} [25 - 5 \times 1 - 5 \times 1] = \frac{15}{25} = \frac{3}{5}$

$N_j = \frac{1}{25} [0 + 5 \times 1 + 0] = \frac{5}{25} = \frac{1}{5}$

$N_k = \frac{1}{25} [0 + 0 + 5*1] = \frac{5}{25} = \frac{1}{5}$ at $(1,1)$

Element Triangle 423

$c_4 = x_2 y_3 - x_3 y_2 \doteq 25, \quad a_2 = 25, \quad a_3 = -25$

$b_4 = y_2 - y_3 = -5, \quad b_2 = 0, \quad b_3 = 5$

$a_4 = x_3 - x_2 = 0, \quad c_2 = -5 \quad c_3 = 5$

$$2 \times (\text{Area of triangle } 423) = \begin{bmatrix} 1 & 0 & 0 \\ 1 & 5 & 0 \\ 1 & 5 & 5 \end{bmatrix} = 25$$

Therefore, $A_2 = 12.5$,

$$[k_2] = \frac{2}{4 \times 12.5} \begin{bmatrix} 25 & 0 & -25 \\ 0 & 0 & 0 \\ -25 & 0 & 25 \end{bmatrix} + \frac{2}{4 \times 12.5} \begin{bmatrix} 0 & 0 & 0 \\ 0 & 25 & -25 \\ 0 & -25 & 25 \end{bmatrix}$$

$$+ \frac{5}{6} \cdot \frac{6}{5} \begin{bmatrix} 2 & 0 & 1 \\ 0 & 0 & 0 \\ 1 & 0 & 2 \end{bmatrix} = \begin{bmatrix} 3 & 0 & 0 \\ 0 & 1 & -1 \\ 0 & -1 & 4 \end{bmatrix}$$

$$\{f_2\} = \frac{6}{5} \cdot \frac{12.5}{3} \begin{Bmatrix} 1 \\ 1 \\ 1 \end{Bmatrix} - 0 + \frac{6}{5} \frac{30 \times 5}{2} \begin{Bmatrix} 1 \\ 0 \\ 1 \end{Bmatrix} + 0$$

$$= 5 \begin{Bmatrix} 1 \\ 1 \\ 1 \end{Bmatrix} + 90 \begin{Bmatrix} 1 \\ 0 \\ 1 \end{Bmatrix} = \begin{Bmatrix} 95 \\ 5 \\ 95 \end{Bmatrix}$$

Assembling the elements we get the global matrix,

$$\begin{bmatrix} 2 & -1 & 0 & -1 \\ -1 & 1+1 & -1 & 0+0 \\ 0 & -1 & 4 & 0 \\ -1 & 0+0 & 0 & 1+3 \end{bmatrix} \begin{Bmatrix} T_1 \\ T_2 \\ T_3 \\ T_4 \end{Bmatrix} = \begin{Bmatrix} -2 \\ 1+5 \\ 95 \\ -4+95 \end{Bmatrix}$$

$$\begin{bmatrix} 2 & 0 & 0 & -1 \\ 0 & 1 & 0 & 0 \\ 0 & 0 & 1 & 0 \\ -1 & 0 & 0 & 4 \end{bmatrix} \begin{Bmatrix} T_1 \\ T_2 \\ T_3 \\ T_4 \end{Bmatrix} = \begin{Bmatrix} -2 \\ 100 \\ 100 \\ 91 \end{Bmatrix}$$

Solving we get,

$$2T_1 - T_4 = 98; \quad T_1 = 60.0°C$$

$$-T_1 + 4T_4 = 92; \quad T_4 = 40.0°C$$

$$T_2 = T_3 = 100°C \quad \text{given}$$

To calculate the temperature at point 6(2.5, 2.5) on 24

$$T_6 = N_{1(6)} T_i + N_{2(6)} T_2 + N_{4(6)} T_4$$

$$N_{1(6)} = \frac{1}{25} [25 - 5 \times 2.5 - 5 \times 2.5] = 0$$

$$N_{2(6)} = \frac{1}{25} [0 + 5 \times 2.5 + 0 \times 2.5] = 0.5$$

$$N_{4(6)} = \frac{1}{25} [0 + 0 \times 2.5 + 5 \times 2.5] = 0.5$$

$$T_6 = 0 \times T_1 + 0.5 + T_2 + 0.5 T_4$$

$$= 0.5 (100 + 4000) = 70°C$$

To calculate the average temperature for elements
For element I.

$$T_{AV} = \frac{T_1 + T_2 + T_4}{3}, = \frac{69.0 + 100 + 40.0}{3}, = 69.67° C$$

$$\Delta T = 69.67 - 30$$

$$= 39.67° \ C \text{ (to be used for calculation of initial strain)}$$

For element II.

$$T_{AV} = \frac{100 + 100 + 40.0}{3}, = 80.0^{\circ} \ C$$
$$\Delta T = 80.0 - 30, = 50.0^{\circ} \ C$$

12.9 Three-Dimensional Heat Conduction

Most of the practical problems are three-dimensional in nature. This section brings out, at the first instance, the enormous amount of computational effort and difficulty that is concentrated for a 3-D problem as compared to the 2-D problem. The three-dimensional heat conduction is illustrated using a linear tetrahedral element. Many of the problems in heat transfer are, in general, axisymmetric cases. The correspondence between the 2-D and axisymmetric cases are brought out so that a slight modification of 2D programs can be used for the solution of axisymmetric heat conduction problems.

12.9.1 Nature of the problem

The equivalent element for triangle in 3D is tetrahedron with 4 nodes. It is very difficult to conceive discretization with tetrahedrons and hence difficult to number the nodes in order. The problem size becomes very large, as can be seen in Table 12.1.

Table 12.1 A comparison of the approximate storage requirements and the Computer time between 2D and 3D analysis

	2D	3D
Mesh	$20 \times 20 = 400$ nodes	$20 \times 20 \times 20$ $= 8000$ nodes
Bandwidth	20 (say)	20×20
Storage	$400 \times 20 = 8000 = SH$	$8000 \times 20 \times 20$ $= 400$ (SH)
Computer time	$1/2 \ (400)(20)^2 = AH$	$1/2(8000)(20 \times 20)^2$ $= 8000$ (AH)
Thermal stress Problem		
Band width	$20 \times 2 = 40$ variables	$20 \times 20 \times 3 =$ 1200 variables
Number of eqns.	$400 \times 2 = 800$	$8000 \times 3 = 24000$
Storage	$800 \times 40 = SS$	$24,000 \times 1200$ $= 900(SS)$
Computer time	$1/2(800)(40)^2 = AS$	$1/2(24000)(1200)^2$ $= 27,000$ (AS)

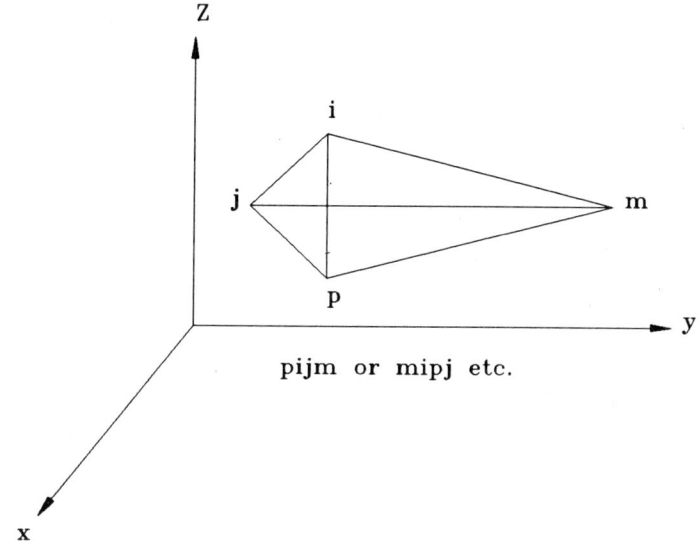

Fig. 12.9 3D-Element-Tetrahedran

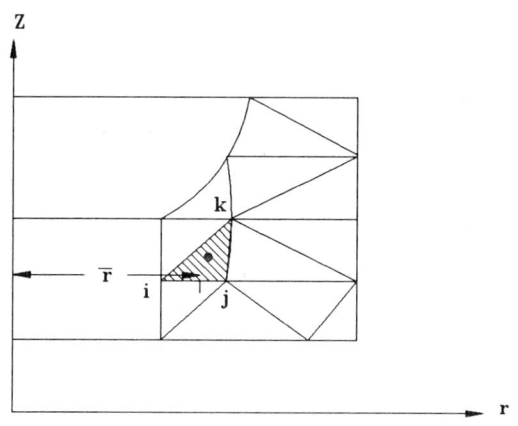

Fig. 12.10 Axisymmetric case

12.9.2 Element Characteristics

Linear Element

$$T = N_i T_i + N_j T_j + N_p T_p + N_m T_m \qquad (12.63)$$

$$N_i = \frac{1}{6V} (A_i + B_i x + C_i y + D_i z) \qquad (12.64)$$

$$A_i = det \begin{vmatrix} x_j & y_j & z_j \\ x_m & y_m & z_m \\ x_p & y_p & z_p \end{vmatrix} \; ; \; B_i = -det \begin{vmatrix} 1 & y_j & z_j \\ 1 & y_m & z_m \\ 1 & y_p & z_p \end{vmatrix}$$

$$C_i = -det \begin{vmatrix} x_j & 1 & z_j \\ x_m & 1 & z_m \\ x_p & 1 & z_p \end{vmatrix} \quad D_i = -det \begin{vmatrix} x_j & y_j & 1 \\ x_m & y_m & 1 \\ x_p & y_p & 1 \end{vmatrix} \qquad (12.65)$$

$$6V = det \begin{vmatrix} 1 & x_i & y_i & z_i \\ 1 & x_j & y_j & z_j \\ 1 & x_m & y_m & z_m \\ 1 & x_p & y_p & z_p \end{vmatrix} \qquad (12.66)$$

Other constants defined by cyclic interchange of the subscripts in the order p, i, j and m.

$$L_i = N_i, \quad L_j = N_j \quad L_p = N_p \quad L_m = N_m$$

$$\int_v L_i^a \, L_j^b \, L_p^c \, L_m^d \, dV = \frac{a! \, b! \, c! \, d!}{(a+b+c+d+3)!} 6V \qquad (12.67)$$

$$[B] = \frac{1}{6V} \begin{bmatrix} B_i & B_j & B_p & B_m \\ C_i & C_j & C_p & C_m \\ D_i & D_j & D_p & D_m \end{bmatrix} \qquad (12.68)$$

$$\int_V [B]^T \, [D] \, [B] \, dV = \frac{k_x}{36V} \begin{bmatrix} B_iB_i & B_iB_j & B_iB_p & B_iB_m \\ & B_jB_j & B_jB_p & B_jB_m \\ & & B_pB_p & B_pB_m \\ \text{sym.} & & & B_mB_m \end{bmatrix}$$

$$= \frac{k_y}{36V} \begin{bmatrix} C_iC_i & C_iC_j & C_iC_p & C_iC_m \\ & C_jC_j & C_jC_p & C_jC_m \\ & & C_pC_p & C_pC_m \\ \text{Sym.} & & & C_mC_m \end{bmatrix}$$

$$+ \frac{k_z}{36V} \begin{bmatrix} D_iD_i & D_iD_j & D_iD_p & D_iD_m \\ & D_jD_j & D_jD_p & D_jD_m \\ & & D_pD_p & D_pD_m \\ \text{Sym.} & & & D_mD_m \end{bmatrix} \qquad (12.69)$$

$$\int_s h\, [N]^T\, [N]\, dS = \frac{hS_{jmp}}{12} \begin{bmatrix} 0 & 0 & 0 & 0 \\ 0 & 2 & 1 & 1 \\ 0 & 1 & 2 & 1 \\ 0 & 1 & 1 & 2 \end{bmatrix} \tag{12.70}$$

$$\int_V G\, [N]^T\, dV = \frac{GV}{4} \begin{Bmatrix} 1 \\ 1 \\ 1 \\ 1 \end{Bmatrix} \text{ and } \int_s hT_\infty\, [N]^T\, dS = \frac{hT_\infty\, S_{jmp}}{3} \begin{Bmatrix} 0 \\ 1 \\ 1 \\ 1 \end{Bmatrix} \tag{12.71}$$

12.9.3 Axisymmetric Field Problems

Many three-dimensional problems possess symmetry about an axis. In addition, they have uniform boundary condition on the periphery, i.e., same for $0 < \theta < 2\pi$. Problems which satisfy the above two conditions can be solved by two-dimensional elements, most of them occur in heat transfer. Functional formulation and evaluation of integrals differ from two-dimensional problems. Heat conduction equation in cylindrical coordinates can be written for steady state as

$$k_r \frac{\partial^2 T}{\partial r^2} + \frac{k_r}{r}\frac{\partial T}{\partial r} + \frac{k_\theta}{r^2}\frac{\partial^2 T}{\partial \theta^2} + k_z \frac{\partial T}{\partial z} + G = 0 \tag{12.72}$$

$$k_r \frac{\partial T}{\partial r}l + \frac{k_\theta}{r}\frac{\partial T}{\partial \theta}m + k_z \frac{\partial T}{\partial z}n + q + h(T - T_\infty) = 0$$

If $T \neq F(\theta)$, then the term containing θ drops out. Thus,

$$\frac{1}{r}\frac{\partial}{\partial r}\left(r\, k_r \frac{\partial T}{\partial r}\right) + k_z \frac{\partial^2 T}{\partial z^2} + G = 0$$

$$\frac{\partial}{\partial r}\left(r\, k_r \frac{\partial T}{\partial r}\right) + r\, k_z \frac{\partial T}{\partial z^2} + rG = 0 \tag{12.72}$$

Compare with two-dimensional equation,

$$\frac{\partial}{\partial x}\left(k_x \frac{\partial T}{\partial x}\right) + \frac{\partial}{\partial y}\left(k_y \frac{\partial T}{\partial y}\right) + G = 0$$

Replace x by r

$\qquad\qquad y \qquad\quad z$

$\qquad\qquad k_x \qquad\ rk_r$

$\qquad\qquad k_y \qquad\ rk_z$

$\qquad\qquad G \qquad\ rG$

Similarity between the axisymmetric and the two-dimensional problem makes the solution of axisymmetric problem quite straight forward.

$$T = N_i (T_i) + N_j (T_j) + N_k (T_k)$$

where

$$N_i = \frac{1}{2A} (A_i + B_i r + C_i z) \tag{12.73}$$

Note that,

$$dV = 2\pi r \, dA \quad \text{and} \quad [D] = \begin{bmatrix} rk_r & 0 \\ 0 & rk_z \end{bmatrix} \tag{12.74}$$

$$\int_V [B]^T [D] [B] \, dV = \frac{2\pi r^2}{4A} \left\{ k_r \begin{bmatrix} b_i b_i & b_i b_j & b_i b_k \\ & b_j b_j & b_j b_k \\ \text{Sym.} & & b_k b_k \end{bmatrix} + \right.$$

$$\left. + k_z \begin{bmatrix} a_i a_i & a_i a_j & a_i a_k \\ & a_j a_j & a_j a_k \\ \text{Sym.} & & a_k a_k \end{bmatrix} \right\} \tag{12.75}$$

where

$$\bar{r} = \frac{r_i + r_j + r_k}{3} \tag{12.76}$$

$$\int h[N]^T [N] \, dS = \int_{l_{jk}} h \begin{bmatrix} N_i \\ N_j \\ N_k \end{bmatrix} [N_i \ N_j \ N_k] \, 2\pi r \, dl_{jk} =$$

$$\frac{2\pi h \, l_{jk}}{12} \times \begin{bmatrix} 0 & 0 & 0 \\ 0 & 3r_j + r_k & r_j + r_k \\ 0 & r_j + r_k & r_j + 3r_k \end{bmatrix} \tag{12.77}$$

$$\int_V rG[N]^T dV = \int_A rG \begin{bmatrix} N_i \\ N_j \\ N_k \end{bmatrix} 2\pi(N_i r_i + N_j r_j + N_k r_k) \, dA$$

$$= \frac{2\pi \, \bar{r} \, G \, A}{12} \begin{bmatrix} 2 & 1 & 1 \\ 1 & 2 & 1 \\ 1 & 1 & 2 \end{bmatrix} \begin{Bmatrix} r_i \\ r_j \\ r_k \end{Bmatrix} \tag{12.78}$$

$$\int h\, T_\infty [N]^T\, dS = \int_{l_{jk}} h\, T_\infty \begin{bmatrix} N_i \\ N_j \\ N_k \end{bmatrix} 2\pi (N_i r_i + N_j r_j + N_k r_k) dl_{jk}$$

$$= \frac{2\pi h\, T_\infty\, l_{jk}}{6} \begin{bmatrix} 0 & 0 & 0 \\ 0 & 2 & 1 \\ 0 & 1 & 2 \end{bmatrix} \begin{Bmatrix} r_i \\ r_j \\ r_k \end{Bmatrix}$$

$$= \frac{2\pi h\, T_\infty\, l_{jk}}{6} \begin{Bmatrix} 0 \\ 2r_j + r_k \\ r_j + 2r_k \end{Bmatrix} \tag{12.79}$$

$$-\int q\, [N]^T\, dS = -\int_{l_{ij}} q \begin{bmatrix} N_i \\ N_j \\ N_k \end{bmatrix} 2\pi r\, l_{ij}$$

$$= \frac{-2\pi q\, l_{ij}}{6} \begin{Bmatrix} 2r_i + r_j \\ r_i + 2r_j \\ 0 \end{Bmatrix} \tag{12.79a}$$

It can be observed that nodes do not receive equal share of thermal loads. A nonsymmetrical temperature distribution in an axisymmetric body does not qualify as an axisymmetric problem. However, using Fourier analysis the asymmetric linear problem can be converted into a number of axisymmetric problems and the method of superposition can be used for obtaining the solution.

12.10   Transient Heat Conduction Problems

In this section, the Galerkin's method of formulation is extended to transient heat conduction problems. The resulting system of ordinary differential equations is solved by using the finite element in time domain. This provides a basis for developing two-level, three-level and multi-level time stepping schemes for transient problems. A numerical example illustrates the method of applying a two-level time-stepping scheme to the transient problem.

12.10.1   Basic Governing Equations with Boundary and Initial Values

The basic differential equation for the transient state includes the time derivative of the temperature on the right hand side.

$$\frac{\partial}{\partial x}\left(k_x \frac{\partial T}{\partial x}\right) + \frac{\partial}{\partial y}\left(k_y \frac{\partial T}{\partial y}\right) + \frac{\partial}{\partial z}\left(k_z \frac{\partial T}{\partial z}\right) + G = \rho c \frac{\partial T}{\partial t} \tag{12.80}$$

$$T = T_B \quad \text{on} \quad S_i$$

$$k_x \frac{\partial T}{\partial x}l + k_y \frac{\partial T}{\partial y}m + k_z \frac{\partial T}{\partial z}n + q = 0 \text{ on } S_2$$

$$k_x \frac{\partial T}{\partial x}l + k_y \frac{\partial T}{\partial y}m + k_z \frac{\partial T}{\partial z}n + h(T - T_\infty) = 0 \text{ on } S_3$$

$$T = T_i \quad \text{when} \quad t = 0$$

ρc = product of density and specific heat

$\rho c \frac{\partial T}{\partial t}$ is the additional term to be treated. Galerkin's method of formulation is used. The temperature in an element is represented by

$$T(x, y, z, t) = \sum N(x, y, z) \, T(t) \tag{12.81}$$

The modification to steady state is, instead of the term $\int_V GN_i \, dV$, we have,

$$\int_V \left(G - \rho c \frac{\partial T}{\partial t}\right) N_i dV = \int GN_i dV - \int N_i \rho c \frac{\partial [N]}{\partial t}\{T(t)\} dV$$

$$= \int G \, N_i \, dV - \int \rho c [N] \, N_i \, dV \, \frac{d\{T\}}{dt} \tag{12.82}$$

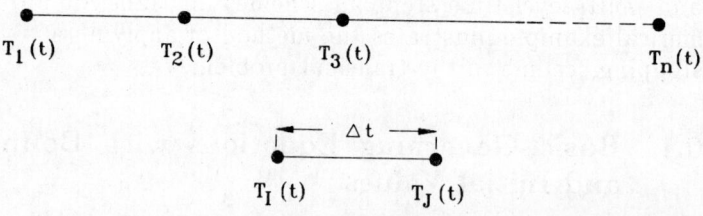

Fig. 12.11 Discretization in time

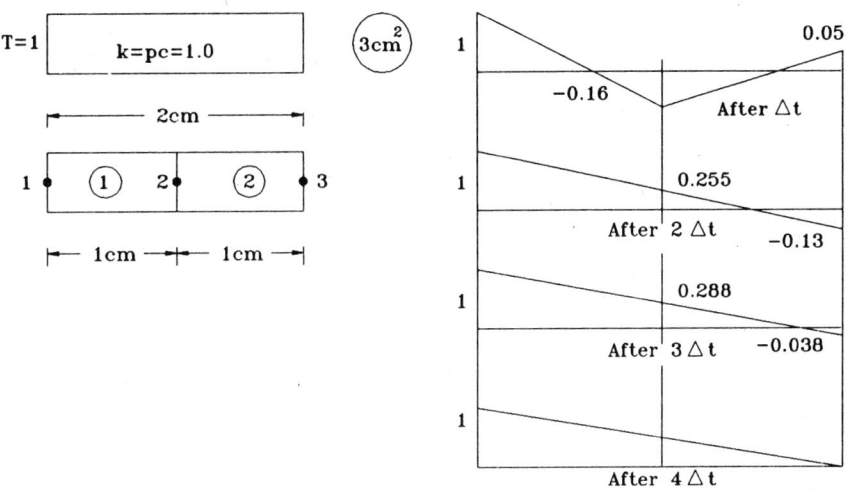

Fig. 12.12 Transient Heat Conduction

Final equation is,

$$[C] \frac{d\{T\}}{dt} + [K]\{T\} = \{f\} \qquad (12.83)$$

where $[C] = \int_V \rho c [N]^T [N] \, dV$

$[K] = \int_V [B]^T [D] [B] \, dV + \int_a h[N]^T [N] \, dS$

$\{f\} = \int_V G [N]^T \, dV - \int q[N]^T \, dS + \int_s hT_\infty [N]^T dS$

We have to solve the ordinary differential equation,

$$[C] \frac{d}{dt} \{T\} + [K] \{T\} = \{f\} \qquad (12.84)$$

Subjected to $T = T_i$ at $t = 0$.

Again Galerkin's method is used in time domain $T(t) = T_i(t) \, N_i + T_j(t) \, N_j$ for an element in time, where $N_i = 1 - \frac{t}{\Delta t}$, $N_j = \frac{t}{\Delta t}$.

Galerkin's Method gives,

$$\int \left([C] \frac{dT}{dt} + [K]\{T\} - f\right) N_i \, dt = 0 \qquad (12.85)$$

$$= \int_0^{t+\Delta t} N_i [C] [-1 \ 1] \begin{Bmatrix} T_i \\ T_j \end{Bmatrix} dt + \int_t^{t+\Delta t} N_i [K] [N_i \ N_j] \begin{Bmatrix} T_i \\ T_j \end{Bmatrix} dt$$

$$- \int_0^t f N_i \, dt = 0$$

$$\int_0^{t+\Delta t} N_i N_i \, dt = \frac{2! \ 0! \ \Delta t}{(2+0+1)!} = \frac{\Delta t}{3} \text{ and } \int_t^{t+\Delta t} N_i \, dt$$

$$= \frac{1! \ 0!}{(1+0+1)!} \Delta t = \frac{\Delta t}{2}$$

$$\int_0^t N_i N_j \, dt = \frac{1! \ 1!}{(1+1+1)!} \Delta t = \frac{\Delta t}{6}$$

$$\frac{[C]}{2} (T_j - T_i) + \frac{\Delta t}{6} [K](2T_i + T_j) - \frac{f\Delta t}{2} = 0$$

Similarly, multiplying by N_j we get,

$$\frac{[C]}{2} (T_j - T_i) + \frac{\Delta t}{6} [K](T_i + 2T_j) - \frac{f\Delta t}{2} = 0$$

Combining the above two,

$$\begin{bmatrix} -\frac{[C]}{2\Delta t} + \frac{[K]}{3} & \frac{[C]}{2\Delta t} + \frac{[K]}{6} \\ -\frac{[C]}{2\Delta t} + \frac{[K]}{6} & \frac{[C]}{2\Delta t} + \frac{[K]}{3} \end{bmatrix} \begin{Bmatrix} T_i \\ T_j \end{Bmatrix} = \begin{Bmatrix} f/2 \\ f/2 \end{Bmatrix} \tag{12.86}$$

If we combine for two such time elements, involving $T_i(t), T_j(t)$ and $T_k(t)$, we get,

$$\begin{bmatrix} -\frac{[C]}{2\Delta t} + \frac{[K]}{3} & \frac{[C]}{2\Delta t} + \frac{[K]}{6} & 0 \\ -\frac{[C]}{2\Delta t} + \frac{[K]}{6} & \frac{[C]}{2\Delta t} + \frac{[K]}{3} \\ & -\frac{[C]}{2\Delta t} + \frac{[K]}{3} & \frac{[C]}{2\Delta t} + \frac{[K]}{6} \\ 0 & -\frac{[C]}{2\Delta t} + \frac{[K]}{6} & \frac{[C]}{2\Delta t} + \frac{[K]}{3} \end{bmatrix} \begin{Bmatrix} T_i(t) \\ T_j(t) \\ T_k(t) \end{Bmatrix} = \begin{Bmatrix} f/2 \\ f \\ f/2 \end{Bmatrix} \tag{12.87}$$

In general,

$$(\frac{-[C]}{2\Delta t} + \frac{[K]}{3})\{T\}_{(n-1)t} + (\frac{[C]}{2\Delta t} + \frac{[K]}{6})\{T\}_{nt} = f/2 \tag{12.88}$$

Eqn. 12.88 is a two level scheme

$$(-\frac{[C]}{2\Delta t} + \frac{[K]}{6})\{T\}_{(n-1)t} + (\frac{2[K]}{3})\{T\}_{nt} + (\frac{[C]}{2\Delta t} + \frac{[K]}{6})\{T\}_{(n+1)} = f \tag{12.89}$$

Eqn. 12.89 is a three level scheme.

12.10.2 Various Time Stepping Schemes

The terms $\frac{dT}{dt}$ and T are written in various forms in different time stepping schemes.

$$[C]\left\{\frac{T(t+\Delta t) - T(t)}{\Delta t}\right\} + [K]\{\theta T(t+\Delta t)+(1-\theta)T(t)\} = \{f\} \quad (12.90)$$

$$
\begin{aligned}
\theta &= 0 && \text{explicit method} \\
\theta &= 1/2 && \text{Crank-Nicholson's method} \\
\theta &= 2/3 && \text{Galerkin's method} \\
\theta &= 1 && \text{implicit method}
\end{aligned}
$$

Lee's three level scheme [9]

$$[C]\left\{\frac{T(t+\Delta t) - T(t-\Delta t)}{2\Delta t}\right\} + [K]\left\{\frac{T(t+\Delta t) + T(t) + T(t-\Delta t)}{3}\right\} = \{f\}$$
$$(12.91)$$

Dupont II Scheme [9]

$$[C]\left\{\frac{T(t+\Delta t) - T(t-\Delta t)}{2\Delta t}\right\} + [K]\left\{\frac{3T(t+\Delta t) + T(t-\Delta t)}{4}\right\} = \{f\}$$
$$(12.92)$$

Many such schemes can be generated and tested for convergence and stability.

12.10.3 Example of Transient Heat Conduction

One-dimensional transient problem
At $t = 0$, $T = 0$, $t > 0$, $T_1 = 1$ and T_2, $T_3 > 0$, Crank-Nicholson's scheme is used.

$$[C]\left\{\frac{T(t+\Delta t) - T(t)}{\Delta t} + \frac{[K]}{2}\right\}\{T(t+\Delta t) + T(t)\} = \{f\}$$

$$\left(\frac{[C]}{\Delta t} + \frac{[K]}{2}\right)\{T(t+\Delta t)\} = \left(\frac{[C]}{\Delta t} - \frac{[K]}{2}\right)T(t)$$

$$\text{i.e.,} \quad [D]\{T(t+\Delta t)\} = [E]\{T_{(t)}\}$$

$$[K] = \frac{Ak}{L}\begin{bmatrix} 1 & -1 \\ -1 & 1 \end{bmatrix} = \begin{bmatrix} 3 & -3 \\ -3 & 3 \end{bmatrix}$$

$$[C] = \frac{\rho c A L}{6} \begin{bmatrix} 2 & 1 \\ 1 & 2 \end{bmatrix} = \begin{bmatrix} 1 & 0.5 \\ 0.5 & 1 \end{bmatrix}$$

Assembly of two element gives,

$$[K] = \begin{bmatrix} 3 & -3 & 0 \\ -3 & 6 & -3 \\ 0 & -3 & 3 \end{bmatrix} \text{ and } [C] = \begin{bmatrix} 1 & 0.5 & 0 \\ 0.5 & 2 & 0.5 \\ 0 & 0.5 & 1 \end{bmatrix}$$

$$[D] = \begin{bmatrix} 10 & 5 & 0 \\ 5 & 20 & 5 \\ 0 & 5 & 10 \end{bmatrix} + \begin{bmatrix} 1.5 & -1.5 & 0 \\ -1.5 & 3 & 1.5 \\ 0 & -1.5 & 1.5 \end{bmatrix} = \begin{bmatrix} 11.5 & 3.5 & 0 \\ 3.5 & 23 & 3.5 \\ 0 & 3.5 & 11.5 \end{bmatrix}$$

$$[E] = \begin{bmatrix} 10 & 5 & 0 \\ 5 & 20 & 5 \\ 0 & 5 & 10 \end{bmatrix} - \begin{bmatrix} 1.5 & -1.5 & 0 \\ -1.5 & 3 & -1.5 \\ 0 & -1.5 & 1.5 \end{bmatrix} = \begin{bmatrix} 8.5 & 6.5 & 0 \\ -1.5 & 17 & 6.5 \\ 0 & 6.5 & 6.5 \end{bmatrix}$$

$$\begin{bmatrix} 11.5 & 3.5 & 0 \\ 3.5 & 23 & 3.5 \\ 0 & 3.5 & 11.5 \end{bmatrix} \begin{Bmatrix} T_1 \\ T_2 \\ T_3 \end{Bmatrix}_{t+\Delta t} \begin{bmatrix} 8.5 & 6.5 & 0 \\ -1.5 & 17 & 6.5 \\ 0 & 6.5 & 6.5 \end{bmatrix} \begin{Bmatrix} T_1 \\ T_2 \\ T_3 \end{Bmatrix}_{t}$$

$$= \begin{bmatrix} 8.5 & 6.5 & 0 \\ 1.5 & 17 & 6.5 \\ 0 & 6.5 & 6.5 \end{bmatrix} \begin{Bmatrix} 0 \\ 0 \\ 0 \end{Bmatrix} = \begin{Bmatrix} 0 \\ 0 \\ 0 \end{Bmatrix}$$

$$\begin{bmatrix} 11.5 & 3.5 & 0 \\ 3.5 & 23 & 3.5 \\ 0 & 3.5 & 11.5 \end{bmatrix} \begin{Bmatrix} T_1 \\ T_2 \\ T_3 \end{Bmatrix} = \begin{Bmatrix} 0 \\ 0 \\ 0 \end{Bmatrix}$$

Since $T_1 = 1$,

$$\begin{bmatrix} 1 & 0 & 0 \\ 0 & 23 & 3.5 \\ 0 & 3.5 & 11.5 \end{bmatrix} \begin{Bmatrix} T_1 \\ T_2 \\ T_3 \end{Bmatrix} = \begin{Bmatrix} 1 \\ -3.5 \\ 0 \end{Bmatrix}$$

Therefore,

$$23T_2 + 3.5\, T_3 = -3.5$$

$$3.5T_2 + 11.5\, T_3 = 0$$

Solving, we get,

$$T_2 = -0.16 \text{ and } T_3 = 0.05$$

To find the temperatures at $(t+2\Delta t)$, we use the values at $(t+\Delta t)$.

$$\begin{bmatrix} 11.5 & 3.5 & 0 \\ 3.5 & 23 & 3.5 \\ 0 & 3.5 & 11.5 \end{bmatrix} \begin{Bmatrix} T_1 \\ T_2 \\ T_3 \end{Bmatrix}_{t+2\Delta t} = \begin{bmatrix} 8.5 & -6.5 & 0 \\ 1.5 & 17 & 6.5 \\ 0 & 6.5 & 6.5 \end{bmatrix} \begin{Bmatrix} 1 \\ -0.16 \\ 0.05 \end{Bmatrix}_{T+\Delta t}$$

$$= \begin{Bmatrix} 7.45 \\ 8.9 - 3.5 \\ -0.625 \end{Bmatrix}$$

Solving we get

$$T_2 = 0.255 \quad T_3 = -0.13$$

Again,

$$\begin{bmatrix} 11.5 & 3.5 & 0 \\ 3.5 & 23 & 3.5 \\ 0 & 3.5 & 11.5 \end{bmatrix} \begin{Bmatrix} T_1 \\ T_2 \\ T_3 \end{Bmatrix}_{t+2\Delta t} = \begin{bmatrix} 8.5 & 6.5 & 0 \\ 1.5 & 17 & 6.5 \\ 0 & 6.5 & 6.5 \end{bmatrix} \begin{Bmatrix} 1 \\ 0.255 \\ -0.13 \end{Bmatrix}_{t+2\Delta t}$$

$$= \begin{Bmatrix} 10.15 \\ 9.99 - 3.5 \\ 0.5525 \end{Bmatrix}$$

Solving we get, $T_2 = 0.288 \quad T_3 = -0.038$
$\Delta t = 0.1$ is large time step. Smaller time step is recommended.

12.11 Modifications to Plane Stress/Strain Program to Determine Temperature Distributions in a Body

By now, the readers have familiarity with the programs which solve plane stress (thin lamina) and plane strain problems in stress analysis. The methodology to be used to convert the above programs to solve heat conduction problems is briefly given here. In heat conduction problems, there is only a scalar quantity temperature. Hence one of the degrees of freedom say v is to be supressed.

The [D] matrix is given by,

$$[D] = \begin{bmatrix} k_x & 0 \\ 0 & k_y \end{bmatrix}$$

As one can observe that the major difference occurs in dealing with heat loss/gain to/from the environment by convection expressed by the relation,

$$q = -k \frac{\partial T}{\partial \eta} = h(T - T_\infty)$$

Since the heat dissipation is dependent on the temperature of the surface which is unknown, we have to add an additional term to the stiffness matrix which leads to the final form for [K] as

$$[K] = \int_V [B]^T [D] [B] \, dV + \int_S h [N]^T [N] \, dS$$

The additional contribution to the loading term is given by

$$\int_s h \, T_\infty [N]^T \, ds$$

The heat flux boundary condition is similar to traction in stress analysis. The heat generation in a body is similar to the body force term in the stress analysis. Thus by a slight modification of the plane stress/strain program, the temperature distribution in a two-dimensional body can be determined. In a similar way the three-dimensional program can also be modified. By multiplying the gradient at any point with the thermal conductivity in a particular direction, we can determine the heat flux at a point in that direction.

REFERENCES

1. Holman, J.P., *Heat Transfer*, McGraw Hill Book Co., Int. Student Edition, 5th Ed., 1981.

2. Segerlind, L.J., *Applied Finite Element Analysis*, John Wiley, New York, 1976.

3. Huebner, K.H., *The Finite Element Method for Engineers*, John Wiley, New York, 1975.

4. Cook, R.D., *Concepts and Applications of Finite Element Analysis*, John Wiley, 2nd Edition, 1981.

5. Reddy, J.N., *An Introduction to the Finite Element Method*, McGraw Hill Book Co., Singapore, International Student Edition, 1985.

6. Zienkiewicz, O.C., *The Finite Element Method*, Tata McGraw Hill Co., 3rd Edition, 1977.

7. Bathe, K.J., *Finite Element Procedures in Engineering Analysis*, Prentice-Hall of India Private Limited, New Delhi, 1990.

8. Lewis, R.W., K.Morgan, H.R.Thomas, and K.N.Seetharamu, *Finite Element Methods in Heat Transfer Analysis*, John Wiley, 1993.

9. Lewis,R.W., K.Morgan, and O.C.Zienkiewicz, *Numerical Heat Transfer*, John Wiley, First Edition, 1981.

EXERCISES

12.1 Calculate the temperature distribution and the heat dissipating capacity of a fin shown in Fig. E12.1. The thermal conductivity of the material is 200 $W/m^{\circ}K$. The surface heat transfer coefficient is 0.5 W/m^2K. The ambient temperature is 30°C. The thickness of the fin is 1 cm.

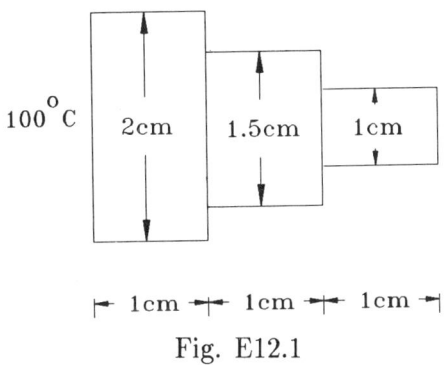

Fig. E12.1

12.2 If in the problem 12.1, the width at the base is 2cm and at the tip 1 cm and varies uniformly along the length, what changes take place in the heat dissipation capacity of the fin?

Fig. E12.2

2.3 A heat sink, as shown in Fig. E12.2, is used to dissipate heat from a transistor whose maximum temperature is limited to 110°C. Each arm has a dimension of 0.5 cm × 1 cm and a length of 3 cm. The base is an annular plate of thickness of 0.5 cm with outer diameter of 2 cm and inner diameter of 1.5 cm. Determine the heat dissipation capacity of an entire sink. Use the data given in problem 12.1.

12.4 The governing differential equation for the case of heat conduction for a body moving in the x direction is given, in general terms, by

$$a\,\frac{d^2T}{dx^2} + b\,\frac{dT}{dx} + cT + d = 0$$

Using the Galerkin's method, derive the element equations using a linear element.

12.5 The heat dissipation in a fin of uniform cross section is given by

$$\frac{d^2T}{dx^2} - \frac{hP(T - T_\infty)}{kA} = 0$$

With $T = T_b$ at $x = 0$ $\frac{dT}{dx} = 0$ at $x = L$. Solve the above equation using (i) collocation method, (ii) sub domain method (iii) Galerkin's method (iv) least squares method.

12.6 For a linear triangular element calculate the nodal load vector if the heat flux on the side ij varies (i) linearly (ii) quadratically.

12.7 For a linear triangular element, calculate the stiffness matrix and the nodal load vector if the surface heat transfer coefficient on the side jk varies (i) linearly (ii) quadratically.

12.8 Calculate the nodal load vector for a six noded quadratic triangular element when the heat flux varies on the side ij (linearly) (ii) quadratically.

12.9 Calculate the stiffness matrix and load vector for a linear triangular element if the thickness at the three nodes are t_i, t_j and t_k.

12.10 Calculate the temperature distribution in a square plate of 30 cm side shown in Fig. E12.10 using (i) 8 linear triangular elements, (ii) 2 six noded quadratic elements (iii) 4 rectangular elements. Compare the centre temperature in all the 3 cases.

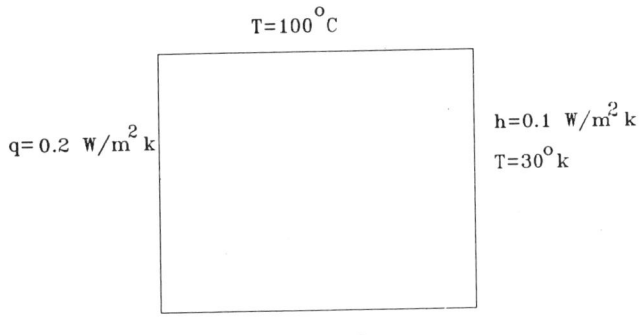

Fig. E12.10

12.11 A long thick walled cylindrical pressure vessel of circular cross section (ID = 20 cm and OD = 40 cm) is subjected to a temperature of 150°C on the inside surface. Determine the temperature distribution in the cylinder thickness if the outside is exposed to ambient. (h = 0.2 W/m² K, T_∞ = 30°C, k = 40 W/m K).

12.12 Determine the temperature distribution in a spherical pressure vessel (ID = 20 cm, and OD = 40 cm) for the conditions given in problem 12.11.

12.13 Determine the temperature distribution in a machine frame with T cross section as shown in Fig. E12.13. h = 0.2 W/m²K, T_∞ = 30°C.

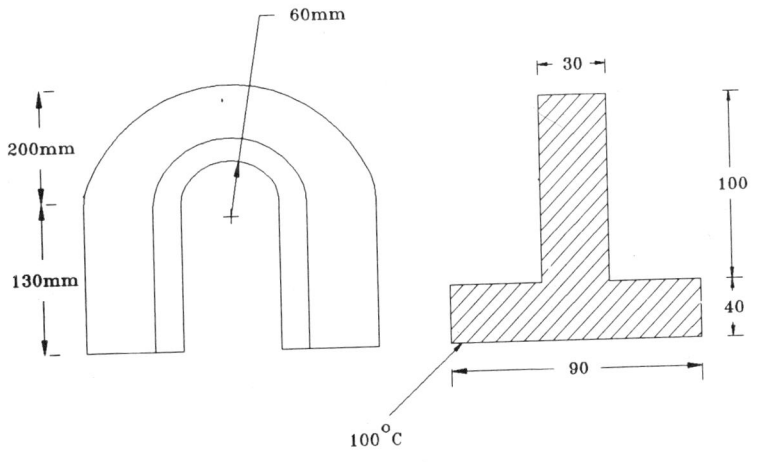

Fig. E12.13

12.14 If the base temperature in problem 12.1 changes suddenly to 200°C and remains constant thereafter, determine the temperature distribution after 15 seconds after the change of base temperature to 200°C takes place.

12.15 In a double pipe heat exchanger, hot fluid flows inside a pipe and cold fluid flows outside in the annular space. The heat exchange between the two fluids is given by the differential equations,

$$(MC) \frac{dT_h}{dA} = -U(T_h - T_c)$$

$$(mc) \frac{dT_c}{dA} = U(T_h - T_c)$$

Assuming both the temperatures of hot fluid and the cold fluid vary linearly along the length, derive the element equations if the inlet energy of the hot fluid is MCT_{h1} and cold fluid is mcT_{c1}. From the above calculations, it is possible to obtain approximately the metal wall temperatures using the relation

$$UA (T_h - T_c) = h_i A_i (T_h - T_m)$$

where T_m is the metal wall temperature and h_i is the inside surface heat transfer coefficient.

Chapter 13

Substructuring Technique

For the analysis of large structural systems substructuring technique has been found to be useful. This technique is also cost effective to incorporate modification in certain parts of structures and also for nonlinear analysis situations. The substructuring technique is presented in basic form and developed into multi-level substructuring or 'superelement' concepts. The computational aspects of the technique are also discussed.

13.1   Multilevel Substructuring Technique

The method of substructuring for static structural analysis is based on subdividing the large structure into smaller parts which are analysed separately to obtain the relationship between forces and displacements at the common interfaces or boundaries [1,2]. These boundary variables are then determined and are used to obtain the unknowns within each substructure. The substructuring technique can be applied to both the displacement and force methods of analysis. But as the finite element method and the computer program presented in the earlier chapters are based on the displacement approach the description here will be limited to displacement method.

The division of the structure into smaller parts is totally left to the analyst. But structures often have convenient internal boundaries which can be easily discerned by the analyst as the interface between the substructures. Also a particular structure can be modelled with different choice of substructures and some amount of insight into the substructuring capability of the program is needed on the user's part to choose the best model amongst these.

An aspect of substructuring which needs clarification at this stage is the structural idealisation and the approximation involved therein.

Using finite element formulation, a continuum is approximated as a group of discrete elements interconnected at the nodal points. It should be noted that application of substructuring to such an ideali-sation does not involve any further approximation, and theoretically substructuring should yield the same results as the basic finite element formulation. But owing to differences in the size of the problems at the solution stage the results are likely to differ slightly. In fact Popov and Peterson [3] report that substructuring method is numerically more accurate than the conventional finite element approach.

Before we take up the computational aspects of the substructuring technique it is necessary to describe the physical concept of the whole process. In this regard, the description given by Przemieniecki in reference [1] is found to be quite useful. But the straight application of this basic approach is not efficient for implementation in finite element program. Instead, the direct approach described in reference [2] is found to be more suitable and leads to the concept of treating a substructure as a complex 'superelement'. Both the approaches are presented in the following and the numerical equivalence of the two approaches is discussed.

13.2 Basic Approach to Substructuring Technique

In the displacement formulation of finite element analysis the basic equation used is the equilibrium equation applied to the structure as a whole and is given by,

$$[K]\,\{r\} \;=\; \{P\} \tag{13.1}$$

Using substructuring the above equilibrium equation is obtained by the assemblage of substructure equations. Treating a typical substructure as a structure, we can write Eq.13.1 for a substructure. To eliminate any ambiguity the substructure stiffness matrix will be denoted by [k].

Now for the substructure, the stiffness matrix, the displacement vector and the load vector are partitioned corresponding to internal and boundary degrees of freedom $\{d_i\}$ and $\{d_b\}$ respectively as,

$$\begin{bmatrix} [k_{ii}] & | & [k_{ib}] \\ --- & + & --- \\ [k_{bi}] & | & [k_{bb}] \end{bmatrix} \begin{Bmatrix} \{d_i\} \\ --- \\ \{d_b\} \end{Bmatrix} = \begin{Bmatrix} \{Q_i\} \\ --- \\ \{Q_b\} \end{Bmatrix} \tag{13.2}$$

In the above equation, a boundary node is defined as a node which is a part of more than one substructure and the degrees of freedom at the boundary nodes are termed as boundary degrees of freedom

Now the analysis can be performed is two stages [1].

(i) Imagine that the degrees of freedom at the boundaries are fixed and analyse each substructure under this condition. This step will be denoted by a superscript α.

(ii) Combine the condensed stiffness of the substructures to get the global structure stiffness matrix and analyse the assemblage by releasing the boundary degrees of freedom. This step will be denoted by the superscript β.

The displacement and load vectors can be expressed as the sum of the two cases as,

$$\left\{ \begin{array}{c} \{d_i\} \\ \hline \{d_b\} \end{array} \right\} = \left\{ \begin{array}{c} \{d_i^\alpha\} \\ \hline \{d_b^\alpha\} \end{array} \right\} + \left\{ \begin{array}{c} \{d_i^\beta\} \\ \hline \{d_b^\beta\} \end{array} \right\} \tag{13.3}$$

and

$$\left\{ \begin{array}{c} \{Q_i\} \\ \hline \{Q_b\} \end{array} \right\} = \left\{ \begin{array}{c} \{Q_i^\alpha\} \\ \hline \{Q_b^\alpha\} \end{array} \right\} + \left\{ \begin{array}{c} \{Q_i^\beta\} \\ \hline \{Q_b^\beta\} \end{array} \right\} \tag{13.4}$$

where the subscripts i and b denote the terms corresponding to the internal and boundary degrees of freedom.

Obviously as $\{d_b^\alpha\}$ is the displacement at the boundary degrees of freedom, when the boundaries are fixed it will become zero. Thus

$$\{d_b^\alpha\} = \{0\} \tag{13.5}$$

Also in the first stage of the analysis, all the forces are applied at the internal nodes of the substructure and hence these forces do not appear at the second stage. Hence,

$$\{Q_i^\beta\} = \{0\} \quad \text{and} \quad \{Q_i^\alpha\} = \{Q_i\} \tag{13.6}$$

Analysis with Fixed Boundaries: Substituting the value of $\{d_b^\alpha\} = \{0\}$ from Eq.13.5 into equilibrium Eq.13.2, we get the set of equations for the first stage of analysis with boundaries of a substructure fixed as,

$$\left[\begin{array}{c|c} [k_{ii}] & [k_{ib}] \\ \hline [k_{bi}] & [k_{bb}] \end{array} \right] \left\{ \begin{array}{c} \{d_i^\alpha\} \\ \hline \{0\} \end{array} \right\} = \left\{ \begin{array}{c} \{Q_i\} \\ \hline \{Q_b^\alpha\} \end{array} \right\} \tag{13.7}$$

Solving the first set of the above equation,

$$\{d_i^\alpha\} = [k_{ii}]^{-1} \{Q_i\} \tag{13.8}$$

Substituting the value of $\{d_i^\alpha\}$ in the second equation we get,

$$\{Q_b^\alpha\} = [k_{bi}] [k_{ii}]^{-1} \{Q_i\} \tag{13.9}$$

Here, the physical interpretations of $\{Q_b^\alpha\}$ is the force required to be applied at the substructure boundaries to keep the boundary displacements equal to zero, i.e., for fixing the boundaries.

The above analysis is to be performed on all the substructures and for the sake of simplicity the subscript denoting the substructure is omitted.

Analysis with Boundaries Released: Again substituting the value of $\{Q_i^\beta\}$ in Eq.13.2 we get the set of equations for the second stage of analysis with boundaries released as,

$$\begin{bmatrix} [k_{ii}] & [k_{ib}] \\ [k_{bi}] & [k_{bb}] \end{bmatrix} \left\{ \begin{array}{c} \{d_i^\beta\} \\ \{d_b^\beta\} \end{array} \right\} = \left\{ \begin{array}{c} \{0\} \\ \{Q_b^\beta\} \end{array} \right\} \tag{13.10}$$

Solving the first set of equation,

$$\{d_i^\beta\} = -[k_{ii}]^{-1} [k_{ib}] \{d_b^\beta\} \tag{13.11}$$

Solving the second set of equation,

$$[k_{bi}] \{d_i^\beta\} + [k_{bb}] \{d_b^\beta\} = \{Q_b^\beta\} \tag{13.12}$$

Substituting from Eq.13.11 for $\{d_i^\beta\}$ into Eq.13.12,

$$-[k_{bi}] [k_{ii}]^{-1} [k_{ib}] \{d_i^\beta\} + [k_{bb}] \{d_b^\beta\} = \{Q_b^\beta\} \tag{13.13}$$

or

$$[[k_{bb}] - [k_{bi}] [k_{ii}]^{-1} [k_{ib}]] \{d_b^\beta\} = \{Q_b^\beta\} \tag{13.14}$$

or

$$[k^*] \{d_b^\beta\} = \{Q_b^\beta\} \tag{13.15}$$

where

$$[k^*] = [k_{bb}] - [k_{bi}] [k_{ii}]^{-1} [k_{ib}] \tag{13.16}$$

The Eq. 13.15 is the equilibrium equation for a substructure in terms of its boundary degrees of freedom and $[k^*]$ is the corresponding stiffness matrix called the condensed stiffness matrix. The above analysis will be carried out for all the substructures and the condensed stiffness matrix for each of the substructure will be obtained. Now treating each substructure as an element, the global structure stiffness matrix can be formed by the usual assembly procedure by direct stiffness method. Thus

$$[K] = \sum_{s=1}^{n} [k^*]_s \qquad (13.17)$$

and

$$\{P\} = \{Q_b\} - \sum_{s=1}^{n} \{Q_b^\alpha\}_s \qquad (13.18)$$

The loads $\{Q_b\}$ are the applied loads along the boundaries but we have already applied loads $\{Q_b^\alpha\}$ at these nodes during the first stage of the analysis. This explains the negative sign in Eq.13.18.

In the above two equations n stands for the number of substructures.

The assemblage of the substructures through Eqs. 13.17 and 13.18 leads to Eq.13.1 where all the degrees of freedom are along the common boundaries of the substructures. Solution of Eq.13.1 gives the global displacements along the boundaries of the substructures. Now picking up the proper displacements, the vector $\{d_b^\beta\}$ can be obtained for each substructure. Then, the values of $\{d_i^\beta\}$ can be determined from Eq.13.11.

From Eq. 13.8, $\{d_i^\alpha\}$ can be calculated for each substructure and it is known that $\{d_b^\alpha\} = \{0\}$. Thus, all the values of $\{d\}$ required in Eq.13.3 are known for each substructure and, the strains and stresses can be calculated following the usual finite element procedure. It may be noted here that though this solution process requires inversion of matrices, actually they can be solved by the Gauss elimination procedure.

13.2.1 'Superelement' Approach

In this approach each substructure is treated as a complex element with a number of finite element and boundary nodes. This approach is described in reference [2].

Consider a typical substructure. Its stiffness matrix and the corresponding force and displacement vectors are partitioned as

$$\begin{bmatrix} [k_{ii}] & [k_{ib}] \\ [k_{bi}] & [k_{bb}] \end{bmatrix} \begin{Bmatrix} \{d_i\} \\ \{d_b\} \end{Bmatrix} = \begin{Bmatrix} \{Q_i\} \\ \{Q_b\} \end{Bmatrix} \qquad (13.19)$$

Here all the notations used are the same as those used in the previous approach. Now the internal degrees of freedom can be eliminated following the same procedure adopted for static condensation of internal nodes of an element described in section 3.8. From the first set of equations,

$$[k_{ii}] \{d_i\} + [k_{ib}] \{d_b\} = \{Q_b\} \qquad (13.20)$$

or

$$\{d_i\} = [k_{ii}]^{-1} \{\{Q_b\} - [k_{ib}]\{d_b\}\} \qquad (13.21)$$

Substituting the value of $\{d_i\}$ in the second set of Eq.13.19 we get,

$$[k_{bi}] [k_{ii}]^{-1} \{\{Q_b\} - [k_{ib}] \{d_b\}\} + [k_{bb}] \{d_b\} = \{Q_i\} \qquad (13.22)$$

Simplifying we have,

$$[[k_{bb}] - [k_{bi}] [k_{ii}]^{-1} [k_{ib}]] \{d_b\} = \{\{Q_i\} - [k_{bi}] [k_{ii}]^{-1} \{Q_b\}\} \qquad (13.23)$$

Equation 13.23 can also be written as,

$$[k^*] \{d_b\} = \{Q^*\} \qquad (13.24)$$

where $[k^*]$ is the condensed substructure stiffness matrix and $\{Q^*\}$ is the corresponding modified load vector.

Following the discussion presented in section 3.8, it can be noted that the condensed stiffness matrix $[k^*]$ is the effective stiffness matrix of a substructure corresponding to the boundary degrees of freedom similar to the effective stiffness matrix of an element corresponding to the external nodal degrees of freedom. Thus, the substructure can be termed as a 'superelement' and the number of such super elements form the actual structural system. Thus Eq.13.24 can be used to assemble structural stiffness matrix and the nodal load vector following the direct stiffness approach described in Chapter 5.

It should be mentioned here again that in the actual implementation of computer program the condensation is carried out by Gauss elimination procedure.

The numerical procedure to be carried out in both the approaches are the same, for Eq.13.16 is the same as the left-hand side of Eq.13.23 and the vector $\{Q_b^\alpha\}$ in Eq.13.18 is the same as the term appearing with a negative sign in Eq.13.23.

13.2.2 Multi-level Substructuring

For the multi-level substructuring, the various levels are defined as follows.

The main structure in the global system is the first level substructure. It may be composed of ordinary finite elements and the second level substructures or only of second level substructure (Fig.13.1)

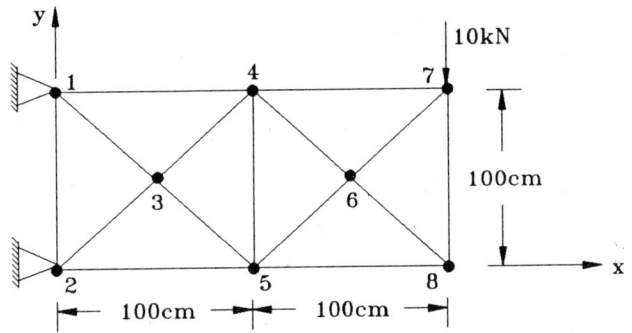

(a) Example problem truss 8 Nodes, 14 Elements

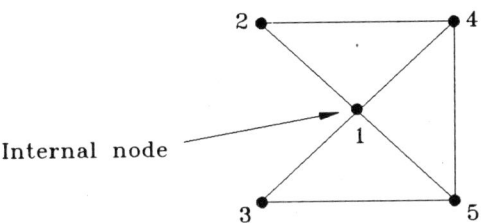

(b) Second level substructure 5 Nodes, 7 Elements

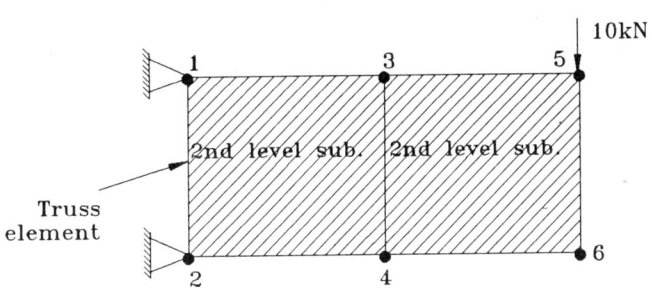

(c) First level (main) substructure
6 Nodes, 1 Element, 2 Superelements

Fig. 13.1 Example Problem 1

A second level substructure is the part of the first level substructure (main structure) and may in turn be composed of ordinary finite elements or third level substructures. Here onwards the successive levels of substructures will be classified as the thrid level, fourth level, etc. It is clear that the highest level substructure must be made up only of ordinary finite elements. A three level substructuring scheme thus indicates that the highest level in the structural idealisation is three. Figure 13.2 illustrates a three level substructuring scheme[4].

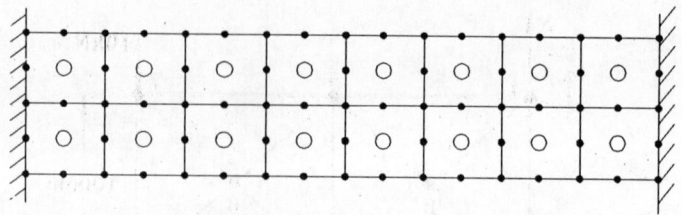

Main Structure (first level sub structure) perforated plate
69 Nodes, 16 Second level substructures

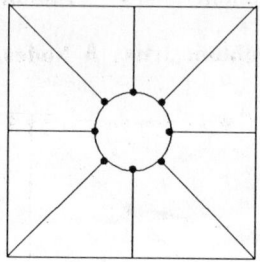

Second level substructure 16 Nodes,
8 Third level substructures

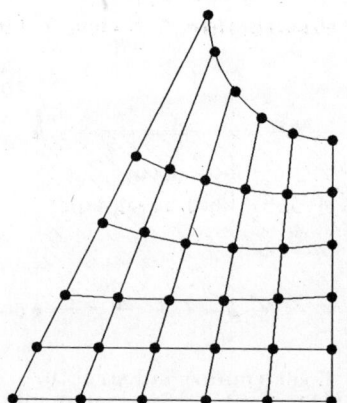

Third level substructure 36 Nodes, 25 Elements

Fig. 13.2 A typical three level substructuring scheme

13.2.3 Examples

Problem 1

In order to illustrate the various aspects of the theory presented above and to help in better understanding of the concept, a simple problem has been worked out here. Figure 13.1 shows an indeterminate truss with eight nodes and fourteen elements. A two level substructuring scheme is chosen and Figs. 13.1b and 13.1c show the first and second level substructures. The problem is solved using substructuring in the following steps.

The material and geometric properties are given below.

$$E = 2 \times 10^4 \ kN/cm^2$$

Area of cross section of vertical members = 10 cm^2

Area of cross section of horizontal members = 10 cm^2

Area of cross section of diagonal members = 14.14 cm^2

The stiffness matrix for a plane truss element is given by Eq.7.12

$$[k] = \frac{EA}{L} \begin{bmatrix} C_x^2 & C_x C_y & -C_x^2 & -C_x C_y \\ C_x C_y & C_y^2 & -C_x C_y & -C_y^2 \\ -C_x^2 & -C_x C_y & C_x^2 & C_x C_y \\ -C_x C_y & -C_y^2 & C_x C_y & C_y^2 \end{bmatrix}$$

where $C_x = cos\alpha$ and $C_y = sin\alpha$ and α is the angle between member and global axes.

The relevant data required for the computation of stiffness matrix of various members is tabulated below.

Member	Length	Area	α	C_x	C_y
	cm	cm^2	degrees		
1	l	a	0	1	0
2	l	a	0	1	0
3	l	a	90	0	1
4	$l/\sqrt{2}$	$\sqrt{2}a$	135	$-1/\sqrt{2}$	$1/\sqrt{2}$
5	$l/\sqrt{2}$	$\sqrt{2}a$	45	$1/\sqrt{2}$	$1/\sqrt{2}$
6	$l/\sqrt{2}$	$\sqrt{2}a$	135	$-1/\sqrt{2}$	$1/\sqrt{2}$
7	$l/\sqrt{2}$	$\sqrt{2}a$	45	$1/\sqrt{2}$	$1/\sqrt{2}$

$l = 100$ cm $a = 10$ cm^2

(iv) With the above information, the stiffness matrix of individual truss element can be worked out and can be assembled to obtain the stiffness matrix of the second level substructure as given below.

$$\frac{EA}{L}\begin{bmatrix}
4 & 0 & -1 & 1 & -1 & -1 & -1 & -1 & -1 & 1 \\
0 & 4 & 1 & -1 & -1 & -1 & -1 & -1 & 1 & -1 \\
-1 & 1 & 2 & -1 & 0 & 0 & -1 & 0 & 0 & 0 \\
1 & -1 & -1 & 1 & 0 & 0 & 0 & 0 & 0 & 0 \\
-1 & -1 & 0 & 0 & 2 & 1 & 0 & 0 & -1 & 0 \\
-1 & -1 & 0 & 0 & 1 & 1 & 0 & 0 & 0 & 0 \\
-1 & -1 & -1 & 0 & 0 & 0 & 2 & 1 & 0 & 0 \\
-1 & -1 & 0 & 0 & 0 & 0 & 1 & 2 & 0 & -1 \\
-1 & 1 & 0 & 0 & -1 & 0 & 0 & 0 & 2 & -1 \\
1 & -1 & 0 & 0 & 0 & 0 & 0 & -1 & -1 & 2
\end{bmatrix} \qquad (a)$$

(ii) By using the Gauss elimination procedure condense out the equations corresponding to the internal degrees of freedom i.e., equation 1 and 2 from the above stiffness matrix. The condensed stiffness matrix is given below.

$$\frac{EA}{L}\left[\begin{array}{cc|cccccccc}
4 & 0 & -1 & 1 & -1 & -1 & -1 & -1 & -1 & 1 \\
0 & 4 & 1 & -1 & -1 & -1 & -1 & -1 & 1 & -1 \\
\hline
0 & 0 & \frac{3}{2} & -\frac{1}{2} & 0 & 0 & -1 & 0 & -\frac{1}{2} & \frac{1}{2} \\
0 & 0 & -\frac{1}{2} & \frac{1}{2} & 0 & 0 & 0 & 0 & \frac{1}{2} & -\frac{1}{2} \\
0 & 0 & 0 & 0 & \frac{3}{2} & \frac{1}{2} & -\frac{1}{2} & -\frac{1}{2} & -1 & 0 \\
0 & 0 & 0 & 0 & \frac{1}{2} & \frac{1}{2} & -\frac{1}{2} & -\frac{1}{2} & 0 & 0 \\
0 & 0 & -1 & 0 & -\frac{1}{2} & -\frac{1}{2} & \frac{3}{2} & \frac{1}{2} & 0 & 0 \\
0 & 0 & 0 & 0 & -\frac{1}{2} & -\frac{1}{2} & \frac{1}{2} & \frac{3}{2} & 0 & -1 \\
0 & 0 & -\frac{1}{2} & \frac{1}{2} & -1 & 0 & 0 & 0 & \frac{3}{2} & -\frac{1}{2} \\
0 & 0 & -\frac{1}{2} & -\frac{1}{2} & 0 & 0 & 0 & -1 & -\frac{1}{2} & \frac{3}{2}
\end{array}\right]$$

$$(b)$$

The 8 × 8 matrix from columns 3 to 10 is the condensed stiffness matrix of the substructure (superelement). The first 2 × 10 matrix will be utilized later to calculate the internal displacements.

(iii) Using the condensed stiffness matrix of the second level substructure and that of the one truss element appearing in the main structure assemble the global stiffness matrix of the first level substructure. The second level substructure appears twice in the first level substructure with its nodal connectivity as 1,2,3,4 and 3,4,5,6 respectively (Fig.13.1c). After assemblage of superelements and the truss element and with application of boundary conditions, the stiffness matrix of the first level (main) substructure can be obtained as,

$$\frac{EA}{L} \begin{bmatrix} 1.5 & -0.5 & -0.5 & 0 & 0 & 0 & 0 & 0 & 0 \\ -0.5 & 3 & 0 & 0 & 0 & -1 & 0 & -0.5 & -0.5 \\ -0.5 & 0 & 2 & 0 & -1 & 0 & 0 & 0.5 & -0.5 \\ 0 & 0 & 0 & 3 & 0 & -0.5 & -0.5 & -1 & 0 \\ 0 & 0 & -1 & 0 & 2 & -0.5 & -0.5 & 0 & 0 \\ 0 & -1 & 0 & -0.5 & -0.5 & 1.5 & 0.5 & 0 & 0 \\ 0 & 0 & 0 & -0.5 & -0.5 & 0.5 & 1.5 & 0 & -1 \\ 0 & -0.5 & 0.5 & -1 & 0 & 0 & 0 & 1.5 & -0.5 \\ 0 & -0.5 & -0.5 & 0 & 0 & 0 & -1 & -0.5 & 1.5 \end{bmatrix}$$

$$(c)$$

The nodal load vector $\{P\}$ can be formed as

$$\{P\}^T = [\,0 \quad 0 \quad 0 \quad 0 \quad 0 \quad -10.0 \quad 0 \quad 0 \quad 0\,] \qquad (d)$$

and

$$[K]\,\{r\} = \{P\} \qquad (e)$$

(iv) Solve the above equilibrium Eq.(d) of the strucuture to obtain the nodal displacements of the first level substructure shown in Fig.13.1c. Then, by substituting the values of the displacements of the boundary nodes of the superelements in the 2 × 10 condensed matrix (Eq.(b)) the displacements at the internal node of each substructure (3,4 of Fig.13.1a) can be found out. The nodal displacement of all the eight nodes (Fig.13.1a) are tabulated below along with the results of the analysis of the whole structure obtained by using PASSFEM and SAP IV programs.

Results of analysis

Node	Substructuring Technique		SAP IV/PASSFEM	
	Vertical displacements cm	Horizontal displacements cm	Vertical displacements cm	Horizontal displacements cm
1	0	0	0	0
2	0	-0.002137	0	-0.002142
3	-0.000710	-0.003560	-0.000714	-0.003571
4	0.007146	-0.013558	0.007143	-0.013570
5	-0.007840	-0.013550	-0.007857	-0.013571
6	-0.007840	-0.013400	-0.007857	-0.013570
7	0.010014	-0.037110	-0.010000	-0.037140
8	-0.009990	-0.034970	-0.010000	-0.035000

It can be seen that in this case the same substructure is used twice and the condensed stiffness matrix of the substructure is formed only once. The number of equations handled at the first and second level is 9 and 10 respectively whereas in the actual original structure (Fig. 13.1a) the number of equations is 13. In a large structural system and in situations where the number of internal nodes is high, the difference in the number of equations between various levels will be significant.

After obtaining the displacements of all the nodes in each substructure, the element strains and stresses can be calculated following the usual procedure.

Problem 2

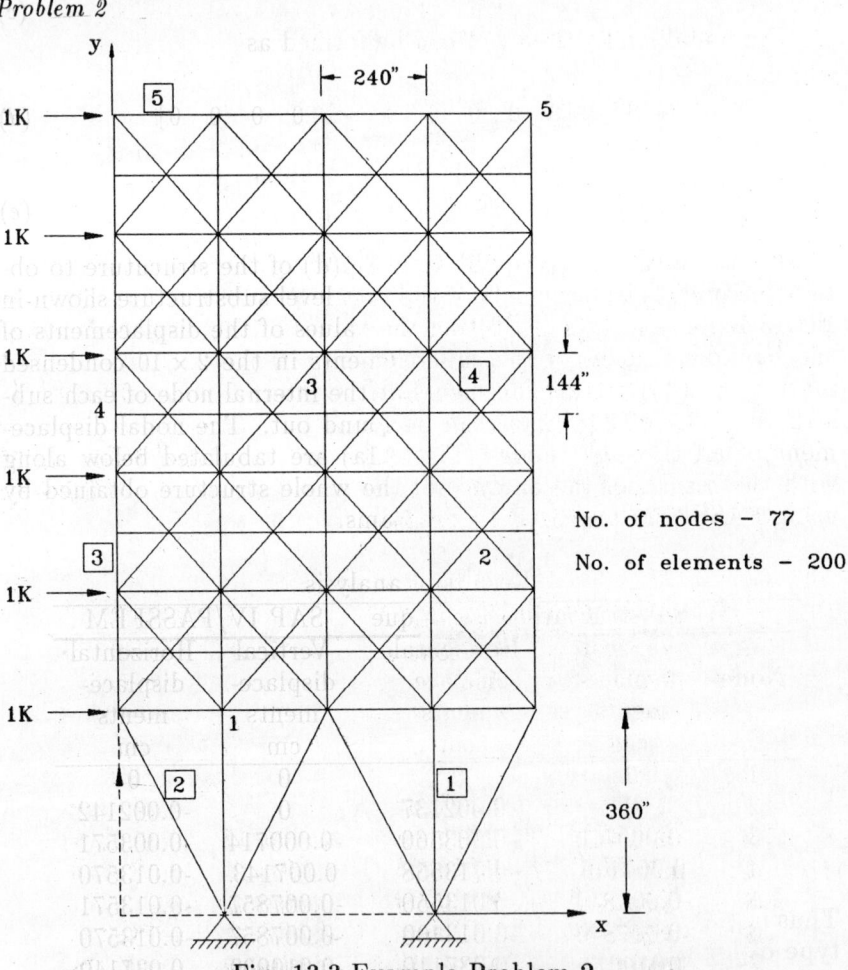

No. of nodes – 77

No. of elements – 200

Fig. 13.3 Example Problem 2

Figure 13.3 shows a plane truss having 77 nodes and 200 elements

[5]. A second level substructure is shown in Fig.13.4(a) and it appears five times in the first level substructure shown in Fig.13.4b.

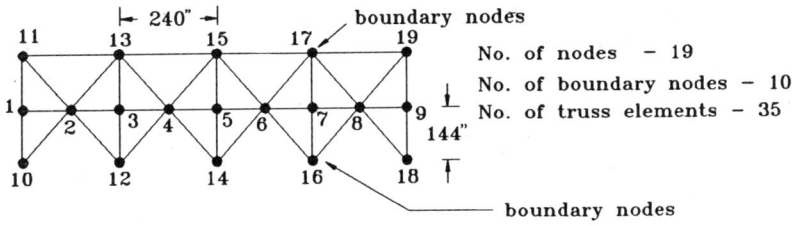

No. of nodes − 19
No. of boundary nodes − 10
No. of truss elements − 35

Fig. 13.4(a) Second level substructure

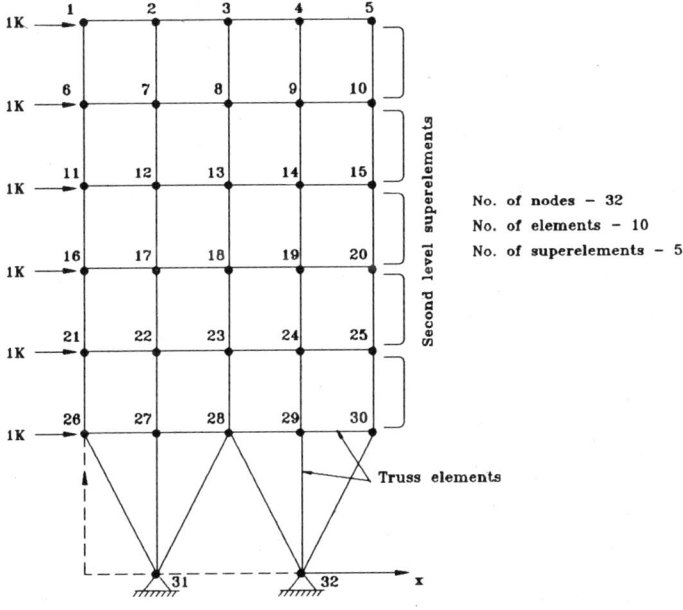

No. of nodes − 32
No. of elements − 10
No. of superelements − 5

Fig. 13.4(b) First level substructure

Thus in the main structure (first level substructure) there is only one type of higher level superelement but the number of superelements of this type is five and in addition it has ten truss elements. It has been observed that the maximum number of equations to be handled using this substructuring scheme is 60 as against 150 for the analysis without substructuring. There is also considerable reduction in the data cards to be prepared because of repetitive use of substructures.

13.2.4 Programming Aspects of Multi-level Substructuring

Having explained the concept of multi-level substructuring in the earlier sections, some of the programming aspects of multi-level substructuring will be discussed in this section. It should be noted that before the formation of global stiffness matrix of the structure, the condensed stiffness matrix of all the second level substructures should be available. In general for getting the stiffness matrix of any nth level substructure, the condensed stiffness matrix of all the $n + 1$th level substructures is required. Therefore, it follows that the formation of stiffness matrix of the substructures should begin at the highest level and proceed towards the first level which is the main structure. The solution of equilibrium equation at the first level gives the nodal displacements of the main structure (first level).

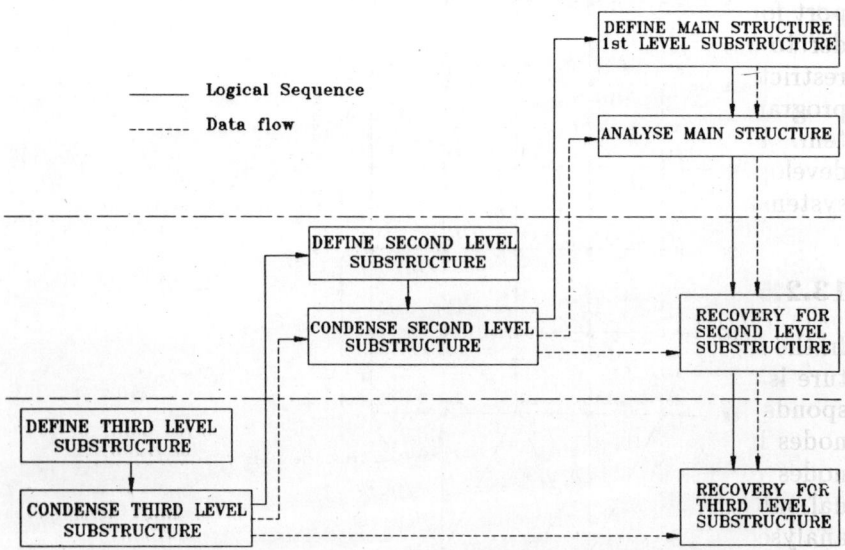

Fig. 13.5 Sequence of operations in a three level substructuring scheme

Again it follows that for getting the internal displacements and stress in the elements, the process should begin at the first level and should proceed towards the higher level. With this background one can now follow the flow of the program control and data transfer for a three level substructuring scheme shown in Fig.13.5

An interesting aspect of the substructuring should be noted here. At the level 3, a substructure termed as a third level substructure, requires input data as a full fledged structure, i.e. nodal data, element connectivity, etc. have to be input and its stiffness matrix has to be assembled and condensed instead of being solved. However, at the second level the same substructure acts as an element and its connectivity and orientation in the second level need be known. Hence, it can be treated as a complex element with number of nodes (boundary nodes). Therefore, a substructure is also aptly called as a 'superelement'. The terms substructure and superelement will be used as synonyms.

Now at any level there can be more than one type of superelement and each type can be repeatedly used in a lower level substructure. Again the data concerning any substructure at any level has to be stored as it will be required during the stress recovering operation. Therefore, for efficient handling, storing and retrieving of data, support for hierarchical data structure should be available. Also in order to give users the flexibility to define the substructure without restrictions on the number of level and numbering of the nodes, the programs should have the support of a good data management system. The package program, for example, POLO-FINITE has been developed using these facilities provided by the software supervisory system POLO[6,7].

13.2.5   Advantages of Multi-level Substructuring

In the substructuring technique the stiffness matrix of each substructure is statically condensed out so that the effective stiffness corresponds to only the boundary degrees of freedom. The number of nodes in the main structure is reduced because only the boundary nodes of the substructure appear there. The size of the individual substructure is obviously less than that of the structure being analysed. Thus at any given instant, the main memory required to process the data corresponding to the main structure or any one substructure is reduced.

Substructuring offers another advantage if the structure can be discretized into identical parts. In such a case, the stiffness matrix of a typical substructure (superelement) can be formed and condensed only once and can be used as many times as the substructure appears in the higher level or main structure.

Another advantage of multi-level substructuring is the reduction of data. But this will be so, only if substructures are identical and are repeatedly used. In such a case, the nodal data and element connectivity need be input only once for a substructure, and its orientation and nodal connectivity only need to be input for higher level substructures. It must be noted here that if the orientation of the substructure with respect to the global coordinate system is different, the transformation matrix can be worked out following the procedure given in subsection 7.4.3.

The full advantage of substructuring can be had in the case of very large sized problems for linear elastic analysis and in the case of non-linear analysis and structural optimization wherein a part of the structure only is modified before the subsequent analysis. These aspects are well illustrated in references [4,6].

REFERENCES

1. Przemieniecki, J.S. *Theory of Matrix Structural Analysis*, McGraw Hill Book Co., New York, 1968.

2. Bathe, K.J. and E.L. Wilson, *Numerical Methods in Finite Element Analysis*, Prentice Hall, Englewood Cliffs, N.J., 1976.

3. Popov, E.P. and H.Peterson, *Substructuring and Equation Solving in Finite Element Analysis*, Int.J.Computers and Structures, Vol.7, pp.197-206, 1977.

4. Noor, A.K., H.A.Kamel and R.E.Fulton, *Substructuring Techniques - Status and Projections*, Computers and Structures, Vol.8, pp.621-632, 1978.

5. Arora, J.S., and A.K.Govil, *An Efficient Method for Optimal Structural Design by Substructuring*, Computers and Structures, Vol.7, pp.507-515, 1977.

6. Dodds, R.H., and L.A.Lopez, *Substructuring in Linear and Non-linear Analysis*, International Journal for Numerical Methods in Engineering, Vol.15, pp.583-597, 1980.

7. Lopez, L.A., R.H.Dodds, D.R.Rehak and J.Urzua, POLO - FINITE, *A Structural Mechanics System for Linear and Non-linear Analysis*, Technical Report, Department of Civil Engineering, University of Illinois at Urbana-Champaign and Department of Civil Engineering and the Academic Computer Center, University of Kansas, Lawrence, Kansas, 1980.

Chapter 14

Finite Element Analysis Software

One of the reasons for wide application of the finite element method is due to the availability of number of package programs. From the computational point of view the five basic steps in finite element analysis, (explained in section 5.2.2) can be well organised in three basic components; pre-processor (step 1), processor (steps 2 to 4) and post-processor (step 5). The requirements of pre- and post-processors are described in the following sections. They are mostly graphics oriented. The processor module involves mathematical computations that require large arithmetic operations, i.e., number crunching operations in a computer.

In order to give an exposure to the students about the finite element analysis software packages, brief description is given on one of the widely used software of moderate size with limited capabilities. The development of a large software package requires a team of persons specialised in finite element analysis, numerical techniques, software technology, and computer graphics. A number of packages are commercially available for wide range of applications. These software packages form integral part of CAD/CAM software in engineering design. An overview of the general capabilities of these software packages is presented in section 14.2.2. The use of alternate programming language, C and object oriented programming technique in future finite element software development are discussed.

Finite element modelling and assessment of error in the solution have been a problem to the analysts. Intensive research is being carried out in the areas of *error estimates* and *adaptive meshing* and these developments are discussed in the last section of the Chapter.

14.1   Pre- and Post-Processors

Finite Element Analysis of practical problems requires handling of
large amount of input data. Manual preparation of input data is a
tedious, time consuming and error-prone task particularly for three-
dimensional stress analysis of solids and shells. In a computer aided
design environment a good and acceptable design is arrived at only
after a few cycles of analysis and design procedure. Hence, it will
be advantageous to have software which will aid in the preparation
of input data and also in the interpretation of analysis results. Such
programs are called pre- and post-processors to finite element analysis package. A pre-processor creates the finite element model and
the input necessary for a finite element analysis program. A post-
processor program accepts the results of the analysis and generates
tables, diagrams/pictures etc. for proper interpretation of results.
The studies conducted by Clerk and Muller [1] show that the cost
of a typical analytical solution can be divided as 80% of engineering cost and 20% of computing cost. The engineering cost consists
of 65% of input data preparation and the rest for interpretation of
results. Due to pre- and post-processing software development and
use, the cost of input data preparation comes down by approximately
20%, that of results interpretation by 10% with a rise in computing
cost by 30%, but results in 40% overall savings.

14.1.1   General Requirements of Pre- and Post- Pro-
cessing

The pre-processor takes minimum input from the user, creates the
finite element mesh and other data required for analysis, and displays
the model for data-check and correction, if any, to be made by the
user in an interactive mode. Graphical post-processing of results
helps to perceive the physical consequences of the analysis. Both the
types of programs should be user oriented, graphics-based and hence
are easy to use with minimum instructions [2,3,4].

A large portion of the input data to finite element analysis is concerned with geometric idealization of the structure and mesh generation. Automatic generation of finite element meshes for two- and
three-dimensional structures not only reduces a good percentage of
man power expenditure of the total analysis time but also minimizes
the errors. Graphics oriented edit facility should be available to edit
and display the finite element model and also enable the user to
input/generate additional data like loads, boundary conditions etc.
The user should have on-line assistance by way of help screens, win-
dowing/zooming, model rotation, change of view direction, element

shrinking, removing hidden lines/surfaces etc.

The nodal values of the field variables are the immediate results of the finite element analysis. In such cases, the post-processor has to work out the higher order derivatives of field variables with reliable accuracy which are of interest in applications. For example, in structural/solid mechanics problems, the field variables are usually displacements, and stress/stress-resultants which are of interest in many design problems. The post-processor helps the user to judge the accuracy of the results by means of various graphical outputs, to identify the critical values and regions for design, and to decide upon the adequacy of modelling/design and the need for remodelling/ re-designing the component/structure/solid.

14.1.2   Methods of Finite Element Model Generation

The construction of a single complex planar or 3-D finite element model usually requires a variety of mesh generation techniques. A single pre-processor must offer many generation options in order to be sufficiently versatile in mesh generation. Several mesh generation techniques which are available in commercial programs are as follows.

1. *Single node and element generation*: This technique is a tedious procedure. However, it can be very effective on models with a few higher order elements.

2. *Digitizing input* : Many users may have scaled drawings of a model. Outlines of the model or nodes and elements can be transferred directly into the program using an electronic digitizing tablet.

3. *Pattern generation*: Often an entire model can be generated by simply repeating a portion of the model.

4. *Duplication*: Portions of an existing model or the entire model can be translated, rotated, mirrored or scaled to generate another portion of the model.

5. *Region generation*: This model generation technique generates nodes and elements within a region bounded by previously defined lines.

6. *Dragging generation*: This mesh generation technique is used to extrude a fixed section of elements. Dragging a line creates 2-D or shell elements. Dragging 2-D elements creates 3-D elements of constant cross section.

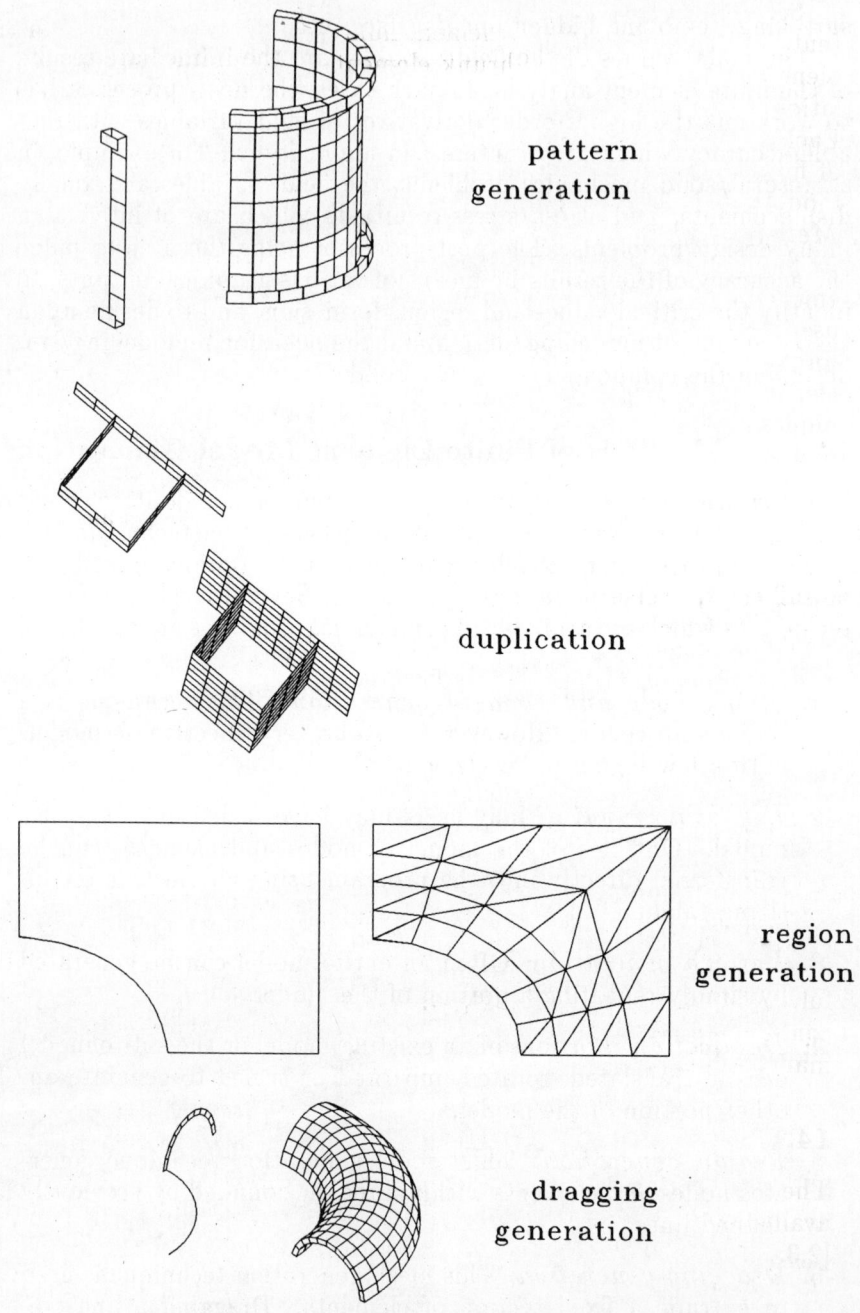

pattern
generation

duplication

region
generation

dragging
generation

Fig. 14.1 Finite element mesh generation techniques

Some of these methods of generation are illustrated in Fig.14.1 Recent developments in automatic mesh generation algorithms include blending functions, coordinate transformations, automatic triangulation and encoding techniques in the form of quadtree and octree. Each of these techniques is designed for more automatic generation of finite element meshes using just the geometric description of the model.

Methods for Specifying Loads and Boundary Conditions

Manual generation of loads and boundary conditions for finite element analysis can be very tedious and time-consuming task. The user normally specifies the nodal point index and the value of each force component for each nodal load. Element loads are specified by element name, face name and load type. The following two techniques are used by interactive pre-processors to reduce the amount of effort required to specify load and boundary conditions:

1. *Menuing:* Nodal couples, nodal loads, element loads and node constraints can be generated by specifying the load or constraint type, the value and the appropriate node or element from the plotted model.

2. *Location and association:* Nodal points and elements have coordinates associated with them. This location can be used as a qualifier to determine if the node or element lies i) on an arbitrary plane ii) within the specified range of the global cartesian or local coordinate system, or iii) on a specified surface. The node or element can be loaded accordingly. The nodal point and element can also be associated with the specific region which generated the node or the elements.

Of these two techniques, the latter is by far the most powerful. With location and association, the user can apply loads and constraints to many nodes or elements with a single interactive command.

14.1.3 Graphical Output Facilities

The common graphical display and output facilities that are generally available in most of the pre- and post-processors are given below: [2,3,4].

1. Two-dimensional and three-dimensional mesh plots:

   The discretized model is shown with options to display element and node numbers, boundary conditions, and loads; to select a portion of a structure either by selecting range of element numbers or by selecting a window on the screen. Shrunk element plots are also available for checking element connectivity.

2. Two-dimensional and three-dimensional node plots:

Here only the nodes are displayed in their respective positions, with options for numbering nodes and selecting range, similar to earlier case.

In addition to the above common facilities, post-processors have the following capabilities.

1. Two-dimensional and three-dimensional deformed meshplots: This plot can be separate or superimposed over the undeformed meshplot to study the structural behaviour.

2. Search and display the maximum and minimum values of user defined variables based on the results of the analysis.

3. Plotting of results such as,

- Force and Moment Diagrams,
- Contours and bands (eg. for stress, strain, temperature, etc.),
- Vectors (for stress, strain etc.),
- Variation of stress or stress resultant along a user defined line/section of model,
- Animation of structural movements,
- Mode shapes in dynamic analysis,
- Load - time - history plots in dynamic analysis,
- Combining the effects of different analyses like different load cases or different load combinations.

Some of the special post-processors can check on satisfaction of specific codal provisions.

14.2　Finite Element Analysis Software

In the field of engineering analysis finite element analysis programs occupy a major computational activity. A large number of general purpose program packages developed in university environments and commercial software houses are now available for stress analysis of complex shapes, material properties and of complicated boundary conditions in several areas of engineering design such as machine components, pressure vessels, boilers, nuclear and off-shore structures, automobile components etc.

If one examines these programs, they seem to fall under two categories; programs developed using the language FORTRAN IV/77 without much of additional software techniques and large sized general purpose packages that are based on sophisticated support software systems including data base management system as explained in section 14.2.2. A brief account of the capabilities of SAPIV will be presented in the next section as a typical example of the widely used software belonging to the first category. A general description of the large sized packages is given in section 14.2.2.

14.2.1 SAPIV (Structural Analysis Program)

SAP IV is a general purpose Finite Element Analysis Program for static and dynamic analysis of linear structural systems. The program was developed by K.J.Bathe, E.L.Wilson and F.E.Peterson at the University of California, Berkeley [5]. The program was coded in standard FORTRAN IV for operation on CDC 6400, 6600 and 7600 computers. SAP IV is a non-proprietary medium sized package and hence is widely used in academic and practical applications. The modifications required to run on IBM 360-370 and on other computer systems were largely undertaken by the SAP Users Group (SUG) at the University of Southern California, Los Angeles. The SUG had also developed enhanced versions of SAP, viz., SAP 6 for linear analysis and SAP 7 for non-linear analysis and were also involved in the development of pre-and-post processors to SAP. There are also several other commercial versions of SAP available around the world.

The program SAP IV has nine element types as shown in Table 14.1. The structural systems to be analysed may be composed of combination of these elements. The capacity of the program depends mainly on the total number of nodal points in the system, the number of eigenvalues needed in the dynamic analysis and the main memory allocation of a master array. There is practically no restriction on the number of elements used, the number of load cases and the bandwidth of the stiffness matrix.

The formation of the structure matrices is carried out in the same way in a static or dynamic analysis. The static analysis is continued by solving the equations of equilibrium followed by the computation of stresses. For the dynamic analysis the following options are available: (i) frequency calculation only, (ii) frequency calculations followed by response history analysis, (iii) frequency calculations followed by response spectrum analysis and (iv) response history analysis by direct integration.

Table 14.1

TYPE	ELEMENT
3–D Truss	
3–D Beam	
Plane stress/strain axisymmetric	
3–D Solid	
Plate/Shell	
Boundary Element	
Thick Shell	
Pipe – tangent and bend	

14.2.2 FEA Software Packages

The rapid advances made in computer hardware and software led to significant developments in finite element analysis software. Finite element programming has emerged as a specialised discipline which requires knowledge and experience in the diverse areas such as finite element technology including foundations of mechanics, and numerical analysis on the one hand and the computational skills in areas of software technology including programming techniques, data structures, data-base management and computer graphics on the other hand. It requires several man years to develop a general purpose finite element analysis software with a processing capability and facility for the user to have a wide choice of several types of elements, analysis for different types of problems - static, dynamic, material and geometric nonlinear, coupled situations, heat transfer, interaction problems etc. and pre- and post-processing features.

As it is not possible here to review the capabilities and compare different commercially available finite element analysis packages only the names of some of the popular packages are given below:

ABAQUS, ADINA, ANSYS, ASKA, COSMOS, GT-STRUDL, NISA, PAFEC, SAP, SESAM-80

The above list is not exhaustive and the reader may refer to publication [6] for more details including the addresses of developers etc. In order to give the readers an overview of the capabilities that are offered in a commercial software package, the major features are listed below:

- Element Library: The element library is a very important component in a package and allows the user to idealise the system and arrive at the finite element model as realistically as possible. The element library may include structural elements such as 3D rods (truss), beams, plane stress/strain and membranes in space, axisymmetric solids, shear panels, plates, thin and thick shells, shells of revolution, 3D solids, discrete stiffeners for plates and shells and pipe elements. The user should have options to use lower or higher order elements in all the above types. Other elements may include matrix elements, substructures, and boundary elements. Other effects which can be modelled include gaps, friction, viscous dampers, cables, plastic hinges, and crack tip.

- Analysis Capabilities and Range of Applications

    - Linear static analysis

- Nonlinear static analysis - nonlinear material, geometry, large deformations, stress stiffening, plasticity, creep, hyperelasticity and rubber like material behaviour, temperature dependent inelastic properties, nonlinear contact problems.
- Dynamic analysis - Eigen-value free vibrations, linear transient dynamics, spectrum analysis due to seismic loading.
- Nonlinear dynamic analysis - material and geometric nonlinear effects, direct integration and different incremental approaches.
- Harmonic response - steady state response of a linear structure subjected to harmonically time-varying loads.
- Stability analysis - calculation of critical loads, bifurcation points, buckled shapes, large deflection analysis to determine the limit load for failure by bifurcation or snap through buckling.
- Heat transfer - temperature distribution and heat flow within a body, linear and nonlinear analysis for temperature dependent material properties and temperature dependent convection boundary conditions, transient heat transfer analysis for time dependent temperature distribution.
- Coupled field analysis - simultaneous solution of interaction of multiple field effect such as structural displacements, temperature and heat flows, solid and fluid flow interactions.

• Types of Loading - Option for analysis of the system due to several types of loading such as concentrated loads, line loads, axisymmetric loads, gravity loads, surface and volume loads, initial stresses, strains or velocities, thermal loading, centrifugal loading, deformation dependent loading, random loading, contact loading.

• Boundary conditions and constraints - Sliding interfaces, prescribed displacements, support at contact points, elastic foundation, multipoint constraints.

• Material Properties and Models - Material properties may be temperature-dependent, isotropic, orthotropic or anistropic and multilayered composites. Nonlinear material behaviour such as plasticity, creep, elastic-strain hardening, visco-elastic or plastic.

• Pre- and post-processing - Commercially available packages offer extensive pre- and post-processing facilities as described in the earlier section.

- Interfaces with CAD/CAM systems - Interfaces to solid modelers and CAD programs are available to form an integrated CAD system.

- Design optimization - Until recently, finite element analysis has been used almost exclusively for the analysis of a user designed model. Some of the packages now offer sophisticated family of computer programs for optimum structural design - the capabilities include minimization of material volume, mass and weight for fixed geometry by changing thickness and cross-sectional dimensions, optimization of structural shapes and optimization of parameters such as area, moment of inertia, thickness etc.

14.2.3 Current Trends in Finite Element Analysis Software

Current trends involve running interactive finite element codes on both larger and smaller computers. There is a strong trend towards implementing total finite element packages, which include interactive pre-processors, equation solvers and post-processors, on stand-alone workstations and personal computers. The application is ideally suited for small and moderate problems because it eliminates the queue time associated with batch computing and opens the door for interactive design and analysis.

Programming language C provides an excellent software tool for development of user interfaces and graphics, in addition it also gives facility to organise the data structures for efficient handling of arrays and multiple load cases etc through pointers [7]. C language is used to develop PASSFEMC to demonstrate the advantages of using the powerful features of C. In the rapidly growing software technology there is an increasing trend to use Object Oriented Programming (OOP). C++ is emerging as a powerful tool for OOP. Besides the several advantages of OOP, it has been found that object oriented techniques facilitate automation of element development and new elements can be added to the element library with very less additional code [8].

Developments are also directed towards merging it with the CAD and CAE technologies. Computer interfaces which allow the geometry from CAD data bases to be transferred directly to CAE data bases are being established [4].

The finite element analysis requires experience and judgement in modelling the solid/structural system, choice of elements, decision on the type of analysis and in the interpretation of results. Users of the finite element packages require assistance in the above aspects and also need expertise to use a particular package. The currently available software systems for finite element analysis are based on the traditional programming languages and do not have facility to incorporate the expert knowledge required to advise the engineer while solving the practical complex problems of engineering design. The recent developments in Knowledge-Based Systems seem to provide a powerful software tool to address the above problems of finite element analysis [9].

A brief description of the expert systems developed for addressing specific problems of finite element analysis is given in reference [10]. Fenves [9] has proposed a KBES framework based on blackboard architecture. Essentially three knowledge modules are required : finite element modelling, analysis and post-processing. They are referred to as Modelling, Analysis and Post-processor Consultants and their main functions are described in reference [10].

14.3　Error Estimates and Adaptive Meshing

Finite Element Method, after three decades of development, has reached that point of maturity at which nobody doubts its effectiveness and power to solve various kinds of engineering problems. But, its results, as one of any other numerical method, can not be relied upon without an assessment of their accuracy. The accuracy of the finite element solution depends on the discretization which is characterized by the finite element mesh and the choice of elements.

Considerable work has been done to investigate the various criteria involved in designing the 'best' mesh. Notable among this are the studies of Melosh, Ebner, Case and Mason, Mallert and Key etc [11]. These studies expose the complexity involved in designing the mesh and provide some guidelines covering problems of idealization and discretization. It is only recently that researchers have focused their attention on the main problems of the 'reliability' of solutions of the finite element method. An increasing number of research workers are now involved in the assessment of errors in the solutions and in improvement of the reliability of the finite element analysis through a procedure known as Adaptive Finite Element Analysis.

Normally finite element analyst designs a mesh based on his previous experience, intuition and guesswork. The results are accepted if they reasonably match with what he has guessed. There is no reliable way of judging the acceptability of the solution. If the analyst's intuition fails while designing the mesh, it is also liable to fail when evaluating the worthiness of the results. Therefore, it is necessary to get a measure of the error in the finite element results as an indicator of their reliability.

Further, if the initial solution is not acceptable, the analyst has to prepare an entirely new set of input data for a finer mesh. This is a very costly and time consuming procedure, and does not guarantee that the new mesh will give sufficiently accurate results.

The main idea behind the adaptive finite element mesh refinement is that given an initial mesh sufficient to define the geometry of the domain, boundary conditions and acceptable level of error, the procedure should indicate whether the solution is accurate enough, on the basis of local information about the error assessed from the first analysis. If it is not accurate enough the procedure should further indicate how best to refine the mesh and automatically re-design it such that the error in the next solution is less than acceptable level. Thus, adaptivity in finite element analysis can be defined as a procedure by which improvement in the accuracy of the approximate solutions is achieved based on the information obtained from the previous solution. The goal of the adaptive procedure is to achieve a prescribed level of accuracy as efficiently as possible.

A finite element software is said to be an 'adaptive' one when it possesses the following features:

1. A local error indicator

2. The capability to upgrade, either automatically or with minimal user interaction, the number of degrees of freedom in those regions where the indicated error is high.

The overall procedure for adaptive mesh refinement is shown in Fig. 14.2. It consists of the following basic modules.

Automatic Mesh Generator : The aim of minimal user interaction in adaptive procedure is achieved in the initial stage of the procedure by coupling it with an automatic mesh generator. It is a technique that can generate subregions of node points and element connectivities for an object given its overall geometry and limited meshing information.

Finite Element Analysis and Error Estimator: It constitutes the main part of procedure i.e., the actual finite element analysis and 'a-posteriori' error estimator. The finite element analysis program should have good library of necessary elements. The error estimator should be such that the cost of computations associated with it is low. It should not be difficult to implement in existing finite element code.

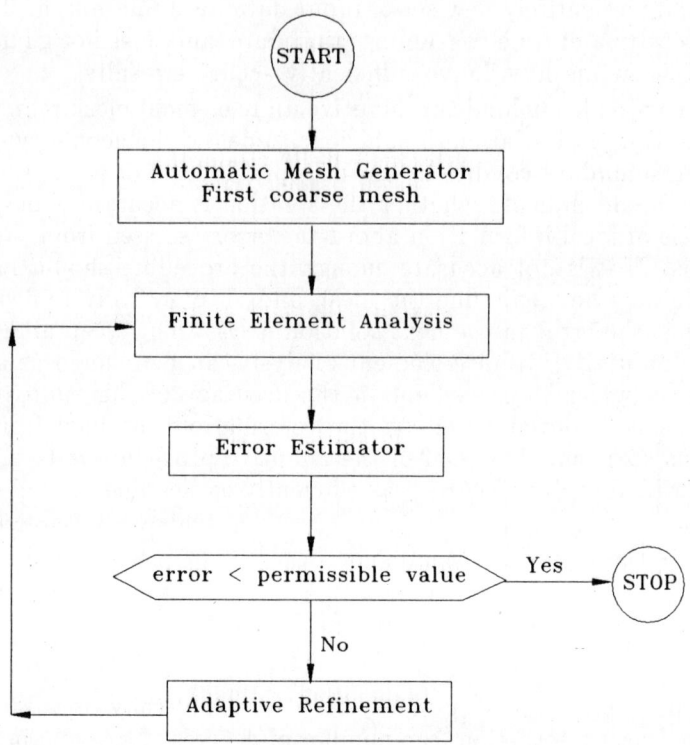

Fig. 14.2 Adaptive finite element analysis

Refinement Algorithm: The refinement strategy depends on the criteria to be adopted to attain the level of accuracy required. For example, if the error in the whole domain is more than the maximum percentage error specified, the mesh should be refined following the criteria such as - the contribution to this total error should be equal from all elements to get the optimal mesh. The refinement scheme and data structure that supports it are crucial since they strongly influence the practical feasibility, storage overheads and related considerations.

After creating the initial mesh for an adaptive procedure using the automatic mesh generator, the first cycle of finite element analysis is carried out. To identify the areas of the mesh that require improvement, an efficient a-posteriori error estimator is needed. Pioneering research has been done on error estimators by Babuska and his co-workers [12]. But these estimators are computationally costly and difficult to implement into an existing code. Recently Zienkiewicz et al [13] have proposed a new error estimator and a refinement strategy that are simple and easy to implement.

The improvement in accuracy can be achieved in several ways. In the conventional approach, called the h-version of the finite element method, the order of polynomial approximation for all elements is kept constant and the number of elements is increased in such a way that the size of the largest element (h_{max}) approaches to zero. The alternative approach is the p-version, in which the number of elements is constant and the order of polynomial approximation (p) of the elements is increased. An adaptive software system can be based either on the h-version or the p-version or a combination of both i.e., $h - p$ version [14].

In order to give the readers an idea about the meshing pattern that emerges in an adaptive finite element analysis, an illustrative example is presented here. Fig. 14.3a shows a cantilever plate having a notch. The plate is analysed for a uniformly distributed load of 10 units per length, acting along the free edge. Zienkiewicz - Zhu error estimator is used [13]. An adaptive mesh generation algorithm for quadrilateral elements has been developed and used to solve the problem. Figures 14.3b, c and d show the successive adaptive meshes generated and they clearly bring out the regions of stress concentration.

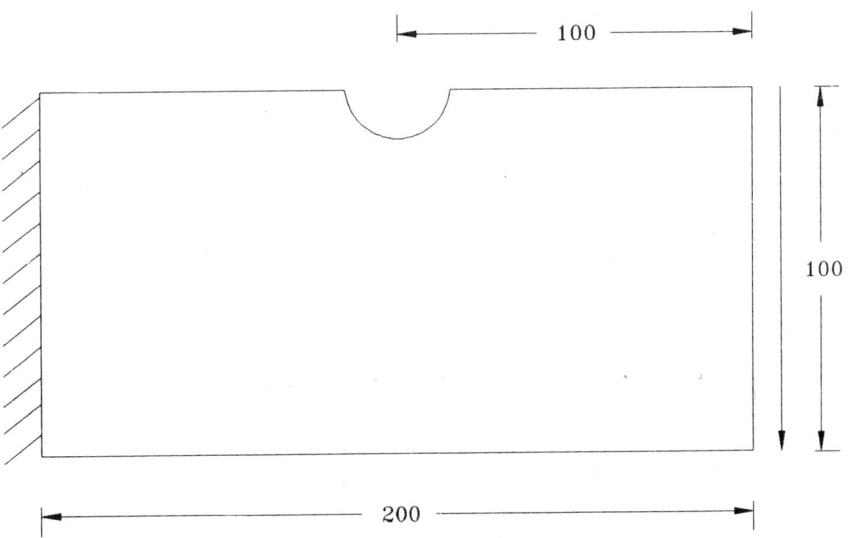

Fig. 14.3a Plate with a notch

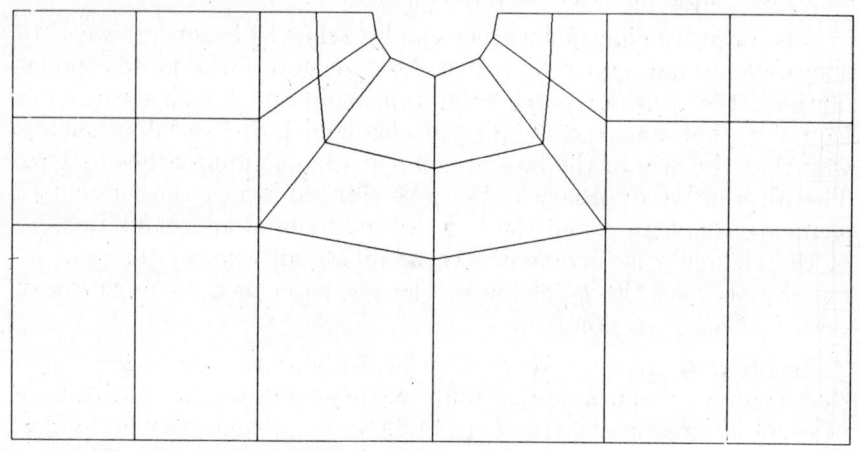

Fig. 14.3b Initial Mesh - percentage error = 23.65 %
No. of nodes = 40, No. of elements = 26

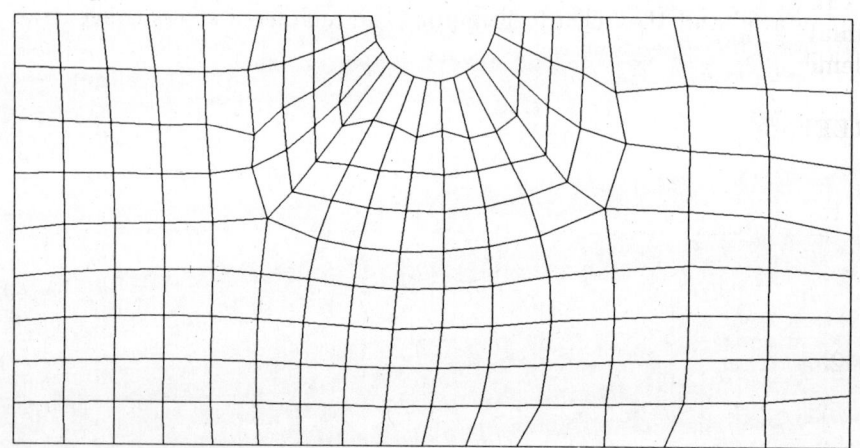

Fig. 14.3c Mesh with percentage error = 13.11 %
No. of nodes = 140, No. of elements = 116

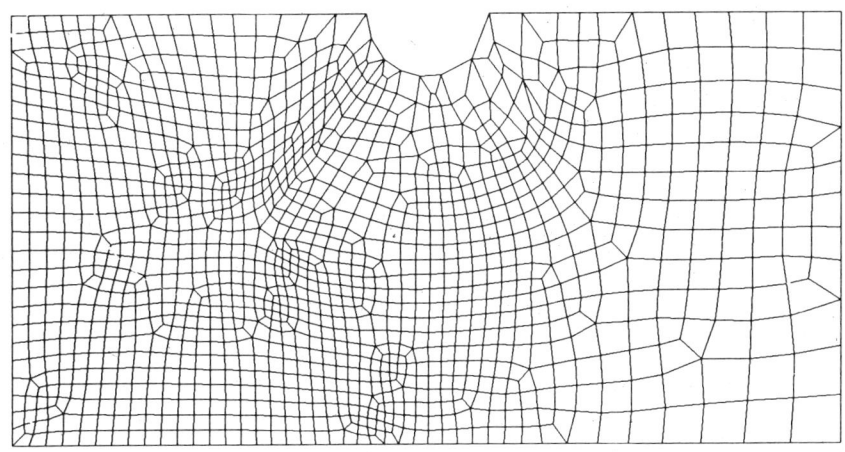

Fig. 14.3d Final Mesh - percentage error = 4.0 %
No. of nodes = 1654, No. of elements = 1555

Intensive research is being carried out in the areas of error estimates and mesh refinement algorithms leading to adaptive finite element analysis for various types of applications.

REFERENCES

1. Clerk, C.V. and R.Muller., *GIFTS-1100: Graphics Oriented Interactive Finite Element Time Sharing System*, In Handbook of Finite Elements, (Ed) C.A.Brebbia, CML Publications, England, 1981.

2. Mackerle, J., *Review of Pre- and Post-Processor Programs in the Major Commercial General Purpose Finite Element Packages*, Advances in Engineering Software, Vol.5, No.1, pp.43-53, 1983.

3. Mackerle, J., *Review of General Purpose Pre- and Post-Processor Programs for the Finite Element Applications*, Advances in Engineering Software, Vol.5, No.1, pp.148-159, 1983

4. Kardestuncer, H.(Ed), *Finite Element Handbook*, McGraw Hill Book Co,, New York, 1987.

5. Bathe, K.J, E.L.Wilson, and F.E.Peterson, *SAPIV - A Structural Analysis Program for Static and Dynamic Response*

of Linear Systems, Earthquake Engineering Research Centre, University of California, Berkeley, Report No.EERC 73-11, 1973.

6. Kardestuncer, H. (Ed), *Survey of Some Finite Element Software Systems*, Chapter 5 in Part 4 - Finite Element Method Computations in Finite Element Handbook, McGraw Hill Book Co., New York, 1987.

7. Krishnamoorthy, C.S. and S.Rajeev and Sooryanarayana Sharma, *A Finite Element Analysis Software in C*, in Engineering Software, (Eds) C.V.Ramakrishnan, A.Varadarajan and C.S. Desai, Narosa Publishing House, pp.697-704, 1989.

8. Benny Raphael and C.S.Krishnamoorthy, *Automating Finite Element Development Using Object Oriented Techniques*, Engineering Computations - An International Journal for Computer Aided Engineering and Software, 1993.

9. Fenves, S.J., *A Framework for Cooperative Development of a Finite Element Modelling Assistant*, Reliability Method for Engineering Analysis, (Ed), K.J.Bathe and D.R.J.Owen, John Wiley, U.K., 1986.

10. Krishnamoorthy, C.S., R.Krishnakumar and S.Rajeev, *Expert System Framework for Finite Element Analysis in Engineering Software*, (Eds). C.V.Ramakrishnan, A.Varadarajan and C.S.Desai, Narosa Publishing House, New Delhi, pp.775-784, 1989.

11. Melosh, K.J. and S.Utku, *Principles for Design of Finite Element Meshes in Finite Element Handbook*, (Ed). H. Kardestuncer, McGraw Hill Book Co., pp.3.425 - 3.441, 1987.

12. Babusaka, I and W.C.Rheinboldt, *A Posteriori Error Estimate for the Finite Element Method*, International Journal for Numerical Methods in Engineering, Vol.11, pp.1597-1618, 1978.

13. Zienkiewicz, O.C. and J.Z.Zhu, *A Simple Error Estimator and Adaptive Procedure for Practical Engineering Analysis*, International Journal for Numerical Methods in Engineering, Vol.24, pp.337-357, 1987.

14. Krishnamoorthy, C.S. and K.Rajeshirke Umesh, *Adaptive Mesh Refinement for Two-Dimensional Finite Element Stress Analysis*, Computers and Structures, An International Journal, Vol.48, No.1, pp.121-133, 1993.

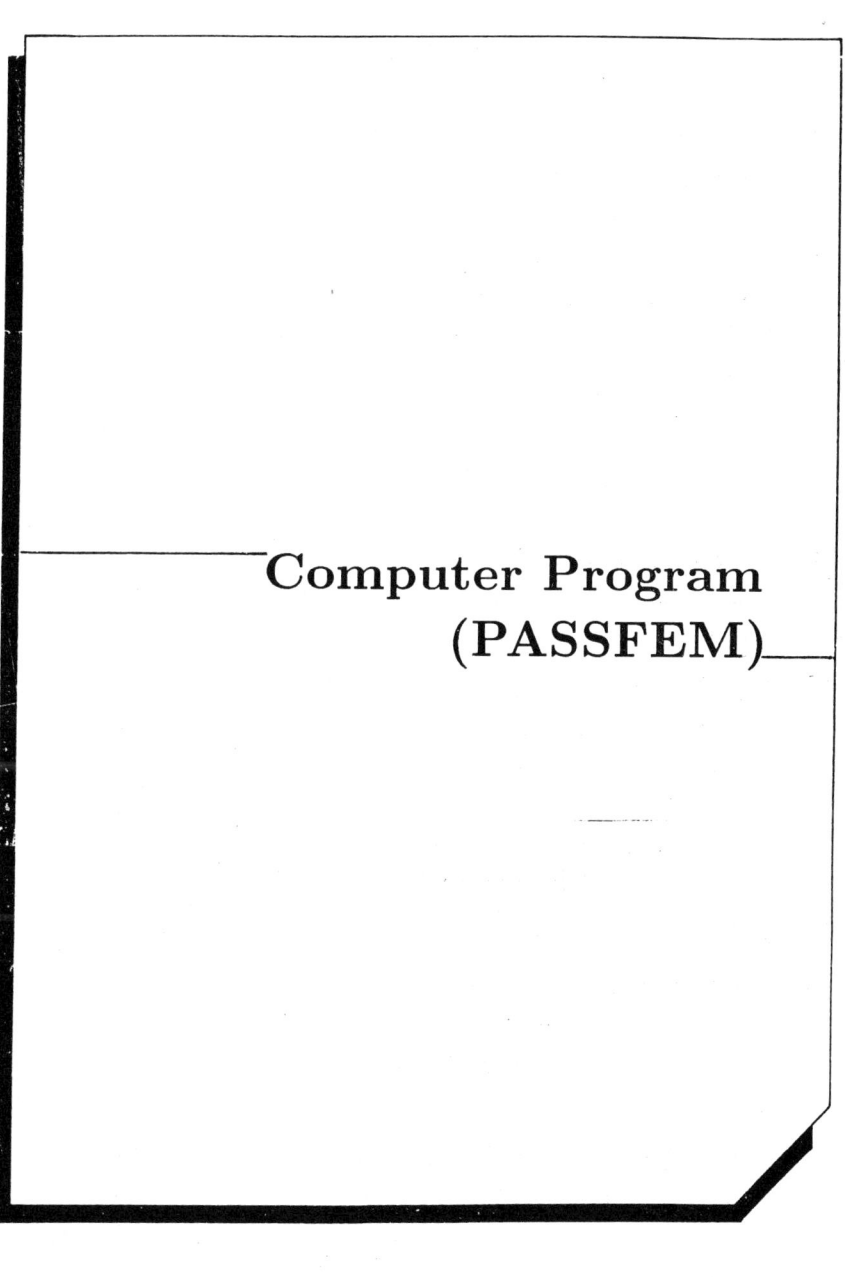

Computer Program
(PASSFEM)

```
C      *******************************************************
C      *                                                     *
C      *                      PASSFEM                         *
C      *                                                      *
C      *       Program for Analysis of Structural Systems     *
C      *                                                      *
C      *                        by                            *
C      *                                                      *
C      *              Finite Element Method                   *
C      *                                                      *
C      *******************************************************
C*************************************************************************
C        MAIN ROUTINE OF THE PROGRAM PASSFEM
C*************************************************************************
C
         IMPLICIT REAL*8(A-H,O-Z)
         REAL*4 S1,S2,S3,S4,S5,S6,S7,S8,SS1,SS2,SS3,SS4
         INTEGER CHT
         COMMON/DIM/N1,N2,N3,N4,N5,N6,N7,N8,N9,N10,N11,N12
        *,N13,N14
         COMMON/PAR/IND,NET,NSN,NMP,NEQ,NSKY,NEQ1,LCOUNT
         COMMON/TAPES/ISTRES,NDARAY,IPR
         COMMON/PRECI/ITWO
         COMMON/MULT/ELMN
         COMMON/ELEM/NDEL(100)
         COMMON/ELOAD/M1,M2,AA1,AA2,AA3,AA4,NSLC,NUM
         DIMENSION SS1(10),SS2(10),SS3(10),SS4(10)
         DIMENSION LLIB(7,3),NLN(10),TITLE(20),ELM(10)
         CHARACTER*15 INFILE,OUTFILE
C
C        THE FOLLOWING TWO CARDS ARE USED TO DETERMINE
C        THE MAIN STORAGE THAT CAN BE USED FOR
C        SOLUTION.TO CHANGE THE MAIN STORAGE
C        AVAILABLE FOR EXECUTION,CHANGE THE VALUE OF MTOT
C        AND CORRESPONDINGLY COMMON A (MTOT)
         DIMENSION A(30000)
         MTOT=30000
         CALL SECOND(S1,0.0)
C
         ITWO=1
C
C        THE FOLLOWING SCRATCH FILES ARE USED
C        ISTRES=FILE STORING STRESS DISPLACEMENT MATRIX
C        NDARAY=FILE STORING ND ARRAY
         ISTRES=1
         NDARAY=2
C
         WRITE (*,*) 'PLEASE ENTER THE INPUT FILENAME:'
         READ (*,1) INFILE
         WRITE (*,*) 'PLEASE ENTER THE OUTPUT FILENAME:'
         READ (*,1) OUTFILE
         OPEN(5,FILE=INFILE ,FORM='FORMATTED',STATUS='OLD',RECL=80)
         OPEN(1,STATUS='SCRATCH',FORM='UNFORMATTED')
         OPEN(2,STATUS='SCRATCH',FORM='UNFORMATTED')
         OPEN(6,FILE=OUTFILE,FORM='FORMATTED',STATUS='NEW',RECL=132)
```

```
C
C        READ AND PRINT STRUCTURE DATA
   100   READ(5,55)(TITLE(I),I=1,20)
         READ(5,11)NSN,NET,NMP,NLC,NEDMAX,MODEX
         IF(NSN.EQ.0)STOP
         READ(5,12)IPR
         WRITE(6,66)(TITLE(I),I=1,20)
         WRITE(6,22) NSN,NET,NMP,NLC
C      READ, GENERATE AND PRINT NODAL DATA &
C      MATERIAL PROPERTIES
         N1=1
         N2=N1+6*NSN
         N3=N2+NSN*ITWO
         N4=N3+NSN*ITWO
         N5=N4+NSN*ITWO
         IF(N5.GT.MTOT) CALL ERROR(N5-MTOT,1)
         IF(N5.GT.MTOT)STOP
         N6=N5+NMP*ITWO
         N7=N6+NMP*ITWO
         N8=N7+NMP*ITWO
         CALL PASSIN (A(N2),A(N3),A(N4),A(N5),
        *A(N6),A(N7),A(N1),NSN,NMP,NEQ)
C      INITIALIZE COLUMN HEIGHTS TO'0'
         N9=N8+NEQ
         DO 10 I=N8,N9
    10   A(I)=0.0
C      READ AND GENERATE ELEMENT DATA
         WRITE(6,33)
         NEQ1=NEQ+1
         REWIND NDARAY
         IND=1
         LCOUNT=0
         CALL FELIB(A,LLIB,MTOT)
         CALL SECOND(S2,S1)
C
C      ADDRESS THE DIAGONAL ELEMENTS OF GLOBAL
C      STIFFNESS MATRIX FOR ASSEMBLY
         N11=N10+NEQ1
         CALL CADNUM(A(N8),A(N10),NEQ,NEQ1,NSKY,MBAND)
C      WRITE DATA OF SOLUTION PHASE
         WRITE(6,88)NEQ,NSKY
C      ALLOT SPACE FOR STIFFNESS MATRIX AND LOAD VECTOR
         N12=N11+NSKY*ITWO
         N13=N12+NEQ*ITWO
         N14=N13+NEQ*ITWO
         IF(N14.GT.MTOT)CALL ERROR(N14-MTOT,3)
         IF(N14.GT.MTOT)STOP
C      INITIALIZE GLOBAL STIFFNESS MATRIX,LOAD VECTOR
C      AND DISPLACEMENT VECTOR
         NSIZE=NSKY+NEQ+NEQ
         CALL INISKP(A(N11),NSIZE)
         IF(MODEX.NE.0) GO TO 222
C
C      CALCULATES LOAD VECTOR & ELEMENT STIFFNESSES AND
C      ASSEMBLES THE STRUCTURE STIFFNESS MATRIX
         REWIND NDARAY
         REWIND ISTRES
         IND=2
```

```
          LCOUNT=0
          CALL FELIB(A,LLIB,MTOT)
          CALL SECOND(S3,S1)
C
C         TRIANGULARIZE STIFFNESS MATRIX
          KTR=1
          CALL PASOLV(A(N11),A(N13),A(N10),NEQ,NEQ1,NSKY,KTR)
222       CONTINUE
          CALL SECOND(S4,S1)
C
C         READ AND PRINT ELEMENT LOAD MULTIPLIERS
C
          READ(5,77)(ELM(J),J=1,NLC)
          WRITE(6,78)
          WRITE(6,79)(L,L=1,NLC)
          WRITE(6,81)(ELM(J),J=1,NLC)
          READ(5,44)(NLN(J),J=1,NLC)
          NN1=N13-1
          CALL SECOND(S5,S1)
          S6 =0.0
          S7 =0.0
          S8 =0.0
          DO 80 L=1,NLC
          CALL SECOND(SS1(L),0.0)
          ELMN=ELM(L)
          DO 144 I=N12,NN1
144       A(I+NEQ)=A(I)*ELMN
          WRITE(6,99) L
          WRITE(6,122)
          NC=NLN(L)
          IF(NC.EQ.0) GO TO 30
C
C         ADD CONCENTRATED LOAD TO CALCULATED LOAD VECTOR
C
          CALL PASLOD (A(N13),A(N1),NC,NSN,NEQ)
C
    30    CONTINUE
          CALL SECOND(SS3(L),SS1(L))
          IF(MODEX.NE.0) GO TO 80
C         CALCULATION OF DISPLACEMENTS
          KTR=2
          CALL PASOLV(A(N11),A(N13),A(N10),NEQ,NEQ1,
         *NSKY,KTR)
C
C         PRINT THE NODAL DISPLACEMENTS
          WRITE(6,111)
          CALL DISP(A(N13),A(N1),NSN,NEQ)
          CALL SECOND(SS3(L),SS1(L))
C
C         CALCULATE STRESSES AT THE SELECTED POINTS OF EACH ELEMENT
          REWIND NDARAY
          REWIND ISTRES
          IND=3
          LCOUNT=0
          CALL FELIB(A,LLIB,MTOT)
          CALL SECOND(SS4(L),SS1(L))
          S6=S6+SS2(L)
          S7=S7+SS3(L)-SS2(L)
```

```
              S8=S8+SS4(L)-SS3(L)
80            CONTINUE
              IF(MODEX.NE.0)GO TO 200
              S1=S2+(S5-S4)+S6
              S2=S3-S2
              S3=(S4-S3)+S7
              S4=S8
              S5=S1+S2+S3+S4
              WRITE(6,130)
              WRITE(6,135)S1,S2,S3,S4,S5
  200         CONTINUE
C
C      READ NEXT ANALYSIS CASE
C
              GO TO 100
  1           FORMAT(A15)
 11           FORMAT(6I5)
 12           FORMAT(I5)
 22           FORMAT(53X,'NUMBER OF STRUCTURE NODES....',
     *I4/53X,'NUMBER OF ELEMENT TYPES......',I4/53X,
     *'NUMBER OF MATERIAL GROUPS....',I4/53X,'NUMBER
     *OF LOADING CONDITIONS.',I3//)
 33           FORMAT(//61X,'ELEMENT DATA',/)
 44           FORMAT(10I5)
 55           FORMAT(20A4)
 66           FORMAT(130('*')//27X,20A4//59X,'STRUCTURE DATA',/)
 77           FORMAT(10F5.2)
 78           FORMAT(//5X,'ELEMENT LOAD MULTIPLIERS'//)
 79           FORMAT(5X,'LOAD CASE NO.',I4,9I8)
 81           FORMAT(15X,10(2X,F6.2))
 88           FORMAT(/51X,'DATA OF SOLUTION PHASE'
     */48X,'NUMBER OF EQUATIONS............(NEQ)=',
     *I7//48X,'NUMBER OF ELEMENTS IN SK..........
     *(NSKY)=',I7//)
 99           FORMAT(//5X,'LOAD CASE NO.....',I4/)
111           FORMAT(///56X,'NODAL DISPLACEMENTS'//5X,
     *'NODE',8X,'X-DISPLACEMENTS',6X,'Y-
     *DISPLACEMENTS',4X,'Z-DISPLACEMENTS',8X,
     *'ROTATION-X',9X,'ROTATION-Y',9X,'ROTATION-Z')
122           FORMAT(/58X,'NODAL LOAD DATA'/)
  130         FORMAT(//5X,'*****O V E R  A L L  T I M E L O G*****'//)
  135         FORMAT(10X,'MODEL INPUT.......................',F8.2/
     *10X,'ELEMENT STIFFNESS FORMULATION'/
     *10X,'AND ASSEMBLAGE....................',F8.2/
     *10X,'EQUATION SOLUTION AND'/
     *10X,'DISPLACEMENT OUTPUT...............',F8.2/
     *10X,'STRESS OUTPUT.....................',F8.2//
     *10X,'TOTAL SOLUTION TIME...............',F8.2//)
              END


C**********************************************************************
C      SUBROUTINE TO DETERMINE THE TIME TAKEN FOR EXECUTION
C      CALLED BY MAIN
C**********************************************************************
              SUBROUTINE SECOND(T2,T1)
C      'SECNDS' IS A SYSTEM DEPENDENT FUNCTION - HERE USE THE
C      APPROPRIATE TIME FUNCTION FOR ANY OTHER SYSTEM
```

```
        T2 = SECNDS(T1)
        RETURN
        END



C**********************************************************************
C     ROUTINE TO PRINT MESSAGES WHEN SPECIFIED MAIN STORAGE
C     IS EXCEEDED
C     CALLED BY : MAIN
C**********************************************************************
        SUBROUTINE ERROR(N,I)
        GO TO (1,2,3),I
1       WRITE(6,11)
        GO TO 10
2       WRITE(6,22)
        GO TO 10
3       WRITE(6,33)
10      WRITE(6,44)N
        RETURN
11      FORMAT(7X,'NOT ENOUGH STORAGE FOR READ-IN OF
       *NDF ARRAY AND NODAL POINT COORDINATES')
22          FORMAT(//////5X,'NOT ENOUGH STORAGE FOR ELEMENTAL
       *NODE COORDINATES')
33          FORMAT(//////'NOT ENOUGH STORAGE FOR ASSEMBLAGE
       *OF GLOBAL STRUCTURE STIFFNESS AND DISPLACEMENT
       *IN STRESS SOLUTION PHASE')
44          FORMAT(////'****ERROR*** STORAGE EXCEEDED BY',I7)
        END



C**********************************************************************
C     ROUTINE READS,GENERATES AND PRINTS NODAL DATA AND
C     MATERIAL PROPERTIES
C     CALLED BY : MAIN
C**********************************************************************
        SUBROUTINE PASSIN(X,Y,Z,E,PR,WD,NDF,NSN,NMP,NEQ)
        IMPLICIT REAL*8(A-H,O-Z)
        COMMON/PRECI/ITWO
        COMMON/TAPES/ISTRES,NDARAY,IPR
        DIMENSION X(NSN),Y(NSN),Z(NSN),E(NMP),PR(NMP),WD(NMP)
        DIMENSION NDF(6,NSN),JF(6)
        DIMENSION NCYLT(1000)
        FLOAT(I)=DBLE(I)
        MN=0
C
100.    READ(5,33)NN,(JF(I),I=1,6),X(NN),Y(NN),Z(NN),
       *NI,NCYL
        NCYLT(NN)=NCYL
        N=MN+NI
        MN=MN+1
110     DO 120 I=1,6
        NDF(I,NN)=JF(I)
120     CONTINUE
        IF (NI.EQ.0) GO TO 130
        IF(NN-MN)130,125,140
125     CONTINUE
        IF(NSN-NN)170,170,100
130     MN=NN
```

```
      GO TO 125
C
C     AUTOMATIC GENERATION OF NODAL DATA
C
140   NX=(NN-N+NI)/NI
      XD=(X(NN)-X(N-NI))/FLOAT(NX)
      YD=(Y(NN)-Y(N-NI))/FLOAT(NX)
      ZD=(Z(NN)-Z(N-NI))/FLOAT(NX)
      MN=NN
150   X(N)=X(N-NI)+XD
      Y(N)=Y(N-NI)+YD
      Z(N)=Z(N-NI)+ZD
      NCYLT(N)=NCYLT(NN)
      DO 160 I=1,6
      NDF(I,N)=JF(I)
160   CONTINUE
      N=N+NI
      IF(N.LT.NN) GO TO 150
      IF(NSN-NN)170,170,100
170   CONTINUE
      DO 180 I=1,NSN
      IF(NCYLT(I).EQ.0) GO TO 180
      THETA=Y(I)*3.14159/180.0
      Y(I)=X(I)*DSIN(THETA)
      X(I)=X(I)*DCOS(THETA)
180   CONTINUE
      IF(IPR.EQ.0)WRITE(6,44)(I,(NDF(J,I),J=1,6),X(I)
     *,Y(I),Z(I),I=1,NSN)
C     CONVERT '0'AND '1' OF 'NDF' ARRAY TO EQUATION
C    *NUMBERS AND '0'S.
      NEQ=0
      DO 30 N=1,NSN
      DO 30 I=1,6
      IF(NDF(I,N))10,20,10
  20  NEQ=NEQ+1
      NDF(I,N)=NEQ
      GO TO 30
  10  NDF(I,N)=0
  30  CONTINUE
      IF(IPR.EQ.0)WRITE(6,77)(I,(NDF(J,I),J=1,6),
     *I=1,NSN)
C
C     READ AND PRINT MATERIAL PROPERTIES
      DO 1 J=1,NMP
1     READ(5,55)I,E(I),PR(I),WD(I)
      WRITE(6,66)(I,E(I),PR(I),WD(I),I=1,NMP)
C
  33  FORMAT(7I5,3F10.3,2I5)
  44  FORMAT(//59X,'NODAL POINT DATA'//5X,'NODE',
     *3X,'NODAL D.O.F.',5X,'X-COORD.',5X,'Y-COORD.',
     *5X,'Z-COORD.',6X,'NODE',3X,'NODAL DOF',
     *5X,'X-COORD',5X,'Y-COORD',
     *5X,'Z-COORD.'//(5X,I4,3X,6I2,3X,F10.4,3X,F10.4,
     *3X,F10.4,3X,3X,I4,3X,6I2,3X,F10.4,3X,F10.4,3X,
     *F10.4/))
  55  FORMAT(I10,E10.3,2F10.4)
  66  FORMAT(//57X,'MATERIAL PROPERTIES'//44X,'GROUP',7
     *X,'YOUNGS',7X,'POISSON',7X,'WEIGHT',/45X,'NO.',
```

```
      *7X,'MODULUS',8X,'RATIO',7X,'DENSITY'//(45X,I3,
      *5X,F10.2,1X,F10.2,6X,F10.5))
   77    FORMAT(/24X,'NODE',11X,'EQUATION NUMBERS',25X,
      *'NODE',12X,'EQUATION NUMBERS'//(24X,I4,3X
      *,6I5,20X,I4,3X,6I5/))
         RETURN
         END


         SUBROUTINE FELIB(A,LLIB,MTOT)
         IMPLICIT REAL*8(A-H,O-Z)
         COMMON/DIM/N1,N2,N3,N4,N5,N6,N7,N8,N9,N10,
      *N11,N12,N13,N14
         COMMON/PAR/IND,NET,NSN,NMP,NEQ,NSKY,NEQ1,LCOUNT
         COMMON/TAPES/ISTRES,NDARAY,IPR
         COMMON/PRECI/ITWO
         COMMON/MULT/ELMN
         DIMENSION A(MTOT)
         DIMENSION LLIB(7,3),LT(7)
         DO 100 I=1,NET
         IF(IND.NE.1) GO TO 5
         READ(5,11)LTYPE,NSHAPE,(LLIB(LTYPE,J),J=1,NSHAPE)
         LT(I)=LTYPE
   5     LA=LT(I)
         GO TO (10,20,30,40,50,60,70),LA
   10    CALL THREDT(A,LLIB,NSHAPE,MTOT)
         GO TO 100
   20    CALL THREDB(A,LLIB,NSHAPE,MTOT)
         GO TO 100
   30    CALL PLANE(A,LLIB,NSHAPE,MTOT)
         GO TO 100
   40    CALL THREDS(A,LLIB,NSHAPE,MTOT)
         GO TO 100
   50    CALL PLATE(A,LLIB,NSHAPE,MTOT)
         GO TO 100
   60    CALL SHELL(A,LLIB,NSHAPE,MTOT)
         GO TO 100
   70    CALL BOUND(A,LLIB,NSHAPE,MTOT)
   100   CONTINUE
   11    FORMAT(5I5)
         RETURN
         END
         SUBROUTINE INISKP(SK,NSIZE)
         IMPLICIT REAL*8(A-H,O-Z)
         DIMENSION SK(NSIZE)
         DO 10 I=1,NSIZE
   10    SK(I)=0.0
         RETURN
         END
         SUBROUTINE COLUMH(CHT,ND,NED,NEQ)
         IMPLICIT REAL*8(A-H,O-Z)
         INTEGER CHT(NEQ),ND(NED)
C     CALCULATES THE COLUMN HEIGHTS OF EACH COLUMN
C     IN THE GLOBAL STIFFNESS MATRIX
C
         LS=100000
         DO 30 K=1,NED
         IF(ND(K)) 10,30,10
```

```
      10    IF(ND(K)-LS)20,30,30
      20    LS=ND(K)
      30    CONTINUE
            DO 40 K=1,NED
            II=ND(K)
            IF(II.EQ.0) GO TO 40
            ME=II-LS
            IF(ME.GT.CHT(II))CHT(II)=ME
      40    CONTINUE
            RETURN
            END


            SUBROUTINE CADNUM(CHT,NDS,NEQ,NEQ1,NSKY,MBAND)
            IMPLICIT REAL*8(A-H,O-Z)
            INTEGER CHT(NEQ),NDS(NEQ1)
C     CALCULATES ADDRESSES OF DIAGONAL ELEMENTS IN
C     BANDED MATRIX WHOSE COLUMN HEIGHTS ARE KNOWN;
C     CALCULATES THE NO.OF ELEMENTS IN THE GLOBAL
C     STIFFNESS MATRIX BELOW THE SKYLINE
C
            DO 10 I=1,NEQ1
      10    NDS(I)=0
            NDS(1)=1
            NDS(2)=2
            MBAND=0
            IF(NEQ.EQ.1) GO TO 30
            DO 20 I=2,NEQ
            IF(CHT(I).GT.MBAND) MBAND=CHT(I)
      20    NDS(I+1)=NDS(I)+CHT(I)+1
      30    MBAND=MBAND+1
            NSKY=NDS(NEQ1)-1
            RETURN
            END


            SUBROUTINE PASSEM (SK,EK,NDS,ND,NED,NEQ1,NSKY,NUED)
            IMPLICIT REAL*8(A-H,O-Z)
            DIMENSION SK(NSKY),NDS(NEQ1),ND(NED),EK(NUED,NUED)
C     ASSEMBLE ELEMENT STIFFNESS INTO COMPACTED GLOBAL STIFFNESS
            DO 70 I=1,NED
            II=ND(I)
            IF(II)70,70,30
      30    CONTINUE
            DO 60 J=1,NED
            JJ=ND(J)
            IF(JJ)60,60,40
      40    CONTINUE
            MI=NDS(JJ)
            IJ=JJ-II
            IF(IJ)60,50,50
      50    KK=MI+IJ
            SK(KK)=SK(KK)+EK(I,J)
      60    CONTINUE
      70    CONTINUE
            RETURN
            END
```

```
              SUBROUTINE PASOLV(SK,P,NDS,NN,NEQ1,NSKY,KKK)
              IMPLICIT REAL*8(A-H,O-Z)
              DIMENSION SK(NSKY),P(NN),NDS(NEQ1)
              IF(KKK-2) 40,150,150
      40      DO 140 N=1,NN
              KN=NDS(N)
              KL=KN+1
              KU=NDS(N+1)-1
              KH=KU-KL
              IF(KH) 110,90,50
      50      K=N-KH
              IC=0
              KLT=KU
              DO 80 J=1,KH
              IC=IC+1
              KLT=KLT-1
              KI=NDS(K)
              ND=NDS(K+1)-KI-1
              IF(ND) 80,80,60
      60      KK=MINO(IC,ND)
              C=0.0
              DO 70 L=1,KK
      70      C=C+SK(KI+L)*SK(KLT+L)
              SK(KLT)=SK(KLT)-C
      80      K=K+1
      90      K=N
              B=0.0
              DO 100 KK=KL,KU
              K=K-1
              KI=NDS(K)
              C=SK(KK)/SK(KI)
              B=B+C*SK(KK)
      100     SK(KK)=C
              SK(KN)=SK(KN)-B
      110     IF(SK(KN)) 120,120,140
      120     WRITE(6,222)N,SK(KN)
              STOP
      140     CONTINUE
              RETURN
C
C       REDUCE RIGHT HAND SIDE LOAD VECTOR
C
      150     DO 180 N=1,NN
              KL=NDS(N)+1
              KU=NDS(N+1)-1
              IF(KU-KL) 180,160,160
      160     K=N
              C=0.0
              DO 170 KK=KL,KU
              K=K-1
      170     C=C+SK(KK)*P(K)
              P(N)=P(N)-C
      180     CONTINUE
C
C       BACK SUBSTITUTION
C
              DO 200 N=1,NN
```

```
         K=NDS(N)
200      P(N)=P(N)/SK(K)
         IF(NN.EQ.1) RETURN
         N=NN
         DO 230 L=2,NN
         KL=NDS(N)+1
         KU=NDS(N+1)-1
         IF(KU-KL) 230,210,210
210      K=N
         DO 220 KK=KL,KU
         K=K-1
220      P(K)=P(K)-SK(KK)*P(N)
230      N=N-1
         RETURN
222      FORMAT(//20X,'STOP-STIFFNESS MATRIX NOT POSITIVE
        *DEFINITE','NON POSITIVE PIVOT FOR EQUATION',I4,//
        *10X,'PIVOT=',E20.12)
         END
         SUBROUTINE PASLOD(P,NDF,NC,NSN,NEQ)
         IMPLICIT REAL*8(A-H,O-Z)
         COMMON/TAPES/ISTRES,NDARAY,IPR
         DIMENSION P(NEQ),NDF(6,NSN)
         DIMENSION CNL(6)
         IF(IPR.EQ.0) WRITE(6,30)
         DO 20 J=1,NC
         READ(5,11) NODE,(CNL(I),I=1,6)
         IF (IPR.EQ.0) WRITE(6,40) NODE,(CNL(I),I=1,6)
         DO 20 I=1,6
         II=NDF(I,NODE)
         IF(II)20,20,10
10       P(II)=P(II)+CNL(I)
20       CONTINUE
         RETURN
11       FORMAT(I10,6F10.4)
30       FORMAT(10X,'NODE',10X,'X-AXIS',10X,'Y-AXIS',10X,
        *'Z-AXIS',10X,'X-AXIS',10X,'Y-AXIS',10X,'Z-AXIS'/
        *25X,'FORCE',10X,'FORCE',10X,'FORCE',10X,'MOMENT',
        *10X,'MOMENT',10X,'MOMENT'/)
40       FORMAT(10X,I4,6F16.3)
         END


         SUBROUTINE DISP(D,NDF,NSN,NEQ)
         IMPLICIT REAL*8(A-H,O-Z)
         DIMENSION D(NEQ),NDF(6,NSN),DISPV(6)
         DO 30 J=1,NSN
         DO 10 I=1,6
10       DISPV(I)=0.0
         DO 20 I=1,6
         KK=NDF(I,J)
20       IF(KK.NE.0) DISPV(I)=D(KK)
30       WRITE(6,22) J,(DISPV(I),I=1,6)
22       FORMAT(5X,I3,4X,E19.7,2E19.7,1X,E20.6,2E18.6/)
         RETURN
         END
```

```
C
C       INTEFRACE ROUTINE FROM DEC 'TO PC FOR GETTING TIME
C
        FUNCTION SECNDS(X)
C
C       RETURN CURRENT TIME - MIDNIGHT - X
C
C
        REAL*4    SECNDS, X
C
C
        INTEGER*2 IHOUR, IMINUT, ISECON, IHUND
        REAL*4    RHOUR, RMINUT, RSECON, RHUND
        REAL*4    X1
C
        CALL GETTIM(IHOUR, IMINUT, ISECON, IHUND)
        RHOUR    = FLOAT( IHOUR )
        RMINUT   = FLOAT( IMINUT)
        RSECON   = FLOAT( ISECON)
        RHUND    = FLOAT( IHUND )
        X1       = RHOUR*3600.0 + RMINUT*60.0 + RSECON +
     &             RHUND/100.0
        SECNDS   = X1 - X
C
        RETURN
        END
```

```
C***********************************************************************
C          CONTROL ROUTINE FOR THE TRUSS ELEMENTS
C***********************************************************************
          SUBROUTINE THREDT(A,LLIB,NSHAPE,MTOT)
          IMPLICIT REAL*8(A-H,O-Z)
          COMMON/DIM/N1,N2,N3,N4,N5,N6,N7,N8,N9,N10,N11,N12,N13,N14
          COMMON/PAR/IND,NET,NSN,NMP,NEQ,NSKY,NEQ1,LCOUNT
          COMMON/TAPES/ISTRES,NDARAY,IPR
          COMMON/PRECI/ITWO
          COMMON/MULT/ELMN
          DIMENSION A(MTOT)
          DIMENSION LLIB(7,3)
          DO 100 J=1,NSHAPE
          LB=LLIB(1,J)
          GO TO (20,40,60),LB
20        NED=6
          IF(IND.NE.1) GO TO 30
          N10=N9+NED
          N11=N10+1*ITWO
          N12=N11+1*ITWO
          N13=N12+1*ITWO
          NSKY=1
30        CALL TRUS1(A(N2),A(N3),A(N4),A(N5),A(N6),A(N7),A(N11),A(
     *    N12),A(N1),A(N8),A(N9),A(N10),A(N13),NED)
          GO TO 100
40        CALL TRUS2
          GO TO 100
60        CALL TRUS3
100       CONTINUE
          RETURN
          END



          SUBROUTINE TRUS2
          RETURN
          END



          SUBROUTINE TRUS3
          RETURN
          END



          SUBROUTINE TRUS1(X,Y,Z,E,PR,WD,SK,PQ,NDF,CHT,ND,NDS,P,NED)
          IMPLICIT REAL*8(A-H,O-Z)
          COMMON/DIM/N1,N2,N3,N4,N5,N6,N7,N8,N9,N10,N11,N12,N13,N14
          COMMON/PAR/IND,NET,NSN,NMP,NEQ,NSKY,NEQ1,LCOUNT
          COMMON/TAPES/ISTRES,NDARAY,IPR
          COMMON/MULT/ELMN
          DIMENSION X(NSN),Y(NSN),Z(NSN),E(NMP),PR(NMP),WD(NMP)
          DIMENSION SK(NSKY),P(NEQ),PQ(NEQ),NDF(6,NSN),NDS(NEQ1),ND(NED)
          INTEGER CHT(NEQ)
          DIMENSION Q(6),XL(2),YL(2),ZL(2),LNC(2),MNC(2),EK(6,6)
          DIMENSION NDEL(20),AREA(100),SWF(3)
C         SQRT(X)=DSQRT(X)
```

```
            GO TO (310,350,400),IND
310         CONTINUE
            READ(5,11)NEL,NGP
            IF(NGP.GT.100)WRITE(6,99)
            IF(NGP.GT.100) STOP
            DO 1 I=1,NGP
1           READ(5,95) J,AREA(J)
            LCOUNT=LCOUNT+1
            NDEL(LCOUNT)=NEL
            WRITE(6,22)NEL
            WRITE(6,96)
            WRITE(6,97)
            WRITE(6,98)(J,AREA(J),J=1,NGP)
            IF(IPR.EQ.0)WRITE(6,32)
            IF(IPR.EQ.0)WRITE(6,33)
            KK=0
            DO 340 I=1,NEL
            READ(5,44)LNUM,MG,(MNC(J),J=1,2),NG,KI
            IF(KI.GT.0) GO TO 110
            DO 120 J=1,2
120         LNC(J)=MNC(J)
110         KK=KK+1
            IF(LNUM-KK) 2,2,3
3           CONTINUE
            DO 5 J=1,2
5           LNC(J)=LNC(J)+KI
2           JJ=0
            DO 330 J=1,2
            K = LNC(J)
            XL(J)=X(K)
            YL(J)=Y(K)
            ZL(J)=Z(K)
            DO 320 L=1,3
            JJ=JJ+1
320         ND(JJ)=NDF(L,K)
330         CONTINUE
            AX=AREA(NG)
            WRITE(NDARAY)(ND(L),L=1,6),(XL(N),YL(N),ZL(N),N=1,2),AX,MG
            CALL COLUMH(CHT,ND,NED,NEQ)
            IF(IPR.EQ.0)WRITE(6,55)KK,MG,(LNC(J),J=1,2),NG
            KK=KK+1
            IF(LNUM.GE.KK)GO TO 3
            KK=KK-1
            IF(LNUM.EQ.NEL)GO TO 341
340         CONTINUE
  341       CONTINUE
            READ(5,66)(SWF(I),I=1,3)
            WRITE(6,89)
            WRITE(6,91)(SWF(I),I=1,3)
            RETURN
C           ********************************************************
350         CONTINUE
            LCOUNT=LCOUNT+1
            NEL=NDEL(LCOUNT)
            DO 200 LNUM=1,NEL
C           INITIALIZE THE ELEMENT STIFFNESS MATRIX
            DO 10 I=1,6
            DO 10 J=1,6
```

```
10        EK(I,J)=0.0
          READ(NDARAY)(ND(L),L=1,6),(XL(N),YL(N),ZL(N),N=1,2),AX,MG
          CX=XL(2)-XL(1)
          CY=YL(2)-YL(1)
          CZ=ZL(2)-ZL(1)
          AL=DSQRT(CX*CX+CY*CY+CZ*CZ)
          CX=CX/AL
          CY=CY/AL
          CZ=CZ/AL
          FQ=AX*E(MG)/AL
          WRITE(ISTRES)CX,CY,CZ,AL
          EK(1,1)=CX*CX*FQ
          EK(1,2)=CX*CY*FQ
          EK(1,3)=CX*CZ*FQ
          EK(2,2)=CY*CY*FQ
          EK(2,3)=CY*CZ*FQ
          EK(3,3)=CZ*CZ*FQ
          EK(2,1)=EK(1,2)
          EK(3,1)=EK(1,3)
          EK(3,2)=EK(2,3)
          DO 30 I=1,3
          DO 30 J=1,3
          EK(I+3,J+3)=EK(I,J)
          EK(I+3,J)=-EK(I,J)
30        EK(I,J+3)=-EK(I,J)
          NUED=NED
          CALL PASSEM(SK,EK,NDS,ND,NED,NEQ1,NSKY,NUED)
C
C         CALCULATION OF LOAD VECTOR'PQ' DUE TO SELF WEIGHT
C
          SWT=0.5*AX*AL*WD(MG)
          JJ=0
          DO 250 J=1,2
          DO 250 K=1,3
          JJ=JJ+1
          M=ND(JJ)
          IF(M.EQ.0)GO TO 250
          PQ(M)=PQ(M)+SWT*SWF(K)
250       CONTINUE
  200     CONTINUE
          RETURN
C****************************************************************
400       CONTINUE
          WRITE(6,77)
          LCOUNT=LCOUNT+1
          NEL=NDEL(LCOUNT)
          DO 410 LNUM=1,NEL
          READ(NDARAY)(ND(L),L=1,6),(XL(N),YL(N),ZL(N),N=1,2),AX,MG
          READ(ISTRES)CX,CY,CZ,AL
          DO 90 I=1,6
90        Q(I)=0.0
          SIGMA=0.0
          DO 80 I=1,6
          K=ND(I)
          IF(K.EQ.0)GO TO 80
          Q(I)=P(K)
80        CONTINUE
          AA=Q(4)-Q(1)
```

```
            BB=Q(5)-Q(2)
            CC=Q(6)-Q(3)
            SIGMA=E(MG)*(CX*AA+CY*BB+CZ*CC)/AL
            FORCE=SIGMA*AX
            WRITE(6,88)LNUM,SIGMA,FORCE
   410      CONTINUE
    11      FORMAT(2I5)
    22      FORMAT(/10X,'NUMBER OF TRUSS ELEMENTS IN THE STRUCTURE',I5/)
    32      FORMAT(/10X,'ELEMENT CONNECTIVITY DATA')
    33      FORMAT(//20X,'ELEMENT NO.',10X,'MATERIAL GROUP',10X,'NODE-I',
        *   10X,'NODE-J',5X,'GEOM.PROP.GROUP'//)
    44      FORMAT(6I5)
    55      FORMAT(25X,I4,18X,I4,12X,I4,12X,I4,14X,I4/)
    66      FORMAT(3F5.2)
    77      FORMAT(//25X,'****AXIAL FORCE/STRESS IN TRUSS ELEMENTS****
        *   '//30X,'ELEMENT NO.',8X,'STRESS',7X,'FORCE'//)
    88      FORMAT(33X,I4,5X,F13.3,2X,F13.3/)
    89      FORMAT(/5X,'SELF WEIGHT FACTORS'/)
    91      FORMAT(/5X,'X-DIRN',F6.2/5X,'Y-DIRN',F6.2/5X,'Z-DIRN',F6.2/)
    95      FORMAT(I5,F10.3)
    96      FORMAT(//5X,'GEOMETRIC PROPERTY TABLE'/)
    97      FORMAT(5X,'GROUP NO.',5X,'AREA OF C/S'/)
    98      FORMAT(5X,I5,10X,F8.3)
    99      FORMAT(//5X,'NO.OF GEOM.PROP.SETS EXCEEDED 100'//5X,'CHANGE THE N
       *UMBERS IN THE CORRESPONDING DIMENSION STATEMENTS'/5X,
       *'IN THE SUBROUTINE TRUS1'//)
            RETURN
            END
```

```
C*******************************************************************
C              CONTROL ROUTINE FOR THE BEAM ELEMENTS
C*******************************************************************
        SUBROUTINE THREDB(A,LLIB,NSHAPE,MTOT)
        IMPLICIT REAL*8(A-H,O-Z)
        COMMON/DIM/N1,N2,N3,N4,N5,N6,N7,N8,N9,N10,
     *  N11,N12,N13,N14
        COMMON/PAR/IND,NET,NSN,NMP,NEQ,NSKY,NEQ1,LCOUNT
        COMMON/TAPES/ISTRES,NDARAY,IPR
        COMMON/PRECI/ITWO
        COMMON/MULT/ELMN
        DIMENSION A(MTOT)
        DIMENSION LLIB(7,3)
        DO 100 J = 1,NSHAPE
        LB=LLIB(2,J)
        GO TO (20,40,60),LB
20      NED=12
        IF(IND.NE.1) GO TO 30
        N10=N9+NED
        N11=N10+1*ITWO
        N12=N11+1*ITWO
        N13=N12+I*ITWO
        NSKY=1
30      CALL BEAM1(A(N2),A(N3),A(N4),A(N5),A(N6),A(N7),A(N11),
     *  A(N12),A(N1),A(N8),A(N9),A(N10),A(N13),NED)
        GO TO 100
40      NED=12
        IF(IND.NE.1) GO TO 50
        N10=N9+NED
        N11=N10+1*ITWO
        N12=N11+1*ITWO
        N13=N12+1*ITWO
        NSKY=1
50      CALL BEAM2(A(N2),A(N3),A(N4),A(N5),A(N6),A(N7),A(N11),
     *  A(N12),A(N1),A(N8),A(N9),A(N10),A(N13),NED)
        GO TO 100
60      CALL BEAM3
100     CONTINUE
        RETURN
        END
        SUBROUTINE BEAM1(X,Y,Z,E,PR,WD,SK,PQ,NDF,CHT,ND,
     *  NDS,P,NED)
        IMPLICIT REAL*8(A-H,O-Z)
        COMMON/PAR/IND,NET,NSN,NMP,NEQ,NSKY,NEQ1,LCOUNT
        COMMON/TAPES/ISTRES,NDARAY,IPR
        COMMON/PRECI/ITWO
        COMMON/MULT/ELMN
        DIMENSION X(NSN),Y(NSN),Z(NSN),E(NMP),PR(NMP),WD(NMP)
C
C       THE MAXIMUM NO. OF GEOMETRIC PROPERTY SETS IS LIMITED
C     * TO 10 AND LOADSETS TO 20. IF THE PROBLEM HAS MORE
C     * SETS CHANGE THE DIMENSION IN THE STATEMENTS GIVEN
C     * BELOW CORRESPONDINGLY
C
        DIMENSION BX(10),BIX(10),BIY(10),BIZ(10),B(12,12)
```

```
          DIMENSION ALIY(20),ADIY(20),ALIZ(20),ADIZ(20),ALJY(20)
          DIMENSION ALJZ(20),ALAX(20)
          DIMENSION LSN(100)
          INTEGER CHT(NEQ)
          DIMENSION Q(12),XL(3),YL(3),ZL(3),LNC(3),MNC(3)
          DIMENSION EK(12,12),NDEL(40),T(12,12)
          DIMENSION FIX(12)
C
          DIMENSION SWF(3)
C         'PQ' IS THE LOAD VECTOR DUE TO ELEMENT FORCES ONLY
C         'P' IS THE FINAL LOAD VECTOR INCLUDING NODAL LOADS
C
          DIMENSION SK(NSKY),P(NEQ),NDF(6,NSN),NDS(NEQ1),ND(NED)
          DIMENSION PQ(NEQ)
C         SQRT(X)=DSQRT(X)
          GO TO (310,350,400),IND
   310    CONTINUE
          READ(5,11) NEL,NGP,NLS
          LCOUNT=LCOUNT+1
          NDEL(LCOUNT)=NEL
          WRITE(6,22)NEL
          DO 1 J =1,NGP
     1    READ(5,55)I,BX(I),BIX(I),BIY(I),BIZ(I)
          WRITE(6,77)
          WRITE(6,88)(I,BX(I),BIX(I),BIY(I),BIZ(I),I=1,NGP)
C         THE FOLLOWING OPTIONS ARE KEPT FOR VARIOUS TYPES OF
C     *   LOADS
C         ALI=CONC.LOAD ON THE BEAM
C         ADI=DISTANCE OF THE LOAD FROM THE I NODE
C         ALJ=UDL ON THE BEAM
C         SUFFICES Y AND Z ARE RESPECTIVELY FOR FORCE IN
C     *   Y AND Z PLANES
          IF(NLS.EQ.0) GO TO 7
          DO 4 J=1,NLS
     4    READ(5,56) I,ALIY(I),ADIY(I),ALIZ(I),ADIZ(I),ALJY(I),
     *    ALJZ(I),ALAX(I)
          WRITE(6,89)
          WRITE(6,91)(I,ALIY(I),ADIY(I),ALIZ(I),ADIZ(I),ALJY(I),
     *    ALJZ(I),ALAX(I),I=1,NLS)
     7    CONTINUE
          IF(IPR.EQ.0)WRITE(6,32)
          IF(IPR.EQ.0)WRITE(6,33)
          KK=0
          DO 340 I=1,NEL
          READ(5,44)LNUM,MG,(MNC(J),J=1,3),NG,LSN(LNUM),KI
          IF(KI.GT.0) GO TO 110
          DO 120 J=1,3
   120    LNC(J)=MNC(J)
          MM=LNC(3)
          XL(3)=X(MM)
          YL(3)=Y(MM)
          ZL(3)=Z(MM)
   110    KK=KK+1
          IF(LNUM-KK)2,2,3
     3    CONTINUE
          DO 5 J=1,2
     5    LNC(J)=LNC(J)+KI
     2    JJ=0
```

```
          DO 330 J=1,2
          K=LNC(J)
          XL(J)=X(K)
          YL(J)=Y(K)
          ZL(J)=Z(K)
          AX=BX(NG)
          AIX=BIX(NG)
          AIY=BIY(NG)
          AIZ=BIZ(NG)
          DO 320 L=1,6
          JJ=JJ+1
   320    ND(JJ)=NDF(L,K)
   330    CONTINUE
          WRITE(NDARAY)(ND(LL),LL=1,12),(XL(NN),YL(NN),
      *   ZL(NN),LNC(NN),NN=1,3),MG,AX,AIX,AIY,AIZ
          CALL COLUMH(CHT,ND,NED,NEQ)
          LSN(KK)=LSN(LNUM)
          IF(IPR.EQ.0) WRITE(6,66)KK,MG,(LNC(J),J=1,3),
      *   NG,LSN(KK)
          KK=KK+1
          IF(LNUM.GE.KK) GO TO 3
          KK=KK-1
          IF(LNUM.GE.NEL) GO TO 341
   340    CONTINUE
   341    READ(5,92)(SWF(I),I=1,3)
69        FORMAT(10X,5I5)
          WRITE(6,94)
          WRITE(6,93)(SWF(I),I=1,3)
          RETURN
   350    CONTINUE
C         INITIALIZE ROTATION TRANSFORMATION AND STIFFNESS
C     *   MATRIX AND LOAD VECTOR
C         ENTER LOOP OVER EACH ELEMENT
          LCOUNT=LCOUNT+1
          NEL=NDEL(LCOUNT)
          DO 200 LNUM=1,NEL
          READ(NDARAY)(ND(LL),LL=1,12),(XL(NN),YL(NN),ZL(NN),LNC(NN),
      *   NN=1,3),MG,AX,AIX,AIY,AIZ
          DO 10 J=1,12
          DO 10 I=1,12
          T(I,J)=0.0
          EK(I,J)=0.0
   10     Q(J)=0.0
C
          CX=XL(2)-XL(1)
          CY=YL(2)-YL(1)
          CZ=ZL(2)-ZL(1)
          AL=DSQRT(CX*CX+CY*CY+CZ*CZ)
          XP=XL(3)-XL(1)
          YP=YL(3)-YL(1)
          ZP=ZL(3)-ZL(1)
C
C         SUBROUTINE'TRANS'COMPUTES ROTATION TRANSFORMATION MATRIX
C
          CALL TRANS(T,CX,CY,CZ,XP,YP,ZP,AL)
C
C         SUBROUTINE'STIFF'COMPUTES THE ELEMENT STIFFNESS MATRIX
C         IN LOCAL COORDINATE SYSTEM
```

```
                F = E(MG)
                AMU = PR(MG)
                CALL STIFF(EK,F,AMU,AL,AX,AIX,AIY,AIZ)
C
C               TRANSFORM STIFFNESS MATRIX IN GLOBAL CO-ORDINATES
C               STORE THE K*T PRODUCT ON 'ISTRES'
                DO 80 I=1,12
                DO 80 J=1,12
                B(I,J)=0.0
                DO 80 K=1,12
     80         B(I,J)=B(I,J)+EK(I,K)*T(K,J)
                WRITE(ISTRES)((B(I,J),I=1,12),J=1,12)
                DO 90 I=1,12
                DO 90 J=1,12
                EK(I,J)=0.0
                DO 90 K=1,12
     90         EK(I,J)=EK(I,J)+T(K,I)*B(K,J)
                NUED=NED
C               ASSEMBLE ELEMENT STIFFNESS INTO COMPACTED GLOBAL
C       *       STIFFNESS
                CALL PASSEM(SK,EK,NDS,ND,NED,NEQ1,NSKY,NUED)
C
C               CALCULATION OF THE LOAD VECTOR DUE TO ELEMENT FORCES
C               CALCULATION OF LOAD VECTOR DUE TO SELF WEIGHT
                CALL SELFWT(Q,NED,AL,AX,MG,WD,SWF,T,NMP)
C
C               CALCULATION OF LOAD VECTOR DUE TO APPLIED ELEMENT LOADS
                L=LSN(LNUM)
                IF(L.EQ.0) GO TO 9
                CALL FIXED(Q,NED,AL,LNUM,LSN,ALIY,ADIY,ALIZ,ADIZ,ALJY,
     *          ALJZ,ALAX)
     9          CONTINUE
C               TRANSFORM THE LOAD VECTOR IN THE GLOBAL COORDINATE SYSTEM
C               ADD THE LOAD VECTOR TO THE 'PQ' VECTOR
                WRITE(ISTRES)(Q(I),I=1,12)
                DO 160 K=1,12
                KK=ND(K)
                IF(KK.EQ.0) GO TO 160
                DO 180 L=1,12
    180         PQ(KK)=PQ(KK)+T(L,K)*Q(L)
    160         CONTINUE
    200         CONTINUE
                RETURN
    400         CONTINUE
                WRITE(6,98)
                WRITE(6,99)
                LCOUNT=LCOUNT+1
                NEL=NDEL(LCOUNT)
                DO 150 LNUM=1,NEL
                DO 130 I=1,12
    130         Q(I)=0.0
                READ(NDARAY)(ND(LL),LL=1,12),(XL(NN),YL(NN),ZL(NN),
     *          LNC(NN),N=1,3),MG,AX,AIX,AIY,AIZ
                READ(ISTRES)((B(I,J),I=1,12),J=1,12)
                READ(ISTRES)(FIX(I),I=1,12)
                DO 140 M=1,12
                K=ND(M)
                IF(K.EQ.0) GO TO 140
```

```
              DO 141 L=1,12
      141     Q(L)=Q(L)+B(L,M)*P(K)
      140     CONTINUE
C
              DO 170 I=1,12
              FIX(I)=FIX(I)*ELMN
      170     Q(I)=Q(I)-FIX(I)
              WRITE(6,132)LNUM,(Q(I),I=1,6)
              WRITE(6,154)(Q(I),I=7,12)
      150     CONTINUE
       11     FORMAT(3I5)
       22     FORMAT(/25X,'NUMBER OF BEAM ELEMENTS.....',I5)
       32     FORMAT(/10X,'ELEMENT CONNECTIVITY DATA'/)
       33     FORMAT(10X,'ELEMENT NO',8X,'MATERIAL GROUP',8X,
              *  'NODE-I',5X,'NODE-J',5X,'NODE-K',8X,'GEO-PROP.GROUP',
              *  8X,'LOAD GROUP'//)
       44     FORMAT(8I5)
       55     FORMAT(I10,4F10.3)
       56     FORMAT(I10,7F10.3)
       66     FORMAT(12X,I4,13X,I4,14X,I4,7X,I4,7X,I4,13X,I4,17X,I4)
       77     FORMAT(/25X,'GEOMETRIC PROPERTY TABLE'//20X,'GROUP NO',
              *  5X,'AREA OF C/S',8X,'TORSIONAL M.I.',8X,'INERTIA',10X,
              *  'INERTIA'/38X,'A',19X,'J',17X,'IYY',14X,'IZZ'/)
       88     FORMAT(20X,I4,4F19.3)
       89     FORMAT(/25X,'ELEMENT LOAD TABLE'//5X,'GROUP',5X,
              *  'CONC.LOAD',10X,'DISTANCE',10X,'CONC.LOAD',10X,
              *  'DISTANCE',10X,'UDL IN',10X,'UDL IN',10X,'AXIAL'/
              *  6X,'NO',6X,'INY-PLANE',9X,'FROM NODE-I',8X,'INZ-PLANE',9X,
              *  'FROM NODE-I',9X,'Y-PLANE',9X,'Z-PLANE',11X,'FORCE'/)
       91     FORMAT(4X,I4,8X,F10.3,8X,F10.3,8X,F10.3,8X,F10.3,6X,
              *  F10.3,6X,F10.3,6X,F10.3)
       92     FORMAT(3F5.2)
       93     FORMAT(5X,'X-DIRN',F6.2/5X,'Y-DIRN',F6.2/5X,'Z-DIRN',
              *  F6.2/)
       94     FORMAT(//5X,'SELF WEIGHT FACTORS'/)
       98     FORMAT(//50X,'****BEAM FORCES AND MOMENTS****'//)
       99     FORMAT(10X,'ELEMENT',5X,'NODE',10X,'AXIAL',10X,'SHEAR',
              *  10X,'SHEAR',10X,'TWISTING',10X,'BENDING',10X,'BENDING'/14X,
              *  'NO',20X,'FORCE',9X,'FORCE-Y',8X,'FORCE-Z',10X,'MOMENT',10X,
              *  'MOMENT-Y',10X,'MOMENT-Z'//)
      132     FORMAT(10X,I5,9X,'I',5X,6(3X,E13.6))
      154     FORMAT(24X,'J',5X,6(3X,E13.6)/)
              RETURN
              END
C*********************************************************************
C        ROUTINE COMPUTES THE ELEMENT LOAD VECTOR DUE TO SELFWEIGHT
C        CALLED BY : BEAM1 AND BEAM2
C*********************************************************************
              SUBROUTINE SELFWT(Q,NED,AL,AX,MG,WD,SWF,T,NMP)
              IMPLICIT REAL*8(A-H,O-Z)
              DIMENSION Q(NED),WD(NMP),SWF(3),T(12,12),GQ(12)
              SWT=AL*AX*WD(MG)
              DO 5 J=1,12
        5     GQ(J)=0.0
              DO 10 I=1,3
              GQ(I)=0.5*SWT*SWF(I)
       10     GQ(I+6)=GQ(I)
              DO 15 I=1,12
```

```
        DO 15 J=1,12
  15    Q(I)=Q(I)+T(I,J)*GQ(J)
        Q(6)=Q(2)*AL/6.0
        Q(12)=-Q(6)
        RETURN
        END


C***********************************************************************
C       ROUTINE COMPUTES THE ELEMENT LOAD VECTOR
C       DUE TO ELEMENT LOADS
C       CALLED BY : BEAM1 AND BEAM2
C***********************************************************************
        SUBROUTINE FIXED(Q,NED,AL,LNUM,LSN,ALIY,ADIY,ALIZ,
     *  ADIZ,ALJY,ALJZ,ALAX)
        IMPLICIT REAL*8(A-H,O-Z)
        REAL*8 LIY,LIZ,LJY,LJZ,LAX
        DIMENSION Q(NED)
        DIMENSION LSN(100)
        DIMENSION ALIY(20),ADIY(20),ALIZ(20),ADIZ(20),ALJY(20)
        DIMENSION ALAX(20),ALJZ(20)
C
        L=LSN(LNUM)
        LIY=ALIY(L)
        DIY=ADIY(L)
        LIZ=ALIZ(L)
        DIZ=ADIZ(L)
        LJY=ALJY(L)
        LJZ=ALJZ(L)
        LAX=ALAX(L)
        IF(LIY.EQ.0) GO TO 20
        CIY=AL-DIY
        AA=LIY*CIY*CIY*(AL+2*DIY)/AL**3
        BB=LIY-AA
        CC=LIY*DIY*CIY*CIY/AL**2
        DD=-LIY*DIY*DIY*CIY/AL**2
        Q(2)=Q(2)+AA
        Q(6)=Q(6)+CC
        Q(8)=Q(8)+BB
        Q(12)=Q(12)+DD
  20    IF(LIZ.EQ.0) GO TO 30
        CIZ=AL-DIZ
        AA=LIZ*CIZ*CIZ*(3*DIZ+CIZ)/AL**3
        BB=LIZ-AA
        CC=-LIZ*DIZ*CIZ*CIZ/AL**2
        DD=LIZ*DIZ*DIZ*CIZ/AL**2
        Q(3)=Q(3)+AA
        Q(9)=Q(9)+BB
        Q(5)=Q(5)+CC
        Q(11)=Q(11)+DD
  30    IF(LJY.EQ.0) GO TO 40
        AA=LJY*AL*0.5
        BB=LJY*AL*AL/12.
        Q(2)=Q(2)+AA
        Q(8)=Q(8)+AA
        Q(6)=Q(6)+BB
        Q(12)=Q(12)-BB
  40    IF(LJZ.EQ.0)GO TO 50
```

```
                    AA=LJZ*AL*0.5
                    BB=LJZ*AL*AL/12.0
                    Q(3)=Q(3)+AA
                    Q(9)=Q(9)+AA
                    Q(5)=Q(5)-BB
                    Q(11)=Q(11)+BB
        50          Q(1)=Q(1)+LAX*0.5
                    Q(7)=Q(7)+LAX*0.5
                    RETURN
                    END


                    SUBROUTINE BEAM3
                    RETURN
                    END


                    SUBROUTINE BEAM2(X,Y,Z,E,PR,WD,SK,PQ,NDF,CHT,ND,NDS,P,NED)
                    IMPLICIT REAL*8(A-H,O-Z)
                    COMMON/DIM/N1,N2,N3,N4,N5,N6,N7,N8,N9,N10,N11,N12,N13,N14
                    COMMON/PAR/IND,NET,NSN,NMP,NEQ,NSKY,NEQ1,LCOUNT
                    COMMON/TAPES/ISTRES,NDARAY,IPR
                    COMMON/PRECI/ITWO
                    COMMON/MULT/ELMN
                    DIMENSION X(NSN),Y(NSN),Z(NSN),E(NMP),PR(NMP),WD(NMP)
                    DIMENSION BX(100),BIX(100),BIY(100),BIZ(100),B(12,12)
                    DIMENSION ALIY(20),ADIY(20),ALIZ(20),ADIZ(20),ALJY(20)
                    DIMENSION ALJZ(20),ALAX(20),LSN(100),NDEL(40),SWF(3)
                    DIMENSION FIX(12),Q(12),XL(3),YL(3),ZL(3),LNC(3),MNC(3)
                    DIMENSION EK(12,12),QS(12),QF(12)
                    DIMENSION T(12,12),TP(12,12),TI(12,12),D(12,12)
                    INTEGER CHT(NEQ)
                    DIMENSION SK(NSKY),PQ(NEQ),P(NEQ),NDF(6,NSN),NDS(NEQ1),ND(NED)
C
C                   'PQ' IS THE LOAD VECTOR DUE TO ELEMENT FORCES ONLY
C                   'P' IS THE LOAD VECTOR INCLUDING NODAL LOADS
C
C                   SQRT(X)=DSQRT(X)
                    GO TO(300,350,400),IND
        300         CONTINUE
                    READ(5,11) NEL,NGP,NLS
                    LCOUNT=LCOUNT+1
                    NDEL(LCOUNT)=NEL
                    WRITE(6,22) NEL
                    DO 1 J=1,NGP
        1           READ(5,55) I,BX(I),BIX(I),BIY(I),BIZ(I)
                    WRITE(6,77)
                    WRITE(6,88) (I,BX(I),BIX(I),BIY(I),BIZ(I),I=1,NGP)
                    IF(NLS.EQ.0) GO TO 7
                    DO 4 J=1,NLS
        4           READ(5,56) I,ALIY(I),ADIY(I),ALIZ(I),ADIZ(I),ALJY(I),
        *           ALJZ(I),ALAX(I)
                    WRITE(6,89)
                    WRITE(6,91) (I,ALIY(I),ADIY(I),ALIZ(I),ADIZ(I),ALJY(I),
        *           ALJZ(I),ALAX(I),I=1,NLS)
        7           CONTINUE
                    IF(IPR.EQ.0) WRITE(6,33)
                    KK = 0
```

```
              DO 340 I=1,NEL
              READ(5,44)LNUM,MG,(MNC(J),J=1,3),NG,LSN(LNUM).KI,EY,EZ
              IF(KI.GT.0) GO TO 120
              DO 110 J=1,3
    110       LNC(J) = MNC(J)
              MM = LNC(3)
              XL(3) = X(MM)
              YL(3) = Y(MM)
              ZL(3) = Z(MM)
    120       KK = KK+1
              IF(LNUM-KK) 2,2,3
      3       CONTINUE
              DO 5 J=1,2
      5       LNC(J) = LNC(J) + KI
      2       JJ = 0
              DO 330 J=1,2
              K = LNC(J)
              XL(J) = X(K)
              YL(J) = Y(K)
              ZL(J) = Z(K)
              AX = BX(NG)
              AIX = BIX(NG)
              AIY = BIY(NG)
              AIZ = BIZ(NG)
              DO 320 L=1,6
              JJ = JJ+1
    320       ND(JJ) = NDF(L,K)
    330       CONTINUE
              WRITE(NDARAY)(ND(LL),LL=1,12),(XL(NN),YL(NN),ZL(NN),LNC(NN),
        *     NN=1,3),MG,AX,AIX,AIY,AIZ,EY,EZ
              CALL COLUMH(CHT,ND,NED,NEQ)
              LSN(KK) = LSN(LNUM)
              IF(IPR.EQ.0)WRITE(6,66)KK,MG,(LNC(J),J=1,3),NG,LSN(KK),
        *     EY,EZ
              KK = KK+1
              IF(LNUM.GE.KK) GO TO 3
              KK = KK-1
              IF(LNUM.GE.NEL) GO TO 341
    340       CONTINUE
    341       CONTINUE
              READ(5,92) (SWF(I),I=1,3)
              WRITE(6,94)
              WRITE(6,93) (SWF(I),I=1,3)
              RETURN
    350       CONTINUE
              LCOUNT = LCOUNT+1
              NEL = NDEL(LCOUNT)
C
C             ENTER LOOP OVER EACH ELEMENT
C
              DO 200 LNUM=1,NEL
              READ(NDARAY)(ND(LL),LL=1,12),(XL(NN),YL(NN),ZL(NN),LNC(NN),
        *     NN=1,3),MG,AX,AIX,AIY,AIZ,EY,EZ
C
C             INITIALIZE STIFFNESS MATRIX,LOAD VECTOR
C             AND TRANSFORMATION MATRICES
C
              DO 10 J=1,12
```

```
        DO 10 I=1,12
        EK(I,J) = 0.0
        T(I,J) = 0.0
        TP(I,J) = 0.0
        TI(I,J) = 0.0
        IF(I.EQ.J) TP(I,J) = 1.0
        IF(I.EQ.J) TI(I,J) = 1.0
        FIX(J) = 0.0
        Q(J) = 0.0
        QF(J) = 0.0
   10   QS(J) = 0.0
        CX = XL(2)-XL(1)
        CY = YL(2)-YL(1)
        CZ = ZL(2)-ZL(1)
        AL = DSQRT(CX*CX+CY*CY+CZ*CZ)
        XP = XL(3)-XL(1)
        YP = YL(3)-YL(1)
        ZP = ZL(3)-ZL(1)
C
C       SUBROUTINE 'TRANS' COMPUTES ROTATION TRANSFORMATION MATRIX
C
        CALL TRANS(T,CX,CY,CZ,XP,YP,ZP,AL)
C
C       COMPUTE PARALLEL AXES TRANSFORMATION MATRICES
C       TP = PARALLEL AXES TRANSFORMATION MATRIX
C       TI = INVERSE OF TRANSPOSE OF TP MATRIX
C
        TP(1,5) = EZ
        TP(1,6) = -EY
        TP(2,4) = -EZ
        TP(3,4) = EY
        TP(7,11) = EZ
        TP(7,12) = -EY
        TP(8,10) = -EZ
        TP(9,10) = EY
C
        TI(4,2) = EZ
        TI(4,3) = -EY
        TI(5,1) = -EZ
        TI(6,1) = EY
        TI(10,8) = EZ
        TI(10,9) = -EY
        TI(11,7) = -EZ
        TI(12,7) = EY
C
C       COMPUTE ELEMENT STIFFNESS MATRIX IN LOCAL COORDINATE SYSTEM
C       BY CALLING THE SUBROUTINE 'STIFF'
        F = E(MG)
        AMU = PR(MG)
        CALL STIFF(EK,F,AMU,AL,AX,AIX,AIY,AIZ)
C
C       TRANSFORM THE STIFFNESS MATRIX TO THE ELEMENT NODE
C       AND THEN TRANSFORM IN GLOBAL COORDINATES
        DO 20 I=1,12
        DO 20 J=1,12
        D(I,J) = 0.0
        DO 20 K=1,12
   20   D(I,J) = D(I,J) + EK(I,K)*TP(K,J)
```

```
          DO 21 I=1,12
          DO 21 J=1,12
          EK(I,J)=0.0
          DO 21 K=1,12
    21    EK(I,J) = EK(I,J) + TP(K,I)*D(K,J)
C         STORE EK*TP*T PRODUCT ON 'ISTRES'
          DO 30 I=1,12
          DO 30 J=1,12
          B(I,J) = 0.0
          DO 30 K=1,12
    30    B(I,J) = B(I,J) + D(I,K)*T(K,J)
          WRITE(ISTRES)((B(I,J),I=1,12),J=1,12)
          DO 40 I=1,12
          DO 40 J=1,12
          D(I,J) = 0.0
          DO 40 K=1,12
    40    D(I,J) = D(I,J) + TP(K,I)*B(K,J)
          DO 50 I=1,12
          DO 50 J=1,12
          EK(I,J) = 0.0
          DO 50 K=1,12
    50    EK(I,J) = EK(I,J) + T(K,I)*D(K,J)
          NUED = NED
C
C         ASSEMBLE ELEMENT STIFFNESS INTO COMPACTED GLOBAL STIFFNESS
C
          CALL PASSEM(SK,EK,NDS,ND,NED,NEQ1,NSKY,NUED)
C         CALCULATION OF LOAD VECTORS DUE TO ELEMENT LOADS
C         CALCULATION OF LOAD VECTOR DUE TO SELF WEIGHT
          IF(WD(MG).EQ.0) GO TO 65
          CALL SELFWT(QS,NED,AL,AX,MG,WD,SWF,T,NMP)
C
    65    CONTINUE
C         CALCULATION OF LOAD VECTOR DUE TO APPLIED ELEMENT LOADS
C
          L = LSN(LNUM)
          IF(L.EQ.0) GO TO 75
          CALL FIXED(QF,NED,AL,LNUM,LSN,ALIY,ADIY,ALIZ,ADIZ,
        * ALJY,ALJZ,ALAX)
    75    CONTINUE
          DO 80 I=1,12
    80    FIX(I) = QS(I) + QF(I)
          WRITE(ISTRES)(FIX(I),I=1,12)
C         TRANSFORM VECTOR'FIX' TO THE ELEMENT NODE
          DO 60 I=1,12
          DO 60 J=1,12
    60    Q(I) = Q(I) + TP(J,I)*FIX(J)
C         TRANSFORM LOAD VECTOR 'Q' IN GLOBAL COORDINATE SYSTEM
C         THEN ADD IT TO THE LOAD VECTOR'PQ'
C
          DO 100 K=1,12
          KK = ND(K)
          IF(KK.EQ.0) GO TO 100
          DO 90 L=1,12
    90    PQ(KK) = PQ(KK) + T(L,K)*Q(L)
   100    CONTINUE
   200    CONTINUE
          RETURN
```

```
      400     CONTINUE
              WRITE(6,98)
              WRITE(6,99)
              LCOUNT = LCOUNT+1
              NEL = NDEL(LCOUNT)
              DO 150 LNUM=1,NEL
              DO 130 I=1,12
      130     Q(I) = 0.0
              READ(NDARAY)(ND(LL),LL=1,12),(XL(NN),YL(NN),ZL(NN),LNC(NN),
         *    NN=1,3),MG,AX,AIX,AIY,AIZ,EY,EZ
              READ(ISTRES)((B(I,J),I=1,12),J=1,12)
              READ(ISTRES)(FIX(I),I=1,12)
              DO 140 M=1,12
              K = ND(M)
              IF(K.EQ.0) GO TO 140
              DO 141 L=1,12
      141     Q(L) = Q(L)+B(L,M)*P(K)
      140     CONTINUE
C             ADD THE MEMBER END FORCES DUE TO FULLY RESTRAINED
C             CONDITION
              DO 170 I=1,12
              FIX(I) = FIX(I)*ELMN
      170     Q(I) = Q(I)-FIX(I)
              WRITE(6,132) LNUM,(Q(I),I=1,6)
              WRITE(6,154)(Q(I),I=7,12)
      150     CONTINUE
      11      FORMAT(3I5)
      22      FORMAT(/25X,'NUMBER OF OFFSET BEAM ELEMENTS.....',I5)
      33      FORMAT(/50X,'ELEMENT CONNECTIVITY DATA'//10X,'ELEMENT',
         *    5X,'MATERIAL',5X,'NODE-I',5X,'NODE-J',5X,'NODE-K',5X,
         *    'GEO.PROP.',5X,'LOAD',5X,'EY',15X,'EZ'/13X,
         *    'NO',7X,'GROUP',42X,'GROUP',8X,'GROUP'//)
      44      FORMAT(8I5,2F10.3)
      55      FORMAT(I10,4F10.3)
      56      FORMAT(I10,7F10.3)
      66      FORMAT(11X,I4,7X,I4,9X,I4,7X,I4,7X,I4,6X,I4,9X,I4,
         *    3X,F10.4,7X,F10.4)
      77      FORMAT(/25X,'GEOMETRIC PROPERTY TABLE'//20X,'GROUP NO',
         *    5X,'AREA OF C/S',8X,'TORSIONAL M.I.',8X,'INERTIA',10X,
         *    'INERTIA'/38X,'A',19X,'J',17X,'IYY',14X,'IZZ'/)
      88      FORMAT(20X,I4,4F19.3)
      89      FORMAT(/25X,'ELMENT LOAD TABLE'//5X,'GROUP',5X,
         *    'CONC.LOAD',10X,'DISTANCE',10X,'CONC.LOAD',10X,'DISTANCE',
         *    10X,'UDL IN',10X,'UDL IN',10X,'AXIAL'/6X,'NO',6X,
         *    'INY-PLANE',9X,'FROM NODE-I',8X,'INZ-PLANE',9X,'FROM NODE-I',
         *    9X,'Y-PLANE',9X,'Z-PLANE',11X,'FORCE'/)
      91      FORMAT(4X,I4,8X,F10.3,8X,F10.3,8X,F10.3,8X,F10.3,6X,
         * /  F10.3,6X,F10.3,6X,F10.3)
      92      FORMAT(3F5.2)
      93      FORMAT(5X,'X-DIRN',F6.2/5X,'Y-DIRN',F6.2/5X,'Z-DIRN',F6.2/)
      94      FORMAT(//5X,'SELF WEIGHT FACTORS'/)
      98      FORMAT(//50X,'****BEAM FORCES AND MOMENTS****'//)
      99      FORMAT(10X,'ELEMENT',5X,'NODE',10X,'AXIAL',10X,'SHEAR',
         *    10X,'SHEAR',10X,'TWISTING',10X,'BENDING',10X,'BENDING'/14X,
         *    'NO',20X,'FORCE',9X,'FORCE-Y',8X,'FORCE-Z',10X,'MOMENT',
         *    10X,'MOMENT-Y',10X,'MOMENT-Z'//)
```

```
 132    FORMAT(10X,I5,9X,'I',5X,6(3X,E13.6)/)
 154    FORMAT(24X,'J',5X,6(3X,E13.6)/)
        RETURN
        END


C**********************************************************************
C       ROUTINE COMPUTES THE ROTATION TRANSFORMATION MATRIX
C       CALLED BY : BEAM1 AND BEAM2
C**********************************************************************
        SUBROUTINE TRANS(T,CX,CY,CZ,XP,YP,ZP,AL)
        IMPLICIT REAL*8(A-H,O-Z)
        DIMENSION T(12,12)
        SQRT(X) = DSQRT(X)
        CX = CX/AL
        CY = CY/AL
        CZ = CZ/AL
        IF(CX.EQ.0.0.AND.CZ.EQ.0.0) GO TO 20
        DEN = DSQRT(CX*CX+CZ*CZ)
        YPY = -1.0*CX*CY*XP/DEN+DEN*YP-CY*CZ*ZP/DEN
        ZPY = CX*ZP/DEN-CZ*XP/DEN
        DIV = DSQRT(YPY*YPY+ZPY*ZPY)
        SI = ZPY/DIV
        CO = YPY/DIV
        T(1,1) = CX
        T(1,2) = CY
        T(1,3) = CZ
        T(2,1) = (-CX*CY*CO-CZ*SI)/DEN
        T(2,2) = DEN*CO
        T(2,3) = (CX*SI-CY*CZ*CO)/DEN
        T(3,1) = (CX*CY*SI-CZ*CO)/DEN
        T(3,2) = -DEN*SI
        T(3,3) = (CY*CZ*SI+CX*CO)/DEN
        GO TO 30
  20    DIV = DSQRT(XP*XP+ZP*ZP)
        SI = ZP/DIV
        CO = -XP*CY/DIV
        T(1,2) = CY
        T(2,1) = -CY*CO
        T(2,3) = SI
        T(3,1) = CY*SI
        T(3,3) = CO
  30    CONTINUE
C       COMPLETE THE TRANSFORMATION MATRIX
        DO 40 I=1,9
        K = I+2
        DO 40 J=I,K
        IF(J.GT.9) GO TO 40
        T(I+3,J+3) = T(I,J)
        T(J+3,I+3) = T(J,I)
  40    CONTINUE
        RETURN
        END
```

```
C***********************************************************************
C          ROUTINE COMPUTES THE ELEMENT STIFFNESS MATRIX IN
C          MEMBER COORDINATE SYSTEM
C          CALLED BY : BEAM1 AND BEAM2
C***********************************************************************
           SUBROUTINE STIFF(EK,F,AMU,AL,AX,X,Y,Z)
           IMPLICIT REAL*8(A-H,O-Z)
           DIMENSION EK(12,12)
           G = 0.5*F/(1.0+AMU)
           EK(1,1) = F*AX/AL
           EK(2,2) = 12.0*F*Z/AL**3
           EK(3,3) = 12.0*F*Y/AL**3
           EK(4,4) = G*X/AL
           EK(5,5) = 4.0*F*Y/AL
           EK(6,6) = 4.0*F*Z/AL
           DO 10 I=1,6
           EK(I+6,I+6) = EK(I,I)
   10      EK(I,I+6) = -EK(I,I)
           EK(5,11) = -0.5*EK(5,11)
           EK(6,12) = -0.5*EK(6,12)
           EK(2,6) = 6.0*F*Z/AL**2
           EK(3,5) = -6.0*F*Y/AL**2
           EK(2,12) = EK(2,6)
           EK(3,11) = EK(3,5)
           EK(5,9) = -EK(3,5)
           EK(9,11) = -EK(3,5)
           EK(8,12) = -EK(2,6)
           EK(6,8) = -EK(2,6)
           DO 20 J=1,9
           M = J+1
           DO 20 I = M,12
   20      EK(I,J) = EK(J,I)
           RETURN
           END
```

```
C*********************************************************************
C          CONTROL ROUTINE FOR THE BOUNDARY ELEMENT
C*********************************************************************
       SUBROUTINE BOUND(A,LLIB,NSHAPE,MTOT)
       IMPLICIT REAL*8(A-H,O-Z)
       COMMON/DIM/N1,N2,N3,N4,N5,N6,N7,N8,N9,N10,N11,N12,
     *  N13,N14
       COMMON/PAR/IND,NET,NSN,NMP,NEQ,NSKY,NEQ1,LCOUNT
       COMMON/TAPES/ISTRES,NDARAY,IPR
       COMMON/PRECI/ITWO
       DIMENSION A(MTOT)
       DIMENSION LLIB(7,3)
       DO 100 J=1,NSHAPE
       LB=LLIB(7,J)
       GO TO (20,40,60),LB
  20   NED=6
       IF(IND.NE.1) GO TO 30
       N10=N9+NED
       N11=N10+1*ITWO
       N12=N11+1*ITWO
       N13=N12+1*ITWO
       NSKY=1
  30   CALL BOUND1 (A(N2),A(N3),A(N4),A(N5),A(N6),A(N7),
     *  A(N11),A(N12),A(N1),A(N8),A(N9),A(N10),A(N13),NED)
       GO TO 100
  40   CALL BOUND2
       GO TO 100
  60   CALL BOUND3
 100   CONTINUE
       RETURN
       END


       SUBROUTINE BOUND2
       RETURN
       END



       SUBROUTINE BOUND3
       RETURN
       END


       SUBROUTINE BOUND1(X,Y,Z,E,PR,WD,SK,PQ,NDF,CHT,
     *  ND,NDS,P,NED)
       IMPLICIT REAL*8(A-H,O-Z)
       COMMON/DIM/N1,N2,N3,N4,N5,N6,N7,N8,N9,N10,N11,N12,
     *  N13,N14
       COMMON/PAR/IND,NET,NSN,NMP,NEQ,NSKY,NEQ1,LCOUNT
       COMMON/TAPES/ISTRES,NDARAY,IPR
       DIMENSION X(NSN),Y(NSN),Z(NSN),E(NMP),PR(NMP),WD(NMP)
       DIMENSION SK(NSKY),P(NEQ),NDF(6,NSN),NDS(NEQ),ND(NED)
       DIMENSION PQ(NEQ)
       INTEGER CHT(NEQ)
       DIMENSION EK(6,6),LNC(2),MNC(2),NDEL(20)
```

```
          DIMENSION Q(6),XL(2),YL(2),ZL(2)
C         SQRT(X)=DSQRT(X)
          MJ=1
          MI=2
          GO TO (310,350,400),IND
   310    CONTINUE
          READ(5,11) NEL
          LCOUNT=LCOUNT+1
          NDEL(LCOUNT)=NEL
          WRITE(6,22)NEL
          IF(IPR.EQ.0)WRITE(6,33)
          KK=0
          DO 340 I=1,NEL
          READ(5,44)LNUM,(MNC(J),J=1,2),KD,KR,KI,SD,SR,STIF
          IF(STIF.EQ.0)STIF=1.0E+10
          IF(KI.GT.0)GO TO 110
          DO 120 J=1,2
   120    LNC(J)=MNC(J)
   110    KK=KK+1
          IF(LNUM-KK)2,2,3
     3    CONTINUE
          DO 5 J=1,2
     5    LNC(J)=LNC(J)+KI
     2    DO 330 J=1,2
          K=LNC(J)
          XL(J)=X(K)
          YL(J)=Y(K)
          ZL(J)=Z(K)
   330    CONTINUE
          K=LNC(1)
          DO 320 LL=1,6
   320    ND(LL)=NDF(LL,K)
          WRITE(MI)(ND(LL),LL=1,6),(XL(NN),YL(NN),ZL(NN),
        * NN=1,2),KD,KR,SD,SR,STIF
          CALL COLUMH(CHT,ND,NED,NEQ)
          IF(IPR.EQ.0)WRITE(6,55)KK,(LNC(J),J=1,2),KD,KR
          KK=KK+1
          IF(LNUM.GT.KK) GO TO 3
          KK=KK-1
          IF(LNUM.EQ.NEL) GO TO 341
   340    CONTINUE
   341    CONTINUE
          RETURN
   350    CONTINUE
          LCOUNT=LCOUNT+1
          NEL=NDEL(LCOUNT)
          DO 200 LNUM=1,NEL
C         INITIALIZE VARIOUS MATRICES
          DO 10 I=1,6
          DO 10 J=1,6
          EK(I,J)=0.0
    10    Q(I)=0.0
          READ(MI)(ND(LL),LL=1,6),(XL(NN),YL(NN),ZL(NN),
        * NN=1,2),KD,KR,SD,SR,STIF
          CX=XL(2)-XL(1)
          CY=YL(2)-YL(1)
          CZ=ZL(2)-ZL(1)
          AL=DSQRT(CX*CX+CY*CY+CZ*CZ)
```

```
          CX=CX/AL
          CY=CY/AL
          CZ=CZ/AL
          IF(KD.EQ.0) GO TO 150
          EK(1,1)=CX*CX*STIF
          EK(1,2)=CX*CY*STIF
          EK(1,3)=CX*CZ*STIF
          EK(2,2)=CY*CY*STIF
          EK(2,3)=CY*CZ*STIF
          EK(3,3)=CZ*CZ*STIF
          PP=SD*STIF
          Q(1)=CX*PP
          Q(2)=CY*PP
          Q(3)=CZ*PP
150       CONTINUE
          IF(KR.EQ.0) GO TO 160
          EK(4,4)=CX*CX*STIF
          EK(4,5)=CX*CY*STIF
          EK(4,6)=CX*CZ*STIF
          EK(5,5)=CY*CY*STIF
          EK(5,6)=CY*CZ*STIF
          EK(6,6)=CZ*CZ*STIF
          PP=SR*STIF
          Q(4)=CX*PP
          Q(5)=CY*PP
          Q(6)=CZ*PP
160       CONTINUE
          DO 170 K=2,6
          KK=K-1
          DO 170 J=1,KK
170       EK(K,J)=EK(J,K)
          DO 180 K=1,6
          KK=ND(K)
          IF(KK.EQ.0) GO TO 180
          PQ(KK)=PQ(KK)+Q(K)
180       CONTINUE
          NUED=NED
          CALL PASSEM(SK,EK,NDS,ND,NED,NEQ1,NSKY,NUED)
200       CONTINUE
          RETURN
400       CONTINUE
          WRITE(6,77)
          LCOUNT=LCOUNT+1
          NEL=NDEL(LCOUNT)
          DO 500 LNUM=1,NEL
          READ(MI)(ND(LL),LL=1,6),(XL(NN),YL(NN),ZL(NN),
     *    NN=1,2),KD,KR,SD,SR,STIF
          DO 410 K=1,6
          Q(K)=0.0
          KK=ND(K)
          IF(KK.EQ.0)GO TO 410
          Q(K)=P(KK)
410       CONTINUE
          FORCE=0.0
          TORQU=0.0
          CX=XL(2)-XL(1)
          CY=YL(2)-YL(1)
          CZ=ZL(2)-ZL(1)
```

```
          AL=DSQRT(CX*CX+CY*CY+CZ*CZ)
          CX=CX/AL
          CY=CY/AL
          CZ=CZ/AL
          FORCE=(Q(1)*CX+Q(2)*CY+Q(3)*CZ)*STIF
          TORQU=(Q(4)*CX+Q(5)*CY+Q(6)*CZ)*STIF
          IF(KD.EQ.0) FORCE=0.0
          IF(KR.EQ.0) TORQU=0.0
          WRITE(6,88)LNUM,FORCE,TORQU
  500     CONTINUE
   11     FORMAT(I5)
   22     FORMAT(20X,'NUMBER OF BOUNDARY ELEMENTS IN THE
      *   STRUCTURE..',I4//)
   33     FORMAT(10X,'ELEMENT NO.',10X,'NODE-I',10X,'NODE-J',
      *   10X,'KODE-D',10X,'KODE-R'//)
   44     FORMAT(6I5,3F10.3)
   55     FORMAT(15X,I4,12X,I4,12X,I4,12X,I4,12X,I4/)
   77     FORMAT(20X,'FORCE IN THE BOUNDARY ELEMENTS'////
      *   20X,'ELEMENT NO.',15X,'FORCE',15X,'TORQUE'//)
   88     FORMAT(25X,I4,10X,F10.3,9X,F10.3/)
          RETURN
          END
```

```
C********************************************************************
C       CONTROL ROUTINE FOR PLANE STRESS, PLANE STRAIN AND
C       AXI SYMMETRIC SOLID ELEMENTS
C********************************************************************
        SUBROUTINE PLANE(A, LLIB, NSHAPE, MTOT)
        IMPLICIT REAL*8(A-H, O-Z)
        COMMON/DIM/N1, N2, N3, N4, N5, N6, N7, N8, N9, N10, N11, N12, N13, N14
        COMMON/PAR/IND, NET, NSN, NMP, NEQ, NSKY, NEQ1, LCOUNT
        COMMON/TAPES/ISTRES, NDARAY, IPR
        COMMON/PRECI/ITWO
        INTEGER CHT
        DIMENSION A(MTOT)
        DIMENSION LLIB(7,3)
        IF(IND.EQ.1) READ(5,1000) NOPT
        DO 100 J=1, NSHAPE
        LB=LLIB(3, J)
        CO TO (60,40,20), LB
20      NED=6
        IF(IND.NE.1)GO TO 30
        N10=N9+NED
        N11=N10+1*ITWO
        N12=N11+1*ITWO
        N13=N12+1*ITWO
        NSKY=1
30      CALL CST(A(N2), A(N3), A(N4), A(N5), A(N6), A(N7), A(N11), A(N12),
     *  A(N1), A(N8), A(N9), A(N10), A(N13), NED, NOPT)
        GO TO 100
40      CALL RECT
        GO TO 100
60      NED=8
        IF(IND.NE.1) GO TO 70
        N10=N9+NED
        N11=N10+1*ITWO
        N12=N11+1*ITWO
        N13=N12+1*ITWO
        NSKY=1
70      CALL QUAD(A(N2), A(N3), A(N4), A(N5), A(N6), A(N7), A(N11), A(N12),
     *  A(N1), A(N8), A(N9), A(N10), A(N13), NED, NOPT)
100     CONTINUE
1000    FORMAT(I5)
        RETURN
        END

        SUBROUTINE RECT
        RETURN
        END


        SUBROUTINE QUAD(X, Y, Z, E, PR, WD, SK, PQ, NDF, CHT, ND, NDS, P, NED, NOPT)
        IMPLICIT REAL*8(A-H, O-Z)
        INTEGER CHT
        COMMON/DIM/N1, N2, N3, N4, N5, N6, N7, N8, N9, N10, N11, N12, N13, N14
        COMMON/PAR/IND, NET, NSN, NMP, NEQ, NSKY, NEQ1, LCOUNT
        COMMON/TAPES/ISTRES, NDARAY, IPR
```

```
        COMMON/JACI/A1,B1,C1,D1,A2,B2,C2,D2
        COMMON/ELEM/NDEL(100)
        DIMENSION X(NSN),Y(NSN),Z(NSN),E(NMP),PR(NMP),WD(NMP),SK(NSKY)
        DIMENSION PQ(NEQ),P(NEQ),NDF(6,NSN),ND(NED),NDS(NEQ1)
        DIMENSION EK(8,8),QK(12,12),XL(4),YL(4),LNC(4),MNC(4),NRAO(4)
        DIMENSION B(4,8),QB(3,12),CB(4,8),BTC(3,12),GP(2),WG(2),Q(8)
        DIMENSION SIG(6),PX1(4),PX2(4),PY1(4),PY2(4),XJAC(2,2),C(4,4)
        DIMENSION LLNUM(50),LNLF(50),LNRAO(50,4)
        DIMENSION PPX1(50,4),PPX2(50,4),PPY1(50,4),PPY2(50,4)
        DIMENSION BBB(12,12),CHT(NEQ)
C
C
C       THE MAXIMUM NUMBER OF LOADED ELEMENTS IS LIMITED TO 50
C       IF THE PROBLEM CONTAINS MORE,CHANGE THE DIMENSIONS IN THE
C       ABOVE STATEMENTS CORRESPONDINGLY
        DIMENSION ELC(100),STR(100,6),TMS(4,4),SWF(3)


C       SQRT(X)=DSQRT(X)
        DATA WG/1.D0,1.D0/
        DATA GP/-0.5773502691896,0.5773502691896/
        GO TO (100,300,700),IND
100     CONTINUE
        NRA=3
        NDEL(1)=0
        NDEL(2)=0
        READ(5,1000)NEL,NLE,KOPT
        IF(NLE.GT.50)WRITE(6,1170)
        IF(NSN.GT.100)WRITE(6,1180)
        IF(NLE.GT.50.OR.NSN.GT.100) STOP
        IF(KOPT.EQ.3)NRA=4
        LCOUNT=LCOUNT+1
        NDEL(LCOUNT)=NEL
        DO 90 I=1,NSN
90      ELC(I)=0.0
        WRITE(6,1010)NEL
        WRITE(6,1020)
        KK=0
        DO 180 I=1,NEL
        READ(5,1040)LNUM,MG,(MNC(J),J=1,4),TH,IOPT,KI
        IF(KI.GT.0)GO TO 120
        DO 110 J=1,4
110     LNC(J)=MNC(J)
120     KK=KK+1
        IF(LNUM-KK)150,150,130
130     CONTINUE
        DO 140 J=1,4
140     LNC(J)=LNC(J)+KI
150     JJ=0
        DO 170 J=1,4
        K=LNC(J)
        XL(J)=X(K)
        YL(J)=Y(K)
        ELC(K)=ELC(K)+1.0
        DO 160 L=1,2
        JJ=JJ+1
        ND(JJ)=NDF(L,K)
160     CONTINUE
```

```
      170    CONTINUE
             WRITE(NDARAY)(ND(L),L=1,8),MG,TH,(XL(N),YL(N),LNC(N),N=1,4),
          *  IOPT
             CALL COLUMH(CHT,ND,NED,NEQ)
             IF(IPR.EQ.0)WRITE(6,1030)KK,MG,(LNC(J),J=1,4),TH,IOPT
             KK=KK+1
             IF(LNUM.GE.KK)GO TO 130
             KK=KK-1
             IF(LNUM.EQ.NEL)GO TO 190
      180    CONTINUE
      190    CONTINUE
C
             READ THE SELF WEIGHT FACTORS
             READ(5,1150)(SWF(I),I=1,3)
C
C            READ THE ELEMENT LOAD DATA
             IF(NLE.EQ.0)GO TO 2
             DO 1 M=1,NLE
             READ(5,1050)LLNUM(M),LNLF(M),(LNRAO(M,I),I=1,4)
             NLF=LNLF(M)
             DO 1 I=1,NLF
             J=LNRAO(M,I)
             READ(5,1060)PPX1(M,J),PPX2(M,J),PPY1(M,J),PPY2(M,J)
        1    CONTINUE
        2    CONTINUE
             WRITE(6,1160)(SWF(I),I=1,3)
             RETURN
      300    CONTINUE
             PI=3.14285714
C            INITIALIZE THE CONSTITUTIVE MATRIX
             DO 310 I=1,NRA
             DO 310 J=1,NRA
      310    C(I,J)=0.0
C            ENTER LOOP OVER EACH ELEMENT
             LCOUNT=LCOUNT+1
             NEL=NDEL(LCOUNT)
             DO 570 LNUM=1,NEL
             READ(NDARAY)(ND(L),L=1,8),MG,TH,(XL(N),YL(N),LNC(N),N=1,4),
          *  IOPT
             IF(NMP.EQ.1.AND.LNUM.GT.1)GO TO 350
             IF(KOPT.EQ.3)GO TO 340
             IF(NOPT.EQ.2)GO TO 320
C
C            CALCULATION OF CONSTITUTIVE MATRIX FOR PLANE STRAIN ANALYSIS
             CF=E(MG)/((1.0+PR(MG))*(1.0-2.0*PR(MG)))
             C(1,1)=CF*(1.0-PR(MG))
             C(1,2)=CF*PR(MG)
             C(2,1)=C(1,2)
             C(2,2)=C(1,1)
             C(3,3)=CF*(1.0-2.0*PR(MG))/2.0
             GO TO 350
C
C            CALCULATION OF CONSTITUTIVE MATRIX FOR PLANE STRESS ANALYSIS
      320    CF=E(MG)/(1.0-PR(MG)*PR(MG))
             C(1,1)=CF
             C(1,2)=CF*PR(MG)
             C(2,1)=C(1,2)
             C(2,2)=C(1,1)
             C(3,3)=CF*(1.0-PR(MG))/2.0
```

```
            GO TO 350
C
C           CALCULATION OF CONSTITUTIVE MATRIX FOR AXI-SYMMETRIC
C           SOLID ANALYSIS
  340       ER=E(MG)/((1.0+PR(MG))*(1.0-2.0*PR(MG)))
            C(1,1)=ER*(1.0-PR(MG))
            C(1,2)=ER*PR(MG)
            C(1,3)=C(1,2)
            C(2,1)=C(1,2)
            C(2,2)=C(1,1)
            C(2,3)=C(2,1)
            C(3,1)=C(2,1)
            C(3,2)=C(2,1)
            C(3,3)=C(2,2)
            C(4,4)=ER*(1.0-2.0*PR(MG))/2.0
C           INITIALIZE STIFFNESS MATRIX,ELEMENT LOAD VECTOR,B MATRIX
  350       DO 360 I=1,8
            Q(I)=0.0
            DO 360 J=1,8
  360       EK(I,J)=0.0
            DO 370 I=1,NRA
            DO 370 J=1,8
  370       B(I,J)=0.0
C
C           COMPUTE MATRIX-B AT CENTROID OF THE ELEMENT
            CALL BMATRX(0.0D0,0.0D0,XL(1),XL(2),XL(3),XL(4),YL(1),YL(2),
     *      YL(3),YL(4),R,DETJ,B,KOPT)
C
C           COMPUTE MATRIX-C*B AT CENTRIOD OF THE ELEMENT
            DO 400 I=1,NRA
            DO 400 J=1,8
            CB(I,J)=0.0
            DO 400 K=1,NRA
  400       CB(I,J)=CB(I,J)+C(I,K)*B(K,J)
            WRITE(ISTRES)((CB(I,J),I=1,NRA),J=1,8)
C
            XBAR=WD(MG)*SWF(1)*TH
            YBAR=WD(MG)*SWF(2)*TH
C
C           FOR IOPT=0,INCOMPATIBLE MODES ARE SUPPRESSED
C           FOR IOPT=1,INCOMPATIBLE MODES ARE INCLUDED
            IF(IOPT.EQ.1) GO TO 460
C           THE DO LOOPS(450) COMPUTE THE MATRIX TRIPLE PRODUCT BT*C*B
C           AT THE FOUR GAUSS POINTS AND ADD THEM TO GET STIFFNESS MATRIX
            DO 450 IX=1,2
            DO 450 IY=1,2
            XJAC(IX,IY)=0.0
            IF(KOPT.EQ.3)GO TO 410
            CALL BMATRX(GP(IX),GP(IY),XL(1),XL(2),XL(3),XL(4),YL(1),YL(2),
     *      YL(3),YL(4),R,DETJ,B,KOPT)
            DA=DETJ*WG(IX)*WG(IY)
            GO TO 420
  410       CALL BMATRX(GP(IX),GP(IY),XL(1),XL(2),XL(3),XL(4),YL(1),YL(2),
     *      YL(3),YL(4),R,DETJ,B,KOPT)
            DA=DETJ*WG(IX)*WG(IY)*2.0*PI*R
C           COMPUTE CB-MATRIX AND WRITE IT ON 'ISTRES'FILE
  420       DO 430 I=1,NRA
            DO 430 J=1,8
```

```
            CB(I,J)=0.0
            DO 430 K=1,NRA
     430    CB(I,J)=CB(I,J)+C(I,K)*B(K,J)
            WRITE(ISTRES)((CB(I,J),I=1,NRA),J=1,8)
C           CALCULATE THE ELEMENT STIFFNESS MATRIX
            DO 440 I=1,8
            DO 440 J=1,8
            DO 440 K=1,NRA
     440    EK(I,J)=EK(I,J)+B(K,I)*CB(K,J)*DA*TH
C           SUBROUTINE 'BODYF' COMPUTES THE ELEMENT LOAD VECTOR
C           DUE TO SELF WEIGHT
            CALL BODYF(GP(IX),GP(IY),XBAR,YBAR,DA,Q)
     450    CONTINUE
            NUED=NED
            CALL PASSEM(SK,EK,NDS,ND,NED,NEQ1,NSKY,NUED)
            GO TO 562
     460    CONTINUE
C
C           COMPUTE JACOBIAN AND INVERSE JACOBIAN MATRIX
C           AT THE CENTROID OF THE ELEMENT
            A1=-YL(1)-YL(2)+YL(3)+YL(4)
            B1=YL(1)-YL(2)-YL(3)+YL(4)
            C1=XL(1)+XL(2)-XL(3)-XL(4)
            D1=-XL(1)+XL(2)+XL(3)-XL(4)
            DETJ1=(A1*D1-B1*C1)/16.0
            TEMP=DETJ1*4.0
            A1=A1/TEMP
            B1=B1/TEMP
            C1=C1/TEMP
            D1=D1/TEMP
C           A1 TO D1 ARE ELEMENTS OF INVERSE JACOBIAN
C           INITIALIZE THE ELEMENT STIFFNESS MATRIX 'QK'
            DO 470 I=1,12
            DO 470 J=1,12
     470    QK(I,J)=0.0
C
C           THE DO LOOPS(520)COMPUTE BT*C*B AT THE FOUR GAUSS POINTS
C           AND ADD THEM TO GET THE STIFFNESS MATRIX 'QK'
            NN=0
            DO 520 II=1,2
            DO 520 JJ=1,2
            S=GP(II)
            T=GP(JJ)
C           COMPUTE JACOBIAN AND INVERSE JACOBIAN AT THE GAUSS POINTS
            AA=0.25*(1.0-T)
            BB=0.25*(1.0+T)
            CC=0.25*(1.0-S)
            DD=0.25*(1.0+S)
C           COMPUTE INVERSE JACOBIAN(A2 TO D2)
            A2=AA*(XL(2)-XL(1))+BB*(XL(3)-XL(4))
            B2=AA*(YL(2)-YL(1))+BB*(YL(3)-YL(4))
            C2=CC*(XL(4)-XL(1))+DD*(XL(3)-XL(2))
            D2=CC*(YL(4)-YL(1))+DD*(YL(3)-YL(2))
            DETJ2=A2*D2-B2*C2
            TEMP=A2
            A2=D2/DETJ2
            B2=-B2/DETJ2
            C2=-C2/DETJ2
```

```
           D2=TEMP/DETJ2
           CALL QBMAT(S,T,QB)
           NN=NN+1
           K3=3*NN
           K2=K3-1
           K1=K2-1
           DO 480 MM=1,12
           BBB(K1,MM)=QB(1,MM)
           BBB(K2,MM)=QB(2,MM)
    480    BBB(K3,MM)=QB(3,MM)
           DO 500 KK=1,3
           DO 500 LL=1,12
           BTC(KK,LL)=0.0
           DO 500 MM=1,3
    500    BTC(KK,LL)=BTC(KK,LL)+C(KK,MM)*QB(MM,LL)
           FQ1=TH*DETJ1*WG(II)*WG(JJ)
           FQ2=TH*DETJ2*WG(II)*WG(JJ)
           DO 510 KK=1,12
           DO 510 LL=1,12
           FQ=FQ2
           IF(LL.GT.8)FQ=FQ1
           IF(KK.GT.8)FQ=FQ1
           DO 510 MM=1,3
    510    QK(KK,LL)=QK(KK,LL)+QB(MM,KK)*BTC(MM,LL)*FQ
C          SUBROUTINE 'BODYF' COMPUTES THE LOAD VECTOR
C          DUE TO SELF WEIGHT
           DA=DETJ2*WG(II)*WG(JJ)
           CALL BODYF(S,T,XBAR,YBAR,DA,Q)
    520    CONTINUE
C          CONDENSE THE STIFFNESS MATRIX FROM 12*12 TO 8*8
C          CONDENSE THE 'B' MATRIX FROM 3*12 TO 3*8
           DO 550 M=1,4
           MN=13-M
           MO=MN-1
C          CONDENSE THE STIFFNESS MATRIX
           DO 530 I=1,MO
    530    QK(MN,I)=QK(I,MN)/QK(MN,MN)
           DO 540 K=1,MO
           SP=QK(MN,K)
           DO 540 J=1,MO
    540    QK(J,K)=QK(J,K)-SP*QK(J,MN)
C          CONDENSE THE 'B' MATRIX
           DO 550 J=1,12
           SP=BBB(J,MN)
           IF(SP.EQ.0)GO TO 550
           DO 545 K=1,MO
           SPQ2=SP*QK(MN,K)
           IF(DABS(SPQ2).LE.1.0D-25)SPQ2=0.0
    545    BBB(J,K)=BBB(J,K)-SPQ2
    550    CONTINUE
           NN=0
           DO 557 M=1,4
           NN=NN+1
           K3=3*NN
           K2=K3-1
           K1=K2-1
           DO 555 N=1,8
           B(1,N)=BBB(K1,N)
```

```
            B(2,N)=BBB(K2,N)
    555     B(3,N)=BBB(K3,N)
            DO 556 KK=1,3
            DO 556 LL=1,8
            CB(KK,LL)=0.0
            DO 556 MM=1,3
    556     CB(KK,LL)=CB(KK,LL)+C(KK,MM)*B(MM,LL)
            WRITE(ISTRES)((CB(I,J),I=1,3),J=1,8)
    557     CONTINUE
            DO 560 I=1,8
            DO 560 J=1,8
    560     EK(I,J)=QK(I,J)
            NUED=NED
            CALL PASSEM(SK,EK,NDS,ND,NED,NEQ1,NSKY,NUED)
    562     CONTINUE
C           ADD THE LOAD VECTOR DUE TO SELF WEIGHT TO GLOBAL LOAD VECTOR
            DO 565 J=1,8
            KK=ND(J)
            IF(KK.EQ.0)GO TO 565
            PQ(KK)=PQ(KK)+Q(J)
    565     CONTINUE
    570     CONTINUE
C           CALCULATE THE LOAD VECTOR FOR THE ELEMENT AND
C           ADD TO THE GLOBAL LOAD VECTOR
            IF(NLE.EQ.0)GO TO 660
            REWIND NDARAY
            LN=0
            DO 640 M=1,NLE
            LNUM=LLNUM(M)
            NLF=LNLF(M)
            DO 3 I=1,4
    3       NRAO(I)=LNRAO(M,I)
            DO 580 I=1,NLF
            J=NRAO(I)
            PX1(J)=PPX1(M,J)
            PX2(J)=PPX2(M,J)
            PY1(J)=PPY1(M,J)
    580     PY2(J)=PPY2(M,J)
            NS1=LNUM-LN-1
            IF(NS1.EQ.0) GO TO 600
            DO 590 J=1,NS1
    590     READ(NDARAY)(ND(L),L=1,8),MG,TH,(XL(N),YL(N),LNC(N),N=1,4),
        *   IOPT
    600     READ(NDARAY)(ND(L),L=1,8),MG,TH,(XL(N),YL(N),LNC(N),N=1,4),
        *   IOPT
            DO 610 J=1,8
    610     Q(J)=0.0
            DO 620 J=1,NLF
            I=NRAO(J)
            IF(I.EQ.4)THH=DSQRT((XL(4)-XL(1))**2+(YL(4)-YL(1))**2)*TH
            IF(I.NE.4)THH=DSQRT((XL(I+1)-XL(I))**2+(YL(I+1)-YL(I))**2)*THH
            AA=(PX1(I)/3.0+PX2(I)/6.0)*THH
            BB=(PY1(I)/3.0+PY2(I)/6.0)*THH
            CC=(PX1(I)/6.0+PX2(I)/3.0)*THH
            DD=(PY1(I)/6.0+PY2(I)/3.0)*THH
            K=2*I-1
            Q(K)=Q(K)+AA
            Q(K+1)=Q(K+1)+BB
```

```
         IF(K.EQ.7)K=-1
         Q(K+2)=Q(K+2)+CC
620      Q(K+3)=Q(K+3)+DD
         DO 630 J=1,8
         KK=ND(J)
         IF(KK.EQ.0) GO TO 630
         PQ(KK)=PQ(KK)+Q(J)
630      CONTINUE
         LN=LNUM
640      CONTINUE
         LK=NEL-LNUM
         IF(LK.EQ.0)GO TO 660
         DO 650 J=1,LK
650      READ(NDARAY)(ND(L),L=1,8),MG,TH,(XL(N),YL(N),LNC(N),N=1,4),
     *   IOPT
660      CONTINUE
         RETURN
700      CONTINUE
C*****************************************************************
C        TRANSFORMATION MATRIX FOR STRESS SMOOTHING
         VAL=3.0
         TERM=DSQRT(VAL)/2.0
         TMS(1,1)=1.0+TERM
         TMS(1,2)=-0.5
         TMS(1,3)=-0.5
         TMS(1,4)=1.0-TERM
         TMS(2,1)=-0.5
         TMS(2,2)=TMS(1,4)
         TMS(2,3)=TMS(1,1)
         TMS(2,4)=-0.5
         TMS(3,1)=TMS(1,4)
         TMS(3,2)=-0.5
         TMS(3,3)=-0.5
         TMS(3,4)=TMS(1,1)
         TMS(4,1)=-0.5
         TMS(4,2)=TMS(1,1)
         TMS(4,3)=TMS(1,4)
         TMS(4,4)=-0.5
         DO 705 I=1,NSN
         DO 705 J=1,6
705      STR(I,J)=0.0
         WRITE(6,1065)
         IF(NRA.EQ.3)WRITE(6,1070)
         IF(NRA.EQ.4)WRITE(6,1080)
         NS=3
         IF(NRA.EQ.4)NS=4
         LCOUNT=LCOUNT+1
         NEL=NDEL(LCOUNT)
         DO 750 LNUM=1,NEL
         READ(NDARAY)(ND(L),L=1,8),MG,TH,(XL(N),YL(N),LNC(N),N=1,4),
     *   IOPT
         DO 738 KK=1,5
         M=KK-1
         READ(ISTRES)((CB(I,J),I=1,NRA),J=1,8)
         DO 710 I=1,6
710      SIG(I)=0.0
         DO 730 I=1,8
         K=ND(I)
```

```
         IF(K.EQ.0)GO TO 730
         DO 720 L=1,NRA
  720    SIG(L)=SIG(L)+CB(L,I)*P(K)
  730    CONTINUE
         IF(NRA.EQ.4)GO TO 732
         IF(KK.GT.1)GO TO 734
         SP=(SIG(1)+SIG(2))/2.0
         SM=(SIG(1)-SIG(2))/2.0
         DS=DSQRT(SM*SM+SIG(3)*SIG(3))
         SIG(4)=SP+DS
         SIG(5)=SP-DS
         IF(SIG(3).NE.0.0.AND.SM.NE.0.0)SIG(6)=28.648*DATAN2(SIG(3),SM)
         WRITE(6,1090)LNUM,(SIG(I),I=1,6)
         GO TO 738
  732    CONTINUE
         IF(KK.EQ.1)GO TO 740
  734    CONTINUE
C        COMPUTE STRESSES AT THE NODES FROM THE STRESSES AT THE
C        GAUSS POINTS BY STRESS SMOOTHING TECHNIQUE
         DO 736 J=1,NS
         DO 736 I=1,4
         NN=LNC(I)
  736    STR(NN,J)=STR(NN,J)+TMS(I,M)*SIG(J)/ELC(NN)
         GO TO 738
  740    WRITE(6,1100)LNUM,(SIG(I),I=1,4)
  738    CONTINUE
  750    CONTINUE
         IF(NRA.EQ.3)WRITE(6,1110)
         IF(NRA.EQ.4)WRITE(6,1120)
         DO 770 I=1,NSN
         IF(NRA.EQ.4)GO TO 760
         SP=(STR(I,1)+STR(I,2))/2.0
         SM=(STR(I,1)-STR(I,2))/2.0
         DS=DSQRT(SM*SM+STR(I,3)*STR(I,3))
         STR(I,4)=SP+DS
         STR(I,5)=SP-DS
         IF(STR(I,3).NE.0.0.AND.SM.NE.0.0)STR(I,6)=28.648*
     *   DATAN2(STR(I,3),SM)
         WRITE(6,1130)I,(STR(I,J),J=1,6)
         GO TO 770
  760    WRITE(6,1140)I,(STR(I,J),J=1,4)
  770    CONTINUE
 1000    FORMAT(3I5)
 1010    FORMAT(10X,'NUMBER OF QUADRILATERAL ELEMENTS IN THE STRUCTURE**'
     *   '***',I5/10X,'FOR IOPT=1 INCLUDE INCOMPATIBLE MODES'/
     *   14X,'IOPT=0 SUPPRESS INCOMPATIBLE MODES'/)
 1020    FORMAT(30X,'ELEMENT CONNECTIVITY DATA'//20X,'ELEMENT NO',
     *   5X,'MATERIAL GROUP',5X,'NODE-I',5X,'NODE-J',5X,'NODE-K',
     *   5X,'NODE-L',5X,'THICKNESS',5X,'IOPT'//)
 1030    FORMAT(25X,I4,13X,I4,7X,I4,7X,I4,7X,I4,7X,I4,5X,F10.3,8X,I2)
 1040    FORMAT(6I5,F10.2,2I5)
 1050    FORMAT(6I5)
 1060    FORMAT(4F10.3)
 1065    FORMAT(/5X,'****QUADRILATERAL ELEMENTS****'/)
 1070    FORMAT(10X,'STRESSES AT CENTROIDS OF ELEMENTS'////5X,
     *   'ELEMENT NO',8X,'SIGMA(X)',8X,'SIGMA(Y)',8X,'TAU(X,Y)',
     *   8X,'SIGMA(1)',8X,'SIGMA(2)',8X,'ANGLE'/)
 1080    FORMAT(10X,'STRESSES AT CENTROIDS OF ELEMENTS'////5X,
```

```
      *    'ELEMENT NO',8X,'SIGMA(R)',8X,'SIGMA(Z)',8X,'SIGMA(T)',
      *    8X,'TAU(R,Z)'/)
 1090    FORMAT(10X,I4,6(4X,E12.4)/)
 1100    FORMAT(10X,I4,4(4X,E12.4)/)
 1110    FORMAT(10X,'STRESSES AT THE NODES'///8X,'NODE NO',8X,'SIGMA(X)',
      *    8X,'SIGMA(Y)',8X,'TAU(X,Y)',8X,'SIGMA(1)',8X,'SIGMA(2)',
      *    8X,'ANGLE'/)
 1120    FORMAT(//10X,'STRESSES AT THE NODES'///8X,'NODE NO',8X,
      *    'SIGMA(R)',8X,'SIGMA(Z)',8X,'SIGMA(T)',8X,'TAU(R,Z)'/)
 1130    FORMAT(10X,I4,6(4X,E12.4)/)
 1140    FORMAT(10X,I4,4(4X,E12.4)/)
 1150    FORMAT(3F5.2)
 1160    FORMAT(//5X,'SELF WEIGHT FACTORS'//5X,'X-DIRN',F6.2/5X,
      *    'Y-DIRN',F6.2/5X,'Z-DIRN',F6.2/)
 1170    FORMAT(//5X,'NO.OF LOADED ELEMENTS EXCEEDED 50'//5X,'CHANGE THE
      * NUMBERS IN THE CORRESPONDING DIMENSION STATEMENTS'/5X,
      * 'IN THE SUBROUTINE QUAD'//)
 1180    FORMAT(//5X,'NO.OF NODES EXCEEDED 100'//5X,'CHANGE THE NUMBERS I
      *N THE CORRESPONDING DIMENSION STATEMENTS IN SUBROUTINE QUAD'//)
         RETURN
         END


C***********************************************************************
C        ROUTINE COMPUTES THE ELEMENT LOAD VECTOR DUE TO SELFWEIGHT
C        CALLED BY : QUAD
C***********************************************************************
         SUBROUTINE BODYF(S,T,XBAR,YBAR,DA,Q)
         IMPLICIT REAL*8(A-H,O-Z)
         DIMENSION Q(8)
         SF1=0.25*(1.0-S)*(1.0-T)
         SF2=0.25*(1.0+S)*(1.0-T)
         SF3=0.25*(1.0+S)*(1.0+T)
         SF4=0.25*(1.0-S)*(1.0+T)
         F1=DA*XBAR
         F2=DA*YBAR
         Q(1)=Q(1)+SF1*F1
         Q(2)=Q(2)+SF1*F2
         Q(3)=Q(3)+SF2*F1
         Q(4)=Q(4)+SF2*F2
         Q(5)=Q(5)+SF3*F1
         Q(6)=Q(6)+SF3*F2
         Q(7)=Q(7)+SF4*F1
         Q(8)=Q(8)+SF4*F2
         RETURN
         END


C***********************************************************************
C        ROUTINE COMPUTES THE STRAIN-DISPLACEMENT MATRIX FOR THE
C        QUADRILATERAL ELEMENT WITH INCOMPATIBLE MODES
C        CALLED BY : QUAD
C***********************************************************************
         SUBROUTINE QBMAT(S,T,QB)
         IMPLICIT REAL*8(A-H,O-Z)
         COMMON/JACI/A1,B1,C1,D1,A2,B2,C2,D2
         DIMENSION QB(3,12)
         DO 10 K=1,3
```

```
        DO 10 L=1,12
10      QB(K,L)=0.0
        QB(1,1)=-A2*(1.0-T)-B2*(1.0-S)
        QB(1,3)=A2*(1.0-T)-B2*(1.0+S)
        QB(1,5)=A2*(1.0+T)+B2*(1.0+S)
        QB(1,7)=-A2*(1.0+T)+B2*(1.0-S)
        QB(1,9)=-8.0*A1*S
        QB(1,10)=-8.0*B1*T
        QB(2,2)=-C2*(1.0-T)-D2*(1.0-S)
        QB(2,4)=C2*(1.0-T)-D2*(1.0+S)
        QB(2,6)=C2*(1.0+T)+D2*(1.0+S)
        QB(2,8)=-C2*(1.0+T)+D2*(1.0-S)
        QB(2,11)=-8.0*C1*S
        QB(2,12)=-8.0*D1*T
        QB(3,1)=QB(2,2)
        QB(3,2)=QB(1,1)
        QB(3,3)=QB(2,4)
        QB(3,4)=QB(1,3)
        QB(3,5)=QB(2,6)
        QB(3,6)=QB(1,5)
        QB(3,7)=QB(2,8)
        QB(3,8)=QB(1,7)
        QB(3,9)=QB(2,11)
        QB(3,10)=QB(2,12)
        QB(3,11)=QB(1,9)
        QB(3,12)=QB(1,10)
        DO 20 K=1,3
        DO 20 L=1,12
20      QB(K,L)=0.25*QB(K,L)
        RETURN
        END



C*************************************************************************
C       ROUTINE COMPUTES THE STRAIN-DISPLACEMENT MATRIX FOR THE
C       QUADRILATERAL ELEMENT WITHOUT INCOMPATIBLE MODES
C       CALLED BY : QUAD
C*************************************************************************
        SUBROUTINE BMATRX(ZT,ET,X1,X2,X3,X4,Y1,Y2,Y3,Y4,R,DETJ,B,KOPT)
        IMPLICIT REAL*8(A-H,O-Z)
        DIMENSION XJAC(2,2),B(4,8)
        AA=0.25*(1.0-ET)
        BB=0.25*(1.0+ET)
        CC=0.25*(1.0-ZT)
        DD=0.25*(1.0+ZT)
        XJAC(1,1)=AA*(X2-X1)+BB*(X3-X4)
        XJAC(1,2)=AA*(Y2-Y1)+BB*(Y3-Y4)
        XJAC(2,1)=CC*(X4-X1)+DD*(X3-X2)
        XJAC(2,2)=CC*(Y4-Y1)+DD*(Y3-Y2)
        DETJ=XJAC(1,1)*XJAC(2,2)-XJAC(1,2)*XJAC(2,1)
        TEMP=XJAC(1,1)
        XJAC(1,1)=XJAC(2,2)/DETJ
        XJAC(2,2)=TEMP/DETJ
        XJAC(1,2)=-XJAC(1,2)/DETJ
        XJAC(2,1)=-XJAC(2,1)/DETJ
        DO 10 I=1,4
        DO 10 J=1,8
10      B(I,J)=0.0
```

```
      B(1,1)=-(XJAC(1,1)*AA+XJAC(1,2)*CC)
      B(1,3)=AA*XJAC(1,1)-DD*XJAC(1,2)
      B(1,5)=BB*XJAC(1,1)+DD*XJAC(1,2)
      B(1,7)=CC*XJAC(1,2)-BB*XJAC(1,1)
      B(2,2)=-(AA*XJAC(2,1)+CC*XJAC(2,2))
      B(2,4)=AA*XJAC(2,1)-DD*XJAC(2,2)
      B(2,6)=BB*XJAC(2,1)+DD*XJAC(2,2)
      B(2,8)=CC*XJAC(2,2)-BB*XJAC(2,1)
      I=4
      IF(KOPT.LE.1)I=3
      B(I,1)=-AA*XJAC(2,1)-CC*XJAC(2,2)
      B(I,2)=-AA*XJAC(1,1)-CC*XJAC(1,2)
      B(I,3)=AA*XJAC(2,1)-DD*XJAC(2,2)
      B(I,4)=AA*XJAC(1,1)-DD*XJAC(1,2)
      B(I,5)=BB*XJAC(2,1)+DD*XJAC(2,2)
      B(I,6)=BB*XJAC(1,1)+DD*XJAC(1,2)
      B(I,7)=-BB*XJAC(2,1)+CC*XJAC(2,2)
      B(I,8)=-BB*XJAC(1,1)+CC*XJAC(1,2)
      IF(KOPT.LE.1)RETURN
      RR=1.0/(AA*(CC*X1+DD*X2)+BB*(DD*X3+CC*X4))
      B(3,1)=AA*CC*RR
      B(3,3)=DD*AA*RR
      B(3,5)=DD*BB*RR
      B(3,7)=CC*BB*RR
      R=4.0/RR
      RETURN
      END


      SUBROUTINE CST(X,Y,Z,E,PR,WD,SK,PQ,NDF,CHT,ND,NDS,P,NED,NOPT)
      IMPLICIT REAL*8(A-H,O-Z)
      INTEGER CHT
      COMMON/PRECI/ITWO
      COMMON/DIM/N1,N2,N3,N4,N5,N6,N7,N8,N9,N10,N11,N12,N13,N14
      COMMON/PAR/IND,NET,NSN,NMP,NEQ,NSKY,NEQ1,LCOUNT
      COMMON/TAPES/ISTRES,NDARAY,IPR
      COMMON/ELEM/NDEL(100)
      COMMON/ELOAD/M1,M2,AA1,AA2,AA3,AA4,NSLC,NUM
      DIMENSION X(NSN),Y(NSN),Z(NSN),E(NMP),PR(NMP),SK(NSKY),P(NEQ)
      DIMENSION NDF(6,NSN),CHT(NEQ),ND(NED),NDS(NEQ1)
      DIMENSION Q(6),XL(3),YL(3),B(3,6),C(3,3),WD(NMP)
      DIMENSION MNC(3)
      DIMENSION EK(6,6),LNC(3),CB(3,6),SIG(6)
      DIMENSION PQ(NEQ)
      DIMENSION M1(50),M2(50),AA1(50),AA2(50),AA3(50),AA4(50),NUM(50)
      DIMENSION SWF(3)
      DIMENSION THI(100)
C     SQRT(X)=DSQRT(X)
      GO TO (100,300,500),IND
100   CONTINUE
      READ(5,1000) NEL
      LCOUNT=LCOUNT+1
      NDEL(LCOUNT)=NEL
      WRITE(6,1020)NEL
      IF(NEL.GT.100) WRITE(6,1110)
      IF(NEL.GT.100) STOP
      WRITE(6,1030)
      KK=0
```

```
            DO 180 I=1,NEL
            READ(5,1010)LNUM,MG,(MNC(J),J=1,3),TH,KI,KGEN
            IF(KI.GT.0)GO TO 120
            DO 110 J=1,3
   110      LNC(J)=MNC(J)
   120      KK=KK+1
            IF(LNUM-KK)150,150,130
   130      CONTINUE
            DO 140 J=1,3
   140      LNC(J)=LNC(J)+KI
   150      JJ=0
            XC=0.0
            YC=0.0
            DO 170 J=1,3
            K=LNC(J)
            XL(J)=X(K)
            YL(J)=Y(K)
            XC=XC+X(K)/3.0
            YC=YC+Y(K)/3.0
            DO 160 L=1,2
            JJ=JJ+1
   160      ND(JJ)=NDF(L,K)
   170      CONTINUE
            WRITE(NDARAY)(ND(LL),LL=1,6),(XL(NN),YL(NN),LNC(NN),NN=1,3),
        *   TH,MG,XC,YC,KGEN
            CALL COLUMH(CHT,ND,NED,NEQ)
            IF(IPR.EQ.0)WRITE(6,1060)KK,MG,(LNC(J),J=1,3),TH
            KK=KK+1
            IF(LNUM.GE.KK)GO TO 130
            KK=KK-1
            IF(LNUM.EQ.NEL)GO TO 190
   180      CONTINUE
   190      CONTINUE
            READ(5,1090)(SWF(I),I=1,3)
            WRITE(6,1100)(SWF(I),I=1,3)
C           READ ELEMENT LOAD DATA
C
            READ(5,1070)NSLC
            IF(NSLC.GT.50)WRITE(6,1120)
            IF(NSLC.GT.50) STOP
            IF(NSLC.EQ.0)GO TO 210
            DO 200 I=1,NSLC
   200      READ(5,1080)NUM(I),M1(I),M2(I),AA1(I),AA2(I),AA3(I),AA4(I)
   210      CONTINUE
            RETURN
   300      CONTINUE
C**************************************************************************
C           INITIALIZE 'C' MATRIX
            DO 310 I=1,3
            DO 310 J=1,3
   310      C(I,J)=0.0
C           ENTER LOOP OVER EACH ELEMENT
            LCOUNT=LCOUNT+1
            NEL=NDEL(LCOUNT)
            DO 420 LNUM=1,NEL
            READ(NDARAY)(ND(LL),LL=1,6),(XL(NN),YL(NN),LNC(NN),NN=1,3),
        *   TH,MG,XC,YC,KGEN
            THI(LNUM)=TH
```

```
          IF(KGEN.GT.0)GO TO 370
C         CONSTRUCT 'C' MATRIX
          IF(NMP.EQ.1.AND.LNUM.GT.1)GO TO 330
          IF(NOPT.EQ.2)GO TO 320
C
C         CONSTITUTIVE MATRIX FOR PLANE STRAIN ANALYSIS
          CF=E(MG)/((1.0+PR(MG))*(1.0-2.0*PR(MG)))
          C(1,1)=CF*(1.0-PR(MG))
          C(1,2)=PR(MG)*CF
          C(2,1)=C(1,2)
          C(2,2)=C(1,1)
          C(3,3)=CF*(1.0-2.0*PR(MG))/2.0
          GO TO 330
C
C         CONSTITUTIVE MATRIX FOR PLANE STRESS ANALYSIS
   320,   CF=E(MG)/(1.0-PR(MG)*PR(MG))
          C(1,1)=CF
          C(1,2)=PR(MG)*CF
          C(2,1)=C(1,2)
          C(2,2)=C(1,1)
          C(3,3)=CF*(1.0-PR(MG))/2.0
   330    CONTINUE
C         INITIALIZE STIFFNESS MATRIX,LOAD VECTOR AND
C         STRAIN-DISPLACEMENT MATRICES
          DO 340 I=1,6
          Q(I)=0.0
          DO 340 J=1,6
   340    EK(I,J)=0.0
          DO 350 I=1,3
          DO 350 J=1,6
   350    B(I,J)=0.0
C         COMPUTE STRAIN DISPLACEMENT MATRIX
          B(1,1)=YL(2)-YL(3)
          B(1,3)=YL(3)-YL(1)
          B(1,5)=YL(1)-YL(2)
          B(2,2)=XL(3)-XL(2)
          B(2,4)=XL(1)-XL(3)
          B(2,6)=XL(2)-XL(1)
          B(3,1)=B(2,2)
          B(3,2)=B(1,1)
          B(3,3)=B(2,4)
          B(3,4)=B(1,3)
          B(3,5)=B(2,6)
          B(3,6)=B(1,5)
          AREA=(B(2,2)*B(1,5)-B(2,6)*B(1,1))/2.0
C         COMPUTE C*B MATRIX
          FB=2.0*AREA
          DO 360 II=1,3
          DO 360 JJ=1,6
          CB(II,JJ)=0.0
          DO 360 KK=1,3
   360    CB(II,JJ)=CB(II,JJ)+C(II,KK)*B(KK,JJ)/FB
   370    WRITE(ISTRES)((CB(I,J),I=1,3),J=1,6)
          IF(KGEN.GT.0)GO TO 410
C         COMPUTE STIFFNESS MATRIX
          DO 380 II=1,6
          DO 380 JJ=1,6
          DO 380 KK=1,3
```

```
 380      EK(II,JJ)=EK(II,JJ)+B(KK,II)*CB(KK,JJ)*TH*0.5
C         COMPUTE ELEMENT LOAD VECTOR DUE TO SELF WEIGHT
C         AND ADD TO THE GLOBAL LOAD VECTOR
          TBODYF=AREA*WD(MG)*TH
          BODYF1=TBODYF*SWF(1)/3.0
          BODYF2=TBODYF*SWF(2)/3.0
          DO 385 I=1,5,2
 385      Q(I)=Q(I)+BODYF1
          DO 390 I=2,6,2
 390      Q(I)=Q(I)+BODYF2
          DO 395 L=1,6
          KK=ND(L)
          IF(KK.EQ.0)GO TO 395
          PQ(KK)=PQ(KK)+Q(L)
 395      CONTINUE
 400      NUED=NED
 410      CALL PASSEM(SK,EK,NDS,ND,NED,NEQ1,NSKY,NUED)
 420      CONTINUE
          IF(NSLC.EQ.0)GO TO 430
          CALL SURFTR(THI,NDF,X,Y,Z,PQ)
 430      CONTINUE
          RETURN
 500      CONTINUE
C***************************************************************
          LCOUNT=LCOUNT+1
          NEL=NDEL(LCOUNT)
          WRITE(6,1045)
          WRITE(6,1050)
          DO 530 LNUM=1,NEL
          READ(NDARAY)(ND(LL),LL=1,6),(XL(NN),YL(NN),LNC(NN),NN=1,3),
     *    TH,MG,XC,YC,KGEN
          READ(ISTRES)((CB(I,J),I=1,3),J=1,6)
          DO 510 I=1,6
          Q(I)=0.0
          NN=ND(I)
          IF(NN.EQ.0)GO TO 510
          Q(I)=P(NN)
 510      CONTINUE
          DO 520 I=1,3
          SIG(I)=0.0
          DO 520 J=1,6
 520      SIG(I)=SIG(I)+CB(I,J)*Q(J)
          SP=(SIG(1)+SIG(2))/2.0
          SM=(SIG(1)-SIG(2))/2.0
          DS=DSQRT(SM*SM+SIG(3)*SIG(3))
          SIG(4)=SP+DS
          SIG(5)=SP-DS
          SIG(6)=0.0
          IF(SIG(3).NE.0.0.AND.SM.NE.0.0)SIG(6)=28.648*DATAN2(SIG(3),SM)
 530      WRITE(6,1040)LNUM,XC,YC,(SIG(I),I=1,6)
1000      FORMAT(I5)
1010      FORMAT(5I5,E10.2,2I5)
1020      FORMAT(10X,'NUMBER OF CST ELEMENTS IN THE STRUCTURE...',
     *    I5)
1030      FORMAT(//25X,'ELEMENT',3X,'MATERIAL',13X,'NODAL CONNECTIVITY',
     *    15X,'THICKNESS'/27X,'NO.',5X,'GROUP NO.'//)
1040      FORMAT(I8,2F10.2,6E12.4/)
1045      FORMAT(/5X,'***STRESSES AT THE CENTROIDS OF CST ELEMENTS*** /)
```

```
1050      FORMAT(/5X,'ELEMENT',9X,'X',9X,'Y',4X,'SIGMA(X)',4X,'SIGMA(Y)',
     *     4X,'TAU(X,Y)',4X,'SIGMA(1)',4X,'SIGMA(2)',7X,'ANGLE'//)
1C60      FORMAT(25X,I5,5X,I5,16X,3I5,18X,F10.4/)
1C70      FORMAT(I5)
1080      FORMAT(3I5,4F10.3)
1090      FORMAT(3F5.2)
1100      FORMAT(//5X,'SELF WEIGHT FACTORS'//5X,'X-DIRN',F6.2/5X,
     *     'Y-DIRN',F6.2/5X,'Z-DIRN',F6.2/)
1110      FORMAT(//5X,'NO.OF CST ELEMENTS EXCEEDED 100'//5X,'CHANGE THE NU
     *MBERS IN THE CORRESPONDING DIMENSION STATEMENTS'/5X,
     *'IN THE SUBROUTINE CST'//)
1120      FORMAT(//5X,'NO.OF LOADED FACES EXCEEDED 50'//5X,'CHANGE THE NUMB
     *ERS IN THE CORRESPONDING DIMENSION STATEMENTS'/5X,
     *'IN THE SUBROUTINE CST'//)
          RETURN
          END




C*************************************************************************
C       ROUTINE COMPUTES THE ELEMENT LOAD VECTOR
C       DUE TO SURFACE TRACTION
C       CALLED BY : CST
C*************************************************************************
          SUBROUTINE SURFTR(THI,NDF,X,Y,Z,P)
          IMPLICIT REAL*8(A-H,O-Z)
          COMMON/PAR/IND,NET,NSN,NMP,NEQ,NSKY,NEQ1,LCOUNT
          COMMON/ELOAD/MM,NN,AA1,AA2,AA3,AA4,NSLC,NUM
          DIMENSION NDF(6,NSN),P(NEQ),X(NSN),Y(NSN),Z(NSN)
          DIMENSION MM(50),NN(50),AA1(50),AA2(50),AA3(50),AA4(50),NUM(50)
          DIMENSION THI(100)
C         SQRT(X)=DSQRT(X)
          DO 4 I=1,NSLC
          LNUM=NUM(I)
          TH=THI(LNUM)
          M=MM(I)
          N=NN(I)
          A1=AA1(I)
          A2=AA2(I)
          A3=AA3(I)
          A4=AA4(I)
          K1=NDF(1,M)
          K2=NDF(2,M)
          K3=NDF(1,N)
          K4=NDF(2,N)
          DX=X(N)-X(M)
          DY=Y(N)-Y(M)
          EL=DSQRT(DX*DX+DY*DY)
          PXI=A1*EL*TH
          PXJ=A2*EL*TH
          PYI=A3*EL*TH
          PYJ=A4*EL*TH
          IF(K1.EQ.0)GO TO 1
          P(K1)=P(K1)+PXI/3.0+PXJ/6.0
    1     IF(K2.EQ.0)GO TO 2
          P(K2)=P(K2)+PYI/3.0+PYJ/6.0
          IF(K3.EQ.0)GO TO 3
          P(K3)=P(K3)+PXJ/3.0+PXI/6.0
    3     IF(K4.EQ.0)GO TO 4
```

```
      P(K4)=P(K4)+PYJ/3.0+PYI/6.0
4     CONTINUE
      RETURN
      END
```

```
C*************************************************************************
C         CONTROL ROUTINE FOR 3D SOLID ELEMENTS
C*************************************************************************
          SUBROUTINE THREDS(A,LLIB,NSHAPE,MTOT)
          IMPLICIT REAL*8(A-H,O-Z)
          COMMON/DIM/N1,N2,N3,N4,N5,N6,N7,N8,N9,N10,N11,N12,N13,N14
          COMMON/PAR/IND,NET,NSN,NMP,NEQ,NSKY,NEQ2,LCOUNT
          COMMON/TAPES/ISTRES,NDARAY,IPR
          COMMON/PRECI/ITWO
          DIMENSION A(MTOT)
          DIMENSION LLIB(7,3)
          DO 100 J=1,NSHAPE
          LB=LLIB(4,J)
          GO TO (20,40,60),LB
  20      NED=24
          IF(IND.NE.1)GO TO 30
          N10=N9+NED
          N11=N10+1*ITWO
          N12=N11+1*ITWO
          N13=N12+1*ITWO
          NSKY=1
  30      CALL THRD08(A(N2),A(N3),A(N4),A(N5),A(N6),A(N7),A(N11),
       *  A(N12),A(N1),A(N8),A(N9),A(N10),A(N13),NED)
          GO TO 100
  40      NED=60
          IF(IND.NE.1) GO TO 50
          N10=N9+NED
          N11=N10+1*ITWO
          N12=N11+1*ITWO
          N13=N12+1*ITWO
          NSKY=1
  50      CALL THRD20(A(N2),A(N3),A(N4),A(N5),A(N6),A(N7),A(N11),
       *  A(N12),A(N1),A(N8),A(N9),A(N10),A(N13),NED)
          GO TO 100
  60      CALL THRED3
 100      CONTINUE
          RETURN
          END


          SUBROUTINE THRD08(X,Y,Z,E,PR,WD,SK,PQ,NDF,CHT,ND,NDS,P,NED)
          IMPLICIT REAL*8(A-H,O-Z)
          COMMON/DIM/N1,N2,N3,N4,N5,N6,N7,N8,N9,N10,N11,N12,N13,N14
          COMMON/PAR/IND,NET,NSN,NMP,NEQ,NSKY,NEQ1,LCOUNT
          COMMON/TAPES/ISTRES,NDARAY,IPR
          COMMON/GASS08/GP(2),WG(2),IPERM(3)
          COMMON/DERI08/E1,E2,E3,DET,XX(8,3),SF(8),XJAC(3,3),
       *  XJACI(3,3),DETC
          INTEGER CHT(NEQ)
          DIMENSION X(NSN),Y(NSN),Z(NSN),E(NMP),PR(NMP),WD(NMP)
          DIMENSION SK(NSKY),PQ(NEQ),P(NEQ),NDF(6,NSN),ND(NED),NDS(NEQ1)
          DIMENSION EK(24,24),QK(33,33),LNC(8),MNC(8)
          DIMENSION SIG(42),SA(11,3),NDEL(21),SPS(21)
          DIMENSION C(6,6),B(42,24),QB(42,33),CB(42,24),Q(24),RF(24)
          DIMENSION LT(1),PN(1),ZREF(1),NF(1)
```

```
            DIMENSION IS(7),ISP(7),LSN(2),KLSN(2),STP(7,3)
            WG(1)=1.D0
            WG(2)=1.D0
            GP(1)=-0.5773502691896
            GP(2)=0.5773502691896
            IPERM(1)=2
            IPERM(2)=3
            IPERM(3)=1
            DATA STP/0.D0,1.D0,-1.D0,0.D0,0.D0,0.D0,0.D0,
      *            0.D0,0.D0,0.D0,1.D0,-1.D0,0.D0,0.D0,
      *            0.D0,0.D0,0.D0,0.D0,0.D0,1.D0,-1.D0/
            GO TO (100,300,500),IND
  100       READ(5,1000) NEL,NDL
            LCOUNT=LCOUNT+1
            NDEL(LCOUNT)=NEL
            WRITE(6,1010)NEL
            IF(NDL)103,103,101
  101       WRITE(6,1025)
            DO 102 I=1,NDL
            READ(5,1026)N,LT(N),PN(N),ZREF(N),NF(N)
  102       WRITE(6,1027)N,LT(N),PN(N),ZREF(N),NF(N)
  103       CONTINUE
            WRITE(6,1023)
  •         WRITE(6,1020)
            KK=0
            DO 180 I=1,NEL
  C         FOR IOPT=1 INCOMPATIBLE MODES ARE INCLUDED
  C         FOR IOPT=0 THESE MODES ARE SUPPRESSED
            READ(5,1030)LNUM,MG,(MNC(J),J=1,8),NINT,KI,LSN(1),LSN(2),
      *     NSS,(ISP(J),J=1,7),IOPT
            KNINT=NINT
            KLSN(1)=LSN(1)
            KLSN(2)=LSN(2)
            KNSS=NSS
            DO 105 J=1,KNSS
  105       IS(J)=ISP(J)
            IF(KI.GT.0)GO TO 120
            DO 110 J=1,8
  110       LNC(J)=MNC(J)
  120       KK=KK+1
            IF(LNUM-KK)150,150,130
  130       CONTINUE
            DO 140 J=1,8
  140       LNC(J)=LNC(J)+KI
  150       JJ=0
            DO 170 J=1,8
            K=LNC(J)
            XX(J,1)=X(K)
            XX(J,2)=Y(K)
            XX(J,3)=Z(K)
            DO 160 L=1,3
            JJ=JJ+1
            ND(JJ)=NDF(L,K)
  160       CONTINUE
  170       CONTINUE
            WRITE(NDARAY)(ND(L),L=1,24),KNINT,MG,(XX(N,1),XX(N,2),XX(N,
      *     LNC(N),N=1,8),KLSN(1),KLSN(2),KNSS,(IS(J),J=1,KNSS),IOPT
            CALL COLUMH(CHT,ND,NFD,NEQ)
```

```
          IF(IPR.EQ.0)WRITE(6,1040)KK,MG,(LNC(J),J=1,8),KNINT,
          KLSN(1),KLSN(2),IOPT
          KK=KK+1
          IF(LNUM.GE.KK)GO TO 130
          KK=KK-1
          IF(LNUM.EQ.NEL)GO TO 190
  180     CONTINUE
  190     CONTINUE
          RETURN
C***************************************************************
  300     CONTINUE
C         INITIALIZE C MATRIX
          DO 310 I=1,6
          DO 310 J=1,6
  310     C(I,J)=0.0
C         ENTER LOOP OVER EACH ELEMENT
          LCOUNT=LCOUNT+1
          NEL=NDEL(LCOUNT)
          DO 490 LNUM=1,NEL
          READ(NDARAY)(ND(L),L=1,24),KNINT,MG,(XX(N,1),XX(N,2),XX(N,3),
        * LNC(N),N=1,8),KLSN(1),KLSN(2),KNSS,(IS(J),J=1,KNSS),IOPT
          IF(NMP.EQ.1.AND.LNUM.GT.1)GO TO 320
          CF=E(MG)/((1.0+PR(MG))*(1.0-2.0*PR(MG)))
          C(1,1)=CF*(1.0-PR(MG))
          C(1,2)=CF*PR(MG)
          C(1,3)=C(1,2)
          C(2,1)=C(1,2)
          C(2,2)=C(1,1)
          C(2,3)=C(1,2)
          C(3,1)=C(1,2)
          C(3,2)=C(1,2)
          C(3,3)=C(1,1)
          C(4,4)=CF*(1.0-2.0*PR(MG))/2.0
          C(5,5)=C(4,4)
          C(6,6)=C(4,4)
          CC1=C(1,1)
          CC2=C(1,2)
          CC3=C(4,4)
  320     CONTINUE
          IF(KNINT.EQ.0)GO TO 425
C         INITIALIZE STIFFNESS MATRIX,LOAD VECTOR AND 'B' MATRIX
          DO 330 I=1,33
          DO 330 J=1,33
  330     QK(I,J)=0.0
          DO 340 I=1,24
  340     Q(I)=0.0
          NS=6*KNSS
          DO 350 I=1,NS
          DO 350 J=1,33
  350     QB(I,J)=0.0
C         COMPUTE STRAIN-DISPLACEMENT MATRIX
          NNN=11
          IF(IOPT.NE.1)NNN=8
          DO 365 L=1,KNSS
          LL=IS(L)+1
          E1=STP(LL,1)
          E2=STP(LL,2)
          E3=STP(LL,3)
```

```
          CALL BMATO8(2,SA)
          L3=6*L-6
          DO 360 K=1,NNN
          K3=3*K
          K2=K3-1
          K1=K2-1
          QB(L3+1,K1)=SA(K,1)
          QB(L3+2,K2)=SA(K,2)
          QB(L3+3,K3)=SA(K,3)
          QB(L3+4,K1)=SA(K,2)
          QB(L3+4,K2)=SA(K,1)
          QB(L3+5,K2)=SA(K,3)
          QB(L3+5,K3)=SA(K,2)
          QB(L3+6,K1)=SA(K,3)
   360    QB(L3+6,K3)=SA(K,1)
   365    CONTINUE
          DO 380 LX=1,2
          E1=GP(LX)
          DO 380 LY=1,2
          E2=GP(LY)
          DO 380 LZ=1,2
          E3=GP(LZ)
          CALL BMATO8(1,SA)
          GT=WG(LX)*WG(LY)*WG(LZ)*DET
          GTC=WG(LX)*WG(LY)*WG(LZ)*DETC
          GG=GT*WD(MG)
          DO 370 I=1,8
          K=3*I
   370    Q(K)=Q(K)+GG*SF(I)
C         ADD CONTRIBUTION TO STIFFNESS MATRIX
          DO 380 I=1,NNN
          K3=3*I
          K2=K3-1
          K1=K2-1
          UI=SA(I,1)
          VI=SA(I,2)
          WI=SA(I,3)
          DO 380 J=1,NNN
          L3=3*J
          L2=L3-1
          L1=L2-1
          UJ=SA(J,1)
          VJ=SA(J,2)
          WJ=SA(J,3)
          UU=UI*UJ
          VV=VI*VJ
          WW=WI*WJ
          UV=UI*VJ
          VU=VI*UJ
          UW=UI*WJ
          WU=WI*UJ
          VW=VI*WJ
          WV=WI*VJ
          IF(I.GT.8.OR.J.GT.8)GO TO 375
          C1=GT*CC1
          C2=GT*CC2
          C3=GT*CC3
          GO TO 376
```

```
        375     C1=GTC*CC1
                C2=GTC*CC2
                C3=GTC*CC3
        376     CONTINUE
                QK(K1,L1)=QK(K1,L1)+C1*UU+C3*(VV+WW)
                QK(K2,L2)=QK(K2,L2)+C1*VV+C3*(WW+UU)
                QK(K3,L3)=QK(K3,L3)+C1*WW+C3*(UU+VV)
                QK(K1,L2)=QK(K1,L2)+C2*UV+C3*VU
                QK(K1,L3)=QK(K1,L3)+C2*UW+C3*WU
                QK(K2,L3)=QK(K2,L3)+C2*VW+C3*WV
                IF(I.EQ.J)GO TO 380
                QK(K2,L1)=QK(K2,L1)+C2*VU+C3*UV
                QK(K3,L1)=QK(K3,L1)+C2*WU+C3*UW
                QK(K3,L2)=QK(K3,L2)+C2*WV+C3*VW
        380     CONTINUE
C               STATIC CONDENSATION
C               WHEN IOPT=0 SKIP THE STATIC CONDENSATION PART
                IF(IOPT.NE.1)GO TO 403
                DO 402 M=1,9
                MN=34-M
                MO=MN-1
C               STIFFNESS MATRIX QK
                SP=QK(MN,MN)
                DO 390 I=1,MO
        390     QK(MN,I)=QK(I,MN)/SP
                DO 400 K=1,MO
                SP=QK(MN,K)
                DO 400 J=1,MO
                SPQ1=SP*QK(J,MN)
                IF(DABS(SPQ1).LE.1.0D-25)SPQ1=0.0
        400     QK(J,K)=QK(J,K)-SPQ1
C                 'B' MATRIX
                DO 402 J=1,NS
                SP=QB(J,MN)
                IF(SP.EQ.0)GO TO 402
                DO 401 K=1,MO
                SPQ2=SP*QK(MN,K)
                IF(DABS(SPQ2).LE.1.0D-25)SPQ2=0.0
        401     QB(J,K)=QB(J,K)-SPQ2
        402     CONTINUE
        403     CONTINUE
                DO 410 I=1,24
                DO 410 J=1,24
                DO 410 K=1,NS
                B(K,I)=QB(K,I)
        410     EK(I,J)=QK(I,J)
                DO 420 I=1,KNSS
                II=6*I-6
                DO 415 J=1,6
                DO 415 K=1,24
                CB(II+J,K)=0.0
                DO 415 L=1,6
        415     CB(II+J,K)=CB(II+J,K)+C(J,L)*B(II+L,K)
                IF(IS(I).LE.0)GO TO 420
                LL=IS(I)+1
                E1=STP(LL,1)
                E2=STP(LL,2)
                E3=STP(LL,3)
```

```
                CALL BMAT08(4,SA)
                CALL FST(IS,XJAC,XJACI,CB,I,24)
     420        CONTINUE
     425        CONTINUE
                WRITE(ISTRES)((CB(I,J),I=1,NS),J=1,24)
                NUED=NED
                CALL PASSEM(SK,EK,NDS,ND,NED,NEQ1,NSKY,NUED)
C               DISTRIBUTED LOAD
                DO 430 I=1,24
     430        RF(I)=0.0
                CALL NDLOAD(LT,PN,ZREF,NF,KLSN(1),KLSN(2),RF)
                DO 440 K=1,24
     440        RF(K)=RF(K)-Q(K)
                DO 450 J=1,24
                KK=ND(J)
                IF(KK.EQ.0)GO TO 450
                PQ(KK)=PQ(KK)+RF(J)
     450        CONTINUE
     490        CONTINUE
                RETURN
C***************************************************************
     500        CONTINUE
                WRITE(6,1070)
                LCOUNT=LCOUNT+1
                NEL=NDEL(LCOUNT)
                DO 550 LNUM=1,NEL
                READ(NDARAY)(ND(L),L=1,24),KNINT,MG,(XX(N,1),XX(N,2),
     *          XX(N,3),LNC(N),N=1,8),KLSN(1),KLSN(2),KNSS,(IS(J),J=1,KNSS)
     *          ,IOPT
                READ(ISTRES)((CB(I,J),I=1,NS),J=1,24)
                NS=6*KNSS
                DO 510 I=1,NS
     510        SIG(I)=0.0
                II=NS-6
                DO 530 I=1,24
                K=ND(I)
                IF(K.EQ.0)GO TO 530
                DO 520 L=1,NS
     520        SIG(L)=SIG(L)+CB(L,I)*P(K)
     530        CONTINUE
                CALL PRSTS(KNSS,NS,IS,SIG,SPS)
                WRITE(6,1080)LNUM,IS(1),(SIG(I),I=1,6),(SPS(I),I=1,3)
                M=7
                KI=1
     540        CONTINUE
                M5=M+5
                M1=(M+1)/2
                M2=(M+5)/2
                KI=KI+1
                NUM=IS(KI)
                WRITE(6,1090)NUM,(SIG(I),I=M,M5),(SPS(I),I=M1,M2)
                M=M+6
                IF(M5.EQ.NS)GO TO 550
                GO TO 540
     550        CONTINUE
    1000        FORMAT(2I5)
    1010        FORMAT(35X,'NUMBER OF 3D-8 NODED ELEMENTS IN THE STRUCTURE='
     *          I5/)
```

```
      1020   FORMAT(/10X,'ELEMENT',4X,'MAT.',3X,25('*'),'ELEMENT CONNEC',
     *       'TIVITY',25('*'),2X,'INT.',2X,'LOAD SET NUMBER',2X,'IOPT'/12X,
     *       'NO.',5X,'GROUP',6X,'1',8X,'2',8X,'3',8X,'4',8X,'5',8X,'6',8X,
     *       '7',8X,'8',3X,'ORDER',4X,'1',6X,'2'//)
      1023   FORMAT(//10X,'FOR IOPT=1,INCOMPATIBLE MODES ARE ACTIVATED'/14X,
     *       'IOPT=0,INCOMPATIBLE MODES ARE SUPPRESSED'//)
      1025   FORMAT(/50X,'ELEMENT DISTRIBUTED LOADS'/12X,'NUMBER',
     *       12X,'TYPE',11X,'PRESSURE',13X,'ZREF',12X,'FACE NO.'/)
      1026   FORMAT(2I5,2F10.3,I5)
      1027   FORMAT(10X,I5,12X,I5,2(10X,F10.3),10X,I5/)
      1030   FORMAT(10I5,13I2)
      1040   FORMAT(5X,13I9,I5/)
      1070   FORMAT(/40X,'STRESSES AT THE CENTROID AND CENTRE OF ELEMENT',
     *       'FACES'/25X,'STRESSES ARE REFERRED TO GLOBAL AXES AT THE',
     *       ' CENTROID AND TO LOCAL AXES AT THE FACES'/4X,'ELEMENT NO.',
     *       2X,'FACE',3X,'SIGMA-XX',4X,'SIGMA-YY',4X,'SIGMA-ZZ',4X,
     *       'SIGMA-XY',4X,'SIGMA-YZ',4X,'SIGMA-XZ',3X,'SIGMA-MAX',
     *       3X,'SIGMA-MIN',4X,'S2/ANGLE'/)
      1080   FORMAT(/2(5X,I5),9(2X,E10.3)/)
      1090   FORMAT(15X,I5,9(2X,E10.3)/)
             RETURN
             END




C***********************************************************************
C       ROUTINE COMPUTES THE STRAIN-DISPLACEMENT MATRIX 'B'
C       CALLED BY : THRD08
C***********************************************************************
        SUBROUTINE BMAT08(KK,SA)
        IMPLICIT REAL*8(A-H,O-Z)
        COMMON/GASS08/GP(2),WG(2),IPERM(3)
        COMMON/DERI08/R,S,T,DET,XX(8,3),SF(8),XJAC(3,3),XJACI(3,3),DETC
        DIMENSION SA(11,3),P(3,11),CXJAC(3,3),CXJACI(3,3)

        RP=(1.0+R)*0.125
        RM=(1.0-R)*0.125
        SP=1.0+S
        SM=1.0-S
        TP=1.0+T
        TM=1.0-T
        IF(KK.EQ.2.OR.KK.EQ.4)GO TO 90
C       SHAPE FUNCTIONS
        SF(1)=RP*SM*TM
        SF(2)=RP*SP*TM
        SF(3)=RM*SP*TM
        SF(4)=RM*SM*TM
        SF(5)=RP*SM*TP
        SF(6)=RP*SP*TP
        SF(7)=RM*SP*TP
        SF(8)=RM*SM*TP
C       DERIVATIVES OF SHAPE FUNCTIONS
   90   P(1,1)=SM*TM*0.125
        P(1,2)=SP*TM*0.125
        P(1,3)=-P(1,2)
        P(1,4)=-P(1,1)
        P(1,5)=SM*TP*0.125
        P(1,6)=SP*TP*0.125
        P(1,7)=-P(1,6)
```

```
                P(1,8)=-P(1,5)
                P(1,9)=-2.0*R
                P(1,10)=0.0
                P(1,11)=0.0
  C
                P(2,1)=-RP*TM
                P(2,2)=-P(2,1)
                P(2,3)=RM*TM
                P(2,4)=-P(2,3)
                P(2,5)=-RP*TP
                P(2,6)=-P(2,5)
                P(2,7)=RM*TP
                P(2,8)=-P(2,7)
                P(2,9)=0.0
                P(2,10)=-2.0*S
                P(2,11)=0.0
  C
                P(3,1)=-RP*SM
                P(3,2)=-RP*SP
                P(3,3)=-RM*SP
                P(3,4)=-RM*SM
                P(3,5)=-P(3,1)
                P(3,6)=-P(3,2)
                P(3,7)=-P(3,3)
                P(3,8)=-P(3,4)
                P(3,9)=0.0
                P(3,10)=0.0
                P(3,11)=-2.0*T
  C             JACOBIAN MATRIX XJAC
                DO 110 I=1,3
                DO 110 J=1,3
                C=0.0
                DO 100 L=1,8
  100           C=C+P(I,L)*XX(L,J)
  110           XJAC(I,J)=C
                IF(KK.EQ.3) GO TO 170
  C             INVERT JACOBIAN MATRIX
                DO 120 I=1,3
                J=IPERM(I)
                K=IPERM(J)
                XJACI(I,I)=XJAC(J,J)*XJAC(K,K)-XJAC(K,J)*XJAC(J,K)
                XJACI(I,J)=XJAC(K,J)*XJAC(I,K)-XJAC(I,J)*XJAC(K,K)
  120           XJACI(J,I)=XJAC(J,K)*XJAC(K,I)-XJAC(J,I)*XJAC(K,K)
                IF(KK.EQ.4)GO TO 160
                DET=XJAC(1,1)*XJACI(1,1)+XJAC(1,2)*XJACI(2,1)+
       *        XJAC(1,3)*XJACI(3,1)
  C             JACOBIAN AT THE CENTROID FOR INCOMPATIBLE MODES
                DO 130 I=1,3
                CXJAC(1,I)=(XX(1,I)+XX(2,I)-XX(3,I)-XX(4,I)+XX(5,I)+XX(6,I)-
       *        XX(7,I)-XX(8,I))*0.125
                CXJAC(2,I)=(-XX(1,I)+XX(2,I)+XX(3,I)-XX(4,I)-XX(5,I)+XX(6,I)+
       *        XX(7,I)-XX(8,I))*0.125
                CXJAC(3,I)=(-XX(1,I)-XX(2,I)-XX(3,I)-XX(4,I)+XX(5,I)+XX(6,I)+
       *        XX(7,I)+XX(8,I))*0.125
  130           CONTINUE
                DO 140 I=1,3
                J=IPERM(I)
                K=IPERM(J)
```

```
              CXJACI(I,I)=CXJAC(J,J)*CXJAC(K,K)-CXJAC(K,J)*CXJAC(J,K)
              CXJACI(I,J)=CXJAC(K,J)*CXJAC(I,K)-CXJAC(I,J)*CXJAC(K,K)
       140    CXJACI(J,I)=CXJAC(J,K)*CXJAC(K,I)-CXJAC(J,I)*CXJAC(K,K)
              DETC=CXJAC(1,1)*CXJACI(1,1)+CXJAC(1,2)*CXJACI(2,1)+
           *  CXJAC(1,3)*CXJACI(3,1)
C             MATRIX OF X-Y-Z DERIVATIVES
              DO 150 I=1,3
              DO 150 J=1,8
              C=0.0
              DO 145 K=1,3
       145    C=C+XJACI(I,K)*P(K,J)
       150    SA(J,I)=C/DET
              DO 160 I=1,3
              DO 160 J=9,11
              C=0.0
              DO 155 K=1,3
       155    C=C+CXJACI(I,K)*P(K,J)
       160    SA(J,I)=C/DETC
       170    CONTINUE
              RETURN
              END



C*********************************************************************
C             ROUTINE CALCULATES THE STRESS-DISPLACEMENT MATRIX 'CB'
C             AT THE CENTRES OF SPECIFIED FACES
C             CALLED BY : THRD08 AND THRD20
C*********************************************************************
              SUBROUTINE FST(IS,XJAC,XJACI,CB,L,NED)
              IMPLICIT REAL*8(A-H,O-Z)
              DIMENSION IS(7),XJAC(3,3),XJACI(3,3),CB(42,NED),NR(6,2)
              DIMENSION TC(6,60),TR(6,6)
              DATA NR/1,1,2,2,3,3,2,2,3,3,1,1/
C
              LL=IS(L)
              I=NR(LL,1)
              TT=XJACI(1,I)*XJACI(1,I)+XJACI(2,I)*XJACI(2,I)+
           *  XJACI(3,I)*XJACI(3,I)
              TT=DSQRT(TT)
              TC(3,1)=XJACI(1,I)/TT
              TC(3,2)=XJACI(2,I)/TT
              TC(3,3)=XJACI(3,I)/TT
              I=NR(LL,2)
              TT=XJAC(I,1)*XJAC(I,1)+XJAC(I,2)*XJAC(I,2)+
           *  XJAC(I,3)*XJAC(I,3)
              TT=DSQRT(TT)
              TC(1,1)=XJAC(I,1)/TT
              TC(1,2)=XJAC(I,2)/TT
              TC(1,3)=XJAC(I,3)/TT
              TC(2,1)=TC(3,2)*TC(1,3)-TC(3,3)*TC(1,2)
              TC(2,2)=TC(3,3)*TC(1,1)-TC(3,1)*TC(1,3)
              TC(2,3)=TC(3,1)*TC(1,2)-TC(3,2)*TC(1,1)
C
              TR(1,1)=TC(1,1)*TC(1,1)
              TR(1,2)=TC(1,2)*TC(1,2)
              TR(1,3)=TC(1,3)*TC(1,3)
              TR(1,4)=TC(1,1)*TC(1,2)*2.0
              TR(1,5)=TC(1,2)*TC(1,3)*2.0
```

```
      TR(1,6)=TC(1,1)*TC(1,3)*2.0
      TR(2,1)=TC(2,1)*TC(2,1)
      TR(2,2)=TC(2,2)*TC(2,2)
      TR(2,3)=TC(2,3)*TC(2,3)
      TR(2,4)=TC(2,1)*TC(2,2)*2.0
      TR(2,5)=TC(2,2)*TC(2,3)*2.0
      TR(2,6)=TC(2,1)*TC(2,3)*2.0
      TR(3,1)=TC(3,1)*TC(3,1)
      TR(3,2)=TC(3,2)*TC(3,2)
      TR(3,3)=TC(3,3)*TC(3,3)
      TR(3,4)=TC(3,1)*TC(3,2)*2.0
      TR(3,5)=TC(3,2)*TC(3,3)*2.0
      TR(3,6)=TC(3,1)*TC(3,3)*2.0
      TR(4,1)=TC(1,1)*TC(2,1)
      TR(4,2)=TC(1,2)*TC(2,2)
      TR(4,3)=TC(1,3)*TC(2,3)
      TR(4,4)=TC(1,1)*TC(2,2)+TC(1,2)*TC(2,1)
      TR(4,5)=TC(1,2)*TC(2,3)+TC(1,3)*TC(2,2)
      TR(4,6)=TC(1,1)*TC(2,3)+TC(1,3)*TC(2,1)
      TR(5,1)=TC(2,1)*TC(3,1)
      TR(5,2)=TC(2,2)*TC(3,2)
      TR(5,3)=TC(2,3)*TC(3,3)
      TR(5,4)=TC(2,1)*TC(3,2)+TC(2,2)*TC(3,1)
      TR(5,5)=TC(2,2)*TC(3,3)+TC(2,3)*TC(3,2)
      TR(5,6)=TC(2,1)*TC(3,3)+TC(2,3)*TC(3,1)
      TR(6,1)=TC(3,1)*TC(1,1)
      TR(6,2)=TC(3,2)*TC(1,2)
      TR(6,3)=TC(3,3)*TC(1,3)
      TR(6,4)=TC(3,1)*TC(1,2)+TC(3,2)*TC(1,1)
      TR(6,5)=TC(3,2)*TC(1,3)+TC(3,3)*TC(1,2)
      TR(6,6)=TC(3,1)*TC(1,3)+TC(3,3)*TC(1,1)
C
      IL=6*(L-1)
      DO 100 I=1,6
      DO 100 J=1,NED
      TC(I,J)=0.0
      DO 100 K=1,6
  100 TC(I,J)=TC(I,J)+TR(I,K)*CB(IL+K,J)
      DO 110 I=1,6
      DO 110 J=1,NED
  110 CB(IL+I,J)=TC(I,J)
      RETURN
      END


C*****************************************************************
C     ROUTINE COMPUTES THE PRINCIPAL STRESSES AT  SPECIFIED POINTS
C     CALLED BY : THRD08 AND THRD20
C*****************************************************************
      SUBROUTINE PRSTS(KNSS,NS,IS,SIG,SP)
      IMPLICIT REAL*8(A-H,O-Z)
      DIMENSION SIG(42),SP(21),IS(7),SG(6)
      DO 60 N=1,KNSS
      K=3*N-3
      II=K*2
      IF(IS(N).EQ.0)GO TO 10
      AA=(SIG(II+1)+SIG(II+2))/2.0
      BB=(SIG(II+1)-SIG(II+2))/2.0
```

```
                CC=DSQRT(BB**2+SIG(II+4)**2)
                SP(K+1)=AA+CC
                SP(K+2)=AA-CC
                SP(K+3)=0.0
                IF(BB.NE.0.0)SP(K+3)=28.648*DATAN2(SIG(II+4),BB)
                GO TO 60
      10        AA=(SIG(II+1)+SIG(II+2)+SIG(II+3))/3.0
                DO 20 I=1,3
                SG(I)=SIG(II+I)-AA
      20        SG(I+3)=SIG(II+I+3)
                C1=(SG(1)**2+SG(2)**2+SG(3)**2)*0.5+SG(4)**2+SG(5)**2+SG(6)**2
                C2=SG(1)*(SG(2)*SG(3)-SG(5)**2)+SG(4)*(SG(5)*SG(6)-
       *        SG(4)*SG(3))+SG(6)*(SG(4)*SG(5)-SG(2)*SG(6))
                T=DSQRT(C1/1.5)
                A=C2*1.414214/T**3
                IF(A.LT.-1.0)A=-1.0
                IF(A.GT.1.0)A=1.0
                A=DACOS(A)/3.0
                T=T*1.414214
                SP(K+1)=T*DCOS(A)
                SP(K+2)=T*DCOS(A+2.0944)
                SP(K+3)=T*DCOS(A-2.0944)
                DO 30 I=2,3
                IF(SP(K+1).GT.SP(K+I))GO TO 30
                C2=SP(K+1)
                SP(K+1)=SP(K+I)
                SP(K+I)=C2
      30        CONTINUE
                IF(SP(K+2).LE.SP(K+3))GO TO 40
                C2=SP(K+2)
                SP(K+2) = SP(K+3)
                SP(K+3)=C2
      40        DO 50 I=1,3
      50        SP(K+I)=SP(K+I)+AA
      60        CONTINUE
                RETURN
                END


C***********************************************************************
C       ROUTINE COMPUTES THE LOAD VECTOR DUE TO ELEMENT APPLIED LOADS
C       CALLED BY : THRD08
C***********************************************************************
                SUBROUTINE NDLOAD(L,P,Z,N,KI,KJ,RF)
                IMPLICIT REAL*8(A-H,O-Z)
                COMMON/GASS08/GP(2),WG(2),IPERM(3)
                COMMON/DERI08/ET(3),DET,XX(8,3),SF(8),XJAC(3,3),XJACI(3,3),DETC
                DIMENSION SA(11,3),KFCON(6,4),KORD(6),FVAL(6),RF(24)
                DIMENSION L(1),P(1),Z(1),N(1),KLSN(2)
                DATA KFCON/1,4,2,1,6,2,2,3,3,4,7,3,6,7,7,8,8,4,5,8,6,5,5,1/
                DATA KORD/1,1,2,2,3,3/
                DATA FVAL/1.D0,-1.D0,1.D0,-1.D0,1.D0,-1.D0/
                KLSN(1)=KI
                KLSN(2)=KJ
                DO 100 KK=1,2
                NN=KLSN(KK)
                IF(NN)100,100,10
      10        LT=L(NN)
```

```
                PR=P(NN)
                ZREF=Z(NN)
                NF=N(NN)
C               INTEGRATE OVER THE SURFACE
                ML=KORD(NF)
                MM=IPERM(ML)
                MN=IPERM(MM)
                ET(ML)=FVAL(NF)
                DO 90 LX=1,2
                ET(MM)=GP(LX)
                DO 90 LY=1,2
                ET(MN)=GP(LY)
                CALL BMATO8(3,SA)
C               COMPUTE DIRECTION COSINES OF NORMAL TO SURFACE AND AREA
                A1=(XJAC(MM,2)*XJAC(MN,3)-XJAC(MM,3)*XJAC(MN,2))
                A2=(XJAC(MM,3)*XJAC(MN,1)-XJAC(MM,1)*XJAC(MN,3))
                A3=(XJAC(MM,1)*XJAC(MN,2)-XJAC(MM,2)*XJAC(MN,1))
                AA=DSQRT(A1*A1+A2*A2+A3*A3)
                A1=A1/AA
                A2=A2/AA
                A3=A3/AA
C               COMPUTE PRESSURE,LOAD COMPONENTS AND STORE IN 'RF'
                IF(LT.EQ.2)GO TO 50
                FORCE=PR
                GO TO 70
        50      ZZ=0.0
                DO 60 I=1,8
        60      ZZ=ZZ+SF(I)*XX(I,3)
                ZZ=ZZ-ZREF
                FORCE=-PR*ZZ
                IF(ZZ.GT.0.0)FORCE=0.0
        70      CONTINUE
                TS=FORCE*WG(LX)*WG(LY)*AA
                DO 80 I=1,4
                NI=KFCON(NF,I)
                QQ=TS*SF(NI)
                K=3*NI
                RF(K-2)=RF(K-2)+QQ*A1
                RF(K-1)=RF(K-1)+QQ*A2
                RF(K)=RF(K)+QQ*A3
        80      CONTINUE
        90      CONTINUE
       100      CONTINUE
                RETURN
                END

C**********************************************************************
C       ROUTINE COMPUTES THE LOAD VECTOR DUE TO ELEMENT APPLIED LOADS
C       CALLED BY : THRD20
C**********************************************************************
                SUBROUTINE DNLOAD(L,P,Z,N,KLSN,RF)
                IMPLICIT REAL*8(A-H,O-Z)
                COMMON/GASS20/DUM(12),GP(4),DDUM(12),WG(4),IPERM(3)
                COMMON/DERI20/ET(3),DET,XX(20,3),SF(20),XJAC(3,3),XJACI(3,3)
                DIMENSION SA(20,3),KFCON(6,8),KORD(6),FVAL(6),RF(60)
                DIMENSION L(1),P(1),Z(1),KLSN(2),N(1)
                DATA KFCON/ 1,4,2,1,6,2,   2,3,3,4,7,3,
              *            6,7,7,8,8,4,   5,8,6,5,5,1,
```

```
     *                    9, 11, 10, 12, 14, 10,    18, 19, 19, 20, 15, 11,
     *                    13, 15, 14, 16, 18, 12,    17, 20, 18, 17, 13, 9/
           DATA KORD/1, 1, 2, 2, 3, 3/
           DATA FVAL/1. D0, -1. D0, 1. D0, -1. D0, 1. D0, -1. D0/
  C
  C

           DO 100 KK=1, 2
           NN=KLSN(KK)
           IF(NN)100, 100, 10
   10      LT=L(NN)
           PR=P(NN)
           ZREF=Z(NN)
           NF=N(NN)
  C        INTEGRATE OVER THE SURFACE
           ML=KORD(NF)
           MM=IPERM(ML)
           MN=IPERM(MM)
           ET(ML)=FVAL(NF)
           DO 90 LX=1, 4
           ET(MM)=GP(LX)
           DO 90 LY=1, 4
           ET(MN)=GP(LY)
           CALL BMAT20(3, SA)
  C        COMPUTE DIRECTION COSINES OF NORMAL TO SURFACE AND AREA
           A1=XJAC(MM, 2)*XJAC(MN, 3)-XJAC(MM, 3)*XJAC(MN, 2)
           A2=XJAC(MM, 3)*XJAC(MN, 1)-XJAC(MM, 1)*XJAC(MN, 3)
           A3=XJAC(MM, 1)*XJAC(MN, 2)-XJAC(MM, 2)*XJAC(MN, 1)
           AA=DSQRT(A1*A1+A2*A2+A3*A3)
           A1=A1/AA
           A2=A2/AA
           A3=A3/AA
  C        COMPUTE PRESSURE, LOAD COMPONENTS AND STORE IN 'RF'
           IF(LT. EQ. 2)GO TO 50
           FORCE=PR
           GO TO 70
   50      ZZ=0. 0
           DO 60 I=1, 20
   60      ZZ=ZZ+SF(I)*XX(I, 3)
           ZZ=ZZ-ZREF
           FORCE=-PR*ZZ
           IF(ZZ. GT. 0. 0)FORCE=0. 0
   70      CONTINUE
           TS=FORCE*WG(LX)*WG(LY)*AA
           DO 80 I=1, 8
           NI=KFCON(NF, I)
           QQ=TS*SF(NI)
           K=3*NI
           RF(K-2)=RF(K-2)+QQ*A1
           RF(K-1)=RF(K-1)+QQ*A2
           RF(K)=RF(K)+QQ*A3
   80      CONTINUE
   90      CONTINUE
  100      CONTINUE
           RETURN
           END
```

```
      SUBROUTINE THRED3
      RETURN
      END



      SUBROUTINE THRD20(X,Y,Z,E,PR,WD,SK,PQ,NDF,CHT,ND,NDS,P,NED)
      IMPLICIT REAL*8(A-H,O-Z)
      COMMON/DIM/N1,N2,N3,N4,N5,N6,N7,N8,N9,N10,N11,N12,N13,N14
      COMMON/PAR/IND,NET,NSN,NMP,NEQ,NSKY,NEQ1,LCOUNT
      COMMON/TAPES/ISTRES,NDARAY,IPR
      COMMON/GASS20/GP(4,4),WG(4,4),IPERM(3)
      COMMON/DERI20/E1,E2,E3,DET,XX(20,3),SF(20),XJAC(3,3),XJACI(3,3)
      INTEGER CHT(NEQ)
      DIMENSION X(NSN),Y(NSN),Z(NSN),E(NMP),PR(NMP),WD(NMP)
      DIMENSION SK(NSKY),PQ(NEQ),P(NEQ),NDF(6,NSN),ND(NED),NDS(NEQ1)
      DIMENSION EK(60,60),LNC(20),MNC(20)
      DIMENSION SIG(42),SA(20,3),NDEL(21),SPS(21)
      DIMENSION C(6,6),B(42,60),CB(42,60),Q(60),RF(60)
      DIMENSION LT(1),PN(1),ZREF(1),NF(1)
      DIMENSION IS(7),ISP(7),LSN(2),KLSN(2),STP(7,3)
      DATA STP/0.D0,1.D0,-1.D0,0.D0,0.D0,0.D0,0.D0,
     *         0.D0,0.D0,0.D0,1.D0,-1.D0,0.D0,0.D0,
     *         0.D0,0.D0,0.D0,0.D0,0.D0,1.D0,-1.D0/
C
      WG(1,1)=2.D0
      WG(2,1)=0.D0
      WG(3,1)=0.D0
      WG(4,1)=0.D0
      WG(1,2)=0.3351800554016
      WG(2,2)=WG(1,2)
      WG(3,2)=0.D0
      WG(4,2)=0.D0
      WG(1,3)=0.5555555555556
      WG(2,3)=WG(1,3)
      WG(3,3)=0.8888888888889
      WG(4,3)=0.D0
      WG(1,4)=0.3478548451375
      WG(2,4)=0.6521451548625
      WG(3,4)=WG(2,4)
      WG(4,4)=WG(1,4)
C
      GP(1,1)=0.D0
      GP(2,1)=0.D0
      GP(3,1)=0.D0
      GP(4,1)=0.D0
      GP(1,2)=-0.7587869106393
      GP(2,2)=-GP(1,2)
      GP(3,2)=0.D0
      GP(4,2)=0.D0
      GP(1,3)=-0.7745966692415
      GP(2,3)=-GP(1,3)
      GP(3,3)=0.D0
      GP(4,3)=0.D0
      GP(1,4)=-0.8611363115941
      GP(2,4)=-0.3399810435849
      GP(3,4)=-GP(2,4)
      GP(4,4)=-GP(1,4)
```

```
C
          IPERM(1)=2
          IPERM(2)=3
          IPERM(3)=1
          GO TO (100,300,500),IND
   100    READ(5,1000)NEL,NDL
          LCOUNT=LCOUNT+1
          NDEL(LCOUNT)=NEL
          WRITE(6,1010)NEL
          IF(NDL)103,103,101
   101    WRITE(6,1025)
          DO 102 I=1,NDL
          READ(5,1026)N,LT(N),PN(N),ZREF(N),NF(N)

   102    WRITE(6,1027)N,LT(N),PN(N),ZREF(N),NF(N)
   103    CONTINUE
          WRITE(6,1020)
          KK=0
          DO 180 I=1,NEL
          READ(5,1030)LNUM,MG,(MNC(J),J=1,20),KI,NINT,LSN,NSS,(ISP(J),
        *   J=1,NSS)
          KNINT=NINT
          KLSN(1)=LSN(1)
          KLSN(2)=LSN(2)
          KNSS=NSS
          DO 105 J=1,KNSS
   105    IS(J)=ISP(J)
          IF(KI.GT.0)GO TO 120
          DO 110 J=1,20
   110    LNC(J)=MNC(J)
   120    KK=KK+1
          IF(LNUM-KK)150,150,130
   130    CONTINUE
          DO 140 J=1,20
   140    LNC(J)=LNC(J)+KI
   150    JJ=0
          DO 170 J=1,20
          K=LNC(J)
          XX(J,1)=X(K)
          XX(J,2)=Y(K)
          XX(J,3)=Z(K)
          DO 160 L=1,3
          JJ=JJ+1
          ND(JJ)=NDF(L,K)
   160    CONTINUE
   170    CONTINUE
          WRITE(NDARAY)(ND(L),L=1,60),KNINT,MG,(XX(N,1),XX(N,2),XX(N,3),
        * LNC(N),N=1,20),KLSN(1),KLSN(2),KNSS,(IS(J),J=1,KNSS)
          CALL COLUMH(CHT,ND,NED,NEQ)
          IF(IPR.EQ.0)WRITE(6,1040)KK,MG,LNC,KNINT,LSN
          KK=KK+1
          IF(LNUM.GE.KK)GO TO 130
          KK=KK-1
          IF(LNUM.EQ.NEL)GO TO 190
   180    CONTINUE
   190    CONTINUE
          RETURN
```

```
C*********************************************************************
   300    CONTINUE
C         INITIALIZE 'C' MATRIX
          DO 310 I=1,6
          DO 310 J=1,6
   310    C(I,J)=0.0
C         ENTER LOOP OVER EACH ELEMENT
          LCOUNT=LCOUNT+1
          NEL=NDEL(LCOUNT)
          DO 490 LNUM=1,NEL
          READ(NDARAY)(ND(L),L=1,60),KNINT,MG,(XX(N,1),XX(N,2),XX(N,3),
      *   LNC(N),N=1,20),KLSN(1),KLSN(2),KNSS,(IS(J),J=1,KNSS)
          IF(NMP.EQ.1.AND.LNUM.GT.1) GO TO 320
          CF=E(MG)/((1.0+PR(MG))*(1.0-2.0*PR(MG)))
          C(1,1)=CF*(1.0-PR(MG))
          C(1,2)=CF*PR(MG)
          C(1,3)=C(1,2)
          C(2,1)=C(1,2)
          C(2,2)=C(1,1)
          C(2,3)=C(1,2)
          C(3,1)=C(1,2)
          C(3,2)=C(1,2)
          C(3,3)=C(1,1)
          C(4,4)=CF*(1.0-2.0*PR(MG))/2.0
          C(5,5)=C(4,4)
          C(6,6)=C(4,4)
          CC1=C(1,1)
          CC2=C(1,2)
          CC3=C(4,4)
   320    CONTINUE
          IF(KNINT.EQ.0)GO TO 440
C         INITIALIZE STIFFNESS MATRIX,LOAD VECTOR AND 'B'MATRIX
          DO 330 I=1,60
          Q(I)=0.0
          DO 330 J=1,60
   330    EK(I,J)=0.0
          NS=6*KNSS
          DO 350 I=1,NS
          DO 350 J=1,60
   350    B(I,J)=0.0
C         COMPUTE STRAIN DISPLACEMENT MATRIX
          DO 365 L=1,KNSS
          LL=IS(L)+1
          E1=STP(LL,1)
          E2=STP(LL,2)
          E3=STP(LL,3)
          CALL BMAT20(2,SA)
          L3=6*L-6
          DO 360 K=1,20
          K3=3*K
          K2=K3-1
          K1=K2-1
          B(L3+1,K1)=SA(K,1)
          B(L3+2,K2)=SA(K,2)
          B(L3+3,K3)=SA(K,3)
          B(L3+4,K1)=SA(K,2)
          B(L3+4,K2)=SA(K,1)
```

```
              B(L3+5,K2)=SA(K,3)
              B(L3+5,K3)=SA(K,2)
              B(L3+6,K1)=SA(K,3)
      360     B(L3+6,K3)=SA(K,1)
      365   CONTINUE
C
              XW=-0.7958224257542
              ICYCLE=0
              MX=0
              DO 420 LX=1,KNINT
              E1=GP(LX,KNINT)
              DO 420 LY=1,KNINT
              E2=GP(LY,KNINT)
              DO 420 LZ=1,KNINT
              E3=GP(LZ,KNINT)
              ICYCLE=ICYCLE+1
              GO TO 390
      380     MX=MX+1
              ICYCLE=ICYCLE+1
              IF(MX.EQ.1)E1=XW
              IF(MX.EQ.2)E1=-XW
              IF(MX.EQ.3)E2=XW
              IF(MX.EQ.4)E2=-XW
              IF(MX.EQ.5)E3=XW
              IF(MX.EQ.6)E3=-XW
              IF(MX.LE.2)E2=0.0
              IF(MX.LE.2)E3=0.0
              IF(MX.GE.3.AND.MX.LE.4)E1=0.0
              IF(MX.GE.3.AND.MX.LE.4)E3=0.0
              IF(MX.GE.5.AND.MX.LE.6)E1=0.0
              IF(MX.GE.5.AND.MX.LE.6)E2=0.0
      390     CALL BMAT20(1,SA)
              GT=WG(1,2)*DET
              IF(KNINT.EQ.3)GT=WG(LX,KNINT)*WG(LY,KNINT)*WG(LZ,KNINT)*DET
              IF(KNINT.EQ.3)GO TO 400
              IF(ICYCLE.GT.8)GT=0.8864265927977*DET
      400     CONTINUE
              GG=GT*WD(MG)
              DO 370 I=1,20
              K=3*I
      370     Q(K)=Q(K)+GG*SF(I)
              C1=GT*CC1
              C2=GT*CC2
              C3=GT*CC3
C         ADD CONTRIBUTION TO STIFFNESS MATRIX
              DO 410 I=1,20
              K3=3*I
              K2=K3-1
              K1=K2-1
              UI=SA(I,1)
              VI=SA(I,2)
              WI=SA(I,3)
              DO 410 J=1,20
              L3=3*J
              L2=L3-1
              L1=L2-1
              UJ=SA(J,1)
              VJ=SA(J,2)
```

```
           WJ=SA(J,3)
           UU=UI*UJ
           VV=VI*VJ
           WW=WI*WJ
           UV=UI*VJ
           VU=VI*UJ
           UW=UI*WJ
           WU=WI*UJ
           VW=VI*WJ
           WV=WI*VJ
C
           EK(K1,L1)=EK(K1,L1)+C1*UU+C3*(VV+WW)
           EK(K2,L2)=EK(K2,L2)+C1*VV+C3*(WW+UU)
           EK(K3,L3)=EK(K3,L3)+C1*WW+C3*(UU+VV)
           EK(K1,L2)=EK(K1,L2)+C2*UV+C3*VU
           EK(K1,L3)=EK(K1,L3)+C2*UW+C3*WU
           EK(K2,L3)=EK(K2,L3)+C2*VW+C3*WV
           IF(I.EQ.J)GO TO 410
           EK(K2,L1)=EK(K2,L1)+C2*VU+C3*UV
           EK(K3,L1)=EK(K3,L1)+C2*WU+C3*UW
           EK(K3,L2)=EK(K3,L2)+C2*WV+C3*VW
  410      CONTINUE
           IF(KNINT.EQ.3)GO TO 420
           IF(ICYCLE.GE.8.AND.ICYCLE.LT.14)GO TO 380
  420      CONTINUE
           DO 430 I=1,60
           DO 430 J=1,60
  430      EK(J,I)=EK(I,J)
           DO 435 I=1,KNSS
           II=6*I-6
           DO 431 J=1,6
           DO 431 K=1,60
           CB(II+J,K)=0.0
           DO 431 L=1,6
  431      CB(II+J,K)=CB(II+J,K)+C(J,L)*B(II+L,K)
           IF(IS(I).LE.0)GO TO 435
           LL=IS(I)+1
           E1=STP(LL,1)
           E2=STP(LL,2)
           E3=STP(LL,3)
           CALL BMAT20(4,SA)
           CALL FST(IS,XJAC,XJACI,CB,I,60)
  435      CONTINUE
  440      CONTINUE
           WRITE(ISTRES)((CB(I,J),I=1,NS),J=1,60)
           NUED=NED
           CALL PASSEM(SK,EK,NDS,ND,NED,NEQ1,NSKY,NUED)
C          DISTRIBUTED LOAD
           DO 450 I=1,60
  450      RF(I)=0.0
           CALL DNLOAD(LT,PN,ZREF,NF,KLSN,RF)
           DO 460 K=1,60
  460      RF(K)=RF(K)-Q(K)
           DO 470 J=1,60
           KK=ND(J)
           IF(KK.EQ.0)GO TO 470
            PQ(KK) = PQ(KK) + RF(J)
```

```
  470    CONTINUE
  490    CONTINUE
         RETURN
C********************************************************************
  500    CONTINUE
         WRITE(6,1070)
         LCOUNT=LCOUNT+1
         NEL=NDEL(LCOUNT)
         DO 550 LNUM=1,NEL
         READ(NDARAY)(ND(L),L=1,60),KNINT,MG,(XX(N,1),XX(N,2),XX(N,3),
     *   LNC(N),N=1,20),KLSN(1),KLSN(2),KNSS,(IS(J),J=1,KNSS)
         READ(ISTRES)((CB(I,J),I=1,NS),J=1,60)
         NS=6*KNSS
         DO 510 I=1,NS
  510    SIG(I)=0.0
         II=NS-6
         DO 530 I=1,60
         K=ND(I)
         IF(K.EQ.0)GO TO 530
         DO 520 L=1,NS
  520    SIG(L)=SIG(L)+CB(L,I)*P(K)
  530    CONTINUE
         CALL PRSTS(KNSS,NS,IS,SIG,SPS)
         WRITE(6,1080)LNUM,IS(1),(SIG(I),I=1,6),(SPS(I),I=1,3)
         M=7
         KI=1
  540    CONTINUE
         M5=M+5
         M1=(M+1)/2
         M2=(M+5)/2
         KI=KI+1
         NUM=IS(KI)
         WRITE(6,1090)NUM,(SIG(I),I=M,M5),(SPS(I),I=M1,M2)
         M=M+6
         IF(M5.EQ.NS)GO TO 550
         GO TO 540
  550    CONTINUE
 1000    FORMAT(2I5)
 1010    FORMAT(35X,'NUMBER OF 3D-20 NODED ELEMENTS IN THE STRUCTURE=',
     *   I5/)
 1020    FORMAT(/8X,'ELEMENT',4X,'MAT',1X,30('*'),'ELEMENT CONNECTIVITY',
     *   30('*'),1X,'INT.',4X,'LOAD SET NUMBER'//10X,'NO.',5X,'GROUP',
     *   1X,'1',3X,'2',3X,'3',3X,'4',3X,'5',3X,'6',3X,'7',3X,'8',3X,'9',
     *   2X,'10',2X,'11',2X,'12',2X,'13',2X,'14',2X,'15',2X,'16',
     *   2X,'17',2X,'18',2X,'19',2X,'20',6X,'ORDER',6X,'1',8X,'2'//)
 1025    FORMAT(/50X,'ELEMENT DISTRIBUTED LOADS'/12X,'NUMBER',12X,
     *   'TYPE',11X,'PRESSURE',13X,'ZREF',12X,'FACE NO.'//)
 1026    FORMAT(2I5,2F10.3,I5)
 1027    FORMAT(10X,I5,12X,I5,2(10X,F10.3),10X,I5/)
 1030    FORMAT(23I3,11I1)
 1040    FORMAT(3X,2I9,20I4,3I9/)
 1070    FORMAT(/40X,'STRESSES AT THE CENTROID AND CENTRE OF ELEMENT',
     *   1X,'FACES'//25X,'STRESSES ARE REFERRED TO GLOBAL AXES AT THE',
     *   1X,'CENTROID AND TO LOCAL AXES AT THE FACES'//4X,'ELEMENT NO.',
     *   2X,'FACE',3X,'SIGMA-XX',4X,'SIGMA-YY',4X,'SIGMA-XY',4X,
     *   'SIGMA-YZ',4X,'SIGMA-XZ',3X,'SIGMA-MAX','SIGMA-MIN',4X,
     *   'S2/ANGLE'//)
```

```
 1080      FORMAT(/2(5X,I5),9(2X,E10.3)/)
 1090      FORMAT(15X,I5,9(2X,E10.3)/)
           RETURN
           END


C******************************************************************
C         ROUTINE COMPUTES THE STRAIN-DISPLACEMENT MATRIX 'B'
C         CALLED BY : THRD20
C******************************************************************
           SUBROUTINE BMAT20(KK,SA)
           IMPLICIT REAL*8(A-H,O-Z)
           COMMON/GASS20/GP(4,4),WG(4,4),IPERM(3)
           COMMON/DERI20/R,S,T,DET,XX(20,3),SF(20),XJAC(3,3),XJACI(3,3)
           DIMENSION SA(20,3),P(3,20)
           RP=(1.0+R)
           RM=(1.0-R)
           SP=1.0+S
           SM=1.0-S
           TP=1.0+T
           TM=1.0-T
           IF(KK.EQ.2.OR.KK.EQ.4)GO TO 90
C         SHAPE FUNCTIONS
           SF(1)=RP*SM*TM*(RP+SM+TM-5.0)*0.125
           SF(2)=RP*SP*TM*(RP+SP+TM-5.0)*0.125
           SF(3)=RM*SP*TM*(RM+SP+TM-5.0)*0.125
           SF(4)=RM*SM*TM*(RM+SM+TM-5.0)*0.125
           SF(5)=RP*SM*TP*(RP+SM+TP-5.0)*0.125
           SF(6)=RP*SP*TP*(RP+SP+TP-5.0)*0.125
           SF(7)=RM*SP*TP*(RM+SP+TP-5.0)*0.125
           SF(8)=RM*SM*TP*(RM+SM+TP-5.0)*0.125
           SF(9)=RP*SP*SM*TM*0.25
           SF(10)=RP*RM*SP*TM*0.25
           SF(11)=RM*SP*SM*TM*0.25
           SF(12)=RP*RM*SM*TM*0.25
           SF(13)=RP*SP*SM*TP*0.25
           SF(14)=RP*RM*SP*TP*0.25
           SF(15)=RM*SP*SM*TP*0.25
           SF(16)=RP*RM*SM*TP*0.25
           SF(17)=RP*SM*TP*TM*0.25
           SF(18)=RP*SP*TP*TM*0.25
           SF(19)=RM*SP*TP*TM*0.25
           SF(20)=RM*SM*TP*TM*0.25
C
C         DERIVATIVES OF SHAPE FUNCTIONS
   90      P(1,1)=SM*TM*(2.0*RP+SM+TM-5.0)*0.125
           P(1,2)=SP*TM*(2.0*RP+SP+TM-5.0)*0.125
           P(1,3)=-SP*TM*(2.0*RM+SP+TM-5.0)*0.125
           P(1,4)=-SM*TM*(2.0*RM+SM+TM-5.0)*0.125
           P(1,5)=SM*TP*(2.0*RP+SM+TP-5.0)*0.125
           P(1,6)=SP*TP*(2.0*RP+SP+TP-5.0)*0.125
           P(1,7)=-SP*TP*(2.0*RM+SP+TP-5.0)*0.125
           P(1,8)=-SM*TP*(2.0*RM+SM+TP-5.0)*0.125
           P(1,9)=SP*SM*TM*0.25
           P(1,10)=-R*SP*TM*0.5
           P(1,11)=-P(1,9)
           P(1,12)=-R*SM*TM*0.5
           P(1,13)=SP*SM*TP*0.25
```

```
            P(1,14)=-R*SP*TP*0.5
            P(1,15)=-P(1,13)
            P(1,16)=-R*SM*TP*0.5
            P(1,17)=SM*TP*TM*0.25
            P(1,18)=SP*TP*TM*0.25
            P(1,19)=-P(1,18)
            P(1,20)=-P(1,17)


            P(2,1)=-RP*TM*(RP+2.0*SM+TM-5.0)*0.125
            P(2,2)=RP*TM*(RP+2.0*SP+TM-5.0)*0.125
            P(2,3)=RM*TM*(RM+2.0*SP+TM-5.0)*0.125
            P(2,4)=-RM*TM*(RM+2.0*SM+TM-5.0)*0.125
            P(2,5)=-RP*TP*(RP+2.0*SM+TP-5.0)*0.125
            P(2,6)=RP*TP*(RP+2.0*SP+TP-5.0)*0.125
            P(2,7)=RM*TP*(RM+2.0*SP+TP-5.0)*0.125
            P(2,8)=-RM*TP*(RM+2.0*SM+TP-5.0)*0.125
            P(2,9)=-RP*S*TM*0.5
            P(2,10)=RP*RM*TM*0.25
            P(2,11)=-RM*S*TM*0.5
            P(2,12)=-P(2,10)
            P(2,13)=-RP*S*TP*0.5
            P(2,14)=RP*RM*TP*0.25
            P(2,15)=-RM*S*TP*0.5
            P(2,16)=-P(2,14)
            P(2,17)=-RP*TP*TM*0.25
            P(2,18)=-P(2,17)
            P(2,19)=RM*TP*TM*0.25
            P(2,20)=-P(2,19)


            P(3,1)=-RP*SM*(RP+SM+2.0*TM-5.0)*0.125
            P(3,2)=-RP*SP*(RP+SP+2.0*TM-5.0)*0.125
            P(3,3)=-RM*SP*(RM+SP+2.0*TM-5.0)*0.125
            P(3,4)=-RM*SM*(RM+SM+2.0*TM-5.0)*0.125
            P(3,5)=RP*SM*(RP+SM+2.0*TP-5.0)*0.125
            P(3,6)=RP*SP*(RP+SP+2.0*TP-5.0)*0.125
            P(3,7)=RM*SP*(RM+SP+2.0*TP-5.0)*0.125
            P(3,8)=RM*SM*(RM+SM+2.0*TP-5.0)*0.125
            P(3,9)=-RP*SP*SM*0.25
            P(3,10)=-RP*RM*SP*0.25
            P(3,11)=-RM*SP*SM*0.25
            P(3,12)=-RP*RM*SM*0.25
            P(3,13)=-P(3,9)
            P(3,14)=-P(3,10)
            P(3,15)=-P(3,11)
            P(3,16)=-P(3,12)
            P(3,17)=-RP*SM*T*0.5
            P(3,18)=-RP*SP*T*0.5
            P(3,19)=-RM*SP*T*0.5
            P(3,20)=-RM*SM*T*0.5
C
C           JACOBIAN MATRIX-XJAC
            DO 110 I=1,3
            DO 110 J=1,3
            C=0.0
            DO 100 L=1,20
  100       C=C+P(I,L)*XX(L,J)
  110       XJAC(I,J)=C
            IF(KK.EQ.3)GO TO 160
```

```
C          INVERT JACOBIAN-XJACI
           DO 120 I=1,3
           J=IPERM(I)
           K=IPERM(J)
           XJACI(I,I)=XJAC(J,J)*XJAC(K,K)-XJAC(K,J)*XJAC(J,K)
           XJACI(I,J)=XJAC(K,J)*XJAC(I,K)-XJAC(I,J)*XJAC(K,K)
    120    XJACI(J,I)=XJAC(J,K)*XJAC(K,I)-XJAC(J,I)*XJAC(K,K)
           IF(KK.EQ.4) GO TO 160
           DET=XJAC(1,1)*XJACI(1,1)+XJAC(1,2)*XJACI(2,1)+XJAC(1,3)*
        *  XJACI(3,1)
C          MATRIX OF X-Y-Z-DERIVATIVES
           DO 150 I=1,3
           DO 150 J=1,20
           C=0.0
           DO 140 K=1,3
    140    C=C+XJACI(I,K)*P(K,J)
    150    SA(J,I)=C/DET
    160    CONTINUE
           RETURN
           END
```

```
C*************************************************************************
C       CONTROL ROUTINE FOR PLATE BENDING ELEMENTS
C*************************************************************************
        SUBROUTINE PLATE(A,LLIB,NSHAPE,MTOT)
        IMPLICIT REAL*8(A-H,O-Z)
        COMMON/DIM/N1,N2,N3,N4,N5,N6,N7,N8,N9,N10,N11,N12,N13,N14
        COMMON/PAR/IND,NET,NSN,NMP,NEQ,NSKY,NEQ1,LCOUNT
        COMMON/TAPES/ISTRES,NDARAY,IPR
        COMMON/PRECI/ITWO
        DIMENSION A(MTOT)
        DIMENSION LLIB(7,3)
        DO 100 J=1,NSHAPE
        LB = LLIB(5,J)
        GO TO (20,40,60),LB
   20   NED = 24
        NSP = 4
        NS = 5
        IF(IND.NE.1) GO TO 30
        N10 = N9+NED
        N11 = N10+1*ITWO
        N12 = N11+1*ITWO
        N13 = N12+1*ITWO
        NSKY = 1
   30   CALL PLATE8(A(N2),A(N3),A(N4),A(N5),A(N6),A(N7),A(N11),
      * A(N12),A(N1),A(N8),A(N9),A(N10),A(N13),NS,NED,NSP)
        GO TO 100
   40   NED = 12
        NSP = 4
        NS = 5
        IF(IND.NE.1) GO TO 50
        N10 = N9+NED
        N11 = N10+1*ITWO
        N12 = N11+1*ITWO
        N13 = N12+1*ITWO
        NSKY = 1
   50   CALL PLATE4(A(N2),A(N3),A(N4),A(N5),A(N6),A(N7),A(N11),
      * A(N12),A(N1),A(N8),A(N9),A(N10),A(N13),NS,NED,NSP)
        GO TO 100
   60   CALL PLATE2
  100   CONTINUE
        RETURN
        END


C*********************************************************************
        SUBROUTINE PLATE4(X,Y,Z,E,PR,WD,SK,PQ,NDF,CHT,ND,NDS,P,
      * NS,NED,NSP)
C
C       A BILINEAR ISOPARAMETRIC PLATE BENDING ELEMENT-A 4 NODED
C       QUADRILATERAL FINITE ELEMENT FORMULATION FOR THE SOLUTION
C       OF PLATE BENDING PROBLEMS.ISOPARAMETRIC CONCEPT AND SELECTIV.
C       INTEGRATION TECHNIQUES ARE ADOPTED
C
        IMPLICIT REAL*8(A-H,O-Z)
        COMMON/DIM/N1,N2,N3,N4,N5,N6,N7,N8,N9,N10,N11,N12,N13,N14
```

```
          COMMON/PAR/IND,NET,NSN,NMP,NEQ,NSKY,NEQ1,LCOUNT
          COMMON/TAPES/ISTRES,NDARAY,IPR
          DIMENSION X(NSN),Y(NSN),Z(NSN),E(NMP),PR(NMP),WD(NMP)
          DIMENSION SK(NSKY),P(NEQ),NDF(6,NSN),ND(NED),NDS(NEQ1)
          DIMENSION ELC(100),STR(100,3)
          INTEGER CHT(NEQ)
          DIMENSION Q(12),XL(4),YL(4),SF(3,4),B(5,12),WG(4),R(4),S(4)
          DIMENSION U(4),V(4),C(5,5),LNC(4),CBS(5,12),CBB(5,12)
          DIMENSION XJACI(2,2),XJAC(2,2),SFG(3,4),CB(5,12),MNC(4)
          DIMENSION EK(12,12),SIG(6),TMS(4,4),NDEL(21),EKB(12,12)
          DIMENSION EKS(12,12),PQ(NEQ)
C         SQRT(X) = DSQRT(X)
          GO TO (310,350,400),IND
   310    READ(5,11) NEL
          LCOUNT = LCOUNT+1
          NDEL(LCOUNT) = NEL
          IF(NSN.GT.100) WRITE(6,133)
          IF(NSN.GT.100) STOP
          DO 314 I=1,NSN
   314    ELC(I) = 0.0
          WRITE(6,33) NEL
          WRITE(6,44)
          KK = 0
          DO 340 I = 1,NEL
          READ(5,22) LNUM,MG,(MNC(J),J=1,4),TH,UDL,KI
          IF(KI.GT.0) GO TO 316
          DO 315 J=1,4
   315    LNC(J) = MNC(J)
   316    KK = KK+1
          IF(LNUM-KK) 2,2,3
     3    CONTINUE
          DO 5 J=1,4
     5    LNC(J) = LNC(J)+KI
     2    JJ = 0
          DO 330 J=1,4
          K = LNC(J)
          ELC(K) = ELC(K)+1.0
          XL(J) = X(K)
          YL(J) = Y(K)
          DO 320 L=3,5
          JJ = JJ+1
   320    ND(JJ) = NDF(L,K)
   330    CONTINUE
          WRITE(NDARAY)(ND(LL),LL=1,12),(XL(NN),YL(NN),LNC(NN),
     *    NN=1,4),TH,MG,UDL
          CALL COLUMH(CHT,ND,NED,NEQ)
          IF(IPR.EQ.0)WRITE(6,55)KK,MG,(LNC(J),J=1,4),TH,UDL
          KK = KK+1
          IF(LNUM.GE.KK) GO TO 3
          KK = KK-1
          IF(LNUM.EQ.NEL) GO TO 341
   340    CONTINUE
   341    CONTINUE
          RETURN
C****************************************************************
C         CALCULATE ELEMENT STIFFNESS MATRIX EK
   350    CONTINUE
          DATA R/-1.D0,1.D0,1.D0,-1.D0/
```

```
          DATA S/-1.D0,-1.D0,1.D0,1.D0/
          DATA U/-1.D0,1.D0,1.D0,-1:D0/
          DATA V/-1.D0,-1.D0,1.D0,1.D0/
          DATA WG/1.D0,1.D0,1.D0,1.D0/
          DO 10 I=1,5
          DO 10 J=1,5
          C(I,J) = 0.0
    10    CONTINUE
C         ENTER LOOP OVER EACH ELEMENT
          LCOUNT = LCOUNT+1
          NEL = NDEL(LCOUNT)
          DO 200 LNUM=1,NEL
          READ(NDARAY)(ND(LL),LL=1,12),(XL(NN),YL(NN),LNC(NN),NN=1,4),
     *    TH,MG,UDL
          IF(NMP.EQ.1.AND.LNUM.GT.1)GO TO 15
C         COMPUTE CONSTITUTIVE MATRIX 'C'
          C(1,1) = E(MG)*TH**3/(12.0*(1.0-PR(MG)**2))
          C(1,2) = C(1,1)*PR(MG)
          C(2,1) = C(1,2)
          C(2,2) = C(1,1)
          C(3,3) = C(1,1)*(1.0-PR(MG))/2.0
          C(4,4) = E(MG)*TH/(2.4*(1.0+PR(MG)))
          C(5,5) = C(4,4)
    15    CONTINUE
C         INITIALIZE ELEMENT STIFFNESS MATRIX'EK' AND LOAD VECTOR'Q'
          DO 20 I=1,12
          Q(I)=0.0
          DO 20 J=1,12
    20    EK(I,J)=0.0
C
C         SHEAR ENERGY TERMS'EKS'(1*1 INTEGRATION IS USED)
C
          DO 175 I=1,12
          DO 175 J=1,12
   175    EKS(I,J) = 0.0
          GPSX = 0.0
          GPSY = 0.0
C         EVALUATE SHAPE FUNCTIONS
          DO 182 I=1,4
          AA = (1.0+R(I)*GPSX)
          BB = (1.0+S(I)*GPSY)
          SF(1,I) = 0.25*R(I)*BB
          SF(2,I) = 0.25*S(I)*AA
          SF(3,I) = 0.25*AA*BB
   182    CONTINUE
C         JACOBIAN MATRIX 'XJAC'
          DO 184 I=1,2
          DO 184 J=1,2
   184    XJAC(I,J) = 0.0
          DO 186 I=1,4
          DO 186 K=1,2
          XJAC(K,1) = XJAC(K,1)+XL(I)*SF(K,I)
          XJAC(K,2) = XJAC(K,2)+YL(I)*SF(K,I)
   186    CONTINUE
C         CALCULATE DETERMINANT JACOBIAN 'DJAC'
          DJAC = XJAC(1,1)*XJAC(2,2)-XJAC(1,2)*XJAC(2,1)
C         COMPUTE JACOBIAN INVERSE 'XJACI'
          XJACI(1,1) = XJAC(2,2)/DJAC
```

```
            XJACI(1,2) = -XJAC(1,2)/DJAC
            XJACI(2,1) = -XJAC(2,1)/DJAC
            XJACI(2,2) = XJAC(1,1)/DJAC
            DB = 4.0*DJAC
      C     FORM GLOBAL DERIVATIVES 'SFG'
            DO 188 I=1,2
            DO 188 J=1,4
      188   SFG(I,J) = 0.0
            DO 190 I=1,2
            DO 190 K=1,4
            DO 190 J=1,2
            SFG(I,K) = SFG(I,K)+XJACI(I,J)*SF(J,K)
      190   CONTINUE
            DO 192 I=1,4
            SFG(3,I) = SF(3,I)
      192   CONTINUE
      C     COMPUTE CBS-THE SHEAR CONTRIBUTION TO STRESS MATRIX'CB'
            DO 197 I=1,NS
            DO 197 J=1,12
      197   B(I,J) = 0.0
            DO 198 I=1,4
            K1 = 3*(I-1)+1
            K2 = K1+1
            K3 = K2+1
            B(4,K1) = SFG(1,I)
            B(4,K3) = SFG(3,I)
            B(5,K1) = SFG(2,I)
            B(5,K2) = -SFG(3,I)
      198   CONTINUE
            DO 205 I=1,3
            DO 205 J=1,12
      205   CBS(I,J) = 0.0
            DO 199 I=4,5
            DO 199 J=1,12
            CBS(I,J) = C(4,4)*B(I,J)
      199   CONTINUE
            II = 0
            DO 196 IA = 1,4
            JJ = 0
            DO 195 IB=1,4
            DO 193 I=1,2
            EKS(II+1,JJ+1)=EKS(II+1,JJ+1)+C(4,4)*SFG(I,IA)*SFG(I,IB)*DB
            EKS(II+I,JJ+3-I)=EKS(II+I,JJ+3-I)-C(4,4)*SFG(I+1,IA)*
          *  SFG(4-I,IB)*DB
      193   EKS(II+I+1,JJ+I+1)=EKS(II+I+1,JJ+I+1)+C(4,4)*SFG(3,IA)*
          *  SFG(3,IB)*DB
            DO 194 I=1,3,2
      194   EKS(II+I,JJ+4-I)=EKS(II+I,JJ+4-I)+C(4,4)*SFG(I,IA)*
          *  SFG(4-I,IB)*DB
      195   JJ = JJ+3
      196   II = II+3
      C
      C     INITIALIZE BENDING STIFFNESS MATRIX'EKB' (2*2 INTEGRATION USED)
            DO 30 I=1,12
            DO 30 J=1,12
      30    EKB(I,J) = 0.0
      C     ENTER LOOPS FOR NUMERICAL INTEGRATION
            NSP=4
```

```
          DO 170 IX=1,NSP
C         EVALUATE SHAPE FUNCTIONS ITS DERIVATIVES,JACOBIAN
C         AND DETERMINANT JACOBIAN
          GP = 0.5773502691896
          DO 50 I=1,4
          AA = (1.0+R(I)*(U(IX)*GP))
          BB = (1.0+S(I)*(V(IX)*GP))
          SF(1,I) = 0.25*R(I)*BB
          SF(2,I) = 0.25*S(I)*AA
          SF(3,I) = 0.25*AA*BB
   50     CONTINUE
C         COMPUTE JACOBIAN MATRIX'XJAC'
          DO 60 I=1,2
          DO 60 J=1,2
   60     XJAC(I,J) = 0.0
          DO 70 I=1,4
          DO 70 K=1,2
          XJAC(K,1) = XJAC(K,1)+XL(I)*SF(K,I)
          XJAC(K,2) = XJAC(K,2)+YL(I)*SF(K,I)
   70     CONTINUE
C         CALCULATE DETERMINANT JACOBIAN'DJAC'
          DJAC = XJAC(1,1)*XJAC(2,2)-XJAC(1,2)*XJAC(2,1)
C         COMPUTE JACOBIAN INVERSE 'XJACI'
          XJACI(1,1) = XJAC(2,2)/DJAC
          XJACI(1,2) = -XJAC(1,2)/DJAC
          XJACI(2,1) = -XJAC(2,1)/DJAC
          XJACI(2,2) = XJAC(1,1)/DJAC
          DA = DJAC*WG(IX)
C         FORM GLOBAL DERIVATIVES'SFG'
          DO 80 I=1,2
          DO 80 J=1,4
   80     SFG(I,J) = 0.0
          DO 90 I=1,2
          DO 90 K=1,4
          DO 90 J=1,2
          SFG(I,K) = SFG(I,K)+XJACI(I,J)*SF(J,K)
   90     CONTINUE
          DO 92 I=1,4
   92     SFG(3,I) = SF(3,I)
C         COMPUTE'CBB'BENDING CONTRIBUTION TO STRESS MATRIX'CB'
C         COMPUTE STRAIN DISPLACEMENT MATRIX'B'
          DO 100 I=1,5
          DO 100 J=1,12
  100     B(I,J) = 0.0
          DO 110 I=1,4
          K1 = 3*(I-1)+1
          K2 = K1+1
          K3 = K2+1
          B(1,K3) = SFG(1,I)
          B(2,K2) = -SFG(2,I)
          B(3,K3) = SFG(2,I)
          B(3,K2) = -SFG(1,I)
  110     CONTINUE

          DO 120 I=1,3
          DO 120 J=1,12
          CBB(I,J) = C(I,I)*B(I,J)
```

```
            IF(I.GT.2) GO TO 120
            II = 3-I
            CBB(I,J) = C(I,II)*B(II,J)+CBB(I,J)
      120   CONTINUE
            DO 113 I=4,5
            DO 113 J=1,12
      113   CBB(I,J) = 0.0
C           COMPUTE THE STRESS MATRIX 'CB'
            DO 123 I=1,NS
            DO 123 J=1,12
      123   CB(I,J) = 0.0
            DO 124 I=1,NS
            DO 124 J=1,12
            IF(IX.EQ.1) GO TO 126
            CB(I,J) = CB(I,J)+CBB(I,J)
            GO TO 124
      126   CB(I,J) = CB(I,J)+CBB(I,J)+CBS(I,J)
      124   CONTINUE
            WRITE(ISTRES)((CB(I,J),I=1,NS),J=1,NED)
            II = 0
            DO 150 IA=1,4
            JJ = 0
            DO 140 IB=1,4
            DO 130 I=2,3
            EKB(II+I,JJ+I)=EKB(II+I,JJ+I)+C(1,1)*SFG(4-I,IA)*SFG(4-I,IB)*DA
     *      +C(3,3)*SFG(I-1,IA)*SFG(I-1,IB)*DA
            EKB(II+I,JJ+5-I)=EKB(II+I,JJ+5-I)-(C(1,2)*SFG(4-I,IA)*
     *      SFG(I-1,IB)+C(3,3)*SFG(I-1,IA)*SFG(4-I,IB))*DA
      130   CONTINUE
      140   JJ = JJ+3
      150   II = II+3
C
C           CALCULATE ELEMENT LOAD VECTOR DUE TO U.D.L
            DO 810 I=1,4
            NN = 3*I-2
            Q(NN) = Q(NN)+SFG(3,I)*UDL*DA
      810   CONTINUE
      170   CONTINUE
C           COMPUTE ELEMENT STIFFNESSMATRIX'EK'
            DO 800 I=1,12
            DO 800 J=1,12
            EK(I,J) = EK(I,J)+EKB(I,J)+EKS(I,J)
      800   CONTINUE
            DO 820 I=2,12
            K= I-1
            DO 820 J=1,K
      820   EK(I,J) = EK(J,I)
            DO 860 J=1,NED
            JJ =ND(J)
            IF(JJ) 860,860,840
      840   PQ(JJ) = PQ(JJ)+Q(J)
      860   CONTINUE
            NUED = NED
            CALL PASSEM(SK,EK,NDS,ND,NED,NEQ1,NSKY,NUED)
      200   CONTINUE
            RETURN
```

```
C***************************************************************
  400     CONTINUE
C         CALCULATION OF BENDING STRESS RESULTANTS AT THE NODES
          LCOUNT = LCOUNT+1
          NEL = NDEL(LCOUNT)
          WRITE(6,84)
          VAL = 3.0
          TERM = DSQRT(VAL)/2.0
          TMS(1,1) = 1.0+TERM
          TMS(1,2) = -0.5
          TMS(1,3) = 1.0-TERM
          TMS(1,4) = -0.5
          TMS(2,1) = -0.5
          TMS(2,2) = TMS(1,1)
          TMS(2,3) = -0.5
          TMS(2,4) = TMS(1,3)
          TMS(3,1) = TMS(1,3)
          TMS(3,2) = -0.5
          TMS(3,3) = TMS(1,1)
          TMS(3,4) = -0.5
          TMS(4,1) = -0.5
          TMS(4,2) = TMS(1,3)
          TMS(4,3) = -0.5
          TMS(4,4) = TMS(1,1)
          DO 430 I=1,NSN
          DO 430 J=1,3
          STR(I,J) = 0.0
  430     CONTINUE
          WRITE(6,77)
          DO 480 LNUM=1,NEL
          READ(NDARAY)(ND(LL),LL=1,12),(XL(NN),YL(NN),LNC(NN),
     *    NN=1,4),TH,MG,UDL
          DO 470 K=1,NSP
          READ(ISTRES)((CB(I,J),I=1,NS),J=1,NED)
          DO 440 I=1,NS
  440     SIG(I) = 0.0
          DO 460 J=1,NED
          JJ =ND(J)
          IF(JJ.EQ.0) GO TO 460
          DO 450 I=1,NS
  450     SIG(I) =CB(I,J)*P(JJ)+SIG(I)
  460     CONTINUE
C         SHEAR STRESS RESULTANTS WITHIN AN ELEMENT AT R=0,S =0
          IF(K.EQ.1)WRITE(6,111)LNUM,(SIG(I),I=4,5)
          DO 470 J=1,3
          DO 470 I=1,4
          NN = LNC(I)
          STR(NN,J) = STR(NN,J)+TMS(I,K)*SIG(J)/ELC(NN)
  470     CONTINUE
  480     CONTINUE
          WRITE(6,122)
          WRITE(6,88)
          WRITE(6,99)(I,(STR(I,J),J=1,3),I=1,NSN)
   11     FORMAT(I5)
   22     FORMAT(6I5,2F10.4,I5)
   33      FORMAT(//35X,'NUMBER OF 4-NODED PLATE ELEMENTS INTHE'
     *    'STRUCTURE.....= ',I5//)
```

```
   44    FORMAT(//25X,'ELEMENT',4X,'MATERIAL',5X,'NODAL CONNECTIVITY',4X,
    *    'THICKNESS',9X,'UDL'/28X,'NO.',5X,'GROUP NO.'//)
   55    FORMAT(25X,I5,5X,I5,5X,4I5,2X,F10.4,5X,F10.4/)
   77    FORMAT(//4X,'ELEMENT NO.',6X,'SHEAR FORCE-QX',6X,
    *    'SHEAR FORCE-QY'//)
   84    FORMAT(//8X,'SHEAR FORCES AT THE CENTRE OF THE ELEMENT'//)
   88    FORMAT(//28X,'BENDING',14X,'BENDING',13X,'TWISTING'/11X,
    *    'NODE-NO.',8X,'MOMENT-MX',12X,'MOMENT-MY',12X,
    *    'MOMENT-MXY'//)
   99    FORMAT(13X,I5,3E20.8/)
  111    FORMAT(7X,I5,3X,2E20.8/)
  122    FORMAT(//36X,'BENDING MOMENTS AT THE NODES'//)
  133    FORMAT(//5X,'NO.OF STRUCTURE NODES EXCCEDE 100'/5X,
    *    'CHANGE THE NUMBERS IN THE CORRESPONDING DIMENSION STATEMENTS'/
    *    5X,'IN THE SUBROUTINE PLATE4'//)
         RETURN
         END


         SUBROUTINE PLATE8(X,Y,Z,E,PR,WD,SK,PQ,NDF,CHT,ND,NDS,P,
    *    NS,NED,NSP)
         IMPLICIT REAL*8(A-H,O-Z)
         COMMON/DIM/N1,N2,N3,N4,N5,N6,N7,N8,N9,N10,N11,N12,N13,N14
         COMMON/PAR/IND,NET,NSN,NMP,NEQ,NSKY,NEQ1,LCOUNT
         COMMON/TAPES/ISTRES,NDARAY,IPR
         DIMENSION X(NSN),Y(NSN),Z(NSN),E(NMP),PR(NMP),WD(NMP)
         DIMENSION SK(NSKY),P(NEQ),PQ(NEQ),NDF(6,NSN),ND(NED),NDS(NEQ1)
         DIMENSION Q(24),XL(8),YL(8),SF(8),SFD(2,8),B(5,24),LNC(8)
         DIMENSION WG(2),GP(2),R(8),S(8),C(5,5),CB(5,24),SFDG(2,8)
         DIMENSION EK(24,24),SIG(6),TMS(4,4),XJAC(2,2),XJACI(2,2)
         DIMENSION ELC(100),STR(100,5),NDEL(20)
         INTEGER CHT(NEQ)
C        SQRT(X)=DSQRT(X)
         GO TO(310,350,400),IND
  310    READ(5,11)NEL
         LCOUNT=LCOUNT+1
         NDEL(LCOUNT)=NEL
         IF(NSN.GT.100)WRITE(6,133)
         IF(NSN.GT.100) STOP
         DO 314 I=1,NSN
  314    ELC(I)=0.0
         WRITE(6,33)NEL
         WRITE(6,44)
         DO 340 I=1,NEL
         READ(5,22)LNUM,MG,(LNC(J),J=1,8),TH,UDL
         DO 315 L=1,4
         NN=LNC(L)
         ELC(NN)=ELC(NN)+1.0
  315    CONTINUE
         JJ=0
         DO 330 J=1,8
         K=LNC(J)
         XL(J)=X(K)
         YL(J)=Y(K)
         DO 320 L=3,5
         JJ=JJ+1
  320    ND(JJ)=NDF(L,K)
  330    CONTINUE
```

```
        WRITE(NDARAY)(ND(L),L=1,24),(XL(N),YL(N),LNC(N),N=1,8),
     *  TH,MG,UDL
        CALL COLUMH(CHT,ND,NED,NEQ)
        IF(IPR.EQ.0)WRITE(6,55)LNUM,MG,(LNC(J),J=1,8),TH,UDL
  340   CONTINUE
        RETURN
C       ****************************************************************
  350   CONTINUE
        DATA WG/1.DO,1.DO/
        DATA GP/-0.5773502691896,0.5773502691896/
        DATA R/-1.DO,1.DO,1.DO,-1.DO,0.DO,1.DO,0.DO,-1.DO/
        DATA S/-1.DO,-1.DO,1.DO,1.DO,-1.DO,0.DO,1.DO,0.DO/
        DO 10 I=1,5
        DO 10 J=1,5
        C(I,J)=0.0
   10   CONTINUE
C       ENTER LOOP OVER EACH ELEMENT
        LCOUNT=LCOUNT+1
        NEL=NDEL(LCOUNT)
        DO 200 LNUM=1,NEL
        READ(NDARAY)(ND(L),L=1,24),(XL(N),YL(N),LNC(N),N=1,8),
     *   TH,MG,UDL
C       COMPUTE THE CONSTITUTIVE MATRIX
        C(1,1)=E(MG)*TH**3/(12.0*(1.0-PR(MG)**2))
        C(1,2)=C(1,1)*PR(MG)
        C(2,1)=C(1,2)
        C(2,2)=C(1,1)
        C(3,3)=C(1,1)*(1.0-PR(MG))/2.0
        C(4,4)=E(MG)*TH/(2.4*(1.0+PR(MG)))
        C(5,5)=C(4,4)
C       INITIALIZE STIFFNESS MATRIX AND LOAD VECTOR
        DO 20 I=1,24
        DO 20 J=1,24
        EK(I,J)=0.0
   20   Q(I)=0.0
C       ENTER LOOPS FOR NUMERICAL INTEGRATION(2*2INTEGRATION IS USED)
        DO 170 IX=1,2
        DO 170 IY=1,2
C       CALCULATE SHAPE FUNCTIONS,THEIR DERIVATIVES,JACOBIAN,
C       DETERMINANT JACOBIAN,JACOBIAN INVERSE
        DO 50 I=1,8
        AA=(1.0+R(I)*GP(IX))
        BB=(1.0+S(I)*GP(IY))
        IF(I.GT.4)GO TO 40
        SF(I)=0.25*AA*BB*(AA+BB-3.0)
        SFD(1,I)=0.25*(2.0*AA+BB-3.0)*BB*R(I)
        SFD(2,I)=0.25*(2.0*BB+AA-3.0)*AA*S(I)
        GO TO 50
   40   AA=AA+(R(I)**2-1.0)*GP(IX)**2
        BB=BB+(S(I)**2-1.0)*GP(IY)**2
        SF(I)=0.5*AA*BB
        SFD(1,I)=0.5*(R(I)+2.0*(R(I)**2-1.0)*GP(IX))*BB
        SFD(2,I)=0.5*(S(I)+2.0*(S(I)**2-1.0)*GP(IY))*AA
   50   CONTINUE
C       COMPUTE THE JACOBIAN MATRIX
        DO 60 I=1,2
        DO 60 J=1,2
   60   XJAC(I,J)=0.0
```

```
          DO 70 I=1,8
          DO 70 K=1,2
          XJAC(K,1)=XJAC(K,1)+XL(I)*SFD(K,I)
          XJAC(K,2)=XJAC(K,2)+YL(I)*SFD(K,I)
   70     CONTINUE
C         CALCULATE DETERMINANT AND INVERSE OF JACOBIAN MATRIX
          DJAC=XJAC(1,1)*XJAC(2,2)-XJAC(1,2)*XJAC(2,1)
          XJACI(1,1)=XJAC(2,2)/DJAC
          XJACI(1,2)=-XJAC(1,2)/DJAC
          XJACI(2,1)=-XJAC(2,1)/DJAC
          XJACI(2,2)=XJAC(1,1)/DJAC
          DA=DJAC*WG(IX)*WG(IY)
C         CALCULATE CARTESIAN DERIVATIVES
          DO 80 I=1,2
          DO 80 J=1,8
   80     SFDG(I,J)=0.0
          DO 90 I=1,2
          DO 90 K=1,8
          DO 90 J=1,2
   90     SFDG(I,K)=SFDG(I,K)+XJACI(I,J)*SFD(J,K)
          DO 100 I=1,5
          DO 100 J=1,24
  100     B(I,J)=0.0
          DO 110 I=1,8
          K1=3*(I-1)+1
          K2=K1+1
          K3=K2+1
          B(1,K3)=SFDG(1,I)
          B(2,K2)=-SFDG(2,I)
          B(3,K3)=SFDG(2,I)
          B(3,K2) = - SFDG(1,I)
          B(4,K1)=-SFDG(1,I)
          B(4,K3)=-SF(I)
          B(5,K1)=-SFDG(2,I)
          B(5,K2)=SF(I)
  110     CONTINUE
          DO 120 I=1,5
          DO 120 J=1,24
          CB(I,J)=C(I,I)*B(I,J)
          IF(I.GT.2)GO TO 120
          II=3-I
          CB(I,J)=CB(I,J)+C(I,II)*B(II,J)
  120     CONTINUE
          WRITE(ISTRES)((CB(I,J),I=1,NS),J=1,NED)
C         CALCULATE THE ELEMENT STIFFNESS MATRIX
          DO 140 I=1,24
          DO 140 J=1,24
          DO 140 K=1,5
  140     EK(I,J)=EK(I,J)+B(K,I)*CB(K,J)*DA
C         CALCULATION OF LOAD VECTOR DUE TO U.D.L
          DO 160 I=1,8
          NN=3*I-2
  160     Q(NN)=Q(NN)+SF(I)*UDL*DA
  170     CONTINUE
          DO 180 I=2,24
          K=I-1
          DO 180 J=1,K
  180     EK(I,J)=EK(J,I)
```

```
             DO 240 J=1,NED
             JJ=ND(J)
             IF(JJ) 240,240,220
   220       PQ(JJ)=PQ(JJ)+Q(J)
   240       CONTINUE
             NUED=NED
             CALL PASSEM(SK,EK,NDS,ND,NED,NEQ1,NSKY,NUED)
   200       CONTINUE
             RETURN
C            ***********************************************************
   400       CONTINUE
             LCOUNT=LCOUNT+1
             NEL=NDEL(LCOUNT)
             WRITE(6,88)
             VAL=3.0
             TERM=DSQRT(VAL)/2.0
             TMS(1,1)=1.0+TERM
             TMS(1,2)=-0.5
             TMS(1,3)=-0.5
             TMS(1,4)=1.0-TERM
             TMS(2,1)=-0.5
             TMS(2,2)=TMS(1,4)
             TMS(2,3)=TMS(1,1)
             TMS(2,4)=-0.5
             TMS(3,1)=TMS(2,2)
             TMS(3,2)=-0.5
             TMS(3,3)=-0.5
             TMS(3,4)=TMS(1,1)
             TMS(4,1)=-0.5
             TMS(4,2)=TMS(1,1)
             TMS(4,3)=TMS(2,2)
             TMS(4,4)=-0.5
             DO 430 I=1,NSN
             DO 430 J=1,NS
   430       STR(I,J)=0.0
             DO 480 LNUM=1,NEL
             WRITE(6,77)LNUM
             READ(NDARAY)(ND(L),L=1,24),(XL(N),YL(N),LNC(N),N=1,8),
     *       TH,MG,UDL
             DO 470 K=1,NSP
             READ(ISTRES)((CB(I,J),I=1,NS),J=1,NED)
             DO 440 I=1,NS
   440       SIG(I)=0.0
             DO 460 J=1,NED
             JJ=ND(J)
             IF(JJ.EQ.0)GO TO 460
             DO 450 I=1,NS
   450       SIG(I)=SIG(I)+CB(I,J)*P(JJ)
   460       CONTINUE
             WRITE(6,111)K,(SIG(I),I=1,NS)
             DO 470 J=1,NS
             DO 470 I=1,4
             NN=LNC(I)
             STR(NN,J)=STR(NN,J)+TMS(I,K)*SIG(J)/ELC(NN)
   470       CONTINUE
   480       CONTINUE
             WRITE(6,122)
             WRITE(6,99)(I,(STR(I,J),J=1,NS),I=1,NSN)
```

```
   11      FORMAT(I5)
   22      FORMAT(10I5,2F10.4)
   33      FORMAT(//35X,'NUMBER OF 8-NODED PLATE ELEMENTS IN THE ST',
    *        'RUCTURE.....=',I5//)
   44      FORMAT(//25X,'ELEMENT',3X,'MATERIAL',13X,'NODAL CONNECTIVITY',
    *        15X,'THICKNESS',10X,'UDL'/27X,'NO',5X,'GROUP NO'//)
   55      FORMAT(25X,I5,5X,I5,5X,8I5,2X,F10.4,5X,F10.4/)
   77      FORMAT(//2X,'ELEMENT NO.',I3/)
   88      FORMAT(//28X,'BENDING',14X,'BENDING',13X,'TWISTING',14X,
    *        'SHEAR',15X,'SHEAR'/11X,'POINT NO.',7X,'MOMENT-MX',12X,
    *        'MOMENT-MY',12X,'MOMENT-MXY',12X,'FORCE-QX',12X,'FORCE-QY'//)
   99      FORMAT(13X,I5,5E20.8/)
  111      FORMAT(11X,I5,2X,5E20.8)
  122      FORMAT(45X,'BENDING MOMENTS AND SHEAR FORCES AT NODES'/)
  133      FORMAT(//5X,'NO.OF STRUCTURE NODES EXCCEDE 100'/5X,
    *        'CHANGE THE NUMBERS IN THE CORRESPONDING DIMENSION STATEMENTS'/
    *        5X,'IN THE SUBROUTINE PLATE4'//)
           RETURN
           END


           SUBROUTINE PLATE2
           RETURN
           END
```

```
C     ****************************************************************
C     CONTROL ROUTINE FOR SHELL
C     ****************************************************************
      SUBROUTINE SHELL (A,LLIB,NSHAPE,MTOT)
      IMPLICIT REAL*8(A-H,O-Z)
      COMMON/DIM/N1,N2,N3,N4,N5,N6,N7,N8,N9,N10,
     * N11,N12,N13,N14
      COMMON/PAR/IND,NET,NSN,NMP,NEQ,NSKY,NEQ1,LCOUNT
      COMMON/TAPES/ISTRES,NDARAY,IPR
      COMMON/PRECI/ITWO
      COMMON/MULT/ELMN
      DIMENSION XB(100),YB(100),ZB(100)
      DIMENSION LLIB(7,3)
      DIMENSION A(MTOT)
      DO 100 J=1,NSHAPE
      LB=LLIB(6,J)
      GO TO (20,40,60),LB
   20 NED=40
      NEN=8
      NSP=8
      NS=5
      IF(IND.NE.1) GO TO 30
      READ(5,11)NADJN
      N10=N9+NED
      N11=N10+1*ITWO
      N12=N11+1*ITWO
      N13=N12+1*ITWO
      NSKY=1
   30 CALL SHELL8(A(N2),A(N3),A(N4),A(N5),A(N6),A(N7),
     *XB,YB,ZB,A(N11),A(N12),A(N1),A(N8),A(N9),A(N10),
     *A(N13),NS,NED,NSP,NADJN)
      GO TO 100
   40 NED=24
      NSP=4
      NS=5
      IF(IND.NE.1) GO TO 50
      N10=N9+NED
      N11=N10+1*ITWO
      N12=N11+1*ITWO
      N13=N12+1*ITWO
      NSKY=1
   50 CALL SHELL4(A(N2),A(N3),A(N4),A(N5),A(N6),A(N7),
     * A(N11),A(N12),A(N1),A(N8),A(N9),A(N10),A(N13),
     * NS,NED,NSP)
      GO TO 100
   60 CALL SHELL2
  100 CONTINUE
      RETURN
   11 FORMAT(I5)
      END


      SUBROUTINE SHELL8 (X,Y,Z,E,PR,WD,XB,YB,ZB,SK,PQ,NDF,
     * CHT,ND,NDS,P,NS,NED,NSP,NADJN)
      IMPLICIT REAL*8(A-H,O-Z)
      COMMON/DIM/N1,N2,N3,N4,N5,N6,N7,N8,N9,N10,N11,N12,N13,N14
```

```
         COMMON/PAR/IND,NET,NSN,NMP,NEQ,NSKY,NEQ1,LCOUNT
         COMMON/TAPES/ISTRES,NDARAY,IPR
         COMMON/PRECI/ITWO
         DIMENSION XM(8),YM(8),ZM(8),XD(8),YD(8),ZD(8),LNC(8),
       * Q(45,3),PRESS(2),VERT(2),NADJ(8),EK(45,45),SIG(6),
       * NDEL(21)
         DIMENSION X(NSN),Y(NSN),Z(NSN),E(NMP),PR(NMP),WD(NMP),
       * SK(NSKY),NDF(6,NSN),P(NEQ),ND(NED),NDS(NEQ1),PQ(NEQ)
         INTEGER CHT(NEQ)
         DIMENSION XB(100),YB(100),ZB(100)
         DIMENSION A1(9),A2(9),A3(9),THICK(9),B1(9),B2(9),
       * B3(9),GP(4,4),WG(4,4),R(8),S(8),SF(9),SFD(2,9),XJAC(3,3),
       * XJACI(3,3),IPERM(3),L1(5),L2(5),M1(6),M2(6),D(3,9),
       * F(3,9),DCOS(3,3),THETA(5,6),T(6,45),B(5,45),ANG(3),
       * C(5,5),CB(5,40)
         DIMENSION M3(8),STRTN(50,5),STRBN(50,5),ELC(50),TMS(4,4)
C        SQRT(X)=DSQRT(X)
C        ABS(X)=DABS(X)
         FLOAT(I)=DBLE(I)
         N=0
         GO TO (10,100,800),IND
   10    READ(5,11)NN,NI,XB(NN),YB(NN),ZB(NN)
         IF(NI.EQ.0)NI=1
         N=N+NI
         IF(NN-N)20,20,30
   20    CONTINUE
         IF(NADJN-NN)50,50,10
   30    NX=(NN-N+NI)/NI
         XN=(XB(NN)-XB(N-NI))/FLOAT(NX)
         YN=(YB(NN)-YB(N-NI))/FLOAT(NX)
         ZN=(ZB(NN)-ZB(N-NI))/FLOAT(NX)
   40    XB(N)=XB(N-NI)+XN
         YB(N)=YB(N-NI)+YN
         ZB(N)=ZB(N-NI)+ZN
         N=N+NI
         IF(N.LT.NN)GO TO 40
         IF(NADJN-NN)50,50,10
   50    CONTINUE
         WRITE(6,55)(NNN,XB(NNN),YB(NNN),ZB(NNN),NNN=1,NADJN)
   60    CONTINUE
         READ(5,44)NEL,ICOUNT
         LCOUNT=LCOUNT+1
         NDEL(LCOUNT)=NEL
         DO 65 I=1,NSN
   65    ELC(I)=0.0
         WRITE(6,66)NEL
         WRITE(6,77)
         DO 90 I=1,NEL
         READ(5,22)LNUM,MG,(LNC(J),J=1,8),(NADJ(K),K=1,8)
         DO 67 L=1,4
         NN=LNC(L)
         ELC(NN)=ELC(NN)+1.0
   67    CONTINUE
         IF(IPR.EQ.0)WRITE(6,88)LNUM,MG,(LNC(J),J=1,8),
       * (NADJ(K),K=1,8)
         JJ=0
         DO 80 J=1,8
         NNT=LNC(J)
```

```
                    NNB=NADJ(J)
                    XM(J)=(X(NNT)+XB(NNB))/2.0
                    YM(J)=(Y(NNT)+YB(NNB))/2.0
                    ZM(J)=(Z(NNT)+ZB(NNB))/2.0
                    XD(J)=(X(NNT)-XB(NNB))/2.0
                    YD(J)=(Y(NNT)-YB(NNB))/2.0
                    ZD(J)=(Z(NNT)-ZB(NNB))/2.0
                    DO 70 L=1,5
                    JJ=JJ+1
       70           ND(JJ)=NDF(L,NNT)
       80           CONTINUE
                    WRITE(NDARAY)(ND(LL),LL=1,40),(XM(N),YM(N),ZM(N),
          *         XD(N),YD(N),ZD(N),LNC(N),N=1,8),MG
                    CALL COLUMH(CHT,ND,NED,NEQ)
       90           CONTINUE
                    RETURN
      100           CONTINUE
                    DATA GP/0.D0,0.D0,0.D0,0.D0,-0.5773502691896,
          *         0.5773502691896,0.D0,0.D0,-0.7745966692415,
          *         0.D0,0.7745966692415,0.D0,-0.8611363115941,
          *         -0.3399810435849,0.3399810435849,0.8611363115941/
                    DATA WG/2.D0,0.D0,0.D0,0.D0,1.D0,1.D0,0.D0,0.D0,
          *         0.5555555555556,0.8888888888889,0.5555555555556,
          *         0.D0,0.3478548451375,0.6521451548625,0.6521451548625,
          *         0.3478548451375/
                    DATA R/-1.D0,1.D0,1.D0,-1.D0,0.D0,1.D0,0.D0,-1.D0/
                    DATA S/-1.D0,-1.D0,1.D0,1.D0,-1.D0,0.D0,1.D0,0.D0/
                    DATA IPERM/2,3,1/
                    DATA L1/1,2,1,2,3/
                    DATA L2/1,2,2,3,1/
                    DATA M1/1,2,3,1,2,3/
                    DATA M2/1,2,3,2,3,1/
                    DATA M3/1,0,3,0,5,0,7,0/
                    READ(5,33)PRESS(1),PRESS(2),VERT(1),VERT(2)
C                   ENTER LOOP OVER EACH ELEMENT
                    LCOUNT=LCOUNT+1
                    NEL=NDEL(LCOUNT)
                    DO 780 LNUM=1,NEL
                    READ(NDARAY)(ND(LL),LL=1,40),(XM(N),YM(N),ZM(N),
          *         XD(N),YD(N),ZD(N),LNC(N),N=1,8),MG
C                   COMPUTE D.C.S OF NORMAL TO MID SURFACE AND
C                   THICKNESSES AT EACH NODE
                    A1(9)=0.0
                    A2(9)=0.0
                    A3(9)=0.0
                    DO 110 I=1,9
                    IF(I.EQ.9)GO TO 110
                    CONST=-0.25
                    IF(I.GT.4)CONST=0.5
                    A1(I)=XD(I)*2.0
                    A2(I)=YD(I)*2.0
                    A3(I)=ZD(I)*2.0
                    A1(9)=A1(9)+CONST*A1(I)
                    A2(9)=A2(9)+CONST*A2(I)
                    A3(9)=A3(9)+CONST*A3(I)
      110           THICK(I)=DSQRT(A1(I)**2+A2(I)**2+A3(I)**2)
C                   NORMALISE D.C.S. OF NORMAL AND 'D.C.S. OF ROTATIONS
C                   'WHICH IS TO BE COMPUTED
```

```
          DO 130 I=1,9
          DL=A1(I)/THICK(I)
          DM=A2(I)/THICK(I)
          DN=A3(I)/THICK(I)
          DIV=DSQRT(DL**2+DM**2)
          IF(DIV.GT.1.E-05)GO TO 120
          A1(I)=0.0
          A2(I)=1.0
          A3(I)=0.0
          B1(I)=1.0
          B2(I)=0.0
          B3(I)=0.0
          GO TO 130
   120    A1(I)=-DM/DIV
          A2(I)=DL/DIV
          A3(I)=0.0
          B1(I)=-DL*DN/DIV
          B2(I)=-DM*DN/DIV
          B3(I)=DIV
   130    CONTINUE
C         COMPUTE CONSTITUTIVE MATRIX 'C' IN LOCAL COORDINATES
          C(1,1)=E(MG)/(1.0-PR(MG)**2.)
          C(2,2)=C(1,1)
          C(1,2)=C(1,1)*PR(MG)
          C(2,1)=C(1,2)
          C(3,3)=E(MG)/(2.0*(1.0+PR(MG)))
          C(4,4)=C(3,3)/1.2
          C(5,5)=C(4,4)
C         INITIALIZE LOAD VECTOR 'Q' AND ELEMENT STIFFNESS
C         MATRIX EK
          DO 150 I=1,45
          DO 140 J=1,3
   140    Q(I,J)=0.0
          DO 150 J=1,I
   150    EK(I,J)=0.0
          NSTRESS=0
C         ENTER LOOPS FOR NUMERICAL INTEGRATION
C         INTEGRATION FOR C12,C21,&C33, ALSO FOR THE GRAVITY LOAD
          IF(ICOUNT.NE.2)GO TO 181
          LOOP=1
          NINTX=2
          NINTY=2
          NINTZ=2
          GO TO 190
C         INTEGRATION FOR C22,&C44
   160    LOOP=2
          NINTX=3
          NINTY=2
          NINTZ=2
          GO TO 190
C         INTEGRATION FOR C11,&C55
   170    LOOP=3
          NINTX=2
          NINTY=3
          NINTZ=2
          GO TO 190
C         INTEGRATION FOR STRESS COMPUTION,SURFACE PRESSURES
C         AND VERTICAL SURFACE LOADS
```

```
180     LOOP=4
        NINTX=2
        NINTY=2
        NINTZ=2
        GO TO 190
181     CONTINUE
        IF(ICOUNT.EQ.3)GO TO 183
182     LOOP=5
        NINTX=3
        NINTY=3
        NINTZ=3
        GO TO 190
183     LOOP=6
        NINTX=2
        NINTY=2
        NINTZ=2
190     DO 640 LX=1,NINTX
        DO 640 LY=1,NINTY
        DO 640 LZ=1,NINTZ
        RG=GP(LX,NINTX)
        SG=GP(LY,NINTY)
        TG=GP(LZ,NINTZ)
        IF(LOOP.EQ.4)TG=(-1)**LZ
        WIJ=WG(LX,NINTX)*WG(LY,NINTY)
        WIJK=WIJ*WG(LZ,NINTZ)
C       COMPUTE SHAPE FUNCTIONS AND IT'S DERIVATIVES W.R.TO
C       LOCAL COORDINATES
        DO 210 I=1,8
        AA=(1.0+R(I)*RG)
        BB=1.0+S(I)*SG
        IF(I.GT.4)GO TO 200
        SF(I)=0.25*AA*BB*(AA+BB-3.0)
        SFD(1,I)=0.25*(2.0*AA+BB-3.0)*BB*R(I)
        SFD(2,I)=.25*(2.0*BB+AA-3.0)*AA*S(I)
        GO TO 210
200     AA=AA+(R(I)**2-1.0)*RG**2
        BB=BB+(S(I)**2-1.0)*SG**2
        SF(I)=.5*AA*BB
        SFD(1,I)=.5*(R(I)+2.0*(R(I)**2-1.0)*RG)*BB
        SFD(2,I)=.5*(S(I)+2.0*(S(I)**2-1.0)*SG)*AA
210     CONTINUE
        AA=1.0-RG**2
        BB=1.0-SG**2
        SF(9)=AA*BB
        SFD(1,9)=-2.0*RG*BB
        SFD(2,9)=-2.0*SG*AA
C       COMPUTE JACOBIAN 'XJAC'
        DO 220 I=1,3
        DO 220 J=1,3
220     XJAC(I,J)=0.0
        DO 240 I=1,8
        DO 230 K=1,2
        XJAC(K,1)=XJAC(K,1)+SFD(K,I)*(XM(I)+TG*XD(I))
        XJAC(K,2)=XJAC(K,2)+SFD(K,I)*(YM(I)+TG*YD(I))
        XJAC(K,3)=XJAC(K,3)+SFD(K,I)*(ZM(I)+TG*ZD(I))
230     CONTINUE
        XJAC(3,1)=XJAC(3,1)+SF(I)*XD(I)
        XJAC(3,2)=XJAC(3,2)+SF(I)*YD(I)
```

```
          XJAC(3,3)=XJAC(3,3)+SF(I)*ZD(I)
    240   CONTINUE
C         INVERT JACOBIAN
          DO 250 I=1,3
          J=IPERM(I)
          K=IPERM(J)
          XJACI(I,I)=XJAC(J,J)*XJAC(K,K)-XJAC(K,J)*XJAC(J,K)
          XJACI(I,J)=XJAC(K,J)*XJAC(I,K)-XJAC(I,J)*XJAC(K,K)
          XJACI(J,I)=XJAC(J,K)*XJAC(K,I)-XJAC(J,I)*XJAC(K,K)
    250   CONTINUE
          DET3=XJAC(1,1)*XJACI(1,1)+XJAC(1,2)*XJACI(2,1)+XJAC(1,3)
      *   *XJACI(3,1)
          DET2=DSQRT(XJACI(1,3)**2+XJACI(2,3)**2+XJACI(3,3)**2)
          DO 260 I=1,3
          DO 260 J=1,3
    260   XJACI(I,J)=XJACI(I,J)/DET3
C         CALCULATE MATRICES D AND F TO FORM B MATRIX
          DO 270 I=1,3
          DO 270 J=1,9
          D(I,J)=0.0
    270   F(I,J)=0.0
          DO 290 I=1,3
          DO 290 K=1,9
          DO 280 J=1,2
          D(I,K)=D(I,K)+XJACI(I,J)*SFD(J,K)
    280   F(I,K)=F(I,K)+XJACI(I,J)*SFD(J,K)*THICK(K)*TG*0.5
    290   F(I,K)=F(I,K)+XJACI(I,3)*SF(K)*THICK(K)*0.5
C         CALCULATE D.C.S. OF ORTHOGONAL LOCAL AXIS SYSTEM,DCOS
          DO 300 I=1,3
          J=IPERM(I)
          K=IPERM(J)
          DCOS(1,I)=XJAC(1,I)
    300   DCOS(3,I)=XJAC(1,J)*XJAC(2,K)-XJAC(1,K)*XJAC(2,J)
          DO 310 I=1,3
          J=IPERM(I)
          K=IPERM(J)
          DCOS(2,I)=DCOS(3,J)*DCOS(1,k)-DCOS(3,K)*DCOS(1,J)
    310   CONTINUE
C         NORMALIZE  DCOS
          DO 320 I=1,3
          DIV=DSQRT(DCOS(I,1)**2+DCOS(I,2)**2+DCOS(I,3)**2)
          DO 320 J=1,3
          DCOS(I,J)=DCOS(I,J)/DIV
    320   CONTINUE
C         COMPUTE MATRIX THETA WHICH TRANSFORMS GLOBAL STRAINS
C         TO LOCAL STRAINS
          DO 340 I=1,5
          I1=L1(I)
          I2=L2(I)
          DO 330 J=1,6
          J1=M1(J)
          J2=M2(J)
    330   THETA(I,J)=DCOS(I1,J1)*DCOS(I2,J2)
          DO 340 J=4,6
          J1=M1(J)
          J2=M2(J)
    340   THETA(I,J)=THETA(I,J)+DCOS(I1,J2)*DCOS(I2,J1)
C         CONVERT THETA TO ENGG.STRAIN TRANSFORMATION MATRIX
```

```
         DO 360 J=1,3
         DO 350 I=3,5
   350   THETA(I,J)=2.0*THETA(I,J)
         K=J+3
         DO 360 I=1,2
   360   THETA(I,K)=0.5*THETA(I,K)
C        MATRIX 'T' TRANSFORMS DISPLACEMENTS TO GLOBAL STRAINS
         DO 370 I=1,6
         DO 370 J=1,45
   370   T(I,J)=0.0
         DO 380 I=1,9
         K1=5*(I-1)+1
         K2=K1+1
         K3=K2+1
         K4=K3+1
         K5=K4+1
         T(1,K1)=D(1,I)
         T(1,K4)=-F(1,I)*B1(I)
         T(1,K5)=F(1,I)*A1(I)
         T(2,K2)=D(2,I)
         T(2,K4)=-F(2,I)*B2(I)
         T(2,K5)=F(2,I)*A2(I)
         T(3,K3)=D(3,I)
         T(3,K4)=-F(3,I)*B3(I)
         T(3,K5)=F(3,I)*A3(I)
         T(4,K1)=D(2,I)
         T(4,K2)=D(1,I)
         T(4,K4)=-F(1,I)*B2(I)-F(2,I)*B1(I)
         T(4,K5)=F(1,I)*A2(I)+F(2,I)*A1(I)
         T(5,K2)=D(3,I)
         T(5,K3)=D(2,I)
         T(5,K4)=-F(2,I)*B3(I)-F(3,I)*B2(I)
         T(5,K5)=F(2,I)*A3(I)+F(3,I)*A2(I)
         T(6,K1)=D(3,I)
         T(6,K3)=D(1,I)
         T(6,K4)=-F(1,I)*B3(I)-F(3,I)*B1(I)
         T(6,K5)=F(1,I)*A3(I)+F(3,I)*A1(I)
   380   CONTINUE
C        TRANSFORMATION MATRIX 'B' FROM DISPLACEMENTS TO
C        LOCAL STRAIN
         IF(LOOP.GT.4)GO TO 411
         DO 400 I=1,2
         IF(I.EQ.LOOP-1)GO TO 400
         DO 390 K=1,45
         B(I,K)=0.0
         DO 390 J=1,6
   390   B(I,K)=B(I,K)+THETA(I,J)*T(J,K)
   400   CONTINUE
         LSTART=LOOP+2
         LFIN=LSTART
         IF(LOOP.EQ.4)LSTART=3
         IF(LOOP.EQ.4)LFIN=5
         DO 410 L=LSTART,LFIN
         DO 410 K=1,45
         B(L,K)=0.0
         DO 410 J=1,6
         B(L,K)=B(L,K)+THETA(L,J)*T(J,K)
   410   CONTINUE
```

```
         GO TO 413
   411   CONTINUE
         DO 412 I=1,5
         DO 412 J=1,45
         B(I,J)=0.0
         DO 412 K=1,6
         B(I,J)=B(I,J)+THETA(I,K)*T(K,J)
   412   CONTINUE
   413   CONTINUE
C        CALCULATE LOCAL STRESS DISPLACEMENTS TRANSFORMATIONS
C        'T' AGAIN
         DV=1.0
         IF(LOOP.NE.4)DV=DET3*WIJK
         IF(LOOP.GT.4)GO TO 445
         DO 430 I=1,2
         IF(I.EQ.LOOP-1)GO TO 430
         II=I
         IF(LOOP.EQ.1.OR.LOOP.EQ.4)II=3-I
         DO 420 K=1,45
   420   T(I,K)=C(I,II)*B(II,K)*DV
   430   CONTINUE
         IF(LOOP.EQ.4)LSTART=1
         IF(LOOP.EQ.4)LFIN=5
         DO 440 K=1,45
         DO 440 L=LSTART,LFIN
         IF(LOOP.NE.4)T(L,K)=0.0
         IF(LOOP.EQ.4.AND.L.GE.3)T(L,K)=0.0
   440   T(L,K)=C(L,L)*B(L,K)*DV+T(L,K)
         GO TO 456
   445   CONTINUE
         DO 446 I=1,5
         DO 446 J=1,45
         T(I,J)=0.0
   446   CONTINUE
         DO 450 I=1,2
         II=3-I
         DO 450 K=1,45
   450   T(I,K)=C(I,II)*B(II,K)*DV
         DO 455 K=1,45
         DO 455 L=1,5
         T(L,K)=C(L,L)*B(L,K)*DV+T(L,K)
   455   CONTINUE
   456   CONTINUE
         IF(LOOP.NE.4)GO TO 570
C        UNIFORM SURFACE PRESSURE
   500   IF(PRESS(LZ).EQ.0.0)GO TO 520
         SURFP=-PRESS(LZ)*(-1)**LZ
         DO 510 I=1,3
   510   ANG(I)=DCOS(3,I)
         NCOL=2
         IDENT=1
         GO TO 530
C        VERTICAL SURFACE LOADS
   520   IF(VERT(LZ).EQ.0.0)GO TO 560
         SURFP=VERT(LZ)
         ANG(1)=0.0
         ANG(2)=0.0
         ANG(3)=-1.0
```

```
                    NCOL=3
                    IDENT=2
C                   LOAD VECTOR FOR THE ABOVE LOADS
        530         DO 550 I=1,9
                    II=5*(I-1)
                    AA=SF(I)*WIJ*DET2*SURFP
                    DO 540 J=1,3
                    Q(II+J,NCOL)=Q(II+J,NCOL)+ANG(J)*AA
        540         CONTINUE
                    Q(II+4,NCOL)=Q(II+4,NCOL)-0.5*THICK(I)*AA*TG*(B1(I)
              *     *ANG(1)+B2(I)*ANG(2)+B3(I)*ANG(3))
                    Q(II+5,NCOL)=Q(II+5,NCOL)+0.5*THICK(I)*AA*TG*(A1(I)
             **ANG(1)+A2(I)*ANG(2)+A3(I)*ANG(3))
        550         CONTINUE
                    GO TO (520,560),IDENT
        560         CONTINUE
                    GO TO 632
C                   ELEMENT BODY LOADS
        570         IF(WD(MG).EQ.0.0)GO TO 585
                    DO 580 I=1,9
                    II=5*(I-1)
                    AA=SF(I)*WIJK*DET3
                    IF(LOOP.EQ.1)AA=-AA
                    Q(II+3,1)=Q(II+3,1)-WD(MG)*AA
                    Q(II+4,1)=Q(II+4,1)+0.5*AA*TG*THICK(I)*B3(I)*WD(MG)
                    Q(II+5,1)=Q(II+5,1)-0.5*AA*TG*THICK(I)*A3(I)*WD(MG)
        580         CONTINUE
        585         CONTINUE
C                   FORM ELEMENT STIFFNESS MATRIX
                    IF(LOOP.LT.5)GO TO 600
                    DO 590 K=1,5
                    DO 590 I=1,45
                    DO 590 J=1,I
        590         EK(I,J)=EK(I,J)+B(K,I)*T(K,J)
                    GO TO 632
        600         DO 620 K=1,2
                    IF(K.EQ.LOOP-1)GO TO 620
                    DO 610 I=1,45
                    DO 610 J=1,I
        610         EK(I,J)=EK(I,J)+B(K,I)*T(K,J)
        620         CONTINUE
                    DO 630 I=1,45
                    DO 630 J=1,I
                    EK(I,J)=EK(I,J)+B(LSTART,I)*T(LSTART,J)
        630         CONTINUE
        632         CONTINUE
                    IF(LOOP.NE.4)GO TO 640
                    DO 635 N=1,5
                    L=45-N
                    M=L+1
                    DO 635 I=1,5
                    AA=T(I,M)/EK(M,M)
                    DO 635 J=1,L
                    T(I,J)=T(I,J)-AA*EK(M,J)
        635         CONTINUE
                    DO 638 I=1,5
                    DO 638 J=1,40
                    CB(I,J)=T(I,J)
```

```
      638     CONTINUE
              WRITE(ISTRES)((CB(I,J),I=1,5),J=1,40)
      640     CONTINUE
              IF(LOOP.GT.4)GO TO 180
              GO TO (160,170,180,650),LOOP
      650     CONTINUE
   C          ELIMINATE CENTER NODE
              DO 710 N=1,5
              L=45-N
              M=L+1
              DO 710 I=1,L
              AA=EK(M,I)/EK(M,M)
              DO 700 K=1,3
      700     Q(I,K)=Q(I,K)-AA*Q(M,K)
              DO 710 J=1,I
              EK(I,J)=EK(I,J)-AA*EK(M,J)
      710     CONTINUE
   C          SYMMETRIZE STIFFNESS MATRIX
              DO 720 I=2,40
              K=I-1
              DO 720 J=1,K
      720     EK(J,I)=EK(I,J)
              DO 750 J=1,NED
              JJ=ND(J)
              IF(JJ)750,750,730
      730     DO 740 I=1,3
      740     PQ(JJ)=PQ(JJ)+Q(J,I)
      750     CONTINUE
              NUED=45
              CALL PASSEM(SK,EK,NDS,ND,NED,NEQ1,NSKY,NUED)
      780     CONTINUE
              WRITE(6,99)
              RETURN
      800     CONTINUE
              LCOUNT=LCOUNT+1
              NEL=NDEL(LCOUNT)
              WRITE(6,122)
              VAL=3.0
              TERM=DSQRT(VAL)/2.0
              TMS(1,1)=1.0+TERM
              TMS(1,2)=-0.5
              TMS(1,3)=-0.5
              TMS(1,4)=1.0-TERM
              TMS(2,1)=-0.5
              TMS(2,2)=TMS(1,4)
              TMS(2,3)=TMS(1,1)
              TMS(2,4)=-0.5
              TMS(3,1)=TMS(2,2)
              TMS(3,2)=-0.5
              TMS(3,3)=-0.5
              TMS(3,4)=TMS(1,1)
              TMS(4,1)=-0.5
              TMS(4,2)=TMS(1,1)
              TMS(4,3)=TMS(2,2)
              TMS(4,4)=-0.5
              DO 810 I=1,NSN
              DO 810 J=1,NS
              STRTN(I,J)=0.0
```

```
              STRBN(I,J)=0.0
      810     CONTINUE
              DO 950 LNUM=1,NEL
              WRITE(6,111)LNUM
              READ(NDARAY)(ND(LL),LL=1,40),(XM(N).YM(N).ZM(N),XD(N),
          *   YD(N),ZD(N),LNC(N),N=1,8),MG
              II=0
              KK=0
              DO 930 K=1,NSP
              IF(K.NE.M3(K))GO TO 820
              II=II+1
              GO TO 830
      820     KK=KK+1
      830     CONTINUE
              READ(ISTRES)((CB(I,J),I=1,5),J=1,40)
              DO 840 I=1,NS
      840     SIG(I)=0.0
              DO 880 J=1,NED
              JJ=ND(J)
              IF(JJ.EQ.0)GO TO 880
              DO 860 I=1,NS
      860     SIG(I)=CB(I,J)*P(JJ)+SIG(I)
      880     CONTINUE
              WRITE(6,144)K,(SIG(I),I=1,5)
              DO 910 J=1,NS
              DO 910 I=1,4
              NN=LNC(I)
              IF(K.NE.M3(K))GO TO 900
              STRBN(NN,J)=STRBN(NN,J)+TMS(I,II)*SIG(J)/ELC(NN)
              GO TO 910
      900     CONTINUE
              STRTN(NN,J)=STRTN(NN,J)+TMS(I,KK)*SIG(J)/ELC(NN)
      910     CONTINUE
      930     CONTINUE
      950     CONTINUE
              WRITE(6,133)(I,(STRBN(I,J),J=1,5),I=1,NSN)
              WRITE(6,155)(I,(STRTN(I,J),J=1,5),I=1,NSN)
       11     FORMAT(2I5,3F10.4)
       22     FORMAT(18I3)
       33     FORMAT(4F10.3)
       44     FORMAT(2I5)
       55     FORMAT(/45X,'COORDINATES OF BOTTOM NODES OF
          *   SHELL8 GROUP '//18X,'NODE',5X,'X-COORD',5X,'Y-COORD',
          *   5X,'Z-COORD',15X,'NODE',5X,'X-COORD',5X,'Y-COORD',
          *   5X,'Z-COORD'//(18X,I4,3F13.4,15X,I4,3F13.4/))
       66     FORMAT(//35X,'NUMBER OF 8-NODED SHELL ELEMENTS IN THE
          *   STRUCTURE=',I5//)
       77     FORMAT(/20X,'ELEMENT',3X,'MATERIAL',6X,'NODAL CONNECTIVITY
          *   OF ELEMENT TOP',14X,'NODAL CONNECTIVITY OF ELEMENT BOTTOM'
          *   /22X,'NO.'5X,'GROUP NO.'//)
       88     FORMAT(20X,I5,5X,I5,5X,8I5,9X,8I5/)
       99     FORMAT(///56X,'NODAL DISPLACEMENTS'//5X,'NODE',8X,
          *   'X-DISPLACEMENTS',6X,'Y-DISPLACEMENTS',4X,'Z-DISPLACEMENTS',
          *   8X,'ROTATION-X',9X,'ROTATION-Y',9X,'ROTATION-Z'/)
      111     FORMAT(//2X,'ELEMENT NO.',I3/)
      122     FORMAT(52X,'STRESSES AT GAUSS POINTS'//25X,'POINT NO.',
          *   3X,'SIGMA-X',8X,'SIGMA-Y',8X,'TAU-XY',9X,'TAU-YZ',9X,
          *   'TAU-ZX'/)
```

```
133    FORMAT(47X,'STRESSES AND MOMENTS AT BOTTOM NODES',//25X,
      * 'NODE NUMBER',4X,'SIGMA-X',8X,'SIGMA-Y',8X,'TAU-XY',9X,
      * 'TAU-YZ',9X,'TAU-ZX'//(25X,I5,2X,5E15.6/))
144    FORMAT(25X,I5,2X,5E15.6/)
155    FORMAT(49X,'STRESSES AND MOMENTS AT TOP NODES',//25X,
      * 'NODE NO.',4X,'SIGMA-X',8X,'SIGMA-Y',8X,'TAU-XY',9X,
      * 'TAU-YZ',9X,'TAU-ZX',8X,//(25X,I5,2X,5E15.6/))
       RETURN
       END

C      SUBROUTINE FOR SHELL2
       SUBROUTINE SHELL2
       RETURN
       END


C      SUBROUTINE FOR ELEMENT SHELL4
       SUBROUTINE SHELL4(X,Y,Z,E,PR,WD,SK,PQ,NDF,CHT,
      * ND,NDS,P,NS,NED,NSP)
       IMPLICIT REAL*8(A-H,O-Z)
C      A BILINEAR DEGENERATED ISOPARAMETRIC SHELL ELEMENT
C      GEOMETRY: 4-NODED QUADRILATERAL IN 3D-SPACE
C      VARIABLES: 6 D.O.F./NODE,3 DISPLACEMENTS&3ROTATIONS
       COMMON/DIM/N1,N2,N3,N4,N5,N6,N7,N8,N9,N10,N11,
      * N12,N13,N14
       COMMON/PAR/IND,NET,NSN,NMP,NEQ,NSKY,NEQ1,LCOUNT
       COMMON/TAPES/ISTRES,NDARAY,IPR
       COMMON/MULT/ELMN
       DIMENSION X(NSN),Y(NSN),Z(NSN),E(NMP),PR(NMP),PQ(NEQ),
      * WD(NMP),SK(NSKY),P(NEQ),NDF(6,NSN),ND(NED),NDS(NEQ1),
      * ELC(300),FORCE(300,3),XMOMNT(300,3),FORCE1(300,3)
       INTEGER CHT(NEQ)
       DIMENSION XL(4),YL(4),ZL(4),SF(3,4),B(3,24),
      * B1(2,24),B2(2,24),B3(2,24),CBB1(3,24),CBB2(3,24),
      * CBS(2,24),CBS1(2,24),CBS2(2,24),CBS3(2,24),LNC(4),
      * MNC(4),DC(3,3),C(3,3),XJAC(3,3),A(4),XJACI(3,3)
       DIMENSION BU(2,4),SFG(3),V3(3,4),VA(3,4),T(4),R(4),
      * S(4),Q(24),LX(3),NDEL(21),TMS(4,4),SIG1(5),SIG2(5),
      * STR(5),EKB(24,24),EKS(24,24),EKT(24,24),EKP(24,24),
      * EKB1(24,24),EK(24,24)
       DIMENSION Q1(24),LOTYPE(5),GSIG(150,3),GMOM(150,3),GSHR(150,3),
      * F(5),TEMPI(4),TEMPF(4),SUMM(150),DIFF(150),TXY(150),PSMX(150),
      * PSMM(150),ANG(150)
       GO TO (10,100,800),IND
10     READ(5,11)NEL
       LCOUNT=LCOUNT+1
       NDEL(LCOUNT)=NEL
       DO 65 I=1,NSN
65     ELC(I)=0.0
       WRITE(6,66)NEL
       WRITE(6,77)
       KK=0
       DO 90 I=1,NEL
       READ(5,44)LNUM,MG,(MNC(J),J=1,4),(T(K),K=1,4),
      * (LOTYPE(I1),I1=1,3),(F(I1),I1=1,3),WT,TORC,IS,MEMB,KI,
      * IC,IP
       IF(KI.GT.0)GO TO 16
       DO 15 J=1,4
15     LNC(J)=MNC(J)
```

```
      16    KK=KK+1
             IF(LNUM-KK)2,2,3
      3      CONTINUE
             DO 5 J=1,4
      5      LNC(J)=LNC(J)+KI
      2      JJ=0
             DO 33 J=1,4
             K=LNC(J)
             ELC(K)=ELC(K)+1.0
             XL(J)=X(K)
             YL(J)=Y(K)
             ZL(J)=Z(K)
             DO 32 L=1,6
             JJ=JJ+1
      32     ND(JJ)=NDF(L,K)
      33     CONTINUE
             WRITE(NDARAY)(ND(LL),LL=1,24),(XL(NN),YL(NN),ZL(NN),
          *  LNC(NN),T(NN),NN=1,4),MG,(LOTYPE(I1),I1=1,3),(F(I1),I1=1,3),
          *  WT,TORC,IS,MEMB,IC,IP
             CALL COLUMH(CHT,ND,NED,NEQ)
             IF(IPR.EQ.0)WRITE(6,55)KK,MG,(LNC(J),J=1,4),
          *  (T(K),K=1,4),(LOTYPE(I1),I1=1,3),(F(I1),I1=1,3),WT
             KK=KK+1
             IF(LNUM.GE.KK) GO TO 3
             KK=KK-1
             IF(LNUM.EQ.NEL) GO TO 34
      90     CONTINUE
      34     RETURN
      100    CONTINUE
             DATA R/-1.D0,1.D0,1.D0,-1.D0/
             DATA S/-1.D0,-1.D0,1.D0,1.D0/
C            ENTER LOOPS OVER EACH ELEMENT
             LCOUNT=LCOUNT+1
             NEL=NDEL(LCOUNT)
             DO 500 LNUM=1,NEL
             READ(NDARAY)(ND(LL),LL=1,24),(XL(NN),YL(NN),ZL(NN),LNC(NN),
          *  T(NN),NN=1,4),MG,(LOTYPE(I1),I1=1,3),(F(I1),I1=1,3),WT,TORC,
          *  IS,MEMB,IC,IP
C            COMPUTE NORMAL AT 4 NODES
             DO 105 I=1,4
             RR=R(I)
             SS=S(I)
             CALL LOCALX(RR,SS,TT,XL,YL,ZL,T,DC,SF,R,S)
             V3(1,I)=DC(1,3)
             V3(2,I)=DC(2,3)
      105    V3(3,I)=DC(3,3)
C            LX(1)=NO OF INTEGRATION POINTS FOR BENDING
C            LX(2)=NO OF INTEGRATION POINTS FOR TRANSVERSE SHEAR
C            LX(3)=NO OF INTEGRATION POINTS FOR INPLANE SHEAR
             LX(1)=IS/100
             LX(2)=MOD(IS,10)
             LX(3)=MOD(IS,100)/10
             IF (LX(1).EQ.0) LX(1)=4
             IF (LX(2).EQ.0) LX(2)=1
             IF (LX(3).EQ.0) LX(3)=LX(1)
C            ENERGY SPLIT LOOP
C            ISP=1,BENDING; ISP=2,TRANSVERSE SHEAR; ISP=3,INPLANE SHEAR
               SPLT=3
```

```
          IF(LX(3).EQ.LX(1))ISPLT=2
          IF(MEMB.EQ.0)ISPLT=1
          DO 350 ISP=1,ISPLT
          LL=LX(ISP)
          LR=1
          IF(LL.EQ.4)LR=2
          LS=2
          IF(LL.EQ.1)LS=1
          AZ=3.0
          GR=(LR/2)/DSQRT(AZ)
          GS=(LS/2)/DSQRT(AZ)
          W=(3.0-LR)*(3.0-LS)
          GO TO (110,210,110),ISP
C         INITIALIZE BENDING STIFFNESS MATRIX :EKB
   110    DO 112 I=1,24
          DO 112 J=1,24
   112    EKB(I,J)=0.0
          IF(ISP.EQ.3) GO TO 133
          GO TO 134
   133    DO 135 I=1,24
          DO 135 J=1,24
   135    EKP(I,J)=EKB(I,J)
C         EKP IS INPLANE SHEAR STIFFNESS MATRIX
C         ENTER LOOPS FOR NUMERICAL INTEGRATION
   134    DO 170 L=1,LL
          RR=R(L)*GR
          SS=S(L)*GS
          IF(ISP.EQ.3) GO TO 151
          GO TO 152
   151    IF(L.EQ.2) SS=S(3)*GS
   152    CALL LOCALX(RR,SS,TT,XL,YL,ZL,T,DC,SF,R,S)
          CALL JACOBN(RR,SS,XL,YL,ZL,T,SF,R,S,XJAC,DJAC,XJACI,V3)
          CALL RIGDTY(E,PR,C,ISP,ISPLT,NMP,MG)
          TOC=0.0
          CALL BMATR1(T,DC,SF,XJACI,V3,B,C,CBB1,CBB2,MEMB,TOC,BU)
C         EVALUATION OF ELEMENT STIFFNESS MATRIX DUE TO BENDING.
   139    WW=DJAC*W*2.0
          DO 123 I1=1,4
          K1=6*(I1-1)+1
          K2=K1+1
          K3=K2+1
          K4=K3+1
          K5=K4+1
          K6=K5+1
          DO 124 I=1,3
          DO 124 J=1,3
          EKB(K1,I)=EKB(K1,I)+B(J,K1)*CBB1(J,I)*WW
          EKB(K2,I)=EKB(K2,I)+B(J,K2)*CBB1(J,I)*WW
          EKB(K3,I)=EKB(K3,I)+B(J,K3)*CBB1(J,I)*WW
          EKB(K4,I+3)=EKB(K4,I+3)+B(J,K4)*CBB2(J,I+3)*WW/3.0
          EKB(K5,I+3)=EKB(K5,I+3)+B(J,K5)*CBB2(J,I+3)*WW/3.0
   124    EKB(K6,I+3)=EKB(K6,I+3)+B(J,K6)*CBB2(J,I+3)*WW/3.0
   123    CONTINUE
          DO 125 I2=2,4
          K1=6*(I2-1)+1
          K2=K1+1
          K3=K2+1
          K4=K3+1
```

```
        K5=K4+1
        K6=K5+1
        DO 126 I=1,3
        DO 126 J=1,3
        EKB(K1,I+6)=EKB(K1,I+6)+B(J,K1)*CBB1(J,I+6)*WW
        EKB(K2,I+6)=EKB(K2,I+6)+B(J,K2)*CBB1(J,I+6)*WW
        EKB(K3,I+6)=EKB(K3,I+6)+B(J,K3)*CBB1(J,I+6)*WW
        EKB(K4,I+9)=EKB(K4,I+9)+B(J,K4)*CBB2(J,I+9)*WW/3.0
        EKB(K5,I+9)=EKB(K5,I+9)+B(J,K5)*CBB2(J,I+9)*WW/3.0
126     EKB(K6,I+9)=EKB(K6,I+9)+B(J,K6)*CBB2(J,I+9)*WW/3.0
125     CONTINUE
        DO 127 I3=3,4
        K1=6*(I3-1)+1
        K2=K1+1
        K3=K2+1
        K4=K3+1
        K5=K4+1
        K6=K5+1
        DO 128 I=1,3
        DO 128 J=1,3
        EKB(K1,I+12)=EKB(K1,I+12)+B(J,K1)*CBB1(J,I+12)*WW
        EKB(K2,I+12)=EKB(K2,I+12)+B(J,K2)*CBB1(J,I+12)*WW
        EKB(K3,I+12)=EKB(K3,I+12)+B(J,K3)*CBB1(J,I+12)*WW
        EKB(K4,I+15)=EKB(K4,I+15)+B(J,K4)*CBB2(J,I+15)*WW/3.0
        EKB(K5,I+15)=EKB(K5,I+15)+B(J,K5)*CBB2(J,I+15)*WW/3.0
128     EKB(K6,I+15)=EKB(K6,I+15)+B(J,K6)*CBB2(J,I+15)*WW/3.0
127     CONTINUE
        DO 129 I=1,3
        DO 129 J=1,3
        EKB(19,I+18)=EKB(19,I+18)+B(J,19)*CBB1(J,I+18)*WW
        EKB(20,I+18)=EKB(20,I+18)+B(J,20)*CBB1(J,I+18)*WW
        EKB(21,I+18)=EKB(21,I+18)+B(J,21)*CBB1(J,I+18)*WW
        EKB(22,I+21)=EKB(22,I+21)+B(J,22)*CBB2(J,I+21)*WW/3.0
        EKB(23,I+21)=EKB(23,I+21)+B(J,23)*CBB2(J,I+21)*WW/3.0
129     EKB(24,I+21)=EKB(24,I+21)+B(J,24)*CBB2(J,I+21)*WW/3.0
170     CONTINUE
        IF(ISPLT.EQ.3) GO TO 131
        IF(MEMB.EQ.0) GO TO 234
        GO TO 350
131     DO 136 I=1,24
        DO 136 J=1,24
136     EKB1(I,J)=EKB(I,J)
        GO TO 350
210     CONTINUE
C       INITIALISE EKS ;THE SHEAR CONTRIBUTION TO STIFFNESS
        DO 211 I=1,24
        DO 211 J=1,24
211     EKS(I,J)=0.0
        RR=0.0
        SS=0.0
        CALL LOCALX(RR,SS,TT,XL,YL,ZL,T,DC,SF,R,S)
        CALL JACOBN(RR,SS,XL,YL,ZL,T,SF,R,S,XJAC,DJAC,XJACI,V3)
        CT=E(MG)/(2.4*(1+PR(MG)))
        CALL BMATR2(T,DC,SF,XJACI,V3,B1,B2,B3,CT,CBS1,CBS2,CBS3,CBS,TOC)
        WW=DJAC*W*2.0
        DO 227 I1=1,4
        K1=6*(I1-1)+1
        K2=K1+1
```

```
          K3=K2+1
          K4=K3+1
          K5=K4+1
          K6=K5+1
          DO 228 I=1,3
          DO 228 J=1,2
          EKS(K1,I)=EKS(K1,I)+B1(J,K1)*CBS1(J,I)*WW
          EKS(K2,I)=EKS(K2,I)+B1(J,K2)*CBS1(J,I)*WW
          EKS(K3,I)=EKS(K3,I)+B1(J,K3)*CBS1(J,I)*WW
          EKS(K4,I)=EKS(K4,I)+B2(J,K4)*CBS1(J,I)*WW
          EKS(K5,I)=EKS(K5,I)+B2(J,K5)*CBS1(J,I)*WW
          EKS(K6,I)=EKS(K6,I)+B2(J,K6)*CBS1(J,I)*WW
          EKS(K1,I+3)=EKS(K1,I+3)+B1(J,K1)*CBS2(J,I+3)*WW
          EKS(K2,I+3)=EKS(K2,I+3)+B1(J,K2)*CBS2(J,I+3)*WW
          EKS(K3,I+3)=EKS(K3,I+3)+B1(J,K3)*CBS2(J,I+3)*WW
          EKS(K4,I+3)=EKS(K4,I+3)+B2(J,K4)*CBS2(J,I+3)*WW
     *    +B3(J,K4)*CBS3(J,I+3)*WW/3.0
          EKS(K5,I+3)=EKS(K5,I+3)+B2(J,K5)*CBS2(J,I+3)*WW
     *    +B3(J,K5)*CBS3(J,I+3)*WW/3.0
228       EKS(K6,I+3)=EKS(K6,I+3)+B2(J,K6)*CBS2(J,I+3)*WW
     *    +B3(J,K6)*CBS3(J,I+3)*WW/3.0
227       CONTINUE
          DO 229 I2=2,4
          K1=6*(I2-1)+1
          K2=K1+1
          K3=K2+1
          K4=K3+1
          K5=K4+1
          K6=K5+1
          DO 230 I=1,3
          DO 230 J=1,2
          EKS(K1,I+6)=EKS(K1,I+6)+B1(J,K1)*CBS1(J,I+6)*WW
          EKS(K2,I+6)=EKS(K2,I+6)+B1(J,K2)*CBS1(J,I+6)*WW
          EKS(K3,I+6)=EKS(K3,I+6)+B1(J,K3)*CBS1(J,I+6)*WW
          EKS(K4,I+6)=EKS(K4,I+6)+B2(J,K4)*CBS1(J,I+6)*WW
          EKS(K5,I+6)=EKS(K5,I+6)+B2(J,K5)*CBS1(J,I+6)*WW
          EKS(K6,I+6)=EKS(K6,I+6)+B2(J,K6)*CBS1(J,I+6)*WW
          EKS(K1,I+9)=EKS(K1,I+9)+B1(J,K1)*CBS2(J,I+9)*WW
          EKS(K2,I+9)=EKS(K2,I+9)+B1(J,K2)*CBS2(J,I+9)*WW
          EKS(K3,I+9)=EKS(K3,I+9)+B1(J,K3)*CBS2(J,I+9)*WW
          EKS(K4,I+9)=EKS(K4,I+9)+B2(J,K4)*CBS2(J,I+9)*WW
     *    +B3(J,K4)*CBS3(J,I+9)*WW/3.0
          EKS(K5,I+9)=EKS(K5,I+9)+B2(J,K5)*CBS2(J,I+9)*WW
     *    +B3(J,K5)*CBS3(J,I+9)*WW/3.0
230       EKS(K6,I+9)=EKS(K6,I+9)+B2(J,K6)*CBS2(J,I+9)*WW
     *    +B3(J,K6)*CBS3(J,I+9)*WW/3.0
229       CONTINUE
          DO 231 I3=3,4
          K1=6*(I3-1)+1
          K2=K1+1
          K3=K2+1
          K4=K3+1
          K5=K4+1
          K6=K5+1
          DO 232 I=1,3
          DO 232 J=1,2
          EKS(K1,I+12)=EKS(K1,I+12)+B1(J,K1)*CBS1(J,I+12)*WW
          EKS(K2,I+12)=EKS(K2,I+12)+B1(J,K2)*CBS1(J,I+12)*WW
```

```
          EKS(K3,I+12)=EKS(K3,I+12)+B1(J,K3)*CBS1(J,I+12)*WW
          EKS(K4,I+12)=EKS(K4,I+12)+B2(J,K4)*CBS1(J,I+12)*WW
          EKS(K5,I+12)=EKS(K5,I+12)+B2(J,K5)*CBS1(J,I+12)*WW
          EKS(K6,I+12)=EKS(K6,I+12)+B2(J,K6)*CBS1(J,I+12)*WW
          EKS(K1,I+15)=EKS(K1,I+15)+B1(J,K1)*CBS2(J,I+15)*WW
          EKS(K2,I+15)=EKS(K2,I+15)+B1(J,K2)*CBS2(J,I+15)*WW
          EKS(K3,I+15)=EKS(K3,I+15)+B1(J,K3)*CBS2(J,I+15)*WW
          EKS(K4,I+15)=EKS(K4,I+15)+B2(J,K4)*CBS2(J,I+15)*WW
     *    +B3(J,K4)*CBS3(J,I+15)*WW/3.0
          EKS(K5,I+15)=EKS(K5,I+15)+B2(J,K5)*CBS2(J,I+15)*WW
     *    +B3(J,K5)*CBS3(J,I+15)*WW/3.0
  232     EKS(K6,I+15)=EKS(K6,I+15)+B2(J,K6)*CBS2(J,I+15)*WW
     *    +B3(J,K6)*CBS3(J,I+15)*WW/3.0
  231     CONTINUE
          DO 233 I=1,3
          DO 233 J=1,2
          EKS(19,I+18)=EKS(19,I+18)+B1(J,19)*CBS1(J,I+18)*WW
          EKS(20,I+18)=EKS(20,I+18)+B1(J,20)*CBS1(J,I+18)*WW
          EKS(21,I+18)=EKS(21,I+18)+B1(J,21)*CBS1(J,I+18)*WW
          EKS(22,I+18)=EKS(22,I+18)+B2(J,22)*CBS1(J,I+18)*WW
          EKS(23,I+18)=EKS(23,I+18)+B2(J,23)*CBS1(J,I+18)*WW
          EKS(24,I+18)=EKS(24,I+18)+B2(J,24)*CBS1(J,I+18)*WW
          EKS(19,I+21)=EKS(19,I+21)+B1(J,19)*CBS2(J,I+21)*WW
          EKS(20,I+21)=EKS(20,I+21)+B1(J,20)*CBS2(J,I+21)*WW
          EKS(21,I+21)=EKS(21,I+21)+B1(J,21)*CBS2(J,I+21)*WW
          EKS(22,I+21)=EKS(22,I+21)+B2(J,22)*CBS2(J,I+21)*WW
     *    +B3(J,22)*CBS3(J,I+21)*WW/3.0
          EKS(23,I+21)=EKS(23,I+21)+B2(J,23)*CBS2(J,I+21)*WW
     *    +B3(J,23)*CBS3(J,I+21)*WW/3.0
  233     EKS(24,I+21)=EKS(24,I+21)+B2(J,24)*CBS2(J,I+21)*WW
     *    +B3(J,24)*CBS3(J,I+21)*WW/3.0
  234     CONTINUE
          DO 220 L=1,4
          AM=3.0
          RR=R(L)/DSQRT(AM)
          SS=S(L)/DSQRT(AM)
          CALL LOCALX(RR,SS,TT,XL,YL,ZL,T,DC,SF,R,S)
          CALL JACOBN(RR,SS,XL,YL,ZL,T,SF,R,S,XJAC,DJAC,XJACI,V3)
          CALL RIGDTY(E,PR,C,1,2,NMP,MG)
          TOC=0.0
          CALL BMATR1(T,DC,SF,XJACI,V3,B,C,CBB1,CBB2,MEMB,TOC,BU)
          IF(MEMB.EQ.0)WRITE(ISTRES)((CBB1(I,J),I=1,3),J=1,24)
          IF(MEMB.EQ.0) GO TO 220
          WRITE(ISTRES)((CBB1(I,J),I=1,3),J=1,24),((CBB2(I,J),
     *    I=1,3),J=1,24)
          CALL BMATR2(T,DC,SF,XJACI,V3,B1,B2,B3,CT,CBS1,CBS2,CBS3,CBS,TOC)
          WRITE(ISTRES)((CBS(I,J),I=1,2),J=1,24)
  220     CONTINUE
          IF (TORC.EQ.0.0) GO TO 350
C         EKT; THE TORSIONAL STIFFNESS CONTRIBUTION
          DO 236 I=1,24
          DO 236 J=1,24
  236     EKT(I,J)=0.0
          RR=0.0
          SS=0.0
          CALL LOCALX(RR,SS,TT,XL,YL,ZL,T,DC,SF,R,S)
          CALL JACOBN(RR,SS,XL,YL,ZL,T,SF,R,S,XJAC,DJAC,XJACI,V3)
          TOC=1.0
```

```
        CALL BMATR1(T,DC,SF,XJACI,V3,B,C,CBB1,CBB2,MEMB,TOC,BU)
        SJAC=DSQRT(XJACI(1,3)**2+XJACI(2,3)**2+XJACI(3,3)**2)*DJAC
        GG=CT*1.2*SJAC*TORC*TT*4.0
        DO 237 IJ=1,4
        DO 237 K=1,3
237     VA(K,IJ)=0.5*(DC(K,1)*BU(2,IJ)-DC(K,2)*BU(1,IJ))
        J1=0
        DO 238 IJ=1,4
        K1=0
        DO 239 IK=1,IJ
        GW=GG*SF(3,IJ)*SF(3,IK)
        DO 240 J=1,3
        DO 240 K=1,3
        EKT(J1+J,K1+K)=EKT(J1+J,K1+K)+VA(J,IJ)*VA(K,IK)*GG
        EKT(J1+3+J,K1+3+K)=EKT(J1+3+J,K1+3+K)+DC(J,3)*DC(K,3)*GW
        EKT(J1+J,K1+3+K)=EKT(J1+J,K1+3+K)+VA(J,IJ)*SF(3,IK)*DC(K,3)*GG
240     EKT(J1+3+J,K1+K)=EKT(J1+3+J,K1+K)+SF(3,IJ)*DC(J,3)*VA(K,IK)*GG
239     K1=K1+6
238     J1=J1+6
350     CONTINUE
C       CALCULATION OF ELEMENT STIFFNESS MATRIX [EK]
        DO 351 I=1,24
        DO 351 J=1,24
351     EK(I,J)=0.0
        IF(MEMB.EQ.0)GO TO 352
        GO TO 353
352     DO 354 I=1,24
        DO 354 J=1,24
354     EK(I,J)=EKB(I,J)
        GO TO 380
353     IF (ISPLT.EQ.2) GO TO 355
        GO TO 360
355     IF(TORC.EQ.0.0) GO TO 356
        GO TO 357
356     DO 358 I=1,24
        DO 358 J=1,24
358     EK(I,J)=EKB(I,J)+EKS(I,J)
        GO TO 380
357     DO 359 I=1,24
        DO 359 J=1,24
359     EK(I,J)=EKB(I,J)+EKS(I,J)+EKT(I,J)
        GO TO 380
360     IF(TORC.EQ.0.0) GO TO 361
        GO TO 364
361     DO 362 I=1,24
        DO 362 J=1,24
362     EK(I,J)=EKB1(I,J)+EKS(I,J)+EKP(I,J)
        GO TO 380
364     DO 365 I=1,24
        DO 365 J=1,24
365     EK(I,J)=EKB1(I,J)+EKS(I,J)+EKP(I,J)+EKT(I,J)
380     CONTINUE
C       ELEMENT GENERATED LOAD
        DO 381 I=1,24
381     Q1(I)=0.0
        DO 386 II=1,3
        LCASE=LOTYPE(II)
        FF=F(II)
```

```
          IF(LCASE.EQ.0) GO TO 386
          DO 391 I=1,24
  391     Q(I)=0.0
          DO 390 L=1,4
          AY=3.0
          RR=R(L)/DSQRT(AY)
          SS=S(L)/DSQRT(AY)
          CALL LOCALX(RR,SS,TT,XL,YL,ZL,T,DC,SF,R,S)
          CALL JACOBN(RR,SS,XL,YL,ZL,T,SF,R,S,XJAC,DJAC,XJACI,V3)
          SJAC=DSQRT(XJACI(1,3)**2+XJACI(2,3)**2+XJACI(3,3)**2)*DJAC
  390     CALL XLOAD(ZL,SF,LCASE,FF,WT,SJAC,DJAC,TT,DC,Q,WD,NMP,MG)
          DO 387 I=1,24
  387     Q1(I)=Q1(I)+Q(I)
  386     CONTINUE
          DO 395 I=1,24
          DO 395 J=1,I
  395     EK(J,I)=EK(I,J)
          DO 405 J=1,NED
          JJ=ND(J)
          IF(JJ)405,405,410
  410     PQ(JJ)=Q1(J)+PQ(JJ)
  405     CONTINUE
          NUED=NED
          CALL PASSEM(SK,EK,NDS,ND,NED,NEQ1,NSKY,NUED)
  500     CONTINUE
          RETURN
C         CALCULATION OF STRESS RESULTANTS
  800     CONTINUE
          LCOUNT=LCOUNT+1
          NEL=NDEL(LCOUNT)
          VAL=3.0
          TERM=DSQRT(VAL)/2.0
          TMS(1,1)=1+TERM
          TMS(1,2)=-0.5
          TMS(1,3)=1-TERM
          TMS(1,4)=-0.5
          TMS(2,1)=-0.5
          TMS(2,2)=TMS(1,1)
          TMS(2,3)=-0.5
          TMS(2,4)=TMS(1,3)
          TMS(3,1)=TMS(1,3)
          TMS(3,2)=-0.5
          TMS(3,3)=TMS(1,1)
          TMS(3,4)=-0.5
          TMS(4,1)=-0.5
          TMS(4,2)=TMS(1,3)
          TMS(4,3)=-0.5
          TMS(4,4)=TMS(1,1)
          DO 806 I=1,NSN
          DO 806 J=1,3
          FORCE1(I,J)=0.0
          FORCE(I,J)=0.0
  806     XMOMNT(I,J)=0.0
          DO 807 I=1,NEL
          SUMM(I)=0.0
          DIFF(I)=0.0
          TXY(I)=0.0
          PSMX(I)=0.0
```

```
          PSMM(I)=0.0
          ANG(I)=0.0
          DO 807 J=1,3
          GSIG(I,J)=0.0
          GMOM(I,J)=0.0
          DO 807 K=1,2
  807     GSHR(I,K)=0.0
C         STRESS RESULTANTS AT GAUSS POINTS
C         SIG1:STRESSES;SIG2:MEMBRANE FORCES;STR:MOMENTS
          DO 810 LNUM=1,NEL
          READ(NDARAY)(ND(LL),LL=1,24),(XL(NN),YL(NN),ZL(NN),
     *    LNC(NN),T(NN),NN=1,4),MG,(LOTYPE(I1),I1=1,3),(F(I1),I1=1,3),WT,
     *    TORC,IS,MEMB,IC,IP
          DO 805 K=1,NSP
          IF(MEMB.EQ.0)READ(ISTRES)((CBB1(I,J),I=1,3),J=1,24)
          IF(MEMB.EQ.0) GO TO 802
          READ(ISTRES)((CBB1(I,J);I=1,3),J=1,24),((CBB2(I,J),
     *    I=1,3),J=1,24)
          READ(ISTRES)((CBS(I,J),I=1,2),J=1,24)
  802     DO 811 I=1,NS
          SIG1(I)=0.0
          SIG2(I)=0.0
  811     STR(I)=0.0
          AX=3.0
          RR=R(K)/DSQRT(AX)
          SS=S(K)/DSQRT(AX)
          CALL LOCALX(RR,SS,TT,XL,YL,ZL,T,DC,SF,R,S)
          DO 817 J=1,NED
          JJ=ND(J)
          IF(JJ.EQ.0) GO TO 817
          DO 818 I=1,3
          SIG2(I)=SIG2(I)+CBB1(I,J)*P(JJ)
          IF(MEMB.EQ.0) GO TO 818
          STR(I)=STR(I)+CBB2(I,J)*P(JJ)
  818     CONTINUE
  817     CONTINUE
          SEMP=TT*TT/6.0
          DO 820 J=1,3
          SIG2(J)=SIG2(J)*TT
          IF(MEMB.EQ.0) GO TO 820
          STR(J)=STR(J)*SEMP
  820     CONTINUE
          IF(MEMB.EQ.0) GO TO 814
          DO 812 J=1,NED
          JJ=ND(J)
          IF(JJ.EQ.0) GO TO 812
          DO 813 I=1,2
  813     SIG1(I)=SIG1(I)+CBS(I,J)*P(JJ)
  812     CONTINUE
          DO 819 I=1,2
  819     SIG1(I)=SIG1(I)*TT
  814     CONTINUE
          IF(IC.EQ.0)GO TO 834
          DO 809 KR=1,3
          GSIG(LNUM,KR)=GSIG(LNUM,KR)+(SIG2(KR)/16.0)
  809     GMOM(LNUM,KR)=GMOM(LNUM,KR)+(STR(KR)/16.0)
          DO 808 KB=1,2
  808     GSHR(LNUM,KB)=GSHR(LNUM,KB)+(SIG1(KB)/16.0)
```

```
      834    CONTINUE
             IF(IP.EQ.0)GO TO 836
             SUMM(LNUM)=SUMM(LNUM)+(SIG2(1)+SIG2(2))/(32.0*T(K))
             DIFF(LNUM)=DIFF(LNUM)+(SIG2(1)-SIG2(2))/(32.0*T(K))
             TXY(LNUM)=TXY(LNUM)+SIG2(3)/(16.0*T(K))
      836    CONTINUE
             IF(MEMB.EQ.0) GO TO 815
             GO TO 816
      815    CONTINUE
             GO TO 821
      816    CONTINUE
             GO TO 822
    C        STRESS SMOOTHENING OF LOCAL STRESS RESULTANTS
      821    DO 823 J=1,3
             DO 823 I=1,4
             NN=LNC(I)
      823    FORCE(NN,J)=FORCE(NN,J)+TMS(I,K)*SIG2(J)/ELC(NN)
             GO TO 805
      822    DO 824 J=1,3
             DO 824 I=1,4
             NN=LNC(I)
             FORCE(NN,J)=FORCE(NN,J)+TMS(I,K)*SIG2(J)/ELC(NN)
      824    XMOMNT(NN,J)=XMOMNT(NN,J)+TMS(I,K)*STR(J)/ELC(NN)
             DO 825 J=1,2
             DO 825 I=1,4
             NN=LNC(I)
      825    FORCE1(NN,J)=FORCE1(NN,J)+TMS(I,K)*SIG1(J)/ELC(NN)
      805    CONTINUE
             IF(IP.EQ.0)GO TO 837
             PSMX(LNUM)=SUMM(LNUM)+DSQRT(DIFF(LNUM)**2+
          *  TXY(LNUM)**2)
             PSMM(LNUM)=SUMM(LNUM)-DSQRT(DIFF(LNUM)**2+
          *  TXY(LNUM)**2)
             DIFP=0.5*(PSMX(LNUM)-PSMM(LNUM))
             IF(DIFP.NE.0.0)ANG(LNUM)=28.648*DATAN2(TXY(LNUM),DIFP)
             IF(LNUM.EQ.1)WRITE(6,890)
             IF(LNUM.EQ.1)WRITE(6,895)
             WRITE(6,896)LNUM,PSMX(LNUM),PSMM(LNUM)
      890    FORMAT(//6X,'PRINCIPAL STRESSES AT THE C.G OF THE ELEMENTS'//)
      895    FORMAT(6X,'ELEMENT NO',5X,'MAX.PRINCIPAL STRESS',5X,
          *  'MIN.PRINCIPAL STRESS',7X,'ANGLE(DEG)'/)
      896    FORMAT(10X,I5,6X,E20.8,5X,E20.8,5X,F10.4)
      837    CONTINUE
      810    CONTINUE
             IF(IC.EQ.0)GO TO 833
             WRITE(6,900)
             WRITE(6,903)
             WRITE(6,904)(I,(GSIG(I,J),J=1,3),I=1,NEL)
             WRITE(6,905)
             WRITE(6,904)(I,(GMOM(I,J),J=1,3),I=1,NEL)
             WRITE(6,906)
             WRITE(6,907)(I,(GSHR(I,J),J=1,2),I=1,NEL)
      900    FORMAT(//6X,'***STRESSES AND MOMENTS AT C.G OF THE ELEMENTS')
      903    FORMAT(//10X,'ELEMENT NO',12X,'N-X',14X,'N-Y',14X,'N-XY'//)
      904    FORMAT(13X,I5,3E20.8)
      905    FORMAT(//10X,'ELEMENT NO',5X,'MOMENT,M-X',8X,'MOMENT,M-Y',
          *  8X,'MOMENT,M-XY'//)
      906    FORMAT(//10X,'ELEMENT NO',10X,'Q-X',10X,'Q-Y'//)
```

```
907     FORMAT(13X,I5,2E20.8)
833     CONTINUE
        WRITE(6,831)
831     FORMAT(//20X,'LOCAL FORCES AND MOMENTS AT NODES'/)
        WRITE(6,832)
832     FORMAT(17X,'*************************************'//)
        WRITE(6,835)
        WRITE(6,840)
        WRITE(6,845)(I,(FORCE(I,J),J=1,3),I=1,NSN)
        WRITE(6,850)
        WRITE(6,855)
        WRITE(6,860)(I,(XMOMNT(I,J),J=1,3),(FORCE1(I,K),K=1,2),I=1,NSN)
11      FORMAT(I5)
44      FORMAT(6I5,4F10.4/3I2,5F10.4,I5,I2,I3,2I2)
66      FORMAT(//35X,'NUMBER OF 4-NODED SHELL ELEMENTS IN THE STRUCTURE
      *  .......=',I5//)
77      FORMAT(//5X,'ELEMENT',4X,'MATERIAL',5X,'NODAL CONNECTIVITY',20X
      * 'THICKNESS',20X,'LOAD',4X,'PRESSURE',5X,'WATER'/8X,'NO.',5X,
      * 'GROUP NO.',7X,'TYPE',4X,'(UNIT WT)',4X,'TABLE'//)
55      FORMAT(5X,I3,2X,I2,4I5,3X,4F10.4,3I2,4X,4F10.4)
835     FORMAT(//25X,'MEMBRANE FORCES AT NODES'//)
840     FORMAT(//8X,'NODE NO.',12X,'N-X',14X,'N-Y',14X,'N-XY'//)
845     FORMAT(13X,I5,3E20.8)
850     FORMAT(//10X,'BENDING MOMENTS AND TRANSVERSE FORCES AT NODES'/)
855     FORMAT(//8X,'NODE NO.',5X,'MOMENT M-X',8X,'MOMENT M-Y',
      * 8X,'MOMENT M-XY',10X,'Q-X',10X,'Q-Y'//)
860     FORMAT(13X,I5,3E20.8,2E20.8)
        RETURN
        END

C**************************************************************
C       ROUTINE FOR DIRECTION COSINE MATRIX
C**************************************************************
        SUBROUTINE LOCALX(RR,SS,TT,XL,YL,ZL,T,DC,SF,R,S)
        IMPLICIT REAL*8(A-H,O-Z)
        DIMENSION XL(4),YL(4),ZL(4),T(4),DC(3,3),SF(3,4),R(4),S(4)
        TT=0.0
        DO 1 I=1,4
        SF(1,I)=0.25*R(I)*(1.0+S(I)*SS)
        SF(2,I)=0.25*S(I)*(1.0+R(I)*RR)
        SF(3,I)=0.25*(1.0+S(I)*SS)*(1.0+R(I)*RR)
1       TT=TT+SF(3,I)*T(I)
C       DIRECTION COSINES (DC)
        DO 10 J=1,3
        DC(J,1)=0.0
10      DC(J,2)=0.0
        DO 11 I=1,4
        DC(1,1)=DC(1,1)+SF(1,I)*XL(I)
        DC(2,1)=DC(2,1)+SF(1,I)*YL(I)
        DC(3,1)=DC(3,1)+SF(1,I)*ZL(I)
        DC(1,2)=DC(1,2)+SF(2,I)*XL(I)
        DC(2,2)=DC(2,2)+SF(2,I)*YL(I)
11      DC(3,2)=DC(3,2)+SF(2,I)*ZL(I)
        CALL CROSSP(DC(1,1),DC(1,2),DC(1,3),0)
        DO 12 J=1,3
12      DC(J,1)=0.0
        DO 13 I=1,4
        DC(1,1)=DC(1,1)+R(I)*0.25*XL(I)
```

```
            DC(2,1)=DC(2,1)+R(I)*0.25*YL(I)
   13       DC(3,1)=DC(3,1)+R(I)*0.25*ZL(I)
            CALL CROSSP(DC(1,3),DC(1,1),DC(1,2),0)
            CALL CROSSP(DC(1,2),DC(1,3),DC(1,1),0)
            RETURN
            END


C****************************************************************
C         ROUTINE FOR JACOBIAN MATRIX
C****************************************************************
            SUBROUTINE JACOBN(RR,SS,XL,YL,ZL,T,SF,R,S,XJAC,DJAC,XJACI,V3)
            IMPLICIT REAL*8(A-H,O-Z)
            DIMENSION XL(4),YL(4),ZL(4),T(4),SF(3,4),R(4),S(4),XJAC(3,3),
         *  XJACI(3,3),V3(3,4),I1(3),I2(3)
            DATA I1/2,3,1/,I2/3,1,2/
C           XJAC-JACOBIAN MATRIX, DJAC-DETERMINANT, XJACI-INVERSE MATRIX
            DO 35 K=1,3
            DO 35 J=1,3
   35       XJAC(K,J)=0.0
            DO 30 J=1,2
            DO 30 I=1,4
            XJAC(1,J)=XJAC(1,J)+SF(J,I)*XL(I)
            XJAC(2,J)=XJAC(2,J)+SF(J,I)*YL(I)
   30       XJAC(3,J)=XJAC(3,J)+SF(J,I)*ZL(I)
            DO 20 I=1,4
            TP=SF(3,I)*T(I)/2.0
            DO 25 K=1,3
   25       XJAC(K,3)=XJAC(K,3)+V3(K,I)*TP
   20       CONTINUE
            DJAC=0.0
            DO 40 J=1,3
            J2=I1(J)
            J3=I2(J)
   40       DJAC=DJAC+XJAC(1,J)*(XJAC(2,J2)*XJAC(3,J3)-XJAC(2,J3)
         *  *XJAC(3,J2))
            CALL CROSSP(XJAC(1,2),XJAC(1,3),XJACI(1,1),1)
            CALL CROSSP(XJAC(1,3),XJAC(1,1),XJACI(1,2),1)
            CALL CROSSP(XJAC(1,1),XJAC(1,2),XJACI(1,3),1)
            DO 50 J=1,3
            DO 50 K=1,3
   50       XJACI(J,K)=XJACI(J,K)/DJAC
            RETURN
            END


C****************************************************************
C         ROUTINE FOR CROSSPRODUCT
C****************************************************************
            SUBROUTINE CROSSP(A,B,C,IN)
            IMPLICIT REAL*8(A-H,O-Z)
            DIMENSION C(3),B(3),A(3)
            C(1)=A(2)*B(3)-A(3)*B(2)
            C(2)=A(3)*B(1)-A(1)*B(3)
            C(3)=A(1)*B(2)-A(2)*B(1)
            IF(IN.NE.0)RETURN
            CNORM=DSQRT(C(1)*C(1)+C(2)*C(2)+C(3)*C(3))
            IF(CNORM.EQ.0.0)RETURN
            DO 1 I=1,3
```

```
   1    C(I)=C(I)/CNORM
        RETURN
        END

C***********************************************************
C        ROUTINE FOR CONSTITUTIVE MATRIX
C***********************************************************
        SUBROUTINE RIGDTY(E,PR,C,ISP,ISPLT,NMP,MG)
        IMPLICIT REAL*8(A-H,O-Z)
        DIMENSION E(NMP),PR(NMP),C(3,3)
        SLAM=E(MG)*PR(MG)/(1.0-PR(MG)*PR(MG))
        TMU=E(MG)/(1.0+PR(MG))
        DO 5 I=1,3
        DO 5 J=1,3
   5    C(I,J)=0.0
        IF(ISPLT.LE.2) GO TO 20
        GO TO (30,10,40),ISP
  30    C(1,1)=2.0*SLAM*TMU/(SLAM+2.0*TMU)
        C(2,2)=C(1,1)
        C(1,2)=C(1,1)
        C(2,1)=C(1,1)
        RETURN
  40    C(1,1)=TMU
        C(2,2)=C(1,1)
        C(3,3)=C(1,1)/2.4
        RETURN
  20    C(1,1)=SLAM+TMU
        C(1,2)=SLAM
        C(2,1)=SLAM
        C(2,2)=C(1,1)
        C(3,3)=TMU/2.0
        RETURN
  10    CONTINUE
        RETURN
        END

C***********************************************************
C        ROUTINE FOR ELEMENT LOAD CALCULATION
C***********************************************************
        SUBROUTINE XLOAD(ZL,SF,LCASE,FF,WT,SJAC,DJAC,TT,DC,Q,WD,NMP,MG)
        IMPLICIT REAL*8(A-H,O-Z)
        DIMENSION ZL(4),DC(3,3),Q(24),SF(3,4),WD(NMP)
        NJ=1
        JJ=0
        IF(LCASE.GE.3) NJ=3
        H=1.0
        IF(LCASE.LE.3) GO TO 20
        H=0.0
        DO 25 I=1,4
  25    H=H+SF(3,I)*ZL(I)
        H=(WT-H)
        IF(H.LT.0.0) H=0.0
  20    DO 32  I=1,4
        DO 30 K=1,NJ
        J=4-K
        A=DC(J,3)*H
        IF(LCASE.EQ.2) A=DC(3,3)
```

```
              IF(LCASE.EQ.1) A=-TT
              IF(LCASE.EQ.1) FF=WD(MG)
      30      Q(JJ+J)=Q(JJ+J)+SF(3,I)*FF*SJAC*A
      32      JJ=JJ+6
              RETURN
              END
```

```
C***************************************************************
C             ROUTINE FOR CBB1,CBB2
C***************************************************************
              SUBROUTINE BMATR1(T,DC,SF,XJACI,V3,B,C,CBB1,CBB2,MEMB,TOC,BU)
              IMPLICIT REAL*8(A-H,O-Z)
              DIMENSION T(4),DC(3,3),SF(3,4),XJACI(3,3),V3(3,4),
      *       B(3,24),C(3,3),CBB1(3,24),CBB2(3,24),BU(2,4),SFG(3)
C             FORM GLOBAL DERIVATIVES SFG & BU
              DO 147 J=1,2
              DO 147 I=1,4
              BU(J,I)=0.0
              DO 148 K=1,3
              SFG(K)=0.0
              DO 149 KA=1,2
      149     SFG(K)=SFG(K)+XJACI(K,KA)*SF(KA,I)
      148     BU(J,I)=BU(J,I)+SFG(K)*DC(K,J)
      147     CONTINUE
              IF(TOC.EQ.1.0) RETURN
C             COMPUTE STRAIN DISPLACEMENT MATRIX ' B '
              DO 150 I=1,3
              DO 150 J=1,24
      150     B(I,J)=0.0
              DO 153 I=1,4
              K1=6*(I-1)+1
              K2=K1+1
              K3=K2+1
              K4=K3+1
              K5=K4+1
              K6=K5+1
              B(1,K1)=BU(1,I)*DC(1,1)
              B(2,K1)=BU(2,I)*DC(1,2)
              B(3,K1)=BU(2,I)*DC(1,1)+BU(1,I)*DC(1,2)
              B(1,K2)=BU(1,I)*DC(2,1)
              B(2,K2)=BU(2,I)*DC(2,2)
              B(3,K2)=BU(2,I)*DC(2,1)+BU(1,I)*DC(2,2)
              B(1,K3)=BU(1,I)*DC(3,1)
              B(2,K3)=BU(2,I)*DC(3,2)
              B(3,K3)=BU(2,I)*DC(3,1)+BU(1,I)*DC(3,2)
              B(1,K4)=(V3(2,I)*DC(3,1)-V3(3,I)*DC(2,1))*T(I)/2.0*BU(1,I)
              B(2,K4)=(V3(2,I)*DC(3,2)-V3(3,I)*DC(2,2))*T(I)/2.0*BU(2,I)
              B(3,K4)=(V3(2,I)*DC(3,1)-V3(3,I)*DC(2,1))*T(I)/2.0*BU(2,I)
      *       +(V3(2,I)*DC(3,2)-V3(3,I)*DC(2,2))*T(I)/2.0*BU(1,I)
              B(1,K5)=(V3(3,I)*DC(1,1)-V3(1,I)*DC(3,1))*T(I)/2.0*BU(1,I)
              B(2,K5)=(V3(3,I)*DC(1,2)-V3(1,I)*DC(3,2))*T(I)/2.0*BU(2,I)
              B(3,K5)=(V3(3,I)*DC(1,1)-V3(1,I)*DC(3,1))*T(I)/2.0*BU(2,I)
      *       +(V3(3,I)*DC(1,2)-V3(1,I)*DC(3,2))*T(I)/2.0*BU(1,I)
              B(1,K6)=(V3(1,I)*DC(2,1)-V3(2,I)*DC(1,1))*T(I)/2.0*BU(1,I)
              B(2,K6)=(V3(1,I)*DC(2,2)-V3(2,I)*DC(1,2))*T(I)/2.0*BU(2,I)
      153     B(3,K6)=(V3(1,I)*DC(2,1)-V3(2,I)*DC(1,1))*T(I)/2.0*BU(2,I)
      *       +(V3(1,I)*DC(2,2)-V3(2,I)*DC(1,2))*T(I)/2.0*BU(1,I)
C             CBB1:BENDING CONTRIBUTION DUE TO U,V,W
```

```
C          CBB2:BENDING CONTRIBUTION DUE TO ROTATIONS
           DO 119 I=1,3
           DO 119 J=1,24
    119    CBB1(I,J)=0.0
           DO 120 I=1,3
           DO 120 J=1,24
    120    CBB2(I,J)=0.0
           DO 121 IK=1,4
           K1=6*(IK-1)+1
           K2=K1+1
           K3=K2+1
           K4=K3+1
           K5=K4+1
           K6=K5+1
           DO 123 I=1,3
           DO 123 J=1,3
           CBB1(I,K1)=CBB1(I,K1)+C(I,J)*B(J,K1)
           CBB1(I,K2)=CBB1(I,K2)+C(I,J)*B(J,K2)
           CBB1(I,K3)=CBB1(I,K3)+C(I,J)*B(J,K3)
           CBB2(I,K4)=CBB2(I,K4)+C(I,J)*B(J,K4)
           CBB2(I,K5)=CBB2(I,K5)+C(I,J)*B(J,K5)
    123    CBB2(I,K6)=CBB2(I,K6)+C(I,J)*B(J,K6)
    121    CONTINUE
           IF(MEMB.EQ.0) GO TO 117
           RETURN
    117    DO 130 I=1,3
           DO 130 J=1,24
    130    CBB2(I,J)=0.0
           RETURN
           END

C****************************************************************
C          ROUTINE FOR CBS;
C****************************************************************
           SUBROUTINE BMATR2(T,DC,SF,XJACI,V3,B1,B2,B3,CT,CBS1
      *    ,CBS2,CBS3,CBS,TOC)
           IMPLICIT REAL*8(A-H,O-Z)
           DIMENSION T(4),DC(3,3),SF(3,4),XJACI(3,3),V3(3,4),B1(2,24),
      *    B2(2,24),B3(2,24),CBS1(2,24),CBS3(2,24),CBS(2,24),
      *    BU(3,4),SFG(3),A(4),CBS2(2,24)
C          FORM GLOBAL DERIVATIVES 'SFG' AND 'BU'
           DO 213 J=1,3
           DO 213 I=1,4
           BU(J,I)=0.0
           DO 214 K=1,3
           SFG(K)=0.0
           DO 215 KA=1,2
    215    SFG(K)=SFG(K)+XJACI(K,KA)*SF(KA,I)
    214    BU(J,I)=BU(J,I)+SFG(K)*DC(K,J)
    213    CONTINUE
C          COMPUTE STRAIN DISPLACEMENT MATRIX 'B'
C          'B1' CONTRIBUTION TO U,V,W ONLY
C          'B2' AND 'B3' CONTRIBUTION TO ROTATIONS ONLY
           DO 216 I=1,2
           DO 216 J=1,24
    216    B1(I,J)=0.0
           DO 217 I=1,2
           DO 217 J=1,24
```

```
217    B2(I,J)=0.0
       DO 218 I=1,2
       DO 218 J=1,24
218    B3(I,J)=0.0
       BN=0.0
       DO 219 I=1,3
219    BN=BN+DC(I,3)*XJACI(I,3)
       DO 220 I=1,4
220    A(I)=T(I)/2.0*BN*SF(3,I)
       DO 221 I=1,4
       D=(V3(2,I)*DC(3,1)-V3(3,I)*DC(2,1))
       E1=(V3(2,I)*DC(3,2)-V3(3,I)*DC(2,2))
       F=(V3(3,I)*DC(1,1)-V3(1,I)*DC(3,1))
       G=(V3(3,I)*DC(1,2)-V3(1,I)*DC(3,2))
       H=(V3(1,I)*DC(2,1)-V3(2,I)*DC(1,1))
       O=(V3(1,I)*DC(2,2)-V3(2,I)*DC(1,2))
       K1=6*(I-1)+1
       K2=K1+1
       K3=K2+1
       K4=K3+1
       K5=K4+1
       K6=K5+1
       B1(1,K1)=BU(3,I)*DC(1,1)+BU(1,I)*DC(1,3)
       B1(2,K1)=BU(3,I)*DC(1,2)+BU(2,I)*DC(1,3)
       B1(1,K2)=BU(3,I)*DC(2,1)+BU(1,I)*DC(2,3)
       B1(2,K2)=BU(3,I)*DC(2,2)+BU(2,I)*DC(2,3)
       B1(1,K3)=BU(3,I)*DC(3,1)+BU(1,I)*DC(3,3)
       B1(2,K3)=BU(3,I)*DC(3,2)+BU(2,I)*DC(3,3)
       B2(1,K4)=D*A(I)
       B2(2,K4)=E1*A(I)
       B2(1,K5)=F*A(I)
       B2(2,K5)=G*A(I)
       B2(1,K6)=H*A(I)
       B2(2,K6)=O*A(I)
       G1=(V3(2,I)*DC(3,3)-V3(3,I)*DC(2,3))
       G2=(V3(3,I)*DC(1,3)-V3(1,I)*DC(3,3))
       G3=(V3(1,I)*DC(2,3)-V3(2,I)*DC(1,3))
       B3(1,K4)=D*BU(3,I)*T(I)/2.0+G1*BU(1,I)*T(I)/2.0
       B3(2,K4)=G1*BU(2,I)*T(I)/2.0+E1*BU(3,I)*T(I)/2.0
       B3(1,K5)=G2*BU(1,I)*T(I)/2.0+F*BU(3,I)*T(I)/2.0
       B3(2,K5)=G2*BU(2,I)*T(I)/2.0+G*BU(3,I)*T(I)/2.0
       B3(1,K6)=G3*BU(1,I)*T(I)/2.0+H*BU(3,I)*T(I)/2.0
221    B3(2,K6)=G3*BU(2,I)*T(I)/2.0+O*BU(3,I)*T(I)/2.0
C      CBS; SHEAR CONTRIBUTION TO STRESS MATRIX
C      CBS1; SHEAR CONTRIBUTION TO STRESS MATRIX
C      CBS2 AND CBS3; SHEAR CONTRIBUTION TO STRESS MATRIX DUE TO THETAS
       DO 222 I=1,2
       DO 222 J=1,24
222    CBS1(I,J)=0.0
       DO 223 I=1,2
       DO 223 J=1,24
223    CBS2(I,J)=0.0
       DO 224 I=1,2
       DO 224 J=1,24
224    CBS3(I,J)=0.0
       DO 225 IK=1,4
       K1=6*(IK-1)+1
       K2=K1+1
```

```
                 K3=K2+1
                 K4=K3+1
                 K5=K4+1
                 K6=K5+1
                 DO 226 I=1,2
                 CBS1(I,K1)=CT*B1(I,K1)
                 CBS1(I,K2)=CT*B1(I,K2)
                 CBS1(I,K3)=CT*B1(I,K3)
                 CBS2(I,K4)=CT*B2(I,K4)
                 CBS2(I,K5)=CT*B2(I,K5)
                 CBS2(I,K6)=CT*B2(I,K6)
                 CBS3(I,K4)=CT*B3(I,K4)
                 CBS3(I,K5)=CT*B3(I,K5)
    226          CBS3(I,K6)=CT*B3(I,K6)
    225      CONTINUE
                 DO 235 I=1,2
                 DO 235 J=1,24
    235      CBS(I,J)=CBS1(I,J)+CBS2(I,J)
                 RETURN
                 END
```

Index

Acoustic vibrations 529
Analytical solution 4
Approximate solution 4
Arch dam
  type1 392
  type5 395
Assemblage of elements 146
Axisymmetric
  problems 34, 546
  stress distribution 33

Band width 152
  minimization 154
Beam
  cantilever 3, 52, 259, 261,
      335, 383
  curved 386
  continuous 146, 228
Beam element
  two-dimensional 214
  three-dimensional 237
Bending stiiffness 258, 419
Bending moment 230, 236
Boundary conditions
  kinematic 2, 51, 154
  natural 52
  static 3
Boundary value problems 1
Box girder 506

Calculus of variations 49
Central difference 8
Circular plates 442, 444, 446

Clamped hyper shell 502
Coefficient of thermal expansion 29
Column height 186
Common block 193
Compatibility conditions 27, 68, 148
Computer program
  PASSFEM 175, 177
Computer graphics 576
Concrete 325
Conduction heat transfer 515
Conduction stiffness matrix 522
Connectivity 149
Convection heat transfer 516
Constitutive
  matrix 28
  relations 28
Convergence criteria 139
Convergence requirements 67
Cooling tower 504
Coordinate
  cartesian 70, 109
  global 112, 223
  local 71
  natural 70, 110
  transformation 110
Curved shell element 467
Cylindrical shell 501
Cylindrical shell element 468

Data base management 581
Data input details
  main program PASSFEM 198

type of analysis   (element type
      and element data)
beams 270
boundary element 276
plane stress/strain 331
plate bending 440
shells 496
three-dimensional solids 379
truss 268
Deep beam 337
Degrees of freedom
external 103
internal 103
global 147
nodal 66
Differential equations 1
Direction cosines 207, 223, 244,
      370, 471
Direct stiffness method 146
Discretization 13
Discretization in time 548
Displacement
function 64
vector 147

Electromagnetic waves 529
Elements
axisymmetric solid 327
beams prismatic 214, 237
boundary 274
constant strain triangle 297
incompatible modes 315
isoparametric
eight-noded
plane stress/strain 314
plate bending 430
shell 486
three-dimensional solid 355
four-noded
plane stress/strain 308
plate bending 420
shell 468
triangular 116
twenty-noded
three-dimensional solid 356,366

Lagrange 87
offset beam 263
one-dimensional line 70
Serendipity 87
three-dimensional tetrahedron 73
two-dimensional triangular 71
variable number of nodes 93
Element
equilibrium condition 99, 100
stiffness matrix 97, 99, 100
strains 96
stresses 96
Element stiffness
fast computation 137
Energy
bending 415
complementary 43
potential 43
shear 415
strain 46
Equation of equilibrium 151
Equilibrium conditions 22
Euler-Lagrange equation 51
Extremum value 49

Fast element stiffness
      computation 137
Field conditions 1
Field variables 1
Film lubrication 528
Finite difference method 8
Finite element analysis
basic steps 168
Finite element method
basic concept of 13
Finite Element
library 182
software 579
Fixed end actions 227
Flexural centre 263
Flow through porous media 528
Folded plate 508
Forces
body 22
centrifugal 22

surface 22
Framed structures 204
  3D beam element 237
  offset beam element 263
  stiffness matrix 238
  transformation matrix 240
  three-dimensional truss
    element 204
  two-dimensional beam
    element 214
  stiffness matrix 219
  transformation matrix 222
  two-dimensional truss
    element 204
  shear deformation in beams 250
Frontal solution method 154
Functional
  approximation 4

Gauss elimination 155
Gauss points 119
  three-dimensional elements 121
  two-dimensional elements 120
Gauss quadrature
  sampling points 120
  weights 120
Geometric invariance 68
Graphical output facilities 579
Gravity loads 370, 484, 494
Green's theorem 35
Grid frame 283

Heat flux 516, 554
Heat transfer 515
  conduction 516
  convection 516
  radiation 517
Heat conduction
  finite element formulation 523
  governing differential eqn. 517
  one-dimensional 530
  three-dimensional 542
  two-dimensional 535
  transient problems 547

weighted residual techniques 524
  Galerkin's method 525
Helmholtz equation 528
Hermitian interpolation 75
Hexahedral elements 354
Hour glass mode 436

Incompatible modes 315
Initial stress 29
Initial strain 29
Input details (*see* Data input)
Integration
  error in 119
  exact 119
  Legendre polynomial 119
  numerical 119
Integration points
  for gauss quadrature 119, 120
Interelement boundary 68
Interelement compatibility 68
Interpolation functions 75
Irrotational flow of fluids 528
Isoparametric elements
  computation of stiffness matrix 123
  convergence criteria 139
Isotropic material 30

Jacobian matrix 113
  determinant 123
  inverse of 113

Kirchhoff's constraints 418
Kirchhoff's theory 403
Knowledge based expert system 586

Lagrange interpolation 87
Legendre polynomial 119
Line element 70
Linear simultaneous equations 145

Main routine 176
  PASSFEM 177
Matrix
  banded 152
  band width of 152
  decomposition 155

diagonal 161
upper triangular 160
Member axes 214, 222, 239, 242
Membrane
contribution 480
effects 480
Mesh
adaptive 575
Methods
Rayleigh-Ritz 4
variational 48
Mindlin's theory 413
Model generation 577

Natural boundary condition
(*see* boundary conditions
natural)
Natural coordinates 70, 110
Nodal degrees of freedom 66
Nodal load vector 99, 100
Nodes
external 103
internal 103
numbering of 153
Numerical integration 119
triangular elements 121

Offset beam element 263
Orthogonality condition 476
Orthotropic material 29

Pascal triangle 69
PASSFEM
CADNUM 186
COLUMH 184
DISP 192
element routines
-guidelines 192

FELIB 182
input details 198
main routine 176
PASLOD 191
PASOLV 190
PASSEM 188
PASSIN 178

Plane frame 231, 282
Plane strain 32, 296
Plane stress 31, 296
Plane truss 209, 278
Plates
bending (see plate bending)
circular 442, 444, 446
continuous 454
shear deformation in 412
square 442, 443, 445, 447
thick 402, 450
thin 403
with circular hole 337
Plate bending
basic theory 402
Polynomial 64
Pre-processing 168
Post-processing 170
Pre- & post-processor 576
Principal plane (axes) 240
Principle of
stationary minimum
potential energy 53
virtual displacement 34, 35

Quadrilateral element 111, 308
Quadratic variations 79
Quadrature (*see* Gauss quadrature)
Quasi-Harmonic equation 528
field of application 528

Rayleigh-Ritz method 4
Reaction 154
Rectangular elements
eight noded 86, 306
four noded 83, 304

Rectangular plates 407, 409
Reduced integration 434
Reinforced concrete element 324
Reynolds equations 528

Sampling points (*see* gauss points)
Shape functions
(*see* also interpolation functions
Shear centre 263

Shells
  analysis of 466
  clamped hyper 503
  cooling tower 504
  cylindrical 501
  folded plate 509
  thin shell theory 466
Shell elements 468
  eight noded 487
  four noded 469
  torsional stiffness 483
Software
  adaptive f.e.a 586
  adaptive meshing 586
  current trends 585
  error estimates 586
  f.e.a. packages 583
  h, h-p versions 589
  model generation 576
  pre- & post-processors 576
Solid elements 350
Space frame 284, 285
Space truss 279
Static condensation 103
Stiffness matrix
  structure 147
Storage scheme
  skyline 152
Strain 26
Strain-displacement relations 26
Strain-energy 45
  bending 256, 427
  shear 256,, 427
Strength of materials 5
Stress 22, 24
Stress resultants 226
Stress smoothing 438
Structural mechanics 22
Subparametric element 111
Subroutine
  BEAM
    description of 270
    listing of 608
  BOUND
    description 274
    listing of 621

Main
  description of 176
  listing of 593
PLANE
  descriptiom of 330
  listing of 625
PLATE
  description of
  listing of 664
SHELL
  description of 496
  listing of 676
THREDS
  description of 377
  listing of 642
TRUSS
  description of 267
  listing of 604
Substructure
  basic approach 560
  multi-level 559, 564
Super-element 563
Superparametric 111

Temperature distribution 515, 554
  stress analysis 554
Tetrahedral element 350, 352
Thermal conductivity of
  the material 516
Thermal load vector 522
Thick cylinder 341
Thick plate 402
Thin plate theory 403
Three-dimensional stress
  analysis 350

Timoshenko beam theory 252
Torsion of noncircular sections 528
Torsional stiffness 482
Transformation matrix
  rotation 222
  strain 376, 491
Trial functions 7, 8
Triangular elements 297
  constant strain 297
  linear strain 297

Unit dummy displacement 39
Unit vector 369, 470
Variational formulation

Variational approach 48
Vertical 3D beam member
Virtual strain 42
Virtual work 34
  external 37
  internal 37
  principle 34

Water pressure 486
Weighted residual technique 524
  Galerkin's method 525
  least square method 525
  point collocation method 524
  subdomain collocation method 524